BRACHIOPODS

PROCEEDINGS OF THE THIRD INTERNATIONAL BRACHIOPOD CONGRESS
SUDBURY/ONTARIO/CANADA/2-5 SEPTEMBER 1995

Brachiopods

Edited by
PAUL COPPER & JISUO JIN
Department of Earth Sciences, Laurentian University, Sudbury, Ontario, Canada

CRC Press
Taylor & Francis Group
Boca Raton London New York

CRC Press is an imprint of the
Taylor & Francis Group, an **informa** business
A BALKEMA BOOK

THIRD INTERNATIONAL BRACHIOPOD CONGRESS

September 2-5, 1995
Laurentian University
Sudbury, Canada P3E 2C6

International
Palaeontological
Association

Cover photo: Slab of Early Silurian (Aeronian) strophomenid brachiopod – rich limestone from the Jupiter Formation, Anticosti Island, Canada

TABLE OF CONTENTS

PREFACE

The study of brachiopods (or 'arm-footers, lamp shells') started more than two hundred years ago. Those of us fortunate enough to have been initiated into the research, rituals and ramifications of brachiopod science, have seen the discipline wax and wane, and wax again in the last century, from a high sparked by the monumental works of Sowerby, Davidson, Barrande, Schnur, Bittner, Hall and others in the last century, to a relatively long pause and recovery in the first half of the twentieth century, and a revitalized burst of effort leading to the 1960 Osnovy Paleontologii, and 1965 Treatise on Invertebrate Paleontology (Brachiopoda), and beyond. Since the 1960's, stratigraphic and monographic descriptions of new faunas, with great diversity of species, genera, and families, have risen almost exponentially, particularly with the discovery of rich faunas in more remote parts of Eurasia, Australia and the Arctic regions of North America. Slightly more than 4830 genera and subgenera have been named to 1995 (see Doescher appendix: A-Z of brachiopods, this volume). Studies once centered around western Europe and North America, have shown that there are fascinating 'new' ancient oceans and evolutionary dramas left to discover and explore.

In 1985, the First International Brachiopod Congress held in Brest, was successfully guided by Patrick Racheboeuf and Christian Emig. This led naturally to the Second Congress in Dunedin in 1990, organized by David MacKinnon, Daphne Lee and June Richardson, and to the Third Congress in Sudbury, Canada. In the ten years elapsed since the first Congress, brachiopodology has seen a number of interesting, parallel developments. The discipline has been enriched by the family nuptial of brachiopod zoology, biochemistry, genetics, ecology and paleontology. We have now at our fingertips information from the mysteries of the gene code providing us with fresh evolutionary scenarios and clues to as far back as the origins of the phylum at the onset of the Cambrian, 545 million years ago or so. We have a wealth of data on the nature and structure of the shell wall, brought to us by the scanning electron microscope, that is providing us with new insight into the evolutionary paths taken by many groups. In the calcite needles of the secondary wall, articulate brachiopods have firmly sealed off the stable isotopes we can now tickle out with a mass spectrometer to give us clues about temperature fluctuations and salinities of the ancient oceans. The fossil record attests to a series of spectacular radiations and mass extinctions in the brachiopods, evidence for global climatic, atmospheric and oceanic events that have sparked the rise and fall of major groups. Most of the brachiopod orders are now extinct, but the past shows that in the Paleozoic and parts of the Mesozoic brachiopods were by far the most abundant shelly elements in the tropical continental shelf regimes. Today articulates are common only in the cooler oceans or deeper waters. Why is this so? Were brachiopods so different from molluscs? No congress can provide answers to all the questions, but this volume is at least a small contribution.

Trying to compile the proceedings of an international meeting, such as the Third International Brachiopod Congress held in Sudbury (September 2-5, 1995), is never an easy task. Manuscripts come in every sort of package and style, and are more commonly written in the second or third language of the authors, and therefore overprinted with all the ethnic and linguistic diversity that this planet has to offer. Advances in word processors have fortunately made our task much easier: papers no longer need to be laboriously re-typed several times.

Diskettes can relatively easily be formatted from one computer language to another. We have tried to stick as faithfully to the submitted, original text as possible, and therefore each paper reflects not only an individual scientific approach, methodology and solution, but also the cultural 'spice' that comes with it. Each paper should thereby be read as one would sit down to order from a menu: commence with your favourite entrées, consume the main course, freshen your palate with dessert, and do not judge too harshly and critically, but wash the meal down with a pleasant linguistic wine, spiced with the reader's own condiments. In translation, mistakes can be blamed on us. We have tried our best to understand the authors' texts, but may have inadvertently misread or misunderstood what was provided. For transliteration of geographic names we have tried to follow the greater Times Atlas. For the cyrillic alphabet we have followed the style of the Geological Society of America. We have re-drawn a number of figures and re-set most tables: those left untouched were either perfect or could not be corrected. Some submitted figures could not be included. Many references were left incomplete by authors, and a number are missing from the list. For missing references we suggest the reader contact the original authors. We did not have the pleasure of providing you with proofs, so that last minute corrections could be made. Some of the authors did not see the reviewers' comments and suggestions, for some could simply not be reached, or the reviews arrived too late to be useful. In the latter instance, we made changes to the text without the author's permission. We apologise and trust you understand our deadlines to get this out on schedule! Some of the submitted papers were not in a publishable state: we regret that they could not be included, but our reviewers were diligent and tough.

The Congress was a lot of fun, as well as hard work, but it has been a beneficial learning process. We saw delegates from some 30 countries, and from all the inhabited continents, but Africa. We thank you for making the effort, and often great personal financial sacrifice, to come to Sudbury and Canada. We dedicate this volume to those who could not be here, the Doyen of brachiopods, 'Gus' Cooper (who celebrates his 94th birthday as this volume is finished), Vladimir Havlicek, and those who contributed greatly to brachiopod science, and whose memory still is strong in our minds, Derek Ager, Jess Johnson and Dick Grant.

Acknowledgments are due. We thank the Natural Sciences and Engineering Research Council of Canada for their generous support in providing part of the travel costs of fourteen delegates from the 'eastern bloc', and China. The Canadian

Geological Foundation provided us with financial aid to assist in publishing these procceedings. The International Paleontological Association gave us seed money to help us mail out the early circulars, and Laurentian University provided us with seed money for mailing and early organizational costs. We thank our spouses, Karin and Bin, for all their help and support, and all the rest of our working team, Laura Pearsall, Darrel Long, Derek Armstrong, Lee Pearsall, Laing Ferguson, Alan Logan, Tristan and Siobhan Long, Jed Day, Mike Koch, and Gao Jianguo for their efforts in helping us to run the Congress facilities, correspondence, sessions and field trips. And finally we thank the many reviewers who helped us steer the manuscripts through their final stages: Neil Archbold, Mike Bassett, Howard Brunton, Sandy Carlson, Jed Day, Keith Dewing, Lars Holmer, Brian Jones, Alan Logan, Willy Norris, Bert Rowell, Mike Sandy, and Derek Walton.

The appendix at the back contains the world list of brachiopod genera or subgenera [to end-1995] collated by Rex Doescher of the Smithsonian Institution, Washington. We thank Rex for making this list available just prior to his early retirement, as this volume goes to press! We have also added the abstracts of those who gave a talk or poster at the Congress, but did not submit a paper for the proceedings.

PAUL COPPER & JISUO JIN

Department of Earth Sciences, Laurentian University, Sudbury, Ontario P3E 2C6, Canada

(The text figures in this volume may not be all at the enlargement or reduction quoted in the explanations. Consult the authors for original scale, where this appears problematic).

TABLE OF TRANSLITERATION FROM CYRILLIC TO ENGLISH

А а	A a
Б б	B b
В в	V v
Г г	G g
Д д	D d
Е е	E e
Ё ё	E e
Ж ж	Zh zh
З з	Z z
И и	I i
Й й	I i
К к	K k
Л л	L l
М м	M m
Н н	N n
О о	O o
П п	P p
Р р	R r
С с	S s
Т т	T t
У у	U u
Ф ф	F f
Х х	Kh kh
Ц ц	Ts ts
Ч ч	Ch ch
Ш ш	Sh sh
Щ щ	Shch shch
Ъ ъ	(omitted)
Ы ы	Y y
Ь ь	(omitted)
Э э	Ee
Ю ю	Yu yu
Я я	Ya ya

Inside photo: Conference photo of the delegates outside the auditorium of the Geological Survey of Ontario, Laurentian University campus, Sudbury, Ontario, September 5, 1995.

X

INTRODUCTION: AMENDED EXCERPTS FROM THE OPENING ADDRESS

ALWYN WILLIAMS

Palaeobiology Unit, 8 Lilybank Gardens, University of Glasgow, Glasgow G12 8QQ, UK

The Third International Brachiopod Congress in Sudbury, Canada, whose proceedings are published herein, was geographically appropriate and scientifically opportune. The Congress was appropriate in that the third of the quinquennial meetings of brachiopodologists, initiated by our French colleagues in one of the cultural bastions of Europe and continued in New Zealand, that paradise for living brachiopods, took place in the heart of North America, the home of so much basic research on the phylum.

The Congress was opportune in that it coincided with the preparation for printing of the first volume of the revision of the brachiopod 'Treatise', three decades after the publication of the first edition of Part (H), Brachiopoda (1965). Of course, 1965 was a momentous year for scientific advances relevant to brachiopod studies as a whole. That first edition just pre-dated by a few months the publication of Hennig's phylogenetic studies (1965), Zuckerkandl and Pauling's proposition that molecules, especially the genome, are 'documents of evolutionary history' (1965), and of the earliest electron microscopy studies of the brachiopod shell (Sass, Munroe and Gerace, 1965; Williams, 1966; Towe and Harper, 1966). Moreover, the Treatise itself contained the first systematic exploration of the degradation of the organic components of living and fossil shells of any marine vertebrates (Jope, 1965).

Such revolutionary thoughts and techniques have had an immense impact on our understanding of brachiopod biology and phylogeny, as was all too evident during the Congress. Yet none has had as much influence on the pace and diversification of recent brachiopod research as the issue of the Treatise itself; not for what the volumes contained but, like Sherlock Holmes's dog that did not bark, for what they did not contain. In effect, the brachiopod edition of the Treatise was an organized display of our ignorance; and the consequences have been remarkable by any measure.

Thus, measuring taxonomic/morphological research by the number of genera created, the growth in these studies has been extraordinary with more taxa erected in the last thirty years than in the two centuries or so since the first brachiopod genus, *Terebratula*, was proposed by Müller in 1776 (Figure 1). Admittedly, many post-1965 genera were erected by relatively few palaeontologists, notably Cooper and Havlícek; but, even so, the great majority were either taxonomic refinements or newly discovered forms, especially from Asia and Australia, which had been recognized as such by using the Treatise as a bench tool for comparative purposes.

The introductory chapters to the 1965 edition of the brachiopod Treatise also revealed the rarity or indeed the absence of any literature on physiology, metabolism and genetics of a phylum which, after all, is unique among marine invertebrates for the continuity and diversity of its geological record. These gaps are now being filled by an

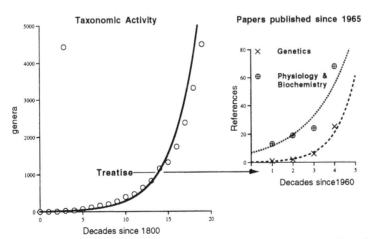

Figure 1. Researches in brachiopod taxonomy and morphology as measured by the number of genera created over the last two centuries (left graph); and in genetics, physiology and biochemistry as indicated by the number of papers published since 1960 (right graph).

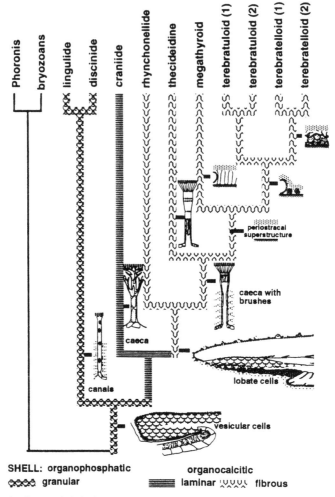

Figure 2. Cladogram showing the genealogical relationships among ten brachiopod species (representing the diversity of the phylum at the present time), based on the differentiation of their integuments compared with those of *Phoronis* and bryozoans.

exponential output of papers with special regard for those data which can serve as indicators of the lifestyle, habits and genealogies of fossil species, like the physiology of growth, musculature and feeding and the elucidation of the genome (Figure 1).

The dramatic impact of these new approaches is well shown by the influence of the electron microscope in the study of the brachiopod integument. In the1965 edition, distinctions were drawn between the primary and secondary layers of apatitic and calcitic shell but it was assumed that the secondary fabric was either fibrous (with needle-like units) or prismatic. In fact, there are at least eight different kinds of secondary shell with three-stratiform, tabular laminar and fibrous - represented in living species. Similarly, even the pioneer studies of Jope (1965) identified no more than two proteins in the *Lingula* shell. Currently at least ten species are now known

(Williams, Cusack & Mackay, 1994), with some covalently attached to glycosaminoglycans (GAGs) and others associated with chitin or fabricated as fibrillar collagens and membranes.

In 1965, the brachiopod shell was described simply as impunctate, punctate or pseudopunctate. Today it can be shown that there are at least three types of pseudopunctae (Williams & Brunton, 1993), and six kinds of punctae with only those of the terebratulides, thecideidines and spiriferides arising from the same stem group characterized by a distal microvillous brush. Yet five of these punctal types accommodated papillose outgrowths of the mantle serving as storage centres. The sixth kind, the canals of organophosphatic shells, could never have contained more than extensions of the secreting plasmalemma of the mantle. In *Lingula* at least, the canals support nets of actin, from which are

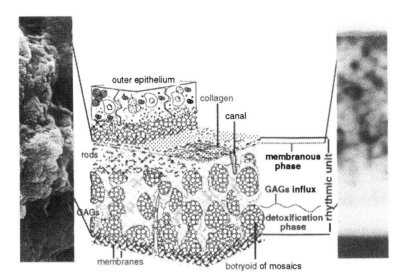

Figure 3. Diagrammatic representation of a laminar set of the shell of Recent *Lingula anatina* showing a complete rhythmic unit correlated with an SEM of a gold-coated, fracture section digested in endoproteinase Glu-C and chitinase (to the left) and a carbon-coated, cut section of a resin-impregnated shell subjected to back scattered electron (BSE) detection (to the right).

suspended mosaics and botryoids of apatite in a gel of GAGs.

In 1965, the outer mantle lobe at the shell margins was seen only as a fold of cells secreting a sheet-like periostracum and part of the primary layer. The periostracum is now known to be a unit membrane normally supporting a complex infrastructure and/or superstructure (Williams and Mackay 1979). Eight or nine different kinds of periostraca characterize living species with infrastructures secreted by vesicular cells and superstructures by either the inner epithelium in inarticulated brachiopods or specialized lobate cells in articulated stocks.

All these recently discovered features of the living integument can be assembled into a phylogenetic frame of the kind that has come into vogue since 1965 and which allow precise comparisons with brachiopod genealogies based on other characters. Such a cladogram (Figure 2) compares reasonably well with the classical interpretation of brachiopod geological history. More importantly, relationships among the brachiopod taxa featured in the cladogram are identical with those described later in this volume by Cohen and Gawthrop, following their ongoing studies of the genome, except for a more distant affinity between thecideidines and terebratuloids.

Given the pace of change in our understanding of the Brachiopoda over the last thirty years, the scope of future research on the phylum inevitably comes in question. My own experience of returning to the electron microscopy of the shell after a break of 12 years is reassuring. When I was asked in front of a geological audience what I intended to do with the facilities being

assembled and I replied that I would be studying the shell of *Lingula*, everybody laughed. The reaction was understandable. *Lingula*, as a fossil, is wholly memorable to everyone with any geological background because it is so featureless.

In fact, the shell of living *Lingula* is a complex sequence of laminae, disposed like lithostratigraphic members of a sedimentary basin with unconformities, oversteps and offlaps and rapid facies changes from predominantly biomineralized to exclusively organic units. The laminae, however, form rhythmic sets (Figure 3) beginning with a compact lamina of apatitic spheroids resting on organic membranes and passing inwardly into a more open network of apatitic bodies suspended in the GAGs gel and then sealed by another set of membranes (Williams, Cusack & Mackay, 1994). The rhythm could be seasonal with the basic apatitic component representing the onset of phosphatic detoxification of tissue.

Despite organic degradation, the chemicostructure of the shell is so well preserved in fossilized *Lingula* that a rhythmic sequence within the skeletal succession of the Carboniferous *L. squamiformis* is virtually indistinguishable from that of living *L. anatina* once allowance is made for an apatitic/clay replacement of GAGs (Figure 4). Even organic strands still survive, traversing canals or, exceptionally, as residual nets on the interfaces between laminar sets. The strands are either actin or collagen (Cusack & Williams, in press).

Notwithstanding the conservatism of this secretory regime, the shell as a whole has undergone a noteworthy anteriomedian change in the proportions of its organic and apatitic constituents. Emig (1981) noted that Recent shells degraded very rapidly, which can be simulated by

L. squamiformis　　　　　　**_L. anatina_**

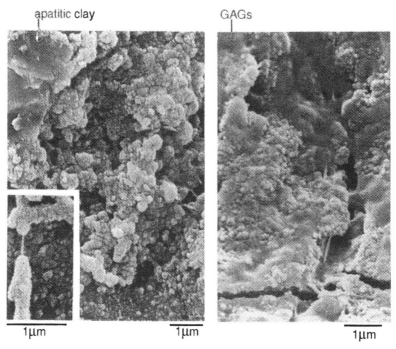

Figure 4. SEMs of gold-coated fracture sections of shells of Lower Carboniferous _Lingula squamiformis_ from Kinghorn, Scotland, and Recent _Lingula anatina_ from Japan showing the essential similarity between virgose laminae of both species.

immersing a valve in 2% sodium hypochlorite for a week. The body platform and parts of the lateral zones survive such bleaching because they are composed of apatitic botryoids and rods in contrast to the anteriomedian sections which are mainly GAGs. Emig concluded that only lingulas, which had been buried catastrophically, could survive as complete shells in the fossilized state. However, the anteriomedian sectors of _L. squamiformis_ consist of well-ordered apatitic laminae indicating that this Palaeozoic species, at least, was more uniformly apatitic in the living state than Recent descendants.

In respect of stock survival, this difference in composition could have affected the relative flexibility of the anteriomedian sectors of the shell. In living _Lingula_, the sectors, being composed mainly of GAGs, readily bend outwards to form a biconvex funnel containing the three setal pseudosiphons (Williams, Cusack and Mackay 1994). The anteriomedian sectors of _squamiformis_, with their comparatively high apatitic content, could have been more brittle and accordingly less favoured by natural selection during the evolution of the _Lingula_ shell.

In conclusion, my experience with _Lingula_, that most modest of brachiopods, unsurprisingly prompts the view that we have only just started to understand the phylum and comprehend its phylogeny. All the sophisticated

techniques now coming into play can only enhance the pleasure arising from our researches in the future. One other thing is certain. There will always be generations of intellectually hungry students coming up with something new.

REFERENCES

CUSACK, M. & A. WILLIAMS. (in press). Chemicostructural degradation of Carboniferous lingulid shells.

EMIG, C. C. 1981. Implications de données récentes sur les Lingules actuelles dans les interprétations paléoécologiques. Lethaia 14: 151-156.

HENNIG, W. 1966. Phylogenetic Systematics. University of Illinois Press, Urbana, Illinois, 263p.

JOPE, H. M. 1965. Composition of brachiopod shell, p.156-163. In R.C. Moore (ed.), Treatise on Invertebrate Paleontology Part (H). Geological Society of America & University of Kansas Press, Lawrence, Kansas.

SASS, D. B., E.A. MONROE & D.T. GERACE. (1965). Shell structure of Recent articulate Brachiopoda. Science 149: 181-182.

TOWE, K.M. & C.W. HARPER. 1966. Pholidostrophiid brachiopods: origin of the nacreous lustre. Science 154:153-5.

WILLIAMS, A. 1966. Growth and structure of the shell of living articulate brachiopods. Nature 211: 1146-1148.

WILLIAMS, A. & C. H. C. BRUNTON. 1993. Role of shell structure in the classification of the orthotetidine brachiopods. Palaeontology 36: 931-966.

WILLIAMS, A., M. CUSACK, M. & S. MACKAY.1994. Collagenous chitinophosphatic shell of the brachiopod *Lingula*. Philosophical Transactions Royal Society London B 346: 223-266.

WILLIAMS, A. & S. MACKAY. 1979. Differentiation of the brachiopod periostracum. Palaeontology 22: 721-736.

ZUCKERKANDL, E. & L.PAULING. 1965. Molecules as documents of evolutionary history. Journal of Theoretical Biology 8: 357-366.

THE GENUS *PSEUDOGIBBIRHYNCHIA* (BRACHIOPODA, RHYNCHONELLACEA) FROM THE TOARCIAN OF PORTUGAL

YVES ALMÉRAS

Centre des Sciences de la Terre et URA 11 [C.N.R.S.],
Université Claude Bernard, Lyon, F - 69622 Villeurbanne Cedex , France

ABSTRACT- Species of *Pseudogibbirhynchia* Ager are similar to those of *Gibbirhynchia* Buckman, but internal characters differ. They may be compared with those of the subfamily Cirpinae and specially with the genus *Cirpa* De Gregorio. *Pseudogibbirhynchia* is known from the Toarcian of the Mid-European Platform and North Tethyan Realm (Catalonia, Iberic Cordillera), and also from the Toarcian and Early Aalenian of the central and eastern High Atlas. It exists also in Portugal (north Lusitanian Sub-basin) where it is associated with *Zeilleria sharpei* Choffat, with terebratulids (the genera *Telothyris* Alméras & Moulan, and *Stroudithyris* Buckman) and rhynchonellids referred to the genera *Soaresirhynchia* Alméras and *Nannirhynchia* Buckman. The present paper describes the characters of three species (*P. moorei, P. jurensis, P. bothenhamptonensis)* and their evolution in the Portuguese Toarcian.

INTRODUCTION

Par leur morphologie et leurs caractères internes, les coquilles décrites dans ce travail se rapportent au genre *Pseudogibbirhynchia* Ager (1962), dont l'espèce-type est *Rhynchonella moorei* Davidson, 1852. Le genre *Pseudogibbirhynchia* (Wellerellidae Likharev, 1956; Cirpinae Ager, 1965) a été créé pour des Rhynchonellidés dont la morphologie est très voisine de celle du genre *Gibbirhynchia* Buckman, 1917 (Ager, 1962). Les espèces attribuées à ce genre (*Rhynchonella moorei* Davidson, 1852; *Terebratula jurensis* Quenstedt, 1858) comprennent des coquilles globuleuses, de taille petite à moyenne, également convexes en vue latérale, de contour subcirculaire, avec une valve brachiale aplatie postérieurement. La commissure frontale est uniplissée, parfois asymétrique, sans toutefois déterminer un pli médian dorsal net ; ce dernier passe progressivement aux parties latérales de la valve brachiale. Le sinus qui lui correspond à la valve pédonculaire est également mal délimité. Les valves sont multicostées et la costulation, de type *tetrahedra* ou *grandis,* prend naissance dès le crochet et l'umbo dorsal. Chaque valve porte de 10 à 20 côtes dont 3 à 6 se situent sur le pli médian dorsal. Le stade lisse développé sur la région postérieure des coquilles de *P. jurensis* résulte d'une usure pendant la vie ou au cours de la fossilisation ou encore d'un phénomène d'épaississement du test par dépôt de calcite sur sa surface externe. Le crochet est petit, quelquefois aigu chez les spécimens bien conservés, légèrement recourbé, sans crêtes latérales nettement marquées. Le foramen est circulaire, submésothyride entre des plaques deltidiales disjointes.

Pseudogibbirhynchia s'éloigne de *Gibbirhynchia* par ses caractères internes qui le rapprochent des *Cirpinae* et en particulier du genre *Cirpa* De Gregorio, 1930 (Ager, 1962). *Pseudogibbirhynchia* se caractérise par des plaques deltidiales épaissies (Ager, 1962, fig. 69) ou dédoublées (ibid., fig. 67), par des dents cardinales droites et présentant des expansions latérales, par des denticules petits et émoussés, par des cruras préfalcifères ainsi que par l'absence de

processus cardinal et de septalium. Les plaques cardinales sont fines et arquées ventralement tandis que le septum médian dorsal est très court, voire à peine esquissé.

Pseudogibbirhynchia a été reconnu dans le Toarcien des Plates-formes de l'Europe moyenne (Ager, 1962), dans le Toarcien du domaine nord-téthysien: Catalogne (Dubar, 1931; Delance, 1969), Cordillère Ibérique (Goy, 1974), ainsi que dans le Toarcien et l'Aalénien inférieur du Haut-Atlas central et oriental (Rousselle, 1973). Dans le Caucase nord-occidental et en Crimée, Kamyschan & Babanova (1973) donnent une acception plus large du genre avec une extension du Toarcien supérieur au Bajocien inférieur et une apogée dans l'Aalénien supérieur. Le genre *Pseudogibbirhynchia* existe également au Portugal, dans le sous-bassin nord-lusitanien où il est associé à *Zeilleria sharpei* Choffat, à des Térébratulidés (représentants des genres *Telothyris* Alméras & Moulan, 1982, et *Stroudithyris* Buckman, 1917) ainsi qu'à des Rhynchonellidés rapportés principalement aux genres *Soaresirhynchia* (Alméras, 1994) et *Nannirhynchia* (Alméras et al., 1995). Ses représentants successifs dans le Toarcien portugais sont *P. moorei* (Davidson), *P. jurensis* (Quenstedt) et *P. bothenhamptonensis* (Buckman). Les caractéristiques de ces espèces et leur évolution sont décrites dans ce travail à partir d'un abondant matériel. Les 124 coquilles rapportées à *Pseudogibbirhynchia* ayant fait l'objet d'une étude biométrique (Table 1: appendix), et 148 exemplaires incomplètement conservés ou partiellement déformés, ont été récoltés dans différents gisements dont la stratigraphie est résumée in Alméras (1994). Pour les coupes de Peniche, de Rabaçal et de Saõ Giao, on se reportera à Alméras (1994, Fig. 2).

EVOLUTION DU *PSEUDOGIBBIRHYNCHIA*

Pseudogibbirhynchia moorei (Davidson, 1852) (Figure 1:1-7) - *P. moorei* marque la première apparition du genre *Pseudogibbirhynchia.* L'espèce est bien représentée au Toarcien inférieur, dans la zone à Polymorphum de Peniche et de Rabaçal où elle est associée au faune à *Koninckella* et à *Nannirhynchia pygmoea.* Les coquilles (Figure 1: 1-3), ressemblent par leur taille et leur morphologie, à l'exemplaire figuré par Davidson (1852, pl. 15, fig. 14). Leur taille est comprise entre 11,3 et 17,8 mm (Figure 1,2). Elles montrent un contour pentagonal arrondi avec une largeur supérieure à la longueur (l/L = 1,10) ; seuls 2 spécimens sur 18 sont un peu plus longs que larges. Les valves sont moyennement convexes (E/L = 0,61), la valve brachiale un peu plus épaisse possède un aplatissement caractéristique sur sa partie postérieure. L'uniplication frontale n'est pas très développée, elle est large et plane, parfois asymétrique (5 coquilles sur 17). Le pli dorsal se différencie peu des parties latérales de la coquille. L'uniplication apparaît entre 9,1 et 11,5 mm à partir du crochet. Un spécimen de 11,4 mm de longueur est encore rectimarginé. Les valves sont ornées de 12 à 20 côtes de type *tetrahedra*, présentes dès le crochet et l'umbo dorsal, dont 5 à 8 se situent sur le pli médian dorsal. Une ou deux côtes intercalaires peuvent être observées sur la moitié des

Table 1. Valeurs moyennes et intervalles de variation des dimensions et du nombre de côtes chez les trois espèces de *Pseudogibbirhynchia* du Toarcien portugais. I = *P. moorei*, zone à Polymorphum, Peniche. II = *P. moorei*, zone à Levisoni, Rabaçal. III = *P. jurensis*, zone à Levisoni *(pars)* à zone à Gradata, Rabaçal. IV = *P. bothenhamptonensis*, ensemble des populations de Saõ Giao(depuis la zone à Bonarellii jusqu'à la sous-zone à Reynesi de la zone à Speciosum). IVa = *P. bothenhamptonensis*, zone à Bonarellii, Saõ Giao. IV b = *P. bothenhamptonensis*, sous-zone à Speciosum, Saõ Giao. IVc = *P. bothenhamptonensis*, sous-zone à Reynesi, Saõ Giao. N = nombre de spécimens. L, I, E = longueur, largeur et épaisseur des coquilles. I/L et E/L = largeur et épaisseur relatives. ED/E = épaisseur de la valve brachiale rapportée à celle de la coquille. Im/L = situation de la largeur maximale, à partir du crochet. hs et ls = hauteur de l'uniplication frontale et largeur du sinus médian ventral. hs/ls et hs/L = hauteur de l'uniplication rapportée à la largeur du sinus et à la longueur des coquilles. Nvd et Nb = nombre de côtes sur la valve brachiale et sur le pli médian dorsal.

THE GENUS *PSEUDOGIBBIRHYNCHIA* (BRACHIOPODA, RHYNCHONELLACEA) FROM THE TOARCIAN OF PORTUGAL

YVES ALMÉRAS

	N	L	I	E	I / L	E / L	E_D / E	Im / L
I	18	15,1 (11,3 - 17,8)	16,4 (11,0 - 20,0)	9,3 (4,8 - 12,0)	1,08 (0,96 - 1,18)	0,61 (0,42 - 0,75)	0,56 (0,51 - 0,63)	0,62 (0,59 - 0,69)
II	7	13,0 (12,0 - 14,3)	14,6 (12,8 - 16,5)	10,0 (8,4 - 11,3)	1,11 (1,02 - 1,19)	0,77 (0,70 - 0,83)	0,57 (0,51 - 0,62)	0,60 (0,54 - 0,66)
III	10	11,8 (9,9 - 13,5)	12,2 (9,5 - 13,7)	7,7 (5,8 - 9,6)	1,03 (0,96 - 1,10)	0,65 (0,51 - 0,76)	0,55 (0,45 - 0,70)	0,59 (0,55 - 0,64)
IV	89	11,5 (6,9 - 16,8)	12,1 (6,9 - 18,8)	8,2 (3,2 - 14,3)	1,05 (0,87 - 1,26)	0,70 (0,46 - ,89)	0,53 (0,43 - 0,65)	0,61 (0,54 - 0,68)
IVa	17	12,5 (8,0 - 15,3)	13,2 (8,0 - 17,4)	9,5 (5,0 - 12,2)	1,05 (0,94 - 1,14)	0,76 (0,62 - 0,87)	0,53 (0,46 - 0,61)	0,61 (0,54 - 0,67)
IVb	17	11,5 (6,9 - 16,5)	12,3 (6,9 - 17,6)	8,4 (3,2 - 13,9)	1,06 (0,99 - 1,17)	0,71 (0,46 - 0,84)	0,57 (0,49 - 0,65)	0,59 (0,54 - 0,63)
IVc	55	11,2 (7,3 - 16,8)	11,7 (7,0 - 18,8)	7,8 (3,7 - 14,3)	1,04 (0,87 - 1,26)	0,68 (0,48 - 0,89)	0,52 (0,43 - 0,61)	0,62 (0,55 - 0,68)

	hs	ls	hs / ls	hs / L	U_1	U_2	Nvd	Nb
I	7,0 (4,0 - 9,6)	11,7 (8,5 - 14,5)	0,60 (0,44 - 0,87)	0,45 (0,34 - 0,60)	10,5 (9,3 - 11,5)	0,68 (0,64 - 0,84)	16,1 (12 - 20)	6,5 (5 - 8)
II	6,8 (5,7 - 8,0)	10,5 (7,9 - 12,6)	0,66 (0,51 - 0,86)	0,52 (0,42 - 0,64)	9,7 (9,1 - 10,5)	0,74 (0,67 - 0,83)	15,0 (13 - 18)	6,6 (6 - 8)
III	4,9 (2,2 - 7,1)	8,3 (5,7 - 10,2)	0,59 (0,32 - 0,81)	0,41 (0,19 - 0,56)	9,3 (8,0 - 10,5)	0,79 (0,72 - 0,89)	13,1 (12 - 14)	5,0 (4 - 6)
IV	5,5 (1,8 - 11,1)	8,8 (5,0 - 13,1)	0,61 (0,24 - 0,98)	0,45 (0,18 - 0,68)	9,8 (6,8 - 13,5)	0,82 (0,73 - 0,93)	11,4 (8 - 18)	4,8 (3 - 8)
IVa	6,3 (3,0 - 8,8)	9,2 (5,9 - 11,7)	0,68 (0,51 - 0,90)	0,50 (0,36 - 0,62)	10,1 (6,8 - 13,2)	0,80 (0,74 - 0,86)	12,3 (10 - 15)	4,8 (3 - 7)
IVb	5,7 (3,5 - 8,1)	8,9 (6,7 - 13,1)	0,64 (0,50 - 0,89)	0,47 (0,35 - 0,63)	9,7 (7,5 - 13,0)	0,81 (0,73 - 0,87)	11,4 (9 - 14)	4,6 (4 - 6)
IVc	5,1 (1,8 - 11,1)	8,6 (5,0 - 12,8)	0,58 (0,24 - 0,98)	0,42 (0,18 - 0,68)	9,7 (7,2 - 13,5)	0,83 (0,75 - 0,93)	11,2 (8 - 18)	4,9 (3 - 8)

coquilles. Le crochet aigu, subdressé à dressé, non crêté, surplombe l'umbo dorsal. Deux crochets seulement montrent de courtes crêtes peu marquées. Foramen submésothyride, ovale (12 ex.) ou circulaire (2 ex.). Plaques deltidiales toujours réunies. Les caractères internes (Figure 2) diffèrent de ceux de *P. moorei* du Dorset (Ager, 1962, fig. 67) par l'absence de collier pédonculaire, par des plaques deltidiales non dédoublées, par des lamelles dentaires parallèles, par la présence d'un court septalium ainsi que par des plaques cardinales horizontales, épaissies et séparées des crêtes internes des fossettes. Les plaques cardinales ne deviennent ventralement arquées que sur leur partie antérieure.

La zone à Polymorphum de Peniche a livré trois exemplaires de même morphologie (Figure 1: 4-6), mais dont l'uniplication frontale davantage développée (hs/L = 0,53 et 0,60) crée un pli médian dorsal large et aplati beaucoup plus élevé que chez *P. moorei*. Les coquilles [Figure 1: 5-6], sont identiques à *Rhynchonella fallax* (Deslonchamps, 1862, pl. 3, fig. 1-5) du Lias moyen (en fait Toarcien inférieur) du Calvados (France). Les caractères internes n'ont pu être recherchés vu le petit nombre d'exemplaires en notre possession. La forme du Carixien (zone à Jamesoni) du Somerset décrite par Ager (1958 et 1962, pl. 8, fig. 1) n'est pas l'espèce de Deslongchamps, mais probablement *Prionorhynchia greppini* (Oppel) (Ager, 1967, p. 163).

Pseudogibbirhynchia moorei se poursuit dans la moitié inférieure de la zone à Levisoni (équivalent téthysien de la zone à Serpentinus) de Rabaçal et de Zambujal de Alcaria où elle est représentée par des coquilles (Figure 1:7) un peu plus petites (L = 12 à 14 mm), plus épaisses (E/L = 0,77 contre 0,61) et à uniplication frontale un peu plus élevée. Leur morphologie correspond à celle des spécimens figurés par Davidson (1852, pl.15, figs.11-13). L'aplatissement postérieur subsiste à la valve brachiale. Certains exemplaires possèdent un sinus ventral bien délimité alors que le pli dorsal n'est guère surélevé par rapport aux parties latérales de la valve brachiale. Les caractères internes se rapprochent de ceux de *P. moorei* figurés par Ager (1962, fig. 67).

Pseudogibbirhynchia jurensis (Quenstedt, 1858) (Figure 1:8-12). *P. jurensis* succède à *P. moorei* vers le milieu de la zone à Levisoni. L'espèce est représentée par des exemplaires isolés. Son extension se poursuit dans la zone à Bifrons et deux coquilles ont aussi été récoltées dans la zone à Gradata de Rabaçal et à la base de la zone à Bonarellii de Zambujal de Alcaria. La morphologie est identique dans ces différents niveaux (Figure 1:8-10). Elle correspond à celle figurée par Quenstedt (1858, pl. 41, fig. 33; 1871, pl. 38, fig. 23) et par Ager (1962, pl. 10, fig. 1-2). Les coquilles sont plus petites (L = 9,9 à 13,5 mm contre 11,3 à 17,8 mm) et plus étroites (I/L = 1,03) que chez *P. moorei* alors qu'on observe une situation

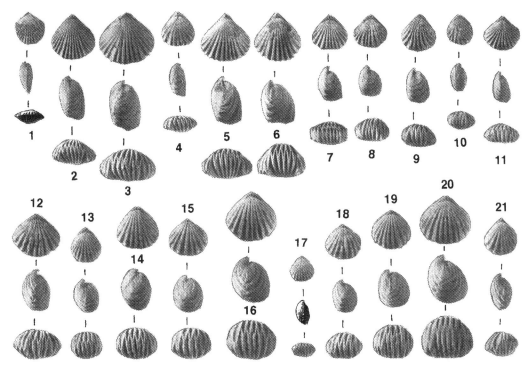

Figure 1. Les *Pseudogibbirhynchia* du Toarcien portugais. **1-3**, *Pseudogibbirhynchia moorei* (Davidson). Toarcien inférieur, zone à Polymorphum. Peniche (16c). Trois exemplaires illustrant la morphogenèse (FSL 307901 à 307903). **4-6**, *Pseudogibbirhynchia fallax* (Deslongchamps). Toarcien inférieur, zone à Polymorphum. Peniche (16c). Trois coquilles illustrant la morphogenèse (FSL 307904 à 307906). **7**, *Pseudogibbirhynchia moorei* (Davidson). Toarcien inférieur, zone à Levisoni. Rabaçal (Ra 19) (FSL 307907). **8-10**, *Pseudogibbirhynchia jurensis* (Quenstedt). 8, Toarcien inférieur, zone à Levisoni. Rabaçal (Ra 20) (FSL 307924). 9, Toarcien moyen, zone à Bifrons, sous-zone à Bifrons. Rabaçal (Ra 51) (FSL 307925). 10, Toarcien moyen, zone à Gradata. Rabaçal (entre Ra 59a et Ra 60) (FSL 307926). **11-12**, *Pseudogibbirhynchia jurensis* (Quenstedt). Toarcien moyen, zone à Bifrons. Zambujal de Alcaria (Z 36). Commissure frontale montrant des caractères ataviques de *P. moorei* (FSL 307927 et 307 928). **13-21**, *Pseudogibbirhynchia bothenhamptonensis* (Walker). 13-14, Toarcien supérieur, zone à Bonarellii. Sao Giao (SG 220 sup') (FSL 307949 et 307950). 15-16, Toarcien supérieur, zone à Speciosum, sous-zone à Speciosum. SaoGiao. 15, SG 221 base. 16, SG 225 sommet (FSL 307951 et 307952). 17-20, Toarcien supérieur, zone à Speciosum, sous-zone à Reynesi. Sao Giao (SG 120). Quatre exemplaires illustrant la morphogenèse (FSL 307953 à 307956). 21, Toarcien supérieur, zone à Speciosum, sous-zone à Reynesi. Sao Giao (SG 120). Coquille à commissure frontale asymétrique (FSL 307957). [scale c. x0.8]

inverse en Angleterre (Ager, 1958). Leur contour est subcirculaire. L'aplatissement postérieur fait défaut sur la valve brachiale. Les valves sont moins densément costées que chez *P. moorei* : 12 à 14 côtes de type *tetrahedra* à *grandis* dont 4 à 6 sur le pli dorsal. Contrairement à *P. moorei* un stade lisse est fréquemment présent au niveau du crochet et de l'umbo dorsal (60 % des spécimens). La naissance de côtes intercalaires s'effectue de la même manière et en mêmes proportions d'individus que chez *P. moorei*. Tous les exemplaires examinés sont uniplissés (hs/L = 0,19 à 0,56). L'uniplication frontale est arrondie et asymétrique dans 60 % des cas. Le crochet petit, non crêté, subdressé à dressé, surplombe l'umbo dorsal. Des crochets droits ont été observés chez deux coquilles longues de 10,5 et de 11,3 mm. Foramen relativement grand pour la taille des coquilles, submésothyride, toujours ovale (il est circulaire chez les exemplaires anglais décrits par Ager). Plaques deltidiales réunies ou séparées en proportions identiques d'individus. Les caractères internes correspondent à ceux de la coquille du Somerset figurés par Ager (1962, fig. 69).

Pseudogibbirhynchia bothenhamptonensis (Walker, 1892)

(Figure 1:13-21). Le genre *Pseudogibbirhynchia* rarement représenté dans la zone à Gradata et à la base de la zone à Bonarellii (quelques exemplaires de *P. jurensis*) va se développer avec *P. bothenhamptonensis* (Figure 1:13-21) au sommet de celle-ci (sous-zone à Fallaciosum) et au cours de la zone à Speciosum. L'apogée de cette espèce se situe dans la sous-zone à Reynesi de la zone à Speciosum (94 exemplaires sur 140 = 67,1 % des peuplements). *P. bothenhamptonensis* n'a pas été trouvé dans les derniers niveaux de la sous-zone à Reynesi. La morphologie des coquilles et la variabilité des caractères dimensionnels sont identiques dans les différents niveaux échantillonnés (Table 1). Dans la sous-zone à Reynesi, la moindre épaisseur des coquilles (E/L = 0,68 contre 0,71 et 0,76) et l'uniplication moins élevée (hs/L = 0,42 contre 0,47 et 0,50) s'expliquent par la plus grande fréquence des petits spécimens de 7 à 9 mm de longueur. La taille des coquilles est celle de *P. jurensis*, elle est inférieure à celle de *P. moorei*. Le contour est dans l'ensemble subcirculaire avec des coquilles un peu plus larges que longues (63 ex.), les coquilles plus longues que larges étant moins fréquentes (21 ex.). L'épaisseur est plus forte que chez *P. jurensis* et chez *P. moorei*, d'où l'aspect relativement gibbeux des spécimens. En

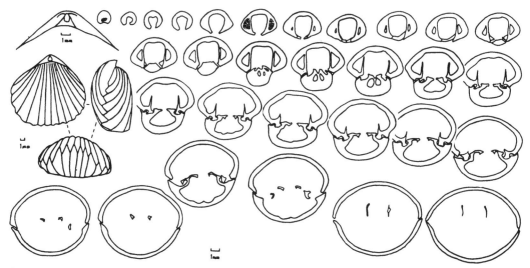

Figure 2. Caractères internes d'une coquille de *Pseudo-gibbirhynchia moorei* (Davidson). Toarcien inférieur, zone à Polymorphum. Peniche (16c). En haut et à gauche: représentation de la coquille sectionnée (FSL 307911) et caractères de son crochet et du foramen [spacing of sections from 0.1 to 0.4mm].

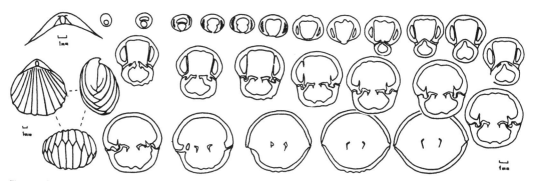

Figure 3. Caractères internes d'une coquille de *Pseudo-gibbirhynchia bothenhamptonensis* (Walker). Toarcien supérieur, zone à Speciosum, sous-zone à Reynesi (base). Saõ Giao (SG 120). En haut et à gauche: représentation de la coquille sectionnée (FSL 307972) et caractères de son crochet et du foramen.

raison de la plus grande épaisseur des coquilles, l'uniplication frontale est plus élevée (hs/L = 0,50 contre 0,41 chez *P. jurensis* et 0,45 chez *P. moorei*). Uniplications symétriques et asymétriques sont également représentées. Les trajectoires ontogénétiques moyennes de la croissance en largeur et en épaisseur des coquilles ainsi que du développement de l'uniplication ont été calculées dans la sous-zone à Reynesi de Saõ Giao. Elles montrent une forte croissance des rapports l/L, E/L et hs/L au cours de l'ontogenèse. Le contour des coquilles est subcirculaire chez les exemplaires jeunes (l/L = 0,98 pour L = 7 à 9 mm et l/L = 1,03 pour L = 9 à 12 mm), les spécimens s'élargissent ensuite (l/L = 1,07 pour L = 12 à 15 mm et l/L = 1,13 pour L = 15 à 17 mm). De même, l'épaisseur est modérée chez les sujets jeunes (E/L = 0,59 pour L = 7 à 9 mm) alors qu'elle augmente par la suite : E/L = 0,66 pour L = 9 à 12 mm ; 0,74 pour L = 12 à 15 mm et 0,84 pour L = 15 à 17 mm. La valve brachiale devient renflée chez les spécimens de 12 à 17 mm. La commissure frontale est rectimarginée chez 10 coquilles dont la longueur est comprise entre 7,3 et 10 mm. L'uniplication

se développe entre 6,8 et 13,5 mm, c'est-à-dire en moyenne au niveau du cinquième antérieur de la longueur. Son développement s'effectue en fonction de la croissance en épaisseur. Le plissement frontal est modéré entre 7 et 12 mm (hs/L = 0,31 et 0,37). Il s'accentue au-dessus de 12 mm de longueur (hs/L = 0,50 pour L = 12 à 15 mm et 0,59 pour L = 15 à 17 mm). Les spécimens, fig. 2. 17-20, illustrent l'ontogenèse de *P. bothenhamptonensis* dans la sous-zone à Reynesi. Le nombre de côtes, de type *tetrahedra* à *grandis*, est compris entre 8 et 18 tandis que 3 à 8 côtes se situent sur le pli dorsal. Par rapport à *P. jurensis*, le stade lisse sur le crochet et sur l'umbo dorsal est moins fréquent (14 ex. sur 89 = 15,7 % contre 60 %). Une ou deux côtes intercalaires ont été observées sur 26 spécimens.

Le crochet petit, aigu, non crêté, est dressé ou légèrement recourbé au-dessus de l'umbo dorsal. Les crochets subdressés sont moins fréquents que chez *P. moorei* et chez

P. jurensis. On observe donc un recourbement du crochet au cours de l'évolution du genre *Pseudogibbirhynchia*. Le foramen est ovale, il existe toutefois quelques foramens circulaires dans la sous-zone à Speciosum. A la différence de *P. jurensis*, les plaques deltidiales sont presque toujours séparées (78 ex.) alors que 7 coquilles seulement montrent des plaques deltidiales réunies.

Les caractères internes de *P. bothenhamptonensis* ont été recherchés chez différents spécimens collectés dans la moitié inférieure de la sous-zone à Reynesi de Saõ Giao. Ils sont concordants avec les coupes sériées (Figures 2-3) obtenues sur deux coquilles prélevées à la base de la sous-zone à Reynesi. Ils correspondent également aux caractères internes de *P. jurensis* figurés par Ager (1962, fig.69). La seule différence concerne le moindre épaississement des plaques deltidiales chez les exemplaires portugais. Les coupes sériées (Figure 3) montrent des épaississements secondaires qui oblitèrent les cavités latérales au niveau du crochet et qui renforcent les lamelles dentaires ainsi que les parois de la coquille.

Dans l'état actuel de nos connaissances, le sommet de la sous-zone à Reynesi marque la fin de l'évolution des *Pseudogibbirhynchia* dans le Toarcien portugais. En effet, exceptions faites de quelques brachiopodes micromorphes à Saõ Giao et de *Soaresirhynchia renzi* (Choffat) à Zambujal de Alcaria (Alméras, 1994), aucun brachiopode n'a été trouvé dans les zones à Meneghinii et à Aalensis du Toarcien supérieur.

RÉPARTITIONS STRATIGRAPHIQUES ET GÉOGRAPHIQUES

Pseudogibbirhynchia moorei (Davidson, 1852)
(Figure 1:1-7)

Rhynchonella moorei Davidson, 1852, p.82 ; Pl.15, figs.11-14. *Pseudogibbirhynchia moorei* (Davidson). AGER, 1962, p.110, fig.67 ; Pl. 9, figs.2-6 *Rhynchonella moorei* Davidson. CHOFFAT, 1880, p. 19, 20.; GARDET ET GÉRARD, 1946, p.23 ; Pl. 2, figs. 7-8. Pars *Rhynchonella moorei* Davidson. CHARLES, 1948, p. 82 ; ? Pl. 4, fig.8. *Pseudogibbirhynchia moorei* (Davidson). DELANCE, 1969, p. 19 ; Pl.B, fig.17.; ROUSSELLE, 1973, p.127; Pl.5, B, C. ? *Pseudogibbirhynchia moorei* (Davidson). GOY, 1974, p.766 ; Pl.110, fig.4; GOY & COMAS-RENGIFO, 1975, p.321; Pl.6, figs.9-10. *Pseudogibbirhynchia moorei* (Davidson). ALMÉRAS & MOULAN, 1982, p.284. Non *Pseudogibbirhynchia moorei* (Davidson). GAKOVIC & TCHOUMATCHENCO, 1994, p.15; Pl.I, figs.1, 4.

Répartition. Angleterre SW : base du Toarcien (Ager, 1962). Provence méridionale : zone à Serpentinus (moitié supérieure) et zone à Bifrons (jusqu'à la sous-zone à Bifrons) (Alméras & Moulan, 1982). Espagne, province de Lérida: Toarcien inférieur (Delance, 1969) et Cordillère ibérique: zone à Bifrons (Goy, 1974; Goy & Comas - Rengifo, 1975). Maroc. Moyen-Atlas septentrional: Toarcien inférieur (Gardet & Gérard, 1946). Haut-Atlas central et oriental: Toarcien moyen à supérieur (Rousselle, 1973). Madagascar (Thévenin, 1908). Portugal-Toarcien inférieur, zone à Polymorphum et moitié inférieure de la zone à Levisoni. Peniche (16c, 16 d), Rabaçal (Ra 8, 9, 11, 14, 19, 20), Porto de Moz (16E, avec *Nannirhynchia pygmoea*), Zambujal de Alcaria (Z 19, 20). Voie aussi Alméras (1994).

Pseudogibbirhynchia jurensis (Quenstedt, 1858)
(Figure 1: 8-12)

Terebratula jurensis QUENSTEDT, 1858, p. 287 ; Pl. 41, fig. 33 seule. Pars *Terebratula jurensis* Quenstedt. QUENSTEDT, 1871,

p.75; Pl.38, fig.23 seule [*non* fig.25= *Praemonticlarella schuleri* (OPPEL)]. *Pseudogibbirhynchia jurensis* (Quenstedt). AGER, 1962, p.112, figs.68-69; Pl.10, figs.1-2. *Rhynchonella jurensis* (Quenstedt). DUBAR, 1931, p.23; Pl.2, figs.13-14. *Pseudogibbirhynchia jurensis* (Quenstedt). ROUSSELLE, 1973, p.124, figs.3-4; GOY, 1974, p.765; Pl.110, figs.1-2.'*Rhynchonella*' *jurensis* (Quenstedt). GOY & COMAS-RENGIFO, p.322; Pl.6, fig.8. *Pseudogibbirhynchia jurensis* (Quenstedt). GAKOVIC & TCHOUMATCHENCO, 1994, p.17; Pl.I, figs. 2-7.

Répartition. Allemagne (Württemberg): Lias Z (Toarcien moyen) (Quenstedt, 1858, 1871). Angleterre (Oxfordshire): Lias supérieur, Communis Zone=Lias supérieur au-dessous de la zone à Jurense et excepté le Toarcien basal (Davidson, 1878 ; Ager, 1962). France: Mont d'Or lyonnais, Toarcien moyen et Toarcien supérieur basal (Dumortier, 1874, et Alméras, sous presse); Jura, zone à Bifrons et peut-être zone à Variabilis (Couches à *Coeloceras crassum* ; Haas, 1889 ; Ardèche et Var, Toarcien moyen (Dumortier, 1874) ; Quercy, zone à Bifrons. Espagne : Catalogne, Toarcien moyen avec *Homoeorhynchia meridionalis* (Dubar, 1931) et Cordillère ibérique, zone à Bifrons (Goy, 1974 ; Goy & Comas-Rengifo, 1975). Maroc: Haut-Atlas central et oriental : depuis le Toarcien à *Hildoceras* jusqu'à l'Aalénien à *Tmetoceras scissum* (Rousselle, 1973). Dinarides (NE Herzegovina): Toarcien inférieur (Gakovic & Tchoumatchenco, 1994). Portugal. - Toarcien inférieur, zone à Levisoni (moitié supérieure) jusqu'au sommet de la zone à Gradata du Toarcien moyen. Rabaçal (Ra 20, 24, 27, 30, 46, 47, 50, 51, 52, 53-54, 58e), Tomar (To 39 = sous-zone à Sublevisoni), Alvaiazere (Az 24 = sous-zone à Bifrons), Alvorge (passage sous-zone à Lusitanicum/sous-zone à Bifrons), Zambujal de Alcaria (Z 36 = zone à Bifrons) et Jamprestes (base de la zone à Bifrons). Un exemplaire à la base de la zone à Bonarellii (Z 39) à Zambujal de Alcaria.

Pseudogibbirhynchia bothenhamptonensis (Walker, 1892)
(Figure 1:13-21)

Rhynchonella jurensis QUENSTEDT var. *bothenhamptonensis* WALKER, 1892, p.442. *Stolmorhynchia bothenhamptonensis* Walker. BUCKMAN, 1917, Pl.13, fig.9. *Rhynchonella jurensis* QUENSTEDT var. *limata* DUBAR, 1931, p.24 ; Pl.2, figs.15-16. *Stolmorhynchia bothenhamptonensis* Walker. AGER, 1962, p. 112. *Stolmorhynchia limata* (Dubar). CALZADA, 1976, p.142, fig.1. *Rhynchonella jurensis* var. *bothenhamptonensis* Walker. ALMÉRAS, 1994, p.300.
Répartition- Angleterre(Dorset): Toarcien supérieur, Yeovilien, horizon à *Striatulum* (Buckman, 1917). Espagne (Catalogne): Toarcien supérieur (= *Rhynchonella limata* DUBAR ; Calzada, 1976). Portugal- Toarcien supérieur, zone à Bonarellii (moitié supérieure= sous-zone à Fallaciosum) à la zone à Speciosum, sous-zone à Reynesi (couches terminales exceptées). Apogée dans la sous-zone à Reynesi. Rabaçal (Ra 64, 65-66), Saõ Giao (SG 217, 220, 221, 224, 225 ; SG 120, 121, 124X, 124 et 1 m au-dessus de SG 124), Alvaiazere (Az 29 = zone à Speciosum), Zambujal de Alcaria (Z 39 = zone à Bonarellii) et Monte Alvão (zone à Speciosum).

Remerciements

La partie stratigraphique de ce travail a bénéficié des travaux, des nombreuses informations et déterminations d'ammonites fournies par R. Mouterde et S. Elmi (Lyon). Je remercie vivement R. Mouterde, C. Ruget ainsi que les Professeurs R.B. Rocha (Lisbonne), A.F. Soares (Coimbra) et tous nos amis portugais M.H. Henriques, J. Marques, J.C. Kullberg et L. Duarte pour leur aide sur le terrain. Cette publication a été facilitée par le personnel du Centre de la Terre, Université Lyon

I : photographies (N. Podevigne), manuscrit (M.T. Boillon), dessins et films (A. Armand et M. Le Hegarat). Elle a été réalisée dans le cadre de la convention d'échanges C.N.R.S./I.N.I.C. (Relations entre cadre stratigraphique, évolution séquentielle et paléotectonique du Jurassique portugais). Le matériel étudié est conservé dans les collections du Centre des Sciences de la Terre, Université Claude Bernard-Lyon, sous les numéros FSL 307901 à 307982.

REFERENCES

AGER, D.V. 1958, 1962, 1967. A monograph of the British Liassic Rhynchonellidae. Part II, III, IV. Palaeontographical Society Monographs, 112, 116, 121: 51-172.

___1965. Mesozoic and Cenozoic Rhynchonellacea, p. 597-H632. In, R.C. Moore (ed.) Treatise on Invertebrate Paleontology, Part H, Brachiopoda. Geological Society of America & University of Kansas Press, Lawrence.

ALMÉRAS, Y. 1994. Le genre Soaresirhynchia (Brachiopoda, Rhynchonellacea, Wellerellidae) dans le Toarcien du sous-bassin nord-lusitanien (Portugal). Documents des Laboratoires de Géologie de Lyon 130: 1-135.

___& MOULAN, G. 1982. Les Térébratulidés liasiques de Provence. Paléontologie, biostratigraphie, paléoécologie, phylogénie. Documents des Laboratoires de Géologie de Lyon 86: 1-365.

___, MOUTERDE, R., S. ELMI & R.B. ROCHA (1995). Le genre Nannirhynchia (Brachiopoda, Rhynchonellacea, Norellidae) dans le Toarcien portugais. Palaeontographica, Stuttgart, (A), 237 (1-4): 1-38.

BUCKMAN, S.S. 1917. The Brachiopoda of the Namyau Beds, Northern Shan States, Burma. Memoirs of the Geological Survey of India, Palaeontologia Indica III, 2: 1-299.

CALZADA, S. 1976. Sobre Stolmorhynchia limata (Dubar, 1931), braquiópodo del Toarciense catalan. Acta Geológica Hispánica, XI(5): 142-144.

CHARLES, R.P. 1948. Le Lias de la Basse Provence Occidentale. Etude paléontologique et paléobiologique. Bulletin du Museum d'Histoire naturelle de Marseille 19: 1-207.

CHOFFAT, P. 1880. Etude stratigraphique et paléontologique des terrains jurassiques du Portugal. 1ère Livraison : Le Lias et le Dogger au Nord du Tage. Travaux Géologiques du Portugal : 72 p.

DAVIDSON, T. 1852. A monograph of British oolitic and Liassic Brachiopoda, Part III, Palaeontographical Society Monographs 6: 65-100.

___1878. A monograph of the British fossil Brachiopoda. Supplement to the Jurassic and Triassic species. Palaeontographical Society Monographs 32 (2): 145-241.

DELANCE, J.H. 1969. Etude de quelques brachiopodes liasiques du Nord-Est de l'Espagne. Annales de Paléontologie, Invertébrés 55 (1): 1-44.

DESLONGCHAMPS, E.E. 1862. Etudes critiques sur des brachiopodes nouveaux ou peu connus. 1. Espèces du Lias. Bulletin de la Société Linéenne de Normandie 3 (7): 248-274.

DUBAR, G. 1931. Brachiopodes liasiques de Catalogne et des régions voisines. Butlleti de Instituto Catalana d'Historia Naturale 2, 31 (4): 103-180.

DUMORTIER, E. 1874. Etudes paléontologiques sur les dépôts jurassiques du bassin du Rhône. 4e partie, Lias supérieur, F, Savy éditions, Paris, 335 p.

GAKOVIC, M. & P. TCHOUMATCHENCO. 1994. Jurassic brachiopods from the Dinarides (NE Herzegovina). Geologica Balkanica 24 (3): 13-29.

GARDET, G. & C. GERARD. 1946. Contribution à l'étude paléontologique du Moyen Atlas septentrional (Lias inférieur à Bathonien). Notes et Mémoires du Service Géologique du Maroc 64: 1-88.

GOY, A. 1974. El Lias de la mitad norte de la Rama Castellana de la Cordillera Iberica. Thèse ès-Sciences Universidad Madrid, 940 p.

GOY, A. & M.J. COMAS-RENGIFO. 1975. Estratigrafia y Paleontologia del Jurásico de Ribarredonda (Guadalajara). Estudios Geologicos 31 (3-4): 297-339.

di GREGORIO, A. 1930. Monografia dei fossili liassici di Monte San Giuliano. Annales de Géologie et Paléontologie de Palerme 53: 40.

KAMYSCHAN, V.P. & L.I. BABANOVA. 1973. Les brachiopodes jurassiques moyens et supérieurs du Caucase nord-occidental et de Crimée. Université de Kharkov, 174 p.

LIKHAREV, B.K. & M.A. RZHONSNITSKAYA. 1956. Nadsemeistvo Rhynchonellacea Gray, 1848. In, Materialii dlya Paleontologii. Trudy Vsesoyuzno Nauchno-Issledovatelskogo Geologicheskii Institut (VSEGEI) 12: 53-61.

QUENSTEDT, F.A. 1858. Der Jura. Laupp, Tübingen, 842 p.

___1868-71. Petrefactenkunde Deutschlands, Band 2, Brachiopoden. Fuess, Tübingen, 748 p.

ROUSSELLE, L. 1973. Le genre Pseudogibbirhynchia (Rhynchonellacea) dans le Toarcien et l'Aalénien inférieur du Haut-Atlas central et oriental. Notes du Service Géologique du Maroc, 34 (254): 121-133.

THEVENIN, A. 1908. Paléontologie de Madagascar, 5, Fossiles liasiques. Annales de Paléontologie 3: 105-143.

WALKER, J.F. 1892. On Liassic Sections near Bridgport, Dorsetshire. Geological Magazine 3 (9): 437-443.

PERMIAN BRACHIOPODS FROM KARAKORUM (PAKISTAN)

LUCIA ANGIOLINI

Dipartimento di Scienze della Terra, Via Mangiagalli 34, 20133 Milano, Italy.

ABSTRACT- Four brachiopod assemblages, and one range zone assemblage, were collected in the Permian successions of northern Karakorum. The oldest association, from the Asselian-early Sakmarian Gircha Fm., has been named the *Trigonotreta lyonsensis-Punctospirifer afghanus* assemblage, and is characterised by low diversity and dominance of spiriferids. The second association, named the *Hunzina electa* assemblage, was collected in the Sakmarian Lupghar Fm., of the upper Hunza valley, and in the Lashkargaz Fm. of Baroghil and Lashkargaz (Chitral), and is characterized by higher diversity. The third, the *Orthotetina convergens-Aldina exilis* assemblage, occurs at the top of the Lashkargaz Fm. at Lashkargaz. The fourth assemblage, typified by *Waagenoconcha macrotuberculata* and *Callytharrella sinensis*, is from the Lashkargaz Fm. at Lashkargaz and Baroghil, and the Panjshah Fm. of the upper Hunza valley. The last assemblage is characterized by high diversity and abundance, and is dominated by productids. The highest, the *Stenocisma armenica - Chapursania tatianae* assemblage, occurs in the Panjshah Fm. These faunas are compared to others from Eurasia and Australia, and their paleogeographic distribution is related to Gondwana glaciation.

INTRODUCTION

Large collections of Permian brachiopods were assembled during four Italian expeditions (1986, 1991, 1992a, 1992b) on the Pakistan side of western Karakorum (Yarkhun valley, Baroghil Pass, and Karambar valley), and central Karakorum (upper Hunza valley, Chapursan valley, Abgarch valley, and Shimshal valley) (Figure 1). Permian brachiopods from the western Karakorum (Baroghil pass) were first collected by Hayden (1915), and described by Reed (1925). A small collection of Permian brachiopods from the upper Hunza valley was described by Fantini Sestini (1965a). Herein, five brachiopod assemblages are described from the Asselian to the Midian of western and central Karakorum, expanding previous knowledge of the faunas (Angiolini, 1994, 1995; Gaetani et al., 1995).

The Permian stratigraphy of Karakorum (Pakistan) has been extensively described by Gaetani et al. (1990), Zanchi & Gaetani (1994), and Gaetani et al. (1995). In the western termination of the Karakorum range, i.e. the Baroghil-Karambar sector, the Early Permian (Asselian-Sakmarian) is represented by a thick quartz-arenitic unit, the Gircha Formation. Over this follows a mixed carbonate-terrigenous succession, several 100m thick, named the Lashkargaz Formation, consisting of four members, spanning Early to Late Permian age (Sakmarian to ? Murgabian). At the top of the Lashkargaz Formation, a gap, roughly corresponding to the Murgabian, sealed by a thin terrigenous unit, the Gharil Formation, occurs. Above this was deposited a thick peritidal dolomitic unit, the Ailak Formation, of Late Permian to Triassic age (Gaetani et al., 1995).

In the central sector of the Karakorum (the upper Hunza Valley), the base of the succession is represented by the Gircha Formation, which consists of arkose to quartzarenites

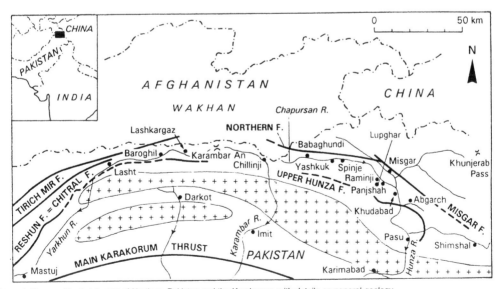

Figure 1. Geographic sketch map of Northern Pakistan and the Karakorum, with details on general geology.

13

with late Asselian and early Sakmarian brachiopods. The overlying Lupghar Formation consists of hybrid sandstones and bioclastic-oolitic limestones, topped by peritidal dolomites. Their age is Sakmarian to ? early Artinskian. At the top, a gap occurs, probably corresponding to the late Artinskian-Bolorian, which is sealed by sandstones, marls and limestones of the Panjshah Formation, of Kubergandian to Midian age (Zanchi & Gaetani, 1994; Gaetani et al., 1995). Above this, the Kundil Formation consists of cherty limestones with late Midian and Dzhulfian conodonts. Finally comes the Wirokhun Formation, consisting of black shales and marly limestones spanning the Early Triassic, Dzhulfian-Spathian, time interval and the Permo-Triassic boundary (Gaetani et al., 1995).

Five brachiopod assemblages were collected from the base to the top of this succession. The **first** sampled the Asselian-early Sakmarian Gircha Fm. of the upper Hunza Valley; the **second** covered the middle and upper part of Sakmarian Lupghar Fm. [Mb. 1] of the upper Hunza Valley, and the Sakmarian Lashkargaz Fm. [Mb. 1] of the Baroghil-Lashkargaz area. The **third** assemblage, Bolorian in age, was collected in two levels at the top of Lashkargaz Fm. [Mb. 2] of Lashkargaz in the upper Yarkhun valley. The **fourth** association, Kubergandian in age, characterizes the Panjshah Fm. [Mb. 1] of the Upper Hunza Valley, and the Lashkargaz Fm. [Mb. 4] of the Baroghil-Lashkargaz area. The **fifth** association was collected in the Panjshah Fm. [Mb. 2] of the upper Hunza Valley and is of Late Murgabian-Midian age.

COMPOSITION OF THE ASSEMBLAGES

The lowest brachiopod assemblage is characterized by the presence of Lyonia, Rhynchopora, Punctospirifer afghanus Termier, Termier, Lapparent & Marin, Trigonotreta lyonsensis Archbold & Thomas, Trigonotreta stokesi Koenig, Spirelytha petaliformis (Pavlova), and Tomiopsis cf. bazardarensis (Grunt). The bivalves Etheripecten , Deltopecten, Eurydesma, and encrusting bryozoans were also present. The assemblage is dominated by the spiriferids Trigonotreta stokesi, T. lyonsensis, and Spirelytha petaliformis. The two species of Trigonotreta represent 80% of the assemblage. The other two spiriferids, Punctospirifer afghanus and Torniopsis cf. bazardarensis are subordinate, whereas rhynchonellids and productids are represented only by the genera Rhynchopora (1.4% of the total assemblages), and Lyonia (3%). This association has been found in fine arkoses (Gaetani et al., 1995) at Spinji and above the Yashkuk Glacier, and in a bioclastic storm lag in fine subarkose in the Ashtigar section. The low diversity of the brachiopod assemblage and the absence of a differentiated macro- and microfauna is probably due to deposition in a prodelta to storm-dominated marine environment during cold climatic conditions, following Gondwanan glaciation (Gaetani et al., 1995).

The second assemblage is dominated by Hunzina electa Angiolini and Trigonotreta paucicostulata (Reed). Specimens of Derbyia cf. baroghilensis Reed, Permochonetes pamiricus Afanaseva, Reticulatia , Globiella cf. rossiae (Fantini Sestini), Costatumulus irwinensis (Archbold), Cleiothyridina ailakensis Reed, Cleiothyridina aff. semiconcava (Waagen), Cleiothyridina , C. nagmargensis (Bion), Hunzina tenuisulcata (Merla) and Gjelispinifera aff. cristata (Schlotheim), are also present. The bivalve Parallelodon desioi Fantini Sestini occurs with the brachiopods. This assemblage is dominated again by spiriferids and athyrids, but chonetids and productids are also present, enhancing the higher diversity of the association. However, the dominating species is from a new genus which represents more than 50% of the assemblage (numbering

about 250 specimens). This new form is absent only in the Shimshal Valley, where it is replaced in the same stratigraphic position by Hunzina tenuisulcata. The lithology is very variable, consisting of shales with limestone lenses in the Baroghil East section, marly limestones in the Lashkargaz, Khudabad East and Abgarch West sections, and arenaceous calcarenites in the Lupghar section. Thus ecological control seems not to affect the range of this assemblage, present over a large area in different depth, substrate and hydrodynamic settings on an open marine shelf. The climate seems to be warmer, or more favourable for carbonate deposition, than for the preceeding fauna, as testified by higher taxonomic diversity of the brachiopod assemblage, and by the increase in limestones, replacing sandstones and shales. The microfacies also shows increasing diversity, of bryozoans, crinoids, gastropods, bivalves, small foraminifers, with rare fragmentary fusulinids at the top (Gaetani et al., 1995).

The third assemblage is characterized by the occurrence of Orthothetina convergens Merla, Orthotichia, Derbyia grandis Waagen, Neochonetes costellata Angiolini, Neochonetes (Sommeriella) baroghilensis (Reed), Paramesolobus sinuosus (Schellwien), Marginifera andreai Angiolini, Retimarginifera praelecta (Reed), Magniplicatina cf. inassueta (Reed), Compressoproductus, and Aldina exilis Angiolini. This assemblage is numerically dominated by the productid Marginifera andreai (53%), and the rhynchonellid Aldina exilis (27%). In particular, Magniplicatina andreai forms significant clusters at the lowest levels, whereas Aldina exilis continues upward for about 45m. The chonetids represent 11% of the association and proliferate only in the uppermost levels (i.e. 45m above the M. andreai cluster). This assemblage has been found in marly limestones only in the Lashkargaz section. The absence of this fauna in the upper Hunza Valley area is probably connected to an erosional gap at the top of the Lupghar Fm. there. The composition of the assemblage, and the host lithofacies, indicate deposition on a muddy, mobile substrate under warm climates. Besides brachiopods, tabulate corals, crinoids, bryozoans, bivalves, rare dasycladaceans, small foraminifers, fusulinids, oncoids and coated grains also occur, suggesting a carbonate ramp environment periodically inundated by clay deposition (Gaetani et al., 1995).

The fourth assemblage (360 specimens) shows the highest taxonomic diversity. It is characterised by Waagenochoncha (Gruntoconcha) macrotuberculata Angiolini, Callytharrella sinensis (Sun), Enteletes, Derbyia grandis Waagen, Orthothetina convergens Merla, Neochonetes costellata Angiolini, N. (Sommeriella) baroghilensis (Reed), N. (S.) vialis (Reed), Paramesolobus sinuosus (Schellwien), Reticulatia chitralis Angiolini, Retimarginifera praelecta (Reed), Transennatia reedi Angiolini, Echinoconchus, Magniplicatina johannis Angiolini, M. vindicata (Reed), and Permophrycodothyris. Thie fauna is dominated by productids (66.7%), especially by big dictyoclostids (30%), attesting, with high taxonomic diversity, to warm climates. Chonetids are also abundant, representing 20% of the assemblage, whereas spiriferids are rare, represented by a single specimen. Transennatia reedi forms a cluster at the top of the unit. Lithic control on this assemblage is limited: it has been collected in cherty bioclastic limestones and in marly limestones in the Karakorum. The associated fauna is very diversified, consisting of corals, crinoids, conodonts, bryozoans, bivalves, gastropods, algae, small foraminifers, abundant fusulinids, poriferans, Tubiphytes, oncoids and coated grains (Gaetani et al., 1995). The diversity of the fauna suggests warm, tropical, equatorial climates. This conclusion is supported by increasing mineralogical stability of detritus (Gaetani et al., 1995).

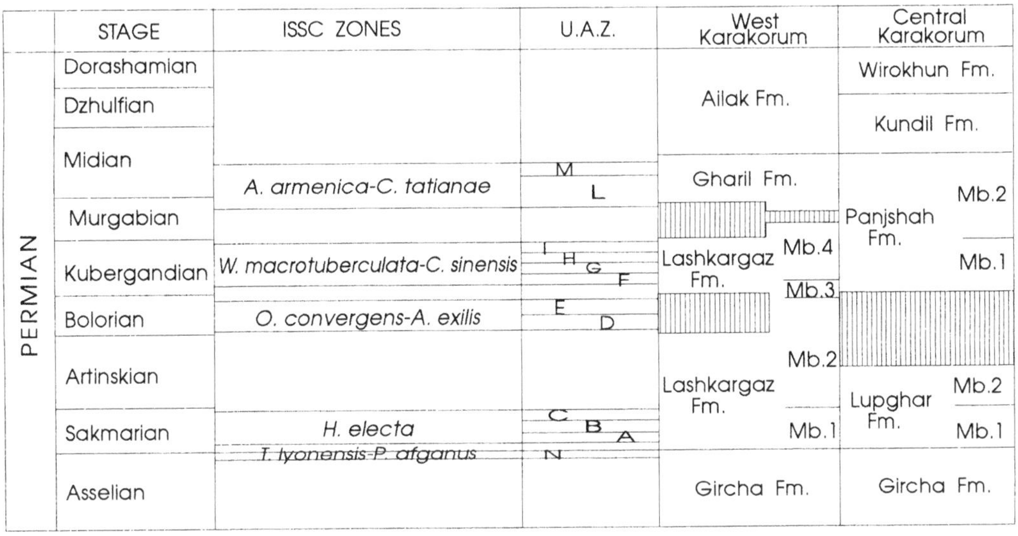

	STAGE	ISSC ZONES	U.A.Z.	West Karakorum	Central Karakorum
P E R M I A N	Dorashamian			Ailak Fm.	Wirokhun Fm.
	Dzhulfian				Kundil Fm.
	Midian	A. armenica-C. tatianae	M / L	Gharil Fm.	Mb.2 (Panjshah Fm.)
	Murgabian			Panjshah Fm.	
	Kubergandian	W. macrotuberculata-C. sinensis	I / H / G / F	Lashkargaz Fm. Mb.4 / Mb.3	Mb.1
	Bolorian	O. convergens-A. exilis	E / D	Mb.2	
	Artinskian			Lashkargaz Fm.	Lupghar Fm. Mb.2
	Sakmarian	H. electa / T. lyonensis-P. afganus	C / B / A / N	Mb.1	Mb.1
	Asselian			Gircha Fm.	Gircha Fm.

Figure 2. Brachiopod biozones in the Permian successions of northern Karakorum. UAZ = biochronozones established using the Unitary Associations Method of Guex (1991). Time scale from Ross et al. (1994).

The highest Permian assemblage of the Pakistani side of the western and central Karakorum is characterised by *Retimarginifera gaetanii* Angiolini, *Magniplicatina*, *Compressoproductus* cf. *mongolicus* (Diener), *Lirellaria*, *Stenoscisma armenica* Sokolskaja, *Martinia*, *Tiramnia tschernyschewi* (Grunt), *Martiniopsis*, and *Chapursania tatianae* Angiolini. It is a low diversity fauna, dominated by new spiriferids (67% of the assemblage). Rhynchonellids are more abundant than in earlier assemblages. This endemic fauna is present in marly limestones only in the upper Hunza valley area, representing the last brachiopod assemblage before 'drowning' of the succession. Its absence in the Baroghil-Lashkargaz sections may be due to a disconformity at the base of the Gharil Fm. Low diversity, together with the appearance of new taxa (one new genus and one new species plus 3 new species here not listed), suggest deposition in muddy basins with cooler and deeper environments. The abundance of crinoids and solitary rugosans supports this concept (Gaetani et al., 1995).

BIOSTRATIGRAPHY

These five brachiopod assemblages from the western and central Karakorum include diagnostic species restricted to each time interval. The successions of faunal assemblages seem to lack local environmental control. The same faunas occur in different lithofacies in the same stratigraphic position, and different consecutive assemblages have been recognized vertically in homogeneous lithofacies. Each assemblage zone is characterised by at least two prominent index species and is bounded by barren intervals.

On this basis, four assemblage biozones and one range zone (Salvador, 1994) have been defined in the Permian successions of Karakorum. Following the Unitary Association Method of Guex (1991) and Savary & Guex (1991), a more detailed biozonation has been attempted, based on 11 biochronozones (Figure 2). The lowest assemblage has been defined as the *T. lyonensis-P. afganus* assemblage zone from the two most characteristic species (Angiolini, 1994, 1995) and this has been recognized in the middle Gircha Fm. in

the Ashtigar valley at Spinje, and above the Yashkuk glacier (Gaetani et al., 1995). The boundaries of this zone coincide with two barren intervals. This assemblage may also be interpreted as a Unitary Association (UA5), which corresponds to biochronozone N.

The age of the *Trigonotreta lyonsensis -Punctospirifer afghanus* Assemblage Zone is late Asselian and early Sakmarian (Angiolini, 1994, 1995), as inferred by the known range of the following forms: *Rhynchopora* and *Lyonia* [genera widespread in the Asselian-early Sakmarian of W Australia: Archbold, 1983, 1992], *P. afghanus* [from the Asselian of Wardak: Termier et al., 1974], *T. lyonsensis* [from the Asselian-early Sakmarian of the Carnarvon Basin: Archbold & Thomas, 1986], and *T. stokesi* [from the Tamarian, i.e. Asselian-early Sakmarian, of Tasmania: Clarke 1979, 1990; Archbold & Dickins, 1989; Archbold, 1992]. Also the bivalve association (*Eurydesma* - like) and *Tomiopsis bazardarensis* attest to this age. The latter species was originally collected from the boulder member at the base of the Tashkazyk Fm. in SE Pamir (Grunt, 1993; Grunt & Novikov, 1993), where the fauna is similar to Karakorum's, except for *Permochonetes pamiricus*, which in the Karakorum occurs higher. The assemblage at the top of the Tashkazyk Fm. is very similar to the *Hunzina electa* fauna (Angiolini, 1995).

The second zone of Karakorum has been defined as the *Hunzina electa* Range Zone (Angiolini, 1995), which occurs in the Lupghar Fm. [Mb. 1] at Lupghar, Khudabad, Abgarch valley and Shimshal valley, and in the Lashkargaz Fm. [Mb. 1] of Lashkargaz and Baroghil (Angiolini, 1995). This 65m thick zone represents strata with *H. electa* Angiolini, and its boundary marks the limits of the species in each section.

Using the method of Guex (1991), four Unitary Associations (UA's) have been recognised with the Biograph 2.02 program (Savary & Guex, 1991). The lowest Unitary Association (UA1) is characterised by *Cyrtella* cf. *nagmargensis*. The second, UA2, is characterised by *Globiella* cf. *rossiae* and *Spirigerella*. UA3 is marked by *Cleiothyridina*, whereas UA4 has *Reticulatia*. *Hunzina electa* and *Trigonotreta paucicostulata* occur throughout the UA's, whereas *Permochonetes pamiricus* and *Cleiothyridina ailakensis* occur in UA2 and UA3. Three biochronozones may thus be recognised (Angiolini, 1995): (1) Zone C = UA4; (2) Zone B = union of UA2 and UA3; (3) Zone A = UA1. The age of the *Hunzina electa* Zone is Sakmarian, as suggested by the occurrence of *Permochonetes pamiricus* and *Cleiothyridina ailakensis* at the top of the Tashkazyk Fm. in Pamir (Sakmarian-early Artinskian in Grunt & Dmitriev, 1973; Afanaseva, 1977; Grunt & Novikov, 1994), *Globiella rossiae* in the late Sakmarian of Central Elburz (Fantini Sestini, 1966). *Trigonotreta paucicostulata* occurs in the Sakmarian of Central Elburz (Fantini Sestini, 1966), in the Sakmarian-early Artinskian of Pamir (Grunt & Dmitriev, 1973; Grunt & Novikov, 1994), and in the Asselian-early Sakmarian of Wardak (Termier et al., 1974). *Hunzina tenuisulcata* occurs in the Early Permian of NE Karakorum (Merla, 1934), and Shaksgam valley (Fantini Sestini, 1965b). *Hunzina* is also present in the Sakmarian middle part of the Mushirebuca Group of Gegyai County, NW Tibet (Sun, 1991). The Sakmarian age of this zone is also supported by the conodont *Anchignathodus paralautus* Orchard (Gaetani et al., 1995), about 4 m below the brachiopod assemblage. This association has three genera (*Hunzina*, *Costatumulus*, and *Cyrtella*) in common with the Sakmarian association of the Callytharra Fm. (W Australia). Finally Sakmarian-early Artinskian fusulinids, dominated by the genus *Pseudofusulina*, have been found at the base of the second member of the Lupghar Fm., above the brachiopod assemblage (Gaetani et al., 1995).

The third Assemblage Zone is the *Orthotetina convergens - Aldina exilis* biozone, occurring at the top of the Lashkargaz Fm. (Gaetani et al., 1995). The lower boundary is placed at the lowermost occurrence of *Magniplicatina inassueta* , whereas the upper boundary is at the highest occurrence of *Aldina exilis* and *Orthotetina convergens*. The thickness of this zone is 45m. Two UA's have been recognized. The lowest, UA6, has *Marginifera andreai* and *Compressoproductus*, whereas UA7 has *A. exilis* with *O. convergens*. These UA's correspond to biochronozones D and E. The age of the *A. exilis* assemblage zone is Bolorian, from the fusulinids *Pseudofusulina norikurensis krafftiformis* and *Darvazites* cf. *zulumartensis* (Gaetani et al., 1995).

The fourth assemblage zone has been named the *Waagenoconcha (Gruntoconcha) macrotuberculata - Callytharrella sinensis* Assemblage Zone. It has been found in the Lashkargaz Fm. [Mb. 4] of Lashkargaz and Baroghil and in the Panjshah Fm. [Mb. 1] of Chapursan valley (fig. 9, 10, 15, in Gaetani et al., 1995). The lower boundary is placed at the lowermost occurrence of *C. sinensis*, *Reticulatia chitralis*, and *Magniplicatina johannis*, whereas the upper boundary is at the highest occurrence of *T. reedi*. The maximum thickness of this zone is 55m. Five UA's have been identified. The lowest, UA8, has *Neochonetes vialis*. The second, UA9, is marked by *Derbyia grandis*. UA10 has *Enteletes*, whereas UA11 has *Waagenoconcha macrotuberculata*. Finally, UA12 sees *Transennatia reedi*. *Magniplicatina johannis* and *Waagenoconcha* occur through UA8 to UA11, whereas *Reticulatia chitralis* occurs in UA8 and UA9. Four biochronozones are recognized (Angiolini, in prep.): (1) Zone I = UA12; (2) Zone H = UA11; (3) Zone G = UA10 ; (4) Zone F = union of UA8 and UA9. The age of the *W. macrotuberculata - C. sinensis* Assemblage Zone is Kubergandian as demonstrated by *Callytharrella sinensis* in the Kubergandian Tunlonggongba Fm. of NW Tibet (Sun, 1983), by the associated fusulinids *Parafusulina* (*Parafusulina*) and *Parafusulina* (*Skinnerella*), and by conodonts (Gaetani et al., 1995).

The highest biozone is the *Stenocisma armenica - Chapursania*

tatianae Assemblage Zone, in the Panjshah Fm. [Mb. 2] of Chapursan valley (Gaetani et al., 1995). The boundaries of this zone coincide with two barren intervals and the thickness is c.55m. Two UA's are evident: the lowest, UA13, has *Tiramnia tschernyschewi* and *Martiniopsis*, whereas in UA14 *Magniplicatina* and *Compressoproductus* cf. *mongolicus* occur. These UA's correspond to biochronozones L and M. The age of this zone is probably Midian, as suggested by *C.* cf. *mongolicus* from Chitichun (Diener, 1897) and the Dzhulfian of SE Pamir (Grunt & Dmitriev, 1973), and *Stenocisma armenica* from the Dzhulfian of SE Pamir (Grunt & Dmitriev, 1973) and Transcaucasia (Ruzhenzev & Sarycheva, 1965). This isconfirmed by the occurrence of Midian Schubertellidae and Murgabian-Midian conodonts (Gaetani et al., 1995).

The reproducibility of the biozones of Karakorum is reasonable, especially the *Hunzina electa - Cleiothyridina ailakensis* and *Waagenoconcha (Gruntoconcha) macrotuberculata - Callytharrella sinensis* zones. Both have been recognized in sections 200 km apart.

CONCLUSIONS

The brachiopod faunas of the Karakorum produce a biozonation, refined by using the Unitary Association method of Guex (1991). Established biozones are local zones recognisable in the Permian successions of western and central Karakorum.

Comparison of the Karakorum faunas with Cimmerian and Gondawanian faunas (Shi & Archbold, 1993; Shi et al., 1995), and the evolution of these assemblages, show that the Asselian-Sakmarian faunas of central Afghanistan, Karakorum, SE Pamir and NW Tibet are similar to the Gondwana faunas of W Australia. During earliest Permian the cold water Gondwana fauna is widespread from W Australia to the Karakorum and central Afghanistan. The wide distribution of the cold Gondwana Province can be explained by Gondwana glaciation (Carboniferous: Tastubian), and by proximity of the mega-Lhasa continent to the Gondwana margin. In Bolorian-Kubergandian time, the faunas from N Iran to NW Tibet and to the Shan-Thai block are characterised by a mixed fauna (the Cimmerian fauna of Archbold, 1983; 'transitional fauna' in Shi et al.,1995), with Gondwana and Tethys, and endemic taxa. This transitional fauna can be explained by climatic warming following Gondwana glaciation and with northward movement of the Cimmerian blocks (mega-Lhasa). In the Late Permian (Midian-Dorashamian), the Cimmerian Province is characterized by Tethyan genera with endemics. This may be due to general warming and to progressive northward movement of the mega-Lhasa plate.

Paleogeographical data suggest proximity of the mega-Lhasa block to the Gondwana margin during the Asselian-Sakmarian, and its north-ward drift from Artinskian-Bolorian time to the Late Permian, in conjunction with climatic warming following Late Carboniferous (Tastubian) glaciation.

REFERENCES

AFANASEVA, G.A. 1977. *Permochonetes* gen. nov. (Brachiopoda) of the Permian of the southeast Pamir. Paleontologicheskii Zhurnal 1977 (1):147-151.
ANGIOLINI, L. 1994. I brachiopodi permiani del Karakorum: sistematica e biostratigrafia. PhD Thesis, Università di Milano, Milano, 204 p.
___(in press). Permian brachiopods from Karakorum: part 1. Rivista Italiano Paleontologia Stratigrafia.
ARCHBOLD, N.W. 1983. Permian marine invertebrate

provinces of the Gondwanan Realm. Alcheringa 7:59-73.

___1992. A zonation of the Permian brachiopod faunas of western Australia. Gondwana 8: Assembly, Evolution and Dispersal. A.A. Balkema, Rotterdam.

___& J.M. DICKINS. 1989. Australian Phanerozoic timescales, 6, Permian. A standard for the Permian System in Australia. Records Bureau Mineral Resources, Geology Geophysics 36:1-17.

CLARKE, M.J. 1979. The Tasmanian Permian spiriferid brachiopods *Trigonotreta stokesi*, Koenig 1825, *Grantonia hobartensis*, Brown, 1953 and *Spirifer tasmaniensis*, Morris 1845. Journal of Paleontology 53(1):197-207.

___1990. Late Palaeozoic (Tamarian; Late Carboniferous-Early Permian) cold water brachiopods from Tasmania. Alcheringa 14: 53-76.

DIENER, C. 1897. The Permo-Carboniferous fauna of Chitichun Palaeontologica Indica, XV (1, 3):1-105.

FANTINI SESTINI, N. 1965a. Permian fossils of the Upper Hunza Valley. In, A. Desio (ed.), Italian expeditions to the Karakorum (K2) and Hindu Kush. IV Paleontology, Zoology, Botany 1 (1):135-148.

___1965b. Permian fossils of Shaksgam Valley. In, A. Desio (ed.), *ibid.*: 149-192.

___1966. The Geology of the Upper Djadjerud and Lar Valleys (N Iran), II Paleontology, Brachiopods from the Geirud Fm., Member D (Lower Permian). Rivista Italiano Paleontologia Stratigrafia 72, 1: 9-50.

GAETANI, M., L. ANGIOLINI, E. GARZANTI, F. JADOUL, E.V. LEVEN, A. NICORA & D. SCIUNNACH. (in press). Permian Stratigraphy in the Northern Karakorum.

GAETANI, M., E. GARZANTI, F. JADOUL, A. NICORA, A. TINTORI, M. PASINI & A.K. SABIR. 1990. The north Karakorum side of the Central Asia geopuzzle. Bulletin Geological Society America 102: 54-62.

GRUNT, T. A. 1993. New spiriferid brachiopods from the Lower Permian of southeastern Pamir. Palaeontologicheskii Zhurnal 1993(4):125-130.

___& V. Yu. DMITRIEV. 1973. Permian Brachiopoda of the Pamir. Akademiya Nauk SSSR, Trudy Palaeontologicheskii Institut 136:1-209.

___& V.P. NOVIKOV. 1994. Biostratigraphy and biogeography of the Early Permian in the southeastern Pamirs. Stratigraphy and Geological Correlation 2 (4): 331-339.

GUEX, J. 1991. Biochronological correlations. Springer Verlag, Stuttgart, 250 p.

HAYDEN, H. 1915. Notes on the geology of Chitral, Gilgit and the Pamirs. Records India Geological Survey 45: 271-335.

MERLA, G. 1934. Fossili antracolitici del Caracorum. Spedizione Italiana de Filippi nell'Himalaya, Caracorum e Turchestan Cinese II (5):99-319.

REED, F.R.C. 1925. Upper Carboniferous fossils from Chitral and the Pamirs. Palaeontologica Indica 6 (4):1-134

RHUZENZEV, V.E. & T.G. SARYCHEVA. 1965. Evolution and succession of marine organisms at the Permo-Triassic boundary. Akademiya Nauk SSSR, Trudy Paleontolo-gicheskii Institut 108: 1-430.

ROSS, C., A. BAUD & M. MENNING. 1994. Tentative project Pangea Time Scale. In, G.D. Klein (ed.). Pangea: paleoclimate, tectonics, and sedimentation during accretion, zenith, and breakup of a supercontinent. Geological Society Amererica Special Paper 288:10.

SALVADOR, A. (ed.) 1994. International Stratigraphic Guide. A. Salvador, Second Edition.

SAVARY, J. & J. GUEX. 1991. Biograph: un nouveau programme de construction des correlations biochrono-logiques basées sur les associations unitaires. Bulletin Société Vaud, Sciences Naturelles 80(3):317-340.

SHI, G.R. & N.W. ARCHBOLD. 1993. Distribution of Asselian to Tastubian (Early Permian) circum-Pacific brachiopod faunas.

Memoirs Association Australasian Palaeontologists 15:343-351.

___,___ & L.P. ZHAN. 1995. Distribution and characteristics of mixed (transitional) mid-Permian (late Artinskian-Ufimian) marine faunas in Asia and general implications. Palaeogeography, Palaeoclimatology. Palaeoecology 114: 241-271.

SUN, TE 1983. Early Permian new genera and species of brachiopod fauna in Rutug Duoma Area, Xizang (Tibet), China. Earth Science Journal Wuhan College Geology 1(19): 119-123.

SUN, DONGLI 1991. Permian (Sakmarian-Artinskian) fauna from Gegyai county, northwestern Xizang (Tibet), and its biogeographic significance. In, Sun, D. et al. (eds.). Stratigraphy and paleontology of Permian, Jurassic and Cretaceous of the Rutog Region, Xizang (Tibet). Nanjing University Press.

TERMIER, G., H. TERMIER, A.F. de LAPPARENT & P. MARIN. 1974. Monographie du Permo-Carbonifère de Wardak (Afghanistan Central). Documents Laboratoires Géologie, Faculté Sciences, Lyon 2:1-167.

ZANCHI, A. & M. GAETANI. 1994. Introduction to the geological map of the North Karakorum Terrain from the Chapursan Valley to the Shimshal Pass (and relative map). Rivista Italiano Paleontologia Stratigrafia 100(1): 125-136.

PALEOBIOGEOGRAPHY OF AUSTRALIAN PERMIAN BRACHIOPOD FAUNAS

N.W. ARCHBOLD

School of Aquatic Science and Natural Resources Management, Rusden Campus,
Deakin University, Clayton, Victoria 3168, Australia

ABSTRACT- Widespread pan-Gondwanan marine transgressions progressively flooded the Australian segment of Gondwana with the receding of Asselian glaciation. Brachiopod faunas invaded the basins of Irian Jaya, western Australia (Perth, Carnarvon, Canning, Bonaparte basins), eastern Australia (Tasmania, Sydney and Bowen Basins) and New Zealand. These Late Asselian-Tastubian faunas were of a wider eastern Gondwanan aspect (Indoralian Province), with widely shared genera such as *Lyonia*, *Bandoproductus*, *Cyrtella*, *Tomiopsis* and *Trigonotreta*. Associated sedimentary sequences are indicative of glaciation or ice rafting. During the Late Sakmarian (Sterlitamakian), two distinct provinces (the Westralian and the Austrazean) became established, separating the faunas of the western basins from those of the east. These provinces persisted for the remainder of the Permian Period. Westralian faunas show significant endemic development of *Neochonetes*, finely spinose *Taeniothaerus*, *Neospirifer*, *Fusispirifer*, *Crassispirifer*, and *Imperiospira*. Cimmerian genera, such as *Demonedys*, *Comuquia*, *Dyschrestia*, *Retimarginifera*, *Spirelytha*, *Spirigerella* and *Stenoscisma*, invaded the Westralian Basins during times of warmer temperatures. Austrazean faunas of eastern Australia and New Zealand exhibit endemic development of *Echinalosia*, *Wyndhamia*, *Pseudostrophalosia*, *Terrakea*, *Magniplicatina* and the spiriferid *Sulciplica*. Genera of the Ingelarellidae show remarkable radiation within the Austrazean Province and include *Tomiopsis*, with over 40 species, *Tabellina*, *Homevalaria*, *Notospirifer*, *Farmerella*, *Glendonia*, *Kelsovia* and *Birchsella*. Only *Tomiopsis* is reliably known in Permian faunas outside the Austrazean Province.

INTRODUCTION

Permian brachiopod faunas from Australia demonstrate a striking contrast in generic and specific composition on either side (western and eastern) of the continent. Australia's palaeogeographical position during the Permian as a large north-eastern extension of Gondwana with warmer water currents affecting the western margin, and cooler water currents affecting the eastern margin throughout most of the Permian (Archbold & Shi, in press), resulted in strikingly different brachiopod faunas on each side of the continent. For this review, the background data for quantitative studies of provincialism is discussed in a series of studies presented by Shi & Archbold (1993; 1995a,b; in press); and Archbold & Shi (1995, in press), and will not be repeated here. Faunal province names (specifically Cimmerian, Indoralian, Westralian and Austrazean) are fully discussed in those studies. This review concentrates on the qualitative data on which those studies rest in terms of the generic evolution and migration history of Australian faunas.

Australian Permian marine faunas are extensively developed in the intracratonic basins of western Australia and the marginal basins of eastern Australia (Figure 1). A discussion of the different tectonic settings of Australian Permian sedimentary basins is provided by Veevers (1984), and Veevers et al. (1994).

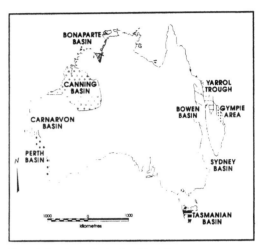

Figure 1. Permian marine basins of Australia.

For the purposes of international correlation, ammonoids provide critical data points for the western Australian faunas (Archbold & Dickins, 1991; Archbold, 1993a), although by no means for all the faunas. Ammonoids are of minor use for the eastern Australian faunas (Archbold & Dickins, 1991). Fusulinid foraminiferans are absent from all Australian marine faunas, and conodonts are known from only two levels in western Australia (Metcalfe & Nicoll, 1995). Reliance for zonation and correlation purposes has therefore been placed on brachiopods and molluscs (Archbold et al., 1993; Archbold, 1993a) for western faunas and brachiopods, molluscs and foraminiferans for eastern faunas (Dickins, 1989; McClung, 1978; Clarke & Farmer, 1976; Parfrey, 1988; Waterhouse, 1987b; Briggs, 1991, Draper et al., 1990), with support from palynological data (Briggs, 1991; Archbold, 1995).

WESTERN AUSTRALIAN PERMIAN BRACHIOPOD FAUNAS

The distinctive brachiopod faunas of the Westralian Province (Archbold, 1983) are found in the marine sequences of the Perth, Carnarvon, Canning and Bonaparte Basins of western Australia. Most species are reviewed by Archbold et al. (1993b) but faunas not covered in that review include those described by Archbold (1991, 1993b, 1995, in press); Archbold & Thomas (1993), and Archbold & Shi (1993). The zonation and international correlations of western Australian faunas are those proposed by Archbold (1993a), with minor modifications (Archbold, 1993b, 1995), as provided herein (Table 1). Migration of genera into, and out of, the Westralian Province has long been suspected (Teichert, 1951). Westralian Permian brachiopod faunas include mixtures of three generic groups.

Genera characteristic of widely distributed Gondwanan faunas include *Arctitreta, Lyonia, Costatumulus, Strophalosia, Taeniothaerus, Cyrtella, Tomiopsis* and *Trigonotreta*, and may indicate cooler water influences. A second group of genera are either endemic to the Westralian Province or demonstrate endemic lineages within the province. This group include *Permorthotetes, Neochonetes (Sommeriella), Gatia, Notolosia, Etherilosia, Coolkilella, Latispirifer, Crassispirifer* and *Imperiospira*. The third group of genera appears to consist of sporadic migrants into the Westralian Province from the warmer Cimmerian Province (Archbold, 1983) of the marginal Gondwanan terranes (Metcalfe,1993). Such genera include *Kiangsiella, Tornquistia, Demonedys, Waagenites, Stictozoster, Dyschrestia, Comuquia, Globiella, Retimarginifera, Callytharrella, Costiferina, Waagenoconcha, Stenoscisma, Elivina, Spiriferella, Spirelytha, Phricodothyris, Gjelispinifera, Callispirina, Hustedia, Rhynchopora, Cleiothyridina* and *Spirigerella*.

Asselian-Tastubian faunas. These faunas include those of the *Lyonia lyoni* and *Trigonotreta occidentalis* Zones. The *Lyonia lyoni* Zone is dominated by cold water genera such as *Lyonia, Cyrtella, Arctitreta, Tomiopsis* and *Trigonotreta*, and is best developed in the Lyons Group of the Carnarvon Basin. The Lyons Group is of glacial origin with a complexity of lithologies reflecting the waxing and waning of glaciation (Dičkins, 1985). The lowest known horizons of the Lyons Group yield the genera *Rhynchopora* and *Kiangsiella*, the latter genus having been used as an indicator of warmer waters (Stehli, 1957). Such genera may represent rare survivors from a warmer preglacial marine fauna occupying the western Australian seas prior to the onset of glaciation (Waterhouse, 1976; Dickins, 1993a). The fauna of the *Trigonotreta occidentalis* Zone, found in the Upper horizons of the Lyons Group and the Carrandibby Formation of the Carnarvon Basin and the Wye Worry Member and the Calytrix Formation of the Grant Group, Canning Basin, demonstrates an increase in diversity. Depending on the locality, the faunas include *Streptorhynchus, Arctitreta, Permorthotretes, Etherilosia,* the earliest *Neochonetes (Sommeriella), Linoproductus, Costatumulus, Callytharrella, Taeniothaerus, Cyrtella, Trigonotreta, ? Martinia, Spiriferellina* and rare *Elivina*. The considerable increase in diversity at the top of the zone indicates amelioration of water temperatures and climate (Thomas, 1976).

Sterlitamakian-Aktastinian faunas. The *Strophalosia irwinensis* Zone of Sterlitamakian age is characterised by one of the most diverse brachiopod western Australian assemblages with over 40 species known from correlative units of the Perth, Carnarvon and Canning basins. The abrupt appearance of Cimmerian genera such as *Tornquistia, Stictozoster, Comuquia, Dyschrestia, Callytharrella, Globiella, Stenoscisma, Elvina, Spirelytha, Phricodothyris, Gjelispinifera, Callispirina, Hustedia,* and *Cleiothyridina* within the faunas indicates significant warming of water temperatures. The earliest species of *Imperiospira* occurs within this zone. A significant drop in diversity and restriction of the fauna to typical Westralian genera are features of the *Neochonetes (Sommeriella)* sp. nov. Zone of early Aktastinian age. This zone, only known from the Perth Basin, includes species of *Neochonetes (Sommerella), Aulosteges, Taeniothaerus, Costatumulus, Neospirifer, Crassispirifer, Cyrtella, Tomiopsis,* rare *Cleiothyridina, Gilledia* and *Hoskingia*. The slightly younger *Strophalosia jimbaensis* Zone from the Carnarvon Basin, with the addition of the genera *Globiella, Callytharrella* and *Spirelytha*, indicates a minor warmer interval. The faunas of this zone are described and illustrated by Archbold (1991) and Archbold & Shi (1993).

Baigendzhinian faunas. Six assemblage zones are recognised within this time interval within the Carnarvon Basin succes-

sions, with perhaps two or three correlative assemblages being known from the Canning Basin. The lower *Echinalosia prideri, Wyndhamia* (or *Arcticalosia*) *colemani* and *Fusispirifer byroensis* zones are characterised by relatively low diversity, with a mix of typical Westralian genera such as *Permorthotetes, Streptorhynchus, Gatia, Neochonetes (Sommeriella), Wyndhamia, Aulosteges, Costatumulus, ?Cyrtella, Neospirifer, Fusispirifer, Crassispirifer* and *Hoskingia*. Rare *Spiriferella, Spirelytha* and *Cleiothyridina* occur in the *Wyndhamia colemani* Zone, and the dominant eastern Australian genus *Echinalosia* occurs in the *Echinalosia prideri* Zone. Cool temperatures are indicated for these lower zones despite the occurrence of *Kiangsiella* within the *Echinalosia colemani* Zone.The higher *Tornquistia magna, Fusispirifer cundlegoensis* and *Fusispirifer wandageensis* Zones show increasing diversity with migrant genera from Cimmerian waters invading the Westralian basins, such as *Tornquistia, Fredericksia?, Demonedys, Quinquenella, Dyschrestia, Retimarginifera, Spiriferella, Spirelytha* and *Hustedia*. The endemic development of distinctive species of *Neochonetes (Sommeriella), Etherilosia, Wyndhamia, Taeniothaerus, Neospirifer, Crassispirifer, Fusispirifer* and *Imperiospira* continued within these zones (Archbold, 1993a). The endemic genera *Lialosia* and *Coolkilella* also appeared during the Baigendzhinian.

Kungurian-?Early Ufimian faunas. Four zones span the Kungurian - ?Early Ufimian time interval (see Archbold, 1993a for the equivocal nature of the latter age determination). The *Neochonetes (Sommeriella) nalbiaensis* Zone includes species of *Wyndhamia, Lialosia, Spirelytha, Fusispirifer* and *Hoskingia* as well as *Svalbardia*, the last genus (in addition to *Lialosia*) also occurs in the minor *Svalbardia thomasi* Zone. Diversity increases with *Derbyia ?, Chonetinella ?, Dyschrestia, Retimarginifera, Spiriferella* and *Spirelytha* returning in the *Neochonetes (Sommeriella) afanasyevae* Zone. This zone also includes species of *Permorthotetes, Streptorhynchus, Aulosteges* and the endemic *Imperiospira* and *Costatispirifer*. The youngest Zone in the Carnarvon Basin succession, the *Fusispirifer coolkilyaensis* Zone, from the top beds of the Coolkilya Sandstone, is characterised by an impoverished fauna of *Fusispirifer* and *Taeniothaerus* and bivalve molluscs (Archbold & Skwarko, 1988).

Ufimian faunas. A subsurface fauna from deep drilling in the Perth Basin appears to be of Ufimian age, on the basis of rare links with eastern Australian faunas (species of *Fusispirifer* and *Sulciplica*), and palynological data (Archbold, 1995). The fauna includes species of *Etherilosia, Spiriferella, Neospirifer, Fusispirifer, Sulciplica, Cleiothyridina* and *Spirigerella*, and hence is a mix of cool water (*Sulciplica*) and warm water (*Spirigerella*) elements.

Dzhulfian faunas. A significant time gap, probably representing most of the Kazanian and Midian Late Permian Zones, from the Canning and Bonaparte Basins, from the earlier Westralian faunas. The *Liveringia magnifica* Zone includes species of *Streptorhynchus, Waagenites,* the endemic *Notolosia, Liveringia, Neospirifer, Tomiopsis* and *Cleiothyridina* and probably requires further collecting to establish its full diversity. The *Waagenoconcha (Wimanoconcha) inperfecta* Zone assemblage is diverse with species of *Streptorhynchus, Derbyia, Neochonetes (Sommeriella), Waagenites, Notolosia, Aulosteges, Taeniothaerus, Megasteges, Waagenoconcha (Wimanoconcha), Latispirifer, Neospirifer, Tomiopsis, Hustedia, Cleiothyridina, Fletcherithyris* and *Hoskingia* all present in the Canning Basin. The Bonaparte Basin correlative horizon also includes species of *Costiferina* and *Leptodus* (Archbold, 1988a). A correlative fauna has also been recorded

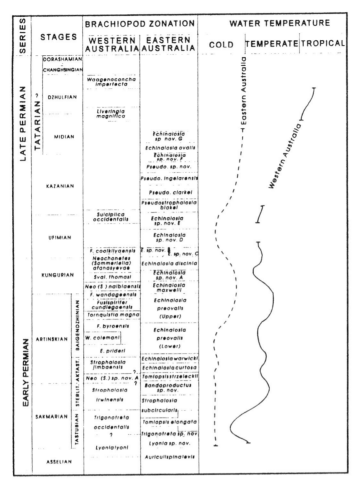

Figure 2. Zonation schemes based on brachiopods for the eastern and western Australia taken from Archbold (1993a) and Briggs (1993), and provisional paleotemperature curves for the Australian Permian.

from deep drilling in the Timor Sea (Archbold, 1988b). Significant links with Djhulfian Himalayan faunas are evident.

EASTERN AUSTRALIAN PERMIAN BRACHIOPOD FAUNAS

The widely developed yet invariably endemic Permian brachiopod faunas of eastern Australia have received sporadic attention for over 170 years. Many species are shared with New Zealand, and collectively these faunas define the Austrazean Province (Archbold, 1983). Broad faunal assemblage zones have been outlined (eg. Dickins, 1970; Runnegar, 1969a; Clarke & Farmer, 1976), and subsequent schemes have proposed named zones (Waterhouse, 1987b; Briggs, 1991, 1993). Considerable debate exists as to precise ages and correlations of the eastern Australian faunas (cf Waterhouse 1987a, 1987b; Archbold & Dickins, 1991; Draper et al. 1990; Briggs, 1993). This review is concerned with the composition of the faunas so that debate over age and correlation will not be discussed herein. The most recent scheme of zones, based on studies of productoids by Briggs (1993) is provided herein (Table 1) for comparison with the western Australian zones. Precise correlation of the zonal schemes should not be inferred from Table 1 because of the problem of extreme provincialism of the faunas. Many faunas require modern descriptions and illustrations, however, the following works provide considerable illustrations of the faunas of eastern Australian basins and additional references to earlier studies. Faunas of the Tasmanian Basin are described by Clarke (1987, 1992a, 1992b) and those of the Sydney Basin have been illustrated more recently by McClung (1978, 1980) and Dickins (1989). Faunas of the Bowen Basin have received considerable attention and more recent studies that illustrate the faunas and include references to earlier studies are those of McClung (1983), Waterhouse (1986, 1987a), Parfrey (1988) and Dickins (1989). Faunas from the Yarrol Trough were described by Maxwell (1964), and those of the Gympie area discussed by Waterhouse and Balfe (1987).

Eastern Australian Permian brachiopod faunas, and those of New Zealand (Waterhouse, 1982), characterise a distinctive faunal province (the Austrazean province of Archbold, 1983) throughout the Permian. Increase in generic diversity, from the Tasmanian faunas in the south to those of the Bowen Basin in the north, is a marked feature of the faunas (Archbold, 1983; Clarke, 1987), those of the Tasmanian Basin representing the most extreme example of cold-water Gondwanan faunas yet known (Clarke, 1987). Core Gondwanan genera of the Austrazean Province include *Arctitreta*, *Strophalosia*, *Trigonotreta*, *Sulciplica*, *Tomiopsis* and allied endemic genera, *Cyrtella* and allies, *Fletcherithyris* and *Gilledia*. Cold water currents were characteristic of the entire marine Permian sequence of the Tasmanian Basin but somewhat warmer influences are noticeable during the Artinskian and Kazanian within the eastern Australian faunas, judging from diversity data, even slightly affecting Tasmanian faunas. All faunas are characterised by evolving lineages of *Tomiopsis* (Asselian to Late Kazanian or Midian) with some species attaining a huge size and *Echinalosia* (Artinskian to ?Midian) as indicated by McClung (1978), Clarke (1987) and Briggs (1991).

Asselian-Sakmarian faunas. Tasmanian faunas, well documented by Clarke (1990, 1992a) include species of *Arctitreta*, *Schuchertella*, *Strophalosia*, the enigmatic *Licharewiella*, *Trigonotreta*, *Sulciplica*, *Tomiopsis*, *Kelsovia*, *Cyrtella*, *Fletcherithyris* and *Gilledia*. Comparable faunas are known from the Sydney Basin (McClung, 1980) and the productoids *Lyonia* and *Bandoproductus* are known from related faunas in the New England area, the Yarrol Trough and the Gympie area (Briggs, 1991).

Artinskian faunas. Faunas of this stage demonstrate an increase in diversity, particularly by the Late Artinskian. Tasmanian faunas include *Orbiculoidea*, *Schuchertella*, *Echinalosia*, *Wyndhamia*, *Taeniothaerus*, *Costatumulus*, *Terrakea*, *Trigonotreta*, *Sulciplica*, *Tomiopsis*, *Homevalaria*, *Notospirifer*, *Glendonia*, *Kelsovia*, *Farmerella*, *Cyrtella*, *Pustulospiriferina*, rare *Cleiothyridina*, *Fletcherithyris* and *Gilledia* (Clarke & Farmer, 1976). Sydney Basin faunas are comparable with additional genera including the chonetid *Tivertonia* and the ambocoelid *Attenuatella*.

The faunas from the Late Artinskian (Baigendzhinian) of the Bowen Basin show a major increase in diversity with additional genera including *Notostrophia*, *Svalbardia*, *Tivertonia*, *Acanthalosia*, *Pseudostrophalosia*, *Wyndhamia*, *Lipanteris*, *Reedoconcha*, *Megasteges*, *Bandoproductus*, *Magniplicatina*, *Paucispinauria*, *Nothokuvelousia*, *Azygidium*, *Permasyrinx*, *Sulcicosta*, *Pustulospiriferina*, *Spiriferellina*, *Pustuloplica*, *Tabellina*, *Fusispirifer*, *Spinomartinia*, *Coledium*, *Stenoscisma*, *Plekonella*, *Plekonina*, rare *Cleiothyridina*, *Maorielasma* and *Marinurnula*. This increase in diversity indicates an increase in temperature for the Bowen Basin during the Baigendzhinian, however, many of the new genera are endemic to the Austrazean Province.

Kungurian-Ufimian faunas. Tasmanian faunas include *Schuchertella*, *Echinalosia*, *Wyndhamia*, *Terrakea*, *Anidanthus*, *Cyrtella*, *Trigonotreta* (or *Aperispirifer*), *Fusispirifer*, *Sulciplica*, *Tomiopsis*, *Notospirifer*, *Glendonia*, *Kelsovia*, *Punctospiriferina*, *Fletcherithyris* and *Gilledia*, indicating a drop in diversity from the Artinskian. Sydney Basin correlative faunas are less well known but also include rare *Neochonetes* and *Magniplicatina*. Bowen Basin faunas include *Capillonia*, *Aulosteges*, *Paucispinauria*, *Lethamia*, rare *Cleiothyridina* and *Marinurnula*. In general, these faunas are of low diversity and indicate a possible cooling of water temperatures.

Kazanian faunas. The low diversity faunas of the Tasmanian and Sydney Basins include *Echinalosia*, *Terrakea*, *Plekonella*, *Fusispirifer*, *Trigonotreta* (*Aperispirifer*), *Sulciplica*, *Tomiopsis*, *Notospirifer*, *Glendonia*, *Birchsella* (only in Tasmania), *Cyrtella*, *Fletcherithyris* and *Gilledia* indicate cool water. However correlative faunas in the Bowen Basin are of higher diversity

with *Capillonia*, *Arctitreta*, *Saetosina*, *Filiconcha*, *Magniplicatina*, *Spiriferellina*, *Cleiothyridina*, *Marinurnula* and *Maorielasma* indicating somewhat warmer waters

Midian? faunas. The youngest faunas of the Bowen and Sydney basins may be of Midian age as suggested by Briggs (1993). These youngest faunas include *Echinalosia*, *Terrakea* and *Tomiopsis* as described by Dickins (1989). *Lethamia* and *Magniplicatina* have also been recorded from these faunas (Briggs, 1991). It should be added that Dickins (1989) has argued that there are no faunas younger than Kazanian in eastern Australia. These youngest faunas are of limited diversity, possibly due more to limited water circulation, rather than a decrease in water temperature (Dickins, 1989).

CONCLUSIONS

An analysis of Permian brachiopod faunas of Australia reveals the following observations:
(1) Two faunal provinces are confirmed by qualitative generic data, a Westralian Province and an Austrazean Province, for most of the Permian.
(2) The Westralian Province is characterised by a mix of Gondwanan, endemic Westralian and warmer water migrant genera from the Cimmerian Province.
(3) The Austrazean Province is characterised by a core group of Gondwanan and endemic genera with a mix of additional endemic genera and warmer water migrant genera in the Bowen Basin faunas. This province is also characterised by evolutionary lineages of *Tomiopsis* and *Echinalosia* and the development of the Ingelarellidae in general.
(4) A provisional temperature curve model can be proposed for the Permian of eastern Australia (Table 1) on the basis of brachiopod generic diversity data and data from bivalve molluscs (Dickins, 1993b). This curve can be compared with a curve proposed for the Westralian Province by Archbold & Shi (1995).

Acknowledgments

Discussion about Australian Permian brachiopod faunas have been held with numerous fellow workers including G.R. Shi, J.M. Dickins, G.A. Thomas and D.J.C. Briggs, whom I thank, although conclusions herein are my own responsibility. M. Grover drafted the figure and table and word processed the manuscript. My work is supported by the Australian Research Council (Grant A39332106).

REFERENCES

ARCHBOLD, N.W. 1983. Permian marine invertebrate provinces of the Gondwanan Realm. Alcheringa 7:59-73.

___1988a. Studies on Western Australian Permian brachiopods, 8, The Late Permian brachiopod fauna of the Kirkby Range Member, Canning Basin. Proceedings Royal Society Victoria 100:21-32

___1988b. Permian Brachiopoda and Bivalvia from Sahul Shoals No. 1, Ashmore Block, Northwestern Australia. ibid. 100:33-38.

___1991. Studies on Western Australian Permian brachiopods 10. Faunas from the Wooramel Group, Carnarvon Basin. ibid. 103:55-66.

___1993a. A zonation of the Permian brachiopod faunas of Western Australia, p.313-321. In, R.H. Findlay, R. Unrug, M.R. Banks & J.J. Veevers (eds.), Gondwana eight assembly, evolution and dispersal. Balkema, Rotterdam.

___1993b. Studies on Western Australian Permian brachiopods, 11, New genera, species and records. Proceedings Royal Society Victoria 105:1-29.

___1995. Ufimian (early Late Permian) brachiopods from the Perth Basin. Memoir Association Australasian Palaeontologists 18:153-163.

___ (in press). Studies in Western Australian Permian brachiopods, 12, Additions to the Asselian-Tastubian faunas. Proceedings Royal Society Victoria.

___& J.M. DICKINS. 1991. Australian Phanerozoic timescales, 6, Permian. Bureau Mineral Resources, Geology Geophysics Australia Record 1989/36, 17p.

___,___& G.A. THOMAS. 1993. Correlation and age of Permian marine faunas in Western Australia. Geological Survey Western Australia Bulletin 136:11-18.

___& G.R. SHI. 1993. Aktastinian (Early Artinskian, Early Permian) brachiopods from the Jimba Jimba Calcarenite, Wooramel Group, Carnarvon Basin, Western Australia. Proceedings Royal Society Victoria 105:187-202.

___&___1995. Permian brachiopod faunas of Western Australia: Gondwanan-Asian relationships and Permian climate. Journal Southeast Asian Earth Sciences 11:207-215.

___&___(in press). Western Pacific Permian marine invertebrate palaeobiogeography. Australian Journal Earth Sciences.

___& S.K. SKWARKO. 1988. Brachiopods and bivalves of Kungurian (late Early Permian) age from the Coolkilya Sandstone, Carnarvon Basin, Western Australia. Geological Survey Western Australia Report 23:1-15.

___& G.A. THOMAS. 1993. *Imperiospira*, a new western Australian Permian Spiriferidae (Brachiopoda). Memoirs Association Australian Palaeontologists 15:313-328.

___,___& S.K. SKWARKO. 1993. Brachiopods. Geological Survey Western Australia Bulletin 136:45-51, 196-264.

BRIGGS, D.J.C. 1991. Stratigraphical distribution and biostratigraphical significance of the Permian Productidina and Strophalosiidina (Order Productida) of eastern Australia, p.367-374. In, D.I. MacKinnon, D.E. Lee & J.D. Campbell (eds.), Brachiopods Through Time. Balkema, Rotterdam.

___1993. Permian depositional sequences of the Sydney Bowen Basin, p.247-254. In, Twentyseventh Newcastle Symposium on 'Advances in the study of the Sydney Basin'. Department of Geology, University of Newcastle, Newcastle.

CLARKE, M.J. 1987. Late Permian (late Lymingtonian= ? Kazanian) brachiopods from Tasmania. Alcheringa 11:261-289.

___1992a. Hellyerian and Tamarian (Late Carboniferous- Lower Permian) invertebrate faunas from Tasmania. Geological Survey of Tasmania Bulletin 69:1-52.

___1992b. A new notospiriferine genus (Spiriferida: Brachiopoda) from the Permian of Tasmania. Papers Proceedings Royal Society Tasmania,126:73-76.

___& D. FARMER. 1976. Biostratigraphic nomenclature for Late Paleozoic rocks in Tasmania. ibid.110:91-109.

DICKINS, J.M. 1985. Late Palaeozoic glaciation. BMR Journal Australian Geology Geophysics 9:163-169.

___1989. Youngest Permian marine macrofossil fauna from the Bowen and Sydney Basins, eastern Australia. ibid. 11:63-79.

___1993a. Climate of the Late Devonian to Triassic. Palaeogeography, Palaeoclimatology Palaeoecology 100:89-94.

___1993b. Permian bivalve faunas- stratigraphical and geographical distribution. Douzième Congres International de la Stratigraphie et Géologie du Carbonifère et Permien, Buenos Aires, 1991, Comptes Rendus 1:523-536.

DRAPER, J.J., V. PALMIERI, P.L. PRICE, D.J.C. BRIGGS & S.M. PARFREY. 1990. A biostratigraphic framework for the Bowen Basin, p.26-35. In, Proceedings of the Bowen Basin Symposium. Geological Society of Australia, Queensland Division, Brisbane.

MAXWELL, W.G.H. 1964. The geology of the Yarrol Region, 1, Biostratigraphy. Papers Department of Geology University of Queensland 5(9):1-79.

McCLUNG, G.R. 1978. Morphology, palaeoecology and biostratigraphy of *Ingelarella* (Brachiopoda: Spiriferida) in the Bowen and Sydney Basins of eastern Australia. Geological Survey Queensland Publication 365, (40):11-87.

___1980. Permian biostratigraphy of northern Sydney Basin. New South Wales Geological Survey Bulletin 26:360-375.

___1983. Faunal sequence and palaeoecology of the marine Permian in GSQ Eddystone 1, Denison Trough, Queensland. Geological Survey Queensland Publication 384: 44-80.

METCALFE, I. 1993. Southeast Asian Terranes: Gondwanaland origins and evolution, p.181-200. In, R.H. Findlay, R. Unrug, M.R. Banks & J.J. Veevers (eds.), Gondwana Eight Assembly Evolution and Dispersal. Balkema, Rotterdam.

___& R.S. NICOLL. 1995. Lower Permian conodonts from Western Australia, and their biogeographic and palaeoclimatological implications. Courier Forschungsinstitut Senckenberg 182:559-560.

PARFREY, S.M. 1988. Biostratigraphy of the Barfield Formation, southeastern Bowen Basin, with a review of the fauna from the Ingelara and lower Peawaddy Formations, southwestern Bowen Basin. Queensland Department Mines Report 1:1-53.

SHI, G.R. & N.W. ARCHBOLD. 1993 Distribution of Asselian to Tastubian (Early Permian) Circum-Pacific brachiopod faunas. Memoirs Association Australasian Palaeontologists 15:343-351.

___&___1995. A quantitative analysis on the distribution of Baigendzhinizn-early Kungurian (early Permian) brachiopod faunas in the western Pacific region. Journal of Southeast Asian Earth Sciences, 11:189-205.

___&___(in press, a). Palaeobiogeography of Kazanian-Midian (Late Permian) western Pacific brachiopod faunas. Journal Southeast Asian Earth Sciences.

___&___(in press, b). A quantitative palaeobiogeographical analysis on the distribution of Sterlitamakian - Aktastinian (Early Permian) western Pacific brachiopod faunas. Historical Biology.

STEHLI, F.G. 1957. Possible Permian climatic zonation and its implications. American Journal Science 255:607-618.

TEICHERT, C. 1951. The marine Permian faunas of Western Australia (an interim review). Paläontologische Zeitschrift 24:76-90.

THOMAS, G.A. 1976. Faunal evidence of climatic change in the early Permian of the Carnarvon Basin. Twentyfifth International Geological Congress Sydney Abstracts 1:318-319.

VEEVERS, J. J. 1984. Phanerozoic earth history of Australia. Clarendon Press, Oxford, 418p.

___,P.J. CONAGHAN & C.McA. POWELL. 1994. Eastern Australia. Geological Society America Memoir 184:11-171.

WATERHOUSE, J.B. 1976. World correlations for Permian marine faunas. Papers Department Geology University of Queensland 7(2):1-232.

___1982. New Zealand Permian brachiopod systematics, zonation and paleoecology. New Zealand Geological Survey Paleontological Bulletin 48:1-158.

___1986. Late Palaeozoic Scyphozoa and Brachiopoda (Inarticulata, Strophomenida, Productida and Rhynchonellida) from the southeast Bowen Basin, Queensland, Australia. Palaeontographica A(193):1-76.

___1987a. Late Palaeozoic Brachiopoda (Athyrida, Spiriferida and Terebratulida) from the southeast Bowen Basin, east Australia. ibid. A(196):1-56.

___1987b. Late Palaeozoic Mollusca and correlations from the south-east Bowen Basin, east Australia. ibid. A(198):129-233.

___& P.E. BALFE. 1987. Stratigraphic and faunal subdivisions of the Permian rocks at Gympie, p.20-33. In, C.G. Murray & J.B. Waterhouse (eds.), 1987 Field Conference Gympie District. Geological Societyf Australia Queensland Division, Brisbane.

BASHKIRIAN AND MOSCOVIAN BRACHIOPOD ASSEMBLAGES FROM NORTHWESTERN SERBIA

N.W. ARCHBOLD & S. STOJANOVIC-KUZENKO

School of Aquatic Sciences and Natural Resources Management, Deakin University, Clayton, Victoria 3168; Department of Geology and Geophysics, University of Adelaide, Adelaide 5000, S Australia

ABSTRACT- Five brachiopod assemblages ranging in age from Early Bashkirian to Podolskian have been identified from the Carboniferous of NW Serbia. The four older assemblages are known from the Likodra Nappe, and the youngest assemblage (Podolskian - the Ivovik assemblage) is from the Jadar region (Krupanj Unit). The oldest faunas, from Vlaska Reka, are dominated by species of *Choristites* typical of the Late Early Bashkirian of the Donets Basin and the Perm Pri-urals such as *C. medovensis*, *C. convexa* and *C. plana*. A small fauna from Ostrikovaca with *Choristites* cf. *medovensis* and *C.* cf. *convexa* as, as well as *Linoproductus* cf. *gasensis*, and *Buxtonia* cf. *mosquensis*, indicates a Late Early to Middle Bashkirian age. A Late Bashkirian fauna is typified by diverse assemblages which include species of *Antiquatonia*, *Brachythyrina* and *Martinia*, as found at Reljaca and Ovcarevac. A distinctive Early Moscovian (Vereian) fauna, represented by assemblages from Obradovici, include such genera as *Meekella*, *Isogramma*, *Megachonetes*, *Karavankina*, *Reticulatia*, *Crurithyris*, *Martinia* and *Stenoscisma*. This assemblage shares genera with the Early Moscovian faunas of Spain to the west, and the contemporaneous faunas of European Russia to the east. The youngest assemblage of Late Moscovian age (Podolskian-age, also constrained by fusulinids, includes *Orthotetes* cf. *plana*, *Neochonetes carboniferus*, and advanced *Choristites* species with fine branching costae.

INTRODUCTION

Bashkirian and Moscovian (Carboniferous) brachiopod faunas have been known from Serbia for some decades (Petronijevic-Kuzenko, 1957; Stojanovic-Kuzenko, 1968) but have invariably been ignored in biogeo-graphical discussions of Carboniferous brachiopods. This review summarises five brachiopod assemblages, ranging in age from Early Bashkirian to Podolskian, known from the Serbian Carboniferous near the Drina River drainage basin. Faunas derive from a series of localities in NW Serbia, as follows (Figure 1): 1, Sabacka Pocerina, Vlaska Reka, at Petkovica village, and Kopljevic and Cerik brooks near Desici; 2, Vlasici Hill area, from Reljaca and Ostrikovaca brooks (Milcinica village), Docki Potok brook (near Sirdija village), and from Ovcarevac Hill, north of Cvetulja church; 3, Obradovici near Stolice; 4, Ivovik, near Krupanj.

Geology- The Carboniferous of NW Serbia has been divided into autochthonous (Jadar region) and alloch-thonous units (Likodra Nappe). A summary of earlier geological work is provided by Filipovic (1974). More recent work by I. Filipovic and D. Jovanovic (field and sedimentological studies), V. Pajic (foraminiferans and algae), and on conodonts by S. Stojanovic - Kuzenko, will be published elsewhere. Four of our brachiopod assemblages are known from the Likodra Nappe which was thrust over the younger autochthonous Carboniferous of the Jadar region during the Early Permian regression and Saalian Orogeny. This allochthonous cover was broken up into three blocks during Alpine tectonic events. All our brachiopod assemblages, except for the Ivovik assemblage, come from

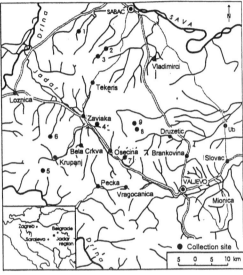

Figure 1. Index map to Carboniferous brachiopod localities in NW Serbia. 1,Vlaska Reka. 2,Cerik. 3,Koplevic. 4,Ovcarevac. 5, Ivovik. 6, Obradovici. 7, Docki potok. 8, Reljaca. 9. Ostrikovaca.

the Stojkovici Formation of the Likodra Nappe, largely thrust over the Ivovik Formation. The Stojkovici Formation varies in thickness from 20 to 60 m, and consists of dark grey siltstone when fresh, weathering to a yellowish or pale brown colour, and minor shales and sandstones.

The Ivovik assemblage was found in the autochthonous Ivovik Formation of the Jadar region, a diverse formation in terms of lithology, age and paleoenvironment; however, brachiopods and foraminiferans found in the section at Ivovik indicate a Podolskian age.

BRACHIOPOD ASSEMBLAGES

The five brachiopod assemblages are informally named the Vlaska Reka, Ostrikovaca, Reljaca, Obradovici and Ivovik assemblages, and are reviewed in terms of their brachiopod fauna, age and paleogeographical relationships. The brachiopod faunas of NW Serbia are shallow water, typical of the Palaeotethys Realm as outlined for other regions by Martinez-Chacon and Winkler Prins (1993). Where species cannot be identified below, only the genus is named.

Vlaska Reka. Brachiopods are abundant at Vlaska Reka and have been investigated previously by Petronijevic-Kuzenko (1957) and Stojanovic-Kuzenko (1968), who illustrated several of the forms present. Revised and new

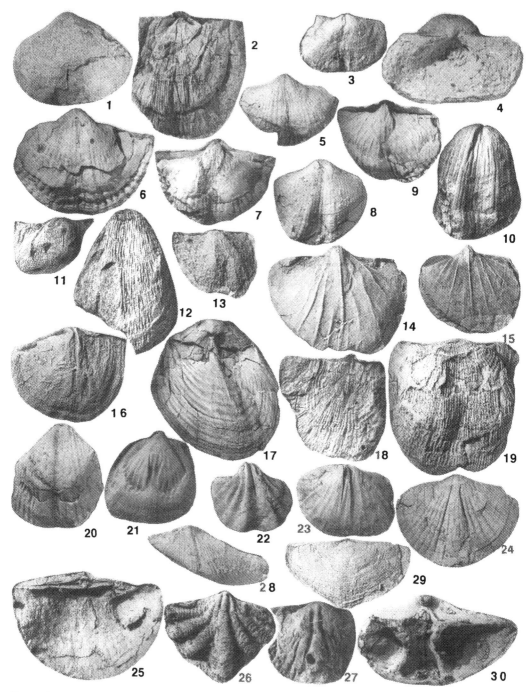

Figure 2. Representative Bashkirian-Moscovian brachiopods from NW Serbia. 1-9, Vlaska Reka; 10-12, Ostrikovaca; 13-15, Reljaca; 16-20, Ovcarevac; 21-24, Kopjevic; 25-28, Obradovici; 29-30, Ivovik. **1**, *Schizophoria* cf. *resupinata* (Martin), VR/14, x2; **2**, *Orthotetes* sp., VR/5, x2; **3**, *Rugosochonetes* sp., VR/15, x4; **4**, *Krotovia* sp., VR/16, x2.5; **5-9**, *Choristites medovensis* Rotai, 5, ventral exterior, VR/27, x1; 6, VR/43, x1.2; 7, ventral interior, VR/35, x1; 8, VR/25, x1; 9, ventral interior, VR/38, x1; **10**, *Choristites* cf. *convexa* Rotai, OS/9, x1; **11**, *Linoproductus* cf. *gasensis* Volgin OS/3, x1; **12**, *Buxtonia* cf. *mosquensis* Ivanov, OS/6, x1.2;**13**, *Rugosochonetes* sp., ventral valve internal mould, specimen RE/10, x3; **14-15** *Martinia* sp., 14, RE/28, x 2; 15, RE/29, x 2; **16**, *Rugosochonetes* sp., OV/1, x3; **17**, *Phricodothyris* sp., OV/28, x2; **18-19**, *Antiquatonia* cf. *hindi* Muir-Wood sensu Semikhatova, 3, OV/27, x2; 4, OV/22, x1; **20**, *Brachythyrina plana* Rotai, OV/34, x2; **21**, *Rhipidomella* sp., KO/4, x1.5; **22**, *Punctospirifer* sp., KO/21, x2.5; **23-24**, *Choristites* cf. *teshevi* Ivanov; 8, KO/16, x2; 9, KO/20, x1.5; **25**, *Megachonetes* sp., OB/8, x2.5; **26**, *Mucrospiriferinella* sp., OB/21, x3; **27**, *Brachythyrina* sp., OB/20, x1.5; **28**, *Isogramma serbica* Stojanovic - Kuzenko, OB/7, x1.5; **29-30**, *Choristites* sp.,14, IV/9, x1;15, IV/14, x1.5.

identifications for this review include *Rhipidomella* cf. *michelini* (Leveille), *Schizophoria resupinata* (Martin), *Orthotetes*, *Schuchertella*, *Rugosochonetes*, *Krotovia*, *Antiquatonia*, *Choristites medovensis* Rotai, *C. convexa* Rotai, *C.* cf. *pseudobisulcatus* Rotai, *C. planus* Lapina [Figure 2].

Stojanovic-Kuzenko (1968) concluded that the age of the Vlaska Reka assemblage is Late Early Bashkirian, a view consistent with subsequent data from microfossil and coral evidence (Kostic and Pajic, 1972). This assemblage represents the oldest Carboniferous brachiopod fauna known from NW Serbia. The assemblage is dominated by *Choristites* species described from the Early Bashkirian of the Donetz Basin (Rotai, 1951) and the Perm Pri-urals (Lapina, 1957). It is a significant fauna within the Paleo-tethyan Realm that stretched from SE Asia into the Mediterranean, and was blocked in the west (Winkler Prins, 1989). The assemblage also demonstrates significant links at the generic level with faunas from Uzkekistan (Sikstel',1975) and the Kuibyshev region of southern Russia (Semenova, 1963). Links with Bashkirian faunas from the eastern part of the Cantabrian Zone, Spain (Sanchez de Posada et al., 1993), were not strong at this time. Detailed analysis of the relationships of Spanish Carboniferous brachiopod faunas is provided by Martinez-Chacon and Winkler Prins (1984,1985).

Ostrikovaca. A small assemblage from Ostrikovaca has proved difficult to date (Stojanovic-Kuzenko, 1968) [Figure 2]. Identified forms include *Schellwienella*, *Rugosochonetes*, *Linoproductus* cf. *gasensis* Volgin, *Buxtonia* cf. *mosquensis* Ivanov, *Choristites* cf. *medovensis* Rotai, and *C.* cf. *convexa* Rotai. The fauna is not diverse. However, the comparison of *Linoproductus*, *Buxtonia* and *Choristites* with species from known Bashkirian faunas of Bashkiria, Fergana and Kirgizia suggests a Late Early to Middle Bashkirian age (Semikhatova, 1964; Volgin, 1965; Galitskaya, 1977). Additional collections will clarify the precise age of this fauna. The assemblage is a typical fauna from the southern border of Laurentia (Winkler Prins, 1989).

Reljaca. This assemblage is typified by faunas from Reljaca and Ovcarevac, localities first recorded by Simic (1934,1938), and Docki Potok. The fauna includes *Rhipidomella*, *Schizophoria* cf. *resupinata* (Martin), *Orthotetes* cf. *radiata* (Fischer de Waldheim), *Schuchertella*, *Rugosochonetes*, *Alexenia* ?, *Antiquatonia* cf. *ivanovi* (Lapina), *Krotovia*, *Linoproductus*, *Podtscheremia* cf. *noviki* (Aizenverg), *Brachythyrina* cf. *plana* Rotai, *Martinia*, *Stenoscisma*, and *Pugnax* ?. Ovcarevac has yielded *Rugosochonetes*, *Fluctuaria* cf. *undata* (Defrance), *Reticulatia* ?, *Alexenia*, *Linoproductus*, *Antiquatonia* cf. *hindi* Muir-Wood (sensu Semikhatova), *Phricodothyris*, *Martinia*, *Brachythyrina plana* Rotai, and *Composita*.

Docki Potok has a smaller fauna which includes *Rhipidomella* cf. *michelini* (Leveille), *Rugosochonetes*, *Schuchertella*, *Linoproductus*, *Brachythyrina* cf. *plana* Rotai, and *Martinia*. They are characterised by distinctive species of *Antiquatonia*, *Brachythyrina*, and *Martinia*. Overall the fauna can be assigned to the Late Bashkirian. Paleogeographical relationships are strongest with the Donetz Basin, Russian Platform and Urals and the Bashkirian region (Sokolskaya, 1954; Lapina, 1957; Semikhatova, 1964). The Kopljevic and Cerik localities, first recorded by Stevanovic (1953), are grouped with the Reljaca Assemblage, but are probably transitional with the Vereian Horizon of the Moscovian Stage. Identified forms from Kopljevic include *Rhipidomella* cf. *michelini* (Leveille), *Rugosochonetes*, *Alexenia* ?, *Linoproductus*, *Reticulatia*,

Echinoconchus punctatus (Martin), *Karavankina*, *Choristites* cf. *teshevi* Ivanov, *Punctospirifer*, and *Martinia* . Cerik has a fauna which includes *Orthotetes*, *Rhipidomella*, *Linoproductus*, *Alexenia* ?, *Karavankina*, *Reticulatia*, *Choristites* cf. *latus* Rotai, *Martinia*, and *Punctospirifer*. Elements such as *Choristites* cf. *teshevi* Ivanov (1937, in Ivanov and Ivanova, 1937), *Karavankina*, and large *Rhipidomella* suggest links with the Vereian Horizon. Nevertheless the Vereian aspect of these localities is not as strongly developed as that of the Obradovici assemblage. The Reljaca assemblage demon-strates closer links with Cantabrian faunas, as well as Alpine and Russian faunas - links that became stronger in the Vereian Obradovici assemblage.

Obradovici. The rich assemblage from Obradovici has already been reported by Stojanovic-Kuzenko (1968), and Archbold & Stojanovic-Kuzenko (1991). It includes *Orthotetes* cf. *radiata* (Fischer de Waldheim), *Meekella eximia* (Eichwald), *Drahanorhynchus*, *Rhipidomella*, *Isogramma serbica* Stojanovic-Kuzenko, *Kozlowskia*, *Megachonetes*, *Karavankina* cf. *rakuszi* Winkler Prins, *Alexenia* ?, *Linoproductus* cf. *tschernyschewi* Ivanov, *Reticulatia* cf *ivanovi* (Lapina), *Choristites*, *Mucrospiriferinella*, *Crurithyris*, *Martinia*, *Cleiothyridina*, and *Stenoscisma*. The character of this assemblage is Early Moscovian (Stojanovic-Kuzenko, 1968), and the age is regarded as being Vereian. It shares many genera with the Early Moscovian faunas of Spain (Rio Garcia & Martinez-Chacon, 1988; Martinez-Chacon, 1990, 1991; Sanchez de Posada et al., 1993). Links are also to be found with European Russian faunas - notably with the genera *Isogramma*, *Linoproductus* and *Reticulatia*. Open links across the Paleo-tethys are indicated by the fauna (Winkler Prins, 1989).

Ivovik. The youngest brachiopod assemblage yet recognised from NW Serbia is represented by Ivovik, within the Krupanj unit. The fauna was regarded as being Visean by Simic (1932), and Upper Carboniferous by Filipovic (1963). Although specimens are grossly distorted by tectonism, the following forms can be recognised: *Orthotetes* cf. *plana* Ivanov, *Neochonetes carboniferus* (Keyserling), and *Choristites*. *Choristites* are of an advanced type, with fine branching costae. The assemblage is of Late Moscovian age (Podolskian), an age also constrained by fusulinids. Brachiopods are comparable with those documented from the Moscow Basin, Russian Platform and the Donetz Basin by Frederiks (1926), Ivanov (1935), Ivanov & Ivanova (1936, 1937), Sokolskaya (1950,1954), Aizenverg (1951), and summarised by Ivanova et al. (1979), indicating open seaways with those areas.

Acknowledgments

We are particularly grateful to our colleague and friend Ivan Filipovic for sharing so much geological knowledge and who also discovered several localities, and patiently, with V. Pajic and the second author built up the collections. Montague Grover word-processed the manuscript. Guang Shi photographed the specimens. Neil Archbold's work on Late Palaeozoic brachiopods is supported by the Australian Research Council. Illustrated specimens are held in the collections of the Museum of Natural History, Belgrade; all specimens are grossly distorted by tectonism.

REFERENCES

AIZENVERG, D.E. 1951. Brakhiopody Kamennougolnykh otlozhenii raiona r. Volchei. Trudy Instituta Geologicheskikh Nauk, Akademiya Nauk Ukrainskoi SSR, Seriya Stratigrafii i Paleontologii 5:1-111.

ARCHBOLD, N.W. & S. STOJANOVIC-KUZENKO. 1991. Carboniferous Brachiopoda of N.W. Serbia, Yugoslavia. Newsletter on Carboniferous Stratigraphy 9:25-26.

FILIPOVIC, I. 1963. Prilog stratigrafiji karbona severozapadne Srbije. Zapisnici Srpskog Geoloskog Drustva, Zbornik 27(3).

___1974. Paleozoik severozapadne Srbije, Geologija. Zavod za geoloska i geofizicka istrazivanja 48:229-252.

FREDERIKS, G. N. 1926. Khoristity i Khoristito-podobnye spirifery iz Myachkova. Izvestiya Akademii Nauk SSSR, VI Seriya 20:253-276.

GALITSKAYA, A. Ya. 1977. Ranne- i Srednekamennougolnye produktidy Severnoi Kirgizii. Izdatelstvo 'ILIM', Frunze, 298p.

IVANOV, A.P. 1935. Fauna brakhiopod srednego i verkhnego Karbona Podmoskovnogo Basseina, Chast 1, Productidae Gray. Trudy Moskovskogo Geologicheskogo Tresta 8:1-161.

___& E.A. IVANOVA. 1936. Fauna brakhiopod srednego i verkhnego Karbona Podmoskovnogo Basseina, Chast I, Vypusk 2. Trudy vsesoyuznogo nauchno-issledovatelskii Instituta Mineralnogo Sirya 108:1-52.

___, & ___1937. Fauna brakhiopod srednego i verkhnego Karbona Podmoskovnogo Basseina (Neospirifer, Choristites). Trudy Paleozoologicheskogo Instituta 6 (2):1-215.

___, M. N. SOLOVIEVA & E. M. SHIK.1979. The Moscovian Stage in the USSR and throughout the world. Yorkshire Geological Society, Occasional Publication 4:117-146.

KOSTIC, V. & V. PAJIC. 1972. Mikrofauna i Korali Baskirskog Kata Sredneg Karbona Zapadne Srbije. Geoloski Anali Balkanskog Poluostrva, 37:101-117.

LAPINA, N.N. 1957. Brakhiopody Kamennougolnykh otlozhenii Permskogo Priuralya. Trudy Vsesoyuznogo Neftyanogo Nauchno-issledovatelskogo Geologorazvedochnogo Instituta (VNIGRI) 108:1-184.

MARTINEZ-CHACON, M.L. 1990. Braquiopodos Carboniferos de la Costa e de Asturias (Espana), I, Orthida, Strophomenida, Rhynchonellida y Athyridida. Revista Espanola de Paleontologia 5:91-110.

___1991. Braquiopodos Carboniferos de la Costa e de Asturias (Espana), 2, Spiriferida y Terebratulida. Revista Espanola de Paleontologia 6:59-88.

___& C. F. WINKLER PRINS. 1984. The brachiopod fauna of the San Emiliano Formation (Cantabrian Mountains, NW Spain) and its connection with other areas. Neuvième Congrès International de Stratigraphie et Géologie du Carbonifère, Washington1979, Comptes Rendus 5:233-244.

___&___1985. Upper Carboniferous (Kasimovian) brachiopods from Asturias (N Spain). Dixième Congrès International de Stratigraphie et Géologie du Carbonifère, Madrid 1983, Comptes Rendus 2:435-448.

___&___1993. Carboniferous brachiopods and the palaeogeographic position of the Iberian Peninsula. Douzième Congrès International de Stratigraphie et Géologie du Carbonifère et Permien, Comptes Rendus 1:573-580.

PETRONIJEVIC-KUZENKO, S. 1957. Gornjokarbonska brakhiopod-ska fauna Vlaske Reke u Gornjoj Pocerini. Zbornik radova Geologskog Instituta 'Jovan Zujovic' 9:163-176.

RIO GARCIA, L.M. & M.L. MARTINEZ-CHACON. 1988. Braquio-podos Moscoviensis del Paquete Levinco (Cuenca Carbonifera Central de Asturias). Trabajos de Geologia Universidad de Oviedo 17:33-56.

ROTAI, A.P.1951. Brakhiopody srednego Karbona Donetskogo Basseina, Chast I, Spiriferidae. Gosudarstvennoe Izdatelstvo, Geologicheskoi Literatury, Moskva, 179 p.

SANCHEZ, DE POSADA L.L., M. L. MARTINEZ CHACON, C. A. MENDEZ, J. R. MENENDEZ ALVAREZ, J. TRUYOLS & E. VILLA. 1993. El Carbonifero de las regiones de picos de Europa y manto del Ponga (zona Cantabria, N de Espana): Fauna y bioestratigrafia. Revista Espanola de Paleontologia, Numero Extraordinario, 89-108.

SEMENOVA, E.G.1963. Stratigraficheskoe rasprostranenie brakhio-pod v kamennougolnykh otlozheniyakh orekhovskoi opornoi, Skvazhiny 1. Trudy Kuibyshevskii Gosudarstvennyi Nauchno-Issledovatelskii Institut Neftyanoi Promyshlennosti 21:66-95.

SEMIKHATOVA, S.V. 1964. Brakhiopody iz opornykh razrezov Bashkirskogo yarusa gornoi Bashkirii. Trudy Vsesoyuznyi Nauchno-Issledovatelskii Geologorazvedochnyi Neftyanoi Instituta (VNIGNI) 43:180-227.

SIKSTEL, T.A. (ed.). 1975. Biostratigrafiya verkhnego paleozoya gornogo Obramleniya Yuzhnoi Fergany. Izdatel'stvo FAN, Tashkent, 166 p.

SIMIC, V. 1932. Prilog geologiji zapadne Srbije. Vesnik Geoloskog Instituta Kraljevine Jugoslavije 1(2):

___1934. Prilog poznavanju gornjokarbonskih i gornjopermskih fauna u Zapadnoj Srbiji. Vesnik Geoloskog Instituta Kraljevine Jugoslavije 3(1):112-128.

___1938. O facijama mladeg paleozoika u Zapadnoj Srbiji. Vesnik Geoloskog Instituta Kraljevine Jugoslavije 6:83-108.

SOKOLSKAYA, A.N. 1950. Chonetidae Russkoi Platformy. Trudy Paleontologicheskogo Instituta Akademiya Nauk SSSR, 27:1-108.

___1954. Strofomenidy Russkoi Platformy. Trudy Paleontologicheskogo Instituta Akademiya Nauk SSSR 51:1-187.

STEVANOVIC, P. 1953. Mladji Paleozoik u gornjem toku Uba i Kladnice (Valjevska Podgorina). Glasnik Srpske Akademije Nauka 5(2):273-275.

___1957. Prikaz geoloske karte liste Valjevo 1 (1:50000). Zapisnici Zapisnici Srpskog Geoloskog Drustva, za 1955, Beograd.

STOJANOVIC-KUZENKO, S. 1968. Biostratigraphy of the Middle Carboniferous of Western Serbia and its correlation with that in Northwest Bosnia, part of Velbit, and Stanisci (Montenegro). Zavod za Geologoska i Geofizicka Istrazivanja A (24/25):221-243.

VOLGIN, V.I. 1965. Brakhiopody Gazskoi Svity Yuzhnoi Fergany. Izdatelstvo Leningradskogo Universiteta, Leningrad, 96 p.

WINKLER PRINS, C.F. 1989. Brachiopod distributions and the palaeogeographic reconstructions for the Carboniferous. Onzième Congrès International de Stratigraphie et Géologie du Carbonifère Beijing 1987, Comptes Rendus 2:382-390.

THE 'AFRO-SOUTH AMERICAN REALM' AND SILURIAN 'CLARKEIA FAUNA'

J.L. BENEDETTO & T.M. SANCHEZ

CONICET, Cátedra de Estratigrafía y Geología Histórica, Universidad Nacional de Córdoba,
Av. Velez Sarsfield 299, 5000 Córdoba, Argentina).

ABSTRACT- On the basis of the spatial distribution of the Silurian endemic brachiopods from South America and Africa, the replacement of the name 'Silurian Malvinokaffric Realm' and 'Clarkeia Fauna', by the name Afro-South American Realm is proposed. From current evidence, the last members of this realm extend to the earliest Devonian (Early Lochkov/ Gedinne). This realm is characterised by the widespread distribution of Heterorthella and other endemic genera, the absence of Pentameroidea, and paucity of Spiriferoidea. Endemism within the Afro-South American Realm shows gradual increase from the Llandovery to Ludlow-Pridoli. Neither geographic distribution nor taxonomic composition of the Afro-South American Realm brachiopods is uniform throughout the Silurian. During the Llandovery, two areas can be differentiated, (1) an infracratonic area characterised by endemics-dominated assemblages, and, (2) an epicratonic Andean area, typified by very low endemism, and presence of several North Silurian Realm taxa. As a result of faunal turnover at the end of the Llandovery, brachiopod faunas were strongly depleted. During the Wenlock-Early Ludlow transgressive event, the open platform was colonised by low-diversity communities dominated by Harringtonina and Australina. By the Ludlow-Pridoli, high-diversity communities including endemic taxa and new immigrants, such as Salopina, Isorthis and Coelospira, flourished in nearshore, well-oxygenated muddy deposits. The end of the Early Lochkov is punctuated by extinction of the last Afro-South American endemics, and the appearance of quite different brachiopod assemblages characteristic of the Devonian 'Malvinokaffric Realm'. Diversity and distribution of brachiopod faunas were controlled primarily by ecologic factors. However, influx of North Atlantic taxa into the Afro-South American Realm in the Silurian and earliest Devonian seems to have been related to relative position of the basins with respect to the continent. Immigrant taxa increased towards the continental margin, especially in the Precordilleran foreland basin.

INTRODUCTION

During the Silurian, a characteristic suite of brachiopods including several endemic genera such as *Australina*, *Harringtonina*, *Anabaia*, *Castellaroina*, and *Amosina* developed along the South American margin of Gondwana, over and around the Brazilian craton and in Northwestern Africa (Figure 1). The only South American brachiopod faunas with a different biogeographic signature are those of the Merida Andes (Venezuela), which belong to the North American Silurian Realm (Boucot et al., 1972). On the basis of the almost complete absence of carbonates and reefs in these Silurian successions, and presence of widespread glaciogenic rocks in the underlying beds, Boucot (1974) considered that the 'Malvinokaffric' brachiopod faunas flourished in a cool to cold water environment. Low diversity of these associations, which contrasts with high diversity North American Silurian Realm faunas, also has been argued as evidence for temperate to peripolar waters (Cocks, 1972; Boucot, 1974), although recently Boucot & Racheboeuf (1993) suggested that it could

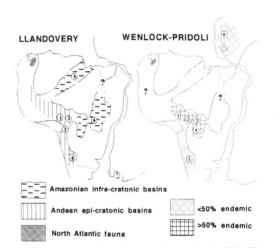

Figure 1. Silurian fossil localities and sedimentary basins of the Afro-South American realm. Key for the Llandovery: 1=Peru; 2=Argentina; 3=Bolivia; 4=Argentina; 5=Brazail; 6=Paraguay; 7=South Africa. For the Wenlock-Pridoli: as before, but 5=Argentina; 6=Senegal; 7=Guinea.

be related locally to low oxygen bottom conditions. Moreover, the distribution of the 'Malvinokaffric' brachiopod associations is consistent with the high latitude placement of these Gondwana regions in all of the paleogeographic reconstructions based on paleomagnetic data (i.e. Cocks & Fortey, 1990; Scotese & McKerrow, 1990), as well as occurrence of glacial diamictitites in the Early Silurian of Brazil (Grahn & Caputo, 1992).

The name 'Malvinokaffric' has usually been applied to Silurian sequences as an extension of the Devonian 'Malvinokaffric Province' of Richter & Richter (1942). As remarked earlier by Cocks & Fortey (1988), despite its generalised use, this name for the Silurian is incorrect because no Silurian sequences are known in the Malvine (Falkland) Islands, and, we add, the name does not reflect the importance of Silurian outcrops in almost the entire South American continent. The name 'Clarkeia Fauna', usually applied to the Llandovery to Pridoli brachiopod faunas of this realm is not completely adequate because, despite the record of one Llandoverian species in Perú (Laudacher et al., 1982), the rhynchonellid genus *Clarkeia* becomes a widespread element in South America and NW Africa only from the Wenlock onwards. Consequently, the name 'Afro-South American Realm' is herein proposed and defined for the African and South American areas characterised by brachiopod assemblages formerly included in the Silurian 'Malvinokaffric' Realm.

Brachiopod species which were classically considered as

Table 1. Geographic distribution of A/SA Realm brachiopod taxa in the Llandovery.

Species	Peru	Bolivia	Argentina Puna	Argentina Prec.	Rarag	Brazil	South Africa
Cordatomyonia umangoensis			*				
Orthostrophella sp.		*					
Calmanella cf. testudinaria				*			
Dalmanella sp.			*				
Dalejina sp.		*					
Heterorthella freitana					*	*	
H. africana							*
H. precordillerana				*			
H. sp.	*						
H.? sp.			*				
Dedzetina? sp.				*			
Plectodonta sp.		*					
Rafinesquina reliqua				*			
Leptaena? sp.				*			
Leptaena "rhomboidalis"		*					
Aphanomena chilcaensis				*			
A. aff. mullochensis				*			
A. sp.			*				
A.? conradii						*	
Eostropheodonta sp.	*						
E. discumbata							*
Fardinia sp.		*					
Protochonetes sp.		*					
Cryptothyrella cf. crassa		*					
C.? sp.	*						
Atrypa "reticularis"		*					
Homoeospira antiqua				*			
H. carinata				*			
Plectothyrella haughtoni							*
Anabaia paraia	*					*	
A.? sp.			*				
Eocoelia paraguayensis						*	
Clarkeia durrelli	*						
Rostricellula sp.				*			
Total species	5	6	5	12	3	2	3
Total genera	5	6	5	10	3	2	3
% endemic taxa	60	0	29	25	33	100	33
Total genera (all localities)	21%						
endemic genera (all localities)	14.3%						

Table 2. Geographic distribution of A/SA brachiopod taxa in the Wenlock-Pridoli.

Species	Peru	Bolivia orien.	Bolivia occi.	Argentina NW	Argentina Prec.	Guinea
Orthostrophella sp.					*	
Heterorthella freitana	*					
H. tacopayana		*	*			
H. zaplensis				*		
H. tondonensis						*
Dedzetina? silurica					*	
Salopina sanjuanensis					*	
S. cf. missendenensis					*	
I. (Protocortezorthis) cuyanum					*	
L. (Leangella) mutabilis					*	
L. (Leangella) sp.				*		
P. (Plectodonta) minima					*	
Costellaroina fascifer					*	
C. sp.				*		
C. telimelensis						*
Leptaena argentina					*	
Leptaena sp.					*	
Amosina fuertensis	*				*	
A. paolae					*	
A. pediculorum					*	
Australina jachalensis			*		*	
Coelospira expansa					*	
C. sp.			*			
Clarkeia antisiensis	*	*	*			
C. ovalis				*		
C. bodenbenderi					*	
C. deflexa					*	
C. tambolarensis					*	
C. alta					*	
C. sp.						*
Howellella? sp.					*	
Harringtonina acutiplicata			*		*	
H. sp.				*		
cf. Anabaia sp.						*
Ancillotoechia cooperensis			*			
Ancillotoechia? sp.						*
Total genera	3	4	3	6	14	5
% endemic genera	100	100	67	67	30	80
Total genera (all localities)	17					
Endemic genera (all localities)	41.2%					

distinctive of this Silurian realm include *Clarkeia antisiensis* (Orbigny), *Castellaroina fascifer* (Conrad), *Leptaena argentina* (Thomas), *Australina jachalensis* Clarke, *Harringtonina acutiplicata* (Kayser), *Amosina fuertensis* (Kayser), and *Heterorthella freitana* (Clarke) (Berry 7 Boucot, 1972; Cocks, 1972; Boucot, 1975). In the last years, numerous taxa have been added, especially from the Argentine Precordillera (Benedetto & Toro, 1989, Benedetto et al., 1992a; Benedetto, 1995, Benedetto et al., in press), the NW and Puna basins of Argentina (Isaacson et al., 1976; Benedetto & Sánchez, 1990; Benedetto, 1991) and NW Africa (Racheboeuf & Villeneuve, 1992. More than 30 genera and species in the Silurian Argentine Precordillera and in the Central Andean basin, exceed those of infracratonic basins of Paraguay and Brazil (Tables 1-2). This, in part, might be due to incomplete studies of these brachiopods, but it seems more likely that low diversity of these assemblages was related to paleogeographic and/or environmental factors (Benedetto et al., 1992b). New taxonomic information indicates that, (1) neither the geographic distribution nor the taxonomic composition of the Afro-South American Realm brachiopods is uniform in the Silurian, and (2), the amount of cosmopolitan or North Atlantic Realm taxa in the Llandovery and in the post-Llandovery successions of the Central Andean basin (western part of the realm) is notably higher than recognised previously .

Considered globally, the Afro-South American Realm is typified by various species of *Heterorthella*, commonly represented in Llandovery, Wenlock and Ludlow/Pridoli beds at many localities in Argentina, Bolivia, Paraguay, Brazil, NW and S

Africa (Tables 1-2). The genus *Clarkeia* is a rare taxon in the middle-late Llandovery, but in younger rocks it becomes useful as a biogeographic indicator, together with several other endemic genera. Another striking feature of this realm, in contrast to the North American Silurian Realm, is the absence of Pentameroidea and Atrypoidea, and the extreme paucity of Spiriferacea, the latter represented only by a few specimens of *Howellella* ? in Wenlock strata of the Precordillera basin. Finally, the Afro-South American Realm covers a time span from early Llandovery to Earliest Devonian (Early Lochkov: Gedinne), although a late Ashgill (Hirnantian) age for the earliest representatives of the Realm cannot be rejected.

TAXONOMY AND STRATIGRAPHY

Inasmuch as no biogeographic assessment is possible without a consistent taxonomic framework, it is necessary to re-evaluate herein some Afro-South American brachiopods as well as their stratigraphic ranges.

Two species of the rhynchonellid genus *Clarkeia* have been recorded in the Llandovery: *C. durreli* Boucot, from the Upper Calapuja Formation of Peru (Laubacher et al., 1982) and *C. antisiensis* (d'Orbigny) from the Ayala Sandstone of Paraguay (Boucot et al.,1991). Paleontological information from the upper Calapuja Fm. is too limited for an accurate age assignment, but according to Laubacher et al. (1982:1143) an early Llandovery age seems to be more consistent. The specimens from Perú undoubtedly belong to the genus *Clarkeia*, but those from the same levels assigned to *Anabaia paraia* (=*Harringtonina paraguayensis*) may correspond to juvenile shells of *C.durrelli*). With regard to *C. antisiensis* from Paraguay, study of a collection made recently by us from the Ayala Fm. indicates that the cardinalium strongly differs from that of the genus *Clarkeia* , and consequently this material needs to be revised. The classic species *C. antisiensis* seems to be restricted to Bolivia. By contrast, in the Silurian of Argentina, several species of this genus have been recognized, including *C. ovalis* Benedetto (1991) from the Lipeón Formation (NW Argentina), and the species *C. bodenbenderi* (Kayser), *C. alta* Benedetto, *C. deflexa* Benedetto, and *C. tambolarensis* Benedetto from the Argentine Precordillera (Benedetto et al., in press). The causes for this speciation are not yet understood (Sánchez & Benedetto, 1993).

Melo & Boucot (1990) pointed out that the genus *Harringtonina* Boucot is a junior synonym of *Anabaia* Clarke, whose type species, *Anabaia paraia*, is from the Trombetas Formation of Brazil. However, the specimens of *Harringtonina acutiplicata* (Kayser) fromt the Argentine Precordillera are very distinct morphologically from the Brazilian species (Benedetto, in Sánchez et al., 1991). Even though both genera are closely related, the differences in the cardinalium, specially in the adult specimens, indicate that *Harringtonina* is a valid genus. The taxonomic panorama of the Afro-South American leptocoeliid rhynchonellids is still more complicated by the fact that *Harringtonina paraguayensis* Boucot, from the Llandovery of Perú (=*Anabaia paraia* according Melo & Boucot, 1990) was probably based on juvenile shells of *Clarkeia durrelli*, present in the same levels from which *H. paraguayensis* was obtained (Benedetto, 1988). Indeed, a comparison between *Harringtonina*, *Anabaia* and *Clarkeia* suggests that these members of the family Leptocoeliidae, as well as other unknown taxa (i.e. '*Clarkeia*' from Paraguay) are phylogenetically related, and members of an endemic stock of diverse rhynchonellids of the Silurian and Early Devonian (Benedetto, in prep.). The earliest representative of this stock seems to be *Anabaia* ? from the Early Llandovery (Rhuddanian) La Chilca Formation of the Precordillera basin (Benedetto, 1995). A younger age (Late Telychian to earliest Wenlock) is suggested by Grahn & Paris (1992) for the fossiliferous levels of the Trombetas Formation (Amazonas basin of Brazil) bearing *A. paraia* and *Heterorthella freitana*.

SPATIAL AND TEMPORAL DISTRIBUTION OF AFRO-SOUTH AMERICAN FAUNAS

The time interval of the proposed Afro-South American Realm is

Table 3. Brachiopods of the Los Espejos Formation in Argentina

Species	Bolivia	Prec.	Seneg.
Orthostrophia meridionalis		*	
cf. *Heterorthella? tacopayana*			*
Salopina sp.			*
I. (Tyersella) megamyaria		*	
Molongella keideli		*	
Leptaena sp.		*	
Amosina tabacucensis	*		
Amosina pediculorum		*	
Australostrophia senegalensis			*
Clarkeia sp.			*
Ancillotoechia? sp.			*
Australocoelia intermedia		*	
Total genera	1	6	5
% endemic genera	100	33	60
Total genera (all localities)	11		
Endemic genera (all localities)	45%		

Silurian. The oldest faunal record in this realm may be that for the South African Cedarberg Formation (Disa Member) faunas, with *Plectothyrella* and *Hirnantia*. Cocks & Fortey (1986) accepted a Hirnantian age for this member, although evidence is inconclusive, with *Heterorthella* suggesting a Llandovery, rather than Ashgill age (Benedetto, 1995). The upper limit of the Afro-South American realm fauna coincides approximately with the Silurian-Devonian boundary, although some taxa have been recorded from the earliest Devonian. In Bolivia, in both the middle Catavi Fm., and the top of the Tarabuco Fm., *Amosina tarabucensis* is associated with palynomorphs which indicate an Early Lochkov/Gedinne age (Lower Devonian) (Racheboeuf & Isaacson, 1993). These levels appear 20-30m above the last record of *Clarkeia antisiensis*. According to Racheboeuf et al. (1993), the palynomorphs associated with *Clarkeia antisiensis* indicate a Pridoli (Upper Silurian) age. The uppermost part of the Los Espejos Formation (Precordillera Basin of Argentina) has yielded a brachiopod fauna including *Amosina* and an *Australocoelia* species which could be an intermediate between *Harringtonina* and the *Australocoelia-Pacificocoelia* stock (Table 3). The brachiopod assemblage indicates and Early Lochkov/Gedinne age (Benedetto et al., 1992a).

The Bafata Group (Bové basin) of East Senegal, NW Africa, contains *Australostrophia senegalensis* associated with *Clarkeia* and *Heterorthella* ? *tacopayana* (Racheboeuf & Villeneuve, 1989: see Table 3). The age of this fauna is poorly constrained, but on the basis of the presence of the Early Devonian genus *Australostrophia* and the tentaculitid *Dicricoconus* (described by Lardeaux, in Drot et al., 1978), an Early Lochkovian age would be accepted. If correct, this is the youngest record of *Clarkeia* and *Heterorthella*. In summary, the evidence from the Argentine Precordillera, Bolivia and NW Africa indicates that scattered members of the Afro-South American Realm extend to the earliest Devonian. The end of the Early Lochkov/Gedinne is marked by the extinction of the last Afro-South American endemics, and the apparition of quite different brachiopod South American and African assemblages which characterise the Devonian 'Malvinokaffric Realm'.

To evaluate biogeographic changes within the Afro-South American Realm through the Silurian, the faunal distribution and composition is analysed for two time intervals, the Llandovery, and post-Llandovery (Tables 1-2). There are marked differences in faunal composition between Early and Late Silurian, and between localities included in the A/SA Realm. One factor utilised to compare brachiopod faunas at each time-interval was the percentage of endemic general. Of the 28 genera recorded in the realm, only *Amosina, Castellaroina, Anabaia,* and *Australina* are considered endemics (though the lissatrypoid *Australina* has been reported both from the Prague Basin and China, and *Anabaia* from Anticosti in Canada). *Clarkeia* was classically considered one of the most characteristic genera, but it is known outside of South America and Africa in beds of Ludlow age from the Prague basin (Havlicek & Storch, 1990), and in the Pridoli of the Armorican Massif (Babin et al., 1979). Likewise, the type species of *Heterorthella* comes from the Wenlock of Nova Scotia (Harper et al., 1969). However, we consider herein the last two genera as 'endemic' taxa because their abundance and diversity in the A/SA realm is more striking than their record outside.

Llandovery. As Benedetto et al., (1992b) pointed out, the Llandovery genera show two major areas of distribution, (1) an intracratonic region, including the Amazonas Basin (Brazil), and the Paraná Basin (Paraguay), characterised mainly by endemic taxa, and, (2) a pericratonic Andean region, which includes the Andes of South America (Perú, Bolivia, W

Argentina), and South Africa, characterized by a mixture of endemic and exotic taxa (Figure 1). In the pericratonic region only two Afro-South American genera have been recorded, *Heterorthella* and *Anabaia* (Table 1). The Andean region contains more diverse assemblages with some local elements such as *Heterorthella* associated with many North Atlantic and cosmopolitan genera, with a low percentage of endemic taxa (14%). The only genus shared by all localities is *Heterorthella*, whereas both *Anabaia* and *Clarkeia* show a more restricted distribution. The remaining genera appear restricted to one locality (Table 1).

Post-Llandovery. Approximately at the Llandovery-Wenlock boundary, an important faunal turnover occurs in the Afro-South American Realm. This event was analysed for the Argentine Precordillera by Sànchez et al. (1993), who suggested that it was the result of abrupt environmental changes related to a relative sea level rise, or regional transgression. This turnover noticeably affected exotic genera which became extinct, or were replaced by other taxa. Immigrants were relatively widespread, occurring in the European and American provinces of the North Atlantic Realm [*Orthostrophella*, *Salopina*, *Isorthis* (*Protocortezorthis*), *Isorthis* (*Tyersella*), *Leangella*, *Coelospira*, *Ancillotoechia*], and some were shared with the Uralian-Cordilleran Region. The Lochkov genus *Molongella* has only been found in Australia (Sino-Australian Province), and the Czech Republic (European Province). For endemic taxa, several features were noted: (1) expansion of *Clarkeia* through the Realm, (2) the spatial distribution of *Heterorthella* appears reduced because it is absent in the post-Llandovery of the Precordillera basin, (3) some endemic taxa show dis-junct distribution, e.g. *Castellaroina*, recorded in NW and Precordillera basins of Argentina and in Guinea (Africa). Taxa with restricted distributions were *Harringtonina* (three localities), and *Australina* and *Amosina* (two localities:Table 2). *Ancillotoechia*, *Coelospira* and *Leangella* (*Leangella*) were found at two localities, but other exotic genera are present at a only a single locality.

DISCUSSION AND CONCLUSIONS

The ranking of the Afro-South American region as a Realm (Malvinokaffric Silurian Realm, Boucot, 1975) is based on low diversity values, rather than on percentage of endemic genera. As Boucot (1975) stated, diversity (total number of taxa) may be considered together with percentage of endemic taxa. In the A/SA Realm, endemic genera form <20% during the Llandovery (Table 1), and c.47% in the post-Llandovery interval (Table 2). Nevertheless, it appears clearly distinguishable from other coeval biogeographic entities. Percentages of endemic genera from all the localities have been plotted for the Llandovery through Devonian (Figure 2). Endemic genera gradually increase, slowly in the Early Silurian (Llandovery), and faster later. Maximum endemism was attained in the Late Silurian (Ludlow-Pridoli). Definition of the Realm was a gradual process, and was achieved only from the Wenlock.

During the Llandovery, the Andean faunas (Puna region, Pojo locality in Bolivia, Precordillera Basin) are characterized by relatively high diversity and high percentage of non-endemics, many of which were shared with the North Atlantic Region. This fact led Isaacson & Boucot (1976), and Boucot (1990), to include the Bolivian-N Argentinian localities in the North Silurian Realm. However, *Heterorthella* in the Salar del Rincón Formation of the Puna region (Benedetto & Sánchez, 1990) strongly suggests that this region belongs to the Afro-South American Realm [though the type is from Nova Scotia. The absence of endemic taxa in the Pojo succession remains unclear, but it may be due to lack of sampling. Nevertheless,

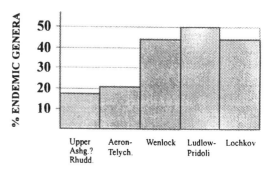

Figure 2. Degrees of endemism with time in South America.

taking into account that this region was physically part of the Central Andean basin, we include it in the Afro-South American Realm. The non-endemic Andean taxa, *Homoeospira*, *Dalmanella*, *Cryptothyrella*, and *Aphanomena*, show more limited migration than endemic taxa, as suggested by their absence in the intracratonic basins. By contrast, the widespread occurrence of *Anabaia* and *Heterorthella* may be due, at least in part, to the fact that these genera were eurytopic. From stratigraphic evidence, these genera presumably originated in extracratonic basins in the earliest Llandovery, and afterwards migrated to intracratonic basins (Amazonas region), where they are recorded in the Late Llandovery.

As a result of Llandovery-Wenlock faunal turnover, brachiopod faunas of the A/SA Realm were strongly depleted. During the regional Wenlock transgressive event, the preceding shallow water brachiopod communities were disrupted and several genera became extinct (Sánchez, et al.,1993). In the Precordillera basin, the open platform was colonised by low-diversity communities dominated by *Harringtonina* and *Australina*. Similar assemblages were found in black mudstones of the Kirusillas Fm. in Bolivia. During the Ludlow-Pridoli, high diversity communities composed of endemic genera and new North Silurian Realm immigrants, such as *Salopina*, *Isorthis*, *Coelospira*, and *Ancillotoechia* flourished in near-shore, well-oxygenated muddy deposits. To explain marked differences in diversity and taxonomic composition within the A/SA Realm, Sánchez et al., (1991) compared the Lipeón Formation (NW Basin of Argentina) with the Los Espejos Formation (Precordillera Basin), and concluded that they probably were related to energy on the shelf environment. *Clarkeia* (Boucot ,1975), seems to have been restricted to high-energy, shallow water sandstones (Benthic Assemblage 2). In the muddy succession of the NW Argentina basin (Lipeón Formation), this genus appears only in a thin sandstone bed (Sánchez et al., 1991), and the partially anoxic mudstones contain low diversity *Heterorthella* -dominated assemblages. We conclude that the diversity and distribution of the brachiopod faunas within the A/SA Realm were controlled primarily by ecologic factors. At continental scale, the distribution of non-endemic dominated brachiopod assemblages throughout the Silurian and earliest Devonian seems to have been related to relative position of the basins with respect to the continental plate. Thus, areas located along, or near the continental margins (i.e. Andean region), received a greater influx of North American Silurian Realm elements than those developed inside the plate (i.e. Amazonas basin).

REFERENCES

BABIN, C., J. DEUNFF, M. MELOU, F. PARIS, A PELHATE, Y. PLUSQUELLEC & P.R. RACHEBOEUF. 1979. La coupe de Porz ar Vouden (Pridoli de la Presqu'île de Crozon) Massif Armoricain, France, Lithologie et stratigraphie. Palaeontographica. A164:52-84.

BENEDETTO, J.L. 1988. Nuevos datos sobre la morfología, ontogenia y distribución estratigráfica de *Harringtonina acutiplicata* (Kayser, 1897) del Silúrico de la Precordillera Argentina. Ameghiniana 25 (2):123-128

___1991. Braquiópodos silúricos de la Formación Lipeón, flanco occidental de la Sierra de Zapla, Provincia de Jujuy, Argentina. Ameghiniana 28(1-2): 111-125.

___(in press). Braquiópodos del silúricos temprano Malvinocáfrico (Formación La Chilca), Precordillera Argentina. Geobios.

___& T.M. SANCHEZ, 1990. Fauna y edad del estratotipo de la Formación Salar des Rincón (Eopaleozoico, Puna Argentina). Ameghiniana 27: 317-326.

___& B.A. TORO, 1989. El género *Coelospira* Hall (Brachiopoda) en el Silúrico de la Precordillera de San Juan, Argentina. Ameghiniana 26: 139-144.

___,P.I. PERALTA & T.M. SANCHEZ (in press). Taxonomía y biometría de las especies del género *Clarkeia* Kozlowski en la Precordillera Argentina. Ameghiniana.

___,P.R. RACHEBOEUF, Z. HERRERA, E. BRUSSA & B.A. TORO, 1992a. Brachiopodes et biostratigraphie de la Formation de Los Espejos, Siluro-Dévonien de la Précordillère (NW Argentine). Geobios 25 (50): 599-637.

___, T.M. SANCHEZ & E. BRUSSA. 1992b. Las Cuencas Silúricas de América Latina, p.119-148. In, J.C.Gutierrez Marco, J. Saavedra & I. Rábano (eds.), Paleozoico Inferior de Ibero-América. Universidad de Extremadura.

BERRY, W.B.N. & A.J. BOUCOT. 1972. Correlation of the South American Silurian rocks. Geological Society America Special Paper 133:1-59.

BOUCOT, A.J. 1974. Silurian and Devonian biogeography. In, C.A. Ross (ed.) Paleogeographic provinces and provinciality. Society Economic Paleontologists Mineralogists Special Publication 21: 165-176.

___1975. Evolution and extinction rate controls. Developments Paleontology Stratigraphy 1:1- 427.

___1990. Silurian biogeography. In, W.S. McKerrow, W.S. & C.R. Scotese (eds.), Palaeozoic palaeogeography and biostratigraphy. Geological Society Memoir 12: 191-196.

___& P.R. RACHEBOEUF. 1993. Biogeographic summary of the Malvinokaffric Realm Silurian and Devonian fossils. In, R. Suarez Soruco (ed.), Fósiles y facies de Bolivia, 2, Invertebrados y paleobotánica. Revista Técnica Y.P.F.B. 13-14:71-75.

___, J.G. JOHNSON & R. SHAGAM, 1972. Braquiópodos del Silúrico de los Andes de Mérida, Venezuela. Actas IV Congreso Geológico Venezolano 2:585-656.

___, J.H. DE MELO, E.V. SANTOS NETO & S. WOLFF, 1991. First *Clarkeia* and *Heterorthella* (Brachiopoda; Lower Silurian) occurrence from the Paraná Basin in Eastern Paraguay. Journal Paleontology 65 (3):512-514.

COCKS, L.R.M.1972. The origin of the Silurian *Clarkeia* shelly fauna of South America, and its extension to West Africa. Palaeontology, 15 (40: 623-630.

___& R.A. FORTEY, 1986. New evidence on the South African Lower Palaeozoic: age of fossils reviewed. Geological Magazine 123 (4): 437-444.

___& ___1988. Lower Palaeozoic facies and faunas around Gondwana. In, M.G. Audley-Charles & A. Hallam (eds.), Gondwana and Tethys. Geological Society Special Publication 37:183-200.

___& ___1990. Biogeography of Ordovician and Silurian faunas. In, W.S. McKerrow & C.R. Scotese (eds.), Palaeozoic palaeogeography and biostratigraphy. Geological Society Memoir 12: 97-104.

DROT, J., H. LARDEAUX & A. LEPAGE, 1978. Sur la découverte de Silurien supérieur au sommet de la série de Youkounkoun au Sénégal oriental : implications paléogéographique et structurale. Bulletin Société Géologique Minéralogique Bretagne. C1:7-30.

GRAHN, Y. & M.V. CAPUTO, 1992. Early Silurian glaciations in Brazil. Palaeogeography, Palaeoclimatology, Palaeoecology 99:9-15.

HARPER, C.W., A.J. BOUCOT & V.G. WALMSLEY, 1969. The rhipidomellid brachiopod subfamilies Heterorthinae. Journal Paleontology 43: 74-92.

HAVLICEK, V. & P. STORCH, 1990. Silurian brachiopods and benthic communities in the Prague Basin (Czechoslovakia). Rozpravy Ustredniho Ustavu Geologickeho 48:1-275.

ISAACSON, P.E., B. ANTELO & A.J. BOUCOT, 1976. Implications of a Llandovery (Early Silurian) brachiopod fauna from Salta Province, Argentina. Journal Paleontology 50:1102-1112.

LAUBACHER, G., A.J. BOUCOT & J. GRAY, 1982. Additions to Silurian stratigraphy, lithofacies, biogeography and paleontology of Bolivia and southern Perú. ibid. 56: 1138-1170.

MELO, GONCALVES de, J.E. & A.J. BOUCOT, 1991. *Harringtonina* is *Anabaia* (Brachiopoda, Silurian, Malvinokaffric Realm). ibid. 64: 363-366.

RACHEBOEUF, P.R. & P.E. ISAACSON, 1993. Los chonetoideos (Braquiópodos) silúricos y devónicos de Bolivia. In: Suarez Soruco, R. (ed.), Fósiles y facies de Bolivia, 2, Invertebrados y paleobotánica. Revisa Técnica Y.P.F.B., 13-14:99-119.

___& M. VILLENEUVE, 1989. *Australostrophia senegalensis* n.sp.: first chonostrophiid brachiopod (Chonetacea) from NW Africa. Implications for the northwestern Gondwanaland margin. Neues Jahrbuch Geologie Paläontologie Monatshefte12:737-748.

___& ___1992. Une faune Malvino-Cafre de Brachiopodes Siluriens du Bassin Bové (Guinée, Ouest de l'Afrique). Geologica Palaeontologica 26: 1-11.

___,A. LEHERISSÉ, F. PARIS, C. BABIN, F. GUILLOCHEAU, M. TRUYOLS-MASSONI & R. SUAREZ SORUCO, 1993. Le Dévonien de Bolivie: biostratigraphie et chronostratigraphie. Comptes-rendus Académie Sciences, Paris 317: 795-802.

RICHTER, R. & E. RICHTER, 1942. Die Trilobiten der Weismes-Schichten am Hohen Venn mit Bemmerkungen über die Malvinocaffrische Provinz. Senckenbergiana 25(1-3): 156-179.

SANCHEZ, T.M. & J.L. BENEDETTO, 1993. Distribución estratigráfica, paleoecología y biogeografía del género *Clarkeia* en el Silúrico Gondwánico. In, R. Suarez Soruco (ed.), Fósiles y Facies de Bolivia, 2, Invertebrados y paleobotánica. Revista Técnica Y.P.F.B., 13-14:93-98.

___,& R.A. ASTINI. 1993. Eventos de recambio faunístico en secuencias deposicionales del Ordovícico tardío-Devónico temprano de la Precordillera de San Juan, Argentina. XII Congreso Geológico Argentino y II Congreso de Exploración de Hidrocarburos 2: 281-288.

___, B.G. WAISFELD & J.L. BENEDETTO, 1991. Lithofacies, taphonomy and benthic assemblages in the Silurian of Western Argentina: A review of the Malvinokaffric Realm communities. Journal South American Earth Sciences 4: 307-329.

SCOTESE, C.R. & W.S. McKERROW, 1990. Revised World maps and introduction. In, W.S. McKerrow & C.R. Scotese (eds.) Palaeozoic palaeogeography and biogeography. Geological Society Memoir 12:1-21.

SILURIAN BRACHIOPOD BIOSTRATIGRAPHY, NE TIMAN-N URALS, RUSSIA

TATYANA M. BEZNOSOVA

Institute of Geology, Russian Academy of Sciences, Pervomaiskaya Per. 54, Syktyvkar 167610, Russia

ABSTRACT- Brachiopods, one of the most abundant benthic elements in the Silurian succession of northeast Timan and the northern (subpolar) Urals, are aligned into a unified biostratigraphic scheme. Dolomites, evaporites, limestones and reefal carbonates were deposited in settings ranging from a western and central (east European) shallow shelf, to eastern (Uralian) slope and basinal environments. Zonal brachiopods include pentamerids, athyrids, atrypids, spiriferids and strophomenids, correlated with sections from the Baltic and southern Urals.

INTRODUCTION

Silurian rocks are widespread in the northeastern Timan-Urals region. They were deposited as shallow water carbonates (mostly dolomites), evaporites and siliciclastics on the Russian Platform, and as coeval slope and basinal facies belonging to the Uralian 'geosyncline' in the east (Figure 1). Four structural and facies zones are characterised by differing rock complexes. In the westernmost sections, studied mainly from subsurface boreholes, Silurian rocks are represented by poorly fossiliferous siliciclastics and carbonates (Zone I - the

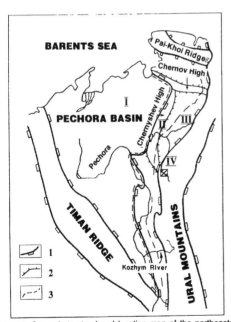

Figure 1. General structural and location map of the northeastern Timan-Pechora and subpolar Urals, Russia. I= shallow siliciclastic-carbonate Pechora Basin; II= Mikhailovskoe-Vaigach area; III= Belsk-Niyayu reefal carbonates; IV= Lemva allochthonous, deep water cherty carbonates, shales.

eastern Pechora Basin). These accumulated within an infracratonic, shallow shelf setting. In the Mikhailovskoe-Vaigach (Zone II) and Belsk-Niyayu (Zone III) zones, carbonates were formed in middle and distal, relatively deeper, shelf environments. Reefal carbonates are characteristic for the Belsk-Niyayu Zone (Zone III). In the Lemva allochtonous zone (Zone IV), sediments consist of limestones, cherts, and shales containing graptolites, usually interpreted as having been deposited in continental slope and basinal settings (Dedeev, 1989; Yudin, 1990).

The geology, stratigraphy and faunal distribution of the Silurian successions of the NE Timan-Urals region have been studied by field mapping, paleontological and lithological studies since the 1920s. These are documented mostly in numerous unpublished accounts, papers, and monographs (Voinovskii-Kriger, 1945, 1956; Pershina,1960;1962;1971). More recently, Silurian stratigraphic data has been summarised by a team of the paleontologists and lithologists, and published as regional, correlative stratigraphic schemes foor the Silurian (Kaljo,1987; Antsygin et al., 1994). The Silurian System of the northern Urals, and adjacent Eastern-European Platform, is divided intoLower and Upper. A lower part comprises the Dzhagal and Sedel groups (Llandovery and Wenlock), with the former including, in ascending order, the Yareney, Lolashor and Filipel horizons, and the latter the Marshrutniy and Ust-Durnayu horizons. The Upper Silurian is composed of the lower Gerdyu Group, in turn including the Padymeytyvis and Sizim Horizons, and the upper Greben Group, composed of the Karpovskiy and Belushya Bay Horizons (Antsygin et al.,1994).

In NE Timan and the N Urals, the Silurian rests on Upper Ordovician evaporites and carbonates (Zone I) and carbonates (Zone II-III). The Ordovician/Silurian boundary is placed at the top of beds with the brachiopod *Proconchidium muensteri*, characteristic for the Ashgill in the area. The boundary is defined within a lithologically homogeneous dolomite with thickness ranging from 40-100m, overlain by dolomites with the brachiopod *Virgiana barrandei* Billings, and conodont *Distomodus kentuckiensis* (Branson & Mehl), characteristic for the early Llandovery (Rhuddanian: Beznosova,1989,1995). The spatial and temporal distribution of brachiopods in this siliciclastic-carbonate and carbonate Silurian succession was studied to develop a Silurian brachiopod zonation scheme. The best Silurian sections are located in the western 'subpolar' Urals of the Kozhym River (Figures 2-3). The area provides a particularly suitable example for study in light of excellent exposure conditions, its richly fossiliferous localities, and most complete stratigraphic successions from the Upper Ordovician through Lower Devonian. The Kozhym River is proposed as type section for the Silurian biostratigraphic subdivision of the Timan-Urals region (Antoshkina et al.,1983, 1987).

BRACHIOPOD ZONATION

Virgiana barrandei- Zygospiraella duboisi Zone. For many years, the Lower Llandoverian deposits of the northern Urals were named the *Virgiana barrandei - Stricklandia lens* Zone. However, in the study area the diagnostic species *Stricklandia lens* (Sowerby) has not yet been found, and therefore the name '*Virgiana barrandei - Zygospiraella duboisi*' Zone was introduced instead (Beznosova,1989). This zone corresponds to most of the Yareney Horizon except for a lower brachiopod-barren unit 30- 50m thick. A late-middle Llandovery age is determined on the basis of characteristic brachiopods, the tabulate corals *Palaeofavosites alveolaris* (Goldfuss), *Catenipora gothlandica* Yabe, the rugose coral *Dalmanophyllum dalmani* (Edwars & Haime), and conodonts *Pedavis vindemus* Melnikov, and *Ozarkodina corona* Melnikov. The lower barren unit of the Yareney Horizon can be correlated as lower Llandovery by *Distomodus kentuckiensis* (Branson & Mehl: in, Antoshina et al., 1987).

The type section of the zone is at Kozhym River (Figure 2; loc.196; Figure 3), and represented by massive dolomites, interbedded with dolomites containing chert nodules and breccias. A monotypic pentamerid community of *Virgiana* is characteristic for the lower part. The upper part of the zone consist of irregularly bedded dolomites with numerous burrows, and interbeds of dolomites holding the atrypid *Zygospiraella duboisi* (Verneuil). The brachiopod zonal assemblage is dominated by *Zygospiraella*, *Protatrypa*, and '*Nalivkinia*', but *Virgiana barrandei* is also rarely found as disarticulated shells and fragments. Numerous tabulate corals and stromatopoids occur throughout the zone. The thickness of this zone ranges from 118m to 200m, and it is recognized in the subpolar Urals, the Chernyshev Rise, Kanin Peninsula and the eastern Pechora Basin.

Pentamerus oblongus Zone. This covers most of the Lolashor Horizon, and is defined as late Llandoverian on the basis of index brachiopods, the typical tabulate coral species *Favosites gothlandicus*, conodonts *Ozarkodina aldzidgei* Uyeno, and ostracodes *Leperditia hisingeri*. Schmidt, and *Hogmochillina maaki* Schmidt. The unit consists of rhythmically alternating limestones and mottled, muddy dolomites containing chert lenses. In its lower part, there are numerous tabulate corals, and in the upper half largely stromatoporoids. The upper boundary of the zone is placed at the top of massive dolomite beds with brachiopods such as *Pentamerus oblongus*. The type section is also at Kozhym River (Figures 2-3: locs108,196), and thickness ranges from 40m in northern Timan, to 265m in the subpolar Urals. In general, the *Pentamerus oblongus* Zone is the part of interregional *Pentamerus oblongus - Harpidium angustum* assemblage (Sapelnikov, 1985). However, additional study should be carried out to refine correlations.

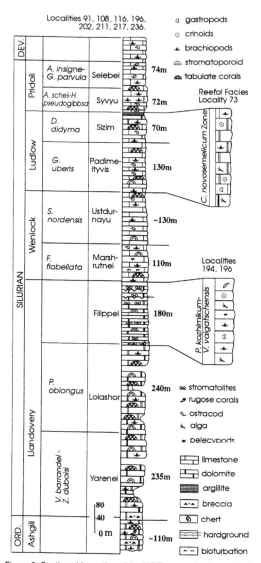

Figure 3. Stratigraphic section of the NE Timan and subpolar Urals region showing the main brachiopod zones.

Virgiana vaigatschensis-Pseudoconchidium kozhimicum Zone. This zone is established only in the reefal type succession in the subpolar and northern Urals, and is defined by the Filippel Horizon, represented by massive, light, 'algal' dolomites holding zonal brachiopod species and ostracodes. A late Llandoverian age was determined by the conodont *Icriodella malvernensis* Aldridge. Shallow-water carbonates of the type section are coeval with thin-bedded dolomites containing stromatolites, stromatoporoids and conodonts. Thickness in the reefal type succession is about 51m, and in other shallow water dolomite sections,180m.

Fardenia flabellata Zone. This corresponds to the Marshrutniy Horizon of Early Wenlockian age, and is also recognised in the Chernyshev High. At the type section it is marked by thickly

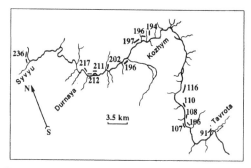

Figure 2. Locality map of the type Kozhym River area, subpolar Urals.

Figure 4. Typical Silurian zonal brachiopods from Timan-Pechora and subpolar Urals areas. **1**, *Holorhynchus giganteus* Kiaer, Kozhym River, loc.198, #198/11088, x1; **2-3,** *Proconchidium muensteri* (St. Joseph), 2, Kozhym River, loc.9, #9/36, x1; 3, Synya, loc.15, #15/204, x1; **4,** *Pentamerus oblongus* Sowerby, Kozhym River, loc.108, #108/25, x2; **5-7,** *Zygospiraella duboisi* (Verneuil): #17/361, x2,5; **8,11,** *Virgiana barrandei* Billings, Povarnitsa River, loc.192, sample 192/702, x1; **9,** *Virgianella vaigatschensis* Nikiforova, Unya River, loc.7, #7/2820, x1; **10,** *Spirinella nordensis* (Lyashenko), Bezymyanniy Creek, loc.2, #2/412, x1; **12,** *Greenfieldia uberis* Modzalevskaya, Padimeytyvis, River, loc.1, #1/59, x1; **13,** *Fardenia flabellata* Beznosova, Marshrutniy Creek, loc.202, #202/522, x2.5; **14,** *Pseudoconchidium kozhimicum* Nikiforova, Unya River, loc.37, #37/2821, x1; **15,23,** *Howellella pseudogibbosa* Nikiforova; 15, Kozhim River, loc. 23, #236/5, x1; 23, borehole 37-Usa, #10/7, x2; **16,** *Didymothyris didyma* (Dalman), Kozhym River, loc.211, #211/62, x1.5; **17,19,** *Collarothyris canaliculata* (Venyukov): borehole 37-Usa, #10/9, x4; **18,** *Atrypoidea scheii* (Holtedahl), Kozhym River, loc.236, #236/8, x1; **20, 21,** *Grebenella parvula* (Chernyshev & Yakovlev), 20 borehole 37-Usa, #10/12, x5; 21, Kozhym, River, loc.236, #236/715, x2.5; **22,** *Atrypoidea insigne* Nikiforova, Kozhym River, loc.236, #236/720, x2.

37

bedded dolomites, with a shelly fauna, alternating with massive coral/stromatoporoid rich limestones and dolomites. Besides *Fardenia*, there are *Protatrypa inflata* Beznosova, and *Leptaena*. Corals and stromatoporoids are often characteristically rock-forming. In the subpolar Urals and the Chernyshev High, brachiopods are associated with the ostracodes *Herrmannina insignis* Abushik, *Giberella prectiosa* Abushik, and *Microcheilinella variolaris*. The latter two species are characteristic only for the lower part of the Wenlock. The lower boundary is underlain by the Filippel Horizon, marked faunally by the first occurrence of index species and lithology (Antoshkina et al., 1987). The upper boundary is placed at the top of dolomites, containing *Fardenia flabellata*, in turn overlain by thinly bedded stromatolitic limestones of the Ust-Dyrnayu Horizon. The type section is near Marshrutnyi Creek (Kozhym River; Figure 2; loc.202), where its thickness is 114.5m.

Spirinella nordensis Zone. This is typified in the Ust-Durnayu Horizon (Beznosova, 1989), except for the lower unit. The lower unit, 60m thick and barren, consists of thinly bedded limestones and dolomites with stromatolite interbeds. The type section is on the Chernov High (Bezymyanniy Creek, loc.2). There, the zone is represented by alternating, thinly bedded, clayey dolomitic limestones, and mottled, massive limestones (Antoshkina & Beznosova, 1988). The lower boundary marks the first occurence of the spiriferid *Spirinella nordensis* (Lyashenko). Other brachiopods include *Atrypoidea lingulata* (Buch), *Atrypa* ex gr. *reticularis* (Linnaeus), and *Hyattidina*, occurring throughout the sequence. They are accompanied by the rugose corals *Phaulactis cyathophylltoides* Ryder, the tabulate corals *Favosites similis* Sokolov, the ostracodes *Eukloedenella grandifabae* Abushik, *Cyrtherellina inornata* Abushik, and *Herrmannina insignis* Abushik. There are also stromatoporoids, gastropods, bivalves, trilobites and crinoids. In the type section, the upper boundary is thought to be at the base of beds with the athyrid *Greenfieldia uberis* Modzalevskaya. Thickness ranges from 60m on the Chernyshev High to 146m on the Chernov High. The *Spirinella nordensis* Zone occurs in the western subpolar Urals (Kozhim, Schugor rivers), northern Urals (Ilych River), Chernov High (Bezymyanniy Creek) and Chernyshev High (Bolshoy Adak, Sharyu rivers).

Greenfieldia uberis Zone. Identified by the Padymeytyvis Horizon, on the Kozhym River (Figure 2; loc.211), the zone consists of limestones and dolomites, alternating with stromatolitic, coral, brachiopod, ostracode and nodular limestones (Antoshkina et al.,1983). The lower boundary of the zone, coinciding with the Wenlock/Ludlow boundary, is placed at the base of a nodular limestone unit with *Greenfieldia uberis* (Antoshkina & Beznosova, 1988), and its upper boundary is placed at the base of beds with *Didymothyris didyma* (Dalman). In the most studied sections of the subpolar Urals, the Chernyshev High, and in the eastern Pechora Basin, brachiopods are only represented by two species *Greenfieldia uberis* and *Lenatoechia clauda* Modzalevskaya. But, in the Chernov High, the brachiopod assemblage comprises *Atrypoidea linguata* (Buch), *A. modesta* (Nikiforova), *Morinorhynchus* cf. *attenuata* (Amsden), *Glassina dissecta* Modzalevskaya, and *Eoreticularia annae* Beznosova. The last species is only known in the stratotype section of the Padymeytyvis Horizon on the Chernov High (Beznosova, 1989). The tabulate corals *Parastriatopora commutabilis* Klaamann, *P. arctica* (Chernyshev), and ostracodes *Signetopsis bicardinata* Aubushik, *Beyrichia parva* Abushik, *Leiocyamus paulus* Zenk, *L. variabilis* Abushik occur with the brachiopods. Thicknesses ranges from 60m in the subpolar Urals, to 150m at the Chernov High. The *Greenfieldia uberis* Zone extends through the western subpolar Urals (Kozhym, Schugor rivers), Chernov High (Padymeytyvis River,

Bezymyanniy Creek), Chernyshev High (Usa, Sharyu, Bolshoy Adak rivers), to subsurface sections in the eastern Pechora Basin.

Didymothyris didyma Zone. This was first discovered by Modzalevskaya (1980) in the upper half of the Luddlow succession, corresponding to the Sizim Horizon (Beznosova, 1989). A Ludlow age is established on the basis of the brachiopods *Didymothyris didyma* (Dalman: type from the lower Hemse Formation, early Ludlow, Gotland), *Conchidium novosemelicum* Nalivkin, the ostracodes *Kiaeria crassa*, *Leperditia quinqueangulata*, *Bingeria infrequense*, and conodonts *Polyganthoides siluricus* (Branson & Mehl), and *Distomodus* ex gr. *dubius* (Rhodes) (Antoshkina et al., 1983; Beznosova, 1984, 1989). The type section is in the subpolar Urals (Kozhym River, locs 211,236; Figure 2). In the European northeast, these deposits are marked by reefal carbonates and carbonates-siliciclastics. Reefs are widespread in the Niyayu River basin (subpolar Urals) and along the Ilych River (northern Urals), and composed of light grey, massive limestones and dolomites, rich in large-shelled brachiopods, tabulate and rugose corals, crinoids and stromatoporoids. The brachiopods are largely pentamerids, e.g.*Conchidium novosemelicum* Nalivkin, *C.* cf. *biloculare* (Hisinger), *C. knighti vogulicum* (Verneuil), *Brooksina conjugula* Khodalevich, *B.* aff. *streisi* Sapelnikov, *Vagranella*, *Didymothyris didyma* (Dalman), *Atrypoidea linguata* (Buch), and *Atrypa* ex gr. *reticularis* (Linnaeus). At present, the upper boundary in the reefal type succession is established only at the Ilych River section, where it is thought to be at the base of a limestone unit with rugose corals characterictic for the Greben Group. In the Niyayu River basin, the boundary is questionable (Antoshkina & Beznosova,1982). Thicknesses reach 600 m.

The carbonate-siliciclastic succession is mainly represented by laminated limestones and dolomites alternating with coquinas, stromatolitic and banded limestones. The assemblage is composed of *Didymothyris didyma*, *Morinorhynchus* cf. *attenuata*, *Conchidium novosemelicum*, and *Atrypoidea linguata*. It also contains tabulate corals, e.g. *Thecia swindernianna* (Goldfuss), *Laceripora cribrosa* Eichwald, *Parastriatopora spinosa* Chernyshev, and the ostracodes named by Abushik *Kiaeria crassa*, *Leiocyamus variabilis*, *L. clausus*, and *L. grandifabae* (Abushik, 1980; Antoshkina, et al., 1983). The brachiopods *Didymothyris didyma*, *Conchidium novosemelicum* and *Atrypoidea linguata* are present in both reefal and carbonate-siliciclastic successions.The upper boundary of the *Didymothyris didyma* Zone coinsides with the Ludlow/Pridoli boundary. In the carbonate succession, it is placed at the base of a shale containing the brachiopods *Hemitoechia distincta* Nikiforova and *Atrypoidea scheii* (Holtedahl). Thickness varies from 50m (Chernyshev High) to 145m (Chernov High). Within the European northeast area, the siliciclastic-carbonate deposits are most widespread. The zonal brachiopod species occur within the northern and subpolar Urals, Chernyshev and Chernov highs, and the eastern Pechora Basin.

Atrypoidea scheii - Howellella pseudogibbosa Zone. Introduced by Beznosova (1989) for the subpolar Urals, where the most complete succession occurs, it corresponds to the Celebey Beds of the Greben Group at the stratotype section (Figure 2), where it consists of rhythmically alternating limestones (nodular, shaly, detrital, with brachiopods, ostracodes), calcareous shales, and shales. An Early Pridolian age is established by the brachiopods *Hemitoechia distincta* Nikiforova, *Pseudohomeospira polaris* Nikiforova, *Collarothyris canaliculata* Modzalevskaya, and ostracodes *Eocloedenella bacata* Abushik, *Calcaribeyrichia grebeni* Abushik, and *Kiaeria*

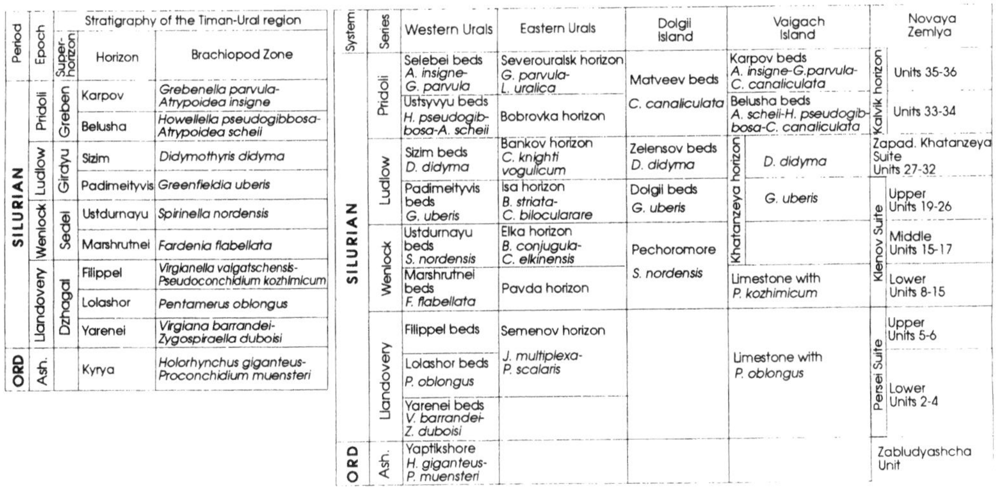

Figure 5. Correlation chart for the Pechora-subpolar Urals region.

Stratigraphy of the Timan-Ural region

Period	Epoch	Super-horizon	Horizon	Brachiopod Zone
SILURIAN	Pridoli	Greben	Karpov	Grebenella parvula-Atrypoidea insigne
SILURIAN	Pridoli	Greben	Belusha	Howellela pseudogibbosa-Atrypoidea scheii
SILURIAN	Ludlow	Girdyu	Sizim	Didymothyris didyma
SILURIAN	Ludlow	Girdyu	Padimeityvis	Greenfieldia uberis
SILURIAN	Wenlock	Sedel	Ustdurnayu	Spirinella nordensis
SILURIAN	Wenlock	Sedel	Marshrutnei	Fardenia flabellata
SILURIAN	Llandovery	Dzhagal	Filippel	Virgianella vaigatschensis-Pseudoconchidium kozhimicum
SILURIAN	Llandovery	Dzhagal	Lolashor	Pentamerus oblongus
SILURIAN	Llandovery	Dzhagal	Yarenei	Virgiana barrandei-Zygospiraella duboisi
ORD	Ash.		Kyrya	Holorhynchus giganteus-Proconchidium muensteri

System	Series	Western Urals	Eastern Urals	Dolgii Island	Vaigach Island	Novaya Zemlya
SILURIAN	Pridoli	Selebei beds A. insigne-G. parvula	Severouralsk horizon G. parvula-L. uralica	Matveev beds	Karpov beds A. insigne-G.parvula-C. canaliculata	Units 35-36 (Kalvik horizon)
SILURIAN	Pridoli	Ustyvyu beds H. pseudogibbosa-A. scheii	Bobrovka horizon	C. canaliculata	Belusha beds A. scheii-H. pseudogibbosa-C. canaliculata	Units 33-34 (Kalvik horizon)
SILURIAN	Ludlow	Sizim beds D. didyma	Bankov horizon C. knighti vogulicum	Zelensov beds D. didyma	D. didyma	Zapad. Khatanzeya Suite Units 27-32
SILURIAN	Ludlow	Padimeityvis beds G. uberis	Isa horizon B. striata-C. bilocularare	Dolgii beds G. uberis	G. uberis (Khatanzeya horizon)	Upper Units 19-26 (Klenov Suite)
SILURIAN	Wenlock	Ustdurnayu beds S. nordensis	Elka horizon B. conjugula-C. elkinensis	Pechoromore		Middle Units 15-17 (Klenov Suite)
SILURIAN	Wenlock	Marshrutnei beds F. flabellata	Pavda horizon	S. nordensis	Limestone with P. kozhimicum	Lower Units 8-15 (Klenov Suite)
SILURIAN	Wenlock	Filippel beds	Semenov horizon			Upper Units 5-6 (Persel Suite)
SILURIAN	Llandovery	Lolashor beds P. oblongus	J. multiplexa-P. scalaris		Limestone with P. oblongus	Lower Units 2-4 (Persel Suite)
SILURIAN	Llandovery	Yarenei beds V. barrandei-Z. duboisi				
ORD	Ash.	Yaptikshore H. giganteus-P. muensteri				Zabludyashcha Unit

lindstroemi (Schmidt). In general, the assemblage is dominated by smooth-shelled brachiopods like *Atrypoidea scheii* (Holtedahl), *A. globa* (Chernyshev), *A. vangirica* Beznosova & Mizens, and *A. pentagonalis* Beznosova & Mizens (Figure 4).

The upper boundary of the *Atrypoidea scheii - Howellella pseudogibbosa* Zone is placed at the top of a dolomite and shale unit with *Grebenella parvula*. Thickness varies from 46m (Chernyshev High) to 130m (Chernov High). The zone occurs in the northern and subpolar Urals, Chernov and Chernyshev highs, and also in numerous boreholes of the Pechora Basin.

Atrypoidea insigna-Grebenella parvula Zone. This was first established in the subpolar Urals along the Kozhym River (Figure 2, loc. 236). The zone corresponds to the Ust--Syvyu Beds of the Greben Group (Beznosova, 1989). A Late Pridoli age is based on the index species, and tabulate corals *Favosites favositiformis*, *Squameofavosites thetidis*, *S. emmonsiaformos*, the rugose corals *Micula simplex* Strelnikov, *Nipponophyllum tardum* (Strelnikov), and ostracodes *Signetopsis michailensis* Zenk, *Leyocyamus limpidus* Gailite. The type section of the *insigne - parvula* Zone includes the Ust-syvyu Beds in the subpolar Urals (Figure 2, loc.236), and consists of massive, platy, shaly, coral, stromatoporoid, brachiopod, and bryozoan-rich limestones, with intercalations of shales. In the upper part of the sequence (the Syvyu Beds, Pershina & Beznosova , 1978), there are laminated dolomites-limestones containing stromatolites. A rich, varied coral fauna is characteristic for the zone (Antoshkina et al., 1983; Beznosova, 1989). The lower part of the zone has brachiopod-coral communities, and the upper brachiopod-ostracode beds. The upper limit of the *Atrypoidea insigne-Grebenella parvula* Zone coincides with the Silurian/Devonian boundary, which is the base of a shale with the Early Devonian brachiopods *Protathyris praecursor* Kozlowski, *Mesodouvillina costatula* (Barrande), and *Howellella angustiplicata* Kozlowski. The zone extends throughout the northern and subpolar Urals, Chernov and Chernyshev highs and Pechora Basin. Silurian brachiopod zones of the Timan-Urals are correlated with the eastern Urals, Dolgiy, Vaigach and Novaya Zemlya (Figure 5).

CONCLUSIONS

On the basis of brachiopods from Silurian strata of the Timan-Urals region nine brachiopod zones have been established. These are, in ascending order, the *Virgiana barrandei - Zygospiraella duboisi*, *Pentamerus oblongus*, *Virgianella vaigatschensis - Pseudoconchidium kozhimicum*, *Fardenia flabellata*, *Spirinella nordensis*, *Greenfieldia uberis*, *Didymothyris didyma*, *Atrypoidea scheii - Howellella pseudogibbosa*, and *Atrypoidea insigne - Grebenella parvula* asssemblage zones. In most cases, the zonal boundaries are defined by a sharp change of brachiopods and mark first and last occurences of diagnostic taxa. Assemblages from the late Ashgill, Llandovery, Ludlow and Pridoli widespread and recognisable in other regions. Wenlock species are largely represented by endemic forms. As a whole, established brachiopods zones correspond to regional stratigraphic 'horizons'.

Acknowledgments

The author thanks the Natural Sciences and Engineering Research Council of Canada (NSERC) for providing me with financial support to attend the Third International Brachiopod Congress, and the Organizing Commettee of the Congress in the persons of Paul Copper and Jin Jisuo. I also wish to thank my colleague Alexandra B. Yudina for helpful rewiews, and preliminary translation of my manuscript into English. I am very grateful to Alexey A. Yudin and Valerii V. Beznosov for computer preparation of the figures, and Vera P. Volkova for taking photographs of brachiopods.

REFERENCES

ANTSYGIN, N.A., K.K. ZOLOEV, M.L. KLUZHINA, V.A. NASEDKINA (eds.). 1994. Ob yasnitelnaya zapiska k stratigraficheskim skhemam Urala (Dokembriy, Paleozoy) Materialy i resheniya IV Ural'skogo Mezhvedomsvennogo Stratigraficheskogo Soveschaniya, Ekaterinburg, 152p.

ANTOSHKINA, A.I., A.F. ABUSHIK & T.M. BEZNOSOVA (eds.) 1983. The key sections of the Silurian/Devonian boundary deposits of the Subpolar Urals, Syktyvkar, 136p.

___, N. Ya. ANTSYGIN & T.M. BEZNOSOVA. 1987. The key sections of the Ordovician and Lower Silurian of the Subpolar Urals, Syktyvkar, 94p.

___& T.M. BEZNOSOVA. 1988, New data on the stratigraphy of the Wenlock deposits of the Bolshaya Zemlya tundra:

Byuletin MOIP, Otdel Geologicheskii 63 (6): 32-39.

BEZNOSOVA, T. M. 1989, Zonal subdivisions of the Upper Ordovician and Lower Silurian of the north European USSR (on the basis of brachiopods), p.5-9. In, Novye dannye po rannim i srednepaleozoiskim brakhiopodam SSSR. Informatsionnie Materialy, URO AN SSSR, Sverdlovsk.

___1994. Biostratigraphy and brachiopods of the Silurian of NE European Russia. Nauka, St Petersburg, 128p.

___1995. Late Ashgillian brachiopod communities of the Subpolar Urals, p.471-472. In, J. Cooper, M. Droeser & S. Finney (eds.). Ordovician Odyssey: short papers for the International Symposium on the Ordovician System. Pacific Section, Society Sedimentary Geology (SEPM) 77.

DEDEEV, V.A., V.V. YUDIN, V.I. BOGATSKII & A.N. SHARDANOV. 1989. Tektonika Timano-Pechorskoi neftegazonosnoi provintsii., Akademiya Nauk SSSR, Syktyvkar , 28p.

KALJO, D.L. (ed.). 1987. Resheniya mezhvedomstvennogo stratigraficheskogo sovetshaniya po ordoviku i siluru vostochno-Evropeiskoi platformy, Leningrad, 115 p., 5 charts.

MODZALEVSKAYA, T. L. 1980. Brachiopods of the Silurian and Lower Devonian of Dolgii Island, and their stratigraphic significance, p.82-106. In, Silur I nizhnii devon ostrova Dolgogo, Akademiya Nauk, Sverdlovsk.

PERSHINA, A.I. 1960. Stratigraphy and facies of the Silurian and Devonian of the Pechora Urals. Trudy Komi Filiala Akademii Nauk SSSR, 10:10-25.

___1962, Silurian and Devonian deposits of the Chernyshev High. Nauka , Moskow, Leningrad, 122p.

___1971, Biostratigraphy of the Silurian and Devonian deposits of the Pechora Urals: Nauka , Leningrad, 129p.

___& T.M. BEZNOSOVA. 1978, Brachiopod complexes of the Silurian of the northeast European part of the USSR. Trudy Instituta Geologii, Komi filial AN SSSR: 25:3-9.

SAPELNIKOV, V.P. 1985. Morphologic and taxonomic evolution of brachiopods (Order Pentamerida). Akademiya Nauk, Sverdlovsk, 231p.

VOINOVSKII-KRIGER, K.G. 1945. Two Paleozoic complexes on the western slopes of the polar Urals. Sovetskaya Geologiya, Moskva 6:27-45.

___1956. Slopes of the polar and subpolar Urals, p.62-64. In, Sovetshanie po unifikatsii stratigraficheskikh skhem Urala i sootnoshenie drevnikh svit Urala i Russkoi Platformy, provodimoe. Unpublished ph.d. thesis, Leningrad.

YUDIN, V.V. 1990. Palinspastic reconstructions of the complex fold systems (on the pattern of the Urals, Pre-Urals and Pai-Khoi): Preprint, Komi Science Centre, Urals Division, Academy Science 33:1-24.

[*We were unable to provide appropriate transliteration of some of the English text, some of which may not follow the style of the remaining volume: editors]

PALEOBIOGEOGRAPHIC - PALEOENVIRONMENTAL SIGNIFICANCE OF THE EOCENE BRACHIOPOD FAUNA, SEYMOUR ISLAND, ANTARCTICA

MARIA ALEKSANDRA BITNER

Institute of Paleobiology, Polish Academy of Sciences,
Al. Zwirki i Wigury 93, 02-089 Warszawa, Poland

ABSTRACT- An abundant and diverse brachiopod fauna from the Eocene La Meseta Formation, Seymour Island, Antarctic Peninsula, comprises 19 genera, including six, *Basiliola*, *Hemithiris*, *Murravia*, *Magella*, *Stethothyris*, and *Macandrevia*, recorded for the first time, and four new, as yet undescribed taxa. Species richness is greatest in the lowermost part of the La Meseta Formation. Some records, e.g. of *Basiliola*, *Hemithiris*, *Notosaria*, *Murravia*, *Stethothyris*, and *Macandrevia*) represent oldest known genus occurrences, thus suggesting an important role for the Antarctic in development and distribution of post-Eocene brachiopod faunas, which may have migrated from this region along shelves and islands to the north. The Seymour Island brachiopod fauna has strong affinities with that from the New Zealand Paleogene. The presence of warm water forms, such as *Lingula* and *Bouchardia*, is in conflict with postulated cooling based on stable oxygen isotopic studies in the uppermost part of the formation.

INTRODUCTION

Very fossiliferous deposits of the La Meseta Formation of Seymour Island, Antarctic Peninsula have been paid special attention by paleontologists since their discovery (Buckman, 1910), and are still providing new information about the composition and biogeographic history of the southern hemi-sphere marine biota. Brachiopods represent an important element of the Seymour Island assemblages, and their presence has been noted since the beginning of Antarctic research, with additional data from more recent studies (Owen, 1980; Wiedman, et al. 1988; Bitner, 1991). Material collected by A. Gazdzicki during the Argentine-Polish field parties in the austral summers of 1987-88, 1991-92 and 1993-94, contains a remarkable, taxonomically diverse brachiopod fauna (Bitner, in press), including many genera recorded for the first time on Seymour Island. The purpose of this paper is to summarize information on diversity, paleogeography, biogeographic affinities and paleoecology of early Tertiary Antarctic brachiopod faunas.

GEOLOGICAL SETTING

The La Meseta Formation is exposed in the northern part of Seymour Island (Figure 1). This several hundred meter thick sequence, which unconformably overlies an upper Cretaceous/Paleocene erosional surface, consists of loosely consolidated sandstones, sandy siltstones, clays, pebbly sandstones and shell beds with an abundant marine fauna (Sadler, 1988; Stilwell & Zinsmeister, 1992). Faunal composition and sedimentary structures indicate a nearshore, shallow water environment (Sadler, 1988; Stilwell & Zinsmeister, 1992). Initially, the La Meseta Formation was divided by

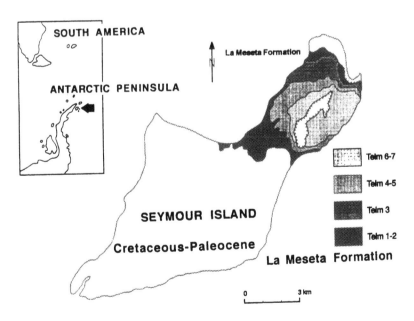

Figure 1. Map of Seymour Island and the Antarctic Peninsula (inset), showing the La Meseta Formation divided into Telm units (from Sadler, 1988; simplified); areas without pattern indicate Cretaceous, Paleocene and Quaternary deposits.

Elliot & Trautman (1982) into three informal lithostratigraphic units. After detailed mapping, Sadler (1988) divided it into seven lithofacies units (Telm1 to Telm7), used in this paper. The age of the La Meseta Formation, first determined as late Eocene to possibly earliest Oligocene (Zinsmeister, 1982a), now is considered to extend down as far as the early ?-middle Eocene, and up to early Oligocene (Cocozza & Clarke, 1992; Stilwell & Zinsmeister, 1992; Tambussi et al., 1994).

THE BRACHIOPOD ASSEMBLAGE

Nineteen genera and 24 species, including several new genera and species, have been identified in the brachiopod collection from the La Meseta Formation of Seymour Island (Bitner, in press). Greatest taxonomic diversity, 18 genera and 23 spp., is observed in the lowermost units, Telm 1 and Telm 2 (early ?-middle Eocene), where all the new taxa were found. Brachiopods from upper units show a considerable decrease in diversity and are represented only by four genera and four species. Of the 19 genera, 9 genera, e.g. *Lingula*, *Notosaria*, *Tegulorhynchia*, *?Plicirhynchia*, *Liothyrella*, *Terebratulina*, *Bouchardia*, *'Terebratella'*, and *Magellania*, have been recorded previously from Eocene strata of Seymour Island (Buckman, 1910; Owen, 1980; Wiedman et al., 1988; Bitner, 1991), and two, *Hemithiris* and *Magella*, have been described

previously from nearby Cockburn Island (Owen, 1980). Four new genera, belonging to the families Hemithyrididae, Terebratulidae and ? Laqueidae, have been recognized in the investigated material. The most interesting occurrences are those of four genera, i.e. *Basiliola*, *Murravia*, *Stethothyris* and *Macandrevia*, known from the Tertiary of other regions, but noted for the first time in the Antarctic.

The most common genera are *Bouchardia*, *Macandrevia* and *Lingula*. While *Bouchardia* occurs in the entire section, *Macandrevia* is known only from the lowermost units, Telm 1 and Telm 2 (early ?-middle Eocene). *Lingula*, in turn, is found solely in the uppermost units, Telm6 and Telm7 (early Oligocene). The genera *Liothyrella*, *Terebratulina* and *Magellania* are also quite common. Other genera are much rarer and represented by up to 16 specimens.

PALEOBIOGEOGRAPHY

The brachiopod fauna of the La Meseta Formation shows several interesting biogeographical and evolutionary aspects. It contains truly cosmopolitan genera, as well as genera restricted only to the southern hemisphere, and forms endemic to Seymour Island. In a number of cases the Seymour Island occurrence is the oldest known record for particular genera.

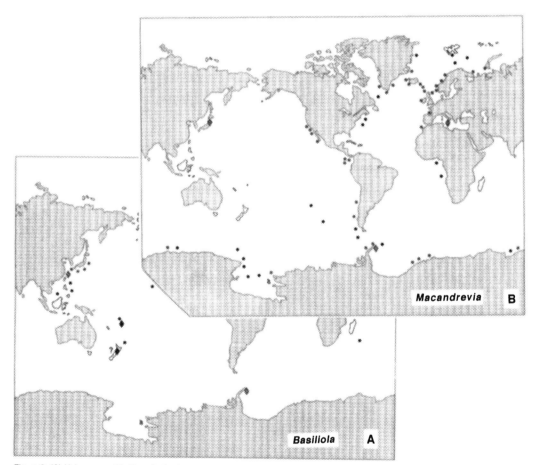

Figure 2. (A) Holocene and Tertiary distribution of *Basiliola* Dall; (B) Holocene and Tertiary distribution of *Macandrevia* King (from Cooper, 1957,1959,1975,1978; Lee,1980a, modified); circles=Recent, diamonds =Tertiary occurrences.

The oldest occurrences of such genera as *Basiliola*, *Hemithiris*, *Notosaria*, *Murravia*, *Stethothyris*, and *Macan-drevia* show that high-latitude heterochroneity, i.e. differential appearance of taxa between high and lower latitudes, is present among Cenozoic Brachiopoda. Early appearances in the Eocene of Seymour Island were first observed among other invertebrates such as molluscs, crinoids, asteroids, and decapods (Zinsmeister & Feldmann, 1984), thus the important role of high southern latitudes in development and evolution of Cenozoic biotas appears supported by brachiopods. Further-more, it suggests an early Tertiary southern high-latitude origin for those Cenozoic brachiopod genera, and rapid northward Paleogene dispersal in the Paleogene before the development of the circum-Antarctic current which made mixing of the fauna impossible. Thus, the data presented in this paper support Model 2 of the evolutionary history of polar regions proposed by Crame (1992).

The genus *Basiliola* Dall, represented by several Holocene species, was hitherto known as fossils from the Pliocene of Okinawa, Ryukyu Islands (Cooper, 1957) and the Miocene and Pliocene sediments of Fiji (Cooper, 1978). Although extant representatives are known in the New Zealand region (Dawson, 1991), *Basiliola* has not been found in the New Zealand and Australian Tertiary, unless specimens from the middle Miocene of New Zealand attributed by Lee (1980a) to the genus *Streptaria* Cooper belong to *Basiliola*. Both genera are very similar externally, but differ in beak structure, and in development of pedicle collars and dental plates. Because interior features are unknown in the New Zealand material, Lee (1980a) placed her specimens in *Streptaria*, using the 'twisted' anterior commissure as the main criterion. However, Cooper (1959) observed specimens of *Streptaria* with and without a twisted commissure. Broad asymmetrical uniplication is also characteristic for *Basiliola* species from Seymour Island. Fossil and Recent distribution of *Basiliola* (Figure 2) suggests that it spread from the Antarctic northwards in post-Eocene time along the New Zealand and Australian shelves, and then along the islands of SE Asia, reaching Okinawa, Japan in the Pliocene.

The distribution pattern of the genus *Tegulorhynchia* Chapman & Crespin is similar to that of *Basiliola*. However, oldest occurrence is not in the Antarctic, but western Australia (McNamara, 1983), *Tegulorhynchia* appears in the Eocene of New Zealand, in the Oligocene of E Australia and Tasmania (Lee, 1980b), and then in the Pliocene of Okinawa (Cooper, 1957). Today it lives in the Pacific from Japan to Borneo (Lee, 1980b), and in the Indian Ocean off St. Paul Island (Cooper, 1981).

In the southern hemisphere the genus *Hemithiris* d'Orbigny is known only as fossils from Cockburn Island (Owen, 1980), and Seymour Island (Bitner, in press), while in the the northern hemisphere it is reported from the Tertiary of Japan, Alaska and Europe (Hatai, 1940; Cooper, 1959). Its absence in the Tertiary of New Zealand and Australia, as well as Recent waters of the southern hemisphere is of interest, however, the distribution pattern is difficult to explain at present.

Notosaria Cooper is known as fossils from Tertiary strata of Seymour Island (Owen, 1980; Wiedman, et al. 1988; Bitner, in press), and from New Zealand (Lee, 1978; Lee & Wilson, 1979), and from the Pliocene of Belgium (Cooper, 1959) in the northern hemisphere. Recent species of *Notosaria* are primarily restricted to New Zealand (Lee, 1978; Lee and Wilson, 1979), but noted as well from subantarctic waters of the Indian Ocean (Foster, 1974; Cooper 1981).

The presence of the genus *Murravia* Thomson in the lowermost part of the La Meseta Formation extends its geographical and stratigraphical range. Hitherto it was known from the Tertiary of New Zealand and Australia (Thomson, 1927; MacKinnon, et al., 1993), and today it lives on the south coast of Australia (Thomson, 1927). Reported for the first time from the Antarctic region, the genus *Stethothyris* Thomson was considered to be restricted to the Tertiary of New Zealand (Lee, 1986), having a late Eocene-late Oligocene stratigraphical range (MacKinnon et al., 1993).

Particularly surprising and interesting was the discovery of a new species of *Macandrevia* King in the lowermost La Meseta Formation. This genus, widely distibuted in modern seas (Figure 3), but is very rare in the fossil record. Earlier it was found only at two fossil localities, in the Pliocene of Sicily (Gaetani & Sacca, 1984) and Miocene of Japan (Hatai, 1940). So far the species from Seymour Island is the oldest representative of the genus. The migration routes of *Macan-drevia* seem to be much more complicated than previously supposed. Its fossil occurrence in the Miocene and Pliocene of the northern hemisphere indicates a northward route of migration similar to that of *Basiliola*, *Hemithiris* and *Tegulo-rhynchia*. *Macandrevia* has not been found to date in the New Zealand and Australian Tertiary. A long history of *Macandrevia*, extending back at least to Eocene, was already suggested by Thomson (1927), based on loop development and wide geographic distribution. Using immunological data, Endo et al. (1994) re-examined the contradiction between traditional and serotaxonomic views of terebratulid classification and assigned the Macandreviidae, together with three other families, Ecnomiosidae, Kraussinidae and Megathyrididae, to one group. They concluded that species of Kraussinidae and Megathyrididae were derived from long-looped Macandreviidae or Ecnomiosidae. This has not yet been confirmed by the fossil record. Megathyridids and kraussinids are known from the Upper Cretaceous and Miocene, respectively, whilst no fossil *Ecnomiosa* is known, and the oldest previous occurrence of *Macandrevia* is from the Miocene. Extending the stratigraphic range of *Macandrevia* to the early ?- middle Eocene corroborates the Endo et al. interpretation.

A single northward Tertiary dispersal pattern for these brachio-pod genera is proposed. Following the generalized oceanic circulation model for the south Pacific during the Early Tertiary (Zinsmeister, 1982b), the suggested dispersal pattern appears to be upcurrent. However, brachiopods probably migrated along the shelf, where their dispersal may have been modified by local gyres and/or a nearshore countercurrents, as postulated by Zinsmeister (1982b, Fig.4b; see also Fig.10, Feldmann & Zinsmeister, 1984; Fig.1, Mancenido & Griffin, 1988). Three genera, *Hemithiris*, *Notosaria*, and *Macandrevia*, are also known from the Pliocene of Europe. Possible routes of dispersal could have been through the Tethys, rather than along the coast of South America, with introdution into the Atlantic basin through the region of Central America, as suggested for crabs by Feldmann (1986). The latter proposition is not confirmed by brachiopods.

At the generic level, the La Meseta brachiopod fauna appears to have closest affinities with that of New Zealand, with 9 genera in common (Lee, 1986). Fewer similarities are observed between brachiopod assemblages of Seymour Island and southern South America. Thus, this may indicate a Tertiary barrier between South America and West Antarctica hampering dispersal of faunas (see also Stilwell & Zinsmeister, 1992), while after separation of New Zealand from Gondwana in the Late Cretaceous, shallow-water connections between New Zealand and Antarctica still existed until the late Eocene (Lee,

1986; Stevens, 1989). This permitted a spreading of faunas before shallow marine links weakened and eventually disappeared, enabling development of the circum-Antarctic current.

PALEOENVIRONMENTS

Until recently, the La Meseta Formation has been interpreted as a nearshore, shallow marine environment, with variable energy, from high to low in protected areas, under warm, temperate climates (Sadler, 1988; Stilwell & Zinsmeister, 1992). Contrary to the most recent interpretation of the La Meseta Formation as an estuarine deposit (Porebski, 1995), the rich and diversified assemblage of brachiopods supports earlier interpretations as normal marine.

Wiedman et al. (1988), plotting temperature, depth, and latitude ranges of six extant brachiopod genera from Seymour Island, concluded that the La Meseta Formation was formed in a shallow-water, temperate lower latitude setting, with temperatures ranging from 10° to 14°C. Brachio-pods support their general conclusions. The presence of such genera as *Lingula, Basiliola, Notosaria, Tegulo-rhynchia, Murravia,* and *Bouchardia,* which live today in warm to tropical waters, clearly indicates higher temperatures than for the present-day Antarctic. This agrees with data provided by bivalves and gastropods (Stilwell & Zinsmeister, 1992).

It is more difficult to estimate the depth of deposition of the La Meseta Formation based on brachiopods, as the bathymetric distribution of most Cenozoic brachiopods is usually extensive. Also, living descendants of the La Meseta brachiopod fauna have remarkable wide depth ranges, although they are most common at depths less than 200m. Sediment associations indicate a shallow water environment (Sadler, 1988). Articulate brachiopod species which inhabit warmer environments today, usually live in deeper waters, as observed for some *Basiliola, Notosaria* and *Macandrevia.* Thus, it seems probable that brachiopods migrating to lower latitudes and warmer waters, have changed bathymetric distribution by moving into deeper and cooler habitats. Such relationships between depth and latitude are corroborated in other southern hemisphere invertebrates (Feldmann & Zinsmeister, 1984).

Another problem involves climatic conditions during deposition of the La Meseta Formation. There is a consensus that it was much warmer in the Eocene Antarctic than today, but considerable climatic cooling at the contact of Telm 5 and Telm 6 is postulated by Gazdzicki et al. (1992) on the basis of stable oxygen isotopic data. However, the presence of such warm water taxa as *Lingula* and *Bouchardia* contradicts the isotopic interpretation (Bitner, in press). A similar conflict between paleontological and stable isotopic data was previously pointed out for the New Zealand Paleogene (Lee, 1986), and at low latitudes during the Tertiary (Adams et al., 1990). This suggests that factors other than climatic might have been involved in producing an isotopic shift of $\delta^{18}O$ values, and that temperatures calculated from stable isotopes must be treated with caution.

CONCLUSIONS

A newly collected Eocene brachiopod fauna from the La Meseta Formation is considerably more diverse than earlier reported. Several taxa which have been previously known from beyond the Antarctic region are especially significant in the paleobiogeographical history of this group. The oldest occurrences of *Basiliola, Hemithiris, Notosaria, Murravia, Stethothyris,* and *Macandrevia* provide evidence that the Antarctic region played an important role in the evolution of

brachiopods and that high-latitude heterochroneity is demonstrable among Cenozoic brachiopods. The observed distribution patterns of the fossil and Recent genera *Basiliola, Hemithiris, Tegulorhynchia, Notosaria, Murravia, Stethothyris,* and *Macandrevia* suggest dispersal patterns of brachiopods from the Antarctic region northward to New Zealand and Australia, and in the case of some, farther to Japan. Brachiopods support earlier interpretations of the La Meseta Formation as shallow, warm-temperate, and normal marine. Brachiopods do not support the dramatic cooling postulated on the basis of stable oxygen isotopes.

Acknowledgments

I am grateful to Rodney M. Feldmann, Kent State University, for critical reviewing and improving this manuscript, and the Natural Sciences and Engineering Research Council of Canada for providing travel support to attend the Third International Brachiopod Congress in Sudbiry, Ontario.

REFERENCES

ADAMS, C.G., D.E. LEE & B.R. ROSEN. 1990. Conflicting isotopic and biotic evidence for tropical sea-surface temperatures during the Tertiary. Palaeogeography, Palaeoclimatology, Palaeoecology 77:289-313.

BITNER, M.A. 1991. A supposedly new brachiopod from the Paleogene of Seymour Island, West Antarctica. Polish Polar Research 12:243-246.

___(in press). Brachiopods from the Eocene La Meseta Formation of Seymour Island, Antarctic Peninsula. In, A. Gazdzicki (ed.), Palaeontological results of the Polish Antarctic expeditions, II, Palaeontologia Polonica 55.

BUCKMAN, S.S. 1910. Antarctic fossil Brachiopoda collec-ted by the Swedish South Polar Expedition, 1901-3. Wissenschaftliche Ergebnisse der schwedischen Sud-polar-Expedition 1901-1903, 3:1-40.

COCOZZA, C.D. & C.M. CLARKE. 1992. Eocene microplankton from La Meseta Formation, northern Seymour Island. Antarctic Science 4:355-362.

COOPER, G.A. 1957. Tertiary and Pleistocene brachiopods of Okinawa, Ryukyu Islands. U.S. Geological Survey Professional Papers 314-A:1-20.

___,1959. Genera of Tertiary and Recent rhynchonelloid brachiopods. Smithsonian Miscellaneous Collections, 139(5):1-90.

___1975. Brachiopods from West African waters with examples of collateral evolution. Journal Paleontology 49:911-927.

___1978. Tertiary and Quaternary brachiopods from the Southwest Pacific. Smithsonian Contributions Paleobio-logy 38:1-23.

___1981. Brachiopods from the Southern Indian Ocean (Recent). ibid. 43:1-93.

CRAME, J. A. 1992. Evolutionary history of the polar regions. Historical Biology 6:37-60.

DAWSON, E.W. 1991. The systematics and biogeography of the living Brachiopoda of New Zealand, p.431-437. In, D.I. MacKinnon, D.E. Lee & J.D. Campbell (eds.), Brachiopods through time. Proceedings Second International Brachiopod Congress, Dunedin, New Zealand.

ELLIOT, D.H. & T.A. TRAUTMAN. 1982. Lower Tertiary strata on Seymour Island, Antarctic Peninsula, p.287-297. In, C. Craddock (ed.), Antarctic Geoscience. Wisconsin University Press.

ENDO, K., G. B. CURRY, R. QUINN, M. J. COLLINS, G. MUYZER & P. WESTBROEK. 1994. Re-interpretation of terebratulide phylogeny based on immunological data. Palaeontology 37:349-373.

FELDMANN, R. M. 1986. Paleobiogeography of two decapod

crustacean taxa in the Southern Hemisphere: global conclusions with sparse data, p.5-19. In, R.H. Gore & K.L. Heck (eds.), Crustacean biogeography. Balkema, Rotterdam.

___ & W.J. ZINSMEISTER 1984. New fossil crabs (Deca-poda: Brachyura) from the La Meseta Formation (Eocene) of Antarctica: paleogeographic and biogeographic implications. Journal Paleontology 58:1046-1061.

FOSTER, M.W. 1974. Recent Antarctic and Subantarctic brachiopods. Antarctic Research Series 21:1-189.

GAETANI, M. & D. SACCA. 1984. Brachiopodi batiali nel Pliocene e Pleistocene de Sicilia e Calabria. Rivista Italiana PaleontologiaStratigrafia 90:407-458.

GAZDZICKI, A., M. GRUSZCZYNSKI, A. HOFFMAN, K. MALKOWSKI, S. MARENSSI, S. HALAS & A. TATUR. 1992. Stable carbon and oxygen isotope record in the Paleogene La Meseta Formation, Seymour Island, Antarctica. Antarctic Science 4:461-468.

HATAI, K.M. 1940. The Cenozoic Brachiopoda of Japan. The Science Reports of the Tohoku Imperial University, Sendai, Japan (Geology) 20:1-413.

LEE, D.E. 1978. Aspects of the ecology and paleoecology of the brachiopod Notosaria nigricans (Sowerby). Journal Royal Society New Zealand 8:395-417.

___ 1980a. Probolarina and Streptaria, two Cenozoic rhynchonellide brachiopods new to New Zealand. ibid. 10:137-144.

___ 1980b. Cenozoic and Recent rhynchonellide brachiopods of New Zealand: systematics and variation in the genus Tegulorhynchia. ibid. 10:223-245.

___ 1986. Paleoecology and biogeography of the New Zealand Paleogene brachiopod fauna, p.477-483. In, P.R. Racheboeuf & C.C. Emig (eds.), Les brachiopodes fossiles et actuels. Biostratigraphie du Paleozoique 4.

___ & J.B. WILSON. 1979. Cenozoic and Recent rhynchonellide brachiopods of New Zealand: systematics and variation in the genus Notosaria. Journal Royal Society New Zealand 9:437-463.

MACKINNON, D.I., S.S. BEUS & D.E. LEE. 1993. Brachio-pod fauna of the Kokoamu Greensand (Oligocene), New Zealand. New Zealand Journal Geology Geophysics 36:327-347.

MANCENIDO, M. O. & M. GRIFFIN. 1988. Distribution and palaeoenvironmental significance of the genus Bouchardia (Brachiopoda, Terebratellidina): its bearing on the Cenozoic evolution of the South Atlantic. Revista Brasileira Geociencias 18:201-211.

MCNAMARA, K.J. 1983. The earliest Tegulorhynchia (Brachiopoda: Rhynchonellida) and its evolutionary significance. Journal Paleontology 57:461-473.

OWEN, E.F. 1980. Tertiary and Cretaceous brachiopods from Seymour, Cockburn and James Ross Islands, Antarctica. Bulletin British Museum (Natural History), Geology 33:123-145.

POREBSKI, S.J. 1995. Facies architecture in a tectonically-controlled incised-valley estuary: La Meseta Formation (Eocene) of Seymour Island, Antarctic Peninsula. Studia Geologica Polonica, 107:7-97.

SADLER, P.M. 1988. Geometry and stratification of uppermost Cretaceous and Paleogene units on Seymour Island, northern Antarctic Peninsula, p.303-320. In, R. M. Feldmann & M. O. Woodburne (eds.), Geology and paleontology of Seymour Island, Antarctic Peninsula. Geological Society America Memoir 169.

STEVENS, G.R. 1989. The nature and timing of biotic links between New Zealand and Antarctica in Mesozoic and early Cenozoic times, p.141-166. In, J. A. Crame (ed.), Origins and evolution of the Antarctic biota. Geological Society London, Special Publication 47.

STILWELL, J.D. & W.J. ZINSMEISTER 1992. Molluscan systematics and biostratigraphy. Lower Tertiary La Meseta Formation, Seymour Island, Antarctic Peninsula. Antarctic Research Series 55:1-192.

TAMBUSSI, C.P., J.I. NORIEGA, A. GAZDZICKI, A. TATUR, M.A. REGUERO & S.F. VIZCAINO. 1994. Ratite bird from the Paleogene La Meseta Formation, Seymour Island, Antarctica. Polish Polar Research 15:15-20.

THOMSON, J.A. 1927. Brachiopod morphology and genera (Recent and Tertiary). New Zealand Board Science Art, Manual 7:1- 338.

WIEDMAN, L.A., R.M. FELDMANN, D.E. LE, & W J. ZINSMEISTER. 1988. Brachiopoda from the La Meseta Formation (Eocene), Seymour Island, Antarctica, p.449-457. In, R.M. Feldmann & M.O. Woodburne (eds.), Geology and Paleontology of Seymour Island, Antarctic Peninsula. Geological Society of America, Memoir 169.

ZINSMEISTER, W.J. 1982a. Review of the Upper Cretaceous-Lower Tertiary sequence on Seymour Island, Antarctica. Journal Geological Society London 139:779-785.

___ 1982b. Late Cretaceous-Early Tertiary molluscan biogeography of the southern circum-Pacific. Journal Paleontology 56:84-102.

___ & R.M. FELDMANN. 1984. Cenozoic high latitude heterochroneity of Southern Hemisphere marine faunas. Science 224:281-283.

THE CLASSIFICATION OF THE BRACHIOPOD ORDER STROPHOMENIDA

C.H.C. BRUNTON & L.R.M. COCKS

Department of Palaeontology, The Natural History Museum,
Cromwell Road, London SW7 5BD, U.K.

ABSTRACT- Changes to the classification of the order Strophomenida in the 1965 Brachiopod Treatise have split the Davidsoniacea, with *Davidsonia* and related genera now placed in the Atrypida, leaving the rest renamed and elevated as the Orthotetidina. Like *Davidsonia*, the Thecospiridae and *Cadomella* have proved to have fibrous shell fabrics and spiralia and are also removed to spire–bearing taxa, leaving no strophomenides after the end of the Permian. At least twelve features are important in the characterisation of the Strophomenida, and their distribution within the order. In the resultant classification the Orthotetidina may stand as a sister group to the more closely related Strophomenidina, Chonetidina, Productidina and Strophalosidina, including aulostegoids, richthofenioids and lyttonioids. The Gemmellaroiidae has been classed with richthofeniids or orthotetids in the past, but the four genera are now grouped with scacchinellids and lyttoniids within the Strophalosiidina, and one genus is possibly an orthotetidine.

INTRODUCTION

In the brachiopod volume of the Treatise on Invertebrate Paleontology (Williams et. al., 1965), the order Strophomenida was divided into four suborders, the Strophomenidina, including the superfamilies Plectambonitacea, Strophomenacea and Davidsoniacea; the Chonetidina, including the superfamilies Chonetacea and Cadomellacea; the Productidina, including the superfamilies Strophalosiacea, Richthofeniacea and Productacea, and the Oldhaminidina, with the sole superfamily Lyttoniacea. Now that a revised edition of that part of the *Treatise* is close to completion, the time is ripe to review this classification and emend it where necessary.

Davidsonia itself has been known to contain a spirolophe since 1973 (Garcia-Alcalde), but at the time this was insufficient evidence to shift the genus from the Strophomenida, which in the 1965 Treatise diagnosis included genera with spiralia. Copper (1979), and Johnson (1982) led debate on the taxonomic position of *Davidsonia*, the former advocating its transfer to the atrypidines. In their study of the Davidsoniacea (sensu 1965) Williams & Brunton (1993) demonstrated that the shell microstructure of *Davidsonia* was not lamellose, but essentially fibrous and closely comparable to that of many spire–bearers. Consequently these authors transferred *Davidsonia* (plus three other closely related genera) to the Atrypida. The remaining genera in the old Davidsoniacea were grouped together in a family named after the genus *Orthotetes* and raised to subordinal rank as the Orthotetidina (Williams & Brunton, 1993), all of which have laminar shells, but range from early impunctate forms to pseudopunctate and, in the Upper Palaeozoic, some extropunctate taxa. *Cadomella* has also proved to possess calcified spiralia (Cowen & Rudwick, 1966), but Brunton & MacKinnon (1974) demonstrated that its shell was fibrous and that morphological features indicated a more accurate assignment to the spire–bearing koninckinaceans. The same was true of *Thecospira*, which had also been in the Davidsoniacea of the 1965 Treatise (MacKinnon, 1974). Thus

the Strophomenida remains as a large and diverse group, but is strengthened in its integrity by the removal of several genera which have proved anomalous and which now more properly reside in the Spiriferida. Two reasons for the earlier assignment of some of these genera to the strophomenides was their cemented habit and strophic hinge lines. Now that they are included in the Spiriferida more weight is given to the suggestion (Williams, 1973; Baker, 1990) that the thecideidines derived from spiriferides. The remaining taxa listed above remain in the broad strophomenide group, but before considering their classification, we will review some key morphological features in turn.

MORPHOLOGICAL FEATURES

Twelve features are outlined describing the main taxonomic characters diagnostic for the Strophomenida.

Supra–apical foramen. This feature occurs throughout the major strophomenide group, at least in the juvenile stage (Figure 1:1–6). It became lost during ontogeny in most groups, but especially early in ontogeny in the Orthotetidina, chonetidines and productidines. A supra–apical foramen is known in some orthoids and was probably inherited from the billingsellids. In orthotetidines the supra–apical foramen of stratigraphically older taxa seems to have been lost in more recent taxa in favour of a cemented habit. It also became widely lost as an adult character within the strophomenides sensu lato, as for instance in chonetidines (Figure 1:1), or productines (Figure 1.2–3), where it was only retained as a juvenile feature holding the brephic shell to the substrate.

The supra–apical foramen has important implications in supra–familial classification and interpretation of larval life. The foramen, commonly associated with a pedicle sheath (Brunton, 1965), is situated within what can be recognised as the protegular node (Figure 1:5). This implies that the late larval stage was already furnished with a small biconvex shell (the protegular node) from which protruded a pedicle ready for attachment to a suitable substrate. This may in turn be suggestive of a more prolonged feeding larval life that is now found in living articulated brachiopods.

Strophic hinge. This normally wide posterior margin provides a subsemicircular outline to representatives of all groups. This wide hinge, associated with interareas, was inherited from orthoids (sensu lato). The shape is lost within some subgroups of the productidines, e.g. lyttoniids, and to a lesser extent richthofeniids and scacchinellids (which are all cemented), but this is considered to be a secondary adaptation and only relevant to family levels of classification.

Interareas. These are present through much of the group although dorsal interareas are short or lacking in some genera; they are largely absent in chonetidines and productidines. Ventral interareas were also lost in some productidines, as in the true productoids, richthofeniids and lyttoniids. In other

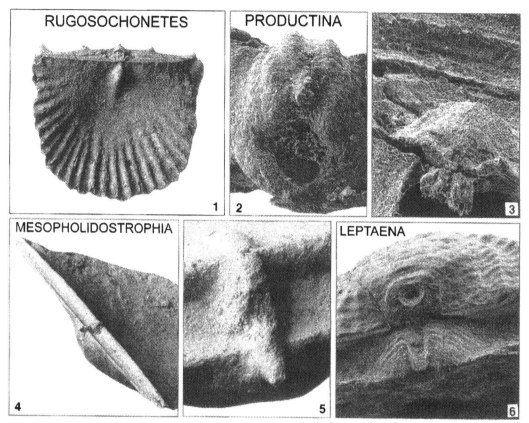

Figure 1. Supra–apical foramina in strophomenide brachiopods. **1**, *Rugosochonetes*, late Visean near Derrygonnelly, County Fermanagh, Ireland, viewed posterodorsally showing spine bases, the pedicle sheath and dorsal protegual node, BB 52768, x30. **2-3**, *Productina* (silicified specimen), late Visean, County Fermanagh, Ireland, viewed posteriorly showing the bilobate ventral protegular node with its central pedicle sheath and the initial pair of clasping spines (below) surrounding a hole representing the initial object of attachment, BB 63843, x150, and posterodorsally (ventral umbo missing) showing the cardinal process and dorsal protegular node, x135. **4-5**, *Mesopholidostrophia*, late Wenlock, Gotland, Sweden, viewed posterolaterally (x20), posteriorly (x80), showing the pedicle sheath and dorsal protegular node. **6**, '*Leptaena*', Early Devonian, Carter Co, Oklahoma, U.S.A., viewed posteriorly with the ventral valve upper-most, showing the pedicle foramen BB 1886 (x35). All views by SEM.

productidines and orthotetidines, the ventral interarea became much elongated in association with the ventral valves becoming conical.

Profile. The common shape of the lateral profile within this otherwise diverse group is concavoconvex, but some of the more extreme shapes of the Productida include planar or conical profiles, while various Strophomenoidea and Plectambonitoidea are resupinate. In general, the initial protegular nodes of brachiopods are biconvex. The node can be preserved on dorsal valves which are concave in adulthood, as in *Mesopholidostrophia* and *Rugosochonetes* (Figure 1:1,4); this initial biconvexity holds for all strophomenide groups. Subsequent growth of the valves commonly led to the loss of the ventral protegular node at the posterior tip of the convex valve through abrasion or attachment in that area. Dorsal valve growth thereafter was commonly concave in all but the Orthotetidina (and rarely in some Strophomenoidea), in which it was planar or gently convex.

Pseudodeltidium and chilidium. These structures have in the past been cited as important characters for the group, but

are now seen as of only lesser importance, for example at generic levels in plectambonitoids, strophomenoids and some strophalosioids. However, the presence of a pair of medially sutured deltidial plates, as compared to a pseudodeltidium, was important in recognising *Davidsonia* as a spiriferide. Although somewhat similar structures, such as stegidia, are present in other orders, the pseudodeltidium remains a key characteristic of strophomenides, although it can be so reduced in size as to be difficult to see at the margins of the delthyrium.

Teeth and sockets. These characterise the group as a whole, but are lost from later representatives of different subgroups. Teeth are always of a deltidiodont nature. In the strophodontids the teeth and sockets gave way to hinge line denticulate articulation; in productidines they are lost in the true productoids (after the Famennian), from aulostegids and in richthofeniids articulation became dorsal knobs fitting into ventral cavities. Teeth and sockets are widespread through all other suborders; only in strophomenides are they so widely lost. The development of teeth within the Strophomenoidea, including the transition between simple teeth through crenulation to strophodontid denticles, is described by Rong & Cocks (1994).

Cardinal process. This morphologically varied structure accommodated the dorsal diductor muscle attachments. It varies from a single–lobed to bilobed structures on which may be set ridges which separated, confined or provided attachment for the muscle bases; these ridges are designated by the termination ....fid, e.g. unifid (a single median ridge), bifid (paired ridges laterally bordering or accommodating the putative muscle bases), trifid (a median ridge plus the previous bordering ridges), quadrifid (paired median ridges plus paired lateral ridges bordering or accommodating the diductor bases). In plectambonitoids the basic type is "single–lobed", the myophore positions being flat or somewhat elevated and commonly separated by a median ridge; the unifid condition. Variations include the trifid condition (when the myophore positions are bordered by ridges) or "undercut" (Cocks & Rong, 1989), in which there is no anterior support for the central process, which is braced laterally by inner socket ridges or their homologues. In strophomenoids, orthotetidines, chonetidines and productidines the basic pattern is bilobed plus a variety of ridging. The diductor myophores are often raised on to the distal ends of a bilobed boss of shell with a variety of ridges bordering and separating the muscle scars. In some extreme examples, as in some aulostegids or meekellids, the cardinal process shaft was greatly extended into somewhat conical ventral valves, but even here the external faces of the shafts retained the tracks of migration of the myophores during growth. These ontogenetic tracks of discrete myophores on shafted cardinal processes seems to be unique to this major group; these tracks are not seen in morphologically similarly–shaped cardinal processes in some athyrididines. Indeed in other non orthide or strophomenide orders there seems to be a widespread tendency for the dorsal diductor scars to have become united (see Williams et al., 1965, fig. 126), at least in early stages of ontogeny (Brunton et al.,1996).

Lophophore supports. In strophomenides there are no structures directly comparable with the brachidia of living brachiopods. However there are a variety of structures in strophomenides which have been interpreted as aiding the support of lophophores, or the anterior body wall in the region of the mouth sector of the lophophore. In the most general terms, many Plectambonitoidea, Strophomenoidea, chonetidines and productidines have indications of schizolophous or simple ptycholophous lophophores attached more or less directly to the inner surface of the dorsal valve. A few Plectambonitoidea have raised dorsal platforms or sets of subparallel ridges probably providing at least partial support to the lophophore. Strophomenoids have sparse structures, but some have paired ridges (termed side septa) probably homologous with the anderidia of many chonetids and early productidines, which have been interpreted as supporting the anterior body wall. In the latter two groups well defined brachial ridges are also found, interpreted as being regions of the dorsal valve interior from which the lophophore hung. In more advanced productids there is direct evidence of a calcified brachidium in support of a folded ptycholophe, seen at its most extreme in lyttoniids (see Grant, 1972). Orthotetidines and some strophomenoids have 'brachiophore' structures more like those of orthoids. These are morphological extensions from the structures originally acting as inner socket ridges, but even when the inner socket ridges lost their original function, for example, in strophodontids in which the teeth are absent in the ventral valve, the inner socket ridges (= 'socket plates') retained their original auxiliary roles as brachiophores. The dorsal valve internal surfaces of these two groups seldom seem to have been involved in supporting the lophophore, unlike those of plectambonitoids, chonetidines and productidines.

Shell fabric. Within these strophomenide groups there are various fabrics: standard fibrous to a mixed crested laminar shell in the plectambonitoids; cross–bladed laminar in strophomenoids and orthotetidines; and crested laminar to fully cross–bladed laminar in chonetoids and productidines. Thus, of itself, shell fabric is not uniquely diagnostic of strophomenides, but the common laminar condition is not found outside the group, other than in some possible early orthoids. The laminate billinselloids were considered to be early 'orthoids', but their shell fabric and other features now places this assignment in doubt.

Pseudopunctae. Pseudopunctation was a feature said to be unique to this major group. However, pseudopunctae may also be present in some gonambonitaceans, and in more recent years it has been recognised that the presence or absence of taleolae in the pseudopunctae also seems to have a significant distribution. Pseudopunctae are seen in all the major strophomenide groups, but they originated independently in both the plectambonitids and the othotetidines. The ortho-tetidines are the only group which seem to lack taleolae and in which the reverse type of pseudopunctation, called extro-punctae (Williams & Brunton, 1993), are also found.

Mode of attachment. Mode of attachment to the substrate is varied and of importance only at family level or below. Many Plectambonitoidea and Strophomenoidea were attached by a functional pedicle in juvenile stages and became 'free' as adults. Chonetoids and productidines respectively lost the pedicle in early stages and lost it completely in stratigraphically younger taxa; they then became free, but stabilised by spines which were introduced monophyletically in the chonetoids and inherited by the productidines; some of which became cemented by the umbo or by spines. Orthotetidines lost the pedicle early in life and their stratigraphical range, with umbonal cementation becoming common; it also occurs in the stropho-menoid family Leptaenoideidae. Cementation by the ventral umbo was commonly accompanied by the development of a conical shape in all groups of brachiopods and is not restricted to strophomenides.

Spines. External hollow, tubular spines are confined to chonetidines and productidines (sensu lato). These are distinct from the long spines of some rhynchonellids, for example, in that the latter first grew as marginal enrolments of the epithelium and shell so as to leave a suture line on the commissural surface of the spines (Brunton & Alvarez, 1989). The spines at the posterior margin of *Barbaestrophia* Havlícek (a Lower Devonian strophodontid) are not hollow and grew as occasional outgrowths at the cardinal extremities, so these too are unlike the tubular chonetid or productide spines. In the plectambonitoid *Eochonetes* there are hollow canals through the ventral interarea in a similar fashion to the canals leading to the hinge spines of chonetoids. The spread of hollow spines to the rest of the ventral valve helps differentiate the early productidines, in which the patterns, types and functions of spines varied enormously.

The unique character of chonetid and productid spines allows their presence on specimens to be taken as good evidence of belonging within these taxa. The four genera assigned to the Gemmellaroiidae by Grant (1993) were placed by him close to *Derbyia* in the Orthotetidina. However, some specimens of *Gemmellaroia* have been shown to possess the stumps of hollow spines and accordingly we classify it (and *Tectaria* Likharev) as a scacchinellid within the Aulostegoidea; *Loczyella* Frech belongs in the Permianellinae (a lyttoniid), and one genus (*Cyndalia* Grant) might remain in the Orthotetidina.

DISCUSSION

All taxa of the old Strophomenidina share, to some extent, the first ten characters (excluding attachment and spines), but none is unique to this group. The supra–apical foramen is important, but shared by some orthides, as is the pseudodeltidium. The shell fabric, varies, but some form of pseudopunctation is common to all strophomenides apart from early orthotetidines. It occurs also in gonambonitids, which may be more closely related to strophomenides than previously thought. In detail, however, taleolae are common to all except the orthotetidines. There is a general trend from fibrous to cross–bladed laminar shell throughout the group. The shell of the Orthotetidines is laminar from their earliest representatives and so may have its origin in laminar billingsellids.

The cardinal process tends to be reported as single–lobed in plectambonitoids and bilobed in the other groups. This distinction may be more apparent than real because in all members of the strophomenide group the evidence is for a paired dorsal diductor attachment set of scars. In plect-ambonitoids these muscle bases were separated by a single ridge and in most other members of the order each muscle base became raised to some extent on their own ridges, giving the bilobed condition. The ancestry of the chonetidines is unresolved. Brunton (1972) suggested that they were derived from the Plectambonitoidea, based on shell structure and some aspects of morphology, but Racheboeuf & Copper (1986) and Cocks & Rong (1989, p. 83) suggested descent from the Strophomenoidea, based on close similarity (apart from the hinge spines and possibly also shell fabric) between *Archeochonetes*, the earliest known chonetidine, which is of Ashgill age, and contemporary strophomenoids such as the eopholidostrophiid *Origostrophia*.

All but the orthotetidines and some strophomenoids seem to have had lophophores attached to the dorsal epithelium of the dorsal valve in such a way as to have resulted in skeletal structures raising the lophophore somewhat above the valve surface. In all but orthotetidines these are series of dorsal ridges or raised platforms giving support, e.g. bema, side septa, anderidia, brachial ridges, and falafer brachidium (of at least some Permian aulostegids). In orthotetidines there are structures derived from the inner socket ridges which extend forward like brachiophores, in a fashion possibly homologous to those of orthoids or early strophomenoids. These brachiophores probably functioned as do crura in Recent brachiopods. Attachment varies throughout the group, as it does within other groups throughout the Brachiopoda. Previously the character was afforded some importance and helped place groups like the davidsoniids, thecospirids and cadomellids within the old Strophomenidina. These are now all assigned to cemented forms of the Spiriferida sensu lato.

The position of the orthotetidines appears less secure than previously thought; the suborder fails to share some characters important to the other groups here considered as being strophomenides. The absence of pseudopunctae in the earliest orthotetidines raises a question as to their origin and Williams & Brunton (1993) concede that the apical pedicle sheath is shared by 'some primitive orthides'. In their discussion of the classification of the orthotetidines, Williams & Brunton (1993) drew attention to the earliest members of the group being impunctate, but strongly laminar in their shell fabric. By late Ordovician times there are traces of the introduction of pseudopunctae in *Fardenia*. Thereafter true pseudopunctation, or an alternative fabric in the Schuchertellidae, which these authors termed extropunctae, is characteristic of all ortho-tetidines. Thus we have to look for a possible ancestry amongst

laminar, impunctate shells. The presence of a pseudodeltidium and supra–apical foramen is quoted as characteristic of strophomenides in general, but Williams & Brunton (1993) also point out that the supra–apical foramen is also present in some primitive orthides. In our study of the cardinal process in taxa of the 'old Treatise' (Moore, 1965) Strophomenida, as noted above, we recognise the early distinction between those taxa in which the diductor muscles are seated directly onto the dorsal valve floor, commonly with a dividing ridge separating the two muscle ends, and those in which the diductor muscle bases became elevated on ridges, forming a bilobed structure, as seen typically in productoids, but also in all other stropho-menide groups other than most plectambonitoids. The earliest plectambonitoids (e.g.*Plectella* and *Akelina*) are known from rocks of Tremadoc age, with the strophomenoids (the undescribed strophomenid from China of Rong & Cocks, 1994, p. 663) appearing in the Arenig. Thus plectambonitoids are likely to have arisen from ancestors with fibrous shell, but lacking a bilobed cardinal process. The earliest known orthotetidines (e.g. *Gacella*) are from the mid-Ordovician (Llandeilo) and can be expected to have evolved from laminar–shelled, impunctate ancestors with a boss–like cardinal process.

Thus the relationship between some Lower Palaeozoic orthides and strophomenides remains unclear and is being studied. The strophomenides can now be more precisely diagnosed than in 1965 since several small groups of cemented and fibrous shelled taxa (having sparse spiralia) have been transferred to the Spiriferida. Thus there are no post–Permian representatives of the Strophomenida. Within the stropho-menides, the Orthotetidina stands more widely separated from the other major taxa than these are to one another. Lineages can be suggested with some confidence between all but the orthotetidines, which evolved from the early Upper Ordovician to the late Permian where, as with other strophomenides, they became extinct.

The following classification is proposed, although taxonomic levels may change:

Order **uncertain** (Orthida or Strophomenida)
  Suborder **Orthotetidina**
Order **Strophomenida**
  Suborder **Strophomenidina**
    Superfamily **Plectambonitoidea**
          Strophomenoidea (incl. strophodontids)
  Suborder **Chonetidina**
  Suborder **Productidina**
    Superfamily **Productoidea**
          Linoproductoidea
          Echinoconconchoidea
  Suborder **Strophalosiidina**
    Superfamily **Strophalosioidea**
        Aulostegoidea
         Richthofenioidea
         Lyttonioidea

REFERENCES

BAKER, P. 1970. The classification, origin and phylogeny of thecideidine brachiopods. Palaeontology 33: 175–191.
BRUNTON, C.H.C.1965. The pedicle sheath of young productacean brachiopods. Palaeontology 7: 703–704.
BRUNTON, C.H.C. 1972. The shell structure of chonetacean brachiopods and their ancestors. Bulletin British Museum (Natural History), Geology 21: 1–26.
___& F. ALVAREZ. 1989. The relationship between lamellae

and epithelial regressions in some articulate brachiopods. Lethaia 22: 247–250.

___, ___& D. MACKINNON. 1972. The systematic position of the Jurassic brachiopod *Cadomella*. Palaeontology 15:405–411.

___, ___ & ___ (in press). Morphological terms used to describe the cardinalia of articulate brachiopods; homo-logies and recommendations. Historical Biology.

COCKS, L.R.M. & JIAYU RONG. 1989. Classification and review of the brachiopod superfamily Plectambonitacea. Bulletin British Museum (Natural History), Geology 45:77–163.

COPPER, P. 1979. Devonian atrypoids from western and northern Canada. Geological Association Canada, Special Paper 18: 289–331.

COWEN, R & M.J.S. RUDWICK. 1966. The spiral brachidium of the Jurassic chonetid brachiopod *Cadomella*. Geological Magazine 103: 403–406.

GARCIA–ALCALDE, J. L. 1973. Braquiopodas Devonicos de la Cordillera Cantabrica, 5, El aparato braquial de *Davidsonia* Bouchard–Chantereaux 1849 (Strophomenida, Davidsonia-cea). Brevoria Geologica Asturica 17:1–5.

RACHEBOEUF, P.R. & P. COPPER. 1986. The oldest chone-tacean brachiopods (Ordovician-Silurian, Anticosti Island, Quebec). Canadian Journal Earth Sciences 23:1297-1308. Canadian Journal of Earth Sciences, 23: 1297-1308.

RONG JIA–YU & L.R.M. COCKS. 1994. True *Strophomena* and a revision of the classification and evolution of stropho-menoid and 'strophodontoid' brachiopods. Palae-ontology 37: 651–694.

WILLIAMS, A. 1973.The secretion and structural evolution of the shell of thecideidine brachiopods. Philosophical Transactions Royal Society, London, B 264:439–478.

___et al. 1965. Brachiopoda. In, R.C. Moore (ed.), Treatise on Invertebrate Paleontology, Part H. Geological Society of America & University of Kansas Press, 927p.

___& C.H.C. BRUNTON. 1993. Role of shell structure in the classification of the orthotetidine brachiopods. Palaeonto-logy 36: 931–966.

Appendix

TREATISE, 1965 CLASSIFICATION

Order **Strophomenida**
Suborder **Strophomenidina**
Superfamily **Plectambonitacea**
          Strophomenacea
              Davidsoniacea
            (incl. Davidsoniidae*, Thecospiridae+)
Suborder **Chonetidina**
Superfamily **Chonetacea**
          Cadomellacea (* *)
Suborder **Productidina**
Superfamily **Strophalosiacea**
          Richthofeniacea
          Productacea
Suborder **Oldhaminidina**
Superfamily **Lyttoniacea** (incl. Bactryniidae +)

(*) Removed to the Atrypida
(+) Removed to the Thecideidina
(* *) Removed to the Koninckinoidea

REVISION AND REVIEW OF THE ORDER PENTAMERIDA

SANDRA J. CARLSON

Department of Geology, University of California, Davis, California 95616, USA

ABSTRACT- Evolution in the Pentamerida is characterized primarily by changes in hinge line length and the degree and nature of development of a spondylium in the ventral valve, and a septalium in the dorsal valve. Two suborders are recognized: the primarily Ordovician, more primitive, paraphyletic Syntrophiidina, and the Siluro-Devonian, derived, monophyletic Pentameridina. Rhynchonellida originates within the Syntrophiidina between two superfamilies, the older Porambonitoidea and the younger Camerelloidea, both paraphyletic. In general, the syntrophiidines transform from primitive orthide-like morphologies to derived, rynchonellide-like morphologies. Pentameridines tend to be larger in size, ventribiconvex, rather than dorsibiconvex, and strongly rostrate. The suborder Pentameridina now includes three superfamilies: Pentameroidea, Stricklandioidea, and Gypiduloidea, that may be distinguished primarily by variation in their dorsal interiors.

INTRODUCTION

In preparation for the revision of the brachiopod volumes (Part H) of the Treatise on Invertebrate Paleontology, I investigated the phylogeny of the Order Pentamerida and, with A.J. Boucot, made relatively minor revisions to the classification of the order. The purpose of this paper is to outline briefly the major features of the phylogenetic analyses and their results, as well as the evolutionary transformations in pentameride character complexes and their effect on the revision of pentameride classification.

Pentamerides comprise a relatively small, but significant group of early and middle Paleozoic brachiopods. The number of pentameride genera recognized has doubled since the publication of the Treatise in 1965, rising from 85 to 169. Evaluating these new genera, many of which occur in Asia, provided an opportunity to re-evaluate the older classification and hypotheses about phylogenetic relationships among pentameride higher taxa. Complexities in discovering phylogenetic relationships among the major groups of pentamerides are understood more completely than before, and underscore the important role that homoplasy (convergence, parallelism, and reversal) has played throughout pentameride evolution.

PHYLOGENY RECONSTRUCTION

The phylogenetic relationships among the pentameride families as they were defined in the 1965 Treatise, were investigated coding each family as a consensus of characters that were present in all genera then assigned to the family (Carlson, 1993). At that time, I stated that a genus-level analysis, not yet completed, would have the potential to affect the results I presented by testing the monophyly of those families. Now that I have completed a genus-level analysis of the Syntrophiidina and revised the generic composition of syntrophiidine families, several differences have emerged that deserve more complete discussion.

Having gathered more extensive character information on each syntrophiidine genus over the past few years and examined specimens of most genera, I selected type genera to represent each revised syntrophiidine and pentameridine family (and most commonly referred to the type species of those genera; two non-type genera, *Clorindella* for Clorindidae and *Conchidium* for Subrianidae). This strategy for selecting terminal taxa has several advantages, not the least of which is reduced dependence on the current generic composition of each family. Two other syntrophiidine genera representing unusual character combinations or whose familial affiliations were initially unclear were also included, as were two rhynchonellide genera (*Drepanorhyncha* and *Rhynchonella*) for a total of 27 genera. The 'skeleton' data matrix then formed the basis for a comprehensive analysis of all syntrophiidine genera. I will discuss only some aspects of the family-level analysis here.

I selected four orthide genera (*Nisusia*, *Eoorthis*, *Billingsella*, and *Finkelnburgia*) as outgroups in these analyses, allowing the character states they possess to polarize character states (determine whether primitive or derived) within the ingroup, the pentamerides. Seventy-one characters concerning valve form, ornament, fold and sulcus, hinge region, ventral and dorsal valve interiors, and shell structure were coded for these taxa; the character list is similar to the one used in Carlson (1993), but some revisions were made. I analyzed the matrix using PAUP 3.1.1 (Swofford, 1993), selecting the heuristic search option with ten replicate analyses, each adding taxa from the matrix to the analysis in a random order.

Twenty-two equally most parsimonious cladograms of length 308 resulted, with a retention index of 0.53 (Figure 1). Variation among the 22 cladograms is manifest only at two poorly resolved regions: among the primitive camerelloideans and pentameridines. Although consistency among the 22 cladograms is fairly high, examining the 706 cladograms just one step longer reveals considerably more variation, primarily among the early syntrophiidines. Certain general patterns do emerge, however. *Cambrotrophia*, one of the earliest syntrophiidines to appear in the fossil record, tends to cluster among the orthide outgroups rather than with the other pentamerides. *Huenella* represents the next most primitive member of the group, with a wide hinge line and pseudospondylium, followed by the syntrophiides and *Triseptata*, an unusual genus from Novaya Zemlya, that has a septalium but is otherwise quite primitive. *Syntrophopsis* and *Tetralobula* consistently form sister groups; *Clarkella* and *Porambonites* are successively more derived syntrophiidines. The rhynchonellides emerge from between the paraphyletic Porambonitoidea and Camerelloidea. *Brevicamera* and *Karakulina* tend to cluster with *Camerella*; and *Parallelasma* and *Parastrophina* are most closely related to the monophyletic Pentameridina. Within the pentameridines, *Clorindella* and *Gypidula* are primitive, and *Stricklandia* and *Enantiosphen* emerge from within the pentamerides.

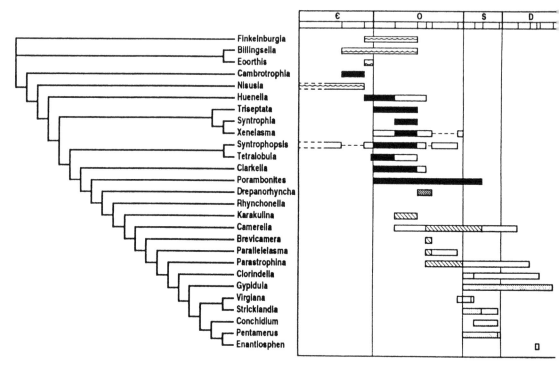

Figure 1. Cladogram #4 of 22 of length 308, retention index = 0.53, illustrating hypothesized phylogenetic relationships among type genera of pentameride families. Stratigraphic ranges of families plotted above; range of type genus shaded: outgroups, scalloped line; porambonitoideans, black; rhynchonellides, dark shading (Jurassic *Rhynchonella* too young to appear on chart); camerelloideans, diagonal lines; pentameridines, light shading.

The earliest pentamerides to appear in the fossil record share a number of similarities with the orthides: wide hinge line and subquadrate outline, poorly developed fold and sulcus, small but distinct cardinal process, pseudospondylium or no spondylium at all, and no septalium. Mosaic and iterative evolution among the (primitive) porambonitoideans make it somewhat difficult to define families within the group unambiguously, while families within the derived taxa - camerelloideans and pentameridines - are more easily discerned. The phylogenetic position of the rhynchonellides is relatively clear. They share a number of primitive characters with the derived porambonitoideans: dorsibiconvexity, uniplicate commissure, and certain features of the dorsal interior. They also share derived characters with the camerelloideans: astrophic hinge line, costate ornament, presence of crura, nature of the socket plates, and occasional presence of septalium. The pentameridines comprise a distinctive, derived and terminal group whose ancestry lies within the camerelloideans.

MORPHOLOGICAL EVOLUTION

Investigating phylogenetic relationships among pentamerides allows the changes in shell morphology that characterize pentameride evolution to be evaluated explicitly in a phylogenetic context. The transformations within character complexes are revealed and the relative pattern of primitive and derived states may be interpreted more clearly.

Overall size and shape.- Size varies considerably, but the largest pentamerides are very large indeed, are among the

most derived in the order, and appear relatively late stratigraphically. Biconvexity increases over time, expressed as dorsibi-convexity in the syntrophiidines, and ventribi-convexity in the pentameridines. Outline transforms from wider than long to longer than wide, and from subquadrate in wide-hinged forms, to subpentameral in short-hinged forms, to subtriangular, round, or elongate oval in astrophic forms. Derived pentamerides are much more rostrate than primitive pentamerides.

Ornament.- Ornament varies considerably among penta-merides. Every family exhibits nearly the full complement of ornament types, from smooth to various types of radial ribs to a range of more unusual kinds of ornament (concentric lamellae, spines, pits, etc.). Attempts to revise syntrophiidine classi-fication based on ornament (Andreeva, 1982) are rejected on the basis of ornament variability when compared with the full suite of morphological characters.

Fold and sulcus.- The earliest and most primitive pentamerides possess a strongly uniplicate commissure, which persists through most of the evolutionary history of the group. A few genera became independently and secondarily rectimarginate; the unisulcate condition evolved at least twice independently among the pentameridines.

Hinge and dentition.- A clear transformation from wide, strophic (straight) hinge lines, to successively shorter and shorter strophic hinge lines, to astrophic (curved) hinge lines is apparent. While hinge line length can, and does vary even intraspecifically, it usually varies within a small and predictable range. Predominantly astrophic species may include individuals with very short, strophic hinge lines, but predominantly strophic species only rarely include astrophic individuals. Most pentamerides have deltidiodont (non-interlocking) dentitions; on taphonomic, but not morphologic, grounds, some porambonitids and camerelloideans appear to have cyrtomatodont (interlocking) dentitions.

Ventral interior.- The dental plates transform over the course of pentameride evolution from small, insignificant features to larger, convergent plates that may form a pseudospondylium or one of three different types of 'true' spondylia. A pseudospondylium lacks a median septum, but possesses a low curved ridge that connects the anterior ends of the dental plates and surrounds the ventral muscle field. A sessile spondylium is characterized by a low median ridge or septum that supports and raises the anterior portion of a spoon- or trough-shaped spondylium formed by the convergence of the dental plates. The posterior portion of a sessile spondylium remains confluent with the valve floor. A median septum supports the entire length of a simplex spondylium, while a duplex spondylium is supported by a 'duplex' septum, apparently formed from the extension and fusion of the anterior portion of the dental plates themselves. It has not been clearly established, without more extensive serial-sectioning, that a simplex and duplex spondylium are strictly homologous. At least three different times in this transformation series the median septum is lost and the dental plates return to a subparallel orientation. The transformation order predicted in a purely structural sense (successive enlarging, converging, and raising of the dental plates above the valve floor) is reflected in the phylogenetic pattern as well.

Dorsal interior.- Features of the dorsal interior are perhaps the most characteristic in pentameridine morphological evolution. Because of their distinctive nature, Amsden (1964) introduced a special set of terms to describe the pentameride 'brachial apparatus' - inner plate, brachial process, outer plate. These features appear to be homologous with the outer hinge plate, crura, and inner hinge plate, respectively, of other, derived articulates; although now entrenched in the literature, the special pentameride terms obscure these homologies, and their use is no longer recommended (Brunton et al., in press). Porambonitoideans lack crura, thus the distinction between outer and inner hinge plates cannot be made, even though these plates in derived porambonitoideans may be quite similar in shape, length, and orientation to outer and inner hinge plates in camerelloideans and pentameridines. Porambonitoidean 'hinge' plates, undivided by a crural base, are referred to simply as socket plates, which can vary from short and subparallel or divergent, to elongate and subparallel, to convergent structures that form a septalium. A true septalium, in which the socket plates converge and unite to form a 'duplex' septum that forms a raised triangular 'pocket' posteriorly, has evolved at least twice, perhaps more, in the pentamerides. Inner socket ridges are distinct, but not elongated or blade-like as are brachiophores, their apparent orthide homologues. A true cardinal process is absent except in some huenellids and pentamerids.

CLASSIFICATION

The revised order Pentamerida comprises two suborders, as before in Amsden (1965) and Biernat (1965): the older, more primitive, paraphyletic Syntrophiidina and younger, more derived, monophyletic Pentameridina. Two syntrophiidine and three pentameridine superfamilies are now recognized: Porambonitoidea and Camerelloidea, and Gypiduloidea, Pentameroidea, and Stricklandioidea, respectively.

One interesting result of this study is the recognition of great morphological 'plasticity' among the older porambonitoideans (reflected in the large number of families recognized by Biernat, 1965), while the younger camerelloideans and pentameridines appear to be more 'stable' morphologically. Several distinguishing characters (width of hinge line, nature of the spondylium, presence or absence of cardinal process or septalium, and, to a lesser extent, ornament and nature of fold and sulcus), appear to be almost randomly distributed in various permutations and combinations among the porambonitoidean families, while the camerelloidean and pentameridine families appear to be morphologically more coherent and clearly defined.

Pentamerida.- Pentamerides range in size from small to large, and may reach extremely large sizes; pentamerides are among the largest brachiopods known. They tend to be strongly biconvex and uniplicate with varied ornament. The hinge line varies from wide and strophic to astrophic, the hinge structures are generally deltidiodont although some appear to be cyrtomatodont, and the delthyrium is generally open. Dental plates commonly, but not invariably, converge to form a spondylium. A cardinal process is commonly absent, socket plates remain discrete and subparallel but may converge to form a septalium, and crura may be absent or present. Shell structure is fibrous (it may contain laminar layers) and impunctate.

Syntrophiidina.- Generally smaller in size than pentameridines, syntrophiidines are wider than long, and their outline varies in shape largely as the hinge line width varies. They are usually strongly dorsibiconvex, and are not commonly rostrate. Inner socket ridges are often well-developed. Socket plates are commonly short, but may continue anteriorly into extensions that, although relatively long, never enclose the adductor muscle field.

Porambonitoidea.- Porambonitoideans have strophic hinge lines of varying width and generally well-developed interareas, particularly in the ventral valve. Dental plates vary in orientation from discrete and subparallel, to convergent and forming a pseudospondylium, or, most commonly, a sessile or simplex spondylium. A cardinal process is usually absent, but may be present as either a low vertical ridge, or as a callosity or narrow shelf at the posterior of the valve. Crura are never present. Socket plates also vary in orientation; they commonly converge and remain discrete, but may unite with a low septum to form a simplex or duplex septalium, may continue anteriorly into extensions of varying length, or may even diverge slightly.

Camerelloidea.- Camerelloideans commonly have astrophic hinge lines and a simplex or duplex spondylium. A cardinal process is never present. Crura are often present as short rod-like or blade-like processes. The outer hinge plates converge and commonly continue anteriorly into long, subparallel inner hinge plates, but may unite to form a simplex, duplex, or even sessile septalium.

Pentameridina.- Pentameridines are often large, strongly ventribiconvex, and strongly rostrate. Most are uniplicate, but some are unisulcate. A spondylium duplex is invariably present, supported, often only posteriorly, by a high median septum. Blade-like or rod-like crura are invariably present. The distinctive 'tripartite' dorsal cardinalia consists of outer hinge plates, crura, and inner hinge plates that either curve smoothly into one another or meet at definite angles. The inner hinge plates are commonly long and enclose the adductor muscle field.

Gypiduloidea.—Gypidoideans are medium to large in size and have astrophic hinge lines. The crura are long and blade-like, and the elements of the dorsal cardinalia meet at distinct angles, forming a 'lyre-shaped brachial apparatus.' The inner hinge plates commonly converge to form a cruralium that houses the adductor muscle field, but may remain subparallel.

Pentameroidea.- Pentameroideans are commonly quite large in size, longer than they are wide, and astrophic. The spondylium duplex tends to be deep and long. A plate-like, bilobed cardinal process or small deltidial plates may be present, but are not common. The crura are long and rod-like, and the dorsal cardinalia form two long, smoothly curved plates. The inner hinge plates commonly remain subparallel, and only rarely converge to form a cruralium.

Stricklandioidea.- The stricklandioideans share a number of features that serve to distinguish them from the other pentamerides. Their size varies but tends to be large, and longer than wide. They are dorsibiconvex, and only moderately so, or planoconvex, and not strongly rostrate. Uniplication is weak, not strong, and they have a strophic hinge line of medium width, with short interareas. The spondylium duplex tends to be quite short; it may be supported only by a short median septum or by none at all. The crura are long and rod-like, and the dorsal cardinalia are somewhat angular. The inner hinge plates are very short or absent, and do not enclose the adductor muscle field.

Comparison with 1965 Treatise classification.- Biernat (1965) reported a single syntrophiidine superfamily comprising 12 families and 40 genera. In the proposed revision, the Alimbellidae and Lycophoriidae have been removed to the Orthida; the Syntrophopsidae and Tetralobulidae have been combined, as have the Brevicameridae and Camerellidae. Amsden (1965) reported a single pentameridine superfamily comprising 5 families and 45 genera. In the revision, Parallelelasmatidae has been removed to the Camerelloidea, and Stricklandiidae and Gypidulinae have been raised to

superfamily rank, with associated shifts in lower level taxonomic ranks.

Most of the genera (and higher taxa) recognized in the 1965 Treatise originated in North America, named by North American brachiopodologists (e.g., Kirk, 1922; Schuchert & Cooper, 1932; Ulrich & Cooper, 1938; Amsden, 1953; Cooper, 1956; Boucot, 1964). However, most of the genera named since 1965 have originated in Eurasia, principally Russia (e.g., Sapelnikov, 1985) and China (e.g., Rong & Yang, 1981) (Figure 2). This increased geographic coverage has greatly expanded our knowledge of the global distribution of the group, both in terms of morphologic diversity and stratigraphic range.

Attempts to establish a phylogenetic classification.- Because the process of evolution generates groups of organisms united by genealogy, it is most informative to recognize and name those groups (taxa) on the basis of their phylogenetic relationships. This is the justification for attempting to establish phylogenetic classifications (see de Queiroz & Gauthier, 1992). The Linnean system of classification was established well before evolution was recognized as the process structuring the diversity of life. Thus, a strictly phylogenetic interpretation of a Linnean classification can be problematic. Attempting to create a phylogenetic classification is also problematic, however, because the phylogeny must be robust to provide the classification with stability. Since phylogenies (of organisms not reared in a controlled, laboratory setting) can only be inferred and never truly 'known', at least some uncertainty is inherent in any phylogenetic reconstruction. Extensive homoplasy (homeomorphy) contributes to this uncertainty, and must be anticipated and evaluated with respect to testable hypotheses of relationship in a group as ancient as the brachiopods, that survives to the present only in highly derived descendants; in such groups, homoplasy should not merely be ignored or dismissed.

Taxon rank and diagnosis differ in Linnean and phylogenetic taxonomy. Linnean taxonomic ranks may lose much of their

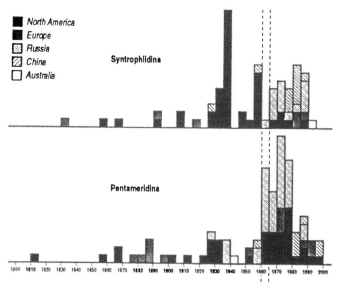

Figure 2. Authorship of new pentameride genera plotted by year and nationality of author.

apparent significance in a phylogenetic hierarchy, because, for example, taxa of the same rank may be nested within one another (e.g., Aves within Reptilia). For this reason, some cladists advocate the abandonment of ranks, opting instead for the adoption of an indented list of unranked names of clades interpreted with reference to a branching phylogenetic diagram (Table 1). Diagnosing monophyletic groups, such as Pentameridina, is relatively easy because clades are recognized by shared, derived characters. Diagnosing paraphyletic groups, such as Syntrophiidina, is more difficult; they are characterized as possessing, primitively, features of the larger clade in which they are nested and lacking derived features of the subclade from which they are being distinguished. These difficulties highlight the advantages of recognizing clades rather than nonmonophyletic taxa, and of adopting a phylogenetic classification, even though the transition from one system of classification to another will likely require a long and difficult transition period.

STRATIGRAPHY AND DIVERSITY

The extent to which the relative stratigraphic appearance of pentameride families is reflected in the hierarchical pattern of relationships (expressed in Figure 1) is quite remarkable. Although the phylogenetic analysis was performed without any explicit inclusion of stratigraphic order, that order is revealed in the cladogram nonetheless. I calculated the stratigraphic consistency index (SCI; Huelsenbeck, 1994) for the cladogram illustrated (in Figure 1), and followed the permutation procedure described therein to estimate the significance of the fit of the

Table 1. (**A**). Phylogenetic classification of pentameride higher taxa; (**B**). Linnean classification of pentameride higher taxa. Monophyletic taxa in bold-faced print.

[A] **PENTAMERIDA**
    RHYNCHONELLIDA
    CAMERELLOIDEA
        PENTAMERIDINA
            PENTAMEROIDEA
            STRICKLANDIODEA

[B] Order **PENTAMERIDA**
    Suborder SYNTROPHIIDINA
        Superfamily PORAMBONITOIDEA
        Superfamily **CAMERELLOIDEA**
    Suborder PENTAMERIDINA
        Superfamily GYPIDULOIDEA
        Superfamily **PENTAMEROIDEA**
        Superfamily **STRICKLANDIOIDEA**
    Order **RHYNCHONELLIDA**

stratigraphic record to the cladogram topology. The SCI value (0.64) was highly significant (p = 0.000), indicating an excellent fit between the two patterns. This provides encouraging reciprocal support for the orthides as useful outgroups for this analysis and for the 'trustworthiness' of the temporal ordering of taxa as they occur in the fossil record. *Tcharella* is the only genus that appears to be more derived than its stratigraphic position as the earliest pentameride would suggest. It is a very

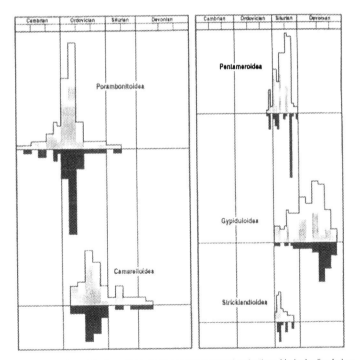

Figure 3. Diversity over time, showing originations (light shading above line) and extinctions (dark shading below line) of syntrophiidine (left) and pentameridine genera (right).

poorly known genus, however, and its assignment to the Syntrophopsidae is tenuous.

Pentameride generic diversity increases to a peak in the Arenig with 32 genera, drops sharply, then peaks again in the Ludlow with 38, drops sharply again and rises somewhat a third and final time in the Emsian with 19 genera. The last four pentameride genera are extinct by the end of the Frasnian stage. The porambonitoideans peak in generic diversity in the Arenig with 27 genera, camerelloideans in the Llandeilo with 14 genera, stricklandioideans in the late Llandovery with 9 genera, pentameroideans in the early Ludlow with 24 genera, and gypiduloideans in the Emsian with 19 genera (Figure 3). In the tallying of genera per superfamily, recall that only the stricklandioideans are strictly monophyletic; these diversity figures should be interpreted accordingly, as representing only portions of clades. Camerelloidea and Stricklandioidea are 'bottom-heavy', while Gypiduloidea and Pentameroidea are 'top-heavy'. Extinctions and originations keep pace with one another in most groups, except in the Gypiduloidea and Pentameroidea, where most extinctions occur substantially later than most originations.

CONCLUSIONS

The phylogenetic relationships among the major groups of pentamerides are better understood than before, but are not unproblematic, particularly at the level of genus and family. The revised classification, although not a true phylogenetic classification, can be interpreted directly with respect to revised hypotheses of pentameride phylogeny. In so doing, it renders evolutionary patterns of character transformation more accessible. Phylogenetic analysis does not solve all the problems of brachiopod classification, but it helps to clarify the nature of the problems, making them easier to test, and it provides the most defensible method available for reconstructing brachiopod evolutionary history.

Acknowledgments

I dedicate this contribution to the memory of R. E. Grant, who seemed to like pentamerides even though they were much transformed long before the Permian. I thank R. Doescher, F. M. Bayer, K. Hill, P. Waterstraat, M. Graziose, N. Buening, and S. M. and Z. A. Rudman for assistance and guidance, and J. Fong for preparing the illustrations. I particularly thank A. J. Boucot for sharing extensive preliminary information on the pentameridines. This research was supported, in part, by a grant from the National Science Foundation (DEB 9221452).

REFERENCES

AMSDEN, T.W. 1953. Some notes on the Pentameracea, including a description of one new genus and one new subfamily. Journal Washington Academy Science 43(5):137-147.

___1964. Brachial plate structure in the brachiopod family Pentameridae. Palaeontology 7:220-239.

___1965. Suborder Pentameridina, p. H536-H552. In, R. C. Moore (ed.), Treatise on Invertebrate Paleontology, Part H, Brachiopoda, 2, Geological Society of America and University of Kansas Press, Lawrence.

ANDREEVA, O.N. 1982. Middle Ordovician brachiopods of Tuva and the Altai region. Paleontologicheski Zhurnal 1982(2):48-59.

BIERNAT, G. 1965. Suborder Syntrophiidina, p. H523-H536. In R. C. Moore (ed.), Treatise on Invertebrate Paleontology, Part H, Brachiopoda, 2, Geological Society of America and University of Kansas Press, Lawrence.

BOUCOT, A.J. 1964. *Callipentamerus*, a new genus of brachiopod from the Silurian of Iowa. Journal of Paleontology 38:887-888.

BRUNTON, C.H.C., F. ALVAREZ & D.I. MACKINNON (in press). Morphological terms used to describe the cardinalia of articulate brachiopods: homologies and recommendations. Proceedings Boucot Symposium, Australia, 1995.

CARLSON, S.J. 1993. Phylogeny and evolution of 'pentameride' brachiopods. Palaeontology 36(4):807-837.

COOPER, G.A. 1956. Chazyan and related brachiopods. Smithsonian Miscellaneous Collections 127(I):1-1024; (II): 1025-1245.

DE QUEIROZ, K. & J. GAUTHIER. 1992. Phylogenetic taxonomy. Annual Review Ecology Systematics 23:449-480.

HUELSENBECK, J.P. 1994. Comparing the stratigraphic record to estimates of phylogeny. Paleobiology 20(4):470-483.

KIRK, E. 1922. *Brooksina*, a new pentameroid genus from the Upper Silurian of southeastern Alaska. Proceedings US National Museum 60(19):1-8.

RONG JIA-YU & XUE-CHANG YANG. 1981. Middle and late Early Silurian brachiopod faunas in southwest China. Memoirs Nanjing Institute Geology Palaeontology, Academia Sinica 13:163-262.

SAPELNIKOV, V.P. 1985. Morphologic and taxonomic evolution of Brachiopoda, the Order Pentamerida. Akademii Nauk SSSR, Uralskii Nauchyi Tsentr, Sverdlovsk, 188 p.

SCHUCHERT, C. & G.A. COOPER. 1932. Brachiopod genera of the suborders Orthoidea and Pentameroidea. Memoirs Peabody Museum Natural History 4:1-259.

SWOFFORD, D.L. 1993. PAUP: Phylogenetic reconstruction using parsimony, Version 3.1.1. Computer program distributed by the Illinois Natural History Survey, Champaign, Illinois.

ULRICH, E.O. & G.A. COOPER. 1936. New genera and species of Ozarkian and Canadian brachiopods. Journal of Paleontology 10:616-631.

GENETIC MODELS OF EARLY DEVONIAN BRACHIOPOD PAVEMENTS, LONGMENSHAN, SICHUAN, CHINA

YUANREN CHEN

Department of Geology, Chengdu Institute of Technology, Chengdu, 610059 Sichuan, China

ABSTRACT- Two kinds of brachiopod shells commonly formed pavement in the Early Devonian of the Longmenshan area, one consisting of flat or concavoconvex shells, and the other of disarticulated biconvex shells. Such pavements have the following features: (1) are commonly one or two valves thick, (2) have convex-up orientation, (3) possess variable sizes, states of preservation, and sorting, dependant mostly on genetic processes, (4) are imbedded in barren muddy siltstones, calcareous mudstones, silty mudstones, and shales, and occasionally bioclastic limestone and siltstone, (5) have pavements parallel to bedding plane, and, (6) show variable frequency of pavements. According to preservation, outcrop, etc., the following genetic pavement models are proposed: (i) opportunistic species explosion, (ii) storm deposit - winnowing, (iii) scouring - transportation, and, (iv) winnowing - diagenetic compaction.

INTRODUCTION

Shell pavement is here defined as a type of thin, usually 1-2 valves thick, virtually two-dimensional bedding-plane shell concentration (following Kidwell et al., 1986). A linear type of localized shell concentration is also referred to as shell pavement (Chen et al., 1992). Shell beds commonly refer to three-dimensional shell concentrations with a thickness of more than two valves, which show gradational change to shell pavements. In the present paper, shell pavements are used for a particular type of shell bed dominated by flat or concavo-convex shells or single valves of biconvex shells. These occur either on bedding plane surfaces, or are parallel to bedding planes.

STRATIGRAPHIC SETTING

The Lower Devonian of the Longmenshan area is divided into five formations nearly 3000m thick, described here in ascending order. Most of the brachiopod pavements discussed in this paper occur in the Bailiuping, Ganxi, and Xiejiawan formations (Figure 1):

Pingyipu Formation. Fine-grained quartz sandstones, siltstones, interbedded with silty mudstone with *Lingula*, *Howellella*, *Orientospirifer* , *Anathyris*, and *Tadschikia* from 600-2390 m thick.

Bailiuping Formation. Siltstones, silty mudstones, with interbedded limestone lenses, bioclastic limestones, with*Protochonetes bailiupingensis*, *Strophochonetes ganxiensis*, *Orientospirifer nakaolingensis*, *O. wangi*, *Acrospirifer primaevus*, *Neoathyrisina typica*, *N. longmenshanensis*, *Pseusoathyrisina fasciata*, *Nikiforovaena ganxiensis*, and *Pseudokymatothyris sinensis*, from10-150 m thick.

Ganxi Formation. Silty-calcareous mudstones, muddy-calcareous siltstones, marls, calcarenites, calcisiltites, intercalated with shell beds containing *Dicoelostrophia punctata*, *Eosophragmophora sinensis*, *E. sichuanensis*, *Xenostrophia yukiangensis*, *Kransia mesodeflecta*, *Rostrospirifer tonkinensis*, *R. papaoensis*, *R. multiplicatus*, *R. ordinaris*, *Howellella incerta*, and*H. yukiangensis* ranging from 30-100 m thick.

Xiejiawan Formation. Silty mudstones, muddy siltstones, interbedded with fine-grained sandstone and bioclastic limestones holding *Euryspirifer paradoxus longmenshanensis*, and *E. paradoxus xiejiawanensis*, from 100-250 m thick.

Yangmaba Formation(lower part). Calcareous sandstones, siltstones, limestones, reef limestones, with *Athyrisina squamosa*, *A. hemi*, and *Otospirifer*. The upper Yangmaba Fm. (*Zdimir* beds) is Eifelian, as dated by conodonts.

CLASSIFICATION OF DEVONIAN BRACHIOPOD PAVEMENTS

There are several kinds of shell beds classifications (Kidwell et al., 1986; Fürsich, 1990). Two of these have shown to be practical for the Devonian shell pavements in the Longmenshan area:

Classification by taxonomic composition

(a) Monotypic- pavement dominated by a single species or genus. This type includes the paucispecific pavement of Kidwell et al. (1986); (b) Polytypic pavement consisting of more than one species and genus, with none of the taxa being dominant.

Classification by shell morphology and preservation

(a) Pavement dominated by flat or concavoconvex shells of disarticulated valves; (b) Pavement dominated by disarticulated valves of biconvex shells; (c) Pavement dominated by fragments of variously shaped shells.

GENERAL FEATURES OF LONGMENSHAN BRACHIOPOD PAVEMENTS

Pavement-forming brachiopods in area studied fall into two shell shape classes: flat or concavo-convex whole shells or single valves, such as those of *Protochonetes*, *Luanquella*, *Dicoelostrophia*, *Eosophragmophora*, *Xenostrophia*, *Nadiastrophia*, and single valves of biconvex shells, such as those of *Orientospirifer*, *Acrospirifer*, *Rostrospirifer*, and *Euryspirifer*. The pavements dominated by these brachiopods are usually 1-2 valves thick, with the valves oriented convex-up, showing varying sizes, states of preservation, and degrees of sorting. The surrounding rocks are commonly muddy siltstones, calcareous mudstones, silty mudstones, shales, generally barren of fossils. The shell pavements usually occur on bedding planes, or parallel to the planes, and their frequency in the stratigraphic section is irregular.

GENETIC MODELS

Three kinds of processes are regarded to be most common and important for the formation of shell beds: biogenic, sedimentologic, and diagenetic processes (Kidwell & Jablonski, 1983; Kidwell et al., 1986; Fürsich, 1990). A shell bed, however, is rarely formed by one process alone, but rather by the combination of two or more processes. The accumulation of skeletal material to form shell beds should meet the following conditions: (1) at any given site, shell input must exceed shell output, (2) rate of net sediment accumulation should be low,

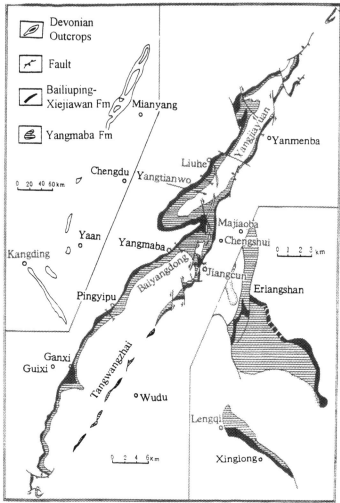

Figure 1. Lower Devonian rock outcrop in the Longmenshan area, Sichuan, South China.

although this does not necessarily mean sediment starvation. Low rate of sediment accumulation can be caused by winnowing and erosion, (3) the depositional environment must be favourable to fossilization and preservation of the shells. Following these general features, state of preservation, and outcrop occurrences of shell pavements, four genetic models are proposed for the Longmenshan area.

Opportunistic species explosion model. Many monotypic pavements dominated by flat of concavo-convex shells are assigned to this type. The *Protochonetes bailiupingensis* pavements, for example, are very common in the lower and middle Bailiuping Fm. The surrounding rocks are typically dark grey to yellowish green, silty mudstone barren of fossils. Small shells of *P. bailiupingensis* are well preserved, with the spines and fine internal structures often intact (Figure 2:2). This indicates that the shells were not subject to long distance transportation before burial, and were probably autochthonous. The thin flat shell beds occur on average at the spacing of 3-4 beds per 1cm of stratal thickness, reaching more than 100 beds in 2m at Qinggangping, near Ganxi. Each pavement

is very thin, with high shell density and good preservation. To form such a pavement, *P. bailiupingensis* probably had high tolerance to environmental stress, high fecundity, great capability to expand its ecotope after invading a new habitat, and was able to regenerate rapidly after short-lived habitat damage or disturbance. These capabilities appear to be characteristic of this opportunistic brachiopod species, which has a small, flat shell (Chen, 1992).

The formation of a typical *P. bailiupingensis* shell pavement is interpreted as follows: (1) A newly created or recently vacated habitat under high environmental stress became available, (2) Numerous larvae of *P. bailiupingensis* invaded the habitat, (3) A population of *P. bailiupingensis* exploded under r-selection, with vertical orientation, (4) Frequent storm disturbance caused burial of populations, or smothering by suspended mud. Most small and thin shells turned convex-up for hydrodynamic stability, (5) Final burial, fossilization, and diagenesis led to the formation of shell pavement (Figure 3).

The thickness of each intercalated muddy bed depended on

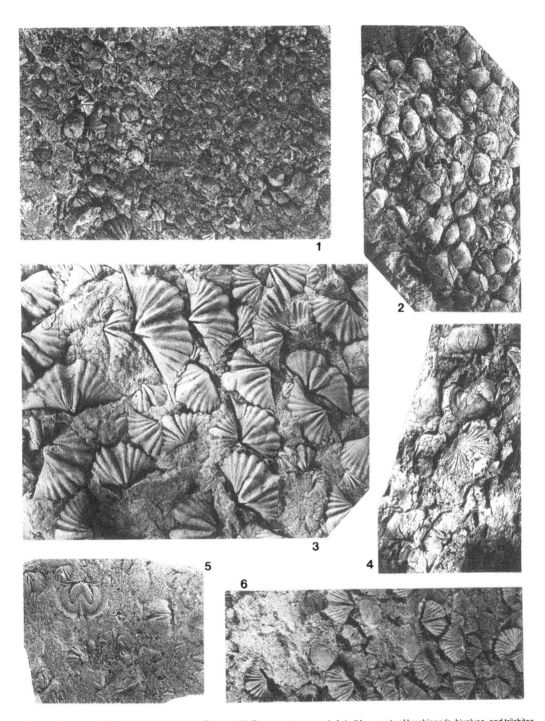

Figure 2. Examples of Devonian shell pavements, Sichuan. (1), Pavement composed of shell fragments of brachiopods, bivalves, and trilobites, Ganxi Fm., Pingyipu; (2) Pavement dominated by shells of *Protochonetes bailiupingensis,* Bailiuping Fm., Yingzuiyan, south of Ganxi; (3) Pavement dominated by *Acrospirifer,* Bailiuping Fm., Yingzuiyan; (4) Pavement dominated by *Xenostrophia yukiangensis, Luanquella kwangsiensis,* and *Dicoelostrophia,* Ganxi Fm., Pingyipu; (5), Pavement dominated by *Dicoelostrophia punctata,* Ganxi Fm.; (6), Pavement of *Orientospirifer nakaolingensis,* Bailiuping Fm., Pingyipu (all x0.75).

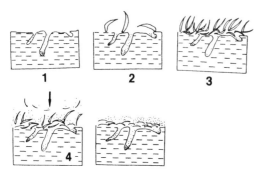

Figure 3. Five stage genetic model of shell pavement dominated by opportunistic species for *Protochonetes bailiupingensis*.

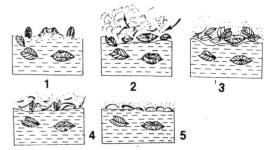

Figure 4. Genetic model of shell pavements formed by storm deposits and winnowing .

the mud input during each storm event, ranging form 2mm to 10mm (prior to compaction). Some shell pavements were covered only by a thin muddy film. The pavements show common vertical bifurcation or convergence along strike. Excellent preservation indicates that the shells of *P. bailiupingensis* were not reworked after their first burial. It is most likely that the pavements were formed at a depth between average storm wave base and maximum storm wave base, where shell accumulation would not be disturbed by small storms. Relatively thick mud cover (>20cm), and low oxygen level in stagnant pore water, would deter settlement of infaunal organisms and bioturbation (Aller & Cochran, 1976; Benninger, 1979; Aller, 1980; McCall et al., 1983). These taphonomic conditions are shown by well preserved shell pavements in dark grey mudstones but poorly preserved pavements in siltstones. According to Johnson (1989), it requires 2-3 years in shallow water and 8-10 years in offshore areas for brachiopods to form a cohort after an event. It is thus possible that each *Protochonetes* pavement represents a 10-year depositional interval, and can be used as an isochronous stratigraphic marker bed.

Storm deposit and winnowing model. Most pavements dominated by single valves of biconvex spiriferids, like *Acrospirifer*, *Rostrospirifer*, *Euryspirifer*, and *Howellella* from the Bailiuping, Ganxi, and Xiejiawan formations, are referred to this type (Figure 2:3,6). The shells are typically well preserved, with the pointed cardinal extremities intact, and predominantly in convex-up orientation. The surrounding rocks consists of mudstone, shale, silty mudstone, and muddy siltstone, and contain sporadic brachiopod shells that are similar to those forming the pavements.

Two kinds of processes led to the formation storm-winnowed shell pavements in Sichuan. The first of these proceeded as follows (Figure 4): (1) under normal sedimentation and environmental stress, brachiopod shells accumulated with varying abundance and diversity; (2) during a storm, strong bottom currents scoured the sea floor and winnowed out fine sediments, which led to concentration of shells; (3) light shells became suspended, or were rolled along the bottom to form a sheet of shells thinning out into deeper water. The thickness, density and abundance of shells would depend on the intensity of the storm; (4) fine sediments covered the shell beds following the storm event, but were winnowed out by waves and bottom currents generated by small storms. Minor disturbances may have been too weak to cause damage or disarticulation to the shells; and, (5) after prolonged winnowing, most single valves turned convex-up.

This type of shell pavement was autochthonous or parautochthonous, as shells were derived from the underlying substrate. The thickness of each pavement and its shell density and diversity are variable from place to place, depending on the initial amount of shell material present on the substrate and the intensity and duration of winnowing. Intercalated beds are commonly much thicker than the pavements themselves. In most pavements, the scouring marks are found on the lower bedding planes. These features indicate that this type of pavement probably was formed near, or slightly above average storm wave base, so that small storms would not scour but simply winnow bottom sediments. Suspended fine sediments in the inner shelf area during small storms could be redeposited here finally to bury the shell pavements.

The second storm-winnowed process is interpreted as follows: (1) under favourable living conditions, rich brachiopod faunas flourished; (2) shells were disturbed repeatedly by storms, waves, and currents. Buried shells remains were re-exposed and then compacted. Thus abundant shell material (from well-preserved shells to shell fragments) accumulated on the sea floor; (3) during violent storms, shell material became suspended and transported together with sediments, and large shells and fragments would be dragged and rolled, breaking into smaller fragments; and, (4) shell material was redeposited in deeper water and was covered by sediments generated by the next storm.

The second type of storm-winnowed shell pavement is similar to the first, but differs in having more fragmented, better sorted shells. The surrounding rock and the interbeds were rapidly deposited, storm-generated, poorly fossiliferous sedi-ments, which show graded bedding. These shell pavements consist of several species, similar or dissimilar to those in the underlying beds. In the Longmenshan area, most shells in the pavements were disarticulated, but relatively well preserved. This suggests that these shells were largely parautoch-thonous. These brachiopods probably lived under relatively shallow water conditions (Benthic Assemblage 2, or upper BA3), but were preserved in slightly deeper water. Storm events and winnowing appear to be the main processes in the production of this type of pavement. Each pavement is thin, and thus shells in pavements could not have been derived from various communities scattered over a large area.

Scouring and transportation model. In the Devonian Bailiuping, Ganxi, Xiejiawan, and to some extent in the Ping-yipu formations, some pavements are composed of very fragmentary, poorly preserved, unidentifiable shells (Figure 2:1). Occasionally, large pieces or single valves of brachiopods are found. The surrounding rocks are mainly poorly fossiliferous

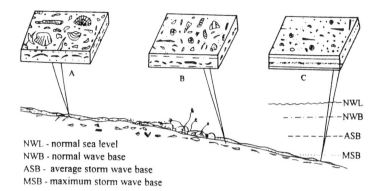

NWL - normal sea level
NWB - normal wave base
ASB - average storm wave base
MSB - maximum storm wave base

Figure 5. Distribution of shell pavements generated by scouring and transport.

siliciclastics, e.g. siltstone, silty mudstone, mudstone, and shale. This type of pavement was probably formed by scouring and transportation during storms, and the process is interpreted as follows: (1) shells in sediments and living on the surface were disturbed repeatedly by storms, waves, and currents. During violent storms, the shell material became suspended and transported together with sediments, and large shells and fragments would be dragged and rolled, breaking into smaller fragments; (2) shell fragments were redeposited in deeper water as thin, fan-sheets and were covered by sediments stirred up and transported by later storms. If the sediment cover were thick enough and undisturbed by subsequent storms, shells would be preserved as sheet-like pavements.

Storm-generated intertidal shell debris deposits of Sanibel Island, Florida (Haas, 1940; Boucot, 1981) are a good example of this type of shell pavement. This type of shell pavement can be found from the intertidal zone to maximum storm wavebase (Figure 5), or even below the maximum storm wavebase, where storms cannot directly affect bottom sediments but storm generated mud and bioclasts can be deposited by storm return flow. This is shown by thin pavements sometimes preserved in dark to black mudstones. Perfectly preserved tentaculitids and ostracodes may also be present. This type of pavement may grade into coquina or shell beds shorewards.

Winnowing and diagenetic compaction model. Some pavements in black shales from the lower Ganxi Formation are assigned to winnowing-diagenetic causes. These are dominated by single valves of the spiriferid *Rostrospirifer tonkinensis*, and whole shells, or single valves of the strophomenid *Xenostrophia*. Some bivalves were also present. All shells were well preserved, parallel to bedding and in convex-up position. Shell density in these pavements and thickness of intercalated beds were locally variable. Some well-preserved bivalve shells were seen to cut across two or three beds. The genetic process for this type of pavements is interpreted as follows: (1) under moderate to high background sedimentation, and low oxygen conditions, various infaunal deposit feeders lived in the sediments; (2) benthic epifaunal shelly faunas gradually became established in this environment, with decreased sedimentation rates and increased oxygen level; (3) most shells were not buried immediately after death, but remained on the surface, underwent prolonged scouring and winnowing, and became disarticulated; (4) shells became buried by storm-generated muds, or increased sedimentation; and, (5) under diagenetic compaction, both surface and within substrate shells formed a single shell pavement (Figure 6).

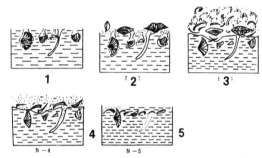

Figure 6. Genetic model of shell pavement by winnowing and diagenetic compaction.

Relatively large and thick shells (mostly infaunal) may cut across several beds (Fürsich, 1978). This type of pavement was probably formed in relatively shallow water conditions (BA2). Well-preserved shells of brachiopods and bivalves indicate a quiet or low water energy setting, probably lagoonal or behind a barrier. Most shells were autochthonous and represent only one or two generations of communities. This type of shell pavement is useful for stratigraphic correlation.

CONCLUSIONS

Four types of shell pavements were generated under different conditions in the Lower Devonian sequence of northwest Sichuan. These genetic models are simplified and generalized, and some pavements may not fit into any model. Such models appear useful for paleoenvironmental reconstructions and stratigraphic correlations by integrating paleontological and sedimentological data (Table 1).

Acknowledgments

I would like to thank Professor Wang Hongfeng for his assistance in this study, and F.T. Fürsich, A.J. Boucot, Paul Copper, and S.M. Kidwell for their constructive discussions. This study was funded by the National Natural Sciences Foundation of China and by Project 96-30 of the Ministry of Geology and Mineral Resources of China. Travel and accommodation support to attend the Third International Brachiopod Congress was provided by the Natural Sciences and Engineering Research Council of Canada

Table 1. OSE = opportunistic species explosion; SDW = storm deposit and winnowing; ST = scour and transport; WDC = winnowing-diagenetic compaction.

models	composition	shells or valves	preservation	orientation	density	area of distribution	process	source	time span
1 OSE	monotypic	flat, concave-convex	well	convex-up	high	large	biogenic	in place	up to 10 yrs
2 SDW	monotypic or polytypic	disjunct biconvex, or flat , concave-convex	well to very well	convex-up	variable	variable	biogenic and sedimentologic	in place	10-100 yrs
3 ST	polytypic	fragments	poor	variable	very high	large	sedimentologic	allochthonous	up to1 yr
4 WDC	polytypic	disjunct or whole shells	variable, mostly well	variable, mostly convex-up	variable	small	sedimentologic and diagenetic	in place	10-100 yrs

REFERENCES

ALLER, R.C.1980. Diagenetic processes near the sediment-water interface of Long Island Sound, I, Decomposition and nutrient element geochemistry (S,N,P). Advances Geophysics 22:237-350.

___& J.K. COCHRAN.1976. Tracking particle associated processes in nearshore environments by use of 234Th/238U disequilibrium. Earth Planetary Sciences Letters 47:161-175.

BENNINGER, L.K. 1979. The increasing influence of biologic activity on sedimentary stratification through the Phanerozoic. Geological Society America, Abstracts with Program 11:38

BOUCOT, A.J. 1981. Principles of benthic marine paleoecology. Academic Press, New York, 463p.

CHEN YUANREN (ed.). 1994. Devonian of the Longmenshan area, Sichuan, China, I, Lower Devonian dynamic stratigraphy and ecostratigraphy. Chengdu University Science Technology Press, Chengdu, 185p.

___, HONGFENG WANG & CHANGJUN ZHANG. 1992. The genetic models and implication of shell beds from the Lower Devonian of the Longmenshan area, Sichuan, China, Mineralogy and Petrology 12:70-78.

FÜRSICH, F.T. 1978. The influence of faunal condensation and mixing on preservation of fossil benthic communities. Lethaia 11:243-250.

___1990. Fossil concentration and life and death assemblages, p.235-239. In, D.E. Briggs & P.R. Crowther (eds), Palaeobiology, a synthesis. Blackwell, Oxford.

HAAS, F. 1940. Ecological observations on the common molluscs of Sanibel Island, Florida. American Midland Naturalist 24:369-378.

JOHNSON, M.E. 1989. Tempestites recorded as variable *Pentamerus* layers in the Lower Silurian of southern Norway. Journal Paleontology 63:195-205.

KIDWELL, S.M., F.T. FÜRSICH & T. AIGNER. 1986. Conceptual framework for the analysis and classification of fossil concentrations. Palaios 1:228-238.

___& D. JABLONSKI. 1983. Taphonomic feedback: ecological consequences of shell accumulation, p.195-248. In, M.J.S. Tevesz & P.L. McCall (eds.), Biotic interactions in Recent and fossil benthic communities. Plenum, New York.

MCCALL, P.L. & M.J.S. TEVESZ. 1983. Soft-bottom succession and fossil record, p.157-193. In, ibid.

COMPETENCE, PRE- AND POST-SETTLEMENT CHOICES OF ARTICULATE BRACHIOPOD LARVAE

S.H. CHUANG

144 Pasir Ris Road, 1851 Republic of Singapore

ABSTRACT- The competent larvae of articulate brachiopods have fully developed larval muscles, a pedicle-forming distal region constricted from the pedicle lobe, and a change in behaviour from photophilous to photophobic. Gliding or creeping movements and frequent contacts of their anterior lobe or ventral surface on the bottom of the laboratory culture dish appear to be exploratory attempts to test suitability of the substratum for attachment or settlement. Settling larvae first glued a small random area at the tip of their pedicle to the substratum with a sticky secretion for tentative attachment. Since these postlarvae may easily be flushed off the substratum with a jet of water from a pipette, or be mechanically separated from substratum without injury to pedicle lobe, presumably during this tentative attachment or settlement, postlarvae could detach themselves or allow themselves to be detached from the substratum by currents or wave action in post-settlement selection for more suitable sites. Subsequently, after a layer of cuticle is secreted between the glued pedicle and substratum to achieve irrevocable permanent attachment, postlarvae cannot be flushed off the substratum nor mechanically removed without injury to the pedicle. Pre- and post-settlement selections of attachment sites or responses of competent larvae to appropriate stimuli from the substratum and the environment, before and during tentative settlemen,t may account for preferences of larvae for certain substrata, and also for natural occurrences of articulate brachiopods in concavities or grooves, and in crypts.

INTRODUCTION

Percival (1944, 1960), Long (1964), Rickwood (1968), Franzen (1969), Doherty (1976), Stricker & Reed (1985), and Peck & Robinson (1994) have previously determined or estimated the developmental time of the embryos and larvae in articulate brachiopods. Then available developmental times, listed in Peck & Robinson (1994), showed large disparity in the time the zygotes took to develop from fertilization to the settlement of the larvae between temperate brachiopods and the Antarctic species *Liothyrella uva* (Broderip). The zygotes of temperate brachiopods took 3-19 days at temperatures around 10˚C, while those of *L. uva* were estimated to require at least 115-160 days at about freezing [0˚C]. However, due to different interpretations of 'embryo' and 'larva', comparison of the precise time, when the embryonic stage ended, and the larval stage began, is difficult.

One confounding feature in articulate development is that the external opening of the blastopore does not close after the invagination of the blastomeres during gastrulation ceases, and the blastopore canal no longer communicates with the coelenteron. From then on, the blastopore becomes a pit and persists into subsequent developmental stages as a blastopore remnant in the following species: *Calloria inconspicua* (Sowerby) [Percival, 1944; personal observation: N.B. *Waltonia* was renamed *Calloria* by Cooper & Lee, 1993], *Notosaria nigricans* (Sowerby) (Percival, 1960), *Pumilus*

antiquatus Atkins (Rickwood, 1968), *Terebratulina retusa* (Linnaeus) (Franzen, 1969), and *Liothyrella neozelanica* Thomson (Chuang, 1994). In *Terebratulina retusa*, the blastopore remnant has an elongated outer opening (Franzen, 1969), while in other articulates, it is irregularly oval or circular. There is difficulty also in determining the hatching of the larva from the embryo, since the loss of the fertilization membrane at hatching has not been documented in articulate brachiopods. The use of the term 'hatching' in articulate embryology should be avoided unless actual loss of the fertilization membrane is witnessed. If the first appearance of cilia is equated with hatching, awkwardness, if not ambiguity, may result, since cilia first appear at different developmental stages from the blastula stage to the wedge larva during the development of different articulate species (Chuang, 1990, 1994). For instance, in *Liothyrella uva* (Broderip) 'larvae' were released from the brooding female at the 'gastrula stage' (Peck & Robinson, 1994). In the present paper the term 'larva' refers to developmental stages beginning with the stage possessing blastopore remnant and with no blastopore canal connecting the blastopore remnant with the coelenteron.

MORPHOLOGICAL AND BEHAVIOURAL CRITERIA OF COMPETENCE

On morphological grounds the larvae of brachiopods fall under three stages, of which the last stage is the pre-settlement stage in *Liothyrella neozelanica* (Chuang, 1994). During the pre-settlement stage the larvae of articulate brachiopods, (1) have a cup-like depression at the apex of the pedicle lobe,perhaps resulting from tonic contractions of longitudinally arranged muscles inserted internally to the tip of the pedicle lobe; (2) have a pedicle-forming distal region constricted from the pedicle lobe; and, (3) have undergone a change in light response from photophilous to photophobic. Articulate larvae with the above attributes were presumed competent. How long competence lasts remains unknown.

PRE-SETTLEMENT SITE CHOICE

In the laboratory culture dish, pre-settlement larvae frequently move along the surface of the dish, touching the bottom of the dish with the anterior lobe or the ventral side of the body. Occasionally, the larvae rest on the bottom of the dish with their anterior lobe touching the dish. This behaviour presumably represents exploratory attempts of competent larvae to test suitability of the substratum for attachment or settlement. This exploratory behaviour lasts from a few hours to several days, especially if the culture dish that holds the larvae is not darkened. Competence of larvae appears to last a few days. During this exploratory behaviour, they apparently respond to environmental cues in finding and choosing a settlement site.

TENTATIVE SETTLEMENT

Tentative settlement of pre-settlement larvae of *Liothyrella neozelanica* occurs some time after the culture dish is covered with foil to exclude light. The competent larvae glue a small random area at the tip of their pedicle to the substratum. This area of attachment is a small fraction of the area defined by the mouth of the cup-like depression. Since only a thin sheet of the sticky secretion of the epithelium of the pedicle tip separates the postlarvae from the substratum, these tentatively settled postlarvae are easily flushed away with a strong jet of water from a pipette or mechanically separated from the substratum with a fine instrument without injury to the pedicle lobe. After detachment, the tentatively settled postlarvae invariably swim to another site to reattach themselves.

PERMANENT SETTLEMENT

After a period of time, tentatively settled postlarvae of *Liothyrella neozelanica* secrete a layer of cuticle to bond a wider area of pedicle to the substratum permanently. With permanent settlement, postlarvae resist flushing with jets of water from a pipette. When forcibly detached from the substratum by mechanical means, postlarvae come off with torn pedicles. Thus postlarvae were firmly attached and irrevocably settled.

DISCUSSION

Articulate brachiopod larvae are presumed competent when they possess the three attributes mentioned earlier. The presence of an apical cup-like depression of the pedicle lobe was already noticed in *Lacazella mediterranea* by Lacaze-Duthiers in 1861, who called it a 'ventouse', presumably implying a sucker-like function for this depression during settlement. This depression is presumably of general occurrence in articulate brachiopod larvae, since it appears in illustrations of the larvae of *Pumilus antiquatus* (in Rickwood, 1968, Fig. 7:8), *Terebratalia transversa* Sowerby (in Gustus & Cloney, 1972, Fig. 4; Stricker & Reed, 1985, Fig. 3). In the pedicle lobe of larvae of the Antarctic *Liothyrella uva*, reared at temperatures between 0˙ and 2˙C place (Peck & Robinson, 1994), the pedicle-forming distal region had not been constricted from the rest of the pedicle lobe 45 days after the release of the brood, but these larvae already congregated under the shell fragments provided in the laboratory container. However, no attachments took (ibid.).

Larvae have previously been reported to swim deliberately along, or creep on the surface, of culture dishes in the laboratory, and to touch the dish bottom with either anterior lobe or ventral surface of the body, suggesting exploratory attempts at testing the suitability of the substratum for attachment or settlement. This has been seen in *Calloria inconspicua* (in Percival, 1944), *Notosaria nigricans* (in Percival, 1960), *Frenulina sanguinolenta* (Gmelin) (in Mano, 1960), and *Terebratalia transversa* (in Long, 1964). Presumably, after coming in contact with the substratum, competent articulate larvae would choose either to settle forthwith, or to swim away. Having made their choice of settlement site, or having responded to appropriate stimuli from the substratum or the environment, the larvae glued a random area of the apex of the pedicle to the substratum with an adhesive substance produced by the pedicle epithelium.

In all the recorded instances of adherence to the substratum during tentative settlements of the larvae of *Liothyrella neozelanica*, the attachment area was invariably smaller than the mouth of the cup-like apical depression. This depression was not used as a sucker in this articulate. Presumably, during this tentative settlement, larvae of articulate brachiopods

could still exercise post-settlement selection of settlement site by vacating the tentative-settlement site, and swimming to a more suitable location on the substratum, because of the ease with which the tentatively settled postlarvae could be flushed away experimentally with jets of water.

In experimental settlements in the confines of the laboratory dishes, the larvae of both *Notosaria nigricans* (in Percival, 1960) and *Pumilus antiquatus* (in Rickwood, 1968) preferred weathered rock surfaces to fresh ones. The larvae of *Frenulina sanguinolenta* showed a preference for attachment to *Lithothamnion* species of coralline algae over other coralline algae, glass, plastic, porcelain, and stones (Mano, 1960). Presumably, these preferences were the result of deliberate pre-settlement choices of competent larvae in response to cues from the substratum and the environment.

The reversal in the behaviour of articulate brachiopod larvae to light, from a photophilous to photophobic state, towards the close of the larval stage was observed in *Calloria inconspicua* by Percival (1944), and in *Frenulina sanguinolenta* by Mano (1960). Articulate choice of a dark place as natural settlement site is presumed to be a consequence of photophobia of pre-settlement larvae. The distribution of subtidal *Argyrotheca johnsoni* Cooper on the underside of foliaceous corals was presumed to be correlated with low light intensity (Jackson et al., 1971). The larvae of *Terebratalia transversa* preferentially settle on the underside of the substratum offered in the laboratory (Long, 1964). The larvae of *Calloria inconspicua* preferentially settle in concavities on the substratum (Wisely, 1969; Doherty, 1976).

CONCLUSIONS

Competent articulate brachiopod larvae typically possess an apical cup-like depression of the pedicle lobe (the 'velouse'), and a distal pedicle-forming region transversely constricted from the pedicle lobe. The behaviour to light of larvae is reversed from photophilous during earlier larval stages, to photophobic during pre-settlement stages. This change, or reversal, results in competent larvae settling in places of subdued light. The presumed pre- and post-settlement choices, or the selections of substratum for settlement in response to appropriate stimuli, or cues from the substratum and from the environment, serve to explain not only larval preference for a particular substratum (experimentally offered for settlement in the laboratory), but also the natural occurrence of articulate brachiopod juveniles in concavities, grooves or crypts in subdued light.

REFERENCES

CHUANG, S. H. 1990. Brachiopoda, p.211-254. In, K. G. Adiyodi & R. G. Adiyodi (eds.), Reproductive Biology of Invertebrates, IV(B), Fertilization, development, and parental care. Oxford & IBH Publishing Co., New Delhi.
___1994. Observations on the reproduction and development of *Liothyrella neozelanica* Thomson 1918 (Terebratulacea, Articulata, Brachiopoda). Journal Royal Society New Zealand 24:209-218.
COOPER, G.A. & D.E. LEE. 1993. *Calloria*, a replacement name for the Recent brachiopod genus *Waltonia* from New Zealand. Journal Royal Society New Zealand 23:257-270.
DOHERTY, P. J. 1976. Aspects of the feeding ecology of the subtidal brachiopod *Terebratella inconspicua* (Sowerby, 1848). M. Sc. thesis, Zoology, University of Auckland, 183p.
FRANZEN, A. 1969. On larval development and metamorphosis in *Terebratulina*, Brachiopoda. Zoologiska Bidrag Fran Uppsala 38:155-174.

GUSTUS, R. M. & R. A. CLONEY. 1972. Ultrastructural similarities between setae of brachiopods and polychaetes. Acta Zoologica, Stockholm 53:229-233.

JACKSON, J. B. C., T. F. GOREAU & W. D. HARTMAN. 1971. Recent brachiopod-coralline sponge communities and their paleoecological significance. Science 173:623-625.

LACAZE-DUTHIERS, H. 1861. Histoire naturelle des brachiopodes vivants de la Mediterranée, I, Histoire naturelle de la Thecidie (*Thecidium mediterraneum*). Annales Sciences Naturelles (Zoologie), Paris 15:259-330.

LONG, J. A. 1964. The embryology of three species representing three superfamilies of articulate brachiopods. Unpublished Ph.D. thesis, University of Washington,185 p.

MANO, R. 1960. On the metamorphosis of a brachiopod, *Frenulina sanguinolenta* (Gmelin). Bulletin Marine Biological Station, Asamushi, Tohoku University 10:171-175.

PECK, L. S. & K. ROBINSON. 1994. Pelagic larval development in the brooding Antarctic brachiopod *Liothyrella uva*. Marine Biology 120:279-286.

PERCIVAL, E. 1944. A contribution to the life-history of the brachiopod,*Terebratella inconspicua* (Sowerby). Transactions Royal Society New Zealand 74:1-23.

___1960. A contribution to the life-history of the brachiopod *Tegulorhynchia nigricans*. Quarterly Journal Microscopical Science 101:439-457.

RICKWOOD, A. E. 1968. A contribution to the life history and biology of the brachiopod*Pumilus antiquatus* Atkins. Transactions Royal Society New Zealand (Zoology) 10:163-182.

STRICKER, S. A. & C. G. REED. 1985. The ontogeny of shell secretion in *Terebratalia transversa* (Brachiopoda, Articulata), I, Development of the mantle. Journal of Morphology 183:233-250.

WISELY, B. 1969. Preferential settlement in concavities (rugophilic behaviour) by larvae of the brachiopod *Waltonia inconspicua* (Sowerby, 1846). New Zealand Journal Marine Freshwater Research 3:273-280.

LOWER PALEOZOIC BRACHIOPOD COMMUNITIES

L.R.M. COCKS

Department of Palaeontology, The Natural History Museum,
Cromwell Road, London, SW7 5BD, U.K.

ABSTRACT- Over the past 30 years a variety of methods have been used to characterise benthic marine communities for comparison with others surrounding them in both space and time. However, because individual brachiopods do not interact with each other, the individual species distribution is a more ecologically natural unit than their grouping together as 'communities'. Nevertheless, the groupings of these species as communities form a recognisable and identifiable series of entities which justifies their formal recognition and naming. From the Cambrian to the Devonian there was a steady spread of brachiopod colonisation from relatively shallow water (BA2) in level–bottom marine communities both shorewards (to BA1) and progressively off–shore (to BA7) so that assemblages thrived from perhaps intertidal to the deeper shelf and slope. A brief summary is presented of key works; there is little published on Cambrian brachiopod–dominated assemblages, but in the early Ordovician there was considerable endemism, which reduced as time went by and the more substantial continents drew closer. By the early Silurian most of the level–bottom faunas were relatively cosmopolitan, apart from those developed in high latitudes, but by the end of the Silurian more endemism had again developed. The absolute depth of each community varied depending on water clarity and other factors but for each benthic assemblage zone was approximately constant as the individual communities replaced each other laterally and over geological time.

GENERAL

Since the dawn of geological studies many workers have noted repetitive assemblages of shelly fossils, which in the Lower Palaeozoic were often dominated by brachiopods. However, it was not until after the classic studies of the marine biologist Thorson (1957), working on modern benthos in the Danish sound, that it was generally realised that marine level–bottom communities were developed in a way that was partly substrate dependent, but also directly related to water depth. The first Lower Paleozoic studies were on the early Silurian of the Welsh Borderland, where Ziegler, Cocks & Bambach (1968) described five communities named the *Lingula*, *Eocoelia*, *Pentamerus*, *Stricklandia* and *Clorinda* Communities, after key brachiopods within them, although the communities were not based on brachiopods alone, but on molluscs, corals and a variety of other associated benthos. This paper provided the numbers of specimens of each species occurring at a single typical locality of each community but, apart from that, did not quantify their results.

Since then, community studies have diverged into two separate ways of describing and assessing the patterns. The first is by identifying recurrent assemblages and naming them after key components; this has usually been done qualitatively. The second way is to put together all of the available quantitative data from many localities and to subject that data to mathematical analyses such as cluster analysis and principal components analysis. The foremost advocate of the first procedure has been Boucot (1975), who took the five

early Silurian assemblages of Ziegler, Cocks & Bambach (1968) and termed them Benthic Assemblages (BA1 to BA5), adding a sixth (BA6) for communities developed in deeper water than the <u>Clorinda</u> community (BA5). This terminology has been followed in various subsequent works, particularly on the Silurian and Devonian, e.g. in an analysis of the late Ordovician to Middle Devonian shelly benthic faunas of China (Wang et al., 1987). The second way has been used for example by Temple (1987), who analysed the early Silurian brachiopods of Wales and who presented his results in a series of ordination diagrams and correlation matrices. Integration and comparison of the results from the two methods is sometimes difficult, but perhaps this stems more from the nature of the communities themselves rather than the methods used to characterise them.

One of the reasons why community analyses of level–bottom shelly faunas are frustrating is that, as different brachiopod genera do not interact with each other, the individual species distributions are the ecological realities rather than their association as 'communities'. Thus to formulate named communities in succession in space or time imposes artificial discontinuities on the natural populations, since the edges of the individual distributions either do not coincide with each other or coincide randomly. This is demonstrated by the data presented by Cocks & McKerrow (1984) for thirty of the hundred or so collections used originally by Ziegler, Cocks & Bambach (1968) to characterise the five named communities from the early Silurian of the Welsh Borderland, in which it can be seen that the divisions between assemblages assigned to adjacent communities are arbitrary, whilst the individual common species have mostly a simple distribution curve. Nevertheless these facts do not negate the usefulness stemming from the recognition of these communities and the pattern of their distributions, since such patterns are invaluable in identifying the positions of old coastlines and their relative movements during periods of transgression, regression and sea level change (Ziegler, Cocks & McKerrow 1968). The discrete species distribution patterns are the ones picked out most clearly in numerical analyses, for example that of Springer & Bambach (1985) for the Ordovician Martinsburg formation of Virginia, although the secondary task of identifying five communities (*Lingula*, bivalve, *Rafinesquina*, *Onniella*, and *Sowerbyella*, in depth descending order), was more easily accomplished by Springer & Bambach (ibid.) by qualitative rather than quantitative means.

Not much has been published on the community aspects of Cambrian faunas, which, as far as brachiopods were concerned, were on the whole less diverse than those of Ordovician and later times. McBride (1976), for example, whilst describing various epifaunal and infaunal outer shelf communities of the late Cambrian of the Great Basin, U.S.A., although able to discriminate various communities based on trilobites, noted that comparable acrotretides and lingulides seemed to occur throughout all of his identified communities, indicating that some brachiopods were perhaps more tolerant then to a wider range of depths than later.

In the Early Ordovician the continents were far enough separated for there to be substantially different faunal realms, each one of which had a separate shelly community structure in the Tremadoc and Arenig. The assemblages in the Iapetus Ocean region, bordering what is now the North Atlantic area, are well known, and communities and their patterns from other areas are also being characterised, for example, the assemblages recognised by Herrera & Benedetto (1990) from Argentina. As the Ordovician progressed, and the oceans narrowed, there were various successful migrations so that by the end of the Period, the Ashgill faunas were more homogeneous, and individually widespread, for example the deeper-water *Foliomena* community (Cocks & Rong, 1988). However, the shallower shelf faunas were still distinct, for example the equatorial communities of Laurentia, such as those described by Bretsky (1970) and Titus (1986) from New York State, and the wider analyses of many recorded Middle and Late Ordovician benthic assemblages from the whole of North America by Potter & Boucot (1992). This last work showed convincingly that most of the craton was flooded only sufficiently deeply to allow BA1 to BA3 communities to be developed, whilst BA4 and BA5 communities were restricted to extracratonic areas. This situation carried on in to the Early Silurian, in which, for example, LoDuca & Brett (1994) have documented BA1 to BA3 communities only in the Llandovery of western New York State; BA1, linguloid; BA2, *Eocoelia* – bivalve, and BA3, *Hyattidina* – crinoid, with *Pentamerus* also developed in BA3 by middle Llandovery time.

On the European side of the Iapetus Ocean during the Ordovician, a different suite of communities occurred, less diverse than the North American group on account of the then higher latitudes. In a series of works, Havlicek (in, Havlicek & Vanek, 1990) has described the 'Mediterranean Province' faunas of Bohemia, southern Europe and North Africa and has identified a variety of shelly communities among which brachiopods dominate BA2 and BA3 depths during the earlier Ordovician and extend to BA5 by late Ordovician times. Lockley (1983) reviewed the Anglo–Welsh communities of

Llanvirn to Caradoc age and demonstrated a variety of assemblages which he links to 'an onshore–offshore palaeoenvironmental transect', although in that relatively restricted area he found that that also broadly corresponds with coarser to finer sedimentary environments. Disentangling the separate factors of sediment grain size and water depth is always difficult, although it is easier in the Early Palaeozoic, where the brachiopods lived largely epifaunally above the substrate, as opposed to more modern situations in which a much higher proportion of the benthos is infaunal.

Near the end of the Ordovician the global cooling clearly affected many brachiopod stocks, although the Ordovician–Silurian turnover was gradual rather than abrupt. The faunas have been analysed, for example by Cocks (1988) and Harper & Rong (1995) who have found that the dominant and widespread *Hirnantia* fauna may be split into a variety of assemblages before being replaced by a less diverse series of faunas in the early Silurian (Baarli & Harper, 1986). The subsequent early Silurian warming was gradual, and it was not until the later Llandovery that the full suite of brachiopod communities already mentioned (Ziegler, Cocks & Bambach, 1968) had become fully differentiated.

Because the Iapetus Ocean had narrowed sufficiently by Silurian times for brachiopod spat to cross it with ease (McKerrow & Cocks 1976), the Silurian assemblages and level bottom communities were almost global in extent, with only the high latitudes providing distinctive assemblages, such as the *Clarkeia* faunas of the southern hemisphere (Cocks & Scotese, 1991) and the *Tuvaella* faunas of the northern hemisphere (Kulkov, 1974; Rong & Zhang, 1982). As the Silurian progressed, the degree of provinciality increased, as has been admirably documented by Havlicek & Storch (1990) for Bohemia, a precursor to the much more segmented situation prevalent in the early Devonian. In the Silurian there was a spread of different brachiopod–dominated community assemblages, from the relatively restricted BA2 environments in which they were chiefly located during the early Cambrian, to

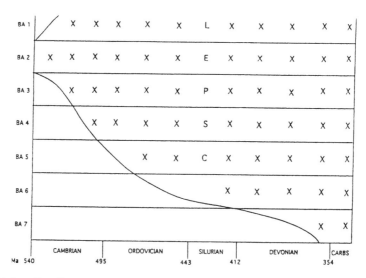

Figure 1. The Benthic Assemblage Zones of Boucot (1975), BA1 (shallow) to BA7 (deep), and their developments through Paleozoic time. The early Silurian communities of Ziegler, Cocks & Bambach (1968) are shown by their letters: L = *Lingula*; E = *Eocoelia*; P = *Pentamerus*; S = *Stricklandia*; and C = *Clorinda*. The Xs in the remaining columns show the presence of brachiopod - dominated benthic assemblages in the various geological periods, spreading both on - and offshore from Early Cambrian times onwards.

the end of the Palaeozoic, when the communities were spread from subtidal environments to the continental slope (Figure 1). This has remained the situation since then, although from Devonian times onwards the niches were often dominated by molluscs and other benthos rather than brachiopods, and in the Mesozoic to Recent deposits the brachiopod–dominated assemblage niches became progressively rarer and more fragmented.

The actual depths inhabited by the benthic associations has been a topic of much debate, particularly for Silurian brachiopod–dominated communities. Ever since the early work (Ziegler, Cocks & Bambach, 1968), there have been a variety of opinions, with some proponents suggesting depths as great as 1500m (Hancock et al., 1974). Most paleontologists have preferred much smaller figures, such as the values of up to 150m postulated by Brett et al. (1993), who compared the Silurian benthos with a range of modern animals and plants with a known depth range. This seems reasonable when weighed against the stratigraphical evidence of eustatic and orogenically caused sea level changes and their rates, but the truth may be that the actual depth range of an individual Lower Palaeozoic animal community may have varied depending on local water clarity and other factors.

To conclude, there is no doubt that the delineation of Lower Palaeozoic brachiopod–dominated animal communities has been an exercise of great help in understanding local geological history. Even though the 'communities' may possess an artifical reality due to the truly heterogeneous nature of their component species, their identification and succession has much assisted in unravelling ancient shorelines, continents and in recognising global changes in sea level.

REFERENCES

BAARLI, B.G. & D.A.T. HARPER. 1986. Relict Ordovician brachiopod faunas in the Lower Silurian of Asker, Oslo Region, Norway. Norsk Geologisk Tidsskrift 66: 88–97.

BOUCOT, A.J. 1975. Evolution and extinction rate controls. Elsevier, Amsterdam, 427p.

BRETSKY, P. 1970. Late Ordovician benthic marine communities in north–central New York. New York State Museum Science Service Bulletin 414: 1–34.

BRETT, C.E., A.J. BOUCOT & B. JONES. 1993. Absolute depths of Silurian benthic assemblages. Lethaia 26: 25–40.

COCKS, L.R.M. 1988. Brachiopods across the Ordovician–Silurian boundary. Bulletin British Museum (Natural History), Geology 43:311–315.

___& W.S. McKERROW. 1984. Review of the distribution of the commoner animals in Lower Silurian marine benthic communities. Palaeontology 27: 663–670.

___& JIAYU RONG. 1988. A review of the late Ordovician Foliomena brachiopod fauna, with new data from China, Wales and Poland. Palaeontology 31: 53–67.

___& C.R. SCOTESE. 1991. The global biogeography of the Silurian period. Special Papers Palaeontology 44:109–122.

HANCOCK, N.J., J.M. HURST & F.T. FÜRSICH. 1974. The depths inhabited by Silurian brachiopod communities. Journal of the Geological Society, London 130:151–156.

HARPER, D.A.T. & JIAYU RONG. 1995. Patterns of change in the brachiopod faunas through the Ordovician–Silurian interface. Modern Geology 20: 83–100.

HAVLICEK, V. & P. STORCH. 1987. Silurian brachiopods and benthic communities in the Prague Basin (Czechoslovakia). Vestnik Ustredniho ustavu geologickeho 48:1–275.

___& J. VANEK. 1990. Ordovician invertebrate communities in black–shale lithofacies (Prague Basin, Czechoslovakia). Vestnik Ustredniho ustavu geologickeho 65: 223–236.

HERRERA, Z.A. & J.L.BENEDETTO. 1990. Early Ordovician brachiopod faunas of the Precordillera Basin, western Argentina: biostratigraphy and paleobiological affinities, p.283–301. In, D.I. MACKINNON, D.E. LEE & J.E. CAMPBELL (eds). Brachiopods through time, Balkema, Rotterdam.

KULKOV, N.P. 1974. Brachiopod communities from the Llandovery of Altai and Tuva and questions of biogeography. Doklady Akademiya Nauk, Sibirskoe Otdelenie, 169–175.

LOCKLEY, M.G. 1983. A review of brachiopod dominated palaeocommunities from the type Ordovician. Palaeontology 26:111–145.

LoDUCA, S.T. & C.E.BRETT. 1994. Revised stratigraphic and facies relationships of the lower part of the Clinton Group (Middle Llandoverian) of western New York State. New York State Museum Bulletin 481:161–182.

McBRIDE, D.J. 1976. Outer shelf communities and trophic groups in the Upper Cambrian of the Great Basin. Brigham Young University Geology Studies 23:139–152.

McKERROW, W.S. & L.R.M. COCKS. 1976. Progressive faunal migration across the Iapetus Ocean. Nature, London 263:304–306.

POTTER, A.W. & A.J. BOUCOT. 1992. Middle and late Ordovician brachiopod benthic assemblages of North America, p.307–323. In, B.D. Webby & J.R. Laurie (eds.), Global perspectives on Ordovician geology. Balkema, Rotterdam.

RONG JIA–YU & ZIXIN ZHANG. 1982. A southward extension of the Silurian Tuvaella brachiopod fauna. Lethaia 15:133–147.

SPRINGER, D.A. & R.K. BAMBACH. 1985. Gradient versus cluster analysis of fossil assemblages: a comparison from the Ordovician of southwestern Virginia. Lethaia 18:181–198.

TEMPLE, J.T. 1987. Early Llandovery brachiopods of Wales. Palaeontographical Society Monographs 139:1–137.

THORSON, G. 1957. Bottom communities (sublittoral or shallow shelf). Geological Society America Memoirs 67:461–534.

TITUS, R. 1986. Fossil Communities of the Upper Trenton Group (Ordovician) of New York State. Journal of Paleontology 60:805–824.

WANG YU, BOUCOT, A.J., JIAYU RONG & XUECHANG YANG. 1987. Community paleoecology as a geologic tool: the Chinese Ashgillian–Eifelian (latest Ordovician through early Middle Devonian) as an example. Geological Society of America Special Paper 211:1–100.

ZIEGLER, A.M., L.R.M. COCKS & R.K. BAMBACH. 1968. The composition and structure of Lower Silurian marine communities. Lethaia 1:1–27.

ZIEGLER, A.M., L.R.M. COCKS & W.S. McKERROW. 1968. The Llandovery transgression of the Welsh Borderland. Palaeontology 11:736–782.

BRACHIOPOD MOLECULAR PHYLOGENY

B. L. COHEN & A. B. GAWTHROP

Division of Molecular Genetics, The University of Glasgow, Glasgow, G11 6NU Scotland

ABSTRACT- Analysis by parsimony, maximum likelihood and distance methods of newly determined nuclear-encoded SSU rRNA gene sequences from 23 species of articulate brachiopods, six inarticulate brachiopods, two phoronids and an ectoproct, together with other sequences from published and unpublished sources show that lophophorates cluster with protostome, not deuterostome metazoa and that phoronids cluster with inarticulate brachiopods. Phoronids, inarticulate, and articulate brachiopods form a monophyletic assemblage. A chiton is the closest known outgroup of brachiopods plus phoronids. Within articulates, separate rhynchonellid and long- and short-looped terebratulid clades are identified and a thecideidine falls within the short-looped articulate clade. Forms with incomplete loops belong either to the short or long-looped clades, thus, a three-fold division of articulate brachiopods suffices to encompass the range of extant diversity so far examined. A perfect correlation was found between clade rank and lineage age rank for five well-dated brachiopod lineages. The important underpinning role of classical brachiopod taxonomy for molecular phylogeny is stressed.

INTRODUCTION

The information bearing molecules of the genome (DNA and its transcript, RNA) are 'the documents of evolutionary history' (Zuckerkandl & Pauling, 1965) and recent developments in DNA sequencing and polymerase chain reaction (PCR) amplification mean that these documents can now be read with reasonable ease. As a result, genealogical relationships are being clarified at an unprecedented rate (Hillis & Moritz, 1990) and our laboratory has been privileged to extend this approach to brachiopods and other lophophorates, using nuclear-encoded small subunit ribosomal RNA (SSU rRNA) gene sequences. To the extent that the genomes of living brachiopods retain phylogenetically useful information, our results and those of our successors will, for the first time, make it possible to justify classification on genealogical grounds independent of morphology and the fossil record.

Previously it was shown that a brachiopod, a phoronid and an ectoproct had nuclear SSU rRNAs of protostome type (Ishikawa, 1977) and this was confirmed by the partial sequence of the *Lingula* RNA (Adoutte & Philippe, 1993; Field, 1988; Ghiselin, 1988; Patterson, 1989). Since then many more nuclear SSU rRNA gene sequences have been determined (Benson et al., 1994; Van de Peer et al., 1994), making this sequence the most useful for wide-ranging phylogenetic studies. Our work extends the range of complete SSU sequences to include all the major extant brachiopod lineages (Cohen & Gawthrop, 1995; Cohen et al., 1995). Two other brachiopod SSU sequences have also been reported (Halanych et al., 1995).

The sequence differences observed in SSU gene comparisons reflect the accumulated results of many evolutionary processes of which we can observe only the end product. The initial events, mutations leading to a nucleotide substitution are, to a first approximation, randomly distributed within and between genes, but we see only those few substitutions that rise to high frequency (effectively to fixation, frequency = 1.0) amongst the multiple genomic copies of rRNA genes and in the population (Coen et al., 1982). Because substitution events are rare and lineage sorting (extinction) is common, extant taxa have an invariant or almost invariant rRNA gene sequence, making it possible to use sequences from only one or a few individuals to exemplify a taxon (Hillis & Dixon, 1991). Since rRNAs play a central structural and functional (even enzymatic) role in protein synthesis with multitudes of interactions within their own structure as well as with ribosomal proteins, their primary and secondary structures are subject to strong selection for conserved function and the majority of mutations that achieve fixation will be neutral or (rarely) advantageous. In consequence, we expect the number of accumulated changes to be roughly proportional to time (a molecular clock hypothesis), though complicating factors exist. An important practical advantage of rRNA sequences is that they combine blocks that are highly conserved (because functionally constrained) interspersed with more variable regions (Hillis & Dixon, 1991). The highly conserved blocks make it possible to be confident that sequences from very diverse taxa have been correctly aligned, with most homologous nucleotide sites in register. The more variable regions are aligned by using the highly conserved segments to anchor their ends, by nucleating the alignment on phylogenetically close taxa and by use of secondary structure information (Huss & Sogin, 1990). Thus, analysis of divergence in complete SSU sequences is possible, providing a wide (but not unlimited) range of phylogenetic resolution. In the absence of evidence for paralogy, the SSU gene phylogeny is assumed to be an honest reporter of organismal phylogeny. Additionally, since metazoan mitochondrial DNA (mtDNA) generally accumulates base substitutions several-fold faster than nuclear DNA, analysis of mitochondrial rRNA genes further increases the divergence-time range over which useful results may be obtained (Adoutte & Philippe, 1993; Hillis & Dixon, 1991; Phillipe et al., 1994).

MATERIALS AND METHODS

Full details of materials and methods will be published elsewhere (Cohen et al., 1995). In brief, nuclear-encoded SSU rRNA gene sequences, each ca. 1,790 nucleotides long were obtained by the direct sequencing of DNA amplification products synthesized by PCR using oligonucleotide primers matching highly conserved terminal regions of the gene. Thus, the sequences are complete except for some short regions and the two large sections missing from the *Lingula reevii* sequence (Field, 1988). In addition to new sequences from 23 species of articulate brachiopods, six inarticulate brachiopods, two phoronids and an ectoproct, sequences from one articulate, two inarticulates, a phoronid and an ectoproct were available elsewhere (Field, 1988; Halanych et al., 1995; Winnepenninckx & De Wachter, 1994). Sequences were

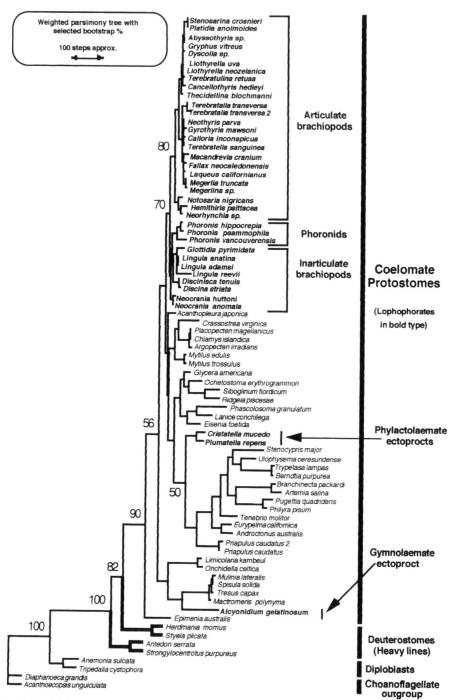

Figure 1. The high-level phylogenetic relationships of brachiopods, phoronids and ectoprocts. Weighted parsimony tree based on an alignment of 80 complete SSU sequences from all available brachiopods, phoronids and ectoprocts together with protostome and other outgroups. The alignment contained 2114 sites of which 804 were parsimony-informative. Heuristic search options (using PAUP, Swofford, 1993) were: collapse zero-length branches, no topological constraints, outgroup rooting, closest addition, no steepest descent, TBR, MULPARS, ACCTRAN. In trial analyses random addition, steepest descent and DELTRAN did not alter tree topology. A search with equally weighted characters found 48 most parsimonious trees of length = 4711 steps, RI = 0.604. After three cycles of character reweighting with the rescaled consistency index (worst fit) followed by heuristic search, the number of trees reduced to 6 of 86565 weighted steps, RI = 0.717. These trees differed only in arrangement of articulate brachiopod terminal taxa on the shortest branches. Bootstrap % are given for key nodes, based on 50 heuristic search replicates with reweighted characters. The limited number of replicates was dictated by computational constraints, which also prevented calculation of support indices.

aligned manually with one another and with protostome and other outgroup sequences (Benson et al., 1994; Maidak et al., 1994; Runnegar et al., 1995; Winnepenninckx et al., 1995; Winnepenninckx & De Wachter, 1994).

DNA was purified, using standard procedures (Sambrook et al., 1989), generally from specimens preserved in alcohol. Sequencing was by the dideoxy termination method using Sequenase 2.0 (USB/Amersham plc) on single-stranded template DNA prepared by asymmetric PCR Allard et al., 1991). Occasionally, a magnetic bead capture method (Dynal, plc) was used (Hultman et al., 1989). Sequence gels and autoradiographs were prepared by standard methods (Sambrook et al., 1989) and sequences were read and recorded manually. With trivial exceptions, every sequence was fully determined from both DNA strands with multiple redundancy. Sequences were aligned by hand Gilbert, 1993; Smith et al., 1994and phylogenetic reconstructions were performed with various methods. The parsimony program PAUP (Swofford, 1993) was used with either equally weighted characters (EP) or with characters reweighted a posteriori on their rescaled consistency index values (WP). Because of the large dataset, parsimony analyses used heuristic searches. Support indices were calculated from the strict consensus trees of non-minimal reconstructions Bremer, 1988; Källersjö et al., 1992). The maximum likelihood method (ML) program fastDNAml Olsen et al., 1994) was used with global branch

exchange. Nucleotide divergence was also estimated with PHYLIP's DNAdist with the Kimura 2-parameter correction and bootstrap samples were prepared with SeqBoot; distance and bootstrap trees were constructed with Neighbor for neighbor-joining trees and Consense for bootstrap trees (Felsenstein, 1993).

RESULTS

Relationships between brachiopods and other phyla-- The WP tree is reconstructed from an alignment of 80 sequences from a wide range of metazoa, rooted on choanoflagellates (Figure 1).This tree and others (Cohen & Gawthrop, 1995) clearly separate diploblasts and deuterostomes from protostomes and the position of the lophophorates within the protostome assemblage is strongly supported. Phoronids clearly tree within the brachiopod clade; they do not represent an independent phylum. These two inferences have been reported elsewhere (Halanych et al., 1995). The ectoprocts appear diphyletic and generally remote from the clade of brachiopods plus phoronids (Cohen & Gawthrop, 1995), but the conclusions which can legitimately be drawn from a result based on so few species are a matter of debate (Conway Morris et al., 1995); additional data are needed. Nodes uniting brachiopods and phoronids in this tree are strongly supported, but the same is not true of nodes connecting many other protostomes: bootstrap values around 50% are common because the SSU

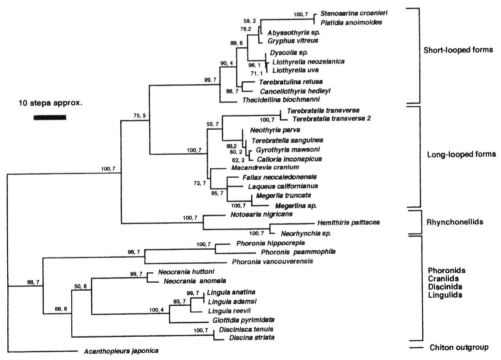

Figure 2. Phylogeny of brachiopods and phoronids based on nuclear-encoded SSU rRNA gene sequences. The alignment was as used for Figure 1 but with unused taxa removed. It contained 1813 sites of which 198 were parsimony-informative. The skewness index of 10,000 random trees was $g_1 = -0.504$, indicating that WP has a high probability of finding the true tree (Hillis et al., 1994). Heuristic search found 36 minimal trees of 495 steps, RI = 0.801. After 3 cycles of reweighting these reduced to 3 of 20355 weighted steps, RI = 0.892, differing in topology only at the unresolved Laqueus, Fallax, Megerlia plus Megerlina node and one of these trees is shown. See caption, Figure 1 for other search details. The numbers adjacent to each node are first, the frequency with which that node appeared amongst 100 bootstrap replicates and second, the support index for that node (Bremer, 1988; Källersjö et al., 1992) on a 7-point scale (Cohen & Gawthrop, 1995).

gene alone cannot resolve the protostome radiation (Adoutte & Philippe, 1993; Phillipe et al., 1994).

<u>The outgroup problem.</u>--Outgroups play a crucial role in phylogenetics (Donoghue & Cantino, 1984; Farris, 1972; Hennig, 1966; Maddison et al., 1984). What is the most appropriate outgroup for the analysis of brachiopod phylogeny? Recent analyses of outgroup rooting (Nixon and Carpenter, 1993; Smith, 1994) stress the dangers of remote outgroups (like choanoflagellates); the closest outgroup is preferred, ideally the ingroup's sister-group. This will minimise homoplasy, facilitate alignment of variable regions and minimise the need to exclude data { ADDIN }(Smith, 1994), in contrast with using evolutionarily remote outgroups (Halanych et al., 1995). The sister-group of the ingroup (brachiopods plus phoronids) would normally be identified by reference to independent evidence such as comparative morphology, but this has led to the lophophorates being regarded as deuterostomes (Brusca & Brusca, 1990; Eernisse et al., 1992), contrary to the molecular results. Since no other, independent evidence exists we must make our outgroup choice recursively, by phylogenetic analysis of the only substantial data-source - nuclear SSU rRNA gene sequences - seeking the protostome sequence phenetically closest to brachiopods plus phoronids. Comparison of branch lengths in unrooted WP, ML and NJ trees and other analyses indicate that the chiton *Acanthopleura* is narrowly the closest outgroup (Cohen & Gawthrop, 1995). This should not be taken to mean that chitons are literally the sister-group of brachiopods plus phoronids; they are simply the phenetically closest outgroup on present evidence, reflecting a combination of available sequences, true phyletic position and similarity of sequence and of nucleotide composition. The next closest taxa are two bivalve molluscs and a polychaete. Thus, *Acanthopleura* will be used as the heuristic outgroup. Subjectivity of outgroup selection has been recognized (Donoghue & Cantino, 1984), and is inescapable in the absence of independent evidence. We have minimised the subjective element by using parameters estimated from the data. The need for a selected outgroup also arises from practical considerations since an alignment of 80 SSU sequences causes computational difficulties.

<u>The phylogeny of brachiopods and phoronids</u> --The WP tree is reconstructed with the selected outgroup (Figure 2). Other close outgroups lead only to alterations in the topology of ambiguously resolved nodes. Comparable ML and NJ trees have been presented elsewhere (Cohen & Gawthrop, 1995). Taking all the evidence together, the conclusions listed below may be drawn:

1. All the new inarticulate brachiopod sequences cluster with the partial sequence from *Lingula reevii* (Field, 1988) and the phylogenetic positions of all other brachiopod sequences are consistent with their reputed brachiopod origin: there is no contamination with DNA from irrelevant organisms.
2. The sister-group of brachiopods plus phoronids probably lies amongst molluscs, annelids and other protostome 'worms' (Winnepenninckx et al., 1995); priapulans and arthropods are more distant. (Priapulans as sister-group of arthropods is unexpected, but the data are unambiguous.)
3. Brachiopods plus phoronids are monophyletic and in WP reconstructions phoronids are basal members of the clade of inarticulate brachiopods. However, in other reconstructions phoronids are apparently diphyletic, with *Phoronis vancouverensis* joining the articulate brachiopods (ML and NJ with low bootstrap support in our trees) as has been reported elsewhere (Halanych et al., 1995). Diphyly of phoronids is biologically implausible and can be explained away by study of

the three phoronid sequences. Those of *Phoronis hippocrepia* and *P. psammophila* show no unusual features when compared with other protostomes, but the *P. vancouverensis* sequence (Halanych et al., 1995) lacks at least 9 nucleotides in otherwise highly conserved sites, suggestive of mis-reading. More importantly, all three phoronid sequences share at least two variable-region motifs that are clear synapomorphies of phoronids alone, and the support index for the node uniting all 3 phoronids is relatively high (Figure 3). Moreover, in a reconstruction based only on the most conserved and hence most reliably aligned nucleotides, phoronids are again a monophyletic sister-group of *inarticulate* brachiopods (Cohen & Gawthrop, 1995). Thus, the suggestion that phoronids are most closely related to *articulate* brachiopods (Halanych et al., 1995) must be erroneous.
4. One reconstruction (Figure 3) joins the craniids and lingulids in a clade that has low bootstrap support. In other reconstructions a clade of phoronids plus craniids occurs and is the sister-group of discinids plus lingulids. Thus, whilst the association of phoronids with inarticulate brachiopods is certain, there is insufficient information in the SSU sequences for unambiguous resolution of inarticulates + phoronids. Other, independent evidence is required.
5. Most reconstructions, but not all, place the origin of discinids before that of lingulids. Also, the long branch and basal position of *Glottidia* amongst lingulids are uncertain because this sequence (Halanych et al., 1995) lacks ca. 14 nucleotides in otherwise highly conserved positions.
6. Rhynchonellids are the sister-group of all other articulate brachiopods.
7. Long-looped and short-looped articulates are sister-groups.
8. Long-looped articulates are monophyletic with at least two sub-clades. *Terebratalia* may be either a basal long-looped form or the sister-group of the New Zealand terebratellids, depending on reconstruction method.
9. The morphological divergence that gave rise to the genus-level diversity of the New Zealand terebratellids has been accompanied by very little change in the SSU gene. *Neothyris* is probably the basal member of this clade.
10. An adequate molecular phylogeny of the long-looped articulates must await results from a wider species sample and a faster-evolving gene.
11. *Macandrevia* is the sister-group of a morphologically diverse clade of long-looped forms that includes kraussinids.
12. Short-looped articulates are clearly monophyletic and at least four sub-clades are recognised.
13. The thecideidine is unambiguously a short-looped articulate, either basal or a sister-group of cancellothyrids. Whilst the probable affinity of these enigmatic brachiopods with short-looped forms has been previously recognized (Baker, 1990; Williams, 1973), the suggestion of a sister-group relationship with cancellothyrids appears to be novel.
14. Subclade relationships within the short-looped forms are not strongly supported, being based on very few sequence differences. An adequate molecular phylogeny of the short-looped articulates must await results from a wider species sample and a faster-evolving gene.
15. Within the short-looped forms the most surprising result is the association of the undoubted short-looped *Stenosarina* with *Platidia*. Although we are confident that neither sequence is an artefact, the *Platidia* sample had an atypical history and this result should be treated with reserve until confirmed. Fortunately we have lately obtained an independent platidiid sample.

<u>Rates of molecular and morphological evolution.</u>--The molecular phylogeny is largely congruent with classical brachiopod systematics represented by the classification used in the Treatise (Williams, 1965). Does congruence extend to

lineage times of origin? Space allows only one preliminary result: Figure 3 shows an analysis of clade rank versus age rank (Norell and Novacek, 1992) for brachiopod lineages with well-established times of origin. In this non-parametric analysis, which depends only on relative age and distance, there is complete agreement between the molecular and stratigraphic rankings.

DISCUSSION

Taking into account both published (Cohen & Gawthrop, 1995) and unpublished analyses, the topology of articulate brachiopod lineages has been stable despite various outgroup(s) and reconstructions, suggesting that the phylogeny is broadly reliable. The topology of the inarticulate plus phoronid clade is less stable, but monophyly of craniids, discinids, lingulids and phoronids is assured, as is monophyly of articulate and inarticulate brachiopods. Thus, our primary aim, to provide a secure, molecular basis for the high-level phylogeny of brachiopods, has been substantially achieved. Remaining uncertain high-level relationships will probably be resolved only by discovery of rare, qualitative evolutionary events such as gene order rearrangements in mtDNA. Such

new data will also be needed to complete reconstruction of the protostome radiation (e.g. Boore et al., 1995). And a more detailed molecular phylogeny of articulate brachiopods below the superfamily or family level will require sequence data from genes that evolve more rapidly than the nuclear-encoded SSU rRNA.

The most important points to emerge from the molecular analysis are:

1) on the evidence of the SSU genes, brachiopods, phoronids and ectoprocts certainly belong in the clade that contains all undoubted protostomes, not in the deuterostomes (Backeljau et al., 1993; Brusca & Brusca, 1990; Eernisse et al., 1992; Field, 1988; Halanych et al., 1995; Irwin, 1991; Nielsen, 1991; Nielsen, 1994; Nielsen, 1995; Schram, 1991). To escape from this conclusion (allowing the lophophorates to continue as deuterostomes) it must be proposed that an ancestor (or ancestors) of all three lophophorate phyla, originally a member of the deuterostome assemblage, received its SSU gene family by horizontal gene transfer from a mollusc-like protostome. There is no precedent for such a hypothetical event, but further

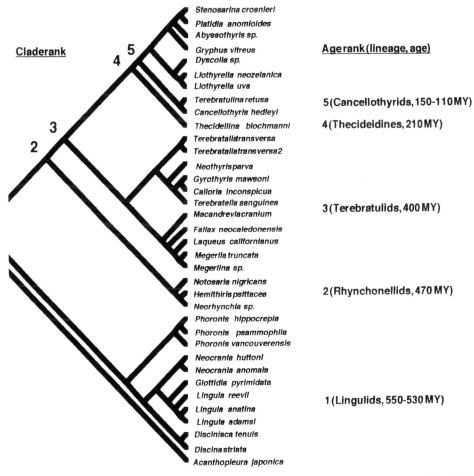

Figure 3. Clade rank compared with lineage age rank. Times of origin of clades are bsed on the origins of the lineages indicated in Benton (1993) and on personal communications (P. Copper, L. Holmer, D. E. Lee, D. I. McKinnon, E. Owen, A. Wliams).

genetic studies are capable of testing predictions that follow from it. In our view, however, this conflict between molecules and morphology (Conway Morris, 1995; Gee, 1995; Patterson, 1985) arises either because traditional histological methods offer too blunt a tool for the proper recognition of homology in dynamic developmental processes or because such processes are more variable than has been appreciated. For example if three distinct methods of embryonic coelom formation occur in brachiopods, and congeneric species differ references in Chuang, 1990), then coelom formation must be a highly plastic character, unsuited to provide evidence of high-level phyletic relationships.

2) articulate and inarticulate brachiopods do form a monophyletic group within which articulates and inarticulates belong to separate clades. Thus, the traditional system of two brachiopod classes is valid, except that separate phylum status of phoronids is excluded; they should be included with articulate and inarticulate brachiopods (but probably not ectoprocts) in a new phylum. If the possible sister-group relationship of phoronids and craniids is verified, a new taxon (craniids + phoronids) may be called for. Alternatively phoronids should be placed as one of three classes within a new phylum;

3) the results exclude the proposed arrangement uniting craniids with articulate brachiopods (Basset et al., 1993; Gorjansky & Popov, 1986; Popov et al., 1993);

4) short-looped and long-looped articulates (as so far analysed) do represent distinct clades, but articulates with atypical, incomplete loops such as *Megerlia* and *Platidia* (subject to confirmation) may belong to either clade. Thus, a three-fold division of the living articulates (rhynchonellids, short-looped forms, long-looped forms) is presently sufficient to encompass extant sequence diversity;

5) thecideidines belong within short-looped articulates and are therefore unlikely to be descendants of spiriferids or strophomenids (Baker, 1990; Williams, 1973);

6) our results clearly exclude the close clustering of *Terebratalia* and *Laqueus* and the grouping of *Macandrevia* relatively close to *Abyssothyris*, *Gryphus* and *Liothyrella* which were proposed following the controversial application of immunological methods to brachiopod taxonomy (Cohen, 1992; Cohen, 1994; Collins et al., 1991a; Collins et al., 1988; Collins et al., 1991b; Curry et al., 1991; Curry et al., 1993; Endo et al., 1994);

7) the sister-group of brachiopods plus phoronids apparently lies amongst molluscs and annelids. This is consistent with various lines of morphological evidence (e.g. Gustus & Cloney, 1972) and focuses attention on fossil groups such as halkieriids as potential common ancestors (Cohen & Gawthrop, 1995; Conway Morris et al., 1995; Conway Morris & Peel, 1995).

ENVOI

A start has been made on brachiopod molecular phylogeny and much has been accomplished since the last Congress, but much remains to be done. Will someone take up the challenge? What will they report in the year 2000? And if, like us, they are not trained in classical taxonomy, who will identify their specimens? Molecular systematics requires both the genome and traditional systematics. Continuity of museum staffing is imperative for progress.

Acknowledgments

The many people who have contributed specimens, identifications, labour and advice are listed elsewhere (Cohen & Gawthrop, in press). Financial support was received from the UK Natural Environment Research Council, the Royal Societies of London and New Zealand, the Natural Sciences and Engineering Research Council of Canada, the Carnegie Trust for the Universities of Scotland, the Nuffield Foundation and the University of Glasgow. We are grateful for permission to the following for permitting the use of unpublished sequences: B. Winnepenninckx and R. DeWachter, University of Antwerp; B. Runnegar, University of California, C. Harrison and C. M. Turbeville, Universities of Indiana and Michigan; K. Halanych, J. Bacheller, A.M. Aquinaldo, S. Liva, D. M. Hillis, University of Texas and J. A. Lake, University of California. Other sequences were obtained from from GenBank (Benson et al., 1994) or from the Ribosomal Database Project (Maidak et al., 1994.).

REFERENCES

ADOUTTE, A. & H. PHILIPPE. 1993. The major lines of meta-zoan evolution: summary of traditional evidence and lessons from ribosomal RNA sequence analysis, p.1-30. In, Y. Pichon (ed.), Comparative Molecular Neurobiology. Birk-häuser, Basel.

ALLARD, M.W., D.L. ELLSWORTH & R.L. HONEYCUTT. 1991. The production of single-stranded DNA suitable for sequencing using the polymerase chain reaction. Bio-Techniques 10:24-26.

BACKELJAU, T., B. WINNEPENNINCKX & L. DE BRUYN. 1993. Cladistic analysis of metazoan relationships: a reap-praisal. Cladistics 9:167-181.

BAKER, P.G. 1990. The classification, origin and phylogeny of thecideidine brachiopods. Palaeontology 33:175-191.

BASSETT, M.G., L.E. HOLMER, L.E. POPOV & J. LAURIE. 1993. Phylogenetic analysis and classification of the brachiopoda - reply and comments. Lethaia 26:385-386.

BENSON, D. A., M. BOGUSKI, D. L. LIPMAN & J. OSTELL. 1994. GenBank. Nucleic Acids Research 22:3441-3444.

BENTON, M. J. 1993. The Fossil Record 2, Chapman & Hall, London, 845p.

BOORE, J. L., T. M. COLLINS, D. STANTON, L. L. DAEHLER & W. M. BROWN. 1995. Deducing the pattern of arthropod phylogeny from mitochondrial DNA rearrangements. Nature 376:163-165.

BREMER, K. 1988. The limits of amino acid sequence data in angiosperm phylogenetic reconstruction. Evolution 42:795-803.

BRUSCA, R. C. & G. J. BRUSCA. 1990. Invertebrates. Sinauer Associates Inc., Sunderland, Mass, 922p.

CHUANG, S.H. 1990. Common and evolutionary features of Recent brachiopods and their bearing on the relationship between, and the monophyletic origin of, the inarticulates and the articulates, p.11-14. In, D.I. McKinnon, D.E. Lee & J.D. Campbell (eds.), Brachiopods through Time. Balkema, Rotterdam.

COEN, E.S., J.M. THODAY & G.A. DOVER. 1982. Rate of turnover of the structural variants in the rDNA gene of *Drosophila melanogaster*. Nature 295:564-568.

COHEN, B.L., 1992. Utility of molecular phylogenetic methods: a critique of immuno-taxonomy. Lethaia 24:441-442.

___1994. Immuno-taxonomy and the reconstruction of bra-chiopod phylogeny. Palaeontology 37:907-911.

___ & A.B. GAWTHROP (in press). The brachiopod genome. In, R.L. Kaesler (ed.), Treatise on invertebrate palaeon-tology, Brachiopoda (revised). Geological Society of America and University of Kansas Press, Lawrence, Kansas

___,___&T. CAVALIER-SMITH (in preparation). Phylogeny of lophophorates, especially brachiopods, based on nuclear-encoded SSU rRNA gene sequences.

COLLINS, M.J., G.B. CURRY, G. MUYZER, R. QUINN, S. XU, P. WESTBROEK & S. EWING. 1991a. Immunological investi-gations of relationships within the terebratulid brachiopods. Palaeontology 34:785-796.

___, G.B. CURRY, R. QUINN, G. MUYZER, T. ZOMERDIJK & P. WESTBROEK. 1988. Sero-taxonomy of skeletal macromolecules in living terebratulid brachiopods. Historical Biology 1:207-224.

___,G. MUYZER, G.B. CURRY, P. SANDBERG & P. WEST-BROEK. 1991b. Macromolecules in brachiopod shells: characterisation and diagenesis. Lethaia 24:387-397.

CONWAY MORRIS, S. 1995. Nailing the lophophorates. Nature 375:365-366.

___, B.L. COHEN, A.B. GAWTHROP, T. CAVALIER-SMITH & B. WINNEPENNINCKX (in press). Lophophorate phylogeny. Science.

___ & J.S. PEEL. 1995. Articulated halkierids from the Lower Cambrian of North Greenland and their role in early protostome evolution. Philosophical Transactions Royal Society, London B (347):305-358.

CURRY, G.B., R. QUINN, M.J. COLLINS, K. ENDO, S. EWING, G. MUYZER & P. WESTBROEK. 1991. Immunological responses from brachiopod skeletal macromolecules; a new technique for assessing taxonomic relationships using shells. Canadian Journal Zoology 24:399-407.

___, R. QUINN, M. J. COLLINS, K. ENDO, S. EWING, G. MUYZER & P. WESTBROEK. 1993. Molecules and morphology - the practical approach. Lethaia, 26:5-6.

DONOGHUE, M.J. & P.D. CANTINO. 1984. The logic and limitations of the outgroup substitution approach to cladistic analysis. Systematic Botany 9:192-202.

EERNISSE, D.J., J.S. ALBERT & F.E. ANDERSON. 1992. Annelida and arthropoda are not sister taxa: a phylogenetic analysis of spiralian metazoan morphology. Systematic Biology 41:305-330.

ENDO, K., G.B. CURRY, M.J. QUINN, M.J. COLLINS, G. MUYZER & P. WESTBROEK. 1994. Re-interpretation of terebratulide phylogeny based on immunological data. Palaeontology 37:349-373.

FARRIS, J. S. 1972. Outgroups and parsimony. Systematic Zoology 31:328-334.

FELSENSTEIN, J. 1993. PHYLIP (Phylogeny Inference Package). Department of Genetics, University of Washington. Distributed by the author, Seattle, Washington.

FIELD, K.G., G.J.OLSEN, D.J. LANE, S.J. GIOVANNONI, M.T. GHISELIN, E.C. RAFF, N.R. PACE & R. RAFF. 1988. Molecular phylogeny of the animal kingdom. Science 239:748-753.

GEE, H. 1995. Lophophorates prove likewise variable. Nature 374:493.

GHISELIN, M.T. 1988. The origin of molluscs in the light of molecular evidence, p.66-95. In, P. H. Harvey & L. Partridge (eds.), Oxford surveys in evolutionary biology, 5.

GILBERT, D. 1993. SeqApp. Available by FTP from Molecular Biology Software Archive, University of Indiana, Bloomington, Ind.

GORJANSKY, W.J. & L.E. POPOV. 1986. On the origin and systematic position of the calcareous-shelled inarticulate brachiopods. Lethaia 19:233-240.

GUSTUS, R.M. & R.A. CLONEY. 1972. Ultrastructural similarities between setae of brachiopods and polychaetes. Acta Zoologica, Stockholm 53:229-233.

HALANYCH, K.M., J.D. BACHELLER, A.M.A. AGUINALDO, S.M. LIVA, D.M. HILLIS & J.A. LAKE. 1995. Evidence from 18S ribosomal DNA that the lophophorates are protostome animals. Science 267:1641-1643.

HENNIG, W. 1966. Phylogenetic systematics. University of Illinois Press, Chicago.

HILLIS, D. M. & M. T. DIXON. 1991. Ribosomal DNA: molecular evolution and phylogenetic inference. Quarterly Review of Biology 66:411-453.

___ J. P. HUELSENBECK & C. W. CUNNINGHAM. 1994. Application and accuracy of molecular phylogenies. Science 264:671-677.

___ & D. MORITZ. 1990. Molecular Systematics. Sinauer, Sunderland, Mass.

HULTMAN, T., S. STAHL, E. HORNES & M. UHLEN. 1989. Direct solid-phase sequencing of genomic and plasmid DNA using magnetic beads as a solid support. Nucleic Acids Research 17:4937-4946.

HUSS, V.A.R. & M.L. SOGIN. 1990. Phylogenetic position of some Chlorella species within the Chlorococcales based upon complete small-subunit ribosomal RNA sequences. Journal Molecular Evolution 31:432-442.

IRWIN, D.H. 1991. Metazoan phylogeny and the Cambrian radiation. Trends Ecology Evolution 6:131-134.

ISHIKAWA, H. 1977. Comparative studies on the thermal stability of animal ribosomal RNA's: V, Tentaculata (Phoronids, moss-animals and lamp-shells). Comparative Biochemistry Physiology 57B:9-14.

KÄLLERSJÖ, M., J.S. FARRIS, A.G. KLUGE & C. BULT. 1992. Skewness and permutation. Cladistics 8:275-287.

MADDISON, W.P., M.J. DONOGHUE & D.R. MADDISON. 1984. Outgroup analysis and parsimony. Systematic Zoology 33:83-103.

MAIDAK, B.L., N. LARSEN, M.J. MCCAUGHEY, R. OVER-BEEK, G.J. OLSEN, K. FOGEL, J. BLANDY & C.R. WOESE. 1994. The Ribosomal Database Project. Nucleic Acids Research 22:3485-3487.

NIELSEN, C., 1991. The development of the brachiopod Crania (Neocrania) anomala (O.F. Müller) and its phylogenetic significance. Acta Zoologica 72:7-28.

___1994. Larval and adult characters in animal phylogeny. American Zoologist 34:492-501.

___1995. Animal evolution: interrelationships of the living phyla. Oxford University Press, Oxford.

NIXON, K.C. & J.M. CARPENTER. 1993. On outgroups. Cladistics 9:413-426.

NORELL, M.A. & M.J. NOVACEK. 1992. The fossil record and evolution: comparing cladistic and paleontological evidence for vertebrate history. Science 255: 1690-1693.

OLSEN, G.J., H. MATSUDA, R. HAGSTROM & R. OVERBEEK. 1994. fastDNAml: a tool for construction of phylogenetic trees of DNA sequences using maximum likelihood. CABIOS 10:41-48.

PATTERSON, C.1985. Introduction, p.1-22. In, C. Patterson (ed.), Molecules and morphology in evolution. Cambridge University Press: Cambridge.

___ 1989. Phylogenetic relations of major groups: conclusions and prospects, p.471-488. In, B. Fernholm, K. Bremer & H. Jörnvall (eds.), The hierarchy of life. Elsevier, Berlin, Dahlem.

PHILIPPE, H., A. CHENUIL & A. ADOUTTE. 1994. Can the Cambrian explosion be inferred through molecular phylogeny?, p.15-25. In, M. Akam, P. Holland, P. Ingham & G. Wray (eds.), The evolution of developmental mechanisms, Development Supplement. Company of Biologists, Cambridge.

POPOV, L.E., M.G. BASSETT, L.E. HOLMER & J. LAURIE. 1993. Phylogenetic analysis of higher taxa of Brachiopoda. Lethaia 26:1-5.

RUNNEGAR, B., A. SCHELTEMA, L. JACKSON, M. JEBB, J. M. TURBEVILLE & C. R. MARSHALL. 1995. A molecular test of the hypothesis that Aplacophora are progenetic Aculifera. [in preparation].

SAMBROOK, J., E. F. FRITSCH & T. MANIATIS. 1989. Molecular cloning, a laboratory manual. Cold Spring Harbor Laboratory, Cold Spring Harbor, NY.

SCHRAM, F. R. 1991. Cladistic analysis of metazoan phyla and the placement of fossil problematica, p.35-46. In, A.M. SIMONETTA & S. CONWAY MORRIS (eds.), The early evolution of the metazoa and the significance of problematic taxa. Cambridge University Press, Cambridge.

SMITH, A.B. 1994. Rooting molecular trees: problems and strategies. Biological Journal Linnean Society 51:279-292.

SMITH, S. W., R. OVERBEEK, C.R. WOESE, W. GILBERT & P.M. GILLEVET. 1994. The Genetic Data Environment, an expandable GUI for multiple sequence analysis. CABIOS 671-675.

SWOFFORD, D.L. 1993. PAUP: phylogenetic analysis using parsimony, Version 3.1. Computer program distributed by Dr D. Swofford. Smithsonian Institution, Washington, D.C., USA.

VAN DE PEER, Y., I. VAN DEN BROECK, P. DE RIJK & R. DE WACHTER. 1994. Database on the structure of small ribosomal subunit RNA. Nucleic Acids Research 22:3488-3494.

WILLIAMS, A. 1965. Brachiopoda, p.524-927. In, R.C. Moore (ed.), Treatise on invertebrate palaeontology, H. Geological Society of America & University of Kansas Press, Kansas.

___1973. The secretion and structural evolution of the shell of thecideidine brachiopods. Philosophical Transactions Royal Society, London B(264):439-478.

WINNEPENNINCKX, B., T. BACKELJAU & R. DE WACHTER. 1995. Phylogeny of protostome worms derived from 18S rRNA sequences. Molecular Biology Evolution 12:641-649.

ZUCKERKANDL, E. & L. PAULING. 1965. Molecules as documents of evolutionary history. Journal Theoretical Biology 8:357-366.

EVOLUTION OF THE SPIRE-BEARING BRACHIOPODS (ORDOVICIAN-JURASSIC)

PAUL COPPER & RÉMY GOURVENNEC

Department of Earth Sciences, Laurentian University, Sudbury, Canada P3E 2C6;
Laboratoire de Paléontologie, Université de Bretagne Occidentale, F-29285 Brest, France

ABSTRACT-- The articulate spire-bearers are presently divided into three (or four) orders, but the monophyletic (from Rhynchonellida), versus polyphyletic origin (from Rhynchonellida and Orthida) of these orders is still in dispute. In terms of their spiral lophophore orientation, they can be divided into two groups, (1) those with medially to dorsomedially-directed spiralia, the Atrypida (Llandeilo-Frasne), and (2), those with laterally to lateroventrally directed spiralia, the Athyridida (Caradoc-Jurassic) and Spiriferida-Spiriferinida (Ashgill-Jurassic), excluding the 'spire-bearing' Thecideidina (Triassic-Recent). In the atrypoids, the oldest taxa have a complete or partial jugum, but by the Silurian nearly all show the jugum split into two separate jugal processes. In the athyroids simple jugate stocks are the ancestral character, but in the Silurian and Devonian highly complex variations of the jugum evolved, including extensions which produced a parallel set of spiral whorls, and those with a jugal web (Mesozoic). In the spiriferoids, the jugum remained essentially very simple throughout the group's history, and was commonly lost in some. The Spiriferida form a large group, usually divided into three suborders [or sometimes split into separate orders, the impunctate Spiriferida, and punctate Spiriferinida]. Atrypoids had an impunctate shell wall, but one family, the Punctatrypidae had a fenestrate ('mega-punctate') shell: the athyroids and spiriferoids (like the Rhynchonellida) comprise both impunctate and punctate stocks, with the impunctate condition always preceding in the fossil record, and punctation repetitively evolved. Deltiodont/cyrtomatodont teeth, and strophic/astrophic hinges, were apparently also repetitive features in all stocks of spire-bearers. This is emphasized by the fact that within atrypids alone, the orthid-like hinge and teeth of Silurian *Tuvaella* and orthid-strophomenoid hinge and teeth of Devonian *Davidsonia* and *Prodavidsonia* evolved independently from different lineages at different times.

INTRODUCTION

More than 110 years ago, both Davidson (1882) and Waagen (1883) suggested a tentative classification of the spire-bearing brachiopods, with the tacit recognition that spiral lophophorate brachiopods probably had a common ancestor. Davidson (1882) recognized four families, the Atrypidae, Spiriferidae, Nucleospiridae, and Athyridae, with the last two united under the Athyridida today. This was generally the approach followed by most later workers. The spire-bearers are not alone in having a double spiral lophophore pump for feeding and respiration: living rhynchonellids also have a double spiral lophophore, but in these it is not supported by calcareous lamellae. Thus spire-bearers provide us with a unique picture of the evolution of a complex lophophore system. The adjectives 'atrypoid', 'athyroid', and 'spiriferoid', and nouns derived therefrom, such as atrypids, athyrids, spiriferids, are used here as common names for shells which are *Atrypa*, *Athyris* and *Spirifer*-like in terms of their brachidia and evolutionary relationships to the three major groups commonly recognized. Approximately 900 genera of spire-bearers have been described; about 210 of these are atrypids, 180 athyrids, and there are more than 500 generic names for spiriferids.

Concepts of the classification of the spire-bearers have fluctuated in the past half century. Roger (1952) in the Traité de Paléontologie, identified three superfamilies of spire-bearers, atrypaceans, spiriferaceans and rostrospiraceans (=athyrids). In 1960 Rudwick analysed the feeding mecha-nisms of the spire-bearing brachiopods, suggesting a common origin for these articulates. In the Osnovy Paleontologii (Sary-cheva, 1960), the spire-bearers were lumped into three separate groups, the Atrypida and Spiriferida, with the 'Athyracea' classified as 'Incerti ordinis'. To the Atrypida, Rzhonsnitskaya (1960) added the leptocoeliids (now removed to the rhynchonellids) and the dayiids, which are athyroids. The Treatise on Invertebrate Paleontology volume on brachiopods (Boucot et al., 1965) grouped the spire-bearers under the Spiriferida, relegating four groups to subordinal status, e.g. the atrypids, retziids, athyrids and spiriferids (the middle two are grouped under Athyrida today). Rudwick (1970) reviewed the development, ontogeny and evolution of the spiral lophophore, and Hoverd (1985, 1986) in particular discussed the spiral lophophore of the living articulate rhynchonellids, showing that feeding currents were directed in from the shell sides and out along the central part of the anterior commissure.

In contrast, Wright (1979a) suggested that the spiriferids (i. e. the *Spirifer* group sensu stricto) were derived independently from the Orthida because of a strophic hinge, and proposed that the remaining two groups, the normally astrophic atrypids and athyrids were derived independently from rhynchonellids. This thus placed primary emphasis on the hinge for evolution and origins, and relegated the nature and orientation of the lophophore to minor considerations. It also assumed that the spiralium repetitively and independently evolved at least three times in the evolution of the lophophore (Wright, 1979b). Some of the oldest atrypids, the anazygids (see for example the Ordovician anazyginids *Zygospira* and Silurian *Tuvaella*, formerly classified with the orthids: Figure 1), possess both an astrophic and strophic hinge system. The oldest athyrids, possessing an astrophic hinge, were described by Fu (1982) from northern China in rocks of late Caradoc to early Ashgill age, but there are also Devonian athyrids with a remarkable *Spirifer*-like shape and hinge. Discovery of the oldest known spiriferid, a deep water inhabiting '*Eospirifer* ' from the middle Ashgill of China (Rong et al., 1994) does not really settle the issue at present, because the specimens have not yet yielded definitive information about the brachidia, and the authors stated that the cardinalia of this earliest form are 'atrypoid' in nature, and lacked crural plates. Illustrations provided, however, show an athyroid-like, astrophic, ovoid shell and cardinalia along the lines of early hindellines, which had laterally directed spiralia. Moreover, the presumed ancestral platystrophids (Wright, 1979a), with a spiriferid appearance, seem to have appeared later in the geologic record than the oldest '*Eospirifer*', throwing doubt on the issue of a platystrophid ancestor for the group.

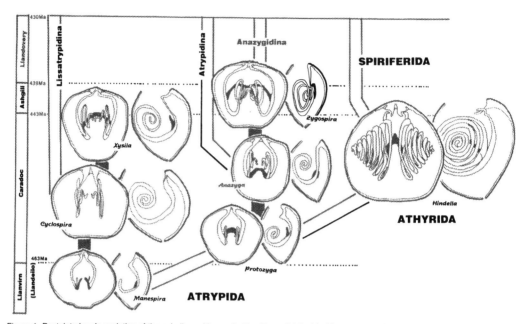

Figure 1. Postulated early evolution of the spiralia and jugum in Atrypida and Athyrida. The ancestry of the middle-late Caradoc Hindellinae (probably in east Asia) may lie with a *Protozyga*-like ancestor which had a simple jugum, and one whorl spiralia oriented in the plane of symmetry, i.e. vertically. In this concept, the oldest '*Eospirifer*', of late Caradoc age from China , may have been derived from a hindelline source by evolving very fine costellae (Copper).

Regardless of the subsequent evolutionary plans for the spire-bearers, the late Ordovician was a period of experimentation and radiation in the group as a whole. By the end of the Ordovician, the major features of all three orders are identifiable in the fossil record. On the one hand one could regard this first 25 million year period of the Caradoc and Ashgill as a period of radiation, diversification, experimentation and innovation in terms of brachidium evolution. The remaining history of the spire-bearers was a period of refinement of shell structure and shell ornamentation, hinge, muscle attachment, evolution of specialized structures, shape modification, and ecological expansion in habitat and abundance. These modes could be visualized as including pictures of both gradualism (stepwise appearance and rise of major clades), as well as periods of stasis, with more conservative, small-scale variation along a common theme. Phases of mass extinction saw the disappearance of the entire order Atrypida, and other spire-bearer families, near the end of the Frasnian (Late Devonian), the almost catastrophic extinction of the athyrids and spiriferids near the end of the Permian, and the end of the Triassic, and the final extinction of the athyrids and true spiriferids in the Early Jurassic (Bathonian ?), a period for which no other mass extinction is generally known. Aberrant thecideid spire-bearers survived into the Recent: thecideidines have been assigned to the spiriferids on the basis of shell structure (Williams, 1973; Baker, 1990), and were derived from Triassic or Jurassic thecospiroid spiriferids under this interpretation. However, analysis of gene sequences of living thecideidines (Cohen & Gawthrop, this volume), proposes that their closest relatives are short-looped terebratuloids, and that the thecideidine loop-bearers are thereby more appropriately assigned to the Terebratulida [accepted herein]. The survival of the Mesozoic thecideidines across the Cretaceous-Paleocene boundary is possibly due to their retreat to a cryptic habitat in reef caves and on the undersides of corals.

Given present divergent opinions, possibly four or more, evolutionary mono- and polyphyletic trends are still open for discussion. In terms of monophyly, the most parsimonious assumption would be that the calcite lamellar spiral lophophore supports evolved only once, and originated with the oldest taxon sometime in Llandeilo time. One monophyletic option would see the atrypoids with a single coil, double spiralium in the plane of shell symmetry, appearing first, giving rise to medially directed spiralia in the earliest Caradoc, then laterally directed spiral athyrids in the middle Caradoc, and with the athyrids in turn providing an ancestral '*Eospirifer*' in the middle Ashgill by losing the jugum (Figure 1). If the oldest '*Eospirifer*' indeed has a complete jugum, and this is not yet demonstrated, it would support this argument. The jugum is unknown in the Paleozoic Spiriferidina studied in detail, with many (if not nearly all) Paleozoic forms apparently lacking one, and the punctate Paleozoic-Mesozoic Spiriferidina possessing a simple jugum. A second option here is the nearly simultaneous independent evolution of both the athyrids and spiriferids in the Caradoc, but their stepwise appearance in the fossil record belies this possibility: athyrids preceded spiriferids.

In terms of polyphyly, at least three possibilities are dictated. The first of these is the separate evolution of atrypids and athyrids in one lineage, with laterally directed spiralia appearing independently from rhynchonellids-atrypids on one side, and orthids on the other (Wright 1979). This would mean having to look for an orthid ancestor for the spiriferids in the Ashgill, and assuming that laterally directed spiralia evolved twice. The second view entertains the possibility that the atrypids with dorsally-medially directed spiralia evolved from rhynchonellids, and that laterally spiralled athyrids, as well as spiriferids, evolved from orthids, thus that the division between laterally and medially directed spiralia was the crucial separation point at a very early, orthoid ancestral stage. This would mean that

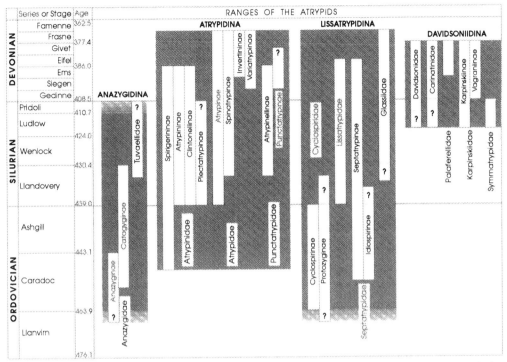

Figure 2. Range chart of the four main suborders of the Atrypida, and the families and subfamilies assigned thereto. The last suborder to appear was the Davidsoniidina, which probably evolved from an *Atrypina*- or *Gracianella*-like stock in the Ludlow (Silurian). Vertical scale by relative time units measured in millions of years (Copper).

the astrophic-hinged, ovoid shape of some ancestral orthid led to athyrids on one hand, and re-evolution of the strophic hinge in some spiriferids on the other. A third, triphyletic alternative was proposed by Grunt (1982), who proposed that athyrids were derived directly from rhynchonellids, without descent via atrypids. This is not an implausible suggestion, but leaves only the problem of finding an alternate smooth-shelled rhynchonellid ancestor in the Caradoc. Phylogenetically the least difficult option, requiring the fewest changes in spiralia and shell shape, would be the monophyletic mode from an early planispiral atrypoid, with successive appearance of the three main orders, assuming the rhynchonellids evolved from orthids.

ANCESTRY OF THE TEREBRATULOID LOOP

Did the spiral lophophore and brachidium of spire-bearers evolve into anything else? In 1961 Williams & Wright suggested that the jugum of the Caradoc genus *Protozyga*, one of the ancestral atrypoid genera, was 'the earliest known [terebratuloid] loop'. This suggestion is not as unlikely as it sounds, though the Ordovician protozyginids were long extinct by the time the first terebratuloids appeared in basal Devonian strata. The loss of spiralia and retention of the jugum in protozyginids could produce a terebratuloid, if punctation were also added to the stock. However, the more than 40 million year stratigraphic gap between such atrypoids and the earliest terebratuloids makes this unlikely. Boucot & Wilson (1994) stated that the Silurian atrypoid *Nalivkinia* was the ancestor of the terebratuloids. This seems also improbable as the genus *Nalivkinia* has separated jugal processes without any loop-like structure, and it would thus have to reverse its evolution in re-

growing a 'jugum loop', or regenerate an entirely new loop system from the crura. Also, *Nalivkinia* has a different hinge plate structure than early terebratuloids, and lacks punctae.

A more likely scenario is evolution of the terebratuloids from an Early Devonian or Late Silurian punctate retziid ancestor (i.e. from the Athyrida). There are solid grounds for this: (1) retziids are already punctate, (2) retziid (and athyrid) shell shape and ornamentation is homeomorphic with the earliest terebratuloids [Boucot & Wilson (1994, p.1018) mentioned that 'athyridoid shells' should be more carefully examined in Early Devonian and Late Silurian strata for possible terebratuloid species], (3), earliest terebratuloids have a dorsal hinge plate virtually identical with that of retziids, (4), retziids, like other athyrids, possess a long, low dorsal median septum, a feature absent in atrypoids, (5) the long narrow dental plates of early terebratuloids are more typical of retziids, and absent in atrypoids (and certainly absent in *Nalivkinia*), (6) only the spiralium would need to be lost, with the retziid jugum becoming the loop, and (7), the stratigraphic ranges of the retziids and earliest terebratuloids overlap. The Mesozoic evolution of thecideidines, assigned to terebratulids, or to a separate order related to terebratulids (Elliott,1965), stresses the possible affinities of spire-bearers in terms of shell structure.

ATRYPIDS (Copper)

The atrypoids are the oldest spire-bearers with a calcite lamellar support, ranging back into late Llandeilo (=latest Llanvirn) time (Figure 2: Copper,1977,1986a,1986b). A spiral lophophore without calcite supporting lamellae was probably already typical of the oldest rhynchonellids, which trace ancestry back into the early or middle Llanvirn [see Jin, this volume]. The order Atrypida is characterised by possessing spiralia which initially evolved as planispiral (a whorl of less than one revolution is automatically planispiral) to medially directed, beginning from a set of crural plates attached to the dorsal hinge. Members of the order later switched the rotation of spiralia into a dorsomedial or dorsal orientation (in most Silurian and Devonian forms). Four suborders are recognised for the Atrypida: the Anazygidina, Lissatrypidina, Atrypidina and Davidsoniidina (previously assigned to the Strophomenida in the 1965 Treatise). In Ordovician time most of the Anazygidina had simple spiralia with less than one complete planispiral whorl, or a few whorls, which were medially directed. This group, a number of which had a strophic hinge, dominated the early stages of spire-bearer evolution. The last of the Anazygidina died out with the Tuvaellidae (assigned in the 1965 Treatise to the Orthida), which survived in the Tuva faunal province of Russia, Mongolia and NE China into late Silurian time (Ludlow-Pridoli). Strophism appears to be a latent feature that developed via iterative evolution in all later independently evolving spire-bearer stocks. The jugum apparently persisted as a single structure throughout the history of the Anazygidina.

The Lissatrypidina, the smooth-shelled atrypids, first appeared in the early Caradoc, but may possibly also have evolved nearly simultaneously with the Anazygidina from a prototype planispiral rhynchonellid. In some of the earliest, 'protozygid' forms only a partially developed, incomplete or complete jugum is known (e.g. as in Caradoc Manespira : Copper, 1988), or there was no jugum altogether, as in Caradoc-Ashgill Cyclospira (ibid.). Normal protozyginids, however, like Protozyga and Xysila, have a complete jugum and medially-directed spiralia. The development of the spiralial apparatus in most Caradoc and Ashgill forms of protozygids is very poorly known because time-consuming, accurate serial sectioning of small shells is difficult and has rarely been carried out (probably less than 5% of species described have been sectioned). Silicification of the spiralia, or post-mortem calcite drusy growth, normally completely obscures the brachidium with secondary crystals. The presence of a jugum is a primitive character in the group. Within the lissatrypinids there is variability in the presence of a jugum, or evolution of the separate jugal processes, unlike the other three suborders of atrypids. The oldest Septatrypidae, e.g. late Caradoc Webbyspira and Idiospira, have a jugum, but all later septatrypids have jugal processes. Within the Lissatrypidae, one Llandovery genus, Cerasina , from eastern North America, has a jugum (Copper, 1995): all others evolved jugal processes terminating in elaborate jugal plates. Another unusual feature for the Silurian and Devonian was the presence of the smooth Glassiidae: this group of shells retained 'primitive' medially directed, barrel-shaped spiralia throughout their evolution, while all other smooth shells moved to a conical dorsally-directed mode. The lissatrypids became minor figures in atrypid evolution and abundance after the Silurian, and even during the Silurian they were frequently confined to specific communities, e.g. the Atrypoidea , Lissatrypa and Glassia assemblages.

The Atrypidina became the dominant figures during the Silurian and Devonian, but even in the Hirnantian they locally were the abundant brachiopods in some shallow tropical shelf settings (e.g. the Ashgill Eospirigerina community). They developed ribbed shells, commonly with frills or spines from the Wenlock on, had a larger shell volume (some reaching widths of 150mm in the Devonian), and had jugal processes nestled between spiralial whorls in a postero-ventral position. During the mid-Silurian (Wenlock), one ribbed group, the family Atrypinidae, appears to have produced a new off-shoot which led to the Davidsoniida: this became a predominantly reef and peri-reef

dwelling stock which led not only to strophomenid and orthid like forms such as Devonian Davidsonia and Prodavidsonia, but also exotic, 'aberrant' lineages such as the Devonian carinatinids, karpinskiinids and palaferellids. These vanished by the end of the Givetian (carinatinids identified from the Frasnian are most likely spinatrypids and atrypids), while the Atrypinida held on into the late Frasnian.

ATHYRIDS (Copper)

The oldest athyrids were smooth shelled forms of hindellines which appeared in the middle to late Caradoc of North China (Fu, 1982). These shells were astrophic, and were characterised internally by laterally directed spiralia held together in the shell centre by a very simple jugum. The first ascending lamella of each spiralial whorl does not appear to have been directly attached to the crura, except via soft, unpreserved muscles and a hook-like clasp (as seen for Hindella : Copper, 1986; see also Figure 1, herein). The crura were directed in the plane of symmetry, with the basal whorl of each spiralium almost lying in the plane of symmetry, on each side. This is in contrast to ancestral atrypoids, where the crura are directed to the sides of the shell, instead of anteriorly, but is similar to the form seen in spiriferids. Grunt (1982) has suggested that athyrids may have evolved directly from rhynchonellids without an intermediate atrypid phase: this may be an alternative to other polyphyletic schemes suggested. However, the oldest athyrids are all smooth, and smooth shells are generally unknown in Middle and Late Ordovician rhynchonellids. The first confirmed, ribbed athyrids are known from middle Llandovery rocks of China (e.g. Metathyrisina), but older taxa may remain to be discovered as internal structures of many ribbed rhynchonellid or atrypid-like shells are undescribed. Ribbed athyrids (retzioids) are also the only athyrids known to be punctate: these first appear in Wenlock time.

The athyrids developed the most complex of jugal structures known in the spire-bearers, and were also the only spire-bearers to develop double spiral lamellae, with the second whorl extending from the jugal stem parallel to the primary whorl right to the apex of the cone. The fossil record suggests that double spiral lamellae may have evolved more than once in unrelated groups. In coelospirids (including Silurian Coelospira, Devonian Anoplotheca and Bifida), the double spiralium is fused as a single spiralium with a trough, but in smooth Devonian Helenathyris and ribbed Kayseria there is a double parallel spiralium right to the apex. Complex jugal and spiralial structures first appeared in the Silurian (e.g. in Wenlock Merista), but were particularly prominent and elaborate in the Devonian athyrids. The phylogenetic significance attached to such structures is still debated (see Modzalevskaya, this volume). If such complex structures are monophyletic, then present classification schemes, based on ribbed versus smooth shells, punctation and cardinalia, require revision. In the Mesozoic, a further complication of the brachidal jugum, a support called a 'jugal web' evolved [see Dagys, this volume].

Athyrids developed frills in the Silurian and some Devonian taxa (plicathyrinids) evolved a spiriferoid shell with a long, strophic hinge. Athyrids, like spiriferids, survived the mass extinctions of the Late Devonian, end-Permian and end-Triassic, possibly because they had a wide geographic range which included cold climate, high latitude siliciclastic shelf settings. The lack of a global mass extinction episode at the end of the Early Jurassic, seeing the disappearance of athyrids and true spiriferids, suggests that these groups probably went extinct from ecologic pressures such as Mesozoic molluscan, arthropod and fish predation, competition for space with bivalves, or less successful reproductive or larval attachment strategies. The sole surviving Recent 'spire-bearers', the thecideinids, are small to minute, cryptic reef inhabitants, and probably belong to the loop-bearing order Terebratulida, having evolved their ventrally-directed spiralia aberrantly from ancestral loop structures.

SPIRIFERIDS (Gourvennec)

Spiriferid brachiopods form a large, diversified group with more than 500, mainly Paleozoic, genera represented by two orders: the generally ajugate impunctate Spiriferida and the normally jugate, punctate Spiriferinida (this number compares with 187 genera, and 76 synonyms, identified in the last edition of the Treatise in 1965). Shell punctation is used as the basic feature in the distinction between the orders. The Spiriferida is the oldest, most prolific and diverse group and flourished during the Siluro-Devonian (Cyrtioidea, Ambocoelioidea, Reticularioidea, Delthyridoidea) and the Carboniferous-Permian (Martinioidea, Spiriferoidea). The earliest punctate Spiriferinida (Cyrtinoidea) derived from the Delthyridoidea in the Early Devonian, but attained their full development in the Late Paleozoic (Syringothyridoidea) and Triassic-Jurassic (Suessioidea, Pennospiriferinoidea, Spiriferinoidea, etc.). Almost all genera were extinct at the end of the Triassic and only a few survived the Early Jurassic (Toarcian) into the Bathonian. Among the diagnostic characters allowing the separation of spiriferids at the order and superfamily levels, shell fabric and growth form are given major importance. Internal structures and micro-ornament are also essential features, but are more useful at the family and subsequent levels.

The order Spiriferida, which comprise two suborders (Spiriferidina and Delthyridina) form the older group, and appear as soon as the Middle Ashgill with the Cyrtioidea: the first representative eospiriferid appears in China, Kazakhstan and Australia (Rong et al, 1994). The suborder saw great expansion during the Devonian (more than 250 genera, 27 families). Towards the end of the Devonian (Frasnian-Famennian), a number of taxa disappeared, but during a Famennian take-over from atrypids, and innovations during the Early Carboniferous, new lineages arose, particularly in the Spiriferidina, which were maintained until the Late Permian. The Delthyridina were strongly affected by the Late Devonian crisis, and only two families (Reticulariidae and Elythynidae) survived through the Permian. No Spiriferida survived the end-Permian double mass extinction.

This is not the case for the Spiriferinida, which comprise two suborders (Cyrtinidina and Spiriferinidina). Their origin is to be found in the Delthyridioidea, probably a cyrtiniopsid, in the Early Devonian. Poorly represented during the Devonian (5 genera, most of them belonging to the Cyrtinoidea), they showed great expansion during the Carboniferous and Permian (more than 60 genera in 12 families). Although strongly affected by the end-Paleozoic mass extinction (loss of c.30 genera, 6 families), they were renewed and still well represented during the Triassic (>60 genera); some genera continued into the late Early Jurassic (Bathonian).

ORIGINS OF SPIRIFERIDS

During the last two decades, the origin of the spiriferids has been debated on morphology of the Llandoverian eospiriferids, e.g. *Eospirifer, Macropleura, Striispirifer, Yingwuspirifer* . Recent work (Rong et al., 1994) described small eospiriferids from the Middle Ashgill Changwu Formation of China, and from the Ulkuntas Horizon of Kazakhstan. Three main theories have been invoked for the origin of spiriferids. The first supposed direct derivation from an older spire-bearer, i. e. a primitive

anazygid atrypid (Boucot et al., 1965; Williams & Hurst, 1977; Rong et al., 1994), implying a common ancestor for all spire-bearers. The second alternative suggested that spiriferids were derived from an impunctate orthid, probably a platystrophid (Wright, 1979a; Rudwick, 1970): in this case the spire-bearers were said to be diphyletic, with the Atrypida (comprising the Athyridida) on one side, and the Spiriferida on the other side. A third solution is to visualize all three as being independently derived, or to have originated separately from a common stock such as rhynchonellids (Grunt, 1989).

Williams & Hurst (1977) implied major changes in the direction of the spiral lophophore and its support (the spiralium), which is inversely directed from that of earliest atrypoids. The lophophore eversion, also seen in athyrids, would have been a fundamental change in the evolution of the group. Thus, the strophic aspect of the shell would have been a secondary effect, in part due to the lateral or postero-lateral orientation of the lophophore, which is a characteristic of spiriferids (and athyrids). Moreover, this orientation of the lophophore is an improvement for the expansion of the lophophore, and separation of inhalant-exhalant currents, which may have induced the development of an anterior, current-separating sulcus and fold (although these structures secondarily disappear in some Spiriferida). The Orthorhynchulidae, a strophic group of Late Ordovician and early Silurian rhynchonellids, suggests the alternate possibility that this group may have given rise to the spiriferids.

In the second hypothesis (Wright, 1979a), the Spiriferida derived from an impunctate orthid stock, from which they are assumed to have inherited their strophic shell (though the earliest spiriferid lacks a strophic hinge during ontogeny). Here a secondary evolutionary stage is the appearance of a lateral spirolophe. In this case, the spirolophe in atrypids and spiriferids is regarded as convergence or parallelism, with the lateral direction of the spirolophe evolving twice in spire-bearers. Thus the spire-bearers would be diphyletic in that model, as visualized by Rzhonsnitskaya (1960), who, however, did not predict an orthid ancestor [and grouped some athyrids with atrypids, despite their inverse spiralium orinetation]. Wright (1979a) took as evidence similarities in the internal features (cardinalia, dental plates) of spiriferids and orthids, particularly the Platystrophiinae, which he cited as the most probable ancestors for the Spiriferida. No Platystro-phiinae have been identified at present in the Ordovician formations of China, Kazakhstan or Australia. Other simila-rities between orthids and spiriferids concern the surface ornament (Gourvennec & Mélou, 1990), nature of the teeth, and disposition of the gonads.

A third viewpoint is Boucot's hypothesis (in Sheehan & Baillie, 1981) suggesting, based on the presence of a pedicle collar and diagonal striae on the ventral valve of *Eospirifer tasmaniensis* , that the ancestor of the eospiriferids could be a meristelloid athyrid. This would support a monophyletic origin for all spire-bearers, or a diphyletic mode if the athyrids> spiriferids evolved from a stock differnt than atrypids.

The discovery of small (c.5mm wide) shells of the deep water species '*Eospirifer' praecursor* in the Middle Ashgill of China (Rong et al, 1994) introduced some contradictory elements in this discussion. Although the presence of crural plates is a characteristic feature of all eospiriferids known, this early species lacks such a structure, and in the early stages of development of *E. praecursor* the shell is astrophic. Basing their arguments on these observations, Rong et al. (1994) suggested an unidentified atrypoid ancestor for *Eospirifer*

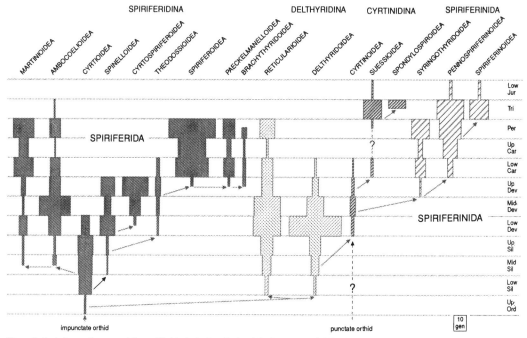

Figure 3. Evolutionary lineages of the spiriferids, including stocks of the impunctate Spiriferida and punctate Spiriferinida (Gourvennec).

rather than an orthid or athyrid. For *E. praecursor* the structure of the spiralium and jugum (if any) remains unknown. The lack of crural plates and the astrophic nature of the hinge suggest a genus other than *Eospirifer*, though the ontogeny of other *Eospirifer* is poorly studied. Johnson & Lenz (1992), following Pitrat's opinion (1965), derive the first ambocoeliid (*Eoplicoplasia*) from an unknown eospiriferid, by loss of crural plates, and re-development of the crural plates in later ambocoeliids (Rhynchospiriferinae). Athyrids with laterally directed spiralia have been described from the late Caradoc of N China by Fu (1982), but these shells, though astrophic and shaped like the Ashgill eospiriferids from E China, are smooth and apparently have a jugum.

The origin of the spiriferids is thus still problematic. If one admits an unknown atrypid ancestor, the strophic character of the hinge appeared independently in rhynchonellids, atrypids and spiriferids, and orthids are not known to have possessed a spiral lophophore. The convergence of strophic hinge and articulation in orthids and spiriferids comprises characters related to articulation (teeth, sockets), to lophophore support (cardinalia), and gonads, ornament, sulcus and fold. Convergent evolution or parallelism is common in brachiopods, and not unique. A polyphyletic origin for spiriferids is not necessarily surprising: the number of morphological possibilities for the development of the lophophore remains limited. Thus one could expect a spiralium, with different orientations, in unrelated stocks. A cladistic analysis made by Carlson (1991), based on the first edition of the Treatise (Moore, 1965), proposed a polyphyletic origin for the spiriferids, using the characters suggested by the original compilers. In the meantime, however, more than four times as many genera have been des-cribed, many new discoveries have extended geologic ranges of spire-bearers, and many alternate phylogenetic lineages unravelled.

EVOLUTION OF THE SPIRIFERIDA

The order Spiriferida comprises impunctate, mostly strophic spire-bearers with laterally directed spiralia (Figure 3). Criteria for subdivision of the Spiriferida relate to specialized internal structures (e.g. ctenophoridium, spondylium, adminicula, syrinx, etc.), the hinge, brachidial attachments, pedicle structures, shell shape, and micro-ornamentation. The oldest genera, the eospiriferids, bear a capillate shell fabric, dental and crural plates, and lack a ctenophoridium. The oldest known eospiriferid lacks the crural plates characteristic of later forms. The suborder Spiriferidina is the most prolific and diverse group (275 genera in 9 superfamilies). The Cyrtioidea gave rise to the Spinelloidea in the Silurian (Wenlock). The oldest spinelloid genus, *Spurispirifer* (Pinguispiriferinae) possess a cyrtiid-type ornament, a smooth cardinal process and short crural plates. Some aspects of its morphology are reminiscent of *Tenellodermis* and the two genera possibly had a common ancestor, in which case the Martinioidea and Spinelloidea were derived from the same stock. The Cyrtospiriferoidea separated from the Spinelloidea by acquisition of a delthyrial plate.

The origin of the Theodossioidea is unclear: they are characterized by plications covering the entire shell and the loss of a well-defined fold and sulcus. The oldest known genus is *Branikia* (Pragian), with short dental plates, lacking crural plates; an acrospiriferid ancestor, as suggested by Havlicek (1959), is rejected. *Theodossia* is the supposed link to younger genera of the Ulbospiriferidae, which are characterized by the presence of an inner prismatic shell layer. Three superfamilies (Spiriferoidea, Paeckelmannelloidea, and Brachythyridoidea) appeared at the same time in the Famennian, following the Frasnian mass extinction, and were derived from the Theodossioidea (probably an ulbospiriferid). This ancestor may have been, at least partially, denticulate, with a capillate ornament and bearing dental adminicula. Furthermore, it may have produced a prismatic shell layer, in as far as this character occurs in some genera of the Spiriferoidea.

The Ambocoelioidea probably derived from an eospiriferid such as *Macropleura*. Johnson & Lenz (1992) pointed out the similarities between *Macropleura* and the earlier ambocoelid (*Eoplicoplasia*) including overall shell shape, lack of ctenophoridium, presence of dental plates, and capillate ornament. *Eoplicoplasia* lacks crural plates, but these are to be found in younger genera of ambocoeliids. The Martinioidea separated from the Ambocoelioidea through acquisition of an inflated dorsal valve and pitted ornament. The earliest family (Tenellodermidae) lacks a ctenophoridium and this suggests close relation to the ambocoeliids. Later, the members of the superfamily developed a ctenophoridium and crural plates. The martinioids were scarce during the Devonian, but their number and diversity increased in the late Paleozoic.

The Delthyridina (Delthyridoidea and Reticularioidea) may have split from the Cyrtioidea during the late Llandovery, but their origin is not clear. Their major evolutionary innovation is fimbriate shell fabric, as illustrated in the oldest known genus, *Howellella*. The latter is already specialized (ctenophoridium, costulation, fimbriate shell fabric, spines), and thus somewhat different from eospiriferids. The ancestor of *Howellella* is unknown. Many delthyridine genera developed a spinose ornament, but some Early Devonian Acrospiriferidae and Cyrtiopsidae (Delthyridoidea) show remnant capillate ornament. The Reticularioidea are characterized by the weakening of fold, sulcus and lateral plications and generally a short hinge line; they derived from the Delthyridoidea in the Early Silurian, perhaps through *Eohowellella*, a minute shelled reticulariid that externally resembles *Howellella*. Both the Delthyridoidea and Reticularioidea flourished during the Devonian, but the former did not survive the Early Carboniferous (Figure 3).

SPIRIFERINIDA

The main feature in the order Spiriferinida is the presence of a punctate shell. In addition, members of the order possessed a well developed ventral area, and the jugum is commonly present. The Spiriferinida comprise two suborders, Cyrtinidina and Spiriferinidina. The earliest cyrtinidine is the genus *Cyrtina* (Cyrtinidina), appearing in the Early Devonian; this genus is highly specialized with its punctate shell and complex internal structure (ventral septum, tichorhinum, dorsal notothyrial platform). No immediate ancestor is known, but some impunctate Silurian Delthyridoidea, e.g. *Kozlowskiellina* or *Cyrtinopsis*, by their overall shape, dental plates and internal septum, remind of *Cyrtina* and both *Cyrtinopsis* and *Kozlowskiellina* possess a high, ventral interarea and conjunct deltidial plates (pierced by a foramen in *Kozlowskiellina*). They also possess a ventral median septum (forming a spondylium in *Cyrtinopsis*). In the latter, the ctenophoridium formed on a raised pad of shell material (Boucot, 1962). Thus *Cyrtina* could have split off from the *Kozlowskiellina-Cyrtinopsis* lineage by the acquisition of a tichorhinum and endopunctae, however, these morphological similarities could be accounted for by convergent or parallel evolution. *Cyrtina* has a cardinal process with few parallel striations located on an elevated platform (Gourvennec, 1989), as in some ambocoeliids (e.g. *Plicoplasia*). Such a structure and the overall form could be an argument for deriving *Cyrtina* from the ambocoeliids, but the ambocoeliids lack well developed dental plates. One cannot exclude separate origins from a punctate orthid (Carter & al., 1994).

The Cyrtinoidea represented the suborder during the Devonian and only one genus crossed the Devono-Carboniferous boundary. They are related to the Suessioidea, represented by the Early Carboniferous Davidsoninidae, but no representatives of the superfamily are recorded in the Late Carboniferous and Early Permian, so that the link between Davidsoninidae and Laballidae is dubious. The Triassic Spondylospiroidea are characterized by a crenulate hinge, differentiated from the denticulate hinge line of the Spiriferoidea because the teeth lack taleolae (? synapomorphy). This group includes widehinged ('spiriferoid') genera with a spondylium (Rastelligeridae) and hemipyramidal ('cyrtinoid') genera with converging dental adminicula (Spondylospiridae). The latter are considered to be more 'primitive' and derived from the laballids. Rastelligeridae seem to have been derived from Spondylospiridae by neotenous loss of the spondylium.

The suborder Spiriferinidina includes the Syringothyridoidea, Pennospiriferinoidea and Spiriferinoidea (Figure 3). The Syringothyridoidea are derived from the Cyrtinoidea (komiellids) and are characterized by either a delthyrial plate or a syrinx in the ventral valve, but in some genera with a low interarea, the delthyrial plate is lost. This superfamily did not survive the Permian mass extinction. The Pennospiriferinoidea appeared approximately at the same period; they possessed a moderately high ventral area, ventral adminicula and median septum. The Spiriferinoidea separated from the Pennospiri-ferinoidea during the Middle Triassic, and judging by micro-ornament and discrete dental adminicula, their ancestor probably was a paraspiriferinid (Figure 3).

Acknowledgments
We thank reviewers for comments on this paper, but are responsible for our own split opinions (Copper for atrypids and athyrids, Gourvennec for spiriferids). Gourvennec supports a polyphyletic origin for the spire-bearers (Option Three), with the spiriferids independently derived from the orthids [and split into two orders, possibly twice], but the athyrids and atrypids related via the rhynchonellids. Copper suggests a monophyletic, three stage, successive development (Figure 1) from Llandeilo through Ashgill time, via the rhynchonellids. New morphological discoveries in the oldest species from each of the three orders will probably dictate the ultimate phylogenetic scenario. Financial support for Copper comes from the Natural Sciences and Engineering Research Council of Canada, and for Gourvennec from the CNRS, France.

REFERENCES

BAKER, P.G. 1990. The classification, origin and phylogeny of the thecideidine brachiopods. Palaeontology 33(1): 175-191.
BOUCOT, A.J., J.G. JOHNSON, C.W. PITRAT & R.D. STATON. 1965, p. 632-728. Spiriferida. In, R.C. Moore (ed.), Treatise on invertebrate paleontology, H, Brachiopoda. Geological Society of America and University of Kansas Press, Lawrence.
___& WILSON, R.A. 1994. Origin and early radiation of terebratuloid brachiopods: thoughts provoked by *Prorensselaeria* and *Nanothyris*. Journal Paleontology 68(5): 1002-1025.
CARLSON, S. 1991. Phylogenetic relationships among brachiopod higher taxa. In, D.I. MacKinnon, D. Lee & J.D. Campbell (eds.), Brachiopods through time, Balkema Press, Rotterdam.
CARTER, J., J.G. JOHNSON, R. GOURVENNEC & HONGFEI HOU. 1994. A revised classification of the spiriferid brachiopods. Annals Carnegie Museum 63(4): 327-374.
COPPER, P. 1977. *Zygospira* and some related Ordovician and Silurian atrypoid brachiopods. Palaeontology 20:295-335.
___1986a. Filter-feeding and evolution in early spire-bearing brachiopods. Biostratigraphie Paléozoique 4:219-230.
___1986b. Evolution of the earliest smooth spire-bearing atrypoids (Brachiopoda: Lissatrypidae, Ordovician-Silurian). Palaeontology 29:827-866.
___1995. Five new genera of Late Ordovician-Early Silurian

brachiopods from Anticosti island, Eastern Canada. Journal Paleontology 69(5): 846-862.

DAVIDSON, T. 1882. Supplement to the British Devonian Brachiopoda. Palaeontographical Society Monographs 5(1):1-134.

ELLIOTT, G.F. 1965. Order uncertain- Thecideidina, p.857-862. In, R.C. Moore (ed.). Treatise on Invertebrate Paleontology H, Brachiopoda. Geological Society of America and University of Kansas Press, Lawrence.

FU, LIPU, 1982. Brachiopoda, p.95-179. In, Paleontological atlas of Northwest China (Shaanxi-Ningxia-Gansu). Geological Publishing House, Beijing.

GOURVENNEC, R. 1989. Brachiopodes Spiriferida du Dévonien inférieur du Massif armoricain. Biostratigraphie Paléozoique 9:1-281.

___ & M. MELOU 1990. Découverte d'un cas d'ornamentation épineuse chez les Orthida (Brachiopoda). Comptes Rendus Académie Sciences, Paris 311(2):1273-1277.

GRUNT, T. 1982. Mikrostruktura rakoviny Brakhiopod Otryada Athyrida. Paleontologicheskii Zhurnal 1982(4): 21-35.

HAVLICEK, V. 1959. Spiriferidae v ceskem Siluru a Devonu (Brachiopoda). Rozpravi Ústredniho Ústavi Geologíckeho 25: 1-275.

HOVERD, W.A. 1985. Histological and ultrastructural observations on the lophophore of the larvae of the brachiopod Notosaria nigricans (Sowerby, 1846). Journal Natural History 19(5): 831-850.

___1986. The adult lophophore of Notosaria nigricans (Brachiopoda). Biostratigraphie Paléozoique 4: 307-312.

JAANUSSON, V. 1971. Evolution of the brachiopod hinge. Smithsonian Contributions Paleobiology 3:33-46.

JOHNSON, J.G. & A.C. LENZ. Eoplicoplasia, a new genus of Silurian-Lower Devonian ambocoeliid. Journal Paleontology 66(3): 763-776.

ROGER, J. 1952. Classe des brachiopodes, p.3-160. In, J. Piveteau (ed.). Traité de paléontologie, 2. Masson & Cie.

RONG, JIAYU, RENBIN ZHAN, & NAIREN HAN, 1994. The oldest known Eospirifer (Brachiopoda) in the Changwu Formation (Late Ordovician) of western Zhejiang, east China, with a review of the earliest spiriferoids. Journal Paleontology 68(4): 763-776.

RUDWICK, M.J.S. 1960. The feeding mechanisms of spire-bearing fossil brachiopods. Geological Magazine 97(5): 369-383.

___1970. Living and fossil brachiopods. Hutchinson & Co., London, 199p.

RZHONSNITSKAYA, M.A. 1960. Otryad Atrypida. In, T.G. Sarycheva (ed.), Osnovy paleontologii, 257-264.

SARYCHEVA, T.G. (ed.) 1960. Osnovy paleontologii, Mshanki, brakhiopody. Izdatelstvo Akademii Nauk, Moskva, 342 p.

SHEEHAN, P.M. & BAILLIE, P.W. 1981. A new species of Eospirifer from Tasmania. Journal Paleontology 55(1):248-256.

WAAGEN, W.H. 1883. Salt range fossils, 4, Brachiopoda. Palaeontologica Indica 1(2): 391-456.

WILLIAMS, A. 1973. The secretion and structural evolution of the shell of thecideidine brachiopods. Philosophical Transactions Royal Society, London B(264): 439-478.

___ & J.M. HURST. 1977. Brachiopod evolution, p.79-121. In, Hallam, A. (ed.). Patterns of evolution as illustrated by the fossil record. Elsevier, Amsterdam.

___& WRIGHT, A.D. 1961. The origin of the loop in articulate brachiopods. Palaeontology 4(2): 149-176.

WRIGHT. A.D. 1979a. The origin of the spiriferidine brachiopods. Lethaia 12: 29-33.

___1979b. Brachiopod radiation, p.235-252. In, M.R. House, (ed.), The origin of major invertebrate groups. Academic Press, London.

ON THE CLASSIFICATION OF THE ORDER ATHYRIDIDA

ALGIRDAS DAGYS

Institute of Ecology, Lithuanian Academy of Sciences, LT-2600 Vilnius, Lithuania

ABSTRACT- Some problems in the classification of the order Athyridida are reviewed, and a new suborder, the Koninckinidina, and new subfamily, the Misoliinae, are added.

INTRODUCTION

Grunt (1986) proposed a revised classification for the Order Athyridida, which was essentially different from previous systematic schemes (Boucot et al., 1964, 1965; Dagys, 1974). I agree with Grunt's basic concept on the evolution and classification of the Athyridida, but discues here some questions, and provide comments on family, and higher taxa.

SUBORDER RETZIIDINA

Boucot et al. (1964, 1965) united two superfamilies in the suborder Retziidina, the costate or plicate, punctate Retzioidea and the impunctate Athyrisinoidea. On the basis of shell structure and cardinalia (spondylium* [sic], or separate hinge plates), Dagys (1974) assigned the family Athyrisinidae to the athyroid superfamily Meristelloidea. Grunt (1986) revived the concept of Boucot et al. (1965) by attributing the Athyrisinoidea to the Retziidina. In Grunt's classification, the Athyrisinoidea contain two ribbed families with essentially different internal structures, the Athyrisinidae, which include genera with a distinct cardinal plate perforated by a 'visceral foramen', and contemporaneous Didymathyrididae and Athyrididae with similar cardinalia.

It is problematic to attribute to the Athyrisinidae such Triassic ribbed genera as *Misolia*, *Stolzenburgella*, and *Anomactinella*, which are separated from the true athyrisinids by a wide stratigraphic gap (Middle Devonian - Middle Triassic). Internal structures of the Triassic genera are poorly known, although at least one genus, *Misolia*, has a cardinal plate lacking a 'visceral foramen*' and distinct cardinal process, as typical of the Mesozoic athyrids [*N.B. the 'visceral foramen' in Russian is the dorsal foramen in English: editor's note]. The family Metathyrisinidae includes ribbed impunctate forms with two different types of cardinalia, one with a distinct septum and septalium (e.g. *Metathyrisina*, *Retziella*), and the other with discrete hinge plates and lacking a prominent septum (*Ornothyrella*, and possibly *Plectothyrella* : see addendum).

The superfamily Athyrisinoidea is a rather artificial taxon, including homeomorphous ribbed genera assignable to different families and superfamilies of the suborder Athyrididina by other criteria. The Athyrisinidae most probably is a member of the superfamily Athyridioidea. Certain Mesozoic genera provisionally assigned to the Athyrisinidae belong to the Diplospirellidae. These genera may be grouped into a new subfamily here proposed as the Misoliinae, characterized by a ribbed diplospirellid shell with short jugal accessory lamellae. Members of the present Metathyrisinidae are related to different groups of the superfamily Meristelloidea. For example, *Ornothyrella*, has cardinalia and brachidia charac-teristic of the Hindellinae, and some genera have a distinct septalium inseparable from the Meristinae [*some assign this genus to the Rhynchonellida: editors].

SUBORDER ATHYRIDIDINA

The majority of the genera in this suborder belong to the relatively conservative superfamily Athyridioidea. Grunt's (1986) classification of this superfamily is based on several evolutionary events postulated for the group. The Athyrididae is characterized by the development of a jugal saddle and lamellose ornament, the Spirigerellidae by the appearance of a massive cardinal plate and cardinal process, and tertiary (prismatic) shell layer in some taxa, and the Diplospirellidae by reduction of the 'visceral foramen' and complex brachidia . On the basis of the internal structures, the Athyrisinidae should be attributed to the superfamily Athyridioidea.

Some nomenclatural problems exist in the family Camelicaniidae, initially established as a subfamily by Merla (1930), The name has not been used for more than fifty years, except by by Waterhouse & Gupta (1986). The Spirigerellidae Grunt, 1986 is effectively a junior synonym of the Camelicaniidae, but in order to avoid nomenclatural changes in the Russian literature, I consider the Camelicaniidae a *nomen oblitum*.

Grunt's classification of the superfamily Meristelloidea is correct, although the Metathyrisidae should also be assigned to this superfamily because of its meristelloid internal structures. The taxonomic rank of nucleospirids is not certain, although Grunt interpreted this group as a monotypic superfamily Nucleospiroidea. Both the external and the internal structures of the nucleospirids are largely athyroid, and their distinct characters consisting only of spinous microornament and simple jugum with a short stem, and without a saddle or bifurcation.

The systematic position of the koninckinids and cadomellids is one the most complicated problems in athyroid taxonomy. In previous classifications, koninckinids were usually assigned to various groups of spear-bearers, and cadomellids to the strophomenids. Schuchert & LeVene (1929) placed the Koninckinidae in the Rostrospiracea. Boucot et al. (1964) interpreted the koninckinids as an independent superfamily within the Athyrididina. Dagys (1974) assigned the Koninck-inoidea to the Order Strophomenida on the basis of their concavoconvex, strophic shells, interareas in both valves, a supra-apical foramen, spiralial orientation, and supposedly pseudopunctate shell structure. Brunton & MacKinnon (1972) demonstrated that koninckinid shells are formed by an impunctate secondary layer, and that the pseudopunctae in some koninckoids, e.g., *Koninckodonta* and *Amphiclinodonta*, are comparable to the tubercles of Recent Terebratulida. These authors also showed that koninckinids lack a strophomenoid supra-apical foramen and assigned the koninckinids and cadomellids to the Order Spiriferida sensu lato.

Taking into account the major differences in shell microstructure between the koninckinids and the strophomenids and the evolutionary conservatism of this feature, the koninckinids are best assigned to the spire-bearers, and particularly to the athyrids. The Koninckinoidea, however, differs from the Athyrididina and the Retziidina in its concavo-convex shell, strophic growth, presence of true interareas in both valves, orientation of the spiralia, and primary lamellae. Some members of the Athyrididina (e.g. *Camelicania*, *Majkopella*) may have a straight hinge line, but no interareas. The distinct morphologic features of the koninckinids warrant their status as an independent suborder within the Athyridida. This new suborder may include the Liassic cadomellids, initially grouped as a superfamily Cadomelloidea by Muir-Wood (1955). Cowen & Rudwick (1966) reported the presence of spiral brachidia in *Cadomella* and thus considered the Cadomelloidea a junior synonym of the Koninckinoidea. The observation of Cowen & Rudwick was based on *Cadomella davidsoni* (Deslongchamps), which is assigned to *Koninckella* in the family Koninckinidae. The type species of *Cadomella*, *Leptaena moorei* Davidson, has a planoconvex, strophic shell with distinct interareas, and differs from the koninckinids in its fine capillae and absence of spiralia. For this reason, the Cadomelloidea are retained as an independent superfamily.

CLASSIFICATION OF THE ATHYRIDIDA

Order Athyridida
Suborder Retziidina Boucot, Johnson & Staton, 1964

Diagnosis. Astrophic, biconvex, ribbed or plicate, ventral palintrope well developed in later taxa; dental plates present only in some early forms; cardinal plate with or without cardinal process, usually supported by septum; spiralia directed laterally; jugum present; accessory lamellae variously developed (may be secondarily connected with spiralia); shell punctate, with simple or branching punctae; tertiary layer present.

Suborder Retziidina Waagen, 1883
Superfamily Retzioidea Waagen, 1883.
[Families Retziidae Waagen, 1883;
Neoretziidae Dagys, 1972]

Suborder Koninckinidina new suborder

Diagnosis. Smooth or plicate, strophic, concavoconvex, with well-developed interareas in both valves; dental plates absent; spiralia directed ventrally; jugum with accessory lamellae continuing between primary volutions of spiralia to their ends; shell impunctate; tertiary layer invariably present.

Superfamily Koninckinoidea Davidson, 1853.
[shell surface smooth]
  Family Koninckinidae Davidson, 1853.
Superfamily Cadomelloidea Schuchert, 1893
[shell surface capillate]
  Family Cadomellidae Schuchert, 1893.

Suborder Athyrididina Boucot, Johnson & Staton, 1964

Diagnosis. Astrophic, biconvex; ventral palintrope usually obsolete; generally smooth, with or without concentric lamellae, rather pauciplicate or ribbe; dental plates commonly absent; cardinalia variable; spiralia directed laterally; jugal accessory lamellae variously developed or absent; shell impunctate; tertiary layer present in some taxa.

Superfamily Athyridoidea M'Coy, 1844
[cardinal plate with or without cardinal process, commonly perforated by visceral foramen; septum absent]
  Families: Didymathyrididae Modzalevskaya, 1977; Athyrididae M'Coy, 1844; Spirigerellidae Grunt, 1965; Diplospirellidae Schuchert, 1894; Athyrisinidae Grabau, 1931.

Superfamily Meristelloidea Waagen, 1883
[hinge plates commonly united with septum forming septalium]
  Families: Meristellidae Waagen, 1883; Meristidae Hall & Clarke, 1895; Metathyrisidae Rong & Yang, 1981*[N.B. Rong et al., 1994, assign the Retziellidae (with taxa from the junior synonym Metathyrisinidae), to the Meristelloidea: *editors].

Superfamily Nucleospiroidea Davidson, 1881
[cardinal plate not perforated; shell surface with numerous spinules]
  Family Nucleospiridae Davidson, 1881

REFERENCES

BOUCOT, A.J., J.G. JOHNSON & R.D. STATON. 1964. On some atrypoid, retzioid, and athyridoid Brachiopoda. Journal Paleontology 38:805-822.
___,___&___1965. Suborders Retziidina and Athyridina, p.649-677. Treatise on Invertebrate Paleontology, H, Brachiopoda. Geological Society of America and University of Kansas Press, Lawrence.
BRUNTON, C.H.C. & D.L. McKINNON. 1972. The systematic position of the Jurassic brachiopod *Cadomella*. Palaeontology 15:405-411.
COWEN, R. & M.J.S. RUDWICK.1966. A spiral brachidium in the Jurassic chonetoid brachiopod *Cadomella*. Geological Magazine 103:403-406.
DAGYS, A.S. 1974. Triassic brachiopods, morphology, systematics, phylogeny, stratigraphic significance, and biogeography. Nauka, Novosibirsk, 387p.
GRUNT, T.A. 1986. Systema brakhiopod otryada atiridida. Trudy Paleontologicheskogo Instituta 215:1-200.
MERLA G. 1930. La fauna del calcare a Bellerophon della regione dolomitica. Memoirs Institute Geological Research, University Padova 9:1- 221.
MUIR-WOOD, H.M. 1955. A history of the classification of the phylum Brachiopoda. British Museum (Natural History), London, 124p.
RONG, JIAYU, D.L.STRUSZ, A.J. BOUCOT, LIPU FU, T.L. MODZALEVSKYA & YANGZHENG SU. 1994. The Retziellidae (Silurian ribbed impunctate athyrididine brachiopods). Acta Palaeontologica Sinica 33(5): 545-574.
SCHUCHERT, C. & C.M. LEVENE. 1929. Brachiopoda (Generum et genetyporum index et bibliographia). Fossilium Catalogus, 1, Animalia, Pars 42. Junk, Berlin, 140 p.
WATERHOUSE, J.B. & V.J. GUPTA. 1986. A significant new Permian athyrid brachiopod from uppermost Gungri Formation (Kuling or *Productus* Shale) at Spiti and its implications for classification and correlation. Bulletin Indian Geological Association19:45-56.

[*Addendum by editors: for another view, consult Modzalevskaya on the Athyridida herein, and see Rong et al., 1994. Some have include the genera *Ornothyrella* and *Plectothyrella* in the Rhynchonellida. The reviewers are of the opinion that the name Camelicaniidae remains valid, and the objective senior synonym of the Spirigerellidae]

REMARKS ON THE CLASSIFICATION OF PUNCTATE SPIRIFERIDS

ALGIRDAS DAGYS

Institute of Ecology, Lithuanian Academy of Sciences, LT-2600 Vilnius, Lithuania

ABSTRACT- The classification of punctate spiriferid brachiopods is re-examined and some systematic emendations for Mesozoic Spiriferinida are proposed. Families and higher taxa of punctate spiriferids are diagnosed, and revised, and a number of genera re-assigned. A new subfamily, the Dinarispirinae, is outlined.

INTRODUCTION

In their revision for the brachiopod Treatise on Invertebrate Paleontology, Carter et al. (1994) proposed a new classification for the spiriferid brachiopods, and raised an entirely new order, the punctate Spiriferinida, distinct from the 'impunctate Spiriferida'. In their classification, orders were based on shell structure such as punctation, suborders and superfamilies on growth form, families and superfamilies on internal structures, and different subfamily-level taxa on micro-ornament. This scheme possibly reflects the partial phylogeny of Paleozoic impunctate spiriferids, but appears questionable for punctate spiriferids, especially those of Mesozoic age.

TAXONOMIC STATUS OF PUNCTATE SPIRIFERIDS

Ivanova (1972) first grouped all punctate spiriferids into the suborder Spiriferidina, including both Paleozoic and Mesozoic forms. Cooper & Grant (1976) earlier proposed to include the retziids (now generally considered to be athyrids), and punctate spiriferids, in a new order, the Spiriferinida, which was very similar to the generally unaccepted Punctospiracea Cooper, 1944. Carter et al. (1994) rasied the suborder Spiriferinida of Ivanova (1972) to order status to include only the punctate spiriferids, but such taxonomic differentiation of spiriferids with different, punctate shell structure at the order level does not seem justified. Punctate taxa are known from several brachiopod orders (Strophomenida, Rhynchonellida, Athyridida, and even some Atrypida have a kind of punctate shell), but have never been grouped as independent orders. At an equivalent level, it would forecast the splitting of four existing orders, and creation of four additional orders, on the sole basis of punctation, thus dramatically increasing the number of existing artculate brachiopod orders. Punctation is regarded as a diagnostic feature at the familial or generic level in the Thecospiroidea (*Thecospira*), at the superfamilial level in the Rhynchonellida (Rhynchoporoidea), or at the suborder or superfamily level in the Retziidina. Punctate spiriferids have all the features characteristic of the Spiriferida and should have a taxonomic rank no higher than suborder within the order Spiriferida.

MAIN GROUPS OF PUNCTATE SPIRIFERIDS

Carter et al. (1994) recognized two suborders (Cyrtinidina and Spiriferinidina) in the punctate Spiriferinida, and included three superfamilies (Cyrtinoidea, Suessioidea, and Spondylospiroidea) in the Cyrtinidina. The relationship between the Devonian-Early Carboniferous Cyrtinoidea and the Middle-Late Triassic Suessioidea and Spondylospiroidea is at best defined as problematic.

The Suessoidea is a rather mixed superfamily including apparently unrelated taxa. The Davidsoninidae, for example, is obviously not spiriferinid and probably not spiriferoid because of its specific punctae (Ivanova, 1971), false spondylium, and lack of spiralia. The Suessiidae is based on the obscure genus *Suessia* Deslongshamps, and should be regarded as a group 'incertae sedis' to avoid taxonomic problems, until it becomes better known. The Laballidae and Bittnerulidae (*sensu* Carter et al., 1994) unite genera with a cyrtiniform shell but different types of ribbing, micro-ornament, and internal structure. Some genera, for example, have a deep spondylium formed by a median septum protruding into the spondylial cavity, and long dental adminicula (Laballinae, Spinolepismatinae, Bittnerulinae), whereas other genera have very short dental adminicula and a shallow spondylium formed by fusion of a long septum that usually does not protrude into the spondylial cavity (Paralepismatininae, Hirsutellinae).

The superfamily Spondylospiroidea is diagnosed by a supposedly synapomorphous character (crenulated hingeline), and includes taxa with variable types of growth forms, internal structure, ribbing, and micro-ornament, i.e. taxa with a cyrtiniform (Spondylospiridae), spiriferiform, or nearly equibiconvex shell (Rastelligerinae), with various kinds of spondylia (Spondylopirinae, Rastelligerinae), or discrete dental adminicula (Dentospiriferinae). For the Mesozoic cyrtiniform Spiriferidina, a truly unique (synapomorphous) character is the special net supporting the jugum. In Carter et al (1994), however, genera with this remarkable feature are scattered in different groups: *Spinolepismatina* in the Laballidae (Suessioidea), *Thecocyrtella* in the Bittnerulidae (Suessioidea), *Zugmayerella* and *Spondylospira* in the Spondylospiridae (Spondylospiroidea). The Triassic cyrtiniform genera with a distinct spondylium, specialized jugal net, and variable hinge-line denticulation should be united instead in one superfamily, the Bittneruloidea, including the families Bittnerulidae Schuchert,1929, Spinolepismatinidae Carter,1994, Spondylospiridae Hoover, 1991, and probably Laballidae Dagys, 1962.

In previous spiriferid classifications (Ivanova, 1960,1972; Pitrat, 1965), Paleozoic and Mesozoic cyrtiniform genera were regarded as belonging to the same phyletic line but separated by a large gap from the Early Carboniferous to Middle Jurassic. Specific morphology of Paleozoic cyrtinids, however, poses problems for the direct relationship between the Cyrtinoidea and Bittneruloidea.

In the suborder Spiriferinidina (sensu Carter et al., 1994), the taxonomic position of the superfamily Syringothyridoidea is questionable. In shell outline, ribbing, micro-ornament, and ventral interior (delthyrial plate and absence of a septum in most taxa), the Syringothyridoidea are spiriferoid. Moreover, the shell structure is impunctate even in late forms of this group (Licharewiidae). On the basis of growth form, true spiriferinids are divided into two superfamilies, the Pennospiriferinoidea (spiriferiform to cyrtiniform) and Spiriferinoidea (reticulariiform). The former includes mainly Paleozoic taxa and is more or less homogenous, but the classification of Mesozoic taxa needs correction. I prefer to restore the original interpretation of the family Pennospiriferinidae to include subequally biconvex, alate, or mucronate forms lacking microornament, with a sessile spondylium and a septum protruding into the spondylial cavity; the hingeline varies from straight (*Pennospiriferina*) to completely denticulate (*Rastelligera*). This group existed only in high paleolatitudes during the Late Triassic.Taxa with reduced dental adminicula (Balatonospiridae) are related to the Mentzeliidae (with weakened dental adminicula), and are excluded from the Pennospiriferinoidea.

The superfamily Spiriferinoidea (*sensu* Carter et al., 1994) unites two different groups of punctate spiriferids: one having a well-developed septum and discrete dental adminicula (Spiriferinidae, Pseudolaballidae, and perhaps Dentospiriferininae), and the other having short dental adminicula united with distal end of the septum to form a shallow spondylium, with the septum not continuing into the spondylial cavity (Mentzeliinae, Tethyspirinae). Several genera with similar internal structures are placed in the Suessioidea (Hirsutellinae, Paralepismatininae). In the classification of Dagys (1974), such taxa were grouped into the subfamily Mentzeliinae, which was later raised to the rank of a superfamily by Sun & Ye (1982). Taxa of the Mentzelioidea belong to a monophyletic group, having a common morphology, limited geographic and chronologic distribution (Tethys, Middle-Late Triassic). In Carter et al. (1994), however, genera of the Mentzelioidea were scattered into three superfamilies (Suessioidea, Pennospiriferinoidea, and Spiriferinoidea), belonging to two suborders.

CLASSIFICATION OF PUNCTATE SPIRIFERIDS

This revised classification of the Paleozoic Spiriferinidina essentially follows Carter et al. (1994), except for reinterpretation of higher taxa, and re-assignment of some genera: diagnoses of Paleozoic families are omitted.

Suborder Spiriferinidina Ivanova, 1972. Punctate.
Superfamily Cyrtinoidea Fredericks, 1911.
Family Cyrtinidae Fredericks, 1911.
? Family Komiellidae Johnson & Blodgett, 1993.
Superfamily Bittneruloidea Schuchert, 1929 (nom. transl. Dagys herein, ex Bittnerulinae Schuchert, 1929, p.21). Cyrtiniform, with distinct spondylium formed by fusion of high septum with long dental adminicula; jugum supported by jugal net. Middle-Upper Jurassic.
Family Bittnerulidae Schuchert, 1929 (nom transl. Carter, 1994, p. 362, *ex* Bittnerulinae Schuchert, 1929, p. 21). (= Thecocyrtellinae Dagys, 1965, p.105; Thecocyrtelloideidae Xu & Liu, 1983, p.821). Fold, sulcus and ribbing of lateral slopes delthyrium closed by deltidium. Middle-Upper Triassic. Genera: *Thecocyrtella* Bittner, 1892 (= *Bittnerulla* Hall & Clarke, 1895); *Klipsteinella* Dagys, 1974; *Thecocyrtelloidea* Yang & Liu, 1966; *Neocyrtina* Yang &Liu, 1966.
Family Spinolepismatinidae Carter, 1994 (nom transl. Dagys herein, ex Spinolepismatininae Carter, 1994, p.362). Fold and sulcus distinct, smooth, lateral slopes plicate; delthyrium open. Upper Triassic. Genera: *Spinolepismatina* Dagys, 1974; *Klipsteinelloidea* Sun, 1981 (possibly =*Spinolepismatina*); *Zugmayerella* Dagys, 1963; ?*Yanospira* Dagys, 1977; ?*Orientospira* Dagys, 1965.
Family Spondylospiridae Hoover, 1991. Entire surface ribbed. Upper Triassic.
Subfamily Spondylospirinae Hoover, 1991. Apex of spondylium not pierced by paired elongate foramina. Genera: *Spondylospira* Cooper, 1942; *Phenacozugmayerella* Hoover, 1991; *Vitimetula* Hoover, 1991.

Subfamily Dagyspiriferinae Hoover, 1991. Apex of spondylium and ventral interarea pierced by paired elongate foramina. Genera: *Dagyspirifer* Hoover, 1991; *Pseudospondylospira* Hoover, 1991.
?Family Laballidae Dagys, 1962 (nom transl. Dagys, 1965, p.91, ex Laballinae Dagys, 1962, p.49). Fold and sulcus variable, lateral slopes smooth; jugal net absent. Upper Jurassic. Genera: *Laballa* Moisseev, 1962; *Pseudolabala* Dagys, 1965; ?*Psioidea* Hector, 1872.
Superfamily Pennospiriferinoidea Dagys, 1972 (nom transl. Carter, 1994, p.336, *ex* Pennospiriferininae Dagys, 1972, p.36). Spiriferiform to cyrtiniform, lateral slopes ribbed; septum and dental adminicula discrete or forming sessile spondylium. Upper Devonian-Lower Jurassic.
Family Punctospiriferidae Waterhouse, 1975.
Family Spiropunctiferidae Carter, 1994.
Family Reticulariinidae Waterhouse, 1975.
Family Paraspiriferinidae Cooper & Grant, 1976.
Family Crenispiriferidae Cooper & Grant, 1976.
Family Sarganostegidae Dagys, 1974.
Family Punctospirellidae Dagys, 1974 (nom transl. herein, ex Punctospirellinae Dagys, 1974, p.135. = Xestotrematidae Cooper & Grant, 1976, = Yangkongidae Xu & Liu, 1983, p.79). Transverse, usually with extended lateral extremities; fold and sulcus smooth, lateral slopes ribbed; dental adminicula discrete; micro-ornament absent. Upper Carboniferous-Middle Triassic. Genera: *Punctospirella* Dagys, 1974 (= *Yangkongia* Xu & Liu, 1983); *Tulungospirifer* Ching & Sun, 1976; (for *Paleozoic* genera see Carter et al., 1994).
Family Lepismatinidae Xu & Liu, 1983. Usually cyrtiniform with distinct smooth fold and sulcus and ribbed lateral lobes. Micro-ornament variable. Middle-Upper Triassic.
Subfamily Lepismatininae Xu & Liu, 1983 (nom transl. Carter, 1994, p.369, ex Lepismatinidae Xu & Liu, 1983, p.82). Micro-ornament consisting of dense spinules. Middle-Upper Triassic. Genera: *Lepismatina* Wang, 1955 (= *Costispiriferina* Ching & Sun, 1974; *Altoplicatella* Xu & Liu, 1983; *Pseudolepismatina* Ching & Sun, 1976; ?*Pseudospiriferina* Yang & Xu, 1966 (= *Lancangjiangia* Jin & Fang, 1976).
Subfamily Pseudocyrtininae Carter, 1994. Micro-ornament absent. Upper Triassic. Genera: *Pseudocyrtina* Dagys, 1962; *Bolilaspirifer* Sun, 1981.
Family Pennospiriferinidae Dagys, 1972 (nom transl. Carter, 1994, p.368, *ex* Pennospiriferininae Dagys, 1972, p.6 = Rastelligerinae Carter, 1994, p.364). Transverse, nearly equibiconvex, usually alate or mucronate; ventral valve with sessile spondylium, with distal end of septum protruding into spondylial cavity; micro-ornament absent. Upper Triassic. Genera: *Pennospiriferina* Dagys, 1965; *Spondylospiriferina* Dagys, 1972; *Rastelligera* Hector, 1879 (= *Boreiospira* Dagys, 1974); *Psioidiella* Campbell, 1969.
?Family Sinucostidae Xu & Liu, 1983 (nom transl. Carter, 1994, p.371, *ex* Sinucostinae Xu & Liu, 1983, p.112). Spiriferiform to reticulariiform, entirely ribbed; dental adminicula discrete. Middle Triassic-Lower Jurassic.
Subfamily Sinucostinae Xu & Liu, 1983 (= Yalongiinae Carter, 1994, p.368, = Dispiriferininae Carter, 1994, p.369). Dorsal septum absent, Middle Triassic-Lower Jurassic. Genera: *Sinucosta* Dagys, 1963; *Guseriplia* Dagys, 1963; *Aequispiriferina* Yang & Yin, 1962 (= *Sinucostella* Xu & Liu, 1983); *Yalongia* Xu & Liu, 1983; *Dispiriferina* Sibilik, 1965.
Subfamily Jiangdaspiriferinae Carter, 1994. Dorsal median septum present. Upper Triassic. Genus:*Jiangdaspirifer* Chen, Rao, Zhou & Pan, 1968.
Superfamily Mentzelioidea Dagys, 1974 (nom transl. Sun & Ye, 1982, p. 162, *ex* Mentzeliinae Dagys, 1974, p. 138). Growth form, ornament, and micro-ornament variable; septum high; dental adminicula short, united with distal end of septum to form shallow spondylium. Middle-Upper Triassic.

Family Mentzeliidae Dagys, 1974 (nom transl. Sun & Ye, 1982, p.162, *ex.* Mentzeliinae Dagys, 1974, p.138. = Hirsutellinae Xu & Liu, 1983, p.82;= Paralepismatininae Carter, 1994, p.362;= Tethyspirinae Carter, 1994, p.370). Features of superfamily. Middle-Upper Triassic. Genera: *Mentzelia* Quenstedt, 1871 (= *Leiolepismatina* Yang & Xu, 1966; = *Paramentzelia* Xu, 1978; *Madoia* Sun & Ye, 1982); *Koeveskallina* Dagys, 1965; *Hirsutella* Cooper & Muir-Wood, 1951 (= *Paralepismatina* Yang & Xu, 1966); *Spiriferinoides* Tokuyama, 1957; *Tethyspira* Siblik, 1991; *Tylospiriferina* Xu, 1978; ? *Quispiriferina* Xu & Liu, 1983.

?Family Balatonospiridae Dagys, 1974 (nom transl. Carter, 1994, p.368, *ex* Balatonospirinae Dagys, 1974, p.137). Shell entirely ribbed; fold and sulcus poorly developed; dental adminicula reduced; septum distinct. Middle-Upper Triassic.

Subfamily Balatonospirinae Dagys, 1974. Micro-ornament absent. Middle-Upper Triassic. Genera: *Balatonospira* Dagys, 1974; ? *Nudispiriferina* Yang & Xu, 1966.

Subfamily Dinarispirinae Dagys, new subfamily. Micro-ornament consisting of short, coarse spines. Middle Triassic. Genera: *Dinarispira* Dagys, 1974; *Quingyenia* Yang & Xu, 1966.

Superfamily Spiriferinoidea Davidson, 1884 (nom transl. Ivanova, 1959, p.57, *ex* Spiriferininae Davidson, 1884, p.354). Reticuliiform or cyrtiniform, cardinal extremities well rounded; fold and sulcus usually weakly developed, with lateral slopes smooth or faintly folded; dental adminicula discrete. Upper Triassic-Lower Jurassic.

Family Spiriferinidae Davidson, 1884 (nom. transl. Ivanova, 1959, p.57, ex Spiriferininae Davidson, 1884). Reticularii-form, lateral slopes smooth or faintly ribbed. Upper Triassic-Lower Jurassic. Genera: *Spiriferina* d'Orbigny, 1847 (= *Leiospiriferina* Rouselle, 1977); *Calyptoria* Cooper, 1989 (= *Cinglospiriferina* Pozza, 1992); *Triadispira* Dagys, 1961; *Viligella* Dagys, 1965; *Mentzelioides* Dagys, 1974; *Mentzeliopsis* Trechman, 1918.

Family Paralaballidae Carter, 1994 (nom. transl. Dagys herein, *ex* Paralaballinae Carter, 1994, p.370). Cyrtiniform, fold and sulcus weakly developed, lateral slopes smooth. Upper Traissic. Genus: *Paralaballa* Sun, 1981.

Family Dentospiriferinidae Carter, 1994 (nom. transl. Dagys herein,ex Dentospiriferininae Carter, 1994, p.364). Reticulariiform to cyrtiniform; fold and sulcus distinct; lateral slopes smooth or weakly folded; hingeline variably denti-culate. Upper Triassic. Genera: *Dentospiriferina* Dagys, 1965; *Canadospira* Dagys,1972

Incertae superfamiliae
Family Suessiidae Waagen, 1883.

REFERENCES

CARTER, J.L., J.G. JOHNSON, R. GOURVENNEC & HOU HONGFEI. 1994. A revised classification of the spiriferid brachiopods. Annals Carnegie Museum 63 (4):327-374.

COOPER, G.A. 1944. Phylum Brachiopoda, p.277-365. In, H.W. Shimer & R.R. Shrock, Index fossils of North America. Wiley & Sons, New York.

___ & R.E. GRANT.1976. Permian brachiopods from West Texas. Smithsonian Contributions Paleobiology 24:2609-3159.

DAGYS, A.S. 1974. Triassic brachiopods, morphology, sys-tematics, phylogeny, stratigraphic significance, and biogeo-graphy. Nauka, Novosibirsk, 387p.

IVANOVA, E.A. 1960. Order Spiriferida, p.264-280. In, Osnovy Paleontologii. Bryozoa, Brachiopoda. Akademiya Nauk SSSR, Moscow.

___1971. An introduction to the study of the spiriferids. Trudy Paleontologicheskogo Instituta 126:1-104.

___1972. The main features of spiriferid evolution (Brachio-poda). Paleontologicheskii Zhurnal 3:28-42.

PITRAT, C.W. 1965. Spiriferidina, p.667-728. In, Treatise on Invertebrate Paleontology, H, Brachiopoda. Geological Society of America & University of Kansas Press, Lawrence.

SUN DONGLI & SONLING YE. 1982. Middle Triassic brachio-pods from the Tosu Lake area, central Qinghai. Acta Palaeontologica Sinica 21:153-173.

TAPHONOMY OF SOME CRETACEOUS AND RECENT BRACHIOPODS

DANIÈLE GASPARD

Université de Paris-Sud Département des Sciences de la Terre,
Bât. 504, 91405 Orsay Cedex, France

ABSTRACT- After death, brachiopod shells are incorporated as live, or in situ (autochthonous), or transported (allochthonous) thanatocoenoses. Shell beds and loose live and dead shells permit observation of partial or nearly complete brachiopod shell degradation through stages of maceration, boring, fragmentation, dissolution, degradation and replacement at the macroscopic and microscopic level.

INTRODUCTION

The term taphonomy was initially proposed by Efremov (1940) as 'the science of the laws of burial', and expanded later to include other concepts (Efremov, 1953, 1958). The philosophy and history of this subdiscipline were later outlined by Olson (1980), then Cadée (1991). In North America it is often called the '3-D science', the science of death, destruction and burial. The term is here used in its widest sense, encompassing physical, chemical, and biological modifications from life through burial. Others have related taphonomy to biostratinomy and fossil diagenesis (Weigelt, 1927). Taphonomy has been used to interpret paleoecology (Behrensmeyer & Kidwell, 1985), with the realization of important losses of information (Lawrence, 1968). Taphonomy includes the analysis of all data playing a role in the formation of shell beds, including orientation (Alexander, 1986), i.e. all processes leading from biosphere to lithosphere.

In this short paper, taphonomic studies which include quantification and mutivariate analyses will not be approached as a tool to discern life and death assemblages, nor to interpret the dynamics of fossil populations, their orientation, and origin (Boucot, 1953; Johnson, 1960; Noble & Logan, 1981; Gaspard & Mullon, 1983; Gaspard, 1988). I will try to pinpoint only the following aspects: infilling of shells by sediments after death, trace fossils of brachiopod pedicle attachment such as *Podichnus*; changes in the shell with biocorrosion and maceration of the organic matrix, internal and external cast making, including phosphatization, and calcitic or siliceous recrystallization, which involve shell modifications during stages of fossilization.

MATERIALS AND METHODS

Cretaceous shells from Europe and North Africa were observed macroscopically, with scanning electron microscopy (SEM) and joint SEM-Cathodoluminescence. Living shells were based on in situ populations from New Zealand and the Mediterranean (Figure 1:1) or dredged during cruises, e.g. the WALDA off the coast of Angola, BIOCAL around New Caledonia, SEAMOUNT 1 at the Madeira Abyssal Plain, BRAPROV and BRACOR in the Mediterranean, and analysed at the same levels.

IDENTIFICATION OF ASSEMBLAGES

In situ assemblages from the Cretaceous (Mid-Cenomanian) of Ciran-Varennes (Indre et Loire, France) are represented by fist-sized clusters of *Sellithyris* of more than 300 specimens representing all age groups. This appears to be a population that died catastrophically and was buried rapidly. Each shell is equivalent to a mini-geode: sections through the shells (Figure 1:8) show valves lined by drusy calcite. The shells buried so rapidly had their valves closed at death. Such rapid burial does not necessarily preclude sediment penetration of the foramen after the loss of the pedicle. Partial or complete infilling by sediment was observed in other Cretaceous shells (Gaspard, 1988, pl. 19, fig.10), and in Recent shells dredged off the Angola coast at 3431m, and from the Madeira Abyssal Plain (Figure 1:10).

Partial infilling by various generations of sediments or calcite has been observed in several Cretaceous specimens of sellithyrids from the Berriasian of Algeria, *Sellithyris* from the Middle Cenomanian of France (Gaspard, 1988, pl. 12, fig.9), and in terebratulids from the Lower Cretaceous of Spain. Several phases of fine grained sediment infilling, by current water movement, show that suspended sediment can settle inside a brachiopod shell when a weak current flows out the shell. One, or several generations of sparry calcite can then fill the remaining space. Such calcite growth appears to depend on water depth, temperature, pH, shell orientation and porosity, which can increase shell thickness and alter shell wall structure. In undersaturated sea water, dissolution can occur, partly stripping off the shell wall (Figure 4:12). This is similar to the phenomena observed by Müller (1979) in fossil echinoids. Processes involving dissolution were also reported by Walter & Morse (1985). Although low Mg-calcite is less soluble than high Mg-calcite, sparry calcite may change the chemical equilibrium inside the shell. Dissolution of biogenic carbonate is also more common in temperate shallow than tropical waters, as seen for bryozoans (Smith et al., 1992).

TRANSPORT, DISARTICULATION, FRAGMENTATION

Transport by currents or removal by storms alters the faunal composition of an assemblage, even in a population in which not all age groups are represented (Figure 1:2), as also reported by Flessa & Brown (1983). Depending on the nature and intensity of the event, fossil brachiopod shell deposits frequently show disarticulation of the two valves, fragmentation, abrasion, bioerosion, encrustation and dissolution. The term 'corrasion' has been introduced by Brett & Baird (1986) for some of these processes. Parsons & Brett (1991) used a series of taphonomic indices to suggest processes which affect shells, but my observations indicate a number of complex interactions for such processes. Transport, disarticulation and fragmentation depend on the composition of the population or assemblage, on the location of the assemblage, on predation, and on shell weakness. In some examples, dorsal valves may be better preserved (Noble & Logan, 1981); on the contrary, ventral valves are particularly well preserved as seen in *Gryphus vitreus* from the Mediterranean. Such variation may be explained by a thickened ventral valve, especially in the posterior two thirds of the valve where the prismatic layer is represented (Gaspard, 1989, pl.3, fig. 5,6). Another explanation may be breakage due to arthropod predators.

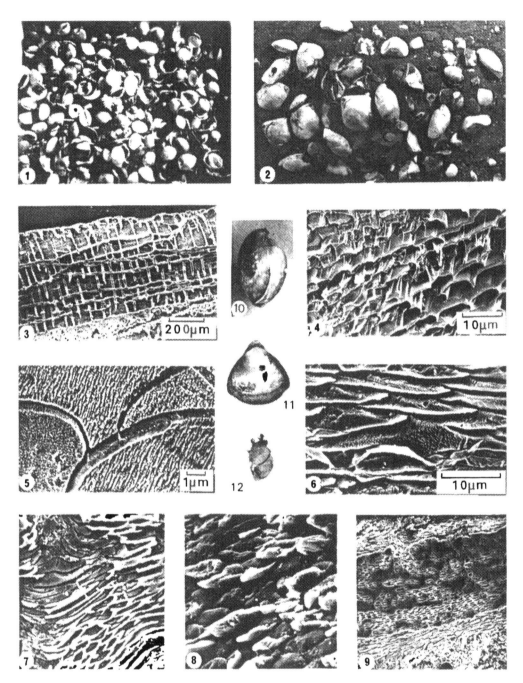

Figure 1. 1. Thanatocoenosis with *Gryphus vitreus* (Born). dredged off Marseille, Mediterranean. 2. Taphocoenosis: *Sellithyris sella* (Sowerby), Aptian, Isle of Wight. 3. Well preserved organic matrix in longitudinal section of *Sellithyris tornacensis* (d'Archiac), Tournai, Belgium. x100. 4. Well preserved organic sheaths in transverse section of *Cyclothyris scaldisensis* (d'Archiac) after demineralization during a few seconds with RDC; Lower Cenomanian, Tournai, Belgium, x 1250. 5. Partial disorganization of organic matrix of *C. scaldisensis*., x5350. 6. Soft remains of organic matrix in *Rhynchonella vespertilio* (d'Orbigny) (Senonian, Indre et Loire, France) exhibiting somewhat recrystallized fibres, x1875. 7. Partial siliceous epigenization of organic matrix in secondary layer of *Gemmarcula menardi* (Sowerby), middle Cenomanian, Sarthe, France, x900. 8. Drusy calcite at each fibre end, internal surface of *Sellithyris cenomanensis* (Gaspard), Middle Cenomanian, Sarthe, France, x500. 9. Invading siliceous nodule in calcitic fibrous layer of *Musculina sanctae-crucis* Catzigras, Hauterivian, Jura, France, x300. 10. Three quarters of shell, *Dallina septigera* Loven, (Seamount 1 cruise, Madeira abyssal plain) filled with fine grained sediment, x1. 11. Iridescence in *Dallina septigera*, indication of softening and maceration, same location, x1. 12. Shell of *Campages furcifera* Hedley (New Caledonia) encrusted by corals, x0.5.

Brachiopods appear to have fewer ways in which to become broken or fragmented, in comparison with molluscs. In the case of *Notosaria* and *Waltonia*, brachiopod shells from the intertidal zone of New Zealand, breakage by birds of mollusc shells as seen in the Wadden sea by Cadée (1995), has not been observed.

BRACHIOPOD TRACE FOSSILS

Whole or isolated brachiopod shells may provide a substrate for other brachiopods or biota. Brachiopod shells may act as sites for successive brachiopod larval settlements, after the free swimming stage ends (Figure 4:10). After death, the pedicle muscles decay, and clustered, circular shallow pedicle borings, i.e. *Podichnus*, become visible (Bromley & Surlyk, 1973). These have been described for some Cretaceous species (Nekvasilova, 1976; Gaspard, 1988). Density of *Podichnus* borings appears to be higher near commissures, as observed in deep-sea populations off Angola (Laurin & Gaspard, 1987). Alexander (1994) measured *Podichnus* around the antero-lateral commissure of Devonian *Leiorhynchus*, indicating a beak down position with the valve erect or inclined to the surface. Borings may provoke a blister-like reaction on the inside of the brachiopod shell as seen in *Macandrevia africana* and *Liothyrella neozelanica* (Figure 2:6; and Gaspard, 1996) as well in cyrtospriferids from China. These blisters indicate that the shells were alive when bored. Brachiopods are frequently encrusted by other biota, e.g. bryozoans, corals (Figure 1:12), and worms (Figure 4: 8). Gastropods are common boring predators of brachiopods (Kowalewski & Flessa, 1994), sometimes also producing blister-like reactions (Smith et al., 1992).

Information loss in the fossil record, due to post-mortem degradation, may be considerable (e.g. Aller, 1982; Archer et al., 1989). Nevertheless, fossilization may also provide certain information gains (Wilson, 1988), as the processes may help in the understanding of the benthic environment. The pteropod muds between Lifou and Ouvea (New Caledonia) provide sediments with fragments of various sized brachiopod shells of different species (Lambert & Roux, 1991; and Figure 4:1). These have been compared with Cretaceous (upper Cenomanian) shelly sands rich in the brachiopod *Phaseolina phaseolina* from the Indre-et-Loire region of France (Figure 4:7), also deposited in nearly tropical latitudes at the time. The ichnia of large borers can accelerate the process of shell fragmentation, as seen in both these examples.

PRESERVATION, DECAY OF ORGANIC MATRIX

When populations of Recent brachiopod shells are dredged, some valves have a glossy iridescence (Figure 1:11). In these, the organic sheaths around shell fibres have disappeared, leading to softening and maceration, and finally to crumbling (Alexandersson, 1978, 1979; and Figure 3). The loss of the organic matrix is due to bacteria, which use this as a food supply, but breakdown may also be the result of a change in pH in deep sea sediments (Archer et al., 1989; and Figure 2:1;. 3:2-4, 7,10). Loss of wall strength on Recent Brachiopods has been tested by Collins (1986), who illustrated crumbling and softening of *Terebratulina retusa* from Scotland under artificially induced fractures.

The organic matrix may also be preserved as soft tissue on fossil material (Gaspard, 1988, pls. 5, 28; 1990, pl.136). This appears to occur after rapid burial under anaerobic conditions. It is mostly the organic sheaths which are preserved (Figure 1:3), but sometimes also the complete caecal epithelium. A

study of Recent specimens (Figure 2:3-4; 4:3-4) and Cretaceous representatives (Figure1:4-6) provides an indication of progressive stages of decay of organic matrix which can be nearly reproduced under experimental conditions in the laboratory, using enzymes (proteinase; Figure 4:2). Partial crumbling may be preserved in fossil shells, but this is restricted to limited areas of the shell with differential hardness as a result of partial recrystallization (Figure 3:8).

OTHER DIAGENETIC ASPECTS

Decay of the organic matrix may have several causes, and the intervention of microborings and bacteria can have unequal effects. Microborings are sometimes present in living brachiopod shells, without having fatal consequences, e.g. as in Mediterranean *Gryphus vitreus* invaded by the borer *Ostreobium queketti* (Fredj & Fa. Iconetti, 1977; and Figure 1:1; 4:10). Such boring algae and fungi have also been noted in taxa such as *Eucalathis tuberata* from the Madeira Abyssal Plain (Gaspard, 1993), and *Frenulina* from the Coral sea (Gaspard, 1989). Complete invasion of microborers may make the shell very fragile and destroy it, e.g. in the case of dead shells of *Neothyris lenticularis* on the seafloor of Patterson Inlet, New Zealand (Figure 3:6). Similar examples are known from the fossil record, observed in *Cyclothyris* (see Figure 3:5; and Gaspard, 1992) and *Gemmarcula menardi* (Figure 2-7).

Micro-organisms may penetrate the shell in several ways: from the external surface, from inside, and via the caecal epithelium. These contribute to destruction or consolidation differently, but the end result is not necessarily the breakdown of the shell. The shell may be entirely destroyed or recrystallized, e.g. silica in the place of vacant space, alteration of the organic matrix or replacement by secondary calcite (Figure 1:7,9). The intervention of various boring organisms produces different results which are not easy to list in sequence. Micro-endolithic associations may be related to water depth, some being associated with the photic zone, like algae and cyanobacteria, while others, like fungi, may operate in shallow to deep environments (Gall et al., 1994).

Algae and fungi may enhance dissolution and diagenetic remobilization, i. e. they may be reponsible for recrystallization (Figures 1:6; 2:4, 7, 8). Sometimes they intervene during early diagenesis, as observed in Recent dead shells dredged with soft parts still enclosed between valves (Figures 2:3, 5; 3:9). In some cases, diagenesis is only partial (Figure 3:9), while in other examples in one and the same shell one can observe shell wall without change, showing even fine calcitic granules preserved (Figure 2:10), or showing attacked zones with blocky calcite (Figure 2:9).

Siliceous epigenetic alteration may be the result of bacterial alteration, which either hardened the organic matrix (Figure 1:7; and Gaspard, 1990, pl. 137), or progressively filled space left by the decay of organic sheaths (Figures 2:11; 3:1), or organic fibres to form siliceous nodules (Figures1:9; 2:11; 4:5). These contributed either to shell strengthening or weakening. Siliceous alteration may also occur in the presence of diatoms located in boring traces on the external surfaces of brachiopods, known from specimens of the New Zealand terebratulid *Waltonia* (Gaspard, 1992), which appears to have incorporated the unstable free silica from dead diatom frustules.

Pyritization in brachiopod shell walls can be effected by sulfobacteria. Framboidal pyrite is found in punctae (Figure 2:2), following the decay of the caecal epithelium (Gaspard &

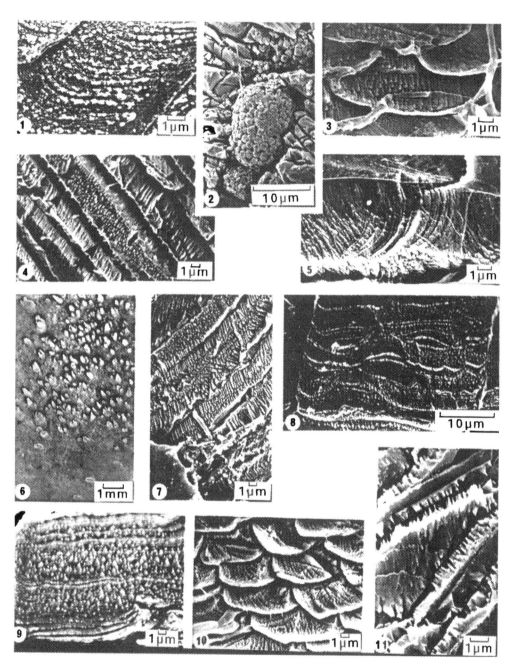

Figure 2. 1. *Aerothyris kerguelenensis*, Crozet Island, Indian Ocean, light dissolved fibres in longitudinal section showing calcitic granules and elementary growth lines, x5500. 2. Framboidal pyrite in punctae of sellithyrid, Berriasian, Algeria, x1850. 3. Section of *Liothyrella notorcadensis* Jackson, Palmer Peninsula, showing soft organic matrix and beginning of recrystallization (blocky calcite) in fibres of shell crossed by microboring organisms, x4500. 4. *Cyclothyris compressa* Lamarck, Cenomanian, Sarthe, France: longitudinal section showing organic sheaths disappearing around partly recrystallized fibres; in place of the organic matrix observe seeds of recrystallization, (arrow), x2500. 5. Microboring organisms inducing calcitic recrystallization in *Eucalathis tuberata* (Jeffreys), Seamount 1 cruise, x4600. 6. Internal surface of *Lyothyrella neozelanica*: Thomson (Rock walls, Doubtful Sound, New Zealand, 20-30 m) showing blisters in reaction to boring predators operating at external), x6. 7. Recrystallized secondary layer in *Gemmarcula menardi* (Lamarck), Jalais (Middle Cenomanian, Sarthe, France), x2400. 8. *Loriolithyris russillensis* (Loriol), Berriasian, Algeria: while growthlines finely identified, recrystallization in structural elements of shell, x1600. 9. *Stenosarina sphenoidea* (Jeffreys), Seamount 1, cruise Gor. CP 11 (805-850 m): observe comparable aspect between this Recent shell and that of Figure 2:8, x2675. 10. Shell (Gor CP 39, 2000 m) of the same species as Figure 2:9, aspect is completely different, with fine, well-preserved granules, x2675. 11. Section in *Magas pumilus*, Campanian, Meudon, showing interlacing siliceous overgrowths at peripheral part of fibres, x5050.

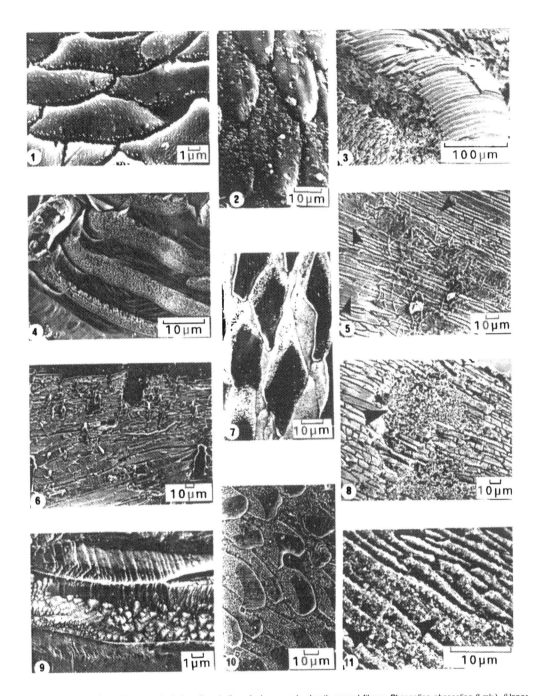

Figure 3. 1. Siliceous `seeds' progressively invading shell, replacing organic sheath around fibres, *Phaseolina phaseolina* (Lmk.), (Upper Cretaceous, Europe, x3500. 2. *Campages furcifera* Hedley, Biocal CP42-01 (380 m): microorganisms, including bacteria present at end of fibres, and on organic sheaths, x1000. 3. *Stenosarina sphenoidea*, Seamount 1 GorCP28 (605-680 m): after disorganisation of organic sheaths, fibres stripped, fragilized, maceration begining, x210. 4. *Dallina septigera* Loven, Gor CP 11 (805-830 m): decay of organic sheaths around fibres, x137. 5. *Cyclothyris sp.*, Cenomanian, SE France: microborers, invading recrystallization, beginning by consolidation of fibres (arrow), x420. 6. *Neothyris lenticularis* Deshayes, dead shells, sea floor (15-25 m), Patterson Inlet, New Zealand: shell invaded by microborers, primary layer lost, x270. 7. Young *Waltonia* sp., compare with Figure 3:4, x750. 8. *Cyclothyris scaldisensis* (d'Archiac): disintegration of fibres in secondary layer after loss of organic matrix, partial crumbling, x357. 9. *Neothyris lenticularis* Desh., dead shell, sea floor, 15-25 m, Patterson Inlet, New Zealand: longitudinal section with recrystallizatio; parts of fibres showing growth lines, others blocky calcite, x357O. 10. Internal surface, dallinacean, Galice bank DW 116 (Seamount 1 cruise), showing last episode of decay of organic sheaths around fibres, after Figure 3:7-8, x600. 11. *Sellithyris cenomanensis* (Gaspard), Middle Cenomanian, Sarthe, France: organic matrix destroyed partial silicification of secondary layer, coalescing some structures (arrow).

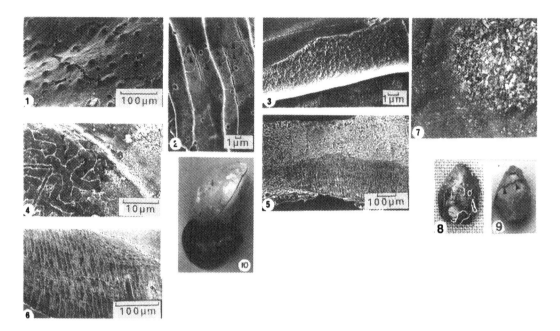

Figure 4. 1. Fragmentary brachiopod shell in pteropod muds, New Caledonia, showing borers, fragmentation of shell, x150. 2. Fragment of *Campages furcifera* Hedley exhibiting fibres, organic sheaths progressively attacked by proteinase for several hours (arrow), x2000. 3. Detail of *Campages*, New Caledonia: natural biocorrosion, x5000. 4. Recent shell of *Campages furcifera* Hedley, New Caledonia: fragment naturally attacked, with periostracum partly preserved (arrows); compare with Figure 4:2, x1500. 5. Longitudinal section, *Crania cenomanensis* d'Orbigny, Middle Cenomanian, Sarthe, France, MNHN Paris B16440, recrystallized secondary layer upper half, myotest lower half, x75. 6. Longitudinal section, *Crania (Isocrania) anomala* (Müller), Scotland (120 m depth): growth bands in myotest, upper valve, x175. 7. Shelly sand rich in fragments of *Phaseolina phaseolina* (Lamarck), Upper Cenomanian, St. Rémy-la-Varenne (France), x0.55. 8. Ventral valve *Gryphus vitreus* (Born), same location, encrusted by worms, predation, x1. 9. Partly dissolved biplicate terebratulid, Gault, Le Frételay, France, x0.5. 10. Adult *Grypheus vitreus*, Mediterranean, -200 m, on a black adult shell, invaded by algae; x0.7. [*Scales may not be exactly as indicated due to reductions]

Roux, 1974). Some of these bacteria have been recognized as 'paleosarcines' (Gaspard, 1988, pl. 10).

Phosphatization has been observed as casts of shells in Cretaceous sediments. Bacterial activity has resulted in phosphate formation, as explained by Lamboy (1990). In Gaspard (1988, pls. 25, 27) , black moulds of *Moutonithyris dutempleana* have been described from the British Cretaceous (Gault) near Cambridge, Gorges du Loup (SW France), and in casts of *Praelongithyris fecampi* from the Gault of Assigny (Cher, France).

CONCLUSIONS

Compared to bivalves, Cretaceous and Mesozoic brachiopods in general, appear to be uncommon in stressed environments. Stenotopic pedunculate brachiopods depend on a relatively stable substrate, stable or normal marine salinity and oxygenation, and low to moderate rates of sedimentation (Aberhan, 1994). Comparisons with brachiopods in Recent environments are tempting, but can be done with circumspection. Athough early diagenesis is evident in live and recently dead shells, it can occur at different rates and earlier than usually suspected. Original mineralogy may not be the determinant in early shell dissolution or recrystallization. In addition to SEM, cathodoluminescence may be accurate in pinpointing diagenesis, but it seems to be useful at first, with fine orange band responses, to highlight some growth rhythms (Gaspard & Barbin, 1993), often less evident in SEM (Figure

4:6). The microprobe was not found to be useful in determining shell zonation or growth because the responsible elements are present in too small quantities. We expect better results from using joint SEM-CL, followed by ionic analysis, in studying Recent as well as Cretaceous shells.

Acknowledgments

I would like to thank the CENTOB laboratory at Brest, and N. Cominardi ; G. Breton at the Museum du Havre, the Institut de Paléontologie MNHN de Paris; the Institut Océanologique de Marseille and Ch. Emig; and D.I. MacKinnon who provided part of the material studied in this paper. The SEM pictures have been realized with the help of D. Guillaumin, CIME, Université de Paris 6 using a JEOL instrument. C'est une contribution aux travaux de l'URA n° 157 du CNRS.

REFERENCES

ABERHAN, M. 1994. Guild-structure and evolution of Mesozoic benthic shelf communities. Palaios 9: 516-545.

ALEXANDER, R.R. 1986. Life orientation and post-mortem reorientation of Chesterian brachiopod shells by paleocurrents. Palaios 1: 303-311.

___ 1994. Distribution of pedicle boring traces and the life habit of late Paleozoic leiorhynchid brachiopods from dysoxic sediments. Lethaia 27: 227-234.

ALEXANDERSSON. E.T. 1978. Destructive diagenesis of

carbonate sediments in the eastern Skagerrak, North Sea. Geology 6(6): 324-327.

___1979. Marine maceration of skeletal carbonates in the Skagerrak Sea. Sedimentology 26(6): 845-852.

ALLER, R.C. 1982. Carbonate dissolution in the nearshore terrigenous muds: the role of physical and biological reworking. Journal of Geology 90(1): 79-95.

ARCHER, D., S. EMERSON & C. REIMERS, 1989. Dissolution of calcite in deep-sea sediments: pH and O_2 microelectrode results. Geochimica Cosmochimica Acta 53(11): 2831-2845.

BEHRENSMEYER, A.K. & S.M. KIDWELL. 1985. Taphonomy's contribution to paleobiology. Paleobiology 11(1):105-119.

BOUCOT, A.J. 1953. Life and death assemblages among fossils. American Journal of Science 261: 25-40.

BRETT, C.E. & G.C. BAIRD. 1986. Comparative taphonomy: a key to paleoenvironmental interpretation based on fossil preservation. Palaios 1(3): 202-227.

BROMLEY, R.G. & F. SURLYK. 1973. Borings produced by brachiopod pedicles fossil and recent. Lethaia 6: 349-365.

CADÉE, G.C. 1991. The history of taphonomy, p.3-21. In, S.K. Donovan (ed.), The processes of fossilization.

___1995. Birds as producers of shell fragments in the Wadden Sea, in particular the role of the Herring gull. First European Palaeontological Congress (Lyon,1993): 'Organism-palaeoenvironment interactions', M. Gayet & B. Courtinat (Coords.). Geobios, Mémoire Special 18: 77-85.

COLLINS, M.J. 1986. Post mortality strength loss in shells of the Recent articulate brachiopod Terebratulina retusa (L.) from the west coast of Scotland. In, Racheboeuf P.R. &C.C. Emig (eds), Les brachiopodes fossiles et actuels. Biostratigraphie du Paléozoïque 4: 209-218

EFREMOV, J.A. 1940. Taphonomy: new branch of paleontology. Pan-American Geologist 74(2):81-93.

___1953. Taphonomie et annales géologiques. Annales du centre d'études et de documentation paléontologiques (transl. S. Ketchian & J. Roger) 4: 1-164.

___1958. Some considerations on biological bases of paleozoology, Vertebrata Palasiatica 2(2/3): 83-98.

FITZGERALD, M.G., C.M. PARMENTER & J.D. MILLIMAN. 1979. Particulate calcium carbonate in New England shelf waters: result of shell degradation and resuspension. Sedimentology 26(6): 853-857.

FLESSA, K.W. & T.J. BROWN. 1983. Selective solution of macroinvertebrate calcareous hard parts: a laboratory study. Lethaia 16(3):193-205.

FREDJ, G. & Cl. FALCONETTI. 1977. Sur la présence d'algues filamenteuses perforantes dans le test de Gryphus vitreus (Born) (Brachiopodes, Térébratulidés) de la limite du plateau continental méditerranéen. Comptes Rendus de l'Académie des Sciences, Paris, D 284(13):1167-1170.

GALL, J-C., P. DURINGER, W. KRUMBEIN & J-C. PAICHELER. 1994. Impact des écosystèmes microbiens sur la sédimentation. Palaeogeography, Palaeoclimatology, Palaeoecology 111: 17-28.

GASPARD, D. 1988. Sellithyridinae Terebratulidae du Crétacé d'Europe Occidentale- Dynamique des populations-Systématique et Evolution. Cahiers de Paléontologie, CNRS (ed.), 243p.

___1989. Quelques aspects de la biodégradation des coquilles de brachiopodes; conséquence sur leur fossilisation. Bulletin Société Géologique, France 8, V, (6):1207-1216.

___1990. Diagenetic evolution of shell microstructure in the Terebratulida (Brachiopoda, Articulata). In, J.G. Carter (ed.), Skeletal biomineralization: Patterns, processes and evolutionary trends, 2 (4): 53-56.

___1992. Evènements environnementaux révélés par la biominéralisation des coquilles de Brachiopodes Articulés. Geobios, Mémoire Special 14:59-70

___1993. Articulate brachiopod shell formation (Terebratulida & Rhynchonellida) and diagenesis evolution, p.21-29. In, Kobayashi, I., H. Mutvei & A. Sahni (eds.). Structure, formation and evolution of hard tissues. Tokyo University Press, Tokyo.

___(in press).. Biomineralized structures of brachiopods and diagenetic changes through time. In, Seventh International Symposium on Biomineralization, Monaco,1993. Bulletin de l'Institut Océanographique, Special issue.

___& M. ROUX. 1974. Quelques aspects de la fossilisation des tests de brachiopodes et de crinoïdes, Relation entre la présence de matière organique et le développement d'agrégats ferrifères. Geobios 7(2):81-89.

___& C. MULLON. 1983. Méthodes de discrimination appliquées à des populations de Sellithyridinae (Brachiopodes) du Cénomanien d'Europe Occidentale. Bulletin Société Géologique France 7, XXV(5): 679-688.

___& V. BARBIN. 1993. Mise en évidence par la cathodo-luminescence des rythmes de croissance liés à la concentration en manganèse dans les coquilles de brachiopodes; implications sur les études géochimiques, p.50. In, Abstracts First European Palaeontological Congress 'Organism-palaeoenvironment interactions', Lyon 1993, Gayet, M. & B. Courtinat (coord.).

JOHNSON, R.G. 1960. Models and methods for analysis of the mode of formation of fossil assemblages. Bulletin Geological Society America 71(7):1075-1086.

KOWALEWSKI, M. & K.W. FLESSA. 1994. A predatory drillhole in Glottidia palmeri Dall (Brachiopoda; Lingulidae) from Recent tidal flats of northeastern Baja California, Mexico. Journal of Paleontology 68(6):1403-1405.

LAMBERT, B. & M. ROUX (coord.) 1991. L'environnement carbonaté bathyal en Nouvelle Calédonie (Programmes envimarges). Documents Travaux IGAL, Paris 15:1-213.

LAMBOY, M. 1990. Microbial mediation in phosphatogenesis: new data from the Cretaceous phosphatic chalks of Northern France, p157-167. In, Notholt, A.J. & I. JARVIS (eds.), Phosphorite research and development. Geological Society, Special publication 52:157-167.

LAURIN, B. & D. GASPARD. 1987. Variations morphologiques et croissance du brachiopode abyssal Macandrevia africana Cooper. Oceanologica Acta 10(4):445-454.

LAWRENCE, D.R. 1968. Taphonomy and information losses in fossil communities. Bulletin Geological Society America 79(10):1315-1330.

MÜLLER, A.H. 1979. Fossilization (Taphonomy), p.2-78. In Moore R.C.; Robinson R.A. &C. Teichert (eds), Treatise on Invertebrate Paleontology, Part A, Introduction. Geological Society of America and University Kansas Press, Lawrence.

NEKVASILOVA, O. 1976. The etching traces produced by pedicles of Lower Cretaceous brachiopods from Stramberk (Czechoslovakia). Casopis pro mineralogii a geologii 21: 405-408.

NOBLE, J.P. & A. LOGAN. 1981. Size-frequency distributions and taphonomy of brachiopods: a recent model. Palaeogeography, Palaeoclimatology, Palaeoecology 36: 87-105.

OLSON, E.C. 1980. Taphonomy: its history and role in community evolution, p.5-19. In, Behrensmeyer A.K. & A.P. Hill (eds.). Fossils in the making. University of Chicago Press, Chicago.

PARSONS, K.M., & C.E. BRETT. 1991. Taphonomic processes and biases in modern marine environments: an actualistic perspective on fossil assemblage preservation,p.22-65. In, S.K. Donovan (ed.), The processes of fossilization.

SMITH, S.A., C.W. THAYER, & C.E. BRETT. 1985. Predation in the Paleozoic: gastropod-like drillholes in Devonian brachiopods. Science 230 : 1033-1035.

SMITH, A.M., C.S. NELSON & P.J. DANAHER. 1992. Dissolution behaviour of bryozoan sediments: taphonomic implications for nontropical shelf carbonates. Palaeogeography, Palaeoclimatology, Palaeoecology 93: 213-226.

WALTER L.M. & J.W. MORSE. 1985. The dissolution kinetics of shallow marine carbonates in seawater: a laboratory study. Geochimica Cosmochimica Acta 49(7):1503-1513.

WEIGELT, F. 1927. Uber Biostratonomie. Eine Betrachtung zu Dollo's siebzigstem Geburstag. Der Geologe 42: 1069-1076.

WILSON, M.V.H. 1988. Taphonomic processes: information loss and information gain. Geoscience Canada 15(2):131-148.

BRACHIOPOD PALEOECOLOGY AND PALEOBIOGEOGRAPHICAL AFFINITIES IN THE EARLY CRETACEOUS OF SOUTHEASTERN ROMANIA

MARIUS-DAN GEORGESCU

Petromar, Port-Dana 34, Laboratory of Paleontology,
8700 Constanta, Romania

ABSTRACT - Early Cretaceous (Late Valanginian-Early Aptian) brachiopods of southeastern Romania show a mixture of boreal and Jura type species, as well as some endemic taxa. These late Valanginian brachiopods inhabited the inner carbonate shelf and lived in a muddy bottom around patch-reefs. The late Barremian - early Aptian populations inhabited the outer shelf. Shell-sediment relationships seem to be the most important factors controlling brachiopod distribution within this region.

INTRODUCTION

Early Cretaceous brachiopods of southeastern Romania usually have been regarded as belonging to the Jura type fauna (Neagu & Barbulescu, 1979), because of the absence of pygopids and the presence of *Loriolithyris russillensis* (Loriol), a characteristic species of the Jura bioprovince. This detailed study, based on Neagu's collection (reposited in the Laboratory of Paleontology of the University of Bucharest), and new collections, provided much more information on the paleoecology and paleobiogeography of these Early Cretaceous brachiopods.

STRATIGRAPHIC SETTING

The brachiopods presented in this paper were collected from the Cernavoda Formation (Alimanu Member) and the Ramadan Formation as defined by Avram et al. (1988) and later refined by Avram et al. (1993). The Alimanu Member consists of light yellowish limestones with interbeds of up to 20 cm thick,

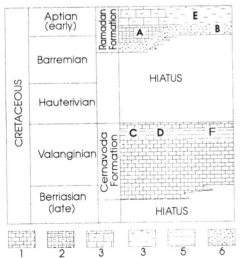

Figure 1. Composite section of the Cernavoda Formation (Alimanu Member) and Ramadan Formation. A-F, brachiopod localities. Key: 1, limestones; 2, marly limestones; 3, sandy limestones; 4, sandstones; 5, claystones; 6, sands.

yellow, greenish or grey claystones (Figure 1).The brachiopods occur in the claystone interbeds (localities C and D) and the muddy limestones (locality F). The foraminifers indicate a Late Berriasian-Valanginian age for the brachiopod-bearing deposits (Georgescu, in Dragomir & Georgescu, 1987).The basal beds of the Ramadan Formation are exposed only in restricted areas. One of these occurrences (locality A) consists of marlstones and light yellowish limestones, containing numerous brachiopods and foraminifers. The foraminiferan assemblage indicates a latest Barremian-early Aptian age. At locality B, brachiopods occur in the basal sandstone unit of the Ramadan Formation. At the top of this lithostratigraphic unit sandstones and quartz sands are the main lithologies. The brachiopods of locality E came from these deposits. An early Aptian age is indicated by the ammonites *Deshayesites flexuosus* Chiriac, and *Cheloniceras ramadanicus* Chiriac (Chiriac, 1981; 1988).

FOSSIL RECORD

Six localities in the western part of southern Dobrogea yielded brachiopods. <u>Locality A</u>: Canlia Village, northern bank of Girlita Lake. Age: Late Barremian-early Aptian. Lithology: muddy limestones. Material: specimens. Taxa: *Lamellaerhynchia*; *Rectithyris* n. sp., *Tamarella tamarindus* (Sowerby), *Gemmarcula aurea* Elliot. <u>Locality B</u>: Adincata Village (Aliman-Urluia Valley). Age: Early Aptian. Lithology: sandstones. Material: 126 well preserved and over 200 crushed specimens.Taxon: *Burrirhynchia romanica* (Barbulescu & Neagu). <u>Locality C</u>. Aliman Village (right side of Aliman-Urluia Valley). Age: Late Valanginian. Lithology: claystones. Material: 87 well preserved and about 100 crushed specimens. Taxa: *Loriolithyris russillensis* (Loriol), *Rugitella villersensis* (Roemer). <u>Locality D</u>: Aliman Village (left side of Aliman-Urluia Valley). Age: Late Valanginian. Lithology: clay-stones. Material: 24 well preserved and over 30 crushed specimens. Taxa: *Loriolithyris russillensis* (Loriol), *Tropaeothyris* n. sp., *Nucleata* n. sp. <u>Locality E</u>: Valea Baciului. Age: Early Aptian. Lithology: muddy and sandy limestones. Material: Over 1000 specimens. Taxa: *Burrirhynchia romanica*; *Cyclothyris antidichotoma* (Bouvignier), *C. latisssima* (Sowerby), *Sellithyris sella* (Sowerby), *Tamarella tamarindus* (Sowerby); *Terebrirostra arduenensis* d'Orbigny. <u>Locality F</u>: Cernavoda Bridge section (right bank of the Danube). Age: Late Valanginian. Lithology: muddy limestones. Material: 45 well preserved and about 30 crushed specimens. Taxa: *Belbekella macoveii* Barbulescu & Neagu; *Lamellaerhynchia desori*, *Loriolithyris* n. sp., *Praelongithyris credneri* (Werth), *Rouillieria tilbyensis* (Davidson), *Sellithyris carteroniana* (d'Orbigny), *S. lindensis* Middlemiss.

Examination of a collection by Avram (Geological Institute at Bucharest) from an unnamed locality of the Ramadan Formation (latest Barremian-early Aptian) showed the presence of *Cyclothyris latissima* (Sowerby) and *Praelongithyris praelongiforma* Middlemiss.

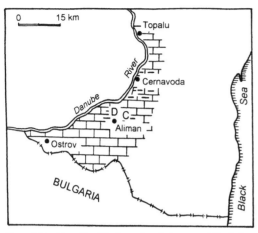

Figure 2. Brachiopod localities in southern Dobrrogea relative to the late Valanginian facies. Key in Figure 1.

PALEOECOLOGY

The late Valanginian brachiopods inhabited the shallow waters spread over southeastern Romania (Figure 2), and were commonly associated with benthic foraminifers and ostracods, branched bryozoans, echinoids, crinoids, ophiuroids, and holoturoids. The terebratulids *Loriolithyris russillensis* and *Tropaeothyris* n.sp. (localities C, D) lived on a muddy bottom, characterized by folded shells, reduced thickness/width ratio and large foramina. Northwards (locality F), where the substrate was mostly limy, *Loriolithris* n.sp. (more weakly folded shells with a rectimarginate to weakly uniplicate anterior commissure) became the dominant species. The limy substrate probably favoured such somewhat flattened shells as *Sellithyris carteroniana*, *S. lindensis*, and *Rouilleria tilbyensis* (Davidson). Most European occurrences of these species are in muddy limestones. The rich dasycladaceans found in adjacent limestones (e.g., *Salpingoporella*, *Likanella*, *Macroporella*, and *Suppililumaella*) indicate clear-water conditions.

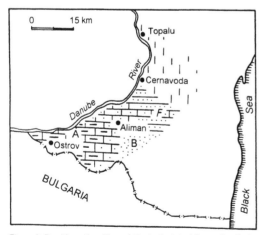

Figure 3. Brachiopod localities relative to the latest Barremian-Early Aptian. Key as in Figure 1.

Latest Barremian-early Aptian paleogeography shows differences with the Berriasian-Valanginian interval (Figure 3). Sandy limestones and sandstones (locality E) were formed in shallow waters with somewhat increased energy. Antidichotomic ornamentation of *Cyclothyris antidichotoma* (Bouvignier) was considered by Ager (1963) as being indicative of a sublittoral environment. *Terebrirostra arduenensis* probably lived with its strongly elongated ventral posterior buried in the sediment. Southwards (locality A), the environ-ment indicates reduced water energy. *Tamarella tamarindus* (Sowerby) has a smooth shell (or with sparse rugae), whereas shells from locality E have numerous rugae on the surface, indicating increased water energy.

PALEOBIOGEOGRAPHY

Boreal and Jura-type taxa - Muddy limestones at locality F yielded three late Valanginian species that were considered by Middlemiss (1979) to be restricted to the boreal faunas: *Praelongithyris credneri* (Weerth), *Rouillieria tilbyensis* (Davidson), and *Sellithyris lindensis* Middlemiss. Most occurrences indicate a very close association with the sediment; usually their shells are found in muddy limestones and marls. In Grenzlerburg northern Germany, these taxa (except for *Rouilleria*) occur in very coarse conglomerates of Valanginian or earliest Hauterivian age (Middlemiss, 1976). Precaution should be given to such occurrences, however, because the shells in conglomerates may have been reworked. It is worth mentioning that the Romanian forms of *R. tilbyensis* are characterized by very large shells (up to 45 mm in length).

Jura-type taxa are the predominant brachiopods in Early Cretaceous rocks of southeastern Romania. Some taxa (e.g., *Loriolithyris russillensis* (Loriol), *Rugitella villersensis* (Loriol), and *Terebrirostra arduenensis* d'Orbigny, are strictly confined to the Jura province (Middlemiss, 1979). Others originated in the Jura province but were spread over the boreal seas (e.g., *Sellithyris carteroniana* (d'Orbigny), *Cyclothyris antidichotoma* (Bouvignier), and *C. latissima* (Sowerby). *Lamellaerhynchia desori* (Loriol) is most abundant in the Jura faunas, whereas *Sellithyris sella* (Sowerby) and *Tamarella tamarindus* (Sowerby) are ubiquitous in both Jura and Boreal faunas.

Endemic taxa - Six taxa are endemic and so far have not been reported outside Romania, being regarded as local offshoots. In the Late Valanginian (locality D), a morphotype with strongly folded valves, an incurved beak and labiate foramen is assigned to *Tropaeothyris*. In the same bed, small forms of *Nucleata* (maximum length 6.9 mm) probably belong to an undescribed species. *Loriolithyris* n.sp. from the upper Valanginian muddy limestones (at locality F), is characterised by reduced folding of the shell and a weakly uniplicate anterior commissure. About two dozen specimens of a small form of *Rectithyris* occur in upper Barremian-lower Aptian limy deposits (locality A). Their external and internal features indicate that they represent a new species. Barbulescu & Neagu (in, Barbulescu et al., 1975) erected two species on the basis of specimens from southern Dobrogea: *Belbekella macoveii* and *Sulcirhynchia romanica*. Later, Sulser (1993) assigned the latter to the genus *Burrirhynchia*. This assignment is supported by my observation of four transversely sectioned specimens (two of them from the type locality).

Five of the six endemic species belong to genera of Jura affinities, *Belbekella*, *Burrirhynchia*, *Loriolithyris*, *Tropaeo-thyris*, and *Rectithyris*. The other, *Nucleata* n.sp., is regarded as characteristic of the Tethys fauna as defined by Middlemiss (1979).

REFERENCES

AGER, D.V. 1963. Principles of Paleoecology. McGraw-Hill, New York, 371 p.

AVRAM, E., A. DRAGANESCU, L. SZASZ & T. NEAGU. 1988. Stratigraphy of the outcropping Cretaceous deposits in southern Dobrogea (SE Romania). Memoriile Institului Geologic 33:5-43.

AVRAM, E., L. SZASZ, E. ANTONESCU, A. BALTRES, M. IVA, M. MELINTE, T. NEAGU, S. RADAN & C. TOMESCU. 1993. Cretaceous terrestrial and shallow marine deposits in northern South Dobrogea (SE Romania). Cretaceous Research 14: 265-305.

BARBULESCU, A., T. NEAGU, V. LAZAROIU & C. VODISLAV. 1975. Brachiopode eocretacice din Dobrogea de Sud (Early Cretaceous brachiopods of South Dobrogea). Studii si cercetari de geologie, geofizica si geografie (Geologie) 20:111-141 [In Romanian with English summary].

CHIRIAC, M. 1981. Amoniti cretacici din Dobrogea de Sud (Cretaceous Ammonites in South Dobrogea). Editura Academiei R.S.R., 143 p. [In Romanian with French summary].

CHIRIAC, M. 1988. Espèces et sous-espèces d'ammonites dans la Crétacé de la Dobrogea méridionale. Memoriile Institutului Geologic 33:45-90.

DRAGOMIR, B.P. & M.D. GEORGESCU. 1987. Aspects of the sedimentary paleoenvironments in the carbonate platform of the Lower Cretaceous in Aliman (southern Dobrogea). Analele Universitatii Bucuresti 36:48-57.

MIDDLEMISS, F.A. 1976. Lower Cretaceous Terebratulidina of northern England and Germany and their geological background. Geologisches Jahrbuch, A30:21-104.

___ 1979. Boreal and Tethyan brachiopods in the European Early and Middle Cretaceos. In Aspekte du Kreide Europas, IUGS Series A, 6:351-361.

NEAGU, T. & A. BARBULESCU. 1979. The paleoecologic and paleobiogeographic values of the Cretaceous Brachiopoda from Dobrudja. Revue Roumaine de Géologie, Géophysique et Géographie, 23:69-75.

SULSER, H. 1993. Brachiopoda: Rhynchonellida Mesozoica. In, F. Westphal (ed.), Fossilium Catalogus I: Animalia, Kugler Publications, Amsterdam, 281 p.

PERMIAN BRACHIOPOD PALEOBIOGEOGRAPHY OF SOUTH AMERICA

REX ALAN HANGER

Department of Geology, George Washington University, Washington, DC 20052, USA

ABSTRACT- A database of 149 Permian South American brachiopod genera from 13 localities were subjected to cluster, probabilistic similarity, and parsimony analysis, in order to detect differential patterns. These were compared to data from North American localities as far north as Axel Heiberg Island, Canada. Results are broadly congruent, showing hetero-geneous distribution patterns paralleling paleolatitudes, supporting further use of multiple analytical methods for paleobiogeographic reconstruction.

INTRODUCTION

Permian brachiopods are abundant and diverse throughout South America. Paleobiogeographic analyses of these faunas have been scant by comparison to other continental areas. The 'traditional' view (R.E. Grant, 1994, pers. commun.) is of a single, Andean Province that encompassed all faunas on the entire continent, distinct from other provinces of North America, as well as other southern hemisphere faunas. Archbold (1982), using endemic chonetid genera, proposed a northern, Andean province, and a southern, Paratinan province. Refined endemism analyses (Shi et al, 1995) have supported the two-fold division of South American Permian faunas, with perhaps a geographically central 'transitional' fauna containing elements of each. All such analyses, of course, assume accurate taxonomic identification.

A common rationale for paleontology is recognizing paleobio-geographic patterns, and then 'explaining' the biogeographical and/or geological processes that may have produced this pattern (Rosen, 1992). This has proved exceedingly difficult for most of the circum-Pacific region, with the probability of long-distance transport of tectonostratigraphic terranes (Jones, 1990). For South America, the accretion of possible allochthonous terranes was largely complete well before the Permian (Ramos, 1986). Thus, the role of tectonic processes in generating any paleobiogeographic patterns for the Permian of South America is minimized. Reduced tectonic effects help to reveal paleoclimatic influences. Coeval, Permian autoch-thonous faunas of North America have been shown to be strongly controlled by temperature, revealing paleolatitude (Yancey, 1975,1979). Detecting a similar, southern hemi-sphere signal for South America should assist in unravelling paleogeographic problems throughout the circum-Pacific region. Before any process or paleoclimatic/paleogeographic interpretations are made for South America, it is necessary to determine its paleobiogeography, preferably in a reproducible, quantified way, which is the purpose of this report. An important ancillary purpose is the comparison of results generated by separate analytical methods.

Table 1. Permian brachiopod localities used in this report.

AR	Argentina	Archangelsky,1987;Polanski,1978
BO	Bolivia	Kozlowski,1914;Samtleben,1971
CA	Chile, loc. A	Minato & Tazawa,1977
CB	Chile, loc. B	unpublished data
CO	Colombia	unpublished data
PE	Peru	Newell et al,1953
VE	Venezuela	Hoover,1981
AH	Axel Heiberg	Stehli & Grant,1971
EL	El Antimonio	Cooper et al,1953
GU	Guatemala	Stehli & Grant,1970
KN	Kansas/Nebras	Dunbar & Condra,1932;Mudge & Yochelson,1962
LD	Las Delicias	King et al,1944
SD	Sierra Diablo	Stehli,1954

DATA AND METHODS

The data base is the faunal lists from published literature on Permian faunas of South America, including 149 brachiopod genera from 13 localities (Figure 1, Table 1), augmented by unpublished data on Colombia (locality CO), and the Antofogasta area of Chile (locality CB). North American localities were included for comparison. To obtain a suitable number of data points, many of the localities were not strict biostratigraphic correlatives, i.e. some are Early Permian or Middle Permian in age, but none are Late Permian. Older publications were updated to modern taxonomic standards.The data matrix was used to calculate a Jaccard similarity matrix, which was subsequently clustered via the UPGMA algorithm of the program, MVSP v.2.1 (Kovach, 1993). A thorough review of multivariate statistics as used in paleontology is given in Shi (1993). The data was also subjected to the probabilistic similarity analysis of Raup & Crick (1979; assumptions and interpretations of the method discussed therein). Finally, the data were subjected to a parsimony analysis of endemicity (=PAE; method discussed in Rosen,1988,1992; Rosen & Smith,1988) using the program, PAUPv.3.0.s (Swofford, 1991).

AXEL HEIBERG

KANSAS/NEBRASKA
SIERRA DIABLO
EL ANTIMONIO

LAS DELICIAS
GUATEMALA
VENEZUELA
COLOMBIA

PERU
BOLIVIA
CHILE, LOC. B
CHILE, LOC. A
ARGENTINA

SELECTED LOCALITIES OF
PERMIAN BRACHIOPODS

Figure 1. Location of Permian brachiopod localities.

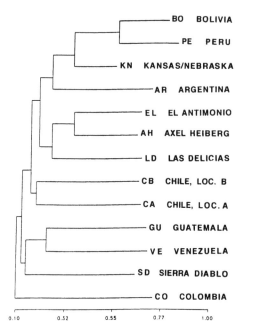

Figure 2. Cluster analysis dendrogram. Locality symbols in Table 1.

RESULTS

Among the clusters produced in the dendrogram (Figure 2) that show relatively high similarity (here arbitrarily defined as Jaccard coefficient >0.25), are Bolivia and Peru, linked with Kansas/Nebraska, then Argentina. El Antimonio, Axel Heiberg and Las Delicias also cluster together with high similarity. Other clusters that have lower Jaccard coefficient values include Chile, locality B with Colombia, and Guatemala, Venezuela and Sierra Diablo. Chile loc. A forms no cluster with any other locality.

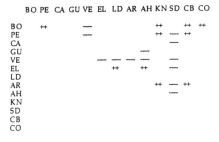

Figure 3. Probabilistic similarity matrix. Locality symbols in Table 1 (++ = significant similarity, ---- = significant dissimilarity).

Results of the probabilistic similarity analysis (Figure 3). show that significantly similar faunal pairs include, (1) Bolivia with Peru, Kansas/Nebraska, Chile loc. B and Colombia, (2) Peru and Kansas/Nebraska and Chile loc. B, (3) Argentina and Kansas/Nebraska and Chile loc. B, and (4) El Antimonio and Las Delicias and Axel Heiberg. The significantly dissimilar faunal pairs include (1) Venezuela and Bolivia, Peru, Axel Heiberg, Sierra Diablo, El Antimonio and Las Delicias, (2) Sierra Diablo and Peru, Argentina, Chile loc. A, El Antimonio and Axel Heiberg, and (3) Guatemala and Axel Heiberg.

Parsimony analysis produced three equally most parsimonious trees of length 296, consistency index of 0.490. A 50% majority-rule consensus tree is shown in (Figure 4) identifies the following groupings, (1) The Mexican localities - El Antimonio and Las Delicias are sister areas, (2) Guatemala and Venezuela are sister areas, (3) Argentina and Chile loc. B are sister areas with Colombia then Kansas/Nebraska as sister areas, and (4) Peru and Chile loc. A are sister areas with each other then with Axel Heiberg. No relationships among these four groups, nor any with Bolivia or Sierra Diablo are resolved.

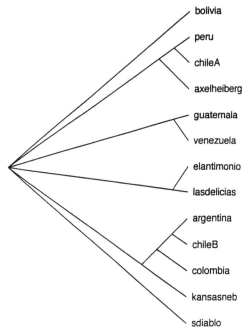

Figure 4. 50% majority-rule consensus of three equally most parsimonious trees [tree-length = 296; consistency index = 0.490].

DISCUSSION

The cluster and probabilistic analyses should produce the same results since they both purport to measure overall similarity. Results from both are broadly congruent, yet differ in several key respects. These include, (1) cluster analysis is equivocal about the relationships of Colombia and Chile loc. B with other localities, though they cluster with each other, while probabilistic analysis groups these with numerous other South American localities, (2) Sierra Diablo and Venezuela cluster together, yet show significant dissimilarity, and (3) Chile loc. A, which clustered with no other locality, shows significant dissimilarity with Sierra Diablo.

Probabilistic methods are independent of sample size (and thus, sampling), while cluster analysis is not. Probabilistic analysis here reveals key relationships not shown by cluster analysis because of differences in sampling of different localities (e.g. Chile loc. B). Therefore, localities that show significant similarity or dissimilarity with other localities may be considered sufficiently sampled, i.e. probabilistic analysis may be considered a rough measure of sampling and should be completed in conjunction with any other statistical analysis. PAE is a relatively new method and the theory behind shared derived taxa from different localities is still immature (Rosen, 1992). The results here again are broadly congruent to the similarity methods. The sister areas of PAE are supposed to reflect recency of biotic contact, a difficult interpretation to justify given the Peru - Chile loc. A - Axel Heiberg grouping, which is not shown by other methods, as well as the overall relationship between North and South American faunal localities.

The grouping of North American localities within South American groupings is most probably a taxonomic artefact. Many of the same genera may be shared by North and South American localities, while comparison of species composition shows very low similarity. Smith (1992) discussed the marked difference in species-level and genus-level results in a study of Cretaceous echinoids. The inconsistent recognition and usage of species in the literature cited here precludes a detailed analysis at the species level. Revision of species, and tighter constraint on time scale, should be completed to refine the paleobiogeographic results.

CONCLUSIONS

(1) Cluster analysis, probabilistic similarity, and PAE, used in conjunction, reveal paleobiogeographic patterns that are interpretable in terms of paleogeography and paleoclimate, key goals of Paleozoic tectonic analysis (Jones,1990).
(2) The traditional view of a single fauna encompassing all of the Permian South America appears to be false. The methods used here reveal a heterogeneous pattern, supporting the results of Shi et al. (1995).
(3) The heterogeneous pattern roughly follows paleolatitudes, with a northern fauna (principally, Venezuela), and a southern fauna for all other South American localities.
(4) The extension of the Grandian Province of Texas into South America is probably not justified. The Venezuelan and Guatemalan faunas group with the Grandian Province using cluster analysis, but not with probabilistic analysis nor PAE. The tropical areas of the American Permian probably had several endemic centers, as did the Tethyan region (Shi et al., 1995).

REFERENCES

AMOS, A.J. 1979. Guia Paleontologia Argentina, I, Paleozoico. Consejo Nacional de Investigaciones Cientificas Tecnicas, Argentina.

ARCHANGELSKY, S. 1987. El Sistema Carbonifero en la Republica Argentina. Academia Nacional Ciencias, Cordoba, 383p.

ARCHBOLD, N.W. 1982. Permian marine invertebrate provinces of the Gondwanan realm. Alcheringa 7:59-73.

DUNBAR, C.O. & G.E. CONDRA.1932. Brachiopoda of the Pennsylvanian System in Nebraska. Nebraska Geological Survey Bulletin 5:1-377.

COOPER, G.A., C.O. DUNBAR, H. DUNCAN, A.K. MILLER & J.B. KNIGHT. 1953. Permian fauna at El Antimonio, western Sonora, Mexico. Smithsonian Miscellaneous Collections 119(2):1-111.

HOOVER, P.R. 1981. Paleontology, taphonomy and paleoecology of the Palmarito Formation (Permian of Venezuela). Bulletin American Paleontology 80(113): 1-138.

JONES, D.L. 1990. Synopsis of Late Palaeozoic and Mesozoic terrane accretion within the Cordillera of western North America. Philosophical Transactions Royal Society London A 331: 479-486.

KING, R.E., C.O. DUNBAR, P.E. CLOUD & A.K. MILLER. 1944. Geology and paleontology of the Permian area northwest of Las Delicias, Southwestern Coahuila, Mexico. Geological Society America Special Paper 52: 3-127.

KOVACH, W.L. 1993. MVSP - A Multivariate Statistical Package for IBM-PC's, version 2.1. Kovach Computing Services, Pentraeth, Wales, U.K.

KOZLOWSKI, R. 1914. Les brachiopodes du Carbonifère supérieur de Bolivie. Annales Paléontologie 9:1-100.

MINATO, M. & JUNICHI TAZAWA. 1977. Fossils of the Huentelaquen Formation at locality F, Coquimbo Province, Chile, p.95-117. In, T. Ishikawa & L. Aguirre (eds.), Comparative studies on the geology of the circum-Pacific orogenic belt of Japan and Chile, First Report Japan Society Promotion Science, Tokyo.

MUDGE, M.R. & E.L. YOCHELSON. 1962. Stratigraphy and paleontology of the uppermost Pennsylvanian and lowermost Permian rocks in Kansas. United States Geological Survey Professional Paper 323:1-213.

NEWELL, N.D., J. CHRONIC & T.G. ROBERTS. 1953. Upper Paleozoic of Peru. Geological Society America Memoir 58:1-276.

POLANSKI, J. 1978. Carbonico Y Permico de la Argentina. EUDEBA, Buenos Aires, 216p.

RAUP, D.M. & R.E. CRICK. 1979. Measurement of faunal similarity in paleontology. Journal Paleontology, 53: 1213-1227.

ROSEN, B.R. 1988. From fossils to earth history: applied historical biogeography, p.437-481. In, A.A. Myers & P. S. Giller (eds.), Analytical biogeography. Chapman and Hall, London.

___1992. Empiricism and the biogeographical black box: concepts and methods in marine paleobiogeography. Palaeogeography, Palaeoclimatology, Palaeoecology 92:171-205.

___& A.B. SMITH. 1988. Tectonics from fossils? Analysis of reef-coral and sea-urchin distributions from Late Cretaceous to Recent, using a new method, p.275-306. In, M.G. Audley-Charles & A. Hallam (eds.), Gondwana and Tethys. Geological Society Special Publication 37.

SAMTLEBEN, C. 1971. Zur Kenntnis der Produktiden und Spiriferiden des bolivianischen Unterperms. Beihefte Geologisches Jahrbuch 111:1- 163.

SHI, G.R. 1993. Multivariate data analysis in palaeoecology and palaeobiogeography - a review. Palaeogeography, Palaeoclimatology, Palaeoecology 105:199-234.

___, N.W. ARCHBOLD & L.P. ZHAN. 1995. Distribution and characteristics of mixed (transitional) mid-Permian (Late Artinskian-Ufimian) marine faunas in Asia and their palaeogeographical implications. Palaeogeography, Palaeoclimatology, Palaeoecology 114:241-271.

SMITH, A.B. 1992. Echinoid distribution in the Cenomanian:. an analytical study in biogeography. Palaeogeography, Palaeoclimatology, Palaeoecology 92:263-276.

STEHLI, F.G. 1954. Lower Leonardian Brachiopoda of the Sierra Diablo. American Museum Natural History Bulletin 105(3):263-358.

___& R.E. GRANT. 1970. Permian brachiopods from Huehuetenango, Guatemala. Journal Paleontology, 44(1):23-36.

___&___1971. Permian brachiopods from Axel Heiberg Island, Canada, and an index of sampling efficiency. Journal Paleontology 45(3):502-521.

SWOFFORD, D.L. 1991. PAUP: Phylogenetic Analysis Using Parsimony, version 3.0s. Illinois Natural History Survey, Champaign, Illinois.

YANCEY, T.E.1975 Permian biotic provinces in North America. Journal Paleontology 49:758-766.

___1979. Permian positions of the northern hemisphere continents as determined from marine biotic provinces, p. 239-247 In, J. Gray & A. J. Boucot (eds.), Historical biogeography, plate tectonics, and the changing environment. Oregon State University Press, Corvallis, Oregon.

THE SEQUENCE OF PERMIAN BRACHIOPOD ASSEMBLAGES IN SOUTH CHINA

XILIN HE & G.R. SHI

Geology Department, China University of Mining and Technology, Xuzhou 221008, China;
School of Aquatic Science - Natural Resources Management, Deakin University, Clayton, Victoria 3168, Australia

ABSTRACT- An integrated three-fold Permian chrono-stratigraphic framework for South China, is based on eleven brachiopod assemblage zones. Regional differences between brachiopod faunas of SW and SE South China are interpreted in terms of facies distribution, paleogeography and paleo-ecology.

INTRODUCTION

The Hercynian Orogeny, especially the Saarlian Phase, conventionally placed at the Lower-Upper Permian boundary, resulted in one of the largest scale regressions in geologic history, ultimately leading to profound changes in paleo-geography at the Paleozoic-Mesozoic boundary. This regression seems to have persisted through the Permian, but was interrupted by localized transgressions and short-term eustatic sealevel changes. The regression produced many incomplete Permian marine sequences, especially for Permo-Triassic boundary strata, continuous sections for which are confined to only a few localities world-wide, such as S China.

The South China area is defined to include almost the entire region of the Yangtze River valley, and most of the provinces to the south. Recent paleogeographic reconstructions locate the South China plate at the paleo-equator in the eastern Tethys throughout the Permian (Huang & Chen, 1987; Scotese, 1994). During the Permian it was covered by an epicontinental sea, divided into two major regions by a prominent, northeast trending paleogeographic high known as the Jiangnan Massif or 'Old Land' (Figure1). The two regions share many stratigraphic characteristics, but are also significantly different. For instance, there is a consistent disconformity between the Middle and Upper Permian in the southwest region, marked by volcanic eruptions (the Omei-shan Basalt), whereas in the southeast region this boundary is conformable. The Permo-Triassic boundary in many areas of the southwest is interrupted, but the same boundary at many localities in the southeast is continuous, with facies changes sharper and more complex. Accordingly, brachiopod faunas and sequences also differ, with brachiopod assemblages in the southwest fairly consistent throughout, in contrast to highly varied brachiopod faunas and community assemblages in the southeast, matching facies complexity. In this paper, we attempt to address these faunal differences and relationships, as well as to establish a general sequence of Permian brachiopod assemblages for the entire South China plate.

Several studies have dealt with Permian brachiopod faunas from South China (Zhan & Li, 1977; Jin & Hu, 1978; Liao, 1980; Wang et al., 1981; Zeng, 1983; Li et al., 1987; Shen & He, 1994; Xu & Grant, 1994). Most of these emphasized either brachiopod assemblages from individual stages or local assemblages. Here we consider the succession of brachiopod assemblages within a three-fold Permian chronostratigraphic

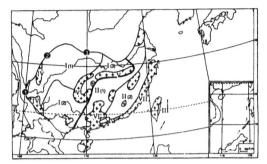

Figure 1. General map of China showing the location of the structural features of the South China plate (*no key provided: editors' note).

framework, and interpret differences among local contemporaneous assemblages as being primarily ecologically controlled. No attempt is made to correlate the succession of brachiopod assemblages of South China with other Permian successions in China, or with brachiopod faunas from elsewhere.

PERMIAN CHRONOSTRATIGRAPHIC SUBDIVISIONS OF CHINA

In China there has been an attempt in recent years to redefine the lower boundary of the Chinese Permian so that it can be correlated with boundaries elsewhere around the world (Wu & Zhao, 1984; Zhou, 1987; Zhou et al., 1987). In addition, there has been a debate whether or not the Chinese Permian System should be divided into three series or maintained as a two-fold system (Huang & Chen, 1987; Zhou et al., 1989; Liang, 1990; Jin et al., 1994; Sheng & Jin, 1994), but the latest seems to favor a three-fold subdivision. Others have proposed different names for Series, Subseries and Stages. The present paper adopts a scheme that is slightly different from that of Sheng & Jin (1994), using brachiopod successions of South China.

BRACHIOPOD FAUNA OF SW SOUTH CHINA

Mapingian Stage. In this paper, we adopt the concept of the Mapingian Stage, as amended by Zhou et al. (1987), corresponding to the Asselian of the Urals. The Mapingian represents the upper part of the traditional Maping Formation of SW China. Grabau (1936) described some 126 brachiopod species from the Maping Limestone in Guizhou and Guangxi, but most of these came from strata older than the type Maping Limestone of Liuzhou (Liao, 1979). Brachiopod faunas from the type section of the Maping Limestone remain poorly known. The best known Mapingian brachiopod fauna was described by Liao (1979) from Longyin in western Guizhou. Here, a continuous section across the uppermost Carboniferous and the lowermost Permian is exposed, divided in ascending order into the Shazitang, Longyin and Baomoshan formations. The Shazitang Fm. contains a rich brachiopod fauna characterised by the abundant, large-sized spiriferid *Choristites jigulensis*, and a new anidanthid called *Protanidanthus*, with large extensive ears and lacking concentric rugae. Liao (1979) named this distinctive fauna the *Choristites jigulensis-Protanidanthus elegans* Assemblage. Other characteristic elements of the assemblage include: *Costachonetes pygmaea, Neochonetes carbonifera, Echinoconchus fasciatus, Juresania juresanensis, Eomarginifera pusilla, Alexenia gratiodentalis, Chaoiella gruenewaldti, Cancrinella cancriniformis, Linoproductus sinensis, Terebratuloidea depressa, Martinia corcula,* and *Phricodothyris extensa.*

Longyinian Stage. Zhan and Li (1984) coined the term Longyinian to represent the strata of both Longyin and Baomoshan formations at Longyin, in western Guizhou. Brachiopods are sparse in the Longyin Fm., but diverse and abundant in the overlying Baomoshan Fm., for which Liao (1979) proposed the *Dictyoclostus uralicus-Neochonetes puanensis* Assemblage. This assemblage is dominated by chonetids such as *Lissochonetes latus,* but *Alexenia mucronata,* '*Undaria*' *longyinensis,* and *Martinia letisulcata* are also very common.

Qixian Stage. We use the Qixian as a stage name in this paper, differing from Zhu & Zhang (1994), who elevated the Qixian to subseries rank, in which they included two new stages, the Luodianian and Xiangboan (in ascending order). In our usage, the Qixian Stage is defined as including three fusulinid zones. the *Misellina claudiae, Nankinella orbicularia* and *Parafusulina multiseptata* zones, from base to top (Wang, 1978). Lithologically, the Qixia Formation is composed of two distinct members, a lower, usually coal-bearing Liangshan Member and an upper Limestone Member. In SW South China, the Liangshan Member has yielded *Chaoina reticulata* Ching, *Monticulifera sinensis* (Frech), *Orthotichia, Spirigerella grandis* Waagen, *Urushtenia* ?, *Orthotichia indica* (Waagen), *Tyloplecta* cf. *nankingensis* (Frech), *Squamularia* cf. *grandis* Huang, *Phricodothyris* cf. *asiatica* (Chao), *Orthotetina, Neoplicatifera lipoensis* (Grabau), *Martinia mongolica* Grabau, *Squamularia. rostrata* (Kutorga), and *Spiriferellina multiplicata* (Sowerby).

The upper Limestone Mbr of the Qixia Fm. contains mainly *Orthotichia, O. chekiangensis* Chao, *O.* cf. *chekiangensis* Chao, *Tyloplecta nankingensis* (Frech), *T. grandicostata* (Chao), *Spirigerella pentagonalis* Chao, *Linoproductus cora* (Orbigny), and *Dictyoclostoidea*. Among these, *Orthotichia chekiangensis* is distributed widely within the member, and has long been regarded as an index fossil for the formation. Similarly, *Tyloplecta nankingensis* (Frech) and *T. grandicostata* (Chao) are also widespread in the Qixia Fm., although they occasionally also occur in the Maokou Fm. Thus, we propose a modified *Orthotichia chekiangensis - Tyloplecta richthofeni-Chaoina reticulata* Assemblage for the Qixian Stage, equivalent to the *Orthotichia chekiangensis - Chaoina reticulata* Assemblage of Zhan & Li (1977).

Maokouan Stage. The Maokou Formation is a sequence of black, calcareous shales-limestones, occasionally intercalated with siliceous beds or cherty nodules. The lower part of

the formation yields abundant and distinctive *Cryptospirifer*, which usually occurs as shell beds at the base of the formation, demarcating the formation clearly from the underlying Qixia Fm. Within this assemblage, the following brachiopods are common, *Cryptospirifer omeishanensis* Huang, *C. striatus* Huang, *C. semiplicata* Huang, *Tyloplecta nankingensis* (Frech), *T. grandicostata* (Chao), *Spirigerella obesa* Huang, *S. grandis* Chao, *S. pentagonalis* Chao, *S. interrupta* Grabau, *Squamularia calori* Gemmellaro, *Marginifera elongata* Huang, *Neophricodothyris asiatica* (Chao), *Martiniopsis omeishanensis* Huang, *Dielasma truncatum* Waagen, and *Meekella ufensis* Chernyshev.

The upper part of the Maokou Formation consists of thick-bedded limestones containing *Tyloplecta nankingensis* (Frech), *Uncisteges crenulata* (Ching), *Neoplicatifera huangi* (Ustritskii), *Martinia triguetra* Gemmellaro, *Asioproductus* aff. *gratiosus* (Waagen), *Cathaysia chonetoides* Chao, and *Chonetes tzunyiensis* Huang. There is an obvious difference between the upper and lower parts of the Maokou Fm., and consequently we recognize two assemblages, with a lower named the *Cryptospirifer omeishanensis,* and an upper the *Uncisteges crenulata-Neoplicatifera huangi* Assemblage. The evolutionary development of brachiopods from the Qixian to Maokouan is continuous, with many species persisting through both stages.

Wujiapingian Stage. A tectonic uplift event called the 'Dongwu Movement', represented by a regional unconformity, occurred between deposition of the Maokou Fm. and overlying Longtan Fm. (equivalent to the Wujiaping Fm.). This unconformity is marked by a continuous sequence of marine strata in SE South China, where a distinct brachiopod fauna, the 'Lengwu Fauna', has been documented. Following uplift, there was a transgression in SW South China during the Wujiapingian, depositing the coal-bearing Longtan Fm. This formation contains abundant, diverse brachiopods, notably,*Orthotetina ruber* (Frech). *O. regularis* Huang, *Orthotichia indica* Waagen, *Steptorhynchus lenticularis* Waagen, *Waagenites barusiensis* (Davidson), *W. soochowensis* (Chao), *W. wongiana* (Chao). *Neochonetes substrophomenoides* Huang, *Cathaysia chonetoides* (Chao), *Meekella abnormalis* Huang, *Asioproductus gratiosus* (Waagen), *A. margaritatus* (Mansuy), *Alatoproductus truncatatus* Ching and Liao, *Spinomarginifera lopingensis* (Kayser), *S. chengyaoyensis* Huang, *Gubleria planata* Ching and Liao, *Leptodus nobilis* (Waagen), *Licharewiella costata* (Waagen), *Haydenella kiangsiensis* (Kayser), *Pugnax pseudoutah* Huang, *Tschernyschewia sinensis* (Chao), *Uncinunellina timorensis* (Beyrich), *Araxathyris timorensis* (Rothpletz), *Punctospirifer multiplicatus* (Sowerby), *P. octapicata* (Davidson), *Squamularia elegantula* (Waagen), *S. indica* (Waagen), *S. grandis* Huang, *Derbyia acutangula* Huang, *Oldhamina grandis* Huang, *O. squamosa* Huang, *O. decipiens* (Koninck), *Edriosteges poyangensis* (Kayser), *Tyloplecta yangtzeensis* (Chao), and *Tschernyschewia* .

Many species characteristic for the Lower and Middle Permian disappeared in the Wujiapingian, while a large number of new forms arose. A discontinuous faunal boundary between Middle and Upper Permian can be recognised. Among the species mentioned above, *Tyloplecta yangtzeensis* Chao, *Asioproductus gratiosus* (Waagen), *Spinomarginifera lopingensis*

(Kayser), *Tschernyschewia sinensis* Chao, *Waagenites soochowensis* (Chao), *Leptodus nobilis* (Waagen), and *Squamularia grandis* Huang are abundant and distributed widely in the Longtan Fm., or equivalent. The Longtan Fm. carries the *Tyloplecta yangtzeensis-Spinomarginifera lopingensis* Assemblage.

Changxingian Stage. The lithology of the Changxing Formation is varied, with three lithofacies recognised (Shen & He, 1994). First is a limestone, well exposed in the Zhongliangshan section at Chongqing, Sichuan. Brachiopods are large and diverse, presumably reflecting an open, normal marine environment. Then comes an alternation of marine and continental siliciclastics, represented by the Wangjiazhai Fm. at Chuanyan in western Sichuan. This facies is dominated by silty to slightly sandy shale, calcareous mudstones and muddy limestones with pyritic nodules and abundant plant fossil fragments, locally interbedded with coal measures. Last are siliceous shales interbedded with clays of the Dalong Fm. at Guiding, in western Guizhou. In some areas, the Dalong Fm. changes laterally to the Changxing Fm. In the Dalong facies, brachiopods are usually less diverse, and mainly represented by small, thin-shelled species, in contrast to abundant ammonoids and radiolarians. Shen & He (1994) recognised three brachiopod assemblages for the Changxingian, in ascending order, an *Oldhamina squamosa - Orthotetina eusarkos* Assemblage, a *Peltichia traversa - Perigeyerella costellata* Assemblage, and a *Waagenites pigmaea - Neochonetes substrophomenoides* Assemblage, and named characteristic brachiopod species for each assemblages. Most species found in the Changxingian continued from the underlying LongtanFm., with only a few restricted to the Changxingian, e.g. *Perigeyerella costellata* Wang, *Semibrachythyrina anshunensis*, and *Peltichia zigzag* Huang.

BRACHIOPODS FROM SE SOUTH CHINA

Early Permian. Brachiopod faunas from the Early Permian Mapingian and Lonyinian stages in this region are represented in the upper and middle Chuanshan Formation, consisting mainly of thick, massive limestones. Brachiopod faunas from this formation are still poorly known and no formal assemblages have yet been proposed.

Qixian Stage. At its type section at Qixia Hill near Nanjing, the Qixia Fm. may be subdivided into, ascending, the Swine Limestone Member, Lower Siliceous Rock Member, Limestone Member, and Upper Siliceous Rock Member. The lower three units contain *Orthotichia chekiangensis* Chao, *O. fushanensis* Liao, *Marginifera obscura* Chao, *Tyloplecta richthofeni* (Chao), *Dictyoclostoidea* sp. *Monticulifera sinensis* (Frech) etc. This *Orthotichia chekiangensis - T. richthofeni* Assemblage is is very similar to the brachiopod fauna of the Qixia Fm. in the SW region. The brachiopod fauna of the uppermost member consists of *Haydenella chiansis* (Chao), *Uncisteges* cf. *crenulata* (Ting), *Urushtenoidea chaoi* (Ching), *Marginifera typica* Waagen, *Cathaysia subpusilla* (Licharew), *Leptodus* cf. *tenuis* (Waagen), *Phricodothyris*, and *Cleiothyridina*. Many of these also occur in the overlying Kufeng Fm. We refer this to the *Haydenella chianensis-Urushtenoidea chaoi - Uncisteges* cf. *crenulata* Assemblage, which may be correlated with the *Uncisteges crenulata-Neoplicatifera huangi* Assemblage of the

Maokou Fm. in the SW region. The Qixia faunas in SE China have been studied by Jin & Hu (1978), Zhao &Tan (1982), and Chang (1987), who proposed slightly different assemblage names, in part reflecting facies.

Kufengian Substage. The Kufeng Formation is widely distributed in Jiangsu and Anhui. Its lower part yields abundant brachiopods, including *Orbicularia anhuiensis* Ching and Hu, *Acosarina indica* (Waagen), *Orthotichia* cf. *chekiangensis* Chao, *O.* aff. *morganiana* (Derby), *Derbyia yangtzeensis* Ching & Hu, *Leptodus tenuis* (Waagen), *Tenuichonetes tenuilirata* (Chao), *T. plicatiformis* (Lee), *Pygmochonetes jingxiangensis* Ching & Hu, *Cathaysia subpusilla* (Licharew), *Haydenella striacosta* (Chan), *H. kiangsiensis* (Kayser), *H. chianensis* (Chao), *Compressoproductus compressus* Waagen, *Anidanthus guichiensis* Ching & Hu, *Cancrinella circularis* Ching & Hu, *Linoproductus* sp., *Dictyoclostoidea kiangsiensis* Wang & Ching, *Neoplicatifera huangi* (Ustritskii), *N. sintanensis* (Chao), *Urushtenoidea chaoi* (Ching), *Uncisteges* sp., *U. crenulata* (Ching), *U. maceus* (Ching), *Alatoproductus trunctus* Ching & Zhu, *Tyloplecta* aff. *nankingensis* (Frech), *Crurithyris lungtanica* Ching & Hu, and *Martinia pentagonalis* Ching & Hu. Among the listed species *Neoplicatifera huangi*, *Tenuichonetes tenuilirata*, *Dictyoclostoidea kiangsiensis* and *Urushtenoidea chaoi* are widespread and abundant within the basal KufengFm. We designate this as the *Uncisteges crenulata - Neoplicatifera huangi - Dictyoclostoidea kiangsiensis* Assemblage. Jin & Hu (1978) correlated this with the *Uncisteges crenulata-Neoplicatifera huangi* fauna of the upper Maokou Fm. in the southwest. Few brachiopods have been discovered in upper Kufeng Fm: its correlation remains problematic, being either equivalent to the Lengwu Fm. in Zhejiang, or the Tongzhiyan Fm. in Fujian.

Lengwuan Substage. The Lengwuan was first proposed by Liang (1990) as a stage name extension of the Lengwu Fm. in western Zhejiang, mainly based on the distinct 'Dipunctella' Fauna (N.B. 'Dipunctella is a junior synonym of *Dicystoconcha* Termier & Termier 1974), transitional from the Maokouan to Wujiapingian faunas. This new stage has since been recognised by Sheng & Jin (1994) as a major chronostratigraphic unit within their Maokouan Series. We prefer to use the Lengwuan at substage rank within the Maokouan Stage. The Lengwu Fm. is characterized by reefal limestones, with a lower member dominated by calcareous siltstones-sandy limestones and an upper member of muddy limestones-bioclastic limestones. Liang (1990) described 235 brachiopod species, and proposed two assemblages corresponding to two lithofacies members. The lower *Semibrachythyrina mucronata-Strophlosiina zhejianensis* Assemblage is dominated by thin-shelled forms with few or no spines. The upper *Monticulifera shizilingensis-Orthotichia gypidularhynchia* Assemblage has abundant thick-shelled, mostly spinose species.

Wujiapingian Stage. The Wujiapingian is marked by the LongtanFm., divisible into a lower coal-bearing argillaceous member containing the *Gigantopteris - Lobatanularia - Pecopteris* flora, a marine sequence with abundant brachiopods, and an upper member of sandstones-carbonaceous shales, interbedded with coal seams. The middle member has *Martinia lopingensis* Chao, *Haydenella kiangsiensis* (Kayser), *Cathaysia chonetoides* (Chao), *Spinomarginifera kueichowensis*

Table 1. Correlation by brachiopod assemblage zones for the southwest and southeast regions of the South China plate during the Permian (*table left as given: editors' note)

Series / Stage	Sequence of brachiopod assemblages	Southwest Region (Yunnan, Guizhou, Guangxi)	Southeast Region — Nanjing Hills	Southeast Region — Zhejiang Province	Southeast Region — Fujian Province
UPPER PERMIAN — Changxing	Waagenites pigmaea-Neochonetes substrophomenoides As. Peltichia tarversa-Perigeyerella costella As. Oldhamina squamularia-Orthotichia cuvarkos As.	*Changxing Fm.:* Perigeyerella costellata. Peltichia zigzag, Waagenites, pigmaea, P. traversa, Spinomarginifera alpha, Permianella liao, Notothyris crassai	*Changxing Fm.:* Brachiopods are rare, containing: Spinomarginifera alpha. Semibrachythyrina anshunensis, Paraphella spp.	*Changxing Fm.:* Orthotichia dowashanensis, Spinomarginifera kueichowensis, Leptodus nobilis, Waagenites soochowensis	*Dalong Fm.:* Oldhamina squamosa, Orthotina ruber, Waagenites soochowensis, Spinomarginifera kueichowensis. Neochonetes substrophomenoides
UPPER PERMIAN — Wujiaping	Tyloplecta yangtzeensis-Spinomarginifera lopingensis Assemblage	*Longtan Fm.:* Havdenella kiangysensis, Tyloplecta yangtzeensis, Eriosteges payangensis, Spinomarginifera granids, Asioproductus gratiosus	*Longtan Fm.:* Tyloplecta yangtzeensis, Eriosteges payangensis, Spinomarginifera lopingensis, Asioproductus gratiosus, Squamularia elegantula	*Lixian Fm.:* Brachiopods rare, but with abundant plant fossils	*Cuipingshan Fm.:* Leptodus nobilis, Orthotina, regularis, O. ruber, Waagenites soochowensis, Oldhamina squamosa. Acosarina indica. Neochonetes substrophomenoides
MIDDLE PERMIAN — Lengwuan	Semibrachythyrina mucronatu-Strophalosiina zhejiangensis; Monticulifera shizilingensis-Orthotichia gypidularhynchia	*Maokou Fm.* [hiatus]	*Gufeng Fm.:* Marine deposits with rare brachiopods	*Lengwu Fm.:* Semibrachythyrina mucronata, Strophalosina, Monticulifera zhejiangensis, Monticulifera shizilingensis	*Tongzhiyan Fm.:* Haydenella, wenganensis, Tyloplecta yangtzeensis, Cathysia chonetoides, Tenuichonetes tenuilirata
MIDDLE PERMIAN — Maokouan / Kufengian	Uncisteges crenulata-Neoplicatifera huangi As. Cryptospirifer omeishanensis Assemblage	*Maokou Fm.:* Uncisteges crenulata, Monticulifera sinensis, Neoplicatifera huangi, Cryptospirifer spp.. etc.	*Gufeng Fm.:* Tenuichonetes tenulirata, Neoplicatifera huangi, Urushtenoidea chaoi, Uncisteges crenulata. Cathaysia subpusilla. Pygmochonetes jinxiangensis	*Unnamed clastic unit:* Neoplicatifera sp.. Uncisteges maceous, Cathysia chonetoides	*Wenbishan Formation:* Uncisteges crenulata. Neoplicatifera cf. sintanensis, Neochonetes substrophomenoides, Linoproductus interruptus. Plichonetes sp.
LOWER PERMIAN — Qixian	Orthotichia chekiangensis-Tyloplecta richthofeni Chaoina reticulata Assemblage	*Qixia Fm.:* Orthotichia chekiangensis, Linoproductus cora, Neoplicatifera lipoensis, Chaoina reticulata, Monticulifera sinensis, Tyloplecta richthofeni, T. nankingensis	*Qixia Fm.:* Orthotichia chekiangensis, Tyloplecta nankingensis, T. richthofeni, Spirigerella pentagonalis, Loczyella nankingensis	*Qixia Fm.:* Orthotichia chekiangensis, Tyloplecta nankiangensis, Spinomarginifera sp.	*Qixia Fm.:* Kiangsiella sp., Chavina. sp.. Neochonetes nantanensis, Neoplicatifera sintanensis, Orthotichia derbyina
LOWER PERMIAN — Longyinian	Dictyoclostus uralicus-Neochonetes puanensis Assemblage	*Longyin & Baomoshan:* Lixsochonetes latus, Neochonetes puanensis, Alexenia mucronata, Dictyoclostus uralicus, "Undaria" logyinensis, Martinia letisulcata	*Middle and Upper Chunshan Formation. Brachiopod faunas poorly known*	Brachiopod faunas poorly known	
LOWER PERMIAN — Mapingian	Chorsiites jigulensis-Protanidanthus elegans Assemblage	*Shazitang Fm.:* Costachonetes pygmaea, Neochonetes carbonifera, Juresania inexsanensis, Alexenia gratioedentalis, Linoproductus simensis, Protanidanthus elegans, Chorstites jigulensis			

114

Huang, *Waagenites soochowensis* (Chao), *Waagenites barusiensis* (Davidson), *Neochonetes substrophomenoides* Huang, *Asioproductus gratiosus* (Waagen), *A. margaritatus* (Mansuy), *Edriosteges poyangensis* (Kayser), *Pugnax pseudoutah* Huang, and *Squamularia elegantula* (Waagen). Almost all also occur in the *Tyloplecta yantzeensis - Spinomarginifera lopingensis* Assemblage of the Longtan Fm. in the southwest.

Changxingian Stage. Liao & Meng (1987) recognised two types of brachiopod biofacies in the Changxingian of SE South China, one associated with limestones, and another in cherts. The limestone facies usually yields a highly diverse brachiopod fauna (Liao & Meng, 1986) at Huatang, Hunan. This fauna is dominated by abundant small, smooth species such as *Uncinunellina tenuis* Liao, *Crurithyris flabilliformis* Liao, *Araxathyris bisulcata* Liao, and *Notothyris crassa* Reed. These limestones may be correlated with the *Peltichia traversa - Perigeyerella costellata* Assemblage of the southwest. In contrast, brachiopods are rare in the cherts, typified by the type Changxingian Fm. at Changxing, N Zhejiang. This assemblage includes *Paryphella sacoformis* Liao, *Paracrurithyris pygmaea* Liao, *Spinomarginifera alpha* Huang, *Semibrachythyrina anshunensis* Liao,and *Huatangia sulcatifera* Liao.

CONCLUSIONS

An integrated scheme of brachiopod assemblages for the whole Permian of South China correlates the SE and SW regions of South China (Table 1). In the southwest, there was generally uplift at the Middle- Upper Permian boundary, resulting in an erosional gap, and faunal discontinuity between the Gufengian and Wujiapingian faunas However, in the southeast region, this gap is represented by the Lengwuan Substage, with a large, diversified '*Dipunctella*' Fauna, containing mixed elements from both the Gufengian and Wujiapingian, indicating Middle to Upper Permian continuity.

REFERENCES

CHANG MEILI. 1987. Fossil Brachiopoda from Chihsia Formation, Anqing, Anhui. Acta Palaeontologica Sinica 26(6):753-766.

GRABAU, A.W. 1936. Early Permian fossils of China, II, Fauna of the Maping Limestone of Kwangsi and Kweichow. Palaeontologia Sinica 8(4):1-441.

HUANG JIQING & BINGWEI CHEN. 1987. The evolution of the Tethys in China and adjacent regions. Geological Pub-lishing House, Beijing, 109p.

JIN YU-GAN, B.F. GLENISTER, G. V. KOTLYAR & SHENG JINZHANG. 1994. An operational scheme of Permian chronostratigraphy. Palaeoworld 4:1-13.

___& SHI-ZHONG HU.1978. Brachiopods of the Kufeng Formation in South Anhui and Nanking Hills. Acta Palaeontologica Sinica 17(2):101-127.

LI LI, DELI YANG & RULIN FENG.1987.The brachiopods and the boundary of the Late Carboniferous-Early Permian in Longlin region, Guangxi. Bulletin Yichang Institute Geology Mineral Resources 11:199-258.

LIANG WENPING. 1990. Lengwu Formation of Permian and its brachiopod fauna in Zhejiang Province. Geological Memoirs, Ministry Geology Mineral Resources 2(10):1-522.

LIAO ZHUOTING.1979. Uppermost Carboniferous brachiopods from western Guizhou. Acta Palaeontologica Sinica 18(6):527-546.

___1980. Brachiopod assemblages from the Upper Permian and Permian-Triassic boundary beds, South China. Canadian Journal Earth Sciences 17(2): 289-295.

___& FENGYUAN MENG. 1987. Late Changxingian brachiopods from Huatang of Chen Xian County, southern Hunan. Memoirs Nanjing Institute Geology Palaeontology 22:71-94.

SCOTESE, C.R. 1994. Permian paleogeographic maps, p.6. In, G.D. Klein (ed.), Pangea: paleoclimate, tectonics, and sedimentation during accretion, zenith, and breakup of a supercontinent. Geological Society of Special Paper 288.

SHEN SHUZHONG & XILIN HE. 1994. Brachiopod assemblages from the Changxingian to lowermost Triassic of southwest China and correlations over the Tethys. Newsletters Stratigraphy18:1-5.

SHENG JINZHANG & YUGAN JIN. 1994. Correlation of Permian deposits in China. Palaeoworld 4:4-113.

___, RUILIN FENG & CHUZHEN CHEN. 1985. Permian and Triassic sedimentary facies and paleogeography of South China, p.59-81. In, K. Nakazawa and & J.M. Dickins (eds), The Tethys, her paleogeography and paleobiogeography from Paleozoic to Mesozoic. Tokyo University Press, Tokyo.

WANG JIANHUA. 1978. The boundaries and zonation of the Qixia Formation near Nanjing. Acta Stratigraphica Sinica 2(1):67-73.

WANG YU, YUGAN JIN, DIYONG LIU et al. 1981. Stratigraphic distribution of Brachiopoda in China. Geological Society America Special Paper 187:97-105.

WU WANGSHI & JIAMING ZHAO.1984. On the biological characteristics and the stratigraphical significance of the Family Kepingothyllidae. Acta Palaeontologica Sinica 23(4):411-419.

XU GUIRONG & R.E.GRANT. 1994. Smithsonian Contributions to Paleobiology 76:1-68.

EARLY PALEOZOIC RADIATION AND CLASSIFICATION
OF ORGANO-PHOSPHATIC BRACHIOPODS

LARS E. HOLMER & LEONID E. POPOV

Institute of Earth Sciences, Historical Geology & Palaeontology, Norbyvägen 22, S-752 36 Uppsala, Sweden;
All-Union Geological Research Institute (VSEGEI), Srednii Prospekt 74, 199026 St Petersburg, Russia

ABSTRACT- The class Lingulata probably represents a monophyletic group within the phylum Brachiopoda, and is redefined here to include also the orders Paterinida, Acrotretida, and Siphonotretida. The origin of various orders within the group is still not fully explicable, but their morphology is here reviewed, and their Paleozoic, particularly Cambro-Ordovician radiation, evaluated using cladistic analysis.

INTRODUCTION

The Class Lingulata, comprising the organo-phosphatic brachiopods, was established recently as a separate clade outside the Brachiopoda (Goryanskii & Popov, 1985), but subsequent studies suggest that it more probably represents a monophyletic group within the phylum (e.g. Popov et al., 1993). Apart from the only extant organo-phosphatic brachiopods, the Lingulida (including the linguloideans and discinoideans), the Lingulata, as defined here, also includes three extinct orders, Paterinida, Acrotretida, and Siphonotretida. This paper provides a brief summary of the morphology of Early Palaeozoic lingulates, and an investigation of the possible patterns in their early radiation by means of a phylogenetic analysis at the family level. The origin of the lingulates is still poorly understood, but the problems of their relationship to other groups of brachiopods is not discussed here [*see also Popov et al.; Ushatinskaya, this volume].

MORPHOLOGY OF EARLY PALAEOZOIC LINGULATES

Linguloidea. The earliest known linguloideans (Obolidae) are from the late Atdabanian (Pelman, 1977). Typical linguloidean characters include well-developed pseudointerareas in both valves, and a well-defined visceral area in the posterior half of the valves. Moreover, the linguloid shells generally have baculate laminae (Holmer, 1989). The muscle patterns of the earliest obolids are more or less identical with those of virtually all younger linguloideans, consisting of six pairs of symmetrically arranged scars (Popov, 1992). Almost all known linguloideans have a well-preserved pair of V-shaped grooves bisecting the ventral visceral area, representing impressions of the pedicle nerve (Holmer, 1991). In most linguloideans (Lingulidae, Pseudolingulidae, Lingulasmatidae, Lingulellotretidae, and most Obolidae), the larval shell is smooth. In contrast, the Zhanatellidae have a larval and post-larval shell with finely pitted micro-ornamentation. The zhanatellids appear at about the same time as the obolids, in the Late Atdabanian. Pitted larval shells also characterize the Eoobolidae, which are first recorded from the Botomian, but here the post-larval shell is finely granulose (Holmer et al., in press). The Elkaniidae are also distinguished by having pitted larval and post-larval shells similar to those of the Zhanatellidae. The elkaniids first appeared during the Late Cambrian. Within this family, there is a tendency towards strongly biconvex shells with high visceral platforms (Popov & Holmer, 1994).

Two families of linguloideans, the Lingulellotretidae and the Dysoristidae, have developed an acrotretide-like pedicle foramen. The oldest of the families, the Lingulellotretidae, arose already in the Botomian, and is characterized by the presence of a pedicle foramen, as well as an internal pedicle tube. The family Dysoristidae, appearing only in the Late Cambrian, also has a pedicle foramen, but unlike in the lingulellotretides, it is placed anterior to the ventral beak, and it was sometimes enlarged through resorbtion of the shell. Also, unlike in the Lingulellotretidae, the larval and post-larval shell of the dysoristids is pitted (Popov & Holmer, 1994).

The Lingulasmatidae, Pseudolingulidae, and Paterulidae appeared early in the Ordovician. The pseudolingulids are characterized by the presence of converging vascula lateralia, poorly developed or absent dorsal vascula media, and reduced pseudointerareas, in addition to a paired umbonal muscle, bisected by the pedicle nerve (Holmer, 1991). Both valves in the paterulids lack pseudointerareas, and have small pitted larval shells much like those of the acrotretoideans; the post-larval shell has a distinctive pitting with rhomboid pits. However, the paterulid interior has a typical linguloidean set of muscle scars, including traces of the pedicle nerve.

Discinoidea. Discinoideans supposedly are younger than the linguloideans, and are first recorded from the Ordovician; however, as yet unstudied discinoidean-like forms possibly occur in the Early Cambrian of China (Holmer et al., unpublished). Recent discinoideans have a musculature with paired posterior and anterior adductors, as well as three pairs of oblique muscles, but the muscle system of most fossil forms is not well known. Discinoideans have a comparatively large, smooth, circular to transversely suboval larval shell, generally about 0.5mm across, which lacks a pedicle notch; the post-larval shells of many fossil discinoideans have a distinctive pitted micro-ornamentation (Holmer, 1989). The shell structure usually includes baculate laminae (Holmer, 1989; Williams et al. 1992). The Trematidae is distinguished from the Discinidae mainly by having a posteriorly unrestricted pedicle opening and larger post-larval pits.

Acrotheloidea. The earliest Botsfordiidae are from the Early Cambrian (Late Atdabanian), and are only slightly older than the Acrothelidae; the superfamily became extinct during the Arenig. The muscle system of the earliest known forms appears to have scars that appear to match a full set of obolid muscles, whilst the number of muscles in the acrothelids was probably reduced (Holmer et al., in press). The botsfordiids have scars of a linguloidean type of V-shaped pedicle nerve. The shell structure of botsfordiids and acrothelids has not yet been studied adequately, but includes baculate laminae (Holmer, 1989; Popov & Holmer, 1994). Both families have pitted larval shells and usually granulose post-larval shells; an enclosed pedicle foramen is present only in the acrothelids, whilst the botsfordiids have a deep pedicle groove (Holmer et al., in press).

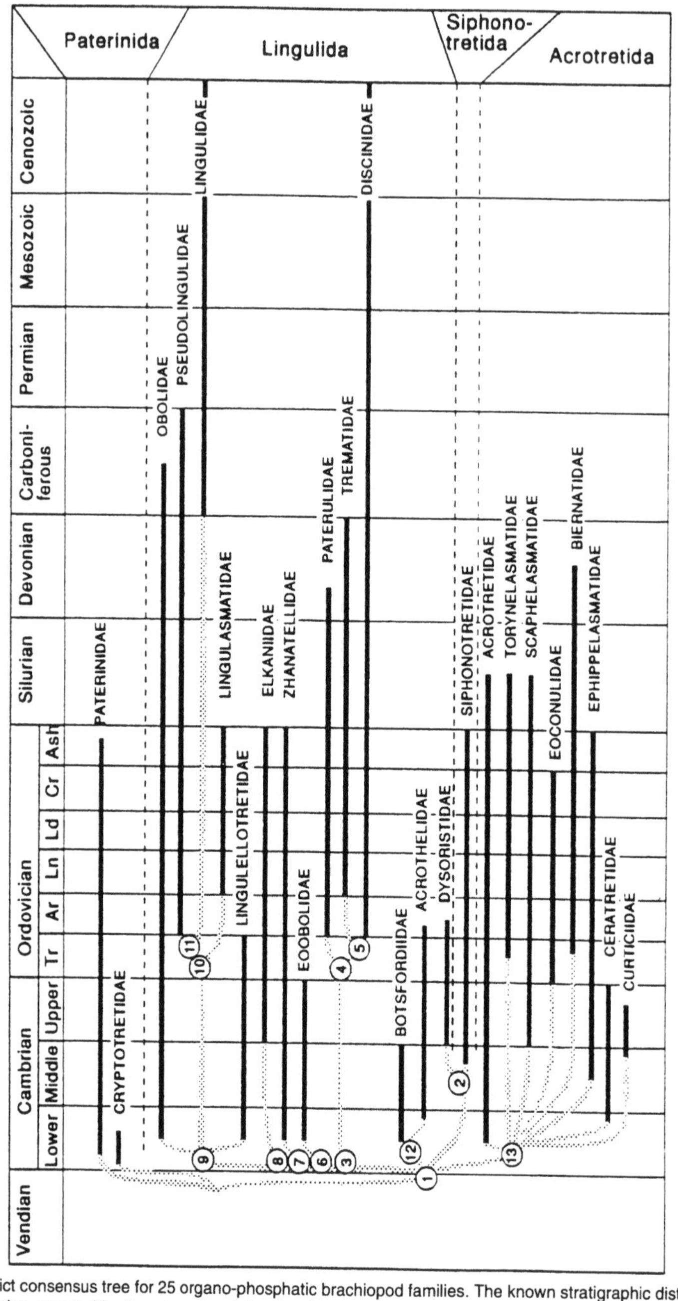

Figure 1. Topology for strict consensus tree for 25 organo-phosphatic brachiopod families. The known stratigraphic distribution is indicated by black bars. Numbered nodes suported by the character states listed in Appendix 1.

Acrotretoidea. The earliest known forms are from the late Atdabanian, and thus of approximately the same age as the earliest acrotheloideans and linguloideans; the group became extinct during the Devonian. The acrotretoidean larval shell is generally small (less than 0.3mm across) and invariably has a pitted micro-ornamentation; the apex is perforated by the pedicle foramen that formed during or shortly after the larval stage (e.g. Holmer, 1989). Acrotretoidean musculature is dis-

tinctive, with large cardinal scars in both valves. The acrotretoidean ventral valve commonly has a distinctive thickened apical process which undoubtedly represents a muscle platform. Popov (1992, fig. 3) proposed that the unique acrotretoidean ventral 'pseudointerarea', having the ventral 'cardinal scars' on its inner side, might not be homologous with that of the linguloideans. Cambrian acrotretoideans are not particularly diverse taxonomically, seemingly belonging mostly to the

Table 1. Character State Matrix used in PAUP analysis of characters listed in Appendix 2. Characters were coded as missing (?) also in the cases where the character is not applicable to the taxon.

	1	2	3	4	5	6	7	8	9	10	11	12	13	14	15	16	17	18	19	20	21	22	23	24	25	26	27	28	29	30	31	32	33	34	35	36	37	38	39	40
	con	rvs	lsh	lsl	lbn	lpo	gra	hsp	vep	apo	lsl	pgr	tms	pmp	hod	dpi	dvu	dp1	pbv	vmc	plm	pad	vam	dvf	amf	apr	dpp	olm	drnc	vme	mbl	dms	daa	shs	shr	plt	vvp	dpm	pnl	lud
PATERINIDAE	2	1	2	0	2	0	0	0	0	0	0	0	?	0	0	0	0	1	0	?	3	?	?	0	0	?	?	?	3	1	0	?	?	?	?	0	1	?	0	0
CRYPTOTRETIDAE	2	1	?	0	2	0	0	0	1	0	0	0	?	0	0	0	0	1	0	?	3	?	?	0	?	?	0	?	3	1	0	?	?	0	?	0	0	?	0	1
OBOLIDAE	1	1	0	0	0	0	0	0	0	1	0	1	0	0	0	0	0	1	1	0	0	0	0	0	0	?	?	0	?	0	0	?	?	0	0	0	0	?	1	0
PSEUDOLINGULIDAE	0	0	0	0	0	0	0	0	2	0	0	1	0	0	0	0	0	1	1	1	0	0	0	0	0	?	0	0	?	0	0	?	1	0	0	0	0	?	1	0
LINGULIDAE	0	0	0	0	0	0	0	0	2	0	0	1	1	1	0	0	0	0	1	0	0	4	0	0	0	?	0	0	1	0	0	?	0	?	0	0	0	?	2	0
LINGULELLOTRETIDAE	1	1	0	0	0	0	0	0	1	1	0	0	0	0	0	0	0	0	0	1	0	0	0	0	1	?	1	0	0	?	0	?	0	1	0	0	1	?	1	0
LINGULASMATIDAE	1	1	0	0	0	0	0	0	2	0	0	1	0	0	0	0	0	1	0	0	0	0	0	0	0	?	0	0	?	0	0	?	0	0	0	0	1	?	1	0
ELKANIIDAE	1	1	0	0	1	0	0	0	1	0	0	0	0	1	0	0	0	0	0	0	0	0	0	0	0	?	1	0	?	1	0	?	1	0	0	1	0	?	0	0
ZHANATELLIDAE	1	1	1	0	1	0	0	0	0	0	0	1	0	0	0	0	0	0	0	0	0	0	0	0	0	?	0	0	1	1	0	?	1	1	1	1	0	?	0	0
DYSORISTIDAE	2	1	1	0	1	0	0	0	1	5	0	0	0	0	0	0	1	0	0	0	0	0	0	0	0	?	0	0	?	?	0	?	1	?	0	1	1	?	1	0
PATERULIDAE	1	1	1	0	2	0	0	0	0	0	0	0	0	0	0	0	0	1	1	0	0	0	0	0	0	?	0	0	?	1	0	?	1	0	1	1	0	?	1	1
EOOBOLIDAE	2	1	1	1	2	0	1	0	1	0	0	1	0	0	0	0	0	0	0	0	0	0	0	0	0	?	0	0	?	0	0	?	1	1	0	0	0	?	0	1
BOTSFORDIIDAE	2	1	1	1	2	0	1	0	2	0	0	0	0	1	0	0	0	1	0	0	?	0	0	0	0	?	0	0	?	?	0	?	0	1	0	0	0	?	0	1
ACCROTHELIDAE	2	1	1	1	2	0	1	0	3	3	0	0	0	0	0	0	0	1	0	0	0	?	0	0	0	?	0	0	?	?	1	?	1	1	1	0	0	?	0	0
SIPHONOTRETIDAE	2	0	0	0	0	0	0	1	1	5	0	0	0	0	0	0	0	1	0	0	0	1	0	0	0	?	0	1	?	?	0	?	1	0	0	0	0	?	0	0
TREMATIDAE	3	0	0	0	?	0	0	0	3	2	1	0	0	0	0	0	1	0	0	1	1	1	0	0	2	?	0	1	1	?	1	?	1	0	0	1	0	?	1	0
DSCINIDAE	?	0	0	0	2	0	0	0	3	2	1	0	2	0	0	0	0	1	0	?	2	2	0	0	2	?	2	2	?	?	1	1	?	2	0	0	1	?	0	0
ACROTRETIDAE	2	1	1	0	2	?	0	0	3	4	0	0	2	0	0	1	0	1	0	0	2	2	0	0	2	1	2	2	?	?	1	1	0	2	0	0	3	2	0	0
TORYNELASMATIDAE	2	1	1	0	2	1	0	0	3	4	0	0	2	0	0	1	0	1	0	?	2	2	0	0	2	6	1	2	?	?	0	2	?	2	0	0	2	1	0	0
SCAPHELASMATIDAE	2	1	1	0	2	0	0	0	3	4	0	0	2	0	0	1	0	1	0	0	2	2	0	0	2	7	2	2	?	?	1	1	0	2	0	0	1	0	0	0
CERATRETIDAE	2	1	1	0	2	0	0	0	3	2	0	0	2	1	0	1	0	0	0	0	2	2	0	0	2	2	1	2	?	?	0	1	0	2	0	0	4	0	0	0
EPHIPPELASMATIDAE	2	1	1	0	2	1	0	0	3	4	0	0	2	0	0	1	0	1	0	0	2	2	0	0	2	6	2	2	?	?	0	?	0	2	0	0	5	1	0	0
EOCONULIDAE	2	1	1	0	2	0	0	0	3	6	0	0	2	0	0	1	?	?	0	0	2	2	0	0	2	4	2	2	?	?	1	?	?	2	1	0	1	?	0	0
CURTICIIDAE	2	1	1	0	2	0	0	0	3	4	0	0	2	0	0	0	1	1	1	?	2	2	0	0	2	2	1	2	2	1	0	1	0	0	0	0	0	1	0	0
BIERNATIDAE	2	1	1	0	2	1	0	0	3	4	0	0	2	0	0	1	0	1	0	?	2	3	0	0	2	6	2	2	2	?	1	0	2	2	0	0	4	1	0	0

119

Acrotretidae, sensu lato, and Ceratretidae. Their main diversification appears to have taken place at about the Late Cambrian to Early Ordovician transition, when most of the acrotretoidean families, such as the Torynelasmatidae, Ephippelasmatidae, Scaphelasmatidae, and others first appeared (e.g. Popov & Holmer, 1994).

Siphonotretoidea. The siphonotretoideans range from the late Middle Cambrian to the late Ashgill, and are distinguished mainly by the presence of hollow spines. They have otherwise a fairly simple shell morphology. The larval and post-larval shell invariably lack pitted micro-ornament. The pedicle foramen commonly extends anteriorly through resorption, producing an elongate triangular pedicle track. Their interior morphology is known inadequately for most taxa; the innermost shell layer is poorly mineralized, and characters like muscle scars and mantle canals are usually not well defined, but are apparently similar to those of linguloideans. Siphonotretoidean shell structure is not well known, but it appears to be made up mostly of micro-granular apatite (Popov & Nõlvak, 1986).

Paterinoidea. This group includes the oldest known organo-phosphatic brachiopods, from the Tommotian. Their morphology is poorly understood and their musculature and mantle canal system seems to differ from those of other organo-phosphatic brachiopods. It appears that some forms have orthide-like musculature and mantle canals (Laurie, 1987).

PHYLOGENETIC ANALYSIS AND PROPOSED CLASSIFICATION

Methods. The data matrix containing 40 unordered and unweighted characters (Table 1, Appendix 2) selected from 25 lingulate family rank taxa was analysed by means of the PAUP programme (Swofford, 1993), using the same methods as described by Holmer et al. (1995) and Popov & Holmer (see this volume). Analyses at the generic level produced inconclusive results as to relationships, in particular for genera of Linguloidea. This problem requires further study, so that here the subdivision at the family level adopted in the previous edition of the Treatise (Rowell, 1965) is largely accepted, with modifications. The origin of the organo-phosphatic brachiopods is still poorly understood. The lingulates are represented by three orders already from the middle part of the Early Cambrian, which suggests that they originated some time earlier, but this is not well documented and potential ancestors are known inadequately. Paterinides are the oldest known bra-chiopods, appearing in the early Tommotian (Pelman, 1977); the families Paterinidae and Cryptotretidae are selected here as the root.

Results and discussion. The analysis produced 354 equally parsimonious trees, each 104 steps long, with a consistency index of 0.712, of which only the highly unresolved strict consensus tree is shown (Figure 1). It is clear from the analysis that the paterinides are not very suitable as reference taxa, because they have a morphology that is not comparable with that of most other organo-phosphatic groups; their phylogenetic position must be studied further. In the absence of other compelling evidence, the Order Paterinida is considered here as a sister stock to all other lingulates. According to the analysis, the non-paterinide lingulates form an unnamed monophyletic group (node 1; Figure 1), which is defined mainly by the absence of a homeodeltidium and in having ventral posterolateral muscle fields anterolateral to the apex.

The resulting consensus tree gives strong support for identifying the Acrotretida (node 13; Figure 1), including only the Acrotretoidea, as a potential monophyletic group defined by 11 derived characters, including the columnar shell struc-ture, changes in the shape of the valves, and the associated simplification of the muscle system, development of the apical process etc. However, the interrelationships beween the included eight families could not be resolved by the present study. The Lingulida and the Linguloidea, as defined here, cannot be confirmed as monophyletic groups based on this analysis; the Linguloidea (including ten families) includes the Paterulidae and Dysoristidae, but excludes the Trematidae, Disciniidae, and Siphonotretidae (Figure 1). The position and relationships of the Dysoristidae and Siphonotretidae are very uncertain. Rowell (1965) placed both groups in the Acrotretida, and the analysis gives weak support for grouping them together as a separate monophyletic stock (node 2; Figure 1), based mainly on the presence of a pedicle foramen anterior to the apex, as well as resorption. However, the Dysoristidae is similar to the lingulids in most other characters, including shell structure and ornamentation, and it is here placed within the Lingulida as proposed by Popov & Ushatinskaya (1992) and Popov & Holmer (1994). The siphonotretids differ from all other lingulates in having hollow spines and are considered here as a separate order (Holmer, 1989).

The discinoideans were included previously in the Acrotretida (Rowell, 1965), or regarded as a separate order (Holmer, 1989). However, this analysis indicates that they might be considered better as a superfamily within the Lingulida, possibly a sister group to the Paterulidae, with which they share some potential derived characters (node 4; Figure 1). Thus, it seems that the discinoideans may have been derived directly from within the Lingulida, but the earliest history of the discinoideans is still poorly known; the Zhanatellidae has also been proposed as a possible ancestral stock for the group (Popov, 1992; Popov & Ushatinskaya, 1992). The botsfordiids and acrothelids were placed previously within the Acrotretida, but the arrangement of muscle scars as well as other features suggest that they may be better classified as a separate superfamily within the Lingulida. The presence of pitting on the larval shell in the acrotheloideans and the acrotretoideans apparently does not represent a synapomorphy for the Acrotretida as proposed by Rowell (1986), and Holmer (1989).

REFERENCES

GORYANSKII, V.Yu & L.E. POPOV. 1985. Morfologiya, sistematicheskoe polozhenie i proiskhozhdenie bezzamkovykh brakhiopod s karbonatnoj rakovinoj. [The morphology, systematic position, and origin of inarticulate brachiopods with carbonate shells] Paleontologicheskii Zhurnal 3 (1985):3-13.

HOLMER, L.E. 1989. Middle Ordovician phosphatic inarticulate brachiopods from Västergötland and Dalarna, Sweden. Fossils Strata 26:1-172.

___1991. The systematic postition of Pseudolingula Mickwitz and related lingulacean brachiopods, p.15-21. In, D.I. MacKinnon, D.E. Lee & J.D. Campbell (eds.), Brachiopods through time. Proceedings of the 2nd International Brachiopod Congress. Balkema, Rotterdam.

___L.E. POPOV, M.G. BASSETT& J. LAURIE (in press). Phylogenetic analysis and ordinal classification of the Brachiopoda. Palaeontology.

___,___& WRONA, R. (in press). Early Cambrian brachiopods from King George Island, West Antarctica. Palaeontologica Polonica.

LAURIE, J. 1987. The musculature and vascular system of two species of Cambrian Paterinida (Brachiopoda). Journal Australian Geology Geophysics 10:261-265.

PELMAN, Yu.L.1977. Ranne i srednekembriiskije bezzam-kovyje brakhiopody Sibirskoj Platformy [Early and Middle Cambrian inarticulate brachioopods of the Siberian Plate].

Trudy Instituta Geologii Geofiziki Sibirskogo Otdelenija 36:1-168.

POPOV, L.E. 1992. The Cambrian radiation of brachiopods, p.399-423. In, J.H. Lipps & P.W. Signor. (eds.). Origin and early evolution of Metazoa. Plenum, New York.

___, M.G. BASSETT, L.E. HOLMER, L.E. & J. LAURIE. 1993. Phyletic analysis of higher taxa of Brachiopoda. Lethaia 26:1-5.

___& L.E. HOLMER. 1994. Cambrian-Ordovician lingulate brachiopods from Scandinavia, Kazakhstan, and South Ural Mountains. Fossils Strata 35:1-156.

___&___1996. Radiation of the earliest calcareous brachiopods [this volume].

___& J. NÕLVAK. 1987. Revision of the morphology and systematic position of the genus *Acanthambonia* (Brachiopoda, Inarticulata). Eesti NSV Teaduste Akadeemia Toimetised, Geoloogia 36:14- 19.

___& G.T. Ushatinskaya. 1992. Lingulidy, proiskhozhdenie discinids, sistematika bysokikh taksonov, p.59-67. [Lingulids, the origin of discinids, systematics of higher taxa]. In, L.N. Repina & A. Yu. Rozanov (eds.), Drevneishie brakhiopody territorii Severnoi Evrazii. Akademiya Nauk, Novosibirsk.

ROWELL, A.J. 1965. Inarticulata, p.260-296. In, R.C. Moore (ed.), Treatise on Invertebrate Paleontology, Part H. Geological Society of America and University of Kansas Press, Lawrence.

___1986. The distribution and inferred larval dispersal of *Rhondellina dorei*: a new Cambrian brachiopod (Acrotretida). Journal Paleontology 60:1056-1065.

SWOFFORD, D.L. 1993. Phylogenetic analysis using parsimony, Version 3.1.1. Computer program, Illinois Natural History Survey, Champaign, Illinois.

WILLIAMS, A., S. MACKAY & M. CUSAK.1992. Structure of the organo-phosphatic shell of the brachiopod *Discinisca*. Philosophical Transactions Royal Society, London B337:83-104.

APPENDIX 1

List of synapomorphies supporting the numbered nodes in Figure 1. Character and characters states numbered as in Appendix 2. Homoplasies are marked by *.

Nodes

(1)ventral pseudointerarea not differentiated (9:3)*; homeodeltidium absent (15:0); ventral posterolateral muscle fields anterolaterally to apex (21:0); ventral valve gently convex (37:0)*.

(2)boundary of larval shell poorly defined (5:0)*; ventral pseudointerarea present, shelf-like (9:1)*; foramen situated anterior to umbo (10:5)*; resorption around pedicle foramen (35:1)*.

(3)shell dorsibiconvex (1:1); pitted post-larval micro-ornamen-tation present (36:1)*; pedicle nerve impression bisecting umbonal muscle scar (39:1)*.

(4)dorsal umbo submarginal (17:1); dorsal pseudointerarea absent (18:0)*; posterior body wall at a distance from posterior margin (19:1)*.

(5)shell convexoplane (1:3); shell equivalved (2:0)*; larval shell smooth (3:0)*; pedicle opening forming elongate track on posterior slope (10:2)*; listrium present (11:1); ventral posterolateral muscle fields posterolateral to apex (21:1); ventral posterior adductor muscle scars paired posterolateral (22:1); outside lateral muscle scars absent (28:1); mantle canals bifurcate (29:1)*.

(6) ventral pseudointerarea present, shelf-like (9:1)*; pedicle groove present (12:1)*.

(7) boundary of larval shell marked by elevated rim (5:1)*.

(8) larval shell smooth (3:0)*; dorsal visceral field forming platform (24:1)*; ventral anterior muscle field (25:1)*.

(9) boundary of larval shell poorly defined (5:0)*; pitted post-larval micro-ornamentation absent (36:0)*.

(10) ventral pseudointerarea vestigial (9:2)*; posterior body wall at a distance from posterior margin (19:1)*; dorsal vascula media absent (30:0).

(11) shell biconvex (1:0); shell equivalved (2:0)*; dorsal vis-ceral field weakly defined (24:0)*; ventral anterior muscle field weakly defined (25:0)*.

(12) ventral larval tubercles present (4:1); granulated orna-mentation (7:1)*; dorsal larval tubercles present (40:1).

(13) small apical foramen (10:4)*; transmedian muscle scars on inner side of acrotretoid 'pseudointerarea' (13:2); acro-tretoid 'pseudointerarea' present (16:1); ventral posterolateral muscle fields situated on inner side of acrotretoid 'pseudo-interarea' (21:2); ventral posterior adductor muscle scars paired subcentral (22:2); ventral anterior muscle field on apical process (25:2)*; posterolateral muscle fields forming cardinal scars (27:2); outside lateral muscle scars combined with middle lateral muscle scars

(28:2); dorsal anterior adductor scars absent (33:0); shell structure columnar (34:2); ventral valve strongly convex to conical (37:1)*.

APPENDIX 2

List of coded characters used in the cladistic analysis (a full discussion of the characters will be published elsewhere).

(1)con-convexity: biconvex (0), dorsibiconvex (1), ven-tribiconvex or planoconvex (2), convexoplane (3).

(2) rvs-relative size of valves: equivalved (0), inequi-valved (1).

(3) lsh-ornamentation of larval shell: smooth (0), pitted (1), pustulose (2).

(4) ls1-ventral larval tubercles: absent (0), present (1).

(5) lbn-boundary of larval shell: poorly defined (0), marked by elevated rim (1), marked by elevated rim and change in the micro-ornamentation (2).

(6) lpo- larval pedicle opening: forming marginal notch in larval shell (0), opening within larval shell (1).

(7) gra-granulation: absent (0), present (1), radial rows of granules (2).

(8) hsp-hollow spines: absent (0), present (1).

(9)vep-ventral pseudointerarea: absent (0), present, shelf-like (1), vestigial (2), not differentiated (3).

(10) apo-adult pedicle opening: delthyrial (0), elongate pedicle foramen on pseudointerarea (1), elongate pedicle track on posterior slope (2), foramen situated posterior to umbo (3), small apical foramen (4), foramen situated anterior to umbo (5), secondary delthyrium (6).

(11) lst-listrium: absent (0), present (1).

(12) pgr-pedicle groove: absent (0), present (1), present in juveniles, transformed into internal pedicle tube in adults (2).

(13)tms-transmedian muscle scars: paired symmetri-cal (0), asymmetrical (1), on inner side on acrotretoid ventral 'pseudointerarea' (2).

(14)pmp-posterolateral muscle platforms: absent (0), present, posterolateral (1), present posteromedian (2).

(15) hod-homeodeltidium: absent (0), present (1), vesti-gial (2).

(16)dpi-acrotretoid ventral 'pseudointerarea': ab-sent (0), present (1).

(17)dvu-dorsal umbo: marginal (0), submarginal (1), sub-central (2).

(18)dp1-dorsal pseudointerarea: absent (0), present (1).

(19) pbv-posterior body wall: close to posterior margin (0), at a distance from posterior margin (1)

(20) vmc-mantle canals: baculate (0), bifurcate (1), pin-nate (2), saccate (3).

(21)plm-ventral posterolateral muscle fields: anterolaterally to apex (0), posterolaterally to apex (1), situ-ated on the inner sides of the acrotretoid 'pseudointerarea' (2), situated on the inner sides of the homeodeltidium (3).

(22)pad-ventral posterior adductor muscle scars: paired posteromedian (0), paired posterolateral (1), paired, subcentral (2), absent (3), forming single scar (4).

(23)vam-ventral vascula media: absent (0), present (1).

(24) dvf-dorsal visceral field: weakly defined or slightly thickened anteriorly (0), forming elevated platform (1).

(25) amf-ventral anterior muscle field: weakly defined or slightly raised (0), forming solid muscle platform (1), on apical process (2).

(26) apr-apical process (if present): low ridge anterior to pedicle tube (0), occluding apex (1), wide ridge perforated by pedicle tube (2), wide, subtriangular, anterior and lateral to foramen (3), wide, subtriangular, anterior to foramen (4), boss-like (5), vestigial to absent (6), high septum (7), high ridge anterior anterior to foramen (8).

(27) dpp-posterolateral muscle fields: weakly defined to slightly raised (0), forming platforms (1), forming cardinal scars (2).

(28) olm-outside lateral muscle scars: present (0), absent (1), combined with middle lateral muscle scars (2).

(29) dmc-mantle canals: baculate (0), bifurcate (1), pin-nate (2), saccate (3).

(30)vme-dorsal vascula media: absent (0), present, short (1), present, long (2).

(31) mbt- median buttress: absent (0), present (1).

(32) dms-dorsal median septum or ridge (in acrotretoideans): absent (0), low triangular (1), high triangular (2).

(33)daa-dorsal anterior adductor scars: absent (0), present (1).

(34)shs-shell structure: baculate (0), granular (1), colum-nar (2).

(35) shr-resorption around pedicle foramen: absent (0), present (1).

(36)pit-pitted post-larval micro-ornamentation: absent (0), present (1).

(37) vvp-profile of ventral valve: gently convex (0), strongly convex to conical (1), low conical, cataline with subcental apex (2), high conical, procline to cataline (3), high conical, apsacline to procline (4), adopting shape of substrate (5).

(38)dpm-shape of acrotretoid pseudointerarea: convex in cross section, well defined laterally (0), convex in cross section, poorly defined laterally (1), flattened, well defined laterally (2).

(39)pni-pedicle nerve impression: absent (0), bisecting umbonal muscle scars (1), passing lateral to umbonal muscle scars (2).

(40)tud-dorsal larval tubercles: absent (0), present (1).

ORDOVICIAN (LLANVIRN-ASHGILL) RHYNCHONELLID BRACHIOPOD BIOGEOGRAPHY

JISUO JIN

Department of Earth Sciences, Laurentian University, Sudbury, Ontario P3E 2C6, Canada

ABSTRACT- The earliest rhynchonellids, *Rostricellula*, *Dorytreta*, *Ancistrorhyncha*, and *Sphenotreta*, evolved during Middle Ordovician (Llanvirn) time and became widespread in shallow marine environments of Laurentia, Siberia, Kazakhstan, and other paleoequatorial continents. Intercontinental migration of these precursor rhynchonellids is interpreted to have been facilitated by dispersal through island faunas as stepping stones during sealevel lowstand, and reduced oceanic deep-water barriers. During the Caradoc, rhynchonellids began to show provincialism. Among the five genera (*Rhynchotrema*, *Oligorhynchia*, *Drepanorhyncha*, *Orthorhynchula*, and *Orthorhynchuloides*) that emerged in Laurentia, only *Rhynchotrema* occurred widely in other paleocontinents. This marked the last successful intercontinental migration of rhynchonellids during the Middle and Late Ordovician. Provincialism became most pronounced in the latest Caradoc and early Ashgill, when three rhynchonellid genera (*Hiscobeccus*, *Lepidocyclus*, and *Hypsiptycha*) evolved and spread rapidly to nearly every part of Laurentia but rarely succeeded in migrating to other continents. Elsewhere, only sporadic and questionable *Lepidocyclus* and *Hypsiptycha* have been reported from Baltica and Kazakhstan. Similarly, the Siberian genera *Lepidocycloides* and *Evenkorhynchia* rarely established themselves in other continents: only one occurrence of *Lepidocycloides* is known from the western margin of Laurentia (Canadian Rocky Mountains). Onset and peak of rhynchonellid provincialism coincided with Caradoc-early Ashgill sealevel rise, which may have caused drowning of island faunas, enhance by further separation of Laurentia and Siberia. By the latest Ordovician, intercontinental mixing of taxa increased, paradoxically as a result of substantial sealevel drop and closure of the lapetus Ocean, as shown by distribution of *Plectothyrella*, which became widespread as a common element of the *Hirnantia* fauna.

INTRODUCTION

There have been several previous studies of Ordovician brachiopod biogeography based on comparison of entire brachiopod faunas (e.g. Williams,1969; Jaanusson, 1971, 1973; Jaanusson & Bergström,1980; Sheehan & Coorough, 1990). Detailed analysis of rhynchonellid biogeography has been rare, despite some puzzling, yet fascinating distribution patterns of these brachiopods in the Middle and Late Ordovician. Why, for example, were the earliest rhynchonellids restricted to the eastern marginal platforms of Laurentia, but globally widespread across the continents in the paleoequatorial regions? Why did some late Caradoc-early Ashgill rhynchonellid genera, such as *Lepidocyclus* Wang, 1949, *Hiscobeccus* Amsden, 1983, and *Hypsiptycha* Wang, 1949, become ubiquitous in the North American epicontinental seas, but rarely succeed in migrating to other continents? Moreover, why did some other genera that co-existed with the *Lepidocyclus* stock, such as *Rhynchotrema* Hall, 1860, and *Rostricellula* Ulrich and Cooper, 1942, still have a cosmopolitan distribution during the Caradoc and Ashgill? The present paper attempts to answer these questions by tracing the dispersal patterns of early rhynchonellids and their descendants.

Rhynchonellids are reliable indicators of intercontinental faunal relationship because these abundant brachiopods first appeared in the Middle Ordovician and their spatial distribution was not complicated by pre-Llanvirn dispersal among continents. Many brachiopod genera have long stratigraphic ranges, and genera spanning several stages are relatively common. In biogeographic studies, these hold-over taxa may create confusion for the time interval under examination because their origin and migration may have taken place at an earlier time. In the present study, investigation of rhynchonellid biogeography is divided into three intervals: Llanvirn, Caradoc, and Ashgill. For each interval, only those taxa that evolved within that time are used for comparison. For example, *Rostricellula* first occurred in the Llanvirn and is taken into consideration only for that time; its subsequent presence is disregarded for the analyses of Caradoc and Ashgill biogeography because intercontinental dispersal of the genus was already achieved in the Llanvirn.

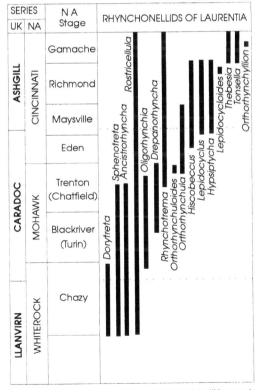

Figure 1. Stratigraphic ranges of Ordovician rhynchonellid genera in Laurentia (North America, Greenland, Scotland, and northern part of Ireland).

123

In the present study, correlation of North American Ordovician series and stages with those of Britain follows the works of Ross & Bergström (1982), Sloan (1987), and Bergström et al. (1988). Following Fortey et al. (1995), the Llandeilo is treated as the last stage of the Llanvirn Series. The new stage terms of Leslie & Bergström (1995) for the Mohawkian Series are not used here because it is not yet easy to assign all the rhynchonellid-bearing Mohawkian stratigraphic units precisely to these stages, and these stage names are difficult to apply outside North America.

RHYNCHONELLID DISTRIBUTION PATTERNS

Llanvirn dispersal. The earliest rhynchonellids are recorded in rocks of Llanvirn (late Abereiddian-early Llandeilian) age (Figure 1). Determination of the exact time and locale for the appearance of the first rhynchonellid species has been hampered by two factors, (1) lack of precise dating and correlation of the rock units containing early rhynchonellids, and difficulty in pinpointing the very earliest rhynchonellid occurrences recorded for every continent or region, and, (2) lack of detailed morphologic and taxonomic study of some of the earliest rhynchonellids, and difficulty in determining precisely which rhynchonellid genus or species appeared first. In general, earliest rhynchonellids appear to be characterised by lack of a dorsal cardinal process, as shown by *Rostricellula* Ulrich & Cooper, 1942; *Ancistrorhyncha* Ulrich & Cooper, 1942; *Dorytreta* Cooper, 1956; *Sphenotreta* Cooper, 1956; and *Lenatoechia* Nikiforova, 1970.

The difficulty in locating 'centres of origin' for characteristic taxa was regarded by Spjeldnaes (1981) as the 'central problem' in his study of Ordovician brachiopod biogeography, because the direction of faunal migration cannot be defined precisely without knowledge of such centres of origin. The problem is further complicated by the hypothesis of 'island faunas' (Neuman, 1972; Fortey, 1984), which suggested that many new taxa in the Ordovician originated around tectonically unstable volcanic islands in the Iapetus Ocean. Such islands with the early island faunas may have disappeared during the destruction of the Iapetus Ocean toward the end of the Ordovician. In the context of this theory, the earliest occurrence of certain taxa in a region does not necessarily indicate that the region was the centre of origin.

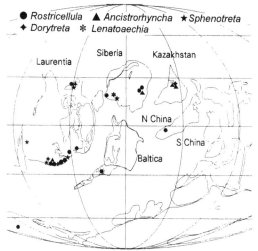

Figure 2. Paleogeographic distribution of the earliest rhynchonellid genera in Llanvirn time.

In Baltica, one of the earliest rhynchonellids, *Rostricellula triangularis* Williams, 1949, occurs in the late Llanvirn (Abereidd-Llandeilo) of Wales (MacGregor, 1961; Cocks, 1978). Most specimens of this species are internal moulds. According to MacGregor (1961), these specimens lack a cardinal process, and thus their assignment to *Rostricellula* is justified. In Laurentia, the earliest rhynchonellids occur predominantly on the eastern margins of the plate (Appalachians; see Figure 2). These include *Sphenotreta acutirostris* (Hall, 1847) from the Crown Point Formation of New York and Vermont, *Dorytreta ovata* Cooper, 1956 from the Lenoir Fm. of Tennessee, *Rostricellula basalaris* Cooper, 1956 from the Mosheim and lower Lenoir formations of Tennessee and Alabama, *Rostricellula plena* (Hall, 1847) from the Valcour Fm. of New York and Vermont, and *Rostricellula orientalis* (Billings, 1859) from the MinganFm. of Quebec (Figure 3:1-4). Cooper (1956, 1976) assigned the Crown Point, Valcour, Mosheim, and Lenoir formations to his late 'Marmorian' (approximately of latest Abereiddian to Llandeilian). On the basis of conodonts and other fossil groups, Bergström et al. (1988) correlated the Mosheim and Lenoir formations in Friendsville, Tennessee with the *friendsvillensis* conodont zone (Abereiddian), noting that the Chazy Group (Day Point, Crown Point, and Valcour formations) spans the *friends-villensis* and the overlying *sweeti* zones (Abereiddian-Llandeilian). More recent international correlations suggest that the *friendsvillensis* conodont zone is largely Llandeilian, and that only its base extends into the uppermost Abereiddian (Webby, 1995). In Laurentia, early rhynchonellids occur in open marine, shallow shelf carbonates along the eastern margin, as indicated by the presence of oolites, cross-laminated calcarenites, and bryozoan-microbial and tabulate coral-sponge reefs in the Day Point, Crown Point, and Valcour formations in Vermont and New York. Ross (1981) postulated that these reefs and carbonate coastal lagoons offered considerable topographic and environmental heterogeneity. Brachiopods were common constituents in a variety of such benthic communities.

The first occurrences of rhynchonellids in Siberia and most other paleocontinents appear to post-date those of Laurentia (Figure 4). By Llandeilian time, rhynchonellids became widespread across all paleoequatorial continents (see Figure 2). *Rostricellula* has been found in Llandeilian rocks of Siberia (Rozman, 1979) and North China (Fu, 1982). From cold-water Gondwana, Havlíček (1961,1989) reported the presence of *Rostricellula ambigena* (Barrande, 1847) in Bohemia, Germany, and North Africa (central Sahara). Other rhynchonellids, which were restricted to the eastern margin of Laurentia during the Abereiddian, also dispersed to other continents by Llandeilian time. *Ancistrorhyncha*, for example, has been found in the Krivoluka beds (Kiren-Kudra horizon, middle Llandeilian) in the Kulyumbe River area of the northwestern Siberian Platform (Yadrenkina, 1974). The genus occurs also in the Tselinograd horizon (Llandeilian) of the Chu-Ili Mountains and Chingiz of Kazakhstan (Nikiforova & Popov, 1981; Nikitin & Popov, 1984). *Sphenotreta* also invaded the Siberian Platform at this time, as documented by Yadrenkina (1989) from the Krivoluka beds of the Lena River area. In the Ordovician, Siberia, like eastern Laurentia, was mostly covered by carbonate platform seas. Shallow water environments prevailed in the Early Ordovician (Tremadoc-Arenig), with wide distribution of stromatolites, dolomites, and minor amount of sandstones in the Tunguska and Lena River regions (Nikiforova & Andreeva, 1961; Kanygin et al., 1989). In the Angara River area, the brachiopod fauna was dominated by lingulid communitiesin the Ustkut and Mamyr shales, siltstones, and sandstones, indicating an intertidal to very shallow subtidal environment (Yadrenkina, 1984). A thick sequence of sandstones developed in the Tunguska region in late Llanvirn time. The Krivoluka beds, of Abereiddian-

124

Figure 3. Some of the characteristic North American rhynchonellids in the Ordovician. 1-4. *Rostricellula orientalis* (Billings, 1859), GSC 102537, Mingan Formation, Llandeilian, Mingan Islands, Québec, x5. 5-8. *Rhynchotrema wisconsinense* Fenton & Fenton, 1923, GSC 99717, Decorah Shale, Trentonian, Kenyon, Minnesota, x4. 9-12. *Hiscobeccus mackenziensis* Jin & Lenz, 1992, holotype, GSC 99747, Whittaker Formation, lowermost Ashgill, Mackenzie Mountains, Canada, x4. 13-16. *Lepidocyclus laddi* Wang, 1949, GSC 99718, Whittaker Formation, x2.5.

Llandeilian age, consist of shales, siltstones, sandstones, and carbonates, with limestones becoming increasingly common upwards. Rhynchonellids first occur in the shales and siltstones (with interbeds of muddy limestones) of the middle Krivoluka beds, such as *Rostricellula transversa* Cooper, 1956 and *R. basalaris* Cooper, 1956 (see Yadrenkina, 1974). In Caradoc and Ashgill time, normal marine carbonate became the main type of deposits, with only minor amount of siltstones and shales (Moskalenko et al., 1978).

So far, there are only five genera described from the upper Abereiddian-lower Llandeilian rocks around the world: *Rostricellula*, *Dorytreta*, *Sphenotreta*, *Ancistrorhyncha*, and *Lenatoechia*. The greatest diversity at both generic and specific levels was reached undoubtedly in the eastern margin of Laurentia. If a Llanvirn age for the bulk of the Marmor and Chazy formations is to be accepted (Bergström et al. 1988), the Appalachian basins contain the earliest known, and taxonomically most diverse, early rhynchonellids, and the

125

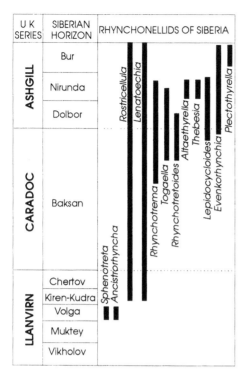

Figure 4. Stratigraphic ranges of Middle and Late Ordovician rhynchonellid genera in Siberia (data mainly from Nikiforova & Andreeva, 1961; Rozman, 1979; Yadrenkina, 1974, 1978, 1984, 1989.)

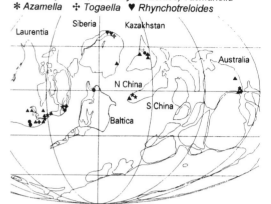

Figure 5. Paleogeographic distribution of characteristic rhynchonellid genera in the Caradoc.

region could well be either the centre of origin for the earliest rhynchonellids, or very close to the centre. The early rhynchonellids in Siberia are perhaps the second most diverse and show great similarity to those in Laurentia, but the earliest representatives of *Rostricellula*, *Sphenotreta*, and *Ancistrorhyncha* did not appear until early Llandeilian time, notably later than those of Laurentia.

In Llanvirn time, rhynchonellid endemism or provincialism at the generic level was minimal. None of the rhynchonellids evolved during this time appear to have been confined to a single continent. *Rostricellula* had a remarkably wide distribution, found not only in the continents in the paleoequatorial regions (except for Kazakhstan), but also in the cold to temperate regions, including southern Europe, North Africa, and South America (Havlíček, 1971, 1989; Havlíček & Branisa, 1980; Leone et al., 1991; Villas, 1992). This indicates that the genus was able to tolerate a wide range of water temperatures and substrate conditions. In view of their ages and diversity, migration of the early rhynchonellids in Llanvirn time was most likely unidirectional -- from Laurentia eastward. N American taxa can be found commonly in Siberia and Kazakhstan, but not vice versa. A similar migration pattern was observed in the Llanvirn cephalopods and, to a lesser extent, trilobites (Spjeldnaes, 1981).

Diversification-endemism (Caradoc). In Caradoc time, the total number of rhynchonellid genera increased considerably, from five at the end of the Llanvirn to fifteen by the end of the Caradoc. This parallels the expansion of the earliest spire-bearing atrypoid brachiopods (Copper, 1986). In rhynchonellid biogeography, the time interval can be regarded as a transition

from the cosmopolitanism to the Ashgill provincialism. Some of the genera that evolved in the early and middle Caradoc became widespread across the continents in the paleoequatorial regions. The best example is *Rhynchotrema*, one of the most common of its group, characterised by a blade-like cardinal process in the dorsal valve. In Laurentia, *Rhynchotrema* first occurs in Trentonian rocks, with species like *R. increbescens* (Hall, 1847) in the Trenton Limestone of New York, *R. wisconsinense* Fenton & Fenton, 1923 from the Decorah Shale of Minnesota (Figure 3:5-8), and other species of *Rhynchotrema* in eastern Laurentia (Figure 3). The oligorhynchiids made their appearances in the early Caradoc of Laurentia, such as *Oligorhynchia angulata*, and *O. subplana* Cooper, 1956 from the Lincolnshire Formation of Tennessee and Virginia and *O. conybearei* (Reed, 1917) from the Balclatchie Formation of Girvan, Scotland (Williams, 1962). By the middle Caradoc, the oligorhynchiids spread to northern Baltica and Kazakhstan, represented, for example, by *O. elegantula* Bondarev, 1968 from the Yugor horizon of Vaigach (Baltica), and *O. mica* Nikitin & Popov (1984) from the Bestamak Formation of Chingiz. *Drepanorhyncha* occurs in Blackriverian strata (middle Caradoc) along the eastern margin of Laurentia (New York, Ontario, Scotland). Outside Laurentia, *Drepanorhyncha* has been found in upper Caradoc rocks of NW China (Fu, 1982).

Differentiation of rhynchonellid faunas of Laurentia from those of Siberia, Kazakhstan, and N China began to manifest itself in late Caradoc time. Rhynchonellid diversification and endemism were apparent in these continents. Among the nine genera present in Laurentia, five appeared during the early-middle Caradoc (*Rhynchotrema*, *Oligorhynchia*, *Drepanorhyncha*, *Orthorhynchula*, *Orthorhynchuloides*). *Orthorhynchula* and *Orthorhynchuloides* (regarded by some as synonymous) are endemic to Laurentia. Among the newly evolved early-middle Caradoc genera, Siberia has little in common with Laurentia, except for *Rhynchotrema*. By late Baksan time (late Caradoc), two genera endemic to Siberia, *Rhynchotretoides* and *Togaella*, developed and were largely confined to the Altai Mountains and Mongolia (Rozman, 1981; Kulkov & Severgina, 1989). Two endemic rhynchonellid genera, *Azamella* and *Tasmanella* Laurie, 1991, are known from eastern Australia. *Tasmanella* from the upper Benjamin Limestone (mid-upper Caradoc) of Tasmania, however, is similar to North American *Orthorhynchula*, and *Azamella* from the lower Benjamin Limestone (lower-mid Caradoc), shows a strong resemblance to the Siberian genus *Lepidocycloides* (late Caradoc) in its imbricating growth lamellae and absence of a cardinal process.

Pronounced provincialism (Ashgill). It has been suggested that North American epicontinental seas contained one of the most endemic brachiopod faunas in the world during the Late Ordovician (Sheehan, 1988; Sheehan & Coorough, 1990). Intensification of rhynchonellid provincialism in Ashgill time is shown most convincingly by the distribution of the *Lepidocyclus-Hypsiptycha* fauna in Laurentia (Figure 6). The fauna is characterised by three distinctive genera, *Lepidocyclus*, *Hypsiptycha*, and *Hiscobeccus*, whose large, commonly globular, and lamellose shells are easy to identify (see Figure 3:9-16). *Hiscobeccus* is the oldest of this stock and the oldest species, *H. mackenziensis* Jin & Lenz, 1992 occurs in the Advance Formation of late Trentonian-Edenian (late Caradoc) age in the northern Canadian Rocky Mountains (Jin & Norford, 1995). This early form of *Hiscobeccus* almost certainly evolved from *Rhynchotrema* because of its moderately biconvex (non-globular) shell that is lamellose only in the anterior two-thirds. Some Trentonian species of *Rhynchotrema*, for example, *R. wisconsinense* from the Decorah Shale of Minnesota, already developed coarse lamellae in the anterior portion of their shells, resembling those of *H. mackenziensis* (Figure 3:9-12). By the Maysvillian and Richmondian (early-mid Ashgill), various species of *Lepidocyclus*, *Hypsiptycha*, and *Hiscobeccus*, all characterised by a completely lamellose shell, spread to nearly every shallow marine, carbonate basin of Laurentia. In the inland basins of Laurentia, these genera have been found in the Maysvillian and Richmondian of the Hudson Bay and Williston basin (Okulitch, 1943; Ross, 1957; Macomber, 1970; Jin et al., in press), Indiana (Conrad, 1842), Tennessee (Foerste, 1909; Howe, 1969), Wisconsin (Whitfield, 1882), Iowa (Wang, 1949), Missouri (Foerste, 1920), Kentucky (Howe, 1979), and Oklahoma (Howe, 1966; Alberstadt, 1973; Amsden, 1983). In the carbonate basins and platforms along the eastern and western paleocontinental margins, these genera have been found on Anticosti Island, Quebec (Billings, 1862; Twenhofel, 1928; Jin, 1989), Baffin Island (Roy, 1941; Miller et al., 1954), northern Ireland (Mitchell, 1977), northern Greenland (Troedsson, 1928), Mackenzie Mountains (Jin & Lenz,1992),

Canadian Rocky Mountains (Wilson, 1926; Norford, 1962; Jin et al., 1989; Jin & Norford, 1995), Nevada (Howe & Reso, 1967), Texas, and New Mexico (White, 1877; Howe,1967). Outside Laurentia, the *Lepidocyclus-Hypsiptycha* fauna is virtually unknown, except for the report of a doubtful species, *Hypsiptycha? procera*, from the Hariuan Series (Caradoc) of Estonia by Roomusoks (1964), and questionable *Lepidocyclus laddi* (internal structures unknown) from the Akdombak Formation (upper Ashgill) of Kazakhstan (Klenina, 1984).

Similarly, two genera characteristic of the Ashgill, *Lepidocycloides* and *Evenkorhynchia*, are very common in Siberia but rarely occur in other continents. *Lepidocycloides* first appearrs in the upper Baksan horizon (latest Caradoc), and becomes abundant in the Dolbor and Nirunda horizons (lower to mid Ashgill) in the Tunguska-Moyero region of Siberia. It may have evolved from the early-mid Caradoc *Azamella* from Australia. *Lepidocycloides* shows a certain degree of homeomorphy with the North American *Lepidocyclus* and *Hiscobeccus* in its globular, lamellose shells. Internally, however, *Lepidocycloides* lacks a cardinal process in the dorsal valve and has been assigned to the family Trigonirhynchiidae. Both *Lepidocycloides* and *Evenkorhynchia* evolved from *Rostricellula*. Rozman (1979) suggested that *Lepidocycloides* branched from *Rostricellula* in late Magazeyan (late Caradoc) time, and *Evenkorhynchia* from *Rostricellula* in Dolborian (early Ashgill) time. Nasedkina (1973) doubtfully assigned to *Lepidocycloides* an unidentified species from the Poluda horizon (uppermost Caradoc) of the central Ural Mountains (western slope), which may have been part of the Baltica paleocontinent in the Ordovician. One species of *Lepidocycloides* (*L. rudicostatus* Jin et al., 1989) occurs in the lower Beaverfoot Formation of the southern Rocky Mountains (western margin of Laurentia), represented by only a few specimens.

Rhynchonellids of Ashgill age are rare in Kazakhstan and Baltica. Only one genus, *Plectothyrella* Temple, 1965, is well documented in northern England (Cocks, 1978). The genus is a common element of the *Hirnantia* fauna of late Ashgill age and has been found in other continents in the paleoequatorial regions. Endemic rhynchonellids are common in NW China, e.g.*Latirhynchia* Fu, 1982 from Shaanxi, and *Wulunguia* Zhang & Zhang, 1981 from Xinjiang.

Contrary to the geographic distribution of most shallow-water rhynchonellids, the deep-water *Foliomena* fauna, dominated by small and thin-shelled orthids and strophomenids, was widespread or even cosmopolitan in early-mid Ashgill time. The fauna apparently evolved in the cold water Mediterranean Province and spread to other continents of warmer-water regions. Sheehan (1979, p.69) postulated that 'the Mediterranean Province developed in high latitudes which became progressively colder during the Ordovician and that the Mediterranean Province communities moved into Sweden following cold water masses'. As recently discussed by Havlicek (1989), however, the late Caradoc and early Ashgill was a relatively warm period, as indicated by the occurrence of marlstones, limestones, and mudmounds in the Mediterranean Province (central Spain and the Carnic Alps). The *Foliomena* fauna possibly formed not in a very cold environment, but simply in moderate- and deep-water conditions. In Bohemia, *Foliomena* and *Kozlowskites* occur in the *Dedzetina* community (*sensu* Havlicek, 1982) found in shales. Similarly, *Foliomena* occurs almost exclusively in siliciclastics in Sweden, Wales (Sheehan, 1973), Ireland (Harper, 1980), Québec (Sheehan & Lespérance, 1978), and northeastern Maine (Neuman, 1994). Based on his observation of the fauna in the calcareous shales and siltstones of the Linxiang Formation (lower Ashgill) in South China, Rong (1984b) inferred that the *Foliomena* fauna lived in a deep-water environment equivalent to Benthic Assemblage 4 or

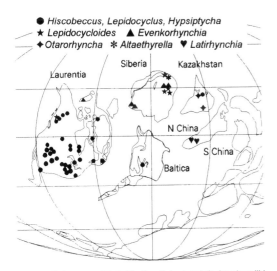

● *Hiscobeccus, Lepidocyclus, Hypsiptycha*
★ *Lepidocycloides*   ▲ *Evenkorhynchia*
◆ *Otarorhyncha*   ✱ *Altaethyrella*   ♥ *Latirhynchia*

Figure 6. Paleogeographic distribution of characteristic rhynchonellid genera in the latest Caradoc and the Ashgill (pre-Hirnantian).

5. Sheehan (1979) assigned *Foliomena* to an even deeper Benthic Assemblage 6, in association with graptolites.

The wide distribution of the *Hirnantia* fauna in siliciclastics was intimately related to late Ashgill glaciation and global fall in sea level. Like the *Foliomena* fauna, typical elements of the *Hirnantia* fauna originated in the cold water Mediterranean Province, although the fauna was first identified in siliciclastics of Bala, North Wales (Cocks & Price, 1975, p.720). In Bohemia, *Hirnantia* appeared early in Berounian (latest Llanvirn-Caradoc) time as a member of the *Aegiromena-Drabovia* fauna, living in a shallow subtidal environment with sandy and silty substrate conditions (Havlíček, 1989). By Kosovian (Hirnantian) time, *Aegiromena* and *Hirnantia*, together with such genera as *Dalmanella*, *Plectothyrella*, *Cliftonia*, *Kinnella*, and *Eostropheodonta*, formed a distinctive *Hirnantia* fauna. In general, the *Hirnantia* fauna differs strikingly from the *Foliomena* fauna in two ways: (1) as pointed out by Rong (1984b), some genera of the *Hirnantia* fauna have large shells (10-35mm width), whereas most genera of the *Foliomena* fauna have small shells (3-8mm width); (2) the *Hirnantia* fauna occurs predominantly in shallow-water sediments, from sandstones and siltstones interbedded with limestones in warmer regions (Wright,1968; Cocks & Price, 1975), to those interbedded with tillites in cold regions (Havlíček, 1989). In the tropical-subtropical regions of Baltica, Laurentia, and South China, the *Hirnantia* fauna has been found most commonly in shales, siltstones, and sandstones interbedded with limestones, indicating that the fauna succeeded in colonizing these regions only in periods of decreased temperature, or in nearshore, shallow siliciclastic settings. On Anticosti Island, Cocks & Copper (1981) reported the presence of a *Hirnantia* fauna in a 'cross-bedded silty sandy quartz limestone' unit in the Ellis Bay Formation, a few meters below the O/S boundary (Hirnantian). Rong (1979) recorded a *Hirnantia* fauna in a limestone unit of the Guanyinqiao Formation of Guizhou, South China. Lithological evidence indicates that the *Hirnantia* fauna had broad temperature tolerance, occurring in circumpolar cold water to subtropical, warm, nearshore settings. There is general consensus that the *Hirnantia* fauna lived in a subtidal environment equivalent to the range of Benthic Assemblage 1 to 3 (Rong, 1984a, 1984b).

DISTRIBUTION CONTROLLING FACTORS

There were many factors that could have affected the patterns of global rhynchonellid distribution in the Ordovician, among which eustatic sealevel changes, relative position and movement of the plates, climatic change, and oceanic currents were probably most important.

Sealevel changes. Ordovician rhynchonellid distribution shows an apparent, counter-intuitive correlation between cosmopolitanism and low sealevel stand, and between provincialism and high sealevel stand. Dispersal of the earliest rhynchonellids during the Llanvirn coincides with a period of relatively cold climate and low sealevel stand (Spjeldnaes,1981; Fortey, 1984; Webby, 1984). Recently, Havlíček (1989) hypothesised the presence of two glacial events in the Gondwana landmasses during the Ordovician, one in the Berounian (late Llandeilian-early Caradoc), and the other in the latest Ashgill. Although precise dating of the first glaciation is not yet available, it is likely that the early Berounian glacial event was a culmination of a cold phase in the late Llandeilian. For bryozoans, Tuckey (1990) recorded peak intercontinental migration at the generic level among Laurentia, Siberia, and Baltica during the Llanvirn-early Caradoc. Most bryozoan taxa were associated with reefs and, therefore, their migration probably responded to sealevel changes (or climate) in a way similar to that for the rhynchonellids.

In discussing provincialism of North American benthic faunas, Neuman (1976) remarked that breeching of a migration barrier in the early Middle Ordovician (Llanvirn) was 'not only because of decreased oceanic distances between continents, but also because of a break-up of the continuity of the trough of deep water at the edge of the continent'. As discussed by Pickering et al. (1988), however, the main, active phase of subduction and closure of the Iapetus Ocean did not occur until latest Ordovician-earliest Silurian time, and thus tectonism probably did not contribute greatly to the breeching of oceanic migration barriers in the Middle Ordovician.

Generic diversification and onset of rhynchonellid endemism in the Caradoc coincided with a major phase of marine transgression. On the basis of lithological and faunal data from eleven regions in North America and Europe, McKerrow (1979) detected a substantial rise in sea level during the Caradoc. According to Fortey (1984), the Caradoc transgression was the most extensive among the transgressive pulses in the Ordovician. Chen (1988) postulated a 300m rise for the same time interval based on his analysis of Chinese faunal data. The transgression was most likely a response to a global warming event, which was discussed recently by Havlíček (1989) on the basis of lithological and faunal data from the cold-water Mediterranean province. In S Europe, limestones and reefal sediments became common in late Berounian-early Kralovdorian (Caradoc-early Ashgill) time, replacing the sandy and silty sediments of pre-Caradoc age. The best example is the limestones and reef-derived rocks of the Urbana Limestone in central Spain. There was also substantial mixing of brachiopods and trilobites from the warm-water Baltic province with those from the cold-water Mediterranean province (Havlíček, 1989). Caradoc transgression was also marked by oceanic graptolitic provinces that went through 'breakdown' and formed a single province (Skevington,1973; Fortey,1984). Trilobites also formed a relatively unified monorakid-remopleuridid province across Laurentia, Siberia, Baltica, and Kazakhstan (Whittington, 1973). The transgression, however, obviously had an opposite effect on the geographic distribution of the shallow-water, fixosessile rhynchonellid faunas. N American rhynchonellids that evolved during the latest Trenton-Richmond (latest Caradoc-mid Ashgill) rarely succeeded in migrating to Siberia, Baltica, and other paleoequitorial continents. Similarly, some 'North American endemic shelf trilobite faunas' (Fortey, 1984) retreated into midcontinental basins (upper Mississippi valley), instead of becoming more widely distributed. These trilobite faunas, like the rhynchonellids, probably had a narrow range of adaptations to shallow-water environments. A deep and wide oceanic barrier preventing migration of shallow-water trilobites was noted also by Whittington (1973).

In contrast to strong rhynchonellid provincialism, the deep-water *Foliomena* fauna had a nearly cosmopolitan distribution during the early and mid Ashgill. This probably can be attributed to the supposition that the *Foliomena* fauna was adapted to relatively cold and deep waters, and that late Caradoc-early Ashgill transgression enabled these species to track their deep water habitat, moving along continental shelves. This clearly illustrates opposing effects of a transgression on biogeographic distribution of deep water versus shallow water faunas.

Intercontinental dispersal of the eurythermal *Hirnantia* fauna was likely to have occurred during sealevel lowstand as a result of latest Ashgill glaciation centred in North Africa. The extent of sea level drop was estimated by Brenchley & Newall (1984) to be about 100m or less, and by Long (1994) as <30m. In the stable cratonic regions of North America and Siberia, the uppermost Ordovician was dominated by shallow-water carbonate sequences with abundant and diverse benthic faunas, punctuated by siliciclastic units. It is thus mostly likely that sealevel fluctuation was closer to 30m (ibid.) so that carbonate environments could be re-established quickly after each cold excursion. During the latest Ordovician, two rhynchonellid genera became widespread. *Plectothyrella* Temple, 1965 has been found commonly as a member of the *Hirnantia* fauna in Baltica, S China (Rong,1984a,1984b), N China (Fu,1982), and eastern Laurentia (Lespérance, 1974). Hirnantian *Thebesia* Amsden, 1974, occurs in both Laurentia and Siberia (Kulkov & Severgina, 1989).

The relationship between global distribution of rhynchonellids and sealevel changes can be explained possibly by the stepstone 'island-fauna hypothesis'. The importance of island faunas as stepping-stones for faunal dispersal has been fully discussed by MacArthur & Wilson (1967, p.133), who suggested that 'dispersal across gaps of more than a few kilometres is by stepping stones wherever habitable stepping stones of even the smallest size exist.' In his study of Ordovician trilobite biogeography, Fortey (1984) hypothesised that oceanic island faunas are probably more effective stepping stones when many oceanic volcanic highs become emergent to form islands during a low sealevel stand. During sealevel rise, submergence of these islands would lead to drowning of the island faunas and dramatically widening the oceanic deep water barriers. Ordovician rhynchonellids were generally shallow water benthos, and their intercontinental migration patterns may have been controlled by the availability of stepping stones during fluctuations of sea level.

Plate tectonics. During the Middle and Late Ordovician, Baltica and Avelonia were moving northward, drifting from a cold to a warm, subtropical climate (Webby, 1984), and Siberia probably drifted somewhat in a northwesterly direction relative to Laurentia. These relative movements may have affected rhynchonellid biogeography, as it resulted in changes in latitudes of the continents as well as distances between continents. Distribution of Ordovician rhynchonellid faunas appears to agree with the placement of the paleocontinents of Scotese & McKerrow (1990). Laurentia and Siberia were close together in the Llanvirn, with a very narrow oceanic barrier between. This would have facilitated intercontinental migra- tions of the shallow-water shelly faunas. By early Ashgill time, the two continents became separated by a wide ocean, which probably acted as a deep water oceanic barrier. Widening of the oceanic barrier and a concomitant marine transgression could explain why the North American rhynchonellid fauna became widespread within the continent, but at the same time isolated from other continents during the late Caradoc-early Ashgill. By the end of the Ashgill, destruction and closure of the Iapetus Ocean brought Laurentia, Siberia, and Baltica very close, which in turn lead to cosmopolitanism of shelly faunas in Silurian time (Pickering et al., 1988).

CONCLUSIONS

Intercontinental migrations of Ordovician rhynchonellids were closely related to changes in sealevel and plate tectonics. An initial wave of migration corresponded to Llandeilian sealevel lowstand, and proximity between Laurentia and Siberia. A growing number of endemic genera marked development of rhynchonellid provincialism in shallow-water brachiopod faunas during major rises in sealevel of the Caradoc and early Ashgill. Provincialism peaked during the latest Caradoc and early Ashgill, when Laurentia and Siberia moved farther away from each other. The onset of shallow water brachiopod cosmo- politanism in the latest Ordovician-early Silurian coincided with Gondwanan glaciation and destruction of the Iapetus Ocean.

Acknowledgments

Dr. R.E. Sloan, Department of Geology and Geophysics, University of Minnesota, Minneapolis, made available the specimen of *Rhynchotrema wisconsinense*, and the US National Museum, Washington, loaned specimens of early rhynchonellids. Research was sponsored by the Natural Sciences and Engineering Research Council of Canada from a grant to P. Copper.

REFERENCES

ALBERSTADT, L.P. 1973. Articulate brachiopods of the Viola Formation (Ordovician) in the Arbuckle Mountains, Okla- homa. Oklahoma Geological Survey Bulletin 117:1- 90.
AMSDEN, T.W. 1974. Late Ordovician and Early Silurian arti- culate brachiopods from Oklahoma, southwestern Illinois, and eastern Missouri. Oklahoma Geological Survey Bulletin 119:1-154.
___1983. Upper Bromide Formation and Viola Group (Middle and Upper Ordovician) in eastern Oklahoma. Oklahoma Geological Survey Bulletin 132(1), Welling-Fite-Corbin Range strata :1-23; (3), The Late Ordovician brachiopod genera *Lepidocyclus* and *Hiscobeccus*, 36-44.
BARRANDE, J. 1847-1848. Über die Brachiopoden der silur- ischen Schichten von Böhmen. Naturwissenschaftliche Ab- handlungen 1 (1847): 357-475; 2 (1848):155-256.
BERGSTRÖM, S.M., J.B. CARNES, J.C. HALL, W. KURAPKAT & B.E. O'NEIL.1988. Conodont biostratigraphy of some Middle Ordovician stratotypes in the southern and central Appalachians. New York State Museum Bulletin 462:20-31.
BILLINGS, E. 1859. Fossils of the Chazy Limestone, with des- criptions of new species. Canadian Naturalist Geologist, 4:426-470.
___1862. New species of fossils from different parts of the Lower, Middle, and Upper Silurian rocks of Canada. Geological Survey Canada, Palaeozoic Fossils 1 (4):96-168.
BONDAREV, V.I. 1968. Stratigrafiya i kharakternye brakhio- pody ordovikskikh otlozhenii yuga Novoi Zemli, ostrova Vaygach i severnogo Pai-Khoia. Trudy Nauchno-Issledo- vatelskogo Instituta Geologii Arktiki, SSSR 157:3-144.
BRENCHLEY, P.J. & G. NEWALL. 1984. Late Ordovician envi- ronmental changes and their effect on faunas. In, D.L. Bruton (ed.), Aspects of the Ordovician System. Palae- ontological Contributions University Oslo 295:65-79.
CHEN JUNYUAN. 1988. Ordovician changes of sea level. New Mexico Bureau Mines Mineral Resources Memoir 44:387- 404.
COCKS, L.R.M. 1978. A review of British Lower Palaeozoic brachiopods, including a synoptic revision of Davidson's Monograph. Monographs Palaeontographical Society, London 131:1-256 .
___& P. COPPER. 1981. The Ordovician-Silurian boundary at the eastern end of Anticosti Island. Canadian Journal Earth Sciences 18:1029-1034.
___& D. PRICE. 1975. The biostratigraphy of the Upper Ordo- vician and Lower Silurian of south-west Dyfed, with com- ments on the *Hirnantia* fauna. Palaeontology 18:703-724.
CONRAD, T. A. 1842. Observations on the Silurian and Devon- ian systems of the U.S., with descriptions of new organic remains. Journal Academy Natural Sciences Philadelphia 8:228-280.
COOPER, G.A.1956. Chazyan and related brachiopods. Smithsonian Miscellaneous Collections 127(1):1-1024; (2):1025-1245.
___1976. Early Middle Ordovician of the United States, p.171- 194. In, M.G. Bassett (ed.), The Ordovician System. Proceedings Palaeontological Association Symposium,

Birmingham, September, 1974. University of Wales Press, Cardiff.

COPPER, P. 1986. Evolution of the earliest smooth spire-bearing atrypoids (Brachiopoda: Lissatrypidae, Ordovician-Silurian). Palaeontology 29:827-867.

FENTON, C.L. & M.A. FENTON. 1923. Some Black River brachiopods from the Mississippi valley. Proceedings Iowa Academy Sciences 29:67-77.

FOERSTE, A.F. 1909. Preliminary notes on Cincinnatian and Lexington fossils. Bulletin Scientific Laboratories Denison University 14:289-324.

___1920. The Kimmswick and Plattin limestones of north-eastern Missouri. ibid. 19:175-224.

FORTEY, R.A. 1984. Global earlier Ordovician transgressions and regressions and their biological implications. In, D.L. Bruton (ed.), Aspects of the Ordovician System, Palaeontological Contributions University Oslo 295:37-50.

___, D.A.T. HARPER, J.K. INGHAM, A.W. OWEN & A.W.A. RUSHTON. 1995. A revision of the Ordovician series and stages from the historical type area. Geological Magazine 132:15-30.

FU LIPU. 1982. Rhynchonellida, p.136-145. In, Xian Institute of Geology and Mineral Resources (eds.), Paleontological Atlas of Northwest China, Xian-Gansu-Ningxia Volume, 1: Precambrian and Early Paleozoic. Beijing, Geological Publishing House.

HALL, J. 1847. Descriptions of the organic remains of the Lower division of the New-York System. New York Geological Survey, Palaeontology New York, 1:1-338 .

___1860. Contributions to palaeontology. Annual Report New York Cabinet Natural History, 13:55-125.

HARPER, D.A.T. 1980. The brachiopod Foliomena fauna in the upper Ordovician Ballyworgal Group of Slieve Bernagh, County Clair. Royal Dublin Society Journal Earth Sciences 2:189-192.

HAVLÍCEK, V.1961. Rhynchonelloidea des böhmischen älteren Paläozoikums (Brachiopoda). Rozpravy Ustredního ústavu geologického 27:1-211.

___1970. Brachiopodes de l'ordovicien de Maroc. Notes Mémoires Service Géologique Maroc 230:1-132.

___1982. Ordovician in Bohemia: development of the Prague Basin and its benthic communities. Sborník Ustredního ústavu geologického 37:103-136.

___1989. Climatic changes and development of benthic communities through the Mediterranean Ordovician. ibid. 44:79-116.

___& L. BRANISA. 1980. Ordovician brachiopods of Bolivia. Rozpravy Ceskoslovenské Akademie 90(1), 54 p.

HOWE, H.J. 1966. The brachiopod genus Lepidocyclus from the Cape (Fernvale) Limestone (Ordovician) of Oklahoma and Missouri. Journal Paleontology 40:258-263.

___1967. Rhynchonellacea from the Montoya Group (Ordovician) of Trans-Pecos, Texas. ibid.41:845-860.

___1969. Rhynchonellacean brachiopods from the Richmondian of Tennessee. ibid. 43:1331-1350.

___1979. Middle and Late Ordovician plectambonitacean, rhynchonellacean, syntrophiacean, trimerellacean, and atrypacean brachiopods. US Geological Survey Professional Paper 1066C:1-18.

___& A. RESO. 1967. Upper Ordovician brachiopods from the Ely Springs Dolomite in southwestern Nevada. Journal Paleontology 41:351-363.

JAANUSSON, V. 1971. Ordovician, p. A136-166. In, R.C. Moore (ed.), Treatise on Invertebrate Paleontology, A, Introduction. Geological Society of America and University of Kansas Press, Lawrence.

___1973. Ordovician articulate brachiopods, p.19-26. In, A. Hallam, (ed.), Atlas of Palaeobiogeography. Elsevier.

___& S. M. BERGSTRÖM. 1980. Middle Ordovician faunal spatial differentiation in Baltoscandia and the Appalachians. Alcheringa 4:89-110.

JIN, J. 1989. Late Ordovician and Early Silurian rhynchonellid brachiopods from Anticosti Island, Quebec. Biostratigraphie Paléozoique 10:1-127.

___,W.G.E. CALDWELL & B.S. NORFORD. 1989. Rhynchonellid brachiopods from the Upper Ordovician-Lower Silurian Beaverfoot and Nonda formations of the Rocky Mountains, British Columbia. Geological Survey Canada Bulletin 396:21-59.

___,___&___ (in press). Late Ordovician brachiopods and biostratigraphy of the Hudson Bay Lowlands, northern Manitoba and Ontario. Geological Survey of Canada, Bulletin.

___& A.C. LENZ. 1992. An Upper Ordovician Lepidocyclus-Hypsiptycha fauna (rhynchonellid Brachiopoda) from the Mackenzie Mountains, Northwest Territories, Canada. Palaeontographica (A) 224:133-158.

___& B.S. NORFORD. 1995. Upper Middle Ordovician (Caradoc) brachiopods from the Advance Formation, northern Rocky Mountains, British Columbia. Geological Survey Canada Bulletin 491:20-77.

JIN YUGAN, SUNGLING YE, HANKUI XU & DONGLI SUN. 1979. Brachiopods, p.60-217. In, Paleontological Atlas of Northwest China, Qinghai Volume,1. Geological Publishing House, Beijing.

KANYGIN, A.V., A.G. YADRENKINA, T. A. MOSKALENKO, G.P. ABAIMOVA, O.V. SYCHEV & A.V. TIMOKHIN. 1989. Opisanie razrezov, p.5-64. In, A.M. Obut (ed.), Ordovik Sibirskoi platformy, fauna i stratigrafiya lenskoi fatzialnoi zony. Trudy Instituta Geologii i Geofiziki, Akademiya Nauk SSSR, Sibirskoie Otdelenie 751.

KLENINA, L.N.1984 Brakhiopody i stratigrafiya srednego i verkhnego ordovika Chingiz-Tarbagataiskogo megantiklinoriya, p.6-125. In, L.N. Klenina, I.F. Nikitin, & L.E. Popov, Brakhiopody i biostratigrafiya srednego i verkhnego ordovika khrebta Chingiz. Alma-Ata, Nauka.

KULKOV, N.P. & L.G. SEVERGINA. 1989. Stratigrafiya i brakhiopody ordovica i nizhnego silura Gornogo Altaya. Trudy Instituta Geologii i Geofiziki, Akademiya Nauk SSSR, Sibirskoye Otdeleniye717:1-224.

LAURIE, J.R. 1991. Articulate brachiopods from the Ordovician and Lower Silurian of Tasmania. Memoirs Association Australasian Palaeontologists 11:1-106.

LEONE, F., W. HAMMANN, R. LASKE, E. SERPAGLI & E. VILLAS. 1991. Lithostratigraphic units and biostratigraphy of the post-sardic Ordovician sequence in south-west Sardinia. Bullettino della Società Paleontologica Italiana 30:190-235.

LESLIE, S.A. & S.M. BERGSTRÖM. 1995. Revision of the North American late Middle Ordovician standard stage classification and timing of the Trenton Transgression based on K-bentonite bed correlation, p. 49-54. In, J.D. Cooper, M.L. Droser & S.C.Finney (eds.), Ordovician Odyssey, Short Papers for the Seventh International Symposium on the Ordovician System, Las Vegas, Nevada, June, 1995. SEPM, Fullerton.

LESPÉRANCE, P. J. 1974. The Hirnantian fauna of the Percé area (Québec) and the Ordovician-Silurian boundary. American Journal Science 274:10-30.

LONG, D.G.F. 1993. Limits on Late Ordovician eustatic sea-level change from carbonate shelf sequences: an example from Anticosti Island, Quebéc. International Association Sedimentologists Special Publication 18:487-499.

MACARTHUR, R.H. & E.O. WILSON. 1967. The theory of island biogeography. Princeton University Press, Princeton, 203p.

MACGREGOR, A.R. 1961. Upper Llandeilo brachiopods from the Berwyn Hills, North Wales. Palaeontology 4:177-209.

MACOMBER, R.W. 1970. Articulate brachiopods from the Upper Bighorn Formation (Late Ordovician) of Wyoming. Journal Paleontology 44:416-450.

MCKERROW, W.S. 1979. Ordovician and Silurian changes in sea level. Journal Geological Society London 136:137-145.

MILLER, A.K., W. YOUNGQUIST & C. COLLINSON. 1954. Ordovician cephalopod fauna of Baffin Island. Memoirs Geological Society of America 62: 1-234.

MITCHELL, W.I. 1977. The Ordovician Brachiopoda from Pomeroy, Co. Tyrone. Palaeontographical Society [London], Monographs 130:1-138.

MOSKALENKO, T.A., A.G. YADRENKINA, V.S. SEMENOVA & A.M. YAROSHINSKAYA. 1978. Ordovik Sibirskoi Platformy, opornie razrezy verkhnego ordovika: Opisanie razrezov. Trudy Instituta Geologii i Geofiziki, Akademiya Nauk SSSR, Sibirskoe Otdelenie 340: 6-59.

NASEDKINA, V.A. 1973. Brakhiopody, p.111-142. In, V.G. Varganov (ed.), Stratigrafiya i fauna ordovika Srednego Urala. Nedra, Moscow.

NETTELROTH, H. 1889. Kentucky fossil shells, a monograph of the fossil shells of the Silurian and Devonian rocks of Kentucky. Kentucky Geological Survey, Frankfort, Kentucky, 245p.

NEUMAN, R.B. 1972. Brachiopods of Early Ordovician volcanic islands. 24th Session International Geological Congress, Montreal, 1972, 7:297-302.

___1976. Ordovician of the eastern United states, p. 195-207. In, M. G. Bassett (ed.), The Ordovician System, Proceedings of the Palaeontological Association Symposium, Birmingham, September 1974.

___1994. Late Ordovician (Ashgill) Foliomena fauna brachiopods from northeastern Maine. Journal Paleontology 68:1218-1234.

NIKIFOROVA, O.I. 1970. Brakhiopody grebenskogo gorizonta Vaigacha (pozdnii silur), p.97-194. In, S.V. Cherkesova (ed.), Stratigrafiya i fauna siluriiskikh otlozhenii Vaigacha. NIIGA, Leningrad.

___& O.N. ANDREEVA. 1961. Stratigrafiya ordovika i silura Sibirskoi Platformy i ee paleontologicheskoe obosnovanie (brakhiopody). Trudy VSEGEI, novaya seriya 56:1-412.

___& L.E. POPOV. 1981. Novye danye ob ordovikskikh rinkhonellidakh Kazakhstana i Srednii Azii. Paleontologicheskii Zhurnal 1:54-67.

NIKITIN, I.F. & L.E. POPOV. 1984. Brakhiopody bestamakskoi i sargaldakskoi svit (srednii ordovik), p.126-195. In, L.N. Klenina, I.F. Nikitin & L.E. Popov (eds.), Brakhiopody i biostratigrafiya srednego i verkhnego ordovika khrebta Chingiz. Alma-Ata, Nauka.

NORFORD, B.S. 1962. Illustrations of Canadian fossils: Cambrian, Ordovician, and Silurian of the Western Cordillera. Geological Survey Canada Paper 62-14:1- 24.

OKULITCH, V.J. 1943. The Stony Mountain Formation of Manitoba. Royal Society Canada Transactions, 3, 37 (4):59-74.

PICKERING, K.T., M.G. BASSETT & D.J. SIVETER. 1988. Late Ordovician-early Silurian destruction of the Iapetus Ocean: Newfoundland, British Isles and Scandinavia--a discussion. Transactions Royal Society Edinburgh, Earth Sciences 79:361-382.

REED, F.R.C. 1917. The Ordovician and Silurian Brachiopoda of the Girvan District. ibid. 51:795-998.

RICE, W.F. 1987. The systematics and biostratigraphy of the brachiopods of the Decorah Shale at St. Paul, Minnesota. Minnesota Geological Survey Report Investigations 35:136-231.

RONG JIAYU. 1979. The Hirnantia fauna in China and the Ordovician/Silurian boundary. Acta Stratigraphia Sinica 3:1-28.

___1984a. Distribution of the Hirnantia fauna and its meaning. In, D.L. Bruton (ed.), Aspects of the Ordovician System, Palaeontological Contributions University Oslo 295:101-112.

___1984b. Ecostratigraphical evidence for a Late Ordovician regression in the upper Yangtze Platform and the impact of glaciation. Journal Stratigraphy, 8:19-29.

ROSS, J.R.P. 1981. Ordovician environmental heterogeneity and community organization, p.1-33. In, J. Gray et al. (eds.), Communities of the past. Hutchinson Ross, Stroudsburg.

ROSS, R.J. 1957. Ordovician fossils from wells in the Williston Basin, eastern Montana. United States Geological Survey Bulletin 1021M:439-506.

___& S.M. BERGSTRÖM. 1982 (eds.). The Ordovician System in the United States, correlation chart and explanation notes. International Union Geological Sciences Publication 12:1-73.

ROY, S.K. 1941. The Upper Ordovician fauna of Frobisher Bay, Baffin Land. Field Museum Natural History Geology Memoir 2:1- 212.

ROZMAN, K.S. 1979. Brakhiopody (Rhynchonellacea) Mangozeyskogo i Dolorskogo Gorizontov In, A.V. Peive (ed.), Fauna ordovika Srednii Sibiri. Trudy Akademiya Nauk SSSR, Geologicheskii Institut 330:37-78.

___1981. Brakhiopody srednego i verkhnego ordovika Mongolii. Atlas fauny ordovika Mongolii. Trudy Akademiya Nauk SSSR, Ordena Trudovogo Krasnogo Znameni Geologicheskii Institut 354:117-176.

ROOMUSOKS, A. 1964. Nekotoryye brakhiopody iz̆ ordovika estonii. Tartu Riikliku Ülikooli Toimetised 153:3-28.

SARDESON, F.W. 1892. The range and distribution of the Lower Silurian fauna of Minnesota, with descriptions of some new species. Minnesota Academy Natural Sciences Bulletin 3:326-334.

SCOTESE, C.R. & W.S. MCKERROW.1990. Revised World maps and introduction. In, W.S. McKerrow & C.R. Scotese (eds.), Paleozoic paleogeography and biogeography. Geological Society Memoir 12:1-21.

SHEEHAN, P.M. 1973. The relation of Late Ordovician glaciation to the Ordovician-Silurian changeover in North American brachiopod faunas. Lethaia 6:147-154

___1979 Swedish Late Ordovician marine benthic assemblages and their bearing on brachiopod zoogeography, p.61-73. In, J. Gray & A.J. Boucot (eds.), Historical biogeography, plate tectonics, and the changing environment. Oregon State University Press.

___1988. Late Ordovician events and the terminal Ordovician extinction. New Mexico Bureau Mines and Mineral Resources Memoir 44:405-415.

___& P.J. COOROUGH. 1990. Brachiopod zoogeography across the Ordovician-Silurian extinction event. In, W.S. McKerrow & C R. Scotese (eds.), Paleozoic paleogeography and biogeography, Geological Society Memoir 12:181-187

___& P.J. LESPÉRANCE.1979. Late Ordovician brachiopods from the Percé region of Québec. Journal Paleontology 53:950-967.

SKEVINGTON, D. 1973. Ordovician graptolites, p. 27-36. In, A. Hallam (ed.), Atlas of Palaeobiogeography. Elsevier.

SLOAN, R.E. 1987. Tectonics, biostratigraphy and lithostratigraphy of the Middle and Late Ordovician of the upper Mississippi valley. Minnesota Geological Survey, Report Investigations 35:7-20.

SPJELDNAES, N. 1981. Lower Palaeozoic palaeoclimatology, p.199-256. In, C.H. Holland (ed.), Lower Palaeozoic of the Middle East, eastern and southern Africa, and Antarctica. John Wiley & Sons, New York.

TEMPLE, J.T.1965. Upper Ordovician brachiopods from Poland and Britain. Acta Palaeontologia Polonica 10:379-427.

TROEDSSON, G.T. 1928. On the Middle and Upper Ordovician faunas of northern Greenland, 2. Meddelelser Grønland 72:1-197.

131

TUCKEY, M.E. 1990. Biogeography of Ordovician bryozoans. Palaeogeography, Palaeoclimatology, Palaeoecology 77:91-126.

TWENHOFEL, W.H. 1928. Geology of Anticosti Island. Geological Survey Canada Memoir 154:1-481.

ULRICH, E.O. & G.A. COOPER. 1942. New genera of Ordovician brachiopods. Journal Paleontology 16:620-626.

VILLAS, E. 1992. New Caradoc brachiopods from the Iberian Chains (northeastern Spain) and their stratigraphic significance. Journal Paleontology 66:772-793.

WANG, Y. 1949. Maquoketa Brachiopoda of Iowa. Geological Society America Memoir 42:1-55.

WEBBY, B.D. 1984. Ordovician reefs and climate: a review. In, D.L. Bruton (ed.), Aspects of the Ordovician System, Palaeontological Contributions University Oslo 295:89-100.

WHITE, C.A. 1877. Report upon the invertebrate fossils collected in portions of Nevada, Utah, Colorado, New Mexico, and Arizona, by parties of the expeditions for 1871, 1872, 1873, and 1874. United States Geological Survey West 100th Meridian 4 (1):3-219.

WHITFIELD, R.P. 1882. Palaeontology. Wisconsin Geological Survey, Geology of Wisconsin 4:161-363.

WHITTINGTON, H.B. 1973. Ordovician trilobites, p. 13-18. In, A. Hallam (ed.), Atlas of palaeobiogeography. Elsevier.

WILLIAMS, A. 1949. New Lower Ordovician brachiopods from the Llandeilo-Llangadock District. Geological Magazine 86:161-174, 226-238.

___1962. The Barr and Lower Ardmillan Series (Caradoc) of the Girvan District, South-west Ayrshire, with descriptions of the Brachiopoda. Geological Society London Memoir 3:1-267.

___1969. Ordovician faunal provinces with reference to brachiopod distribution, p.117-150. In, A. Wood (ed.), The Pre-Cambrian and Lower Palaeozoic Rocks of Wales. University of Wales Press, Cardiff.

WILSON, A.E. 1926. An Upper Ordovician fauna from the Rocky Mountains, British Columbia. Geological Survey of Canada, Bulletin 44:1-34.

WRIGHT, A.D. 1968. A westward extension of the Upper Ashgill Hirnantia fauna. Lethaia 1:352-367.

YADRENKINA, A.G. 1974. Brakhiopody verkhnego kembriya i ordovika Severo-Zapada Sibirskoi Platformy. Trudy Ministerstvo Geologii SSSR, SNIIGGIMS 151:1-164.

___1984. Tip Brachiopoda. In, T.A. Moskalenko (ed.), Ordovik Sibirskoi Platformy, Paleontologicheskii Atlas. Trudy Instituta Geologii i Geofiziki, Akademiya Nauk SSSR, Sibirskoe Otdelenie 590:32-57.

___1989. Opisanie fauny. Brakhiopody. In, A.M. Obut (ed.), Ordovik Sibirskoi Platformy, fauna i stratigrafiya lenskoi fatzialnoi zony. Trudy Instituta Geologii i Geofiziki, Akademiya Nauk SSSR, Sibirskoe Otdelenie 751:65-81.

ZHANG CHUAN & ZIXIN ZHANG. 1981. Brachiopoda, p.75-106. Paleontological Atlas of Northwest China, Xinjiang Weiwur Autonomous Region, 1: Late Proterozoic-Early Paleozoic. Beijing, Geological Publishing House.

KANINOSPIRIFERINAE, A NEW SUBFAMILY OF THE SPIRIFERIDAE (BRACHIOPODA)

N.V. KALASHNIKOV

Institute of Geology, Komi Scientific Centre, Russian Academy of Sciences,
Syktyvkar 167000, Russia

ABSTRACT- The new subfamily Kaninospiriferinae (family Spiriferidae) is proposed for the genera *Kaninospirifer* and *Imperiospira*, characterized by an absence or weak development of dental plates in the ventral valve. The genus *Kaninospirifer* is reviewed.

INTRODUCTION

Kaninospirifer was described by Kulikov & Stepanov (in Stepanov et al.,1975), and based on the Late Permian species *Spirifer kaninensis* Likharev (1943) from the eastern coast of the Kanin Peninsula, north European Russia. The genus has been widely reported from the Ufimian-Kazanian stages of the Kanin Peninsula (Kalash-nikov, 1986,1988; Molin et. al.,1983), the Pechora Basin (Kalashnikov et al.,1990), the Russian Far East, Mongolia, North China (Pavlova, 1991), and the northern part of the Permian boreal zoogeographic realm.

SYSTEMATIC PALEONTOLOGY

Family Spiriferidae King, 1846
Subfamily Kaninospiriferinae Kalashnikov, new subfam.

Diagnosis. Transverse shells of Spiriferidae with fine, branching costae, weakly fasciculate close to umbones; growth lines developed toward shell anterior, rarely lamellose; sulcus and fold costate; interareas distinct with transverse growth striations; delthyrium prominent. Ventral dental plates absent or weakly developed, delthyrial ridges prominent; cardinal process striate; socket plates distinct.

Genera included. *Kaninospirifer* Kulikov & Stepanov 1975, type species from the Ufimian and Kazanian of northern European Russia; *Imperiospira* Archbold & Thomas, 1994, type species from the Late Artinskian of western Australia.

Discussion. The new subfamily is distinguished from the Neospiriferinae Waterhouse 1968 and other subfamilies of the Spiriferidae by the absence or weak development of dental plates. Fasciculation of costae is often restricted to the umbonal area of the shell.

Age and distribution. Late Permian of northern European Russia, the Arctic (Spitzbergen, Greenland and Arctic Canada), the Russian Far East, Mongolia, North China, and Early Permian (Sakmarian-Kungurian) of Western Australia.

Kaninospirifer Kulikov &Stepanov, 1975

Type species.*Spirifer kaninensis* Likharev, 1943, p.279, figs.14.

Diagnosis. Transverse shells with fine costae, fasciculate only near umbones. Dental plates absent or very weakly developed.

Discussion. The type species was described by Likharev (1943) from the Kanin Peninsula. A second species,

Kaninospirifer borealis, was described by Kulikov & Stepanov (in Stepanov et al., 1975) from Kazanian strata of the Kanin Peninsula, and was also figured by Kalashnikov (1961, 1986). Pavlova (1991) described *K. incertiplicatus* from the lower Upper Permian of Mongolia and assigned to the genus other species from coeval rocks of the Russian Far East and North China. *Kaninospirifer* has also been recorded form the Late Permian of the Pechora Basin (Kalashnikov et al., 1990). Late Permian Arctic species, such as *Spirifer striatoplicatus* Gobbett (1964) from Spitzbergen, and *Spirifer striatoparadoxus* Toula (1873) from Greenland (Dunbar, 1955), Spitzbergen, and Arctic Canada (Harker & Thorsteinsson, 1960), are referred to the genus, as are specimens assigned to *Spirifer marcoui* and *Spirifer ravana* by Wiman (1914) from Bear Island, Spitzbergen. *Imperiospira* Archbold & Thomas (1993) differs from *Kaninospirifer* in possessing weakly developed dental plates and anterolateral extensions to the shell margin.

Age and distribution. Ufirmian and Kazanian of the Arctic, northern European Russia, the Russian Far East, Mongolia, and North China.

Kaninospirifer kaninensis (Likharev, 1943)
Figure 1

Spirifer kaninensis Likharev, 1943, p.279, figs.1-4; *Kaninospirifer kaninensis* (Likharev), Stepanov et. al., 1975, p.63; Molin et al., 1983, p.22, p1.3, figs.1-3, pl.4, fig.1; Kalashnikov, 1986, p.98, pl.125, fig.9, pl.126, figs.5-6.

Holotype. A complete shell (10/5969) reposited in the T.N. Chernyshev Museum (VSEGEI), St.Petersburg, Russia. Age and distribution: Ufimian and Kazanian of the Kanin Peninsula.

Remarks. Likharev (1943) described the species in detail and additional material has been figured by subsequent authors (synonymy above). Illustrations of key specific features of the species are from the Fish River, section 5061, Kanin Peninsula and are reposited in the Geological Institute.

REFERENCES

ARCHBOLD, N.W. & G.A. THOMAS. 1993. *Imperiospira*, a new Western Australian Permian Spiriferidae (Brachiopoda). Memoirs Association Australasian Palaeontologists 15:313-328.
DUNBAR, C.O. 1995. Permian brachiopod faunas of central east Greenland. Meddelelser øm Grönland 110(3):1-169.
GOBBETT, D.J. 1964. Carboniferous and Permian brachiopods of Svalbard. Norsk Polarinstitut Skrifter, 127:1-201.
HARKER, P. & R. THORSTEINSSON. 1960. Permian rocks and faunas of Grinnell Peninsula, Arctic Archipelago. Memoirs Geological Survey Canada 309:1-89.
KALASHNIKOV, N.V.1961. Fauna Permskikh otlozhenii yugo-vostochnoi chasti Poluostrova Kanin, p.44-61. In, Materialy po geologii i petrografii Timana i Poluostrova Kanin. Institut Geologii, Komi Filiala Akademiya Nauk SSSR. Izdatelstvo

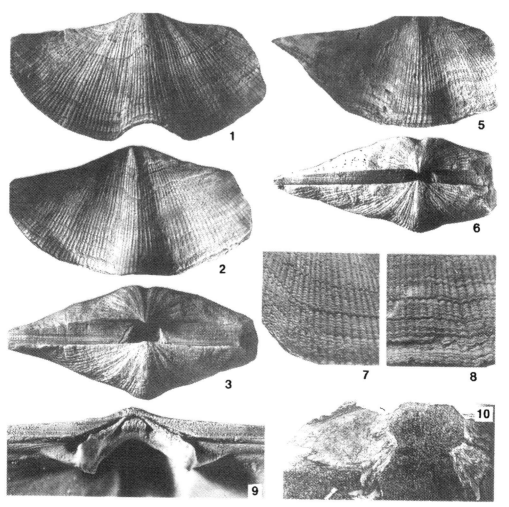

Figure 1. *Kaninospirifer kaninensis* (Likharev, 1943). 1-3. Dorsal, ventral, posterior views, 280/33, x1; 4. Dorsal interior showing cardinalia, 280/34, x3.5; 5-6. Ventral, posterior views, 280/199, x1; 7. Anterior dorsal valve showing micro-ornament, 280/200, x3.5; 8. Anterior ventral valve, showing micro-ornament, 280/33, x3.5; 9. Polished section, ventral posterior indicating lack of dental plates extending to valve floor, 280/201, x2.

Akademiya Nauk SSSR, Moskva.

___1986. Tip Brachiopoda. Trudy VSEGEI, Novaya Seriya 331:29-30, 89-94.

___1988. Razvitie Permskikh brakhiopod na severe Borealnoi (Biarmiiskoi) selasti, p.38-46. In, Permskaya sistema: voprosy stratigrafii i razvitiya organicheskogo Mira. Izdatelstvo Kazanskogo Universiteta, Kazan.

___, G.P. KANEV, N.A. KOLODA, G.G. MANAEVA, E.I. POLETAEVA & S.K. PUKHONTO. 1990. Paleontologicheskaya kharakteristika uglenosnykh svit, p.28-58. In, V.A. Dedeev (ed.), Uglenosnaya Formatsiya Pechorskogo Basseina. Nauka, Leningrad.

LIKHAREV, B.K. 1943. O novom Permskom*Spirifer*, priblizhayushchemsya k sp. *striatus* Sow. Izvestiya Akademii Nauk SSSR, Otdelenie Biologich Nauk, 1943 (5):279-285.

MOLIN, V.A., N.V. KALASHNIKOV, N.A. KOMODA & S.O. MELNIKOVA.1983. Novye dannye po paleontologicheskoi kharakteristike verkhnepermskikh otlozhenii Poluostrova Kanin. Trudy Institut Geologii, Komi filial, Akademiya Nauk SSSR 43:7-24.

PAVLOVA, E.E. 1991. Family Spiriferidae. Trudy, Sovmestnaya Sovetsko-Mongolskaya Palaeontologicheskaya Ekspeditsiya 40:124-134.

STEPANOV, D.L., M.V. KULIKOV & A.A. SULTANAEV. 1995. Stratigrafiya: brakhiopody verkhnepermskikh otlozhenii Poluostrova Kanin. Vestnik Leningradskogo Universiteta 1975 (6):51-65.

TOULA, F. 1873. Kohlenkalk Fossilien von der Südspitze von Spitzbergen. Sitzungsberichte Kaiserlichen Akademie Wissenschaften Wien, Mathematische-Naturwissenschaftliche Klasse 68:267-291.

WIMAN, C. 1914. Über die Karbonbrachiopoden Spitzbergens und Beeren Eilands. Nova Acta Regiae Societatis Scientiarum Uppsaliensis, Serie 4, 3 (8):1-91.

WATERHOUSE, J.B. 1968. The classification and descriptions of Permian Spiriferida (Brachiopoda) from New Zealand. Palaeontographica (A), 129:1-94.

LATE EIFELIAN-EARLY GIVETIAN (MIDDLE DEVONIAN) BRACHIOPOD PALEOBIOGEOGRAPHY OF EASTERN AND CENTRAL NORTH AMERICA

WILLIAM F. KOCH & JED DAY

Department of Zoology, Oregon State University, Corvallis, Oregon 97331;
Department of Geography - Geology, Illinois State University, Normal, Illinois 61790-4400, USA

ABSTRACT- A distinct brachiopod fauna, composed of a mixture of Cordilleran Old World Realm (OWR) endemic genera, cosmopolitan and formerly endemic genera of the Appohimchi Subprovince of the Eastern Americas Realm (EAR), spread across central and eastern North America duringthe late Eifelian (upper part *kockelianus* Zone). The breakdown of North American Devonian brachiopod provincialism in the late Eifelian-earliest Givetian, as evidenced by mixing of EAR and OWR faunas in the US Midcontinent Devonian carbonate platforms (Iowa and Michigan basins) and Appalachian and Moose River basins, was a direct consequence of widespread transgression of the North American craton at this time. This late Eifelian sea level rise resulted in profound expansion of cratonic seaways that provided avenues for migrations of elements of formerly isolated North American brachiopod faunas. Rocks deposited during this transgression can be correlated in most successions in North America, which corresponds to the sea level rise of Devonian eustatic Transgressive-Regressive (T-R) cycle **1e**. OWR migrants found in U.S. midcontinent and eastern U.S. faunas include *Variatrypa* (*V.*) *arctica*, *Spinatrypa borealis*, *Spinatrypina*, *Carinatrypa*, *Warrenella*, *W.* (*Warrenellina*), *Gypidula*, *Kayserella americana*, shallow water species of *Emanuella* (*E. sublineata*, *E. meristoides*), *Subrensselandia*, and possibly *Stringocephalus*. Some of these, as well as the OWR *Independatrypa* occur with coeval faunas in the Moose River Basin of eastern Canada. Elements of this paleobiogeographically mixed fauna are now known to occur in coeval deposits of: the Murray Island Formation in the Moose River Basin in Ontario; the Bakoven (upper part), Stony Hollow, and Hurley members of the Union Springs Formation in eastern New York; the Delaware Limestone of Ohio; the Rogers City Limestone of Michigan; the Lake Church Formation of eastern Wisconsin; the Spillville and Otis formations in northern and eastern Iowa; and the Elm Point Formation of southern Manitoba.

INTRODUCTION

Temperate and subtropical endemic brachiopod taxa characterized shelf faunas of the Eastern Americas Realm (EAR) in the US midcontinent and eastern N America, which were isolated for much of the Middle Devonian from warm-water tropical Old World Realm brachiopod faunas of western N America (Johnson 1970, 1971; Johnson & Boucot, 1973). N American Middle Devonian provincial shelly invertebrate faunas were replaced by cosmopolitan faunas as a direct result of a series of major marine transgressions that began in the late Givetian (Taghanic Onlap of Johnson, 1970; = Devon-ian Transgressive-Regressive [T-R] cycle IIa of Johnson et al., 1985) which allowed for migrations and mixing of formerly isolated mid-Givetian faunas.

Preliminary comparisons of older N American Middle Devonian brachiopod faunas by Koch (1983) and Day & Koch (1994a) indicate that a significant pre-Taghanic breakdown of brachio-

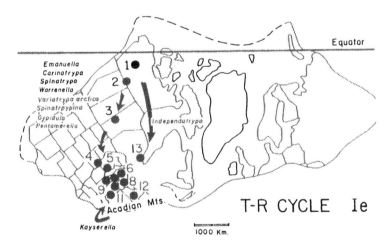

Figure 1. Middle Devonian (late Eifelian) modified paleogeographic base map of Euramerica (Witzke & Heckel, 1989) showing locations of successions with strata deposited during Devonian T-R cycle Ie of Johnsonet al. (1985), Johnson & Klapper (1992). Genera listed in upper right are Old World Realm (OWR) carbonate platform taxa that migrated into central and eastern North America from western and northern Canada; *Kayserella* shown in the lower part of map migrated from OWR shelf habitats in north Africa. Large arrow on right side labelled *Independatrypa* shows possible migration route from western Canada. Locality key: 1=Lower Mackenzie River Valley, NWT; 2=south shore Great Slave Lake, NWT, and NE Alberta; 3=SW Manitoba; 4=Iowa Basin (Iowa, SE Minnesota, W Illinois); 5=SE Wisconsin; 6=NE Michigan; 7=SE Michigan; 8=SW Ontario; 9=central Indiana; 10=north-central Ohio; 11=central Ohio; 12=New York; 13=Hudson Bay Lowlands.

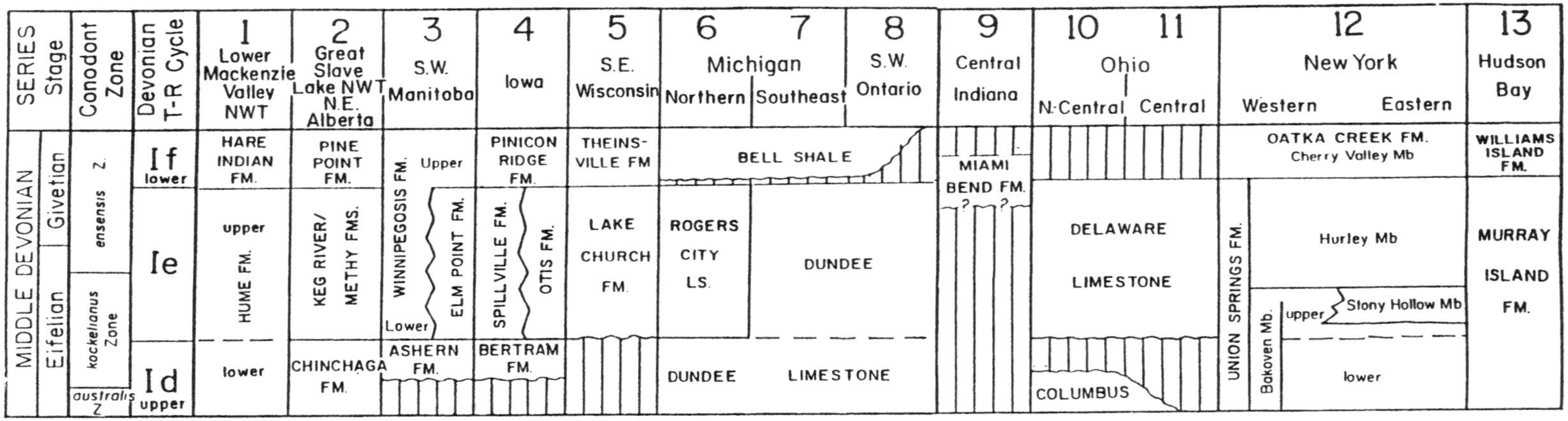

Figure 2. Correlation of late Eifelian-early Givetian strata at localities of Figure 1. Conodont biostratigraphy and Devonian Transgressive-Regressive (T-R) cycles after Johnson et al.(1985), and Johnson & Klapper (1992).

SERIES	Stage	Conodont Zone	Devonian T-R Cycle	1 Lower Mackenzie Valley NWT	2 Great Slave Lake NWT N.E. Alberta	3 S.W. Manitoba	4 Iowa	5 S.E. Wisconsin	6 Michigan Northern	7 Michigan Southeast	8 S.W. Ontario	9 Central Indiana	10 11 Ohio N-Central Central	12 New York Western Eastern	13 Hudson Bay
MIDDLE DEVONIAN	Givetian	ensensis Z	If lower					THEINS-VILLE FM.	BELL SHALE			MIAMI BEND FM.	DELAWARE LIMESTONE	OATKA CREEK FM. Cherry Valley Mb / Hurley Mb / upper Stony Hollow Mb / lower	WILLIAMS ISLAND FM.
	Eifelian	kockelianus Zone	Ie	HARE INDIAN FM.	PINE POINT FM.	Upper WINNIPEGOSIS FM. ELM POINT FM.	PINICON RIDGE FM. OTIS FM. SPILLVILLE FM.	LAKE CHURCH FM.	ROGERS CITY LS.	DUNDEE LIMESTONE	DUNDEE		COLUMBUS	Bakoven Mb UNION SPRINGS FM.	MURRAY ISLAND FM.
		australis Z	Id upper	upper HUME FM. lower	KEG RIVER/METHY FMS. CHINCHAGA FM.	Lower ASHERN FM.	BERTRAM FM.		DUNDEE						

pod provincialism occurred in the late Eifelian, marked by incursions of tropical OWR genera into central and eastern N American shelf areas. Koch & Boucot (1982), and Koch (1983) observed that this apparent faunal realm boundary shift or breakdown in provincialism corresponded to changes in seaway circulation patterns, water temperatures, and destruction of barriers to brachiopod migrations associated with a major Late Eifelian transgression of the N American craton. This late Eifelian (upper part *kockelianus* Zone) sealevel event was later designated as Devonian eustatic T-R cycle 1e by Johnson et al. (1985), and Johnson & Klapper (1992). Fossiliferous rocks deposited during Devonian T-R cycle 1e are now widely correlated in most cratonic and continental margin basins in N American (Figures1-2), based on revised and refined conodont correlations of N American Middle Devonian rocks outlined by Johnson et al. (1985), Braun et al. (1989), Witzke et al. (1989), Johnson & Klapper (1992), Day et al. (1994,1996), Norris (in press), and Uyeno (in press).

New information from our study of Middle Devonian brachiopod faunas in the central and eastern US (Figure 3) makes it possible to reconstruct and analyse faunal distribution patterns accurately, and gauge the effects of Middle Devonian sealevel events on brachiopod paleobiogeography. Comparisons of late Eifelian and early Givetian faunas from deposits in various N American shelf successions in this study, document the nature of faunal migrations, and breakdown of brachiopod provincialism that was the direct consequence of profound changes in paleogeography associated with major transgression of Devonian T-R cycle 1e noted earlier by Koch (1983), Johnson & Klapper (1992), and more recently by Day & Koch (1994a). The sealevel rise of Devonian T-R cycle 1e initiated deposition of carbonate platform sediments across vast areas of central and western N America (Figures 1-2). This late Eifelian sealevel rise breached the Transcontinental Arch and established open marine connections between formerly isolated shelves of the Cordilleran Province of the Old World Realm (OWR) and the Eastern Americas Realm (EAR). Equatorial and tropical carbonate platforms in western N America served as sources of OWR genera that mixed with EAR and endemic taxa in the Iowa, Michigan, Appalachian, and Moose River basins. No cool water EAR taxa appear to have migrated successfully into tropical carbonate platforms of western and arctic N America (Koch & Boucot, 1982), where none has been noted in Devonian faunas of the Great Basin (Johnson, 1990, Great Basin Devonian Faunal Intervals 17-18), and western and arctic Canada.

Movement of OWR taxa into eastern N America appears to have taken place along two main routes. Most OWR genera arrived in eastern N American shelves by migrations through the ESE Elk Point Basin of central Canada (Figures1-3, loc. 3) across the submerged Transcontinental Arch into the Iowa and Michigan basins. Evidence for this is the occurrence of OWR migrants common to western Canadian and US midcontinent faunas such as *Carinatrypa*, *Spinatrypina*, and *Emanuella*, which are absent in eastern N American faunas. US midcontinent carbonate platform faunas were dominated by OWR and cosmopolitan genera with smaller numbers of EAR taxa (Figure 3). The incursion of warm-water OWR taxa into the Appalachian Basin is marked by arrivals of stenothermal tropical gypidulinids such as *Gypidula* and *Pentamerella*, and the atrypids *Variatrypa*, and *Spinatrypa* (*Isospinatypa*). Another seaway connection probably developed between western Canadian shelves and the Moose River Basin of Ontario. The occurrence of OWR *Independatrypa* in the Murray Island Formation, and its absence in New York and Iowa Basin faunas, indicate development of a direct seaway connection with western Canadian platforms, as shown by Ziegler (1989),

and suggested by Johnson & Klapper (1992). Older comparisons of faunas known from these rocks have clearly indicated western Canada as the principle source of OWR migrants seen in late Eifelian and early Givetian deposits in central and eastern N America.

WESTERN AND CENTRAL CANADIAN FAUNAS

Late Eifelian and early Givetian brachiopod faunas are described or listed from the upper Hume Formation in the Lower Mackenzie River Valley and the Keg River and Methy formations from the south shore of the Great Slave Lake in the Northwest Territories (NWT) and NE Alberta, and the Elm Point Formation of SW Manitoba (Figures 1-3, loc. 1-3).

Upper Hume Fauna. Crickmay (1960) originally described '*Spinatrypa*' *dysmorphostrota* from the upper part of the Hume Fm. and named this interval the '*verrilli* zone'. Norris (1968, 1985), and Caldwell (1971), referred to the brachiopod fauna in the middle-upper part of the Hume as the *Spinulicosta stainbrooki* fauna. This interval was later designated as the *dysmorphostrota* Zone by Pedder (1975). Brachiopods from the *dysmorphostrota* Zone of the upper Hume were illustrated in Meek (1868), Warren (1944), Warren & Stelck (1956), Crickmay (1960), McLaren et al.(1962), Caldwell (1968,1971), and Copper (1973, 1978). Important elements of the upper Hume fauna with OWR affinities seen in US midcontinent and eastern N American T-R cycle 1e faunas include *Carinatrypa dysmorphostrota*, *Variatrypa* (*V.*) *arctica*, *Independatrypa aperanta*, *Spinatrypa* (*Isospinatrypa*) *borealis*, *Spinatrypina* (*S.*) *edmundsi*, *Emanuella* sp.II of Caldwell *E. sublineata* (1968), *E. meristoides*, and *E. lineata*.. North Tropical OWR genera restricted to western N America during Devonian T-R cycle 1e include the atrypids *Carinatina*, *Invertrypa*, and possibly *Planatrypa*. The upper Hume spiriferid '*Plectospirifer*' *compactus*, illustrated in Caldwell (1971), has weakly developed plicae in the ventral sulcus, lacks plicae on the dorsal fold, and displays micro-ornament suggesting assignment of this form to *Warrenella* (*Warrenellina*) Brice (1982). Species of *W.* (*Warrenellina*) also occur in US midcontinent faunas of the same age (Figure 3).

Keg River/Methy Fauna. Correlations of the Keg River and Methy formations (Figures1-2, loc. 2) are outlined in Braun et al. (1989), Norris & Uyeno (in Day et al., 1996), Norris (in press), and Uyeno (in press). Widespread elements of the lower Methy brachiopod fauna include *Variatrypa* (*V.*) *arctica*, *Emanuella sublineata*, and *Independatrypa aperanta*. The first two occur in the Spillville or Otis formations in the Iowa Basin (Figure 3), and the latter is listed from Murray Island Fm. in the Moose River Basin of Ontario by Norris (in Sanford & Norris, 1975). Widespread elements of the Keg River brachiopod fauna include *Variatrypa* (*V.*) cf. *arctica*, *Independatrypa perfimbriata*, and *Emanuella* cf. *sublineata*.

Elm Point Fauna. The Elm Point and lower part of the Winnipegosis formations of the lower Elk Point Group in SW Manitoba are correlated with the interval of T-R cycle 1e (Figures 1-3, loc. 3) on the basis of conodont and brachiopod faunas discussed by Norris et al. (1982), Witzke et al. (1989), Braun et al. (1989), and Norris et al. (in Day et al., 1986). According to Norris (in Norris et al., 1982), the Elm Point fauna contains widespread OWR taxa such as *Variatrypa arctica*, *Gypidula*, and *Spinulicosta spinulicosta*.

137

BRACHIOPOD TAXA	Biogeographic Affinity: Old World Realm	Cosmopolitan	Eastern Americas R	Europe	3 S.W. Manitoba	4 Iowa Basin	5 S.E. Wisconsin	6 N. Michigan	9 C. Indiana	12 New York	13 Hudson Bay
Atlanticocoelia acutiplicata			■							●	
Carinatrypa sinuata	■						●				
C. dysmorphostrota	■					●		●			
Spinatrypa (S.) n.sp.	■						●				
S. sp.	■									●	
S. (Isospinatrypa) borealis	■					●	●				
S. cf. borealis	■									●	
Spinatrypina (S.) edmundsi	■					●					
Variatrypa (V.) arctica	■				●	●	●	●		●	●
Fimbrispirifer cf. grieri			■				●				
Orthospirifer ? n. sp.			■	■		●	●	●			
Warrenella maia	■									●	
W. (Warrenellina) extensa	■					●	●				
Emanuella meristoides	■					●					
E. sublineata	■					●		●			
E. sp. II of Caldwell	■					●	●				
indet. ambocoelid										●	
indet. rostrospiroid										●	
Nucleospira ? sp.		■								●	
Cyrtina sp.		■							●	●	
C. cf. hamiltonensis		■				●	●	●			
Cranaena n.sp.			■			●	●				
Subrensselandia n.sp.	■							●			
S. subpyriformis	■								●		
Stringocephalus sp. A	■								●		
S. sp. B	■								●		
Kayserella americana	■									●	
Schizophoria macfarlanei		■					●				
S. cf. macfarlanei		■								●	
S. n.sp. A		■						●			
S. sp.		■				●					
Skenidium sp.		■								●	
Gypidula n.sp. A	■					●					
G. n.sp. B	■						●	●			
G. sp.	■				●					●	
Pentamerella cf. winterei	■									●	
Camarotoechia limitaris			■							●	
indet. unciniellid	■					●					
Arcuaminetes scitulus			■			●	●	●			
indet. chonetid										●	
Longispina sp.			■							●	
Schuchertella n.sp. A		■				●	●	●			
S. or Eoschuchertella sp.										●	
Leptaena sp.			■	■						●	
Pholidostrophia sp.		■					●	●			
Strophodonta (S.) musculosa			■	■		●	●	●			
Productella sp.		■								●	
Spinulicosta spinulocosta		■			●	●	●	●			
productellid n.gen.						●					

Figure 3. Chart comparing compositions and paleobiogeographic affinities of elements of late Eifelian-early Givetian brachiopod faunas known from selected Devonian T-R cycle Ie deposits in central and eastern North America. Numbers in uppermost row correspond to localities in Figs. 1 and 2. Where no affinity indicated, paleobiogeographic affinities of those genera either unknown due to uncertainty of identifications, or endemic.

US MIDCONTINENT FAUNAS

Our study of US Midcontinent Middle Devonian brachiopods from the Iowa and Michigan basins (Figure 3) demonstrate their close similarities at the generic and species levels, and dominance of OWR genera in these faunas. The common occurrence of *OWR Variatrypa, Carinatrypa, Spinatrypa* (*Isospinatrypa*), *Emanuella*, and *Gypidula*, in western Canadian and US Midcontinent faunas indicates that midcontinent Iowa and Michigan basin areas constituted a broad, contiguous carbonate platform that was directly connected by open marine seaways to western N America through the Elk Point Basin of Manitoba during T-R cycle Ie (Figure 1). T-R cycle Ie rocks in this region are included in the Spillville and Otis formations in the Iowa Basin, the Lake Church Fm. of SE Wisconsin, the Rogers City Limestone of N Michigan, and the upper Dundee Fm. of SE Michigan and SW Ontario. Some authors (Ehlers & Kesling,1970) have corre-lated the Miami Bend Fm. (Cooper & Phelan,1967) of central Indiana with the Rogers City Limestone.The available evidence suggests that the Miami Bend is younger than the Rogers City and its equivalents in the western Michigan Basin and adjacent Iowa Basin. The upper Dundee brachiopod faunas of Ontario and SE Michigan are poorly known. These rocks yield *kockelianus* and *ensensis* Zone conodonts (Uyeno, in Uyeno et al., 1982; Sparling, 1988), and were assigned to T-R cycle 1e by Klapper & Johnson (1992).

Spillville-Otis Fauna. In the Iowa Basin, rocks of the Spillville and Otis formations were deposited during Devonian T-R cycle 1e (Witzke et al., 1989; Bunker & Witzke, 1992; Johnson & Klapper, 1992; Day et al., 1994, 1996). A diverse open-marine brachiopod fauna was recently reported from the Spillville by Day & Koch (1994a, 1994b). The Otis yields brachiopod assemblages included in the *Emanuella*, defined to include assemblages consisting of *E. sublineata* in the Coggin and Cedar Rapids members of east-central Iowa, and *E.* cf. *E. sublineata* in the undifferentiated Otis in SE Iowa and NW Illinois (Figure 4). The 'Lake Meyer Member' of Bunker et al. (1983) does not yield brachiopods. Day & Koch (1994a, 1994b) noted a succession of two major faunas, reflecting arrivals of two major waves of migrants in the Spillville shelf. The oldest assemblages from the lower 4-5m of the Spillville comprise the *Spinulicosta-Spinatrypa* Fauna, which is succeeded by elements of the *Spinatrypina-Brachyspirifer* ? Fauna in the remainder of the Spillville succession (Figure 4). The older *Spinulicosta-Spinatrypa* Fauna is comprised of *Spinulicosta spinulicosta, Emanuella* sp. II (Caldwell, 1968), *Gypidula* n.sp.A, *Spinatrypa* (*Isospinatrypa*) cf. *borealis, Carinatrypa dysmorphostrota*, and *Warrenella* (*Warrenellina*) cf. *extensa*.

In Michigan, *Carinatrypa dysmorphostrota* is restricted to the lower Rogers City Limestone (Ehlers & Kesling,1970), and is closely related to *C. sinuata* in the Lake Church Fm. of Wisconsin (Cleland, 1911; Griesemer, 1965). The *Spinatrypina-Brachyspirifer* ? Fauna of the upper Spillville (Figure 4) is renamed the *Spinatrypina-Orthospirifer* Fauna, because the alate spiriferid identified earlier as *Brachyspirifer* ? (in Day & Koch, 1994a), is considered to be *Orthospirifer*. This fauna includes *Spinatrypina* (*S.*) *edmundsi, Gypidula* n.sp.A, *Schizophoria, Spinatrypa* (*I.*) *borealis, Schuchertella, Arcuaminetes scitulus, Plicanoplia, Emanuella* sp.II of Caldwell (1968), *E. meristoides, Strophodonta* (*S.*) *musculosa, Cyrtina, Cranaena, Variatrypa* (*V.*) *arctica*, a new genus of productellid similar to *Productella*, and *Orthospirifer* n.sp.A [identical to *Brachyspirifer* from the Rogers City Limestone of Michigan in Ehlers & Kesling (1970, Pl.12, figs. 44-48), in the Lake Church Formation of SE Wisconsin, and a closely related, but more alate form from the St. Laurent Limestone of E Missouri named

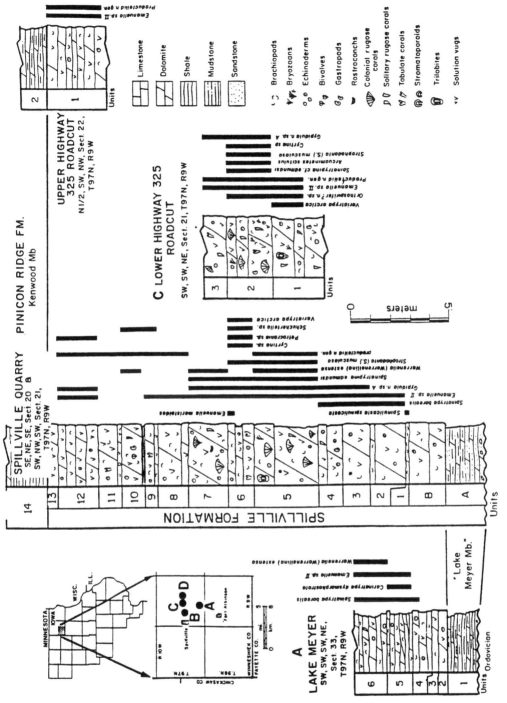

Figure 4. Distribution of brachiopod species and stratigraphy of the Spillville Formation of the Wapsipinicon Group in its type area in NE Iowa (locality 4 of figures 1-3).

139

B. ventroplicatus by Cooper (1945)]. Although the Spillville-Otis fauna is dominated by OWR genera, it also features EAR genera such as *Orthospirifer*, *Arcuaminetes*, *Strophodonta*, and *Cranaena*.

Lake Church Fauna. The Lake Church Fm. of SE Wisconsin yields a brachiopod fauna with many elements common to the Spillville fauna of Iowa and the Rogers City Limestone in northern Michigan (Figures1-3:loc.5). Lake Church brachiopods were described in Cleland (1911), and Griesemer (1965), and general aspects of the brachiopod sequence were outlined in Raasch's (1935) original description of the formation. Raasch (1935) reported *Atrypa reticularis* (= *Variatrypa*), *A. hystrix* (=*Spinatrypa* ?), *Spirifer* (= *Mediospirifer*), *Chonetes* '*vicinus*', *Gypidula*, and *Pholidostrophia iowensis* from the basal Belgium Mbr. He also reported *Stropheodonta musculosa* and a 'large species of *Productella*' from the overlying Ozaukee Mbr. Our study yielded *Carinatrypa sinuata*, *Variatrypa* (*V.*) *arctica*, *Spinatrypa* (*I.*) *borealis*, *S.* (*S.*) n.sp., *Cyrtina* cf. *hamiltonensis*, *Emanuella* sp.II (Caldwell,1968), *Fimbrispirifer*, *Mediospirifer* n.sp. A, *Warrenella* (*Warrenellina*) *extensa*, *Strophodonta* (*S.*) *musculosa*, *Spinulicosta spinulicosta*, *Pholidostrophia* (*P.*) n.sp., *Arcuaminetes scitulus*, *Schuchertella* n.sp., *Schizophoria macfarlanei*, and *Cranaena*. Paleobiogeographic affinities are dominantly OWR or cosmopolitan, except for the EAR taxa *Fimbrispirifer*, *Orthospirifer*, *Cranaena*, *Arcuaminetes*, and *Strophodonta* (*S.*), the latter also occurring in European faunas (Figure 3).

Rogers City Fauna. Brachiopods from the Rogers City Fm. were illustrated by Ehlers & Kesling (1970), who provided data on taxon ranges within two faunal subdivisions named the *Emanuella* and *Omphalocirrus* zones (Figure 5). Linsley (1973) later summarized stratigraphic distribution of the fauna. Current correlations appear in Norris et al. (1982), Johnson & Klapper (1992), Day & Koch (1994a), Day (1996), and Day et al.(1996). Restudy of Ehlers and Kesling's type material and new collections confirm similarities with the Lake Church and Spillville faunas. The fauna of the *Emanuella* Zone (units1-2) consists of *Carinatrypa dysmorphostrota*, *Orthospirifer* n.sp.A, *Emanuella sublineata*, *Cyrtina*, *Spinatrypa* (*I.*) *borealis*, *Independatrypa*, *Spinulicosta spinulicosta*, *Schuchertella* n.sp, and *Schizophoria* n.sp. A. The brachiopod fauna of the *Omphalocirrus* Zone (units 3-6) is most diverse abundant in the *Atrypa* Zonule, and is succeeded by a low diversity fauna in the interval of the *Gypidula* Zonule (units 3-4). Brachiopods are rare in the remainder of the overlying strata (units 5-6; Linsley, 1973). The fauna of the *Atrypa* Zonule includes *Variatrypa* (*V.*) *arctica*, *Spinatrypa* (*I.*) *borealis*, rare *Carinatrypa dysmorphostrota*, *Orthospirifer* n.sp. A, *Gypidula* n.sp.B, *Strophodonta* (*S.*) *musculosa*, *Pholidostrophia* (*P.*) n.sp., and distinctive *Subrensselandia*. Brachiopods are abundant in the *Gypidula* Zonule, and assemblages are dominated by *G.* n.sp.B, with smaller numbers of *V.* (*V.*) *arctica* and *Pholidostrophia* n.sp.

Miami Bend Fauna. The fauna of the Miami Bend Fm. of north-central Indiana (Figures 1-3, loc.9) was first defined by Cooper & Phelan (1966), and correlated it with the Rogers City using *Subrensselandia*, and with part of the Winnipegosis Fm. of Manitoba, using *Stringocephalus* and gastropod faunas. Three of four brachiopod species known from the Miami Bend (Figure 3) were described by Cooper & Phelan (1966). We recovered abundant *Subrensselandia subpyriformis* and *Cyrtina* , but did not find *Stringocephalus*. The Miami Bend yielded a sparse conodont fauna of *Polygnathus linguiformis linguiformis* (indet. morphotype), *Dvorakia*, and *Icriodus* cf. *orri*, insufficient to make an unequivocal zonal correlation. In western N America and central Canada, *Stringocephalus* and *Subrensselandia* is are not observed together any lower than the *castanea* Zone and its equivalents in the basal part of T-R cycle 1f. In the

Great Basin, species of *Stringocephalus* first occur in shallow platform deposits no older than the *castanea* Zone (Devonian Interval 19 of Johnson, 1990). Norris (in press) and Uyeno (in press) will outline new faunal data and refined correlations of the Middle Devonian in western Canada, demonstrating that the oldest *Stringocephalus* occur in the upper Methy Fm. in NE Alberta, and in the upper Keg River Fm. of the Great Slave Lake area, NWT. These forms apparently first appear below the lowest *Eliorhynchus castanea*, and consequently might fall within the upper Devonian T-R cycle 1e. If so, the migration of *Stringocephalus* into the US midcontinent happened during the later phase of T-R cycle Ie (Figure 2).

EASTERN NORTH AMERICAN FAUNAS

Late Eifelian and early Givetian strata yielding diagnostic conodont or brachiopod faunas of Devonian T-R cycle Ie in eastern N America include: the Delaware Limestone of Ohio; rocks now included in the upper part of the Union Springs Fm. of New York; and the Murray Island Fm. of the Hudson Bay Lowlands (Figures1-3, loc.10-13). The OWR spiriferid *Warrenella* occurs in Ohio and New York, and distinctive OWR atrypids are found in the Union Springs and Murray Island faunas.

Delaware Fauna. The Delaware Limestone of Ohio yields conodont faunas of the *kockelianus* and lower *ensensis* zones described in reports by Sparling (1983,1988), and discussed by Klapper (in Johnson & Klapper,1992), who placed the Delaware in the interval of T-R cycle Ie (Figure 2). The Delaware brachiopod fauna is largely undescribed, or in need of revision. Stauffer (1910) listed *Warrenella maia* (Billings, 1860), illustrated earlier by Hall (1867), and Hall & Clarke (1894).

Union Springs Fauna. Deposits correlated with Devonian T-R cycle Ie yielding significant OWR genera were recently recovered from the upper Union Springs Fm. of New York (see Ver Straeten et al., 1994; Brett & Baird,1994), in central and western N America (Figures 1-3, loc. 12). The part of the Union Springs yielding OWR genera such as *Spinatrypa* (*Isospinatrypa*), *Variatrypa*, *Warrenella*, *Gypidula*, *Pentamerella*, and *Kayserella* are included in the upper Bakoven, Stony Hollow and Hurley members. In N America, the OWR orthid *Kayserella* is only known in late Eifelian strata of New York. Its absence in coeval western N American and US midcontinent faunas indicates that migration of *Kayserella* must have been possible by seaway from northern Gondwana (Figure 1), where it is known in the Eifelian strata of N Africa.

Cooper (1943) identified important elements of the Stony Hollow Member fauna, including '*Kayserella*, *Pentamerella*, *Leptaena*, *Productella*, *Chonetes*, a fine ribbed *Atrypa*, and a small *Atrypa* of the *spinosa* stock'. The latter two taxa correspond to *Variatrypa* and *Spinatrypa* recently found in Ver Straeten's Union Springs samples by one of us (WFK) with Art Boucot and Robert Blodgett. Koch, Blodgett, and Boucot identified a low diversity fauna (unpublished) from the upper Bakoven Mbr that includes *Camarotoechia limitaris*, *Warrenella* cf. *maia*, *Atlanticocoelia acutiplicata*, and an indeterminate schuchertellid. The Stony Hollow Mbr yields *Spinatrypa* cf. *borealis*, *Variatrypa* (*V.*) *arctica*, *Cyrtina*, *Warrenella* cf. *maia*, *Atlanticocoelia acutiplicata* ?, *Nucleospira* ?, *Gypidula*, '*Productella*' (n.gen.), *Spinulicosta*, *Schizophoria* cf. *macfarlanei*, and an indeterminate chonetid, ambocoeliid, and rostrospiroid. The youngest Union Springs fauna occurs in the Hurley Mbr, which yielded '*Productella*' (n.gen.), *Longispina*, *Kayserella americana*, *Skenidium*, *Atlanticocoelia acutiplicata*, *Nucleospira* ?, *Spinatrypa* cf. *borealis*, *Warrenella* cf. *maia*, *Cyrtina*, a leiorhynchid, and schuchertellid.

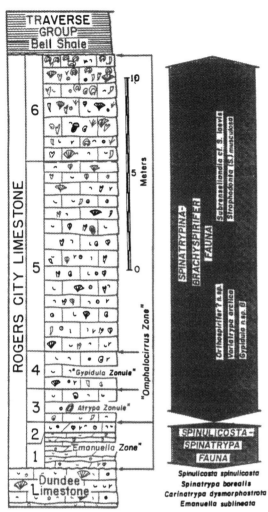

Figure 5. Faunal subdivisions and stratigraphy of the Rogers City Limestone in its type area, Michigan. Correlations of the *Emanuella* and *Omphalocirrus* zones (Ehlers & Kesling, 1970) and the *Spinulicosta-Spinatrypa* and *Spinatrypina-Orthospirifer* ? faunas of the Spillville.

existence of such a connection. Our discovery of specimens of *Independatrypa* in the Rogers City Limestone of N Michigan point to a migration of OWR *I. aperanta* and *I. perfimbriata* from W Canada via the Michigan Basin into the Moose River Basin. Alternatively, migration from the Moose River Basin into the Michigan Basin from the north, is equally likely. *Variatrypa* could also have taken a more direct northerly migration route into eastern N America with *Independatrypa* (Figure 1).

SUMMARY

The late Eifelian sea level rise of Devonian T-R cycle 1e (Johnson et al. (1985) and Johnson & Klapper (1992) was sufficient to breach barriers to benthic migrations and resulted in a breakdown of Middle Devonian brachiopod provincialism. This is indicated by the widespread occurrence of tropical warm water OWR genera in US midcontinent and eastern N American faunas, derived from W Canada. The most wide-spread OWR genera in western, central, and eastern N American T-R cycle 1e faunas include *Variatrypa*, *Spinatrypa* (*S.*), *S. (Isospinatrypa)*, *Warrenella*, and *Gypidula*. US midcon-tinent T-R cycle 1e faunas in the Iowa and Michigan basins are dominated by OWR genera including *Variatrypa* (*V.*), *Spin-atrypa* (*Isospinatrypa*), *S. (S.)*, *Spinatrypina* (*S.*), *Carinatrypa*, *Emanuella*, *Gypidula*, and *Warrenella* (*Warrenellina*). These occur in assemblages with EAR taxa such as *Fimbrispirifer*, *Orthospirifer* ?, *Arcuaminetes*, *Strophodonta* (*S.*), *Plicanoplia*, and *Cranaena*. Migrations of OWR genera into the US midcontinent occurred through the SE Elk Point Basin of Manitoba. In eastern N America, OWR taxa are known from T-R cycle 1e deposits in Ohio (*Warrenella*), New York, and the Hudson Bay Lowlands of Ontario. New York faunas of the upper Union Springs Formation yield species of OWR taxa such as *Varia-trypa* (*V.*), *Spinatrypa* (*I.*), *Spinatrypa* (*S*)., *Warrenella*, *Gypi-dula*, *Pentamerella*, and *Kayserella*. All but the last moved from western N American shelves, into the eastern US, through midcontinent basins. *Kayserella*, found in Union Springs strata of New York, probably migrated from northern Gondwana (north Africa). These OWR taxa are associated with important EAR endemics including: *Atlanticocoelia*, *Leptaena*, *Arcuaminetes*, and *Camarotoechia*. Strata of the Hudson Bay Lowlands contain species of the OWR genera *Variatrypa* (*V.*), *Spinatrypa* (*S.*), and *Independatrypa*. The presence of the latter OWR genus suggests that a direct migration route may have existed between western Canadian shelf areas and the Moose River basin of Ontario.

Acknowledgments

This paper is dedicated to the memory of Dr. Jess Johnson who made momumental contributions towards the understanding of North American Devonian brachiopod systematics, paleobiogeography, and biostratigraphy. The authors would like to thank Dr. Willy Norris (Emeritus, Geological Survey of Canada, Calgary) for his careful and thorough review of this study, and providing prepublished copies of his systematic and biostratigraphic study of the brachiopod faunas of the Middle-Upper Devonian rocks in the Great Slave Lake area. We thank Dr. Dan Fisher and Dr. Gregg Gunnel of the University of Michigan Museum of Paleontology for their assistance and hospitality during our visit to the museum and providing loans of type specimens and stratigraphic collections of brachiopods from the Rogers City Limestone of Michigan. We also thank Dr. Rodney Watkins of the Milwaukee Public Museum (Milwaukee, Wisconsin) for his hospitality during our stay in his home and visit to the museum, and for providing loans of collections from the Lake Church Formation of SE Wisconsin.

Murray Island Fauna. Correlations of the Murray Island Fm. of the Hudson Bay Lowlands, based on brachiopod and conodont faunas, were outlined by Norris (in Sanford & Norris, 1975), which cited older faunas recovered from the subsurface Murray Island by Wilson (1953). They noted the unusual aspects of the fauna comprised of a mix of OWR atrypids and EAR endemics. Important OWR taxa identified by Norris include *Variatrypa* (*V.*) *arctica*, *Independatrypa* cf. *aperanta*, *I.* cf. *perfimbriata*, *Spinatrypa* cf. *costata*, and *S. ehlersi* [the spinatrypids also occur in the Dundee Fm. of Ontario and SE Michigan (Basset, 1935)]. Johnson & Klapper (1992) noted the occurrence of OWR atrypids in the Murray Island fauna from the report by Norris (in Sanford & Norris,1975) and postulated movement of these taxa into Ontario along a direct route from W Canada, based on Ziegler's (1989,fig.5) paleogeographic reconstructions of Laurussia. No intervening Devonian rocks (outliers) are known which could provide direct evidence for the

REFERENCES

BASSET, C.F. 1935, Stratigraphy and paleontology of the Dundee Limestone of southeastern Michigan. Geological Society America Bulletin 46: 425-462.

BILLINGS, E.1860. Canadian Journal Industry, Science Art 31:276.

BIZZARRO,M.1995 The Middle Devonian chonetoidean brachiopods from the Hamilton Group of New York. Documents Laboratoire Géologie, Lyon 136:149-190.

BRAUN, W.K., A.W. NORRIS & T.T.UYENO.1989. Late Givetian to early Frasnian biostratigraphy of western Canada: the Slave Point-Waterways boundary and related events. Canadian Society Petroleum Geologists Memoir 14(3):93-112.

BRETT, C.E. & G.C. BAIRD 1994. Depositional sequences, cycles,and foreland basin dynamics in the Middle Devonian (Givetian) of the Genesee Valley and Finger Lakes region, p.505-586. In, C.E. Brett, & J.Scatterday, (eds.), Field Trip Guidebook, 66th Annual Meeting New York State Geological Association.

BRICE, D. 1982. Brachiopodes du dévonien Inferieur et moyen des formations de Blue Fiord des îles arctiques canadiennes. Geological Survey Canada Bulletin 326:1-175.

BUNKER, B.J. & B.J. WITZKE. 1992. An upper Middle through lower Upper Devonian lithostratigraphic and conodont biostratigraphic framework of the Midcontinent Carbonate Shelf area, Iowa. Iowa Department of Natural Resources, Guidebook Series16:3-26.

___, G. KLAPPER & B.J. WITZKE.1983. New stratigraphic interpretations of the Middle Devonian rocks of Winneshiek and Fayette counties, northeastern Iowa. Geological Society Iowa, Guidebook 39: 1-38.

CALDWELL, W.G.E., 1968. Ambocoeliid brachiopods from the Middle Devonian rocks of northern Canada, p.601-616. In, D.H. Oswald (ed.), Proceedings First International Symposium on the Devonian System 2.

___1971. The biostratigraphy of some Middle and Upper Devonian rocks in the Northwest Territories: an historical review. Musk-Ox 9:1-20.

CLELAND, H.F.1911. The fossils and stratigraphy of the Middle Devonic of Wisconsin. Wisconsin Geological Natural History Survey Bulletin 21 (6):1-222.

COOPER, G.A. 1943. Geology of the Coxsackie Quadrangle, New York. New York State Museum Bulletin 332:247-248.

___1945. New species of brachiopods from the Devonian of Illinois and Missouri. Journal Paleontology 19(5):479-489.

___& D. PHELAN. 1966. Stringocephalus in the Devonian of Indiana. Smithsonian Institution Miscellaneous Collections 157(1):1-20.

COPPER, P.1973 New Siluro-Devonian atrypoid brachiopods. Journal Paleontology 47 (3): 484-500.

___1978. Devonian atrypoids from western and northern Canada. Geological Association Canada Special Paper 18:289-331.

CRICKMAY, C.H. 1960. The older Devonian faunas of the Northwest Territories. Evelyn deMille Books, Calgary, 20p.

DAY, J. 1989. The brachiopod succession of the late Givetian-Frasnian of Iowa. Canadian Society of Petroleum Geologists Memoir 14(3):303-326.

___1992. Middle-Upper Devonian (late Givetian-early Frasnian) brachiopod sequence in the Cedar Valley Group of central and eastern Iowa. Iowa Department Natural Resources, Guidebook 16:53-105.

___1994. The late Givetian and early Frasnian (Middle- Upper Devonian) brachiopod fauna of the Cedar Valley Group of the Iowa Basin p.65-84. In, W. Hickerson (ed.), Geological Society Iowa Guidebook.

___(in press). Faunal signatures of Middle-Upper Devonian depositional sequences and sea level fluctuations in the Iowa Basin: U.S. Midcontinent. In, B.J. Witzke, G.A. Ludvigson & J. Day (eds.), Paleozoic sequence stratigraphy, perspectives from the North American craton. Geological Society America Special Paper 306.

___& W.F. KOCH. 1994a. The Middle Devonian (late Eifelian-early Givetian) brachiopod fauna of the Wapsipinicon Group ofeastern Iowa and northwestern Illinois. Geological Society Iowa Guidebook 59:31-44.

___&___1994b. The previously undescribed Middle Devonian (late Eifelian-Givetian) brachiopod fauna of the Spillville Formation of the Iowa Basin. Geological Society America, Program Abstracts.

___,T.T.UYENO, A.W. NORRIS, B.J. WITZKE & B.J. BUNKER.1994. Middle-Upper Devonian relative sea level histories of central and western North American interior basins. ibid. 26(5):12.

___,___,___,&___(in press). Middle-Upper Devonian relative sea level histories of central and western North American interior basins. In, B.J. Witzke, G.A. Ludvigson & J. Day (eds.), Paleozoic sequence stratigraphy, perspectives from the North American craton. Geological Society America Special Paper 306.

EHLERS, G.M. & R.V. KESLING. 1970. Devonian strata of Alpena and Presque Isle counties, Michigan. Michigan Basin. Geological Society Guide Book North-Central Section Meeting Geological Society America, 130 p.

GRIESEMER, A.D., 1965, Middle Devonian brachiopods of southeastern Wisconsin. Ohio Journal Science 65:241-294.

HALL, J. 1867. Descriptions and figures of the fossil Brachiopoda of the upper Helderberg, Hamilton, Portage and Chemung groups. New York State Geological Survey, Paleontology New York 4:1-428 .

___& J.M. CLARKE. 1894. Introduction to the study of the Brachiopoda. Annual Report New York State Geologist, Paleontology 13(2):1-394.

JOHNSON, J.G. 1970. Taghanic onlap and the end of North American provinciality. Geological Society America Bulletin 81:2077-2106.

___1971. A quantitative approach to faunal province analysis. American Journal Science 270:257-280.

___1978. Devonian, Givetian age brachiopods and biostratigraphy, central Nevada. Geologica Palaeontologica 12:117-150.

___1990. Lower and Middle Devonian brachiopod-dominated communities of Nevada, and their position in a biofacies-province-realm model. Journal Paleontology 64(6):902-941.

___& A.J. BOUCOT. 1973, Devonian brachiopods, p.89-96. In, A. Hallam (ed.), Atlas of palaeobiogeography, Elsevier, Amsterdam.

___& G. KLAPPER.1992. North American midcontinent T-R cycles. Oklahoma Geological Survey Bulletin 145:127-135.

___,___& C.A. Sandberg. 1985. Devonian eustatic fluctuations in Euramerica. Geological Society America Bulletin 96:567-587.

KLAPPER, G. & J.E. BARRICK. 1983. Middle Devonian (Eifelian) conodonts from the Spillville Formation in northern Iowa and southern Minnesota. Journal Paleontology 57:1212-1243.

KOCH, W.F. 1983. Late Eifelian paleobiogeographic boundary fluctuations in North America. Geological Society of America, Programs Abstracts15:616.

___& A.J. BOUCOT. 1982. Temperature fluctuations in the of the Devonian Eastern Americas Realm.Journal Paleontology 56(1):240-243.

LINSLEY, R.M., 1973. Paleoecological interpretation of the Rogers City Limestone (Middle Devonian, northeastern Michigan). Contributions University Michigan Museum Paleontology 24(11):119-123.

MCLAREN, D.J., A.W. NORRIS & D.C. MCGREGOR. 1962. Illustrations of Canadian fossils, Devonian of western Canada. Geological Survey Canada Paper 62-4:1-35.

MEEK, F.B. 1868. Remarks on the geology of the valley of the Mackenzie River with figures and description of fossils from that region, in the museum of the Smithsonian Institution, chiefly collected by the late Robert Kennicott. Chicago Academy Sciences1 (1):61-113.

NORRIS, A.W. 1968. Devonian of northern Yukon Territory and adjacent District of Mackenzie. In, D.H. Oswald (ed.), First International Symposium on the Devonian System, Calgary. Alberta Society Petroleum Geologists1:753-780.

___1985. Stratigraphy of Devonian outcrop belts in northern Yukon Territory and northwestern District of Mackenzie (Operation Porcupine area). Geological Survey Canada Memoir 410:1-81.

___(in press). Brachiopod faunas of the Bituminous Limestone Member of the Pine Point Formation and Sulphur Point Formation (Middle Devonian), Pine Point area on south side of Great Slave Lake, District of Mackenzie; and a discussion of related Devonian stratigraphy and biostratigraphy of adjacent areas. Geological Survey of Canada Bulletin.

___, T.T. UYENO & H.R. McCABE. 1982. Devonian rocks of the Lake Winnipegosis-Lake Manitoba outcrop belt, Manitoba. Geological Survey of Canada, Memoir 392:1-280.

ORR, R.W. 1971. Conodonts from Middle Devonian strata of the Michigan Basin. Indiana Geological Survey Bulletin 45:1-110.

PEDDER, A.E.H. 1975. Revised megafossil zonation of Middle and lowest Upper Devonian strata, central Mackenzie Valley. Geological Survey Canada Paper 75-1(A):571-576.

RAASCH, G.O., 1935, Devonian of Wisconsin, p.261-267. Kansas Geological Society Guidebook, Ninth Annual Field Conference.

SANFORD, B.V. & A.W. NORRIS. 1975. Devonian stratigraphy of the Hudson Platform, 1, Stratigraphy and economic geology,124p., 2, Outcrop and subsurface sections, 248 p. Geological Survey Canada Memoir 379.

SPARLING, D.R. 1983. Conodont biostratigraphy and biofacies of lower Middle Devonian limestones, north-central Ohio. Journal Paleontology 57(4):825-864.

___1988. Middle Devonian stratigraphy and conodont biostratigraphy, north-central Ohio. Ohio Journal Science 88:2-18.

STAUFFER, C.R. 1910. The Middle Devonian of Ohio. Geological Survey Ohio Bulletin 10:341-533.

UYENO, T.T., P.G. TELFORD & B.V.SANFORD. 1982. Devonian conodonts and stratigraphy of southwestern Ontario. Geological Survey Canada Bulletin 332:1-55.

___(in press). Biostratigraphy and taxonomy of Middle Devonian conodonts from south side of Great Slave Lake, District of Mackenzie. Geological Survey Canada Bulletin.

VER STRAETEN, C.A., D.H. GRIFFING & C.E. BRETT. 1994. The lower part of the Marcellus "Shale", central to western New York State: stratigraphy and depositional history p.271-324. In, C.E. Brett & J. Scatterday (eds.), Field Trip Guidebook, 66th Annual Meeting New York State Geological Association.

WARREN, P.S., 1944, Index brachiopods of the Mackenzie River Devonian. Transactions Royal Society of Canada 4(3), 38:105-135.

___& C.R. Stelck. 1956. Devonian faunas of western Canada. Geological Association Canada, Special Paper1:1-75.

WILSON, A.E. 1953. Report of fossil collections from the James Bay Lowland. Ontario Department Mines 61(6):59-81.

WITZKE, B.J., B.J. BUNKER & F.S. ROGERS.1989. Eifelian through lower Frasnian stratigraphy and deposition in the Iowa area, midcontinent, USA. Canadian Society Petroleum Geologists Memoir 14(1):221-250.

___& P.H. Heckel. 1989. Paleoclimatic indicators and inferred Devonian paleolatitudes of Euramerica. ibid. 14(1): 49-66

ZIEGLER, P.A. 1989. Laurussia-the Old Red Continent. ibid. 14(1):15-48.

TAPHONOMIC MEGABIAS IN THE FOSSIL RECORD
OF LINGULIDE BRACHIOPODS

MICHAL KOWALEWSKI

Department of Geosciences, University of Arizona, Tucson, AZ 85721

ABSTRACT- The published literature on the taphonomy of Recent and fossil lingulides (those infaunal brachiopods with a thin organo-phosphatic shell) was surveyed to evaluate changes in the group's fossilization potential through the Phanerozoic. The available lines of evidence include: (1) quality of shell preservation; (2) presence of original shell material; (3) mode of shell occurrence; and (4) taphonomic distribution. They consistently suggest that the fossilization potential of lingulides increased slightly by the end of the early Paleozoic, and was at a relative peak in the late Paleozoic, Triassic, and Jurassic. It was considerably lower in the Cretaceous and especially in the Tertiary when it was comparable with the very low fossilization potential of Recent lingulides. Causes for the changes in fossilization potential may have included changes in shell thickness, shell mineralization, habitat and behavior, and the intensity of extrinsic taphonomic agents. The decrease in the fossilization potential of lingulides may have generated an important megabias in their fossil record. The Phanerozoic record of taxonomic diversity, morphologic disparity, and ecological importance of lingulides may, to some extent, reflect changes in the quality of their fossil record.

INTRODUCTION

Among other things, reconstructions from the fossil record hinge upon a crucial, but rarely stated assumption, that the quality of the record is constant through time. However, this is not necessarily true. The fossilization potential of organisms - - and hence, the quality of their fossil record - may change through geological time. Such changes may result in taphonomic megabias: the large-scale distortion of the fossil record. Changes in the fossilization potential of organisms may be common. First, changes in morphology, behavior, or ecology of an organism are likely to alter its fossilization potential. Second, temporal trends in external environments (e.g., Brandt & Elias, 1989; Crimes & Droser, 1992) often cause changes in the intensity of taphonomic agents, and thus, the percentage of depositional environments that favour preservation may vary through time (see also Allison & Briggs, 1993). Not surprisingly, therefore, a number of taphonomic megabiases has been suggested at various scales and for various organisms (e.g., Hewitt, 1988; Greenstein, 1992; Allison & Briggs, 1993; Kidwell & Brench-ley, 1994).

In this report, I discuss a recent hypothesis (Kowalewski, in press, a) that the fossil record of lingulide brachiopods is affected by a taphonomic megabias. Firstly, I review the taphonomic characteristics of Recent lingulides and discuss their fossilization potential. Secondly, I analyze taphonomic characteristics of fossil lingulides to assess variation in their fossilization potential through the Phanerozoic. Finally, I discuss the causes and consequences of the megabias.

TAPHONOMY AND FOSSILIZATION POTENTIAL OF RECENT LINGULIDES

In the first rigorous taphonomic experiments on an extant lingulide, Worcester (1969) observed that shells of the Hawaiian *Lingula reevei* from Kaneohe Bay are fragile and decompose rapidly due to biological and chemical degradation. Worcester (1969, p.39) placed freshly killed specimens on the Bay bottom. After 90 days only fragile fragments remained. Because no shell remains could be found deeper than 5 cm below sediment-water interface, Worcester (1969, p.45) concluded that fossilization of *L. reevei* is not occurring in Kaneohe Bay and suggested that fossilization of lingulides requires catastrophic environmental changes. Similarly, in a series of studies, Emig (1981, 1983, 1986, 1990) pointed out that shells of Recent *Lingula* are thin, fragile, and due to their high organic content (see Iwata, 1981), susceptible to rapid hydrolysis. Emig (1981, 1983) estimated that lingulide shells are completely destroyed 2-3 weeks after death and also concluded that catastrophic events are necessary for the fossilization of lingulides.

Recent taphonomic studies on *Glottidia palmeri* Dall (Kowalewski, in press, a) from the tidal flats of Baja California, Mexico indicate that the fossilization potential of the other extant lingulide genus is also very low. *Glottidia* shells are fragile, susceptible to decay, and unlikely to survive reworking. The rate of destruction of *G. palmeri* shell varies, but even in the most favorable circumstances (in situ preservation), shells do not last for more than a few years -- consequently, as with *Lingula*, rapid to catastrophic burial is needed for the fossilization of *Glottidia*. Indeed, despite the fact that *G. palmeri* populations are a common component of the intertidal fauna, the beach ridges, formed through gradual reworking of the intertidal flats, are dominated by mollusks (Kowalewski et al., 1994) and contain no lingulide shells.

In sum, the two extant genera of lingulide have an extremely low fossilization potential and require favorable taphonomic circumstances to be preserved in the fossil record. Due to their fragility and susceptibility to decay, Recent lingulide shells are unlikely to be substantially transported or significantly time-averaged (see Kowalewski, in press, b).

HOW DOES ONE ASSESS THE FOSSILIZATION POTENTIAL OF FOSSILS?

It may seem a bit odd to ask about the preservational potential of shells as fossils: after all, to become fossils they must be preserved. However, not all shells are created equally as fossils: some, such as thick-shelled oysters found in reworked lag-deposits, are preserved because their remains were resistant to destruction, whereas others, such as Burgess Shale-type fossils, are preserved because they are lucky enough to die under favorable taphonomic

Table 1. Empirical lines of evidence used to assess the preservational potential of fossil lingulides.

Taphonomic Indicator	Low fossilization potential	High fossilization potential
Quality of shell preservation	Fragmented, fragile, and rare shells. May be well preserved in deposits that favour preservation (i.e., *Fossil Lagerstätten* deposits)	Shells are often well preserved and complete. May be abundant, well-preserved, and not confined to deposits that favour preservation
Preservation of shell material	Casts or molds, with the original shell material removed. Original shell material is found occasionally in deposits that favour preservation	Original shell material is often preserved
Mode of occurrence	Shells typically preserved paired and oriented perpendicularly to the bedding plane (i.e., *in situ* preserved specimens that were not removed from their burrows after death)	Shells often reworked and represented by single valves that are oriented horizontally to the bedding plane
Taphonomic distribution	Fossils occur predominantly in taphofacies that favor preservation (e.g., in obrution assemblages preserved through rapid burial) and are rarely found in deposits that do not favor preservation of fragile fossils	Fossils occur in a wide range of taphofacies and may be abundant in deposits that do not favour preservation of fragile fossils

circumstances. In other words, preserved fossils vary in their initial preservational potential and this variation, if sufficient, generates a taphonomic megabias.

Variation in fossilization potential can be deciphered from various taphonomic indicators. In general, organisms with lower fossilization potential should be found in deposits and circumstances that favoured preservation. Organisms with higher fossilization potential should be found in a wider range of taphofacies, may carry evidence of post-mortem alteration, and should be better preserved (everything else being equal). Specific taphonomic indicators differ among organisms depending on their anatomy, ecology, and other characteristics. The indicators employed here (Table 1) are based both on actualistic observations (see above) and on published literature. They include: (1) quality of shell preservation; (2) presence of original shell material; (3) mode of shell occurrence; and (4) taphonomic distribution. To some extent, each of these indicators reflects the initial fossilization potential.

I surveyed 105 publications to evaluate variation in lingulide taphonomic indicators throughout the Phanerozoic [using the criteria of Table 1],. The papers were selected from GEOREF, UNCOVER, the Brachiopod Reference Database of Rex Doesher (Smithsonian Institution), and other sources. I do not claim that all (or even the majority) of the publications dealing with lingulides are included in this study. However, the large number of publications considered here probably provides a reasonable and representative sample.

This study concerns all infaunal brachiopods that possess an inarticulate, organo-phosphatic, linguliform shell. This tapho-nomically and ecologically consistent group corresponds roughly to the family Lingulidae (*Apsilingula, Barroisella, Glottidia, Langella, Lingula, Lingularia, Lingulipora*(?), *Semilingula*(?), *Trigonoglossa*), but can be extended also to many infaunal obolides (e.g., *Ectenoglossa, Lingullela*). The taxonomy of lingulides, especially those from the Early *Semilingula*(?), *Trigonoglossa*), but can be extended also to many infaunal obolides (e.g., *Ectenoglossa, Lingullela*). The taxonomy of lingulides, especially those from the Early Paleozoic, is disputable and the oldest unquestionable lingulides date back to the Late Devonian (Williams, 1977). Many organo-phosphatic linguliform taxa (e.g., *Pseudo-*

lingula, Holmer, 1991) may belong to other families of lingulids, and many are not preserved in sufficient detail. As a result, the taxonomy of lingulides is unsettled and *Lingula,* sensu lato, is described far more commonly than the unquestioned forms. The questionable forms are included here. Addition of these 'lingulides' extends the taphonomic discussion to a substantially larger group of fossils and expands this study's temporal scope to the entire Phanerozoic. The addition is unlikely to introduce any substantial bias because the anatomic features that raise taxonomic issues are of little taphonomic importance. Finally, some lingulides, especially in the early Paleozoic, may not have been infaunal (e.g., Rudwick, 1970; Percival, 1978), although many certainly were (see especially Pemberton & Kobluk, 1978; Pickerill et al., 1984; Over, 1988). Only infaunal lingulides were included in this analysis because the taphonomic characteristics and initial fossilization potential of non-infaunal lingulides must have been different.

TAPHONOMY OF PHANEROZOIC LINGULIDES

Of 105 papers that were searched, 71 contained useful data. Only a few papers gave detailed explicit taphonomic information, and thus, I was frequently forced to infer taphonomic indicators indirectly from the text and/or figures. For example, if specimens represented in figures were exclusively molds or impressions, I assumed that no original shell material was preserved, and if an associated fauna was well preserved and in place, I assumed that lingulides also were preserved *in situ*. Some of the taphonomic indicators could not be inferred. Thus, the amount and accuracy of information varied among the taphonomic indicators discussed below.

Quality of shell preservation. In the Cenozoic fossil record, lingulide shells are generally poorly preserved. Specimens are typically limited to a few brittle fragments (e.g., Owen, 1980; Cooper, 1988; Popiel-Barczyk & Barczyk, 1990). Complete, well-preserved specimens are found almost exclusively in unusually favorable preservational circumstances. For example, in rapid-burial obrution deposits (e.g., Ohara, 1969), in exceptionally rich assemblages (e.g., Hertlein, 1966; Bullivant, 1969), or in unique, preservation-favouring circumstances (e.g., a few complete specimens of *Glottidia inexpectans* found between the articulated valves of *Pecten --* Olsson, 1914). In the Cretaceous, specimens are typically rare, fragile -- often so fragile that they cannot survive sample processing (e.g., Surlyk, 1972) -- and fragmented (e.g., Scott, 1970; Thomson & Owen, 1979). In the Triassic and Jurassic, the average quality of shell preservation is much better. Complete, well-preserved specimens are found even among transported horizontal valves (e.g., Rowell, 1970; Fürsich, 1984). In the Paleozoic, complete, well-preserved valves are not restricted to favorable preservational conditions and are common in reworked assemblages (e.g., Alexandrowicz & Jarosz, 1971; Archbold, 1981; Craig, 1954; Ferguson, 1963; Williams, 1977).

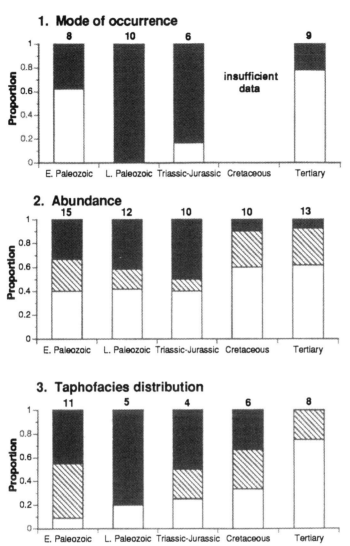

1. Mode of occurrence

2. Abundance

3. Taphofacies distribution

Figure 1. Some taphonomic indicators of lingulide fossilization potential. 1. Mode of occurrence. Black bars represent assemblages that include horizontal/ transported shells and white bars assemblages of exclusively *in situ* preserved lingulides; 2. Relative abundance of lingulides in fossil assemblages. Black bars = abundant, striped = common, white = rare; 3. Lingulide taphofacies distribution. Black bars = widely distributed in a variety of taphofacies, striped = confined primarily to assemblages favoring preservation, white = found exclusively in Fossil Lagerstätten-type deposits.

Preservation of shell material. In the Cenozoic and the Cretaceous, original shell material is often preserved. This is mostly because, for that time interval, lingulides are predominantly found in favourable preservational conditions (see below). In assemblages created under less favorable taphonomic conditions, lingulides occur as moulds or impressions (Thomson & Owen, 1979; Scott, 1970). Although moulds and impressions predominate in many Triassic and Jurassic lingulide assemblages (e.g., Biernat & Emig, 1993; Campbell, 1987), the original shell material is frequently preserved, even when preservational conditions were not particularly favourable (e.g., Rowell, 1970; Muir-Wood, 1959). Original shell material is also commonly preserved in the Paleozoic, even when assemblages underwent reworking (e.g., Alexandrowicz & Jarosz, 1971; Craig, 1954).

Mode of occurrence. In the Cenozoic, lingulides are either preserved in situ or almost in situ. This can be inferred from the nature of the associated fauna, which is typically articulated and well-preserved. Out of ten assemblages, eight contained lingulides that were most likely preserved in situ (Figure 1:1). The Tertiary lingulides of New Zealand, usually found to have been transported, are an interesting exception (Lee & Campbell, 1987). No sufficient information on the modes of occurrence of Cretaceous lingulides was available in the surveyed literature. In the Triassic and Jurassic, the proportion of transported horizontal valves was considerably higher (Figure 1:1). Five out of six assemblages that were described in sufficient detail included transported horizontal shells. Assemblages often consist of convex-up and directionally oriented valves (e.g., Fürsich, 1984), which is good evidence for post-mortem transport (see Kowalewski, in press, a). In the late Paleozoic, all surveyed assemblages included horizontal valves (Figure 1:1), which often predominated (e.g., Graham, 1970). Interestingly, in the early Paleozoic, Silurian, and especially Ordovician, the majority of assemblages consisted exclusively of in situ shells (e.g., Pickerill, 1973; Over, 1988).

Abundance and taphonomic distribution. In the Cenozoic, lingulides are rare in fossil assemblages (Figure 1:2; see also Hertlein & Grant, 1944; Cooper, 1988). They are sometimes locally abundant in the assemblages formed under very favourable taphonomic conditions (e.g., Ohara, 1969; Bullivant, 1969). A similar pattern is observed in the Cretaceous (Figure 1:2). In contrast, pre-Cretaceous lingulides (Figure 1:2), are often common or even abundant. Most notably, they are common in deposits that were not formed under favourable taphonomic conditions (e.g., Craig, 1954; Graham, 1970; Williams, 1977). The striking difference in abundance between post-Jurassic and pre-Cretaceous lingulides is well illustrated by the following comparison. Graham (1970), when studying Scottish Carboniferous lingulides, obtained 5,000 well-preserved specimens. In contrast, studies of Tertiary lingulides are often based on only few fragments (e.g., Chuang 1964; Owen, 1980; Popiel-Barczy & Barczyk, 1990).

The taphofacies distribution of Cenozoic lingulides (Figure 1:3) is confined primarily to preservation-favoring deposits (e.g., Ohara, 1969; Marquet, 1984). In the Cretaceous, however, lingulides are relatively more common in taphofacies that do not favor preservation (Figure 1:3). In the pre-Cretaceous fossil record, lingulides are commonly found in a variety of taphofacies, many of which were not particularly favorable for preservation, and Fossil-Lagerstätten type occurrences constitute a minor proportion of all occurrences (Figure 1:3).

Principal component analysis [PCA] was used to quantify the magnitude and nature of changes in the fossilization potential of lingulides (Figure 2). This analysis included three taphonomic variables (mode of occurrence, abundance, taphonomic distribution) and five time intervals (as defined on Figure 1). To estimate a missing value ('mode of occurrence' for Cretaceous), three iterative, optimal-transformation methods (spline, linear, and montone) were used (see SAS Institute, 1990, PROC PRINQUAL, p. 1265-1323). PCA was consistent in all three transformed datasets: all taphonomic indicators displayed high positive correlations with the first principal component [PC1], which accounted for most of the variation in the data (from 80% to 90%). This suggests that PC1 is a reasonable estimator of changes in the fossilization potential. The resulting PC score curves (Figure 2), indicate that the highest fossilization potential was in the late Paleozoic and early-to-mid Mesozoic. This was followed by a considerable decrease in the Cretaceous and Tertiary. Although the three curves reveal the same general trend they differ in details (Figure 2), pointing to a need for further data collection. In sum, the PCA identified the direction and approximate timing of changes, but the magnitude of these changes (i.e., the shapes of the curves) cannot be trusted fully.

It is suggested (Kowalewski, in press, a) that the greatest change in the fossilization potential of lingulides occurred sometime at the beginning of the Mesozoic. The more rigorous estimates presented here, which are based on substantially larger dataset, suggest that it occurred in the Late Jurassic or Early Cretaceous. An additional, smaller-magnitude change, may have occurred near the end of the early Paleozoic (Figure 2). Note that the change in fossilization potential was most likely gradual, as Cretaceous lingulides may have had a slightly higher fossilization potential than the Tertiary ones (Figure 1:3, and above). Also, changes may not have affected

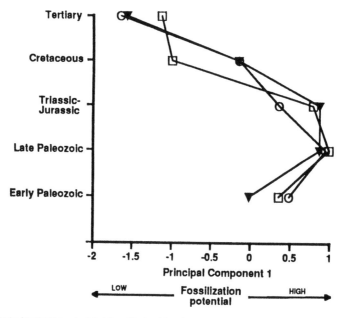

Figure 2. Changes in the fossilization potential of lingu-lide brachiopods throughout the Phanerozoic, as estimated by scores derived from a principal component analysis of the three taphonomic indicators (Figure 1). The three curves correspond to three transformation techniques used to fill in missing values. Symbols are as follows: Empty circles = spline method; black triangles = linear method; empty squares = monotone method.

all lingulide species equally. For instance, the Tertiary lingulides of New Zealand may have had a relatively higher preservational potential than a typical post-Jurassic lingulide. Thus, we should understand the curve (Figure 2) as depicting average fossilization potential of taxa rather than uniform trends that affected all lingulides.

CAUSES AND CONSEQUENCES OF THE TAPHONOMIC MEGABIAS

A number of factors may have generated the observed pattern. They include: (1) shell thickness -- a decrease in shell thickness may result in a decrease in preservational potential; (2) shell mineralization - shells of pre-Cretaceous lingulides may have had a lower organic content, and thus, may have been less prone to decay; (3) ecology -- changes in bathymetric distribution or burrowing behavior may have affected chances for preservation; (4) extrinsic taphonomic factors -- changes in external conditions such as the intensity of bioturbation (Crimes & Droser, 1992) may have affected chances for shell preservation; and (5) some other, as yet unknown factors. Unfortunately, the data needed to evaluate these rival hypotheses are not readily available. For instance, very few authors report shell thickness or composition. Thus, these possible causes are hypotheses that can be tested rigorously only after more data are assembled. It seems more likely that intrinsic characteristics of shells were responsible for observed shifts. Taphonomic indicators such as 'mode of occurrence' point to a change in shell strength rather than to external or ecological factors.

Regardless of causes, postulated changes in preservational potential are likely to have introduced a significant taphonomic megabias. Most importantly, changes in taxonomic diversity of lingulides may reflect variation in fossilization potential. Thus, a much lower diversity of lingulides in the post-Paleozoic fossil record (Kowalewski, in press, a, and unpubl. data) may be partly attributed to the taphonomic megabias (especially in the post-Jurassic record). This does not necessarily mean that the actual diversity of post-Jurassic lingulides was comparable with the Paleozoic level, but only that the post-Jurassic diversity may be more severely underestimated. Changes in the morphological disparity within lingulides may be affected in a similar manner.

Given the low fossilization potential of lingulides, their paleoecological importance in respect to other shelly organisms is likely to be underestimated (see also Patzkowsky, 1995). Again, this bias may be much more substantial in the post-Jurassic fossil record. On the other hand, outcrop-scale fidelity may be higher for post-Jurassic lingulides. Due to their extremely low fossilization potential, they are typically preserved in their original habitats and are unlikely to undergo significant time-averaging.

FINAL REMARKS

This study shows that the fossilization potential of the lingulides may have varied substantially throughout the Phanerozoic. Consequently, their fossil record may be affected by taphonomic megabias. However, it is clear that the data assembled thus far are insufficient to quantify the causes of the megabias rigorously. Unwanted taphonomic distortion of the Phanerozoic history of lingulide brachiopods will only be eliminated when these issues are evaluated. This progress report is important primarily in providing strong evidence that preservational megabias exists. More generally, this study illustrates the importance of, and suggests a methodology for, research on large-scale, macro-taphonomic patterns.

Acknowledgments

I thank Karl Flessa for support, encouragement, and many helpful comments on the manuscript. I am grateful to John Alroy and Peter Holterhoff for stimulating discussions, helpful comments, and suggestions. I thank Rex Doescher, who let me use his brachiopod reference database, and Jack Sepkopski, who let me use his generic database. This study was supported by NSF grant EAR 9417054 (to Karl W. Flessa). This is publication number 15 of C.E.A.M. [Centro de Estudios de Almejas Muertas].

REFERENCES

ALEXANDROWICZ, S.W. & J. JAROSZ. 1971. Palaeoecology of the Zechstein *Lingula* sandstone from Lubin (West Poland). Bulletin Polish Academy Science, Earth Science Series 29 (3): 183-191.

ALLISON, P.A. & D.E.G. BRIGGS. 1993. Exceptional fossil record: Distribution of soft-tissue-preservation through the Phanerozoic. Geology 21: 527-530.

ARCHBOLD, N.W. 1981. *Lingula* (Lingulidae, Brachiopoda) from the late Artinskian (Permian), Carnarvon basin, western Australia. Proceedings Royal Society Victoria 92:169-180.

BIERNAT, G. & C.C. EMIG. 1993. Anatomical distinctions of the Mesozoic lingulide brachiopods. Acta Palaeontologica Polonica 38 (1/2): 1-20.

BRANDT, D.S. & R.J. ELIAS. 1989. Temporal variation in tempestite thickness may be a geological record of atmospheric CO_2. Geology 17: 951-952.

BULLIVANT, J.S. 1969. The Bathhouse Beach assemblage. Southern California Academy of Sciences Bulletin, Los Angeles 68: 86-95.

CAMPBELL, J.D. 1987. Triassic records of the genus *Lingula* (Brachiopoda: Inarticulata) in New Zealand. Journal Royal Society New Zealand 17 (1): 9-16.

CHUANG, S.H. 1964. *Glottidia glauca* n.sp. from the Lower Claiborne of Texas. Journal of Paleontology 38(1): 157-159.

COOPER, G.A. 1988. Some Tertiary brachiopods of the East Coast of the United States. Smithsonian Contributions Paleobiology 64: 1-29.

CRAIG, G.Y. 1954. The palaeoecology of the Top Hosie Shale (Lower Carboniferous) at a locality near Kilsyth. Quarterly Journal Geological Society, London 438:103-119.

CRIMES, T.P. & M.L. DROSER. 1992. Trace fossils and bioturbation: The other fossil record. Annual Reviews Ecology Systematics 23: 339-360.

EMIG, C.C. 1981. Observations sur l'écologie de *Lingula reevei* Davidson (Brachiopoda: Inarticulata). Journal Experimental Marine Biology Ecology 52: 47-61.

___1983. Comportement expérimental de *Lingula anatina* (Brachiopodes, Inarticulés) dans diverses substrats meubles (Baie de Mutsu, Japan). Marine Biology 75:207-213.

___1986. Conditions de fossilization du genre *Lingula* (Brachiopoda) et implications paléoécologiques. Palaeogeography, Palaeoclimatology, Palaeoecology 53: 245-453.

___1990. Examples of post-mortality alteration in Recent brachiopod shells and (paleo)ecological consequences. Marine Biology 104: 233-238.

FERGUSON, L. 1963. The paleoecology of *Lingula squamiformis* Philips during a Scottish Mississippian marine transgression. Journal of Paleontology 37: 669-681.

FÜRSICH, F.T. 1984. Paleoecology of boreal invertebrate faunas from the Upper Jurassic of central East Greenland. Palaeogeography, Palaeoclimatology, Palaeoecology 48: 309-364.

GRAHAM, D.K. 1960. Scottish Carboniferous Lingulacea. Bulletin Geological Survey Great Britain 31: 139-184.

GREENSTEIN, B.J. 1992. Taphonomic bias and the evolutionary history of the family Cidaridae (Echinodermata: Echinoidea). Paleobiology 10: 50-79.

HERTLEIN, L.G. 1966. Pliocene fossils from Rancho El Refugio, Baja California, and Cerralvo Island, Mexico. California Academy Sciences, Proceedings, San Francisco (Ser.4) 30: 265-284.

HERTLEIN, L.G. & U.S. GRANT, IV. 1944. The Cenozoic Brachiopoda of western North America. Publications UniversityCalifornia, Los Angeles, Mathematical Physical Sciences 3:1-24.

HEWITT, R.A. 1988. Nautiloid shell taphonomy: interpretation based on water pressure. Palaeogeography, Palaeoclimatology, Palaeoecology 63 (1/3): 15-26.

HOLMER, L.E. 1991. The systematic position of *Pseudolingula* Mickwitz and related lingulacean brachiopods, p. 3-14. In, D.I. MacKinnon, D.E. Lee & J.D. Campbell (eds.), Brachiopods through time, Balkema, Rotterdam.

IWATA, K. 1981. Ultrastructure and mineralization of the shell of *Lingula unguis* Linne (Inarticulate, Brachiopoda). Journal Faculty Science, Hokkaido University (Ser. 4) 20: 35-65.

KIDWELL, S.M. & P.J. BRENCHLEY. 1994. Patterns in bioclastic accumulations through the Phanerozoic: Changes in input or in destruction? Geology 22: 1139-1143.

KOWALEWSKI, M. (in press,a). Taphonomy of a living fossil: the lingulide brachiopod *Glottidia palmeri* Dall from Baja California, Mexico. Palaios.

___ (in press, b). Time-averaging, overcompleteness, and the geological record. Journal of Geology.

KOWALEWSKI, M, K.W. FLESSA & J.A. AGGEN. 1994. Taphofacies analysis of Recent shelly cheniers (beach ridges), northeastern Baja California, Mexico. Facies 31: 209-242.

LEE, D.E. & J.D. CAMPBELL.1987. Cenozoic records of the genus *Lingula* (Brachiopoda: Inarticulata) in New Zealand. Journal Royal Society New Zealand 17(1): 17-30.

MARQUET, R. 1984. A remarkable Molluscan fauna from the Kattendijk Formation (Lower Pliocene) at Kallo (Oost-Vlaanderen, Belgium). Bulletin de la Société belge de Géologie, 93(4): 335-345.

MUIR-WOOD, H.M. 1959. Report on the Brachiopoda of the John Murray Expedition. John Murray Expedition 1933-1934, Scientific Reports 10: 283-317.

OHARA, S. 1969. Discovery of *Lingula* from the Oligocene Shimokine Formation of the Uryu Coal-field in Central Hokkaido. Journal Geological Society Japan 75(7): 387-388.

OLSSON, A.A. 1914. New and interesting Neogene fossils from the Atlantic Coastal Plain. Bulletin American Paleontology 5(24): 41-72.

OVER, D.J. 1988. Lingulid brachiopods and *Lingulichnus* from a Silurian shelf-slope carbonate sequence, Delorme Group, Mackenzie Mountains, Northwest Territories. Canadian Journal Earth Sciences 25(3): 465-471.

OWEN, E.F. 1980. Tertiary and Cretaceous brachiopods from Seymour, Cockburn and James Ross Islands, Antarctica. Bulletin British Museum Natural History, Geology 33(2):123-145.

PATZKOWSKY, M.E. 1995. Gradient analysis of Middle Ordovician brachiopod biofacies: Biostratigraphic, biogeographic, and macroevolutionary implications. Palaios 10: 154-179.

PEMBERTON, G.E. & D.R. KOBLUK. 1978. Oldest known brachiopod burrow: the Lower Cambrian of Labrador. Canadian Journal Earth Sciences 15: 1385-1389.

PERCIVAL, I.G. 1978. Inarticulate brachiopods from the Late Ordovician of New South Wales, and their paleoecological significance. Alcheringa 2: 117-141.

PICKERILL, R.K. 1973. *Lingulasma tenuigranulata* - Palaeoecology of a large Ordovician linguloid that lived within strophomenid-trilobite community. Palaeogeography, Palaeoclimatology, Palaeoecology 13: 143-156.

___, T.L. HARLAND & D. FILLION. 1984. *In situ* lingulids from deep-water carbonates of the Middle Ordovician Table Head Group of Newfoundland and the Trenton Group. Canadian Journal Earth Sciences 21: 194-199.

POPIEL-BARCZYK, E. & W. BARCZYK. 1990. Middle Miocene (Badenian) brachiopods from the southern slopes of the Holy Cross Mountains, Central Poland. Acta Geologia Polonica 40(3/4): 159-181.

ROWELL, A.J. 1970. *Lingula* from the basal Triassic Kathwai Member, Mianwali Formation, Salt Range and Surghar Range, West Pakistan. p. 111-116. In, B. Kummel & C. Teichert, Stratigraphic boundary problems: Permian and Triassic of West Pakistan. University Kansas, Department Geology, Special Publication 4.

RUDWICK, M.J.S. 1970. Living and fossil brachiopods. Hutchison, London, 199p.

SCOTT, R.W. 1970. Paleoecology and paleontology of the Lower Cretaceous Kiowa Formation, Kansas. University Kansas Paleontological Contributions 52: 5-94.

SAS INSTITUTE. 1990. SAS/STAT® User's Guide: Version 6. 4th ed., Sas Institute, Cary, 943 p. (Vol.1), 846 p. (Vol.2).

SURLYK, F. 1972. Morphological adaptations and population structures of the Danish Chalk brachiopods. Kongelige Danske Videnskabernes Selskab, Biol. Skrifter 19 (2):1-57.

THOMSON, M.R.A. & E.F. OWEN. 1979. Lower Cretaceous Brachiopoda from south-eastern Alexander Island. British Antarctic Survey Bulletin 48: 15-36.

WILLIAMS, A. J. 1977. Insight into lingulid evolution from the Late Devonian. Alcheringa 1: 401-406.

WORCESTER, W. 1969. On *Lingula reevei*. Ph.D. Thesis. University of Hawaii, 49 p.

INTRACRYSTALLINE PROTEINS FROM *TEREBRATULINA RETUSA* (LINNAEUS) AND THEIR ROLE IN BIOMINERALISATION

J. H. LAING

School of Environmental and Applied Sciences, University of Derby, Derby DE22 1GB, England

ABSTRACT- In assessing the role of the intracrystalline proteins from *Terebratulina retusa* in biomineralisation, results indicate that with a number of known protein species, there are previously unidentified, low molecular weight protein species (<10 kDa). Using Stains All, a cationic carbocyanine dye which specifically stains acidic and/or calcium binding proteins blue and all other proteins red, pink or not at, has revealed that three of the main protein species stain blue (positive), suggesting that these proteins possess a calcium binding function. Data from hplc profiles indicate the acidic rather than hydrophobic nature of these proteins.

INTRODUCTION

Biomineralisation, the precipitation and manipulation of inorganic minerals by organisms occurs throughout a wide range of phyla (Lowenstam, 1981; Mann, 1993). Organic macromolecules, principally proteins (Hudson, 1967), but also polysaccharides (Berman, 1993; Weiner, 1979; 1985), which are intimately associated with the mineral phase, play an essential, but as yet poorly understood role in biomineralisation. Initial studies of brachiopods (Jope, 1979) focused mainly on the intracrystalline material between the biocrystals, but this has been shown to be extensively degraded biochemically while appearing physically well-preserved (Towe, 1980). For this reason much recent work on fossiliferous and extant systems (Berman et al., 1988; Curry et al., 1991a; b; Cusack et al., 1992; Walton et al., 1993) has been concerned with the intracrystalline fraction, i.e. macromolecules entombed within the biocrystals.

The fashioning of minerals under atmospheric conditions into complex and often elaborate morphologies, quite unlike their inorganic counterparts, has intrigued scientists for the last two decades. Despite intensive research, however, the underlying mechanisms of biomineralisation remain poorly understood, principally due to a lack of coherence between studies throughout different mineralising phyla. Lowenstam (1981) outlined two processes of biomineralisation according to the degree of biological control: firstly, biologically induced mineralisation, commonly found in bacteria and algae, which involves bulk intra and /or extracellular mineral formation without the elaboration of organic matrices, e.g. precipitation simply arising through the interaction of biogenically formed gases and metal ions present in the external medium. Here, crystal habits are similar to those seen inorganically precipitated minerals. As this process is under minimal biological control, it may represent a more primitive stage in the evolution of biomineralisation. Secondly, organic matrix mediated (or biologically controlled) mineralisation, common among the Animalia, which involves the construction of an organic framework with three proposed functions: nucleation of crystal growth; determination of the shape and volume of the biocrystal and control of crystal development through termination of growth. The mineral type, crystallographic axes and morphology are therefore all under genetic control. One of the most important considerations in biomineralisation is the interaction between the organic macromolecules and the mineral phase of the biocrystal. The organic phase is princi-pally composed of proteins (Hudson, 1967) but also poly-saccharides (Berman, 1993; Weiner, 1979; 1985), carbohy-drates (Collins et al., 1991) and lesser amounts of lipids (Curry et al., 1991a).

Terebratulina retusa has already been the focus of extensive biological studies (Endo & Curry, 1991; James et al., 1991a; b; c; Cohen et al., 1991; 1993) including work on the intracrystalline proteins (Collins et al.,1988; 1991; Curry et al., 1991a,b), and this study both supports and supplements the previous work. Amino acid analysis should confirm the acidic nature of these proteins with their prevalence of Asp and Gly residues both in brachiopods (Walton et al., 1993) and other marine invertebrates (Weiner, 1979; Worms & Weiner 1986; Cariolou & Morse, 1988). Protein sequence data may reveal potentially active sites in the primary sequence such as the $(Asp-X-Asp)_n-$ sequence (where X is a neutral residue) reported by Weiner & Hood (1975), and suggested by Mann (1988) to be a key component of calcium binding in vivo. Staining with the dye Stains All, specific for calcium binding proteins (Campbell *et al.*, 1983b) may provide a further clue as to the function of these proteins.

The complex, heterogenous mix of macromolecules that comprise the organic matrix can be separated into two main fractions according to their solubility in ethylene diamine tetra acetic acid (EDTA) or weak acid. The insoluble fraction consists of a mix of proteins and carbohydrates (Weiner & Traub, 1980), some of which is relatively hydrophobic and crosslinked, which may account for the insolubility of this fraction (Meenakshi et al., 1971). The soluble fraction is mainly composed of acidic proteins and polysaccharides (Weiner,1979). Weiner (1982) pointed out the numerous technical difficulties involved with these proteins; their tendency to aggregate in solution, anomalous chromatographic behaviour and relative lability under mild conditions. Despite this, Worms & Weiner (1986) resolved the soluble matrix from three classes of molluscs into two quite different categories of proteins by reverse phase hplc according to the methods of Weiner (1982, 1983). The first class of proteins are characteristically rich in aspartic acid and glycine and are associated with small amounts of poly-saccharide. The second class are rich in serine and glutamic acid and are associated with larger amounts of polysaccharide. The acidic nature of the soluble matrix, its intimate association with the mineral phase and its ubiquitous presence in diverse phyla all indicate that it plays an important role in the regulation of crystal growth (Weiner, 1982; Worms & Weiner, 1986). Soluble matrix is thought to possess both stimulatory and inhibitory properties towards crystal growth depending on whether it is bound to a substrate or free in solution (Mann, 1983; Wheeler & Sikes, 1984; Weiner, 1984; 1986). This dual functionality would allow the same class of soluble macromolecules both to initiate and manipulate crystal growth (Addadi & Weiner, 1985).

Alternatively, there may be different classes of soluble matrix macromolecules responsible for this function (Wheeler & Sikes, 1984). When adsorbed to a substrate (e.g., the insoluble matrix) soluble matrix macromolecules can nucleate crystal growth by binding to, and stabilising clusters of ions or nascent crystallites. In contrast, when in solution, they can inhibit growth, perhaps by adsorbing to specific growth sites on the crystal faces and blocking further growth in that direction (Addadi & Weiner, 1985; Wheeler et al., 1987; Berman et al., 1990). Cessation of crystal growth may be total or only from specific faces, in which case the morphology of the growing crystal can be altered (Addadi & Weiner, 1985; Mann, 1983; Wheeler & Sikes, 1984). A system of mineralisation may be envisaged in which the insoluble matrix core serves as a template for shaping the growing crystal, while nucleation is achieved through soluble matrix macromolecules adsorbed onto the insoluble core at specific sites. Crystals may then increase in size laterally and their growth terminated or manipulated when soluble matrix macromolecules in solution adsorb onto the growing faces.

Intracrystalline material may arise through macromolecules, particularly those of the soluble matrix, becoming entombed within the crystal as a result of overgrowth or overlap between adjacent crystals or between cycles of deposition (Towe, 1980; Wheeler & Sikes, 1984). The location of these intracrystalline macromolecules may provide important clues to their precise effects in vivo, e.g. they may affect cleavage along certain planes (Berman et al., 1990), or they may confer strength to the biomineral (Mann, 1993). The mineral phase offers excellent protection to the entombed molecule from various diagenetic processes such as hydrolysis and leaching by percolating ground waters or bacterial contamination. Intracrystalline molecules may still be susceptible to breakdown by endogenously trapped water (Towe, 1980), but analysis of fossil (and extant) degradatory products from within this closed crystal system should allow likely pathways of decomposition to be reconstructed. Drawing comparisons without a knowledge of Recent biomineralisation processes is rather like trying to complete a jigsaw puzzle without having an overall picture to which one can refer. Knowledge of the primary sequence and 3–D conformation of the extant protein will allow one to work back-wards, following likely pathways of diagenesis to arrive at the fossiliferous intracrystalline end-products. An understanding of the biomineralisation processes in extant systems will pro-vide an accurate database, containing information on a reper-toire of extant intracrystalline macromolecules and their various diagenetic products, necessary to compare the molecular breakdown products of fossiliferous intracrystalline systems.

MATERIALS AND METHODS

Extraction of intracrystalline molecules and concentration of extract. Samples of Terebratulina retusa were obtained by dredging in Kerrera Sound off the NW coast of Scotland (Figure 1). The protocol employed here was a modification from Collins et al. (1988) and Cusack et al. (1992). Extensively bored or otherwise damaged shells were discarded. Shell valves were disarticulated and the body tissue removed. Surfaces were thoroughly cleaned of sediment by scrubbing, and encrusting epifauna scraped off. Remaining organic matter, including the periostracum, was removed by incubation in an aqueous solution of sodium hypochlorite (1 % v/v) for 2–3 hours at room temperature. After rinsing in double de-ionised water and airdrying, the shells were crushed to a fine powder in a ceramic mortar and pestle. Intracrystalline material was oxidised by incubation in an aqueous solution of sodium hypochlorite (1 % v/v) for 24 hours at 4°C. The bleach was removed by extensive washing with double de-ionised water followed by centrifugation

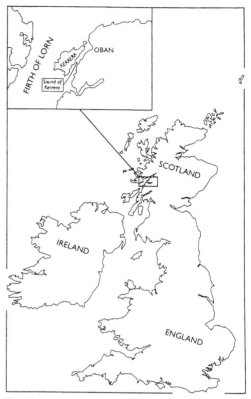

Figure 1. Map of Britain detailing Kerrera Sound, Oban on the west coast of Scotland (inset), the collection site of Terebratulina retusa.

(8g.h.). The intracrystalline molecules were released by demineralising the mineral phase in ethylene diamine tetra–acetic acid (EDTA), 20 %w/v, pH 8, at a ratio of 23 ml/g powder (Table 1). This mixture was incubated with agitation at 4°C for 2–3 days, or until all the inorganic phase had solubilised. After centrifugation (20g.h.) the supernatent was concentrated, washed and the EDTA removed using tangential flow ultrafiltration (Millipore). Virtually all the EDTA is removed by this method (Curry et al., 1991a). The extract was further concentrated and desalted, in three separate batches, using both centriprep^{TM} (Amicon) and ultraspin^{TM} (Lida) filter units either singly or in combination (Table 1). In all cases filter membranes of nominal pore size 3, 5 or 10 kDa were employed.

Sodium dodecyl sulphate polyacrylamide gel electrophoresis (SDS–PAGE). Polyacrylamide minigels (8cm x 7cm x 0.75mm) were prepared according to the method of Schägger & Von Jagow (1987), where tricine replaces glycine as the trailing ion. Aliquots from all three batch preparations were analysed by SDS–PAGE. Samples for electrophoresis were mixed with an equal volume of (2x) sample buffer to contain final concentrations of Tris base (50mM, pH 6.8), SDS (4%w/v), glycerol (12%v/v), mercaptoethanol (2%v/v), Coomassie Brilliant Blue– G (CBB–G), 0.01%w/v, and heated at 100°C for 4 minutes. Protein standards of known molecular weight were run on every gel and the apparent molecular weights of the sample proteins calculated by plotting relative mobility against log molecular weights of the standard proteins. Samples were electrophoresed (Hoefer mighty small^{TM}) at a constant voltage

Table 1. Final sample volumes and concentration factors of individual shell protein extracts from *Terebratulina retusa*.

Weight of shell powder	174 grams		
Volume of EDTA used	4 litres		
Volume after tangential flow ultrafiltration	80 ml		
Batch volume	30 ml		50 ml
Batch number (corresponding gel and blot)	Batch 1 (gels 1, blot A)	Batch 2 (gel 2, blot B)	Batch 3 (blot C)
3 k Da Centriprep concentration	0.5 ml	0.5 ml	-
10 kDa Centriprep concentration	-	-	2 ml
10 kDa Ultraspin concentration	-	30 μl	200 μl
concentration from original volume	8000 fold	1.3×10^8	2×10^7

of 50V for 4 hours with continuous cooling. Proteins were visualised by staining with CBB followed by destaining with an aqueous solution of methanol (50%v/v).

Electroblotting (Western Blotting). A constant voltage of 55V for 30 minutes was applied to transfer the proteins from the polyacrylamide gel onto Problott^{TM} membrane in an aqueous solution of 3–[cyclohexylamino]–1–propane sulfonic acid (CAPS) buffer (10 mM, pH 11) with methanol (10%v/v). After washing the membrane in double de-ionised water and 100% methanol, the immobilised proteins were visualised by staining in CBB (0.1%w/v) for one minute and background staining was reduced with an aqueous solution of methanol (50%v/v).

Stains All. Stains All is a cationic carbocyanine dye which specifically stains calcium-binding proteins blue, and all other proteins red, pink or not at all (Campbell et al., 1983b). Aliquots from Batch 2 were electrophoresed. Before staining, the gel was fixed with an aqueous solution of TCA (12.5%w/v) and ethanol (50%v/v). All traces of SDS were removed by incubating the gel in an aqueous solution of acetic acid (10%v/v) and methanol (30%v/v). After washing with an aqueous solution of isopropanol (25%v/v), any cations were removed by washing with EDTA (0.31%w/v) in water:isopropanol (3:1). The gel was further washed with isopropanol (25&v/v) and then Tris buffer (10mM, pH 8.8) before staining with Stains All solution. Protein bands were visualised after destaining with double de-ionised water.

Silver Stain. Silver staining is 100x more sensitive than CBB staining (Switzer et al., 1979) and is routinely used to visualise lower concentrations of proteins. Aliquots from Batch 1 were electrophoresed as described above. The polyacrylamide gels were rinsed with double de-ionised water before incubating in an aqueous solution of dithiothreitol (5mg/litre). The gels were then stained by incubating in freshly prepared silver nitrate solution (0.2%w/v). After rinsing with double de-ionised water, protein bands were developed using an aqueous solution of sodium carbonate (decahydrate), 8%w/v, and formaldehyde (35%), 0.05%v/v. Staining was stopped with an aqueous solution of acetic acid (1%v/v).

RESULTS

SDS–PAGE and Electroblotting. The soluble intracrystalline protein fraction, was extracted from *Terebratulina retusa*, and resolved by SDS–PAGE and electroblotting (Figures 2-3). As SDS–PAGE only provides an approximation of the true molecular mass, all molecular masses given are average values. Four main protein bands of apparent molecular mass 71kDa, 46kDa, 40kDa and 13kDa are observed (Figure 2). The

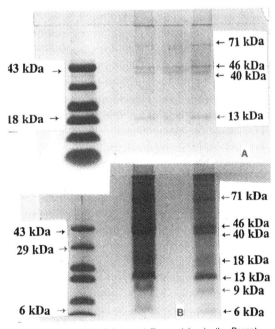

Figure 2. SDS-PAGE of intracrystalline proteins in the Recent brachiopod *T. retusa* from Kerrera Sound, Oban, Scotland. Proteins are visualised with Coomassie Brilliant Blue. (a) Batch 1, gel 1, sample volume per well = 7.5 ml; (b) Batch 2, gel 2, sample volume per well = 10 ml.

153

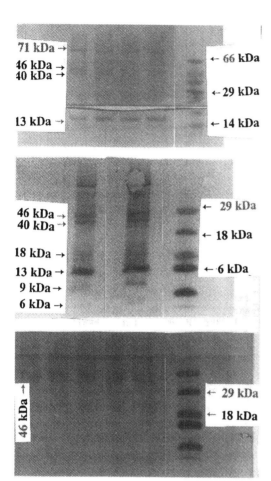

Figure 3. Western Blots of intracrystalline proteins from *Terebratulina retusa*, Kerrera Sound, Oban. Proteins were separated by SDS-PAGE according to Schägger & Von Jagow (1987), and electroblotted; (a) Blot A, Batch 1, sample volume per well = 10 ml; (b) Blot B; Batch 2, sample volume per well = 5 ml; (c) Blot C, Batch 3, sample volume per well = 7.5 ml.

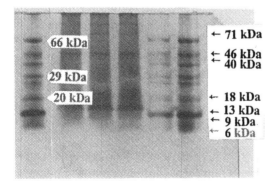

Figure 4. SDS-PAGE of intracrystalline proteins in *Terebratulina retusa*, Kerrera Sound, Oban. Proteins are visualised with silver staining. Sample volume per well = 7.5 ml.

major 30-35kDa protein band identified previously in the intracrystalline extracts of *T. retusa* (Curry et al., 1991a) is observed here as one of a pair of proteins of molecular mass 46 and 40kDa respectively (Figure 2a). It is likely that the 40 kDa band and the previously identified 30–35 kDa band represent the same protein. SDS–PAGE has also revealed previously unidentified protein species in the intracrystalline fraction of *T. retusa* of molecular mass 18, 13, 9 and 6kDa. These lower molecular weight bands are visible in the highly concentrated extract after staining with CBB–G (Figures 2b, 3 b) but are only visible in the less concentrated extract when stained with silver (Figure 4).

Stains All. Preliminary results indicate that the 46, 40 and 13kDa bands, i.e. 3 of the 4 main protein bands stain blue (positive) with Stains All, suggesting that they are acidic in nature and/or possess a calcium-binding function. These results have yet to be verified.

DISCUSSION

EDTA contamination. Weiner (1983, 1985), Worms & Weiner (1986), and Albeck et al. (1993) all voice concern regarding EDTA contamination. This potent metal ion chelator interacts with proteins with the resulting EDTA–protein complexes probably being stabilised by a calcium ion 'bridge' between the anionic EDTA and the soluble matrix molecules (Wheeler et al., 1987). EDTA–calcium–protein complexes have proven notoriously difficult to remove by conventional methods such as dialysis and ultrafiltration (Worms & Weiner, 1986) and may have been responsible for all (or nearly all) of the high affinity calcium-binding observed in whole oyster or clam soluble matrix extract (Wheeler et al., 1987). In addition, EDTA must be removed from the protein solution, as it leads to spurious patterns in polyacrylamide gel electrophoresis and interferes with subsequent amino acid analyses by co–eluting with the amino acids Glu, Ser, Pro and Gly (Walton & Curry, 1994). Tangential flow ultrafiltration (Millipore) efficiently reduces the concentration of such EDTA complexes to less than picomolar (10^{-12}M) levels, (Curry *et al.*, 1991a) while simultaneously concentrating and washing proteins in solution.

Comparisons of Gels and Blots. Batches 1 and 2 (gels 1 and 2) were further concentrated using a 3kDa centriprep[TM] filter unit (Amicon), while Batch 2 (gel 2) was additionally concentrated using a 10kDa ultraspin[TM] filter unit (Lida). In each case four main bands of approximately 71kDa, 46kDa, 40kDa and 13kDa in size are observed (Figure 2). The additional higher and lower molecular weight bands reported in this study will necessitate a revision of previously published work on the intracrystalline proteins from *T. retusa* (Curry et al., 1991a). Gel 2 (Batch 2) contains three additional protein bands of apparent molecular mass 18kDa, 9kDa and 6kDa (Figure 2b). The staining intensity of all the bands in gel 2 is stronger, reflecting the greater concentration of protein in Batch 2. This difference in staining intensities between the two gels is so marked that it is unlikely to be solely due to there being a greater volume (and therefore quantity) of protein extract loaded per well in gel 2 (10ml cf. 7.5ml in gel 1), but is probably a result of the extra concentration step employed in Batch 2. Staining with silver (Figure 4), a more sensitive protein stain, shows that the 9kDa and 6kDa proteins visible in gel 2 (Batch 2) are visible in gel 1(Batch 1). After initial purification using membranes of nominal pore sizes 5, and then 3kDa, the low molecular weight proteins would be retained in both batch 1 and 2. The nominal pore size of filtration membranes is not absolute, however, as passage through the membrane depends on both the molecular size and physical conformation of the protein. Therefore, the extra

154

concentration step in Batch 2, with a 10kDa membrane, must have retained some of these low molecular weight proteins and concentrated them to a level where they could be visualised by staining with CBB–G (Figure 2b).

Batches 1-2 (blots A and B) were further concentrated using a 3 kDa centriprep[TM] filter unit (Amicon), while Batch 2 (blot B) was additionally concentrated using a 10kDa ultraspin[TM] unit (Lida). The four main bands of 71kDa, 46kDa, 40kDa and 13kDa are present in both blots (Figures 3a,b), however blot B contains an increased overall protein concentration with a greater number of strongly stained, distinct bands (cf. gel 2, Figure 2b) despite there being only half the sample loaded (5ml/well cf. 10ml/well in blot A). This effect is most likely due to the extra concentration of the sample used in blot B as described above. The additional protein bands of 18kDa, 9kDa and 6kDa are clearly seen in blot B, but not in blot A, reflecting the increased concentration of the sample in blot B, sufficient to allow visualisation of these low abundance proteins.

Batch 3 (blot C) was further concentrated with a 10kDa centriprep[TM] filter unit (Amicon) followed by microfiltration using a 10kDa ultraspin[TM] filter unit (Lida). It is useful there-fore to compare blots B and C (Figures 3b,c) which differ only in the initial extra concentration step. Individual band strength is stronger overall and the lower molecular weight components are more evident in blot B. One can therefore conclude that in this case further concentration across a 3kDa membrane rather than a 10kDa membrane increases the yield of protein, in particular, of low molecular weight protein species.

Stains All. Provisional results indicate that the 46, 40 and 13kDa bands, i.e. 3 of the 4 main bands stain blue (positive) with Stains All, suggesting, in common with data from hplc profiles (Laing, unpublished), that these proteins are acidic in nature. Additionally, the positive staining suggests that these proteins possess a calcium binding quality which may have implications for their role in biomineralisation, in particular the sequestration of calcium ions. Positive (blue) staining, however, can also be a direct consequence of interactions between dye molecules and any anionic satellite groups, e.g. carbohydrate moieites (or sulphate groups) attached to the protein chain. Glycosylated proteins (glycoproteins) do not readily stain with CBB. However, the bands electrophoresed here are sharp and readily stained and therefore it is likely that these are proteins and not glycoproteins.

In order to eliminate the possibility of interfering reactions, enzymic analysis, specifically to digest any attached anionic groups, must be performed. The occurrence of a positive stain after enzymic treatment will confirm the calcium binding nature of these proteins. There are a number of binding states with different absorption maxima associated with the Stains All dye-protein complex (Bean et al., 1965). These states vary with the nature and conformation of the interacting macromolecule and it is the interaction of the dye with anionic sites in the protein that is responsible for the J (blue) state absorbing maximally between 610 and 650nm (Green et al.,1973; Green & Pastewka, 1974). Various dye-salt aggregates also absorb in this range (Kay et al., 1964), and therefore care must be taken when interpreting results of dye-protein interactions. It is likely that the blue staining observed in this study is due to dye-protein and not dye-salt interactions, as one of the four main bands stained red and not blue. Had the sample been contaminated with salt, e.g. SDS, then all four bands would have stained blue.

Campbell et al. (1983b) found that a number of well known calcium binding proteins form the J (blue) state on binding the dye, leading him to conclude that Stains All is specific for, and therefore can be used to identify, this class of proteins. The positive result in this study demonstrates that three of the four main intracrystalline proteins from *T. retusa* share an affinity with other well known calcium binding proteins. As intra-crystalline proteins are so intimately associated with the mineral phase (being released only upon its dissolution), this result represents another positive step forward in assigning a calcium binding role to these proteins. Campbell et al. (1983b) found that the extent of the blue staining may be proportionate to the number of calcium binding sites in the protein. Calsequestrin, with one binding site/9 amino acids and troponin C, with one binding site/40 amino acids both stain blue but the $(Ca^{2+} + Mg^{2+})$–ATPase from rabbit skeletal muscle with one binding site/500 amino acid residues stains red. Therefore the red staining intracrystalline proteins (71, 18, 9, and 6kDa) in this study may either have very low calcium binding site ratios or no calcium binding function.

A distinction should be made between calcium modulating proteins (in Campbell et al., 1983b), and other calcium binding proteins, e.g. calsequestrin, a calcium storage protein. Most calcium-modulating proteins, e.g. calmodulin, troponin C and members of the S–100 protein family, have EF hands (Kretsinger, 1980). These are recurring helix-loop-helix motifs, approximately 30 amino acids in length, each optimally coordinating one calcium ion. Calcium-modulating proteins normally bind between 1 and 4 calcium ions in order to exert their effects on cellular processes. Sequestering calcium ions through EF hands would seem to be an improbable means of initiating shell deposition via epitaxy or the stereochemical effect. Although the largest of the three positively staining intracrystalline proteins reported here could contain a maximum of 15 EF hands (one EF hand/30 amino acid residues), in practice this is unknown with most known calcium modulating proteins binding 1–4 calcium ions through a corresponding number of functional EF hand sites (Kretsinger, 1980; Heizmann & Hunziker, 1991). Furthermore, as the EF hand motif is 'ancient and designed to provide extremely tight and specific binding of the biologically important cation' (Vyas et al., 1987, p.638), the calcium ion, optimally coordinated by water molecule(s) and residue side chains within the EF hand, is likely to be shielded, both physically and chemically, from incoming anions. The bound calcium ion may therefore be effectively restricted from participating in epitaxial or stereochemical processes.

Other calcium binding proteins such as calsequestrin, bind calcium ions without the facility of EF hands. Calsequestrin is a very acidic protein with aspartate and glutamic acid together constituting approximately 32 mole % (Campbell et al., 1983a). The high number of calcium binding sites (one site per 9 amino acid residues along the chain) do not possess a sufficiently high affinity for calcium to be EF hands (ibid.), and it is likely that the calcium ion binds to specific acidic motifs in the primary sequence.

Binding of the Stains All dye to the protein is not necessarily a function of acidity as proteins eluting at the same acidity as calsequestrin (which stains blue) are found to stain red (Campbell et al., 1983b). This suggests there is a feature, common among calcium sensitive proteins which encourages the formation of the J (blue) state. Stains All binds to both native and denatured proteins (ibid.), suggesting that the association is not due to a conformational motif such as the EF hand, but is through a sequence motif, accessible to the dye in both native and denatured forms of the protein. The

155

conformational motif may be composed wholly or in part by this sequence motif, e.g. the calcium-binding loop in the EF hand but it plays little or no part in the binding of the dye. There may be an anionic amino acid sequence, similar in or common to both calcium-modulating and calcium-binding proteins, that is responsible for the binding of the dye Stains All.

EDTA–free oyster and clam soluble matrix extracts bind very little calcium under ionic compositions similar to the extrapallial fluid of these molluscs (Wheeler et al.,1987). This is curious, as it is generally accepted that for matrix to nucleate growth it must first bind calcium ions, e.g. by epitaxy, in which the lattice dimensions of the interacting surfaces match (Weiner & Hood, 1975), or by the 'Stereochemical Effect' where the matrix proteins interact specifically with a certain face of the developing crystal (Addadi & Weiner, 1985). Wheeler et al. (1987) therefore suggest that the immobilised soluble matrix perhaps initiates crystal growth not by binding calcium ions, but by binding to, and stabilising clusters of ions or nascent crystallites which continually form and dissolve in such metastable physiological solutions. Once precursors are stabilised, crystal growth may proceed.

The SDS–PAGE and Western Blotting results clearly advocate the continued use of both centriprep^{TM} and ultraspin^{TM} filter units with membranes of a nominal pore size of 3 (or 5) kDa at all stages throughout the sample preparation, including tangential flow ultrafiltration (Millipore). Evidently, employing membranes of such a small pore size permits retention of the lower molecular weight components (cf. Curry et al., 1991a in which filtration membranes of 10kDa were used throughout preparation). Improved results are obtained when the filtration steps are carried out in sequence, as this minimises protein loss (especially of the low molecular weight species) and maximises protein concentration. A number of the protein bands observed here may be developmentally expressed. Carioulou et al.(1988) found that the (combined inter- and intracrystalline) protein profile from adult, juvenile and post–larval shells of the mollusc Haliotis rufescens varied, with juvenile shell extracts yielding a pattern intermediate between that of adult and post–larval shells. They point to developmentally regulated differential gene expression in which individual genes of a multi-gene family are expressed at different stages of development, thereby producing a developmentally regulated pattern of protein expression. There was no size or developmental census carried out in this study and therefore it is possible that some of the observed proteins are due to differential gene expression. Further work will be required to determine if there are different protein profiles in T. retusa shells of different ages.

Shell matrix proteins, like acidic calsequestrin, are similar to those found in the intracrystalline proteins from a variety of Recent brachiopod species (Walton et al, 1993; Laing, unpublished). These acidic residues are presumably involved, directly or indirectly, in the binding of calcium ions. In calcium-modulated proteins, e.g. calmodulin, the carboxylate (and hydroxyl) groups of residue side chains are instrumental in coordinating the calcium ion within the EF hand (Kretsinger, 1980). Canine cardiac calsequestrin, with high levels of aspartic acid, binds 300 nm calcium per mg protein (Campbell et al., 1983a) and yet contains no EF hand motif, indicating the binding site is probably a primary sequence or secondary structural motif. In addition to the high levels of acidic residues, uniquely high levels of glycine are reported in various shell matrix proteins (Weiner, 1985; Cariolou, 1988), including brachiopod shell proteins (Walton et al., 1993). Cariolou et al. (1988) reported that the proportion of aspartic acid and glycine residues in the total shell protein fraction of the gastropod Haliotis rufescens increases with increasing age and size of the shell. This clearly suggests that these residues play an important role in the binding of calcium ions during mineral formation.

Weiner & Hood (1975) demonstrated that a putative calcium binding sequence of–(–Asp–X–Asp–)$_n$– can be isolated from the soluble organic matrices of mollusc shells. Preferential cleavage of aspartic acid residues released significant levels of glycine and serine indicating that X may be one or both of these amino acids. Rusenko et al. (1991), in an analysis of the major shell matrix protein of the oyster Crassostrea virginica, found lower levels of glycine and serine released through preferential cleavage of aspartic acid residues. They warned that the above sequence may not exist, or at least may be present at lower levels than those reported by Weiner & Hood (1975). Rusenko et al. (1991) suggested that there are domains of polyaspartic acid and polyphosphoserine in common with the mammalian dentin phosphophoprotein, phosphoryn (Veis et al., 1991). Rusenko et al. (1991) further proposed that the high levels of glycine may represent regions of polyglycine which may offer flexibility to the protein chain. These polyglycine regions may allow isolated acidic residues (or sequences of polyaspartic acid) spread out in the primary sequence, to be brought into an optimal orientation for calcium binding when the protein folds into its 3–D conformation. The calcium binding site might only be functional in the folded protein, the 3–D structure of which may be partially stabilised through electrostatic interactions with calcium ions. Worms & Weiner (1986) determined, by infrared spectroscopy, that aspartic-rich proteins isolated from three classes of molluscs adopt the b–sheet conformation upon binding calcium.

Crystal growth is thought to proceed upon a structural template composed of matrix macromolecules. X–ray diffraction (Weiner & Traub, 1980) of insoluble matrix proteins from three classes of molluscs have shown that some adopt an antiparallel b–sheet conformation not unlike that of silk–fibroin proteins. This resemblance is supported in that levels of Gly, Ala and Ser are similar in both classes of proteins (Keith et al., 1993). Weiner (1984) proposed that the b–sheet could be composed, at least in part, of stretches of the putative calcium binding sequence, –(–Asp–X–Asp–)$_n$–. The principle of epitaxial crystal growth is that the distance between the anionic (carboxylate) groups of the aspartic acid residues in the b–sheet should geometrically match the distance between the anionic (carbonate) groups in the interacting face of the crystal lattice. However, this ionic matching is insufficient to explain why a particular polymorph may be favoured, e.g. aragonite instead of calcite, when the spacing of the calcium ions in the crystal lattice is almost identical in both forms. This specifity may lie in a combination of epitaxy and the stereochemical effect, where the carboxylate groups of the aspartic acid residues in the b–sheet not only match crystal lattice dimensions but also coordinate the calcium ions in the interacting face of the crystal with the correct stereochemistry of a particular polymorph. After nucleation, successive layers of calcium and then carbonate are built up, each retaining the confirmation of the original layer. In this way, aspartic acid rich matrix proteins direct orientated growth of the crystal. Crystal growth may be modulated or terminated at any point by the release, from mantle epithelial cells, of soluble matrix macromolecules. These may bind to specific growth faces inhibiting further growth in that direction or may blanket the entire crystal preventing growth in any direction.

N–terminal sequence analyses of comparable intracrystalline proteins from four extant brachiopod genera (Cusack et al., 1992; Curry et al.,1991b; Laing, unpublished) reveal a high degree of sequence homology between the first twenty amino acids of the sequence. Searching of protein databases did not reveal any homology between these intracrystalline proteins and any other known protein suggesting that these represent an entirely new class of protein. In contrast, there is surprisingly little primary sequence homology among the EF hand proteins. Even in the calcium binding loops, where the overall conformation of the loop is identical in different proteins, the sequence identity is only in the order of 25% (Heizmann & Hunziker, 1991).

Macromolecules associated with mineral deposition are acidic in character with a predominance of Asp, lesser amounts of Glu, and significantly high levels of Gly and Ser (Weiner, 1979). An $-(Asp-X-Asp-)_n-$ sequence, where X= Gly or Ser, is reported in EDTA soluble proteins from a number of different mollusc shells (Weiner & Hood, 1975), and is thought to bind calcium ions directly. This sequence may form part of an antiparallel b-sheet conformation which may nucleate and direct oriented crystal growth through a combination of epitaxy and stereochemistry.

CONCLUSIONS

This study has revealed a number of different proteins from the intracrystalline fraction of *Terebratulina retusa*, some of them previously unreported. It has also shown that some of the major intracrystalline proteins from *T. retusa* share an affinity with known calcium-binding proteins in that they stain positive (blue) with the dye Stains All. This result, along with data from hplc profiles (Laing, unpublished), also indicates their acidic nature. Further characterisation through reverse phase hplc, amino acid analysis and protein sequencing is underway, which will lead to a greater understanding of the role that these special proteins play in biomineralisation.

Acknowledgments
This research is funded by a bursary from the University of Derby, England. I should like to thank D. Walton for his instructive comments and endless patience while reviewing this paper and A. Ansell at the Dunstaffnage Marine Labora-tory, Oban, Scotland, for supplying samples of *Terebratulina retusa*.

REFERENCES

ADDADI, L. & S. WEINER. 1985. Interactions between acidic proteins and crystals: stereochemical requirements in bio-mineralisation. Proceedings National Academy Science USA 82:4110-4114.

ALBECK, S., L. AIZENBERG, L. ADDADI & S. WEINER. 1993. Interactions of various skeletal intracrystalline components with calcite crystals. Journal American Chemical Society 115:11691-11697.

BEAN, R.C., W.C. SHEPHERD, R.E. KAY & E.R. WALWICK. 1965. Spectral changes in a cationic dye due to interactions with macromolecules, III, stoichiometry and mechanism of the complexing reaction. Journal Physical Chemistry 69:4368-4379.

BERMAN, A., L. ADDADI, A. KVICK, L. LEISEROWITZ, M. NELSON & S. WEINER. 1990. Intercalation of sea urchin proteins in calcite: study of a crystalline composite material Science 250:664–667.

___,___& S. WEINER. 1988. Interactions of sea urchin skeleton macromolecules with growing calcite crystals - a study of intracrystalline proteins. Nature 331: 546-548.

___, J. HANSON, L. LEISEROWITZ, T.F. KOETZLE, S. WEINER & L. ADDADI. 1993. Biological control of crystal texture: a widespread strategy for adapting crystal properties to function. Science 259:776-779.

CAMPBELL, K.P., D.H. MACLENNAN & A.O. JORGENSEN 1983b. Staining of the Ca^{2+} – binding proteins, calsequestrin, calmodulin, troponin C, and S–100, with the cationic carbocyanine dye "Stains All", Journal Biological Chemistry, 258:11267–11273.

___,___,___, & M.C. MINTZER. 1983a. Purification and characterisation of calsequestrin from canine cardiac sarcoplasmic reticulum and identification of the 53,000 dalton glycoprotein. ibid. 258:1197–1204.

CARIOLOU, M.A. & D.E. MORSE. 1988. Purification and characterisation of calcium binding conchiolin shell peptides from the mollusc, *Haliotis rufescens*, as a function of development. Journal Comparative Physiology, B 157:717–729.

COHEN, B.L., B. BALFE, M. COHEN & G.B. CURRY. 1991. Molecular evolution and morphological speciation in North Atlantic brachiopods (*Terebratulina* spp.). Canadian Journal Zoology 69: 2903-2911.

___,___,___. 1993. Molecular and morphometric varia-tions in European populations of the articulate brachiopod *Terebratulina retusa*. Marine Biology 115:105-111.

COLLINS, M.J., G.B. CURRY, R. QUINN, G. MUYZER, T. ZOMERDIJK & P. WESTBROEK. Sero-taxonomy of skeletal macromolecules in living terebratulid brachiopods. Historical Biology 1:207-224.

___, G. MUYZER, G.B. CURRY, P. SANDBERG & P. WEST-BROEK. 1991. Macromolecules in brachiopod shells: characterization and diagenesis. Lethaia 24:387-397.

CURRY, G.B., M. CUSACK, D. WALTON, K. ENDO, H. CLEGG, G. ABBOTT & H. ARMSTRONG. 1991a. Biogeochemistry of brachiopod intracrystalline molecules. Philosophical Trans-actions Royal Society London 333B:359–366.

___, K. ENDO, D. WALTON & R. QUINN. 1991b. Intracrystalline molecules from brachiopod shells, p.35–40. In, S. Suga and H. Nakahara (eds.), Mechanisms and phylogeny of mineralisation in biological systems. Springer Verlag, Tokyo.

___,___, H. CLEGG & G. ABBOTT. 1992. An intracrystalline chromoprotein from red brachiopod shells: implications for the process of biomineralisation. Comparative Biochemistry Physiology 102B: 93-95.

ENDO, K. & G.B. CURRY. 1991. Molecular and morphological taxonomy of a Recent brachiopod genus *Terebratulina*, p.101–107. In, D.I. MacKinnon, D.E. Lee & J. D. Campbell (eds.), Brachiopods through time. Balkema, Rotterdam.

GREEN, M.R., J.V. PASTEWKA 1974. Simultaneous differential staining by a cationic carbocyanine dye of nucleic acids, proteins and conjugated proteins, II, Carbo-hydrate and sulfated carbohydrate containing proteins. Journal Histochemistry Cytochemistry 22:774-781.

___,___ & A.C. PEACOCK. 1973. Differential staining of phosphoproteins on polyacrylamide gels with a cationic carbocyanine dye. Analytical Biochemistry 56:43-51.

HEIZMANN, C. W. & W. HUNZIKER.1991. Intracellular calcium binding proteins: more sites than insights. Trends Biochemical Sciences 16:9-103.

HUDSON, J. D. 1967. The elemental composition of the organic fraction, and the water content of some Recent and fossil mollusc shells. Geochimica Cosmochimica Acta 31:2361-2378.

JAMES, M.A., A.D. ANSELL & G.B. CURRY.1991. Functional morphology of the gonads of the articulate brachiopod *Terebratulina retusa*. Marine Biology 111:401-410.

___,___&___. 1991a. Oogenesis in the articulate brachiopod *Terebratulina retusa*. ibid.111:411-423.

___.___&.1991b. Reproductive cycle of the brachiopod *Terebratulina retusa* on the west coast of Scotland. ibid.111:441-451.

JOPE, M. 1979. The protein of brachiopod shell,VI, C-terminal end groups and sodium dodecylsulphate-polyacrylamide gel electrophoresis: molecular constitution and structure of the protein. Comparative Biochemistry Physiology 63B:163-173.

KAY, R.E., E.R. WALWICK & C.K. GIFFORD. 1964. Spectral changes in a cationic dye due to interaction with macromolecules, I, Behavior of dye alone and in solution and the effect of added macromolecules. Journal Physical Chemistry 68:1896-1906.

KEITH, J., S. STOCKWELL, D. BALL, K. REMILLARD, D. KAPLAN, T. THANNHAUSER & R. SHERWOOD. 1993. Comparative analysis of macromolecules in mollusc shells. Comparative Biochemistry Physiology, 105B:487-496.

KRETSINGER, R.H.1980. Structure and evolution of calcium modulated proteins. CRC Critical Reviews Biochemistry ?: 119–174.

LOWENSTAM, H.A. 1981. Minerals formed by organisms. Science 211:1126-1131.

MANN, S. 1983. Mineralisation in biological systems. Structure Bonding 54:125-174.

___1988. Molecular recognition in biomineralisation. Nature 332:119-124.

___1993. Biomineralisation: the hard part of bioinorganic chemistry! Journal Chemical Society, Dalton Transactions 1:1-9.

MEENAKSHI, V.R., P.E. HARE & K.M. WILBUR. 1971. Amino acids of the organic matrix of neogastropod shells. Comparative Biochemistry Physiology 40B:1037-1043.

RUSENKO, K.W., J.E. DONACHY & A.P. WHEELER. 1991. Purification and characterisation of a shell matrix phosphoprotein from the American Oyster, p.107-124. In, C.S. Sikes & A.P. Wheeler (eds.). Surface reactive peptides and polymers: discovery and commercialization. American Chemical Society, Washington.

SCHÄGGER, H. & G. VON JAGOW. 1987. Tricine-sodium dodecyl sulfate-polyacrylamide gel electrophoresis for the separation of proteins in the range from 1 to 100 kDa. ? 166:368-379.

SWITZER, R.C., C.R. MERRIL & S. SHIFRIN. 1979. A highly sensitive silver stain for detecting proteins and peptides in polyacrylamide gels. Analytical Biochemistry 98:231-237.

TOWE, K.M. 1980. Preserved organic ultrastructure: an unreliable indicator for Paleozoic amino acid biogeochemistry, p. 65–74. In, P.E. Hare, T.C. Hoering & K. J. King (eds.). Biogeochemistry of amino acids. John Wiley, New York.

VEIS, A., B. SABSAY & C.B. WU. 1991. Phosphoproteins as mediators of biomineralisation, p.1-12. In, C.S. Sikes & A.P. Wheeler (eds.), Surface reactive peptides and polymers: discovery and commercialization. American Chemical Society, Washington.

VYAS, N.K., N.V. MEENAKSHI & F.A. QUIOCHO. 1987. A novel calcium binding site in the galactose-binding protein of bacterial transport and chemotaxis. Nature 327:635-638.

WALTON, D & G.B. CURRY. 1994. Extraction, analysis and interpretation of intracrystalline amino acids from fossils. Lethaia 27:179 -184.

___, M. CUSACK & G. B. CURRY. 1993. Implications of the amino acid composition of Recent New Zealand brachiopods. Palaeontology 36:883-896.

WEINER, S. 1979. Aspartic acid rich proteins: major components of the soluble organic matrix of mollusc shells. Calcified Tissue International 29:163-167.

___1982. Separation of acidic proteins from mineralized tissues by reversed-phase high-performance liquid chromatography ? 245: 148 -154.

___1983. Mollusc shell formation: isolation of two organic matrix proteins associated with calcite deposition in the bivalve Mytilus californianus. Biochemistry 22:4139-4145.

___1984. Organization of organic matrix components in mineralized tissues. American Zoology 24:945-951.

___1985. Organic matrix like macromolecules associated with the mineral phase of sea urchin skeletal plates and teeth. Journal Experimental Zoology 234:7-15.

___1986. Organisation of extracellularly mineralised tissues: a comparative study of biological crystal growth. CRC Critical Reviews Biochemistry 20:365-408.

___& L. HOOD. 1975. Soluble protein of the organic matrix of mollusc shells: a potential template for shell formation Science 190: 987-989.

___ & W. TRAUB. 1980. X-ray diffraction study of the insoluble organic matrix of mollusc shells. FEBS Letters, Elsevier 111:311-316.

WHEELER, A.P., K.W. RUSENKO, J.W. GEORGE & C.S. SIKES. 1987. Evaluation of calcium binding by molluscan shell organic matrix and its relevance to biomineralisation, Comparative Biochemistry Physiology 87B:953-960.

WORMS, D. & S. WEINER. 1986. Mollusc shell organic matrix: Fourier transform infra-red study of the acidic macromolecule. Journal Experimental Zoology 237:11-20.

THE PROBLEM OF COMPARABILITY IN TAXONOMIC RANKS FOR BRACHIOPOD SYSTEMATICS

STANISLAV S. LAZAREV

Paleontological Institute, Russian Academy of Sciences, Moscow 117647, Russia

ABSTRACT- A philosophical approach is taken to the problem of rank in systematic studies of brachiopods. Criteria are suggested for separating different levels of taxonomic hierarchies. A critique of existing taxonomic approaches, including cladistics, is outlined. The concepts of 'adjacent' and 'iterative parallelism' are elaborated.

CRITERIA FOR TAXONOMY

In assigning a taxonomic rank, brachiopodologists, as well as specialists on other groups, usually rely on their intuition and common sense, which, unfortunately, varies from person to person. In this respect, the cladistic approach to the assessment of taxonomic rank is preferable to traditional systematics. In my opinion, it is the cladistic approach to the relative assessment of a taxonomic rank that deserves special attention and development. I understand the difficulty in finding an overall mutual understanding of this problem. Probably, this obstacle has prevented putting the problem on the agenda in the preparation of the Treatise. The problem has two aspects: (1), to find common basic criteria (not characters!) for the taxa of different ranks; (2), to create, on the basis of common criteria, a structure within brachiopod systematics that would make taxa of corresponding rank in different orders as comparable to each other as possible. Only the first aspect is discussed below.

It is obvious that criteria for taxonomic units at different hierarchical levels will be different. One may discriminate between groups of criteria that determine different levels in a hierarchy. But before doing this, it is useful to separate species level distinctions from those of higher taxa. Species, unlike higher taxa, are based mainly on the pattern of distribution of characters not on the characters proper (Smirnov, 1938). Thus, here I shall discuss further only the hierarchy of supraspecific higher taxa.

All supraspecific taxa are based on characters of a different nature. In other words, the problem of hierarchy may be formulated as the problem of arrangement of taxa according to their taxonomic significance. Supraspecific taxa reflect differences in the scale of the evolutionary process, so the analysis of evolutionary processes may provide grounds for taxonomic rank. A very general statement about the taxonomic significance of characters may be expressed as follows: 'taxonomic weight of characters is determined by its evolutionary role, namely the degree of propagation and phylogenetic stability'. Consequently, composition or revision of a system of classifying any group (the stage of 'synthesis') should be preceeded by an analytical ('meronomic' *sic*) stage of research into the sequence of appearing characters (Lazarev, 1993). In other words, taxonomic composition (separation of the taxa of different ranks) should be preceeded by phylogenetic reconstruction (dendrograms). Such an arrangement of systematic work is typical of cladists, but unfortunately it is not obligatory in traditional systematics studies. If one accepts the peculiarity of cladistic methodology, it does not necessarily mean the acceptance of the methodology of reconstructing a phylogeny as a whole. Since the probability of appearance of homoplasies in evolution, as paleontological data show, is substantially more than 0.5, the cladistic method of phylogenetic reconstruction, based on the principle of parsimony, seems to be inacceptable. The results obtained by these methods are less reliable by far than those received when paleontological data are taken into consideration.

The use of data on the sequence of appearance of apomorphies as the basis for arrangement (attributing ranks to) of features is most important. If purely typological methods are used in phylogenetic reconstructions, one may expect a totally different picture of the sequence of apomorphies. Consequently, we can juxtapose the level of superior taxa of brachiopods (classes and higher), the history of which extends far back into the Precambrian. A level of superior taxa should be reconstructed exclusively by typological methods, and lower taxa (genus to order) should be based on phylogenetic reconstruction, which, in turn, may be based on paleontological data (Lazarev, 1993). To reconstruct superior level taxa, the deductive (typological) approach should be used almost exclusively, based on taxonomic criteria such as architectonics, a 'Bauplan', or embyological data (strongly distorted after more than 500 Ma of brachiopod history).

The basic criterion for determining taxonomic weight of a character, for taxa ranging from genus to order, is the evolutionary history of the character, i.e. the earlier it appears, the longer its uninterrupted history, the more elementary its taxa (species, genera) embraced, the greater the taxonomic signiificance of the character. In this connection one may reconsider the method of 'equal level' used by neontologists, and some paleontologists. This method has restricted significance for assessment of ranks of groups with an already known phylogeny. The phylogenetic basis for this method is essentially provided by the principle of the 'main link' outlined by Ruzhentsev (1960). More precisely, the phylogenetic aspect of the 'main link' principle, states that close (adjacent) phylogenetic branches are characterised by similar evolutionary trends. These are the subfamilies which constitute families, whereas genera appear to be taxa reflecting stages of development in a trend. Unfortunately, studies of brachiopods show that longterm (parallel) trends embracing all genera and families can be identified only occasionally. Usually such trends are based on two manifestations of a character only, and a researcher seldom employs the principle of the main link' in this connection. The method of 'equal level' and its phylogenetic analogue, the principle of the 'main link' is seldom used in paleontological work for two perceptions. The first perception, less important, is the appearance of groups based on occasional or unique features. The second perception, more important, is the abundance of parallelism manifested at different phylogenetic stages in any diversified group. I am sure that the refinement of the systematics of productids, as for any other group of brachiopods, will essentially be determined by the discovery of more and more parallelisms.

It is useful to distinguish two kind of parallelism according to the order of appearance (Lazarev, 1990). First is 'adjacent parallelism' one which appears in the same succession in parallel lineages. This kind of parallelism corresponds to the principle of the 'main link'. It promotes construction of good (clear) systematics, where the place of any taxon is determined easily by a dichotomous key. The second is the kind of parallelism which has beenn named as iterative ('unnexted', *sic*) parallelism. This kind of parallelism appears by iteration and breaks the common succession of appearance of characters. Any point of divergence corresponding to 'adjacent parallelism' is the 'sliding down' along a dendrogram never to cross the analogic point of divergence. On the contrary, the upper point of divergence of the appearance of 'iterative parallelism' is the 'sliding down' along a dendrogram to intersect the analogic point of a more early appearing character at least once. In practice, we always have a mixture of adjacent and iterative parallelism. The more common the iterative parallelisms in evolution, the more difficult it is to construct good (clear) systematics with good diagnoses of taxa. Experience in the study of productid brachiopods exposes characters which appeared repeatedly and independently in large groups (superfamilies), i.e. evolutionary destiny was very different in close lineages. One good example is the character of dorsal spines. In the new systematics for the suborder Productidina, this character has different taxonomic rank from superfamily to species. It is clear that the history of dorsal spinosity was connected with iterative parallelism.

For the present, only genera, providing the first level of species integration, may claim comparability among different orders (provided we have more or less the same degree of knowledge about the genera). Higher taxa, among different orders of brachiopods, are constructed like the Tower of Babel. This can be explained by the lack of a solid methodological approach towards the assessment of taxonomic rank. Much depends on prejudice among brachiopod workers, i.e. that internal characters are more important than external, or that this character in the group (order) studied is considered to be a familial feature. As an editor of the Paleontologicheskii Zhurnal, I can only comment on these statements in the following manner: (1) higher taxa of brachiopods (classes, orders) are easily recognised exactly on the basis of external characters, that is different external characters (as well as internal characters) have different taxonomic significance, and, (2), persistence of taxonomic rank of the character across one order phylogenetically means that the character is conditioned by manifestation of adjacent parallelisms. This is probable within one family, much less probable between two families, and highly improbable between all families of one order.

Functional significance of a morphological feature also seems to be inapplicable towards assessment of taxonomic rank of a given character. That is why arguments about relative importance of a structure for mode of life, and, consequently, for the assessment of taxonomic rank, e.g. the structure of dental plates or cardinal process, are absolutely useless. Assessments of taxonomic ranks based on functional statements give rice to skepticism in relation to objectivity of taxa higher than species and, particularly, taxa higher than the genus level. Of course, this does not mean that all work dedicated to the construction of systematics in different orders, with more or less comparable taxa, should be recognised as useless.

It is very difficult, if not impossible, to equalise taxa at the rank of genus, family, and order across groups with such a different evolutionary history, as, let us say, brachiopods and vertebrates, or brachiopods and insects. Cladists usually

ignore this problem. To find taxonomic matches (conformities) in groups with a similar evolutionary destiny seems to be a more realistic objective. I refer here to groups in whose evolution, as in the case of brachiopods, no substantial apomorphic changes have occurred. Bryozoans, corals, and bivalves are also among such groups. When these groups are compared, analogies are striking. For example, almost all orders of brachiopods, bryozoans and bivalves appeared in the Early Palaeozoic. Specific groups, such as rudists among the Bivalvia, and oldhaminides among the Brachiopoda, originated much later. The rank of order, attached to the bivalves and brachiopods today, is more doubtful. If evolutionary criteria are to be applied, that is when we consider the early partition of the largest branches within classes and subclasses as orders (or ?supraorders), we should re-establish the chonetids and productids within one order (or ?superorder), and the taxonomic rank of oldhaminids should not be higher than suborder, or even superfamily. The problem of comparability of higher taxa of brachiopods (as well as other organisms) in relation to refining systematics deserves further consideration for these taxa are used to solve problems of stratigraphy and biogeography, and are used in the analyis of biodiversity dynamics, ecological crises, etc.

REFERENCES

LAZAREV, S.S. 1990. Taxonomicheskie criterii ot otrjada do roda na primere brachiopod, p. 99-103. In, Sistematika i filogenia bespozvonochnych. Nauka, Moskva.
___1993. On the methods of phylogenetics. Paleontologicheskii Zhurnal 27(2): 5-19.
RUZHENTSEV, V.E. 1960. Printsipy sistematiki, sistema i filogenia paleozoyakikh ammonoidei. Trudy Paleontologicheskii Institut 83:1-331.
SMIRNOV, E.S. 1938. Konstruktsiya vida s taksonomicheskoy tochki zrenia. Zoologicheskii Zhurnal 17(3):387-418.

* Footnote (eds.): We presume that the 'term meronomic' refers to merosystematic, or fractional classification, i.e. as the opposite to 'holosystematic'. We include this submitted article on the philosophy of systematics basically as it stands so that the reader can provide their own judgment. We are not familiar with the method of 'equal level', nor with the principle of the 'main link'. The term 'succession' as used in the manuscript may presumably be visualised either as 'stratigraphic succession', or a 'succession of taxa'.

FUNCTIONAL AND TAPHONOMIC IMPLICATIONS OF ORDOVICIAN STROPHOMENID BRACHIOPOD VALVE MORPHOLOGY

LINDSEY R. LEIGHTON & MICHAEL SAVARESE

Department of Geological Sciences, Indiana University, Bloomington, 47405, Indiana, USA

ABSTRACT- Functional hypotheses for strophomenid shape were evaluated, inclusive of potential taphonomic bias. Results from experimental biomechanical experiments (e.g. flume measurement of flow-induced forces; sediment scour around shells on mobile, sandy substrates; stability of muds with varying pore-water content; wind-tunnel pressure mapping over shells) were used to predict paleoenvironmental occurrence, and then corroborated in the field. *Rafinesquina alternata*, a ubiquitous Upper Ordovician strophomenid from the Illinois Basin, was used as a standard for similarly shaped brachiopods. Three presumed functional hypotheses were tested: (1) strophomenids lived semi-infaunally with the convex valve down; (2) geniculation placed the commissure above the sediment-water interface to prevent fouling; and (3) flow through the valves for suspension feeding was not lateral-medially separated. Geniculated brachiopods with convex-down orientations generated greater drag, induced greater scour, and experienced larger pressure differentials than arcuate or convex-up specimens, supporting the claim that convex-up orientations should predominate in shell beds. A convex-up life position appears inconsistent with results from muddy substrate experiments; the commissure quickly sank below the sediment-water interface, ostensibly rendering the lophophore ineffective. Geniculate convex-down specimens probably maintained their entire commissure above the substrate. These patterns were tested on substrates with 25% and 50% water contents. On sandy substrates, arcuate, convex-down brachiopods induced little scour and were less likely to become buried than geniculate forms. The paleoenvironmental distribution of *R. alternata* in Cincinnatian (Ashgill) strata from the Illinois Basin appears predictable, given these interpretations for life position and shell shape. Wind tunnel pressure measurements along the commissure of convex-down *Rafinesquina* were inconsistent with a laterally incurrent, medially excurrent flow, suggesting that *Rafinesquina* may not have had a spirolophe.

INTRODUCTION

Functional significance of strophomenid shape (geniculate commissure, concavoconvex valves, broad planar area) is highly speculative. Taphonomic overprinting has complicated functional interpretation, with life position inferred from epibiont distribution and shell orientation, when both may be due to post-mortem influences. Many hypotheses concerning the paleobiology and taphonomy of benthic fossil organisms are founded upon life orientations that are assumed to be correct, but are often not adequately tested. Functional morphological interpretations may be incorrect if life orientation is misinterpreted. Moreover, once a life orientation is proposed, whenever the fossil is found in a different position, it is commonly explained as a consequence of taphonomic processes.

The concavoconvex geometry common to most strophomenid brachiopods is a classic illustration of the problems associated with inferring life orientation. In 1934, Lamont suggested that strophomenids lived with their convex valves down, and that this orientation enabled the organism to live on soft, muddy substrates. Subsequently, Rudwick (1970) and Richards (1972) also supported this interpretation. The latter used taphonomic evidence, and epibiont distribution patterns to support the convex-down model. Functional studies of geniculation, i.e. abrupt change in growth and profile curvature, have generally assumed that the convex-down paradigm is correct. Alexander (1975), utilizing biomechanics and paleoenvironmental distribution, concluded that geniculation in the Upper Ordovician strophomenid *Rafinesquina alternata* was probably ecophenotypic, following the Lamont paradigm. Fossil strophomenids are observed in many orientations, including convex-up, convex-down, and, rarely, even upright. Savarese (1994) observed that the convex-up orientation appeared more stable hydrodynamically, but this was not necessarily used to imply life orientation. Lescinsky (1993) suggested that the traditional interpretation has been incorrect, and that concavoconvex brachiopods lived with the convex valve up. Richards (1972) and Lescinsky (1993) reached different conclusions for orientation using epibiont distribution.

This study used an alternative approach to interpreting life orientation and functional morphology of strophomenids. We examined the validity of the convex-up and convex-down life orientations and the function of geniculation in four sets of

Figure 1. Six *Rafinesquina alternata* morphotypesused in soft substrate experiments. 1, prolate, strongly geniculate; 2, prolate, moderately geniculate; 3, prolate and arcuate; 4, oblate, moderately geniculate; 5, oblate, alate, moderately geniculate; 6, oblate, arcuate.

biomechanics experiments on strophomenid models and real fossils. Two experiments were designed to determine the relationship between passive strophomenid morphology and disturbance due to moving water. The third examined strophomenid stability on soft substrates, and the last experiment analysed the effect of orientation on feeding efficiency. All experiments used the Late Ordovician brachiopod *Rafinesquina alternata*, chosen as representative of strophomenids because of its abundance, wide geographic and environmental range, and morphological variability (Figure 1). In addition, three other Ordovician strophomenid species were used in the soft substrate experiment, in order to broaden the morphological range under examination: *Rafinesquina nasuta*, similar in shape to *R. alternata*, but with a medial projection of the geniculate, anterior margin; *Leptaena richmondensis*, which is small, oblate, always geniculate, with pronounced, rugose growth lines; and *Strophomena planumbona*, which is small, oblate, and never geniculate, but varies in valve convexity.

As well as examining possible life orientations, all four experiments also compared data from both geniculate and arcuate morphotypes of *Rafinesquina alternata*. If these morphotypes are ecophenotypic, as suggested by Alexander (1975), then these should be functional. Therefore, bio-mechanical experiments may demonstrate the respective function of each morphotype, also providing corroborative evidence for life orientation. Results from biomechanics experiments were compared to paleoenvironmental distribution of *Rafinesquina alternata* from SE Indiana (Illinois Basin), to determine if functional implications of experiments would be consistent with the environment in which the brachiopod lived. Specimens used were qualitatively described as either geniculate or arcuate. More than 650 specimens were collected from nine localities exposing limestones and mudstones of the Arnheim, Waynesville, and Liberty Members of the Dillsboro and Whitewater formations (Upper Ordovician, Ashgill, type Richmondian: Figure 2). Environmental facies were defined by a weighted moving average of the relative abundance of limestone to mudstone (explained in Leighton, 1995). The White-water Fm. is interpreted as the shallowest paleoenvironment, based on sedimentary analyses and its stratigraphic position below the Saluda Dolomite, inferred to be a shallow, saline lagoon (Hay, 1981). Although the interpretation of Dillsboro Fm.

paleodepth is still controversial, our interpretation of the Whitewater as the shallowest paleoenvironment is consistent with previous work (Martin, 1977; Oldroyd, 1978; Hay, 1981; Holland, 1993).

On biomechanical principles, we hypothesize the following for *Rafinesquina*, and similarly shaped strophomenids: (1) Although specimens oriented convex-up would be apparently more hydrodynamically stable, this orientation may not have been employed by living strophomenids, because these could not maintain their commissure (and lophophore) clear of sediment. We furthermore predict that specimens oriented convex-down, on both sandy and muddy substrates, would be less prone to burial of the commissure than specimens oriented convex-up; (2) An arcuate morphotype would experience less drag than a geniculate morphotype, and arcuate specimens would be less susceptible to disturbance by sediment scour. Consequently, arcuate morphotype would be expected to prevail in shallow, higher energy environments, such as interpreted for the Whitewater Fm.; (3) Specimens with a high planar surface area to volume (herein SA/V) ratio would be less prone to sink in soft substrates because body mass of the organism is more evenly distributed (i.e., 'snowshoe' strategy, Thayer, 1975); (4) Given the function of geniculation to be elevation of the commissure above the sediment-water interface, so as reduce risk of lophophore fouling (Alexander, 1975), then, when oriented convex-down on soft substrates, the greater body volume of a geniculate brachiopod would sink below the interface, but the commissure would be kept well above the sediment (i.e., 'iceberg' strategy, Thayer, 1975). As a corollary, the geniculate morphotype would be more abundant in muddier environments, such as determined for the Dillsboro Fm.; (5) Flow through the mantle cavity should be consistent with environmentally induced pressure gradients. The pressure gradient should be distri-buted in concert, with low pressures at the excurrent region, and high at the incurrent region. Because a lateral-medial flow through the mantle cavity appears to be impractical for a semi-infaunal organism, we predict that environmentally induced pressure gradients along the surface of the specimens would be inconsistent with a lateral-medial flow, i.e. if the brachiopod were utilizing a lateral-medial flow, the organism would be pumping against pressure gradients.

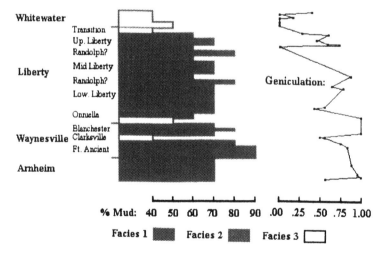

Figure 2. Geniculation decrease of *Rafinesquina* through the Richmondian (Ashgill: Late Ordovician), with stratigraphy, lithofacies, and bed-by-bed percentage of geniculate *Rafinesquina alternata* (out of total *R. alternata* population). Lithofacies defined by relative abundance of limestone to mudstone. The Whitewater Fm. is inferred to represent shallowest, highest energy.

EXPERIMENTAL METHODS

Measurement of flow-induced drag during flume experiments.
Details concerning model making techniques and flume
experimentation are described in Savarese (1994). A general
overview of the methods is presented below.   Data are
presented for four models of *Rafinesquina alternata*, two
arcuate and two geniculate, one of each postured convex up
and convex down, and in three orientations: hingeline
upstream, hingeline parallel to flow direction, and hingeline
downstream. Results for other concavoconvex brachiopod
species (*Leptaena richmondensis*, *Strophodonta demissa*, and
Tropidoleptus carinatus, a Devonian orthid, but similarly
shaped) and for other orientations are published in Savarese
(1994).

A high speed recirculatory flume (following design of Vogel &
LaBarbera, 1978) at Hopkins Marine Station (Stanford
University Marine Laboratory, Pacific Grove, California) was
used for testing. The flume has a square cross-section with
10cm long sides and can generate speeds ranging from 0.4-
4m/s. Identical experiments were conducted in a low speed
recirculatory flume housed at Indiana University (capable of
flows between 2-50cm/s). Those data are not here repeated,
but are compared to the high speed results [see 'Discussion'].
Force due to drag was measured with a beam-bending force
transducer following the design of Denny (1982, 1989), and
data were collected by a chart recorder (Savarese, 1994).

Pressure drag is a force experienced by an object in a moving
fluid that is directed in a downstream direction, and is a
function of the pressure differential that exists fore and aft.
Drag can have two ill-effects for a living benthic organism. If
the force were great enough, the organism may become
entrained. More commonly, however, turbulence created in the
wake of the organism may cause sediment scour and
eventually bury the organism [provided the organism has no
capacity to clean itself]. Often sediment is left downstream and
adjacent to the object, causing a depositional 'shadow' in its
wake. The pressure drag generated by a brachiopod is a
function of flow characteristics (fluid density, viscosity,
velocity), and brachiopod size and shape (for a review of flow-
induced forces and their controls see Denny, 1988,1993;
Vogel, 1994). Consequently, a brachiopod's shape in profile
(i.e., its cross-sectional shape in a plane perpendicular to the

flow direction), and its orientation, significantly affect entrainment and sediment scour. Comparing force due to drag and the
drag coefficient (Denny, 1989; Savarese, 1994), for different
brachiopod shapes and orientations within flows of varying
velocities, permits a qualitative assessment of a brachiopod's
relative susceptibility to disturbance.

Mobile substrate experiments. The effects of shape and
orientation on drag were investigated directly by flume testing
brachiopods on mobile, sandy substrates in a low speed,
recirculatory flume. Presumably *Rafinesquina* morphologies
and orientations generating greatest drag in experiments
described above should be the ones generating greatest
sediment scour and burial during life. Two specimens of *Rafinesquina alternata*, one arcuate and one highly geniculate,
were placed on fine-grained (50 mesh size), quartz sand in a
low speed, recirculating flume (dimensions: channel length
2.5m; width 30cm; depth 40cm). Experiments were conducted
at free stream speeds of 20 and 30cm/s. For each experimental trial a single specimen was positioned either convex-up
or convex-down, and in one of two orientations, hingeline
upstream or hingeline downstream. Scour and burial patterns
were observed after 5 minutes.

Soft substrate experiments. Two soft sedimentary substrates
were constructed, one with 50% water and 50% carbonate
mud, and another with 25% water and 75% mud (by volume).
Fixed volumes of lime aggregate mud and sea water were mixed
in a blender to create desired porosities. The mixtures were
decanted in 1000ml beakers, and enough sea water was added
above the substrate to create a water column of 200ml depth.
Twelve experiments were conducted concurrently in 6 beakers
for each of the two sediment types.

Six specimens of *Rafinesquina alternata* with morphologies
varying in arcuate vs geniculate shape, in prolation, and in
alation, were chosen for study (Figure 1:1-4; Table 1). In addition, two specimens of *Strophomena planumbona*, and one
specimen each of *Rafinesquina nasuta*, and *Leptaena
richmondensis*, were also used in this experiment (Table 1).
Although fossil shell material probably would be denser than
that of a living organism, soft tissue volume of concavoconvex
strophomenids was probably small enough that the difference
in density would be minimal. Moreover, the experimental design
is assumed to be conservative, i.e., if a fossil specimen does

Table 1. Physical description, measurements, and settling behavior of strophomenid specimens on muddy substrates. Described behavior is
for specimens oriented convex down on mud that is 50% water by volume. All convex-down specimens experienced minimal settling
on firmer mud (25% water by volume). All convex-up specimens sank sufficiently into sediment to bury commissure, regardless of substrate
consistency. `Snowshoe' indicates that brachiopod did not sink significantly after placement; `Iceberg' indicates that much of the body volume
sank below the sediment-water interface, but commissure remained elevated above interface. `Sank' indicates commissure buried by sediment
due to settling.

SPECIMEN	MORPHOLOGY	MASS (g)	VOLUME (cc)	DENSITY (g/cc)	S. AREA (cm²)	SA / V (cm⁻¹)	BEHAVIOR
R. alternata # 1	Prolate, Strongly geniculate	17.76	6.83	2.60	11.20	1.64	Iceberg
R. alternata #2	Prolate, Geniculate	8.34	3.11	2.68	9.11	2.93	Sank
R. alternata #3	Prolate, Arcuate	7.74	2.91	2.66	7.92	2.72	Snowshoe
R. alternata #4	Oblate, Geniculate	9.16	3.39	2.70	14.21	4.19	Snowshoe
R. alternata #5	Alate, Geniculate	11.01	4.13	2.67	11.03	2.67	Sank
R. alternata #6	Oblate, Arcuate	3.94	1.47	2.68	6.35	4.32	Sank
R. alternata #7	Prolate, Arcuate	1.65	0.62	2.66	3.33	5.37	Snowshoe
R. alternata #8	Prolate, Geniculate	14.54	5.38	2.70	16.45	3.06	Iceberg
R. nasuta	Prolate, Strongly geniculate	15.24	5.85	2.61	9.04	1.55	Iceberg
L. richmondensis	Oblate, Strongly geniculate	2.36	0.88	2.68	3.20	3.64	Snowshoe
S. planumbona #1	Oblate, Weakly convex	1.88	0.70	2.69	3.11	4.44	Snowshoe
S. planumbona #2	Oblate, Strongly convex	2.30	0.86	2.67	3.00	3.49	Snowshoe

not sink in mud, then the live organism would be even less likely to do so. Planar area of each brachiopod (area in contact with the sediment) was measured using computer image analysis. Brachiopod mass was measured on a pan balance, and volume was obtained by measuring the mass of water displaced when the brachiopod was suspended in water (Table 1).

Brachiopods were gently placed on the muds in either a convex-down or convex-up orientation. Nail polish was used to define a plane through the brachiopod parallel to the substrate at time of initial placement. Experiments were monitored over a 48 hour period. Three characteristics were noted: degree of sinking, degree of rotation (rotation of the horizontal plane defined by the nail polish), and degree of and extent of sediment fouling of the commissural edge and hingeline. Orientation of each brachiopod was photographed at the end of the 48 hour interval.

Pressure mapping during wind tunnel experiments. A wind tunnel (Purdue University Hydromechanics Laboratory) was used to conduct pressure mapping experiments. The tunnel has a 2 x 2m cross section and is capable of generating wind velocities up to 10m/s. Experiments were conducted at four wind speeds equivalent to water flows of 0.25, 0.32, 0.53, and 0.63m/s (data reported herein for only the 0.25m/s flow). Because seawater and air have different densities and dynamic viscosities, air speeds ranging from 4 to 12m/s were used to maintain dynamic similarity with (see Vogel, 1994). Unlike water, air density and viscosity are affected significantly by temperature. Consequently, air speed during testing was adjusted regularly as temperature fluctuated.

Due to time limitations and difficulties in model construction, only four shell models of *Rafinesquina alternata* were tested: convex-down arcuate, convex-up arcuate, convex-down

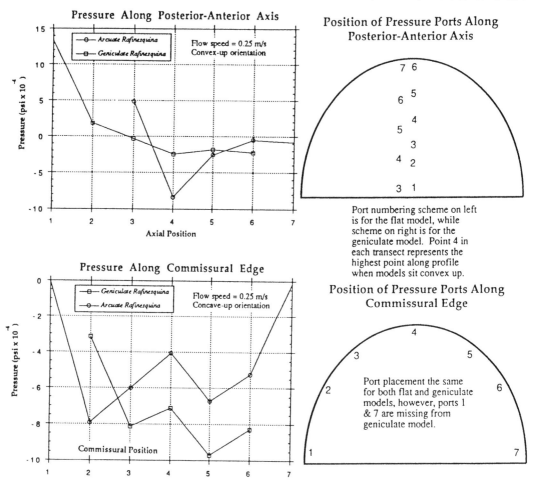

Figure 3. Results from wind tunnel, pressure mapping experiments with free stream flow speed equivalent to a water current of 0.25 m/s; relative patterns identical for experiments at 0.32, 0.53, and 0.63 m/s (not shown), though pressure magnitudes progressively more extreme (i.e., larger positive and negative pressures); pressure reported in 10^{-4} pounds per square inch, and relative to tunnel ambient pressure; positive values higher than ambient, and negative values lower than ambient; position 4 = highest point on each profile. Diagrams at right show relative port positions for both arcuate (flat) and geniculate models. 1. Pressure measured along posterior-anterior axis of both arcuate and geniculate, convex-up *Rafinesquina alternata*.; models oriented with hingeline upstream; arcuate shell behaves as streamlined airfoil, whereas geniculation disrupts streamlining and moves point of flow separation in upstream direction, causing greater drag and turbulence in wake. 2. Pressure measured along commissure; pressure maxima consistently occurring at left and right lateral margins and at anterior margin, though pressures all negative relative to ambient; minima at medial-lateral points.

geniculate, and convex-up geniculate (the same four shells used in the force transducer experiments). We assumed that the patterns generated for these models are representative of other arcuate and geniculate taxa, and plan to conduct further testing on other species. Each of the four models were tested in two orientations: hingeline oriented upstream; and hingeline oriented downstream [data herein are reported only for the hingeline upstream orientation]. Plaster replicas were constructed at 2x original size using an iterative mold and cast process. Latex molds were enlarged by soaking them in kerosene, producing up to a 50% increase in size. The enlarged mold was cast and the procedure was repeated until the desired size was obtained (see Wilson, 1989 for methodologic details). Larger-than-life models were necessary to permit the drilling of numerous pressure ports across the shell surface.

Pressure ports were drilled in 20 or 21 places in the geniculate and arcuate models respectively (Figure 3). A 1/8" [c.3mm] diameter catheter tube was threaded through each port, glued in place, and cut flush with the model's exterior surface. Each model was individually attached to a steel plate suspended 20cm above the tunnel floor. The catheter tubes passed through the model and plate base and exited the tunnel through a pipe perforating the floor. Tubes were attached to a multi-channeled scanivalve and then to a pressure transducer with a 5 volt resolution. Analog data were then delivered via an analog to digital converter to an IBM-compatible personal computer. A program written in QuickBasic calculated the desired wind speed to maintain dynamic similarity, switched through each of the scanivalve channels in sequence, and recorded pressure measurements at each of the ports. Ten consecutive pressure measurements were made at each port at a frequency of 0.1Htz; 3 rounds of ten measurements were made for each port, and the 30 measurements per port were averaged. Pressure at each port was compared to the tunnel's ambient pressure, measured with a pitot tube positioned in the free

stream flow. A value less than zero at a port indicates a negative pressure relative to tunnel pressure (model experiences a force pulling it off the substrate at this position); a value greater than zero indicates a relative positive pressure (force pushing the model into the substrate at this point: see Olivera ,1995). Though pressure was measured at all these points, data are reported herein for 2 transects, a posterior-anterior transect (from umbo to anterior commissure) running along the axis of bilateral symmetry, and a transect following the commissural margin running from the left lateral margin, through the anterior margin, and ending at the right lateral margin (Figure 3).

RESULTS

Flow-induced drag. Patterns in flow-induced drag of the *Rafinesquina alternata* models are intuitively predictable. Drag increased for all models and orientations as free stream velocity in the flume increased (Figure 4). Models and orientations that projected larger cross-sectional areas to the oncoming flow experienced greater drag. For example, geniculate models generated greater drag than similarly oriented arcuate models, and drag for models oriented with the hingeline parallel to the current was less than that for the same model oriented with either the hingeline upstream or downstream (with the hingeline parallel to the current, a given model exposes a smaller profile area). Convex-up oriented models experienced significantly lower drag than the same morphology oriented convex-down. *Rafinesquina* oriented convex-down, especially the geniculate model, behaved like 'bluff bodies', while convex-up *Rafinesquina*, in particular the arcuate model, functioned like stream-lined objects. Although the data are not reported herein, relative patterns seen during high speed flume testing were identical to those of lower speed flume experiments (speeds <0.5m/s).

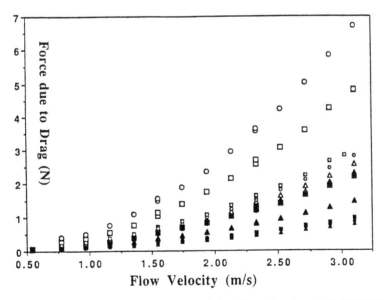

Figure 4. Force due to drag, in Newtons, as function of flow velocity, in m/s, for both arcuate and geniculate *Rafinesquina alternata* models measured during high flow regime flume experiments. Results from low flow regime flume experiments, at flow speeds <0.5 m/s (not shown), exhibit similar relative patterns; ordering in drag for various shapes and orientations remaining identical; plotted values = mean from 15 seconds of continuous measurements; large symbols = geniculate models; small symbols = arcuate models. Open circle = convex-down, hingeline upstream; open triangle = convex-down, hingeline parallel to flow; open square = convex-down, hingeline downstream. Closed circle = convex-up, hingeline upstream; closed triangle = convex-up, hingeline parallel to flow; closed square = convex-up, hingeline downstream.

Figure 5. Geniculate specimen of *Rafinesquina alternata* on fine sandy substrate in low speed, recirculatory flum; upstream to lef; specimen oriented convex (ventral) valve down, hingeline upstream; left arrow pointing to scoured depression at upstream margin; middle arrow pointing to sedimentary deposition against downstream margin; right arrow pointing to scoured depression a short distance downstream, marking edge of vortex in current.

Models and orientations with higher magnitude drags are also those which generate greater turbulence in the model's wake and would consequently induce greater sediment scour. This supposition is supported by results from the mobile substrate experiments (Figure 5). Similarly, greater drag implies greater ease of entrainment of the brachiopod, assuming all models had equal immersed weights and experienced comparable forces due to friction with the substrate. Force due to lift was not measured (see Savarese, 1994); lift could act in concert with or oppose drag and would therefore contribute to the brachiopod's entrainment potential.

Drag coefficients, which are obtained empirically from drag measurement, permit a normalized comparison of drag effects irrespective of brachiopod size (Denny, 1989; Savarese, 1994). Therefore, they better reflect the relative ease of *Rafinesquina* entrainment with different morphologies and orientations (Savarese, 1994, fig.5). The convex-down geniculate *Rafinesquina*, oriented with its hingeline upstream, exhibited the highest drag coefficient at a given flow velocity (ranging from 1.1-1.4), and the convex-up arcuate model in all three orientations had the lowest (< 0.4). Drag coefficients do not change drastically when models are tested at flow velocities under 0.5m/s in the low speed flume (Savarese, 1994). For example, coefficients for geniculate, convex-down *Rafinesquina* oriented hingeline upstream range between 1.0 and 1.4 for flow speeds between 5 - 30cm/s.

Mobile substrate experiments. Flume experiments with a mobile, sandy substrate demonstrated, as predicted by force measurement, that a passive geniculate shell would generate greater sediment transport and scour than an arcuate specimen. Regardless of orientation relative to current or to substrate, the passive geniculate shell generated scour on its upstream margin, and sediment was transported downstream along the flanks. Two counter-rotating vortices formed immediately downstream of the geniculate shell; sediment captured in these vortices formed paired small sand piles that migrated medially and upstream, eventually burying the downstream margin of the specimen (Figure 5). These results were observed at both free-stream velocities of 20 and 30cm/s. Sediment scour and burial occurred more rapidly at 30cm/s, but, regardless of free-stream velocity, final scour and burial pattern were the same for any given specimen and orientation. Sediment scour and burial occurred particularly rapidly if the geniculate

specimen were oriented convex-up with the hingeline upstream. In this position, the downstream margin (commissure) was completely buried in less than two minutes (flow=30cm/s), compared to four to five minutes for other orientations.

When oriented convex-down with the hingeline downstream, the geniculate specimen was sometimes observed to topple into the cavity formed by upstream scour. If the convex-down geniculate specimen were oriented with the hingeline upstream, sediment accumulated against the geniculate, downstream margin, forming a bulwark that stabilized the specimen (Figure 5). However, any sediment on the surface of the dorsal, concave valve was eventually stripped away by the current, exposing the specimen, even if the specimen was originally placed in a semi-infaunal position, with only the commissure above the substrate. In contrast, the arcuate specimen produced neither significant upstream scour, nor downstream burial, regardless of orientation. Scour might occur around the arcuate specimen at greater flow velocities, but presumably for any given current velocity, the rate of sediment scour and burial around an arcuate specimen would be less than that for a geniculate specimen. Clearly, the onset of scour around arcuate specimens occurs at a higher free-stream velocity than that required to generate scour around a geniculate specimen.

Soft substrate experiments. Results of the soft substrate experiments showed several apparent trends (Table 1): (1) All shells oriented convex-down rested on the firmer substrate (25% water by volume) with minimal sinking; (2) When oriented convex-down, shells with small volumes (<1cm3) did not sink in the more fluid-rich mud (50% water by volume). These specimens had surface area to volume (SA/V) ratios greater than 3.45 cm/-1. With one exception (*R. alternata* specimen #6), larger (volume >1cm3) specimens with SA/V ratios greater than 3.45 cm/-1 also maintained a position such that their commissures were clear of the substrate; (3) When oriented convex-down, geniculate specimens 'floated' in the more fluid substrate. Although much of the peripheral margin sank below the sediment-water interface, the commissure was always elevated well above the substrate (Figure 6). The two most strongly geniculate specimens had SA/V ratios less than 2cm/-1, implying that the organism maintained position in the substrate in spite of its SA/V ratio, rather than because of it. Arcuate specimens with low SA/V ratios also 'floated' in the sediment, but usually sank to a depth sufficient for mud to bury the commissure and hingeline; (4) When specimens were oriented convex-up, the anterior margin (commissure) sank below the sediment-water interface within a few seconds after placement, even on the firmer mud. This effect was less

Figure 6. Geniculate specimen of *Rafinesquina alternata* `floating' convex valve down in mud; mud 50% water by volume. This is an example of `ceberg' strategy (Thayer, 1975); much of body volume of shell below the sediment-water interface, but geniculate profile maintaining commissure well above sediment (shell = specimen 1 in Figure 1).

pronounced for smaller (volume <1cm3) specimens, but in all cases, more than half of the anterior margin was buried. The anterior margin of larger specimens immediately sank to depths greater than 1cm. Some specimens also displayed a tendency to rotate laterally.

Paleoenvironmental distribution. The geniculate morphotype of Richmondian *Rafinesquina alternata* is least abundant in Lithofacies 3 (Mud < 50%), based on t-test, one-way analysis of variance (ANOVA), Tukey's pairwise comparison, and the Chi-squared test of homogeneity (all at 95% level of confidence: Figure 2). This lithofacies is inferred to be the shallowest, highest energy facies in the study. Although the geniculation percentage experiences minor fluctuations through time, overall the geniculation percentage decreases upsection with the lowest percentages in the Whitewater Fm., the coarsest-grained unit in the section. The geniculation percentage for the entire Whitewater Fm. is only 16% compared to 70%, 66%, and 51% respectively for the Arnheim, Waynesville, and Liberty Members of the Dillsboro Fm. The percentage of geniculate *R. alternata* from the Whitewater is significantly different from that of all other units (t-test, ANOVA, Tukey's, Chi-square, all at 95% level of confidence).

Pressure mapping. The observations made concerning the shape effects on drag are confirmed by wind tunnel pressure measurements made along the models' posterior-anterior axes (Figure 3). The arcuate convex-up *Rafinesquina* exhibited a posterior-anterior pressure distribution characteristic of streamlined bodies. Pressure dropped as flow passed over the leading edge of the shell, reached a minimum at the shell's highest point (port 4), and then increased until the point of flow separation occurs in the aft region (between ports 5 and 6). The pressure in the model's wake never regained the maximal level seen at the leading edge. The geniculate, convex-up *Rafinesquina* behaved like a 'bluff body'. Pressure dropped more gradually as the high point was reached, but flow separation occurred immediately beyond this point, and the pressure quickly leveled off. These patterns were identical at the higher free stream velocities (0.32, 0.53, 0.63m/s; not recorded herein), however, the absolute value of the pressure magnitudes increased. For example, the pressure minimum measured at port 4 for the arcuate convex-up *Rafinesquina* during the 0.63m/s experiment was -100 x 10^{-4}psi, an order of magnitude smaller than that measured at 0.25 m/s.

Pressure measurements made along the commissure (i.e., where the two valves would gape to permit the passage of water in and out of the shell) indicate where brachiopod-generated currents for feeding, respiration, and waste removal would be operating relative to environmentally induced pressure gradients. For example, points of excurrent flow, if located at pressure maxima, would require that the brachiopod work against the environmental pressure gradient. Conversely, if the excurrent region were located at a pressure minimum, the brachiopod would be able to conserve metabolic energy while maintaining its internal flow. Pressure maxima occurred at the anterior margin and at the left- and rightmost margins of the shell for *Rafinesquina* oriented concave-up (Figure 3:1-2). Minima were located at left-medial and right-medial positions. As with the posterior-anterior transects, the pattern was identical regardless of the free stream flow speed; the pattern was maintained during the 0.32, 0.53, and 0.63 m/s experiments, but the pressure magnitudes were augmented.

DISCUSSION

Biomechanical experiments uphold the hypothesis that concavo-convex strophomenid brachiopods lived in the convex-down position. Moreover, the convex-up hypothesis is not supported. These results hold for both sandy and muddy substrates.

Strophomenid brachiopods that lived on firm substrates such as sand would presumably have been subjected to relatively high current velocities, yet even at free stream current velocities as low as 20cm/s, a geniculate specimen experiences rapid burial of the downstream margin. For an organism oriented convex-up, this is a serious problem, as the commissure will be buried at least partly, unless the anterior commissure is always oriented upstream. If the brachiopod is oriented convex-down, the commissure is elevated above the sediment by geniculation, but the brachiopod may still experience difficulties due to sediment scour. Scouring may undermine the specimen, or remove sediment covering much of the brachiopod, consequently increasing the risk of entrainment. As demonstrated by the flow-induced drag experiments, an exposed, geniculate strophomenid positioned convex-down experiences greater drag than other morphologies or orientations, thereby increasing the likelihood of entrainment and scour.

An arcuate specimen presents a lower, more streamlined profile to currents than does a geniculate specimen. Consequently, arcuate morphology experiences less drag, less turbulence, and less sedimentary scour and is less likely to be disturbed in a high energy environment. The paleoenvironmental data also are consistent with the hypothesis that arcuate specimens are more suited to relatively higher energy regimes. The percentage of geniculate *Rafinesquina alternata* was lowest in Lithofacies 3, the least muddy of the facies. The Whitewater Fm., inferred to have been deposited in shallow water, dominated by Lithofacies 3, has a significantly lower geniculation percentage than any other lithostratigraphic unit in the study.

Differences in biomechanical effects between convex-up and convex-down orientations are more apparent on softer substrates. Even on the firmer mud (25% water by volume), no specimen maintained its commissure above the sediment-water interface when the specimen was positioned convex-up. In contrast, when oriented convex-down on the firmer mud, all specimens had commissures free of sediment. This is strong evidence supporting the convex-down paradigm in strophomenids, and suggests that the function of their concavo-convex geometry is to support the organism on soft substrates. Concavoconvex brachiopods apparently adopted one of two strategies (originally suggested by Thayer, 1975) for maintaining stability on highly liquid muds. Geniculate specimens functioned as 'icebergs'; most of the body volume sank below the sediment-water interface, but the commissure remained elevated above the sediment. The results support the hypothesis that the function of geniculation is to keep the commissure above the substrate and clear of sediment. The alternative strategy is to distribute the body mass over a large surface area. With the single exception mentioned previously, specimens with a high surface area to volume ratio experienced minimal sinking within the highly fluid mud (50% water by volume), demonstrating that the 'snowshoe' strategy is an adaptation to living on soft substrates.

Environmental distribution appears consistent with biomechanical experiments. Overall, the Arnheim, Waynes-ville, and Liberty members of the Dillsboro Fm. are muddier than the Whitewater Fm. (Figure 2), and strophomenid geniculation percentage for each of the three members is significantly greater than that for the Whitewater. Geniculate specimens do not completely dominate the muddier units, perhaps because the arcuate individuals present in muddier intervals adopted the snowshoe strategy. These results suggest that morphotypes of *Rafinesquina alternata* had functional significance. Although arcuate individuals with high surface area to volume ratios may have survived on soft substrates, arcuate geometry appears most successful on harder substrates and in a higher energy regime. The geniculate morphotype functions best as an 'iceberg', floating in highly fluid muds. If the organism is oriented convex-up, the geniculate profile is ineffective, regardless of substrate consistency. This corroborates the hypothesis that strophomenids lived convex-down. Given the above evidence, strophomenids observed in the convex-up position probably experienced post-mortem reorientation, and convex-up orientation may be a taphonomic indicator of disturbance. However, transport may have been minimal after flipping, because a convex-up individual would experience greater hydrodynamic stability.

Pressure mapping experiments demonstrated that pressure gradients along margins of *Rafinesquina alternata* are not consistent with lateral-medial flow. This is compatible with the inferred life orientation of strophomenids, because a lateral-medial flow through the mantle cavity would be detrimental to a semi-infaunal brachiopod, as the incurrent regions of the commissure would be below the substrate. Although external morphology of *R. alternata* varies for different environments, presumably the lophophore and incurrent-excurrent directions remained constant for the species. If strophomenids did not have lateral-medial flow, they possibly did not have a spirolophe. *Rafinesquina alternata* has simple brachiophores that do not indicate the type of lophophore present, but Grant (1972), Rudwick (1970), and Brunton (1972) suggested that some strophomenids may have had a 'ptycholophe', based on an interpretation of internal structures. Racheboeuf & Copper (1990) proposed that Ordovician chonetids, a group derived from, and morphologically similar to strophomenids, possessed a 'mesolophe', with water flow derived posteriorly. Possession of the two-dimensional ptycholophe implies a ventral-dorsal flow through the mantle cavity. The observed pressure gradients for *R. alternata* do not conclusively support this interpretation, but the inferred semi-infaunal life mode of some strophomenids would not interfere with a ventral-dorsal flow. No data is avalaible for posterior-anterior flow in passive shells.

Acknowledgments
We thank R. Alexander and S. Carlson for their insightful comments. We also thank W. Wood, A. Olivera, and T. Cooper at Purdue's Hydromechanics Laboratory for use of the wind tunnel, and M. Denny at Hopkins Marine Station (Stanford University) for use of the high speed recirculatory flume. T. Bonafair and B. Zeigler provided assistance in the collection of data. Research was supported in part by funding from the American Chemical Society's Petroleum Research Fund and the Galloway-Perry Fund from Indiana University's Department of Geological Sciences. The senior author also wishes to acknowledge financial support provided by the Lerner-Gray Fund for Marine Research, American Museum of Natural History. We also wish to thank our wives, Leslie Rapp and Mona Wright, for their love and support.

REFERENCES

ALEXANDER, R.R. 1975. Phenotypic lability of the brachiopod *Rafinesquina alternata* (Ordovician) and its correlation with the sedimentologic regime. Journal Paleontology 49:607-618.

BRUNTON, C.H.C. 1972. The shell structure of chonetacean brachiopods and their ancestors. Bulletin British Museum Natural History 21:1-26.

DENNY, M.W. 1982. Forces on intertidal organisms due to breaking ocean waves: design and application of a telemetry system. Limnology Oceanography 27:178-183.

___1988. Biology and the mechanics of the wave-swept environment. Princeton University Press, Princeton, 329p.

___1989. A limpet shell shape that reduces drag: laboratory demonstration of a hydrodynamic mechanism and an explanation of its effectiveness in nature. Canadian Journal Zoology 67:2098-2106.

------1993. Air and water: the biology and physics of life's media. Princeton University Press, Princeton, 341p.

GRANT, R.E. 1972. The lophophore and feeding mechanism of the Productidina, Brachiopoda. Journal Paleontology 46:213-248.

HAY, H.B. 1981. Lithofacies and formations of a carbonate-clastic ramp: the Cincinnatian Series (Upper Ordovician), southeastern Indiana and southwestern Ohio. Ph.D. thesis, Miami University, Miami 236p.

HOLLAND, S.M. 1993. Sequence stratigraphy of a carbonate-clastic ramp: the Cincinnatian Series (Upper Ordovician) in its type area. Geological Society America Bulletin 105:306-322.

LAMONT, A. 1934. Lower Palaeozoic brachiopods of the Girvan District, with suggestions on morphology in relation to environment. Annals Magazine Natural History, London 10 (14):161-184.

LEIGHTON, L.R. 1995. Factors influencing morphological variation in the articulate brachiopod *Rafinesquina alternata*, Upper Ordovician (Richmondian), southeastern Indiana. M.Sc. thesis, University of Indiana, Bloomington, 96p.

LESCINSKY, H.L. 1993. Life orientation of strophomenid brachiopods: overturning the concavo-convex paradigm. Geological Society America Abstracts Programs 25(6):A51.

MARTIN, W.D. 1977. Petrology of the Cincinnatian series limestones (upper Ordovician) of Indiana and Ohio, p.II:I-26. In, J.K. Pope and W.D. Martin (eds.), Biostratigraphy and paleoenvironments of the Cincinnatian Series, southeastern Indiana. Seventh Annual Field Conference Guidebook, SEPM, Great Lakes Section.

OLDROYD, J.D. 1978. Paleocommunities and environments of the Dillsboro and Saluda formations (Upper Ordovician), Madison, Indiana. Ph.D. thesis, University of Michigan, Ann Arbor, 329p.

OLIVERA, A. 1995. Hydrodynamics of bivalve shell entrainment and transport, and their applications to paleoenvironmental reconstruction. Unpublished Ph.D. thesis, Purdue University, West Lafayette, Indiana, 202p.

RACHEBOEUF, P.R. & P. COPPER. 1990. The mesolophe, a new lophophore type for chonetacean brachiopods. Lethaia 23: 341-346.

RICHARDS, R.P. 1972. Autecology of Richmondian brachiopods (Late Ordovician of Indiana and Ohio). Journal Paleontology 46:386-405.

RUDWICK, M.J.S. 1970. Living and fossil brachiopods. Hutchinson, London, 199p.

SAVARESE, M. 1994. Taphonomic and paleoecologic implications of flow-induced forces on concavo-convex articulate brachiopods: an experimental approach. Lethaia 27:301-312.

THAYER, C.W.1975. Morphological adaptations of benthic invertebrates to soft substrata. Journal Marine Research 33:177-189.

VOGEL, S. 1994. Life in moving fluids. Princeton University Press, Princeton, NJ, 467p.

------ & M. LABARBERA. 1978. Simple flow tanks for research and teaching. Bioscience 28:638-643.

WILSON, M. A. 1989. Enlarging latex molds and casts, p.282-283. In, R.M. Feldmann, R.E. Chapman, and J.T.Hannibal (eds.), Paleotechniques. Paleontological Society Special Publications 4.

EARLY BRACHIOPOD ASSOCIATES: EPIBIONTS ON MIDDLE ORDOVICIAN BRACHIOPODS

HALARD L. LESCINSKY

Department of Geology, University of California, Davis 95616-8605

ABSTRACT- The origination of Paleozoic epibiont communities is temporally constrained through an investigation of encrusters and borers on brachiopods (and some trilobites) older than the Late Ordovician. Well preserved brachiopods from the Arenigian (Kanosh Shale, Utah) and Llandeilan (Bromide Formation, Oklahoma; Benbolt Formation, Virginia) were examined for epibionts. Kanosh brachiopods contained no epibionts; Bromide and Benbolt specimens had numerous encrusters but few borings. For each encrusted host, the percent cover of various epibionts was computed and ecological interactions (i.e., overgrowth relationships) were noted. Bromide and Benbolt epibiont communities were closely similar to each other and to previously described Late Ordovician communities in relative abundance of taxa and ecological structure. They were spatially dominated by several species of mat trepostome bryozoans and the cyclostome *Corynotrypa*. Other encrusters include craniids, cornulitids, crinoids and edrioasteroids. Mat trepostomes were most common on convex surfaces of host *Strophomena* and *Corynotrypa* was most common on concave surfaces and trilobite exuviae. Epibiont growth patterns suggest these assemblages encrusted the upper and lower surfaces, respectively, of live strophomenids. Similarity between Llandeilan and later Ordovician brachiopod epibionts, and the paucity of epibionts on Arenigian specimens suggest that typical epibiont faunas, like many other Paleozoic faunas, arose through rapid diversification of various taxonomic groups during the Arenigian-Llanvirnian.

INTRODUCTION

Communities of encrusting and boring organisms have inhabited brachiopods and other shells throughout most of the Phanerozoic, and biological associations between host brachiopods and epibionts are an important aspect of brachiopod ecology. For example, brachiopod hosts can either benefit from (e.g., camouflage), or be harmed (e.g., parasitism), by encrusters and borers. This paper focuses on the epibiont communities in an effort to understand the origination and early history of brachiopod associates.

The earliest descriptions of communities of epibionts are from the Caradocian-Ashgillian (Upper Ordovician) where epibionts were diverse and widespread (e.g., Flower, 1961; Havlicek, 1972; Richards, 1972; McNamara, 1978; Alexander & Scharpf, 1990; Morris & Felton, 1993). Older epibionts are known primarily through individual taxa; little is known of the ecological structure of older encrusting assemblages. Waddington (1980) described epibiontic edrioasteroids, and Mayoral et al. (1994) described epibiontic ctenostome bryozoan borings on Middle Ordovician brachiopods; both noted, but did not discuss the presence of other encrusters. Early Ordovician and Cambrian epibiont records are limited primarily to cryptic reef species (e.g., Kobluk & James, 1979; Riding and Zhuravlev, 1995), and a few other examples (e.g., Sprinkle, 1973; Taylor, 1984; Conway Morris, 1985; Bengtson & Urbanek, 1986; Hesselbo, 1989). In general, encrusters on Cambrian shells were apparently extremely rare or absent.

Epibionts occur on most available host skeletons, but brachiopod shells are ideal substrates for epibiont studies and they are the primary hosts used in this investigation. Brachiopods are usually the largest and most numerous hosts available in Paleozoic environments, and uniform host shell shape and size presents a standard unit of measurement. For example, epibiont assemblages on similar host brachiopods can be compared between areas and across geological time to examine patterns of community change (e.g. Lescinsky, in press). In this study, to increase sample size, large and robust trilobite exuviae from the Benbolt Formation were also examined.

METHODS

Three collections of Llandeilan brachiopods were investigated for encrusting and boring epibionts: the Mountain Lake Member, Bromide Formation, Oklahoma; Poolville Member, Bromide Formation; and the Benbolt Formation, Virginia (Table 1). A large collection of Arenigian brachiopods from the Kanosh Shale, Utah was also examined, but contained no epibionts.

The majority of the Bromide specimens belong to the genus *Strophomena*, collected from the Lower Echinoderm Zone of the Mountain Lake Member at Cornell Ranch (section 3, Sprinkle, 1982: Oklahoma Geol. Surv. Colln.). Other Bromide *Strophomena* are from the Poolville Member at the Criner Hills (Sprinkle, 1982, Univ. Kansas Colln.). Specimens are articulated and unabraded. They were collected from greenish-gray to light brown, thinly bedded shales, interbedded with limestones that were deposited during intervals of high sealevel in a shallow continental sea (Sprinkle, 1982).

Benbolt samples were collected from thinly bedded, argillaceous skeletal packstones, interbedded with calcareous shales 300m NW of Rye Cove School, Rye Cove Virginia (Steinhauff and Walker 1995). Specimens are disarticulated and fragmented. Epizoans were examined on both internal and external brachiopod shell surfaces and on trilobite exuviae. Brachiopods from the Kanosh Shale, Millard County Utah, were collected from calcareous shale beds, interbedded with locally developed fossiliferous hardgrounds (Fossil Mountain locality, Wilson et al., 1992).

Each host brachiopod or trilobite was examined under a microscope at 10X magnification and epibionts were identified. Mat-forming bryozoans were usually poorly preserved and immature, and could not be accurately distinguished. Percent cover of each epibiont was calculated by capturing a video image of each host shell surface, and by outlining and digitizing the area of each encruster on the surface using Image 1.5 software from the National Institute of Health. Overgrowth relationships between adjacent epibionts, and epibiontic growth patterns providing clues to host vitality, were also noted.

RESULTS

Epibionts did not occur on Arenigian brachiopods from the Kanosh Shale, but were common on Middle Ordovician shells

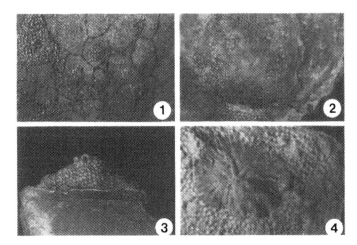

Figure 1. Middle Ordovician epibionts. 1. *Corynotrypa delicatula* on trilobite fragment, Benbolt Fm; 2. Epibiontic inarticulate brachiopod (*Petrocrania*?) entirely encrusted by bryozoan, except across commissure. Inarticulate is overgrowing another bryozoan (bottom center), Bromide Fm.; 3. Bryozoan growing on convex valve of *Strophomena*, does not cross the commissure indicating that brachiopod was probably alive, Bromide Fm.; 4. Crinoid holdfast with numerous outer plates missing, attributed to *Apadasmocrinus* by Lewis (1982). Width of field of view for each photograph is 1.5 cm.

from the Bromide and Benbolt Formations (Figure 1). Total epibiont cover ranged from eight to 44 percent of the surface area, and taxon richness (number of distinguishable morphospecies) was between four and eight (Table 1). Most areal cover was attributed to either mat bryozoans (each formation was dominated by a single (different) morphospecies) and the chain cyclostome *Corynotrypa*. *Corynotrypa delicatula* was more common in the Bromide while *C. inflata* was more common in the Benbolt (Taylor & Wilson 1994). Other encrusters included *Cornulites* (up to 10.0 mm in length on some Benbolt fossils), a craniid inarticulate brachiopod and several echinoderms. Because samples are relatively small, rare taxa will be under-represented in richness counts, but would little change patterns of relative cover. For example, Lewis (1982) describes eight types of crinoid holdfasts from the Bromide Formation and only three of these occurred on shells in this study. Similarly, Bell (1982) described six edrioasteroid species from the Bromide, but only the robust *Cyathocystis americanus* was encountered. Larger sample sizes would probably provide additional epibiontic echinoderm taxa.

Epibiont cover was not random among shell surfaces. On *Strophomena*, convex valves had greater encrustation in 31 of 35 instances where comparisons could be made ($p<.001$, binomial test). The relative abundance of some epibionts also varied significantly with surface type. Mat trepostome bryozoans were the dominant encruster on convex valves of Bromide *Strophomena* and on shell exteriors from the Benbolt (Figure 2). On concave surfaces of *Strophomena* and interior brachiopod surfaces and trilobite exuviae, the chain bryozoan *Corynotrypa* was the spatial dominant. Overgrowth relationships between the bryozoan *Corynotrypa* and mat bryozoans suggest that mat bryozoans were the competitive dominants (Wilson, 1985). In 14 of 18 observed encounters mat morphologies overgrew *Corynotrypa* ($p < .05$). A craniid inarticulate brachiopod overgrew a mat trepostome in the only observed encounter (Figure 1:4).

Table 1. Study collections. N = number of encrusted fossils, or, for the Kanosh Fm., the number inspected for epibionts. External shell surfaces (Bromide Fm.), or internal and external surfaces (Benbolt Fm.) were examined. % cover is total cover of all epibionts with respect to total available shell area. Richness refers to number of observed morphospecies.

Ordovi-cian	Formation	Host	N	Valve	% cover	Rich-ness
Early	Kanosh	Orthids	150	Both	0	0
Middle	Bromide (Mountain Lake Mbr)	*Stropho-mena*	54	Convex Concave	44 25	8 5
Middle	Bromide (Poolville Mbr)	*Stropho-mena*	11	Both	8	7
Middle	Benbolt	Brachio-pods Trilobites	21 20	Exterior Interior All-surfaces	31 23 41	6 4 7

DISCUSSION

Life Orientation of Host. Epibiontic organisms, if they encrusted live hosts, can provide information on host orientation. For example, much debate has centered on the life-orientation of concavo-convex brachiopods such as *Strophomena* (e.g., Alexander, 1975; Alexander & Scharpf, 1990; Bordeaux & Brett, 1990; Lescinsky, 1995). Recently, I presented epibiontic evidence suggesting that concavo-convex brachiopods lived convex valve up (Lescinsky, 1995). Data from Bromide *Strophomena* further support a convex-up orientation. Greater cover on convex than concave valves, and the differentiation of epibionts into assemblages dominated by mat bryozoans and those dominated by early successional, cryptic *Corynotrypa* (Wilson, 1985) suggest that convex valves of

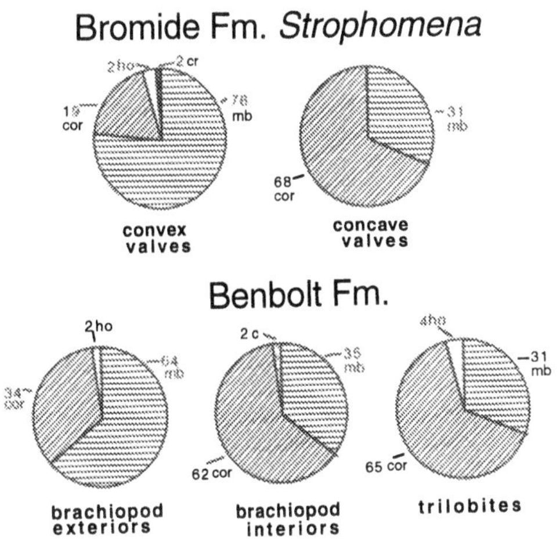

Bromide Fm. *Strophomena*

convex valves

concave valves

Benbolt Fm.

brachiopod exteriors   brachiopod interiors   trilobites

Figure 2. Epibiont composition on host surfaces as % cover by epibionts. Two epibionts formed most epibiont cover, (1) *Corynotrypa*, (2) mat bryozoans. Mat bryozoans dominated on convex valves of *Strophomena* and on exterior shell surfaces; *Corynotrypa* dominated on concave (lower) surfaces of *Strophomena*, and surfaces encrusted post-mortally, e.g. shell interiors, trilobite fragments. cor = *Corynotrypa*, cr = *Petrocrania*, c = *Cornulites*, ho = crinoid holdfasts, mb = mat bryozoans.

these *Strophomena* were uppermost during epibiont encrustation. The early successional, cryptic role of *Corynotrypa* also explains encrustation on Benbolt specimens where internal surfaces of brachiopod shells and all surfaces of trilobite exuviae have high *Corynotrypa* cover, whereas external brachiopod surfaces, which would have been exposed to epibiont settlement throughout the host's life are dominated by mat bryozoans (Figure 2). Careful observation of epibiont growth patterns also suggest that the *Strophomena* were encrusted while alive (Table 2; Lescinsky, 1995). Epibiont colonies that arched over but did not cross the commissure probably indicate a functioning commissure and epibiont colonies that ceased growth linearly along the host hinge probably indicate the flexing hinge of a live host. For articulated *Strophomena*, I recorded 16 growth relationships suggesting live hosts, 23 suggesting probable live hosts and only 5 suggesting postmortem encrustation.

Similarity With Late Ordovician Communities. The Llandeilan epibiont communities examined here are very similar to better known communities from later in the Ordovician (e.g., Richards, 1972; Alexander & Scharpf, 1990; Lescinsky, 1995). For example, Caradocian *Strophomena* epibiont communities (Lescinsky, 1995) are spatially dominated by similar taxa with similar ecological patterns. Convex surfaces of *Strophomena* in both cases are spatially dominated by mat bryozoans (76% and 55% of cover for Llandeilan and Caradocian respectively; Figure 2; Lescinsky, 1995, fig. 2). On concave surfaces chain bryozoans predominate (68& and 60% cover, respectively) and mat bryozoans are less common (31 and 9%, respec-tively). Less abundant epibionts common to both assemblages include *Cornulites*, *Petrocrania*, and different crinoid holdfasts and edrioasteroid species. Caradocian communities in addition have epibiontic corals, attached pedunculate articulates (e.g., *Zygospira*), a greater number of encrusting bryozoans and a variety of shell boring organisms for overall higher taxon richness. High Caradocian diversity is probably not simply the result of better sampling because many of the additional taxa

Table 2. Epibiontic indicators of host vitality. Values refer to number of hosts observed with epibiont growth patterns indicating live, probable live, and dead hosts. Observations attributed to live hosts include epibionts that elevated their margin adjacent to the commissure but did not cross it (avoiding valve movement), and colonies that ceased growth linearly along the hinge (avoiding hinge movement). Probable live hosts exhibited cessation of epibionts at the shell margins, and postmortem indicators include epibiontic colonies that successfully crossed the commissure or hinge

Epibionts adjacent to commissure		Epibionts adjacent to hingeline	
Live host	11	Live host	5
Probably live	17	Probably live	5
Dead host	5	Dead host	0

Table 3. Epibionts from the Bromide and Benbolt Formations. Crinoids are identified by holdfast type following Lewis (1982).

Bromide Formation	Benbolt Formation

Bryozoans
  Corynotrypa delicatula
  Corynotrypa inflata
  Diplotrypa sp.
  trepostomes
	Bryozoans
	  Corynotrypa delicatula
	  Corynotrypa inflata
	  trepostomes (at least 3 species)

Brachiopods *Petrocrania* ?	

Echinoderms
  Holdfast 1C (*Reteocrinus*)
  1D (*Apodasmocrinus*)
  4B (*Bromidocystus*)
  Cyathocystis americanus
	Echinoderms
	  1A (*Palaeocrinus*)

Others
  Cornulites
  pyrite outlines
	Others
	  Cornulites

contribute significantly to the total areal cover. However, additional sampling of communities of both ages is needed to support a general pattern of increasing richness within epibiont communities during the later Ordovician.

Origination of Paleozoic Epibiont Communities. Epibiont communities, like other encrusting faunas, apparently had two periods of relative stability in their geological history, corresponding to Paleozoic and modern 'Great Evolutionary Faunas' (Sepkoski ,1981; Wilson & Palmer, 1992, Taylor & Michalik, 1991). This study suggests that diverse epibiont communities were founded during a short interval prior to the Llandeilan. Profound change in community composition and structure then also occurred during the Permian/Triassic transition, with more gradual taxonomic and ecological change occurring throughout the Paleozoic and during the Mesozoic-Cenozoic.

It is impossible to characterize an epibiont contribution to Sepkoski's (1981) Cambrian fauna. Prior to the rise of diverse epibiont communities during the Ordovician (bracketed here between the Arenigian and Llandeilan), there are few reported encrusting organisms. Guensburg & Sprinkle (1992) attributed the paucity of Cambrian encrusters (eocrinoids are a notable exception, e.g., Brett et al., 1983) to a shortage of appropriate hard settlement sites (specifically hardgrounds). However, shells existed throughout most of the Cambrian and could have served as suitable substrates. Signor & Vermeij (1994) suggested that the planktic food source for filter feeding organisms may have not yet evolved in the Cambrian, and this could explain the rarity of epibiont occurrences and/or their delayed evolution. A further search for Cambrian encrusters is needed before either suggestion can be accepted.

Diverse encrusting assemblages, including those with encrusting bryozoans, appear for the first time on Arenigian hardgrounds from the Kanosh Shale, Utah (Wilson et al., 1992). However, no epibionts were found on brachiopods from the same localities, suggesting that the evolution of shell encrusters was delayed relative to encrusters of more permanent and extensive hardgrounds. Additional samples of brachiopods are clearly needed to document the establishment of Ordovician multispecies epibiont assemblages. It would be surprising, however, if epibiont abundance and diversity did not initially increase during the Arenigian-Llanvirnian with the onset of rapid diversification among many invertebrate groups, including shell encrusters such as bryozoans (Sepkoski & Sheehan, 1983).

CONCLUSIONS

(1). Epibiont communities are well developed on Llandeilan (Middle Ordovician) brachiopods from the Bromide and Benbolt Formations. Common encrusting taxa include mat trepostome bryozoans, *Corynotrypa*, and cornulitids.
(2). Llandeilan epibiont communities are similar in taxonomic composition and structure to Late Ordovician communities, but less diverse.
(3). The origin of Paleozoic epibiont communities must be sought in assemblages older than the Llandeilan, probably during the Arenigian-Llanvirnian.

Acknowledgments

I thank P. Sutherland, University of Oklahoma, and the Museum of Invertebrate Paleontology, University of Kansas for the loan of Bromide Formation brachiopods, and F. McKinney, Department of Geology, Appalachian State University, for loan of Benbolt material.

REFERENCES

ALEXANDER, R.R. 1975. Phenotypic lability of the brachiopod *Rafinesquina alternata* (Ordovician) and its correlation with the sedimentologic regime. Journal of Paleontology 49:607-618.

___ & C.D. SCHARPF. 1990. Epizoans on Late Ordovician brachiopods from southeastern Indiana. Historical Geology 4:179-202.

BELL, B.M. 1982. Edrioasteroids. University of Kansas Paleontological Contributions, Monograph 1:57-64.

BENGTSON, S. & A. URBANEK. 1986. *Rhabdotubus,* a Middle Cambrian rhabdopleurid hemichordate. Lethaia 19:293-308.

BORDEAUX, Y.L. & C.E. BRETT. 1990. Substrate specific associations of epibionts on Middle Devonian brachiopods: implications for paleoecology. Historical Biology 4:221-224.

BRETT, C.E., M.E. BROOKFIELD & K.L. DERSTLER. 1983. Late Cambrian hard substrate communities from Montana/Wyoming: the oldest known hardground encrusters. Lethaia 16:281-289

CONWAY MORRIS, S. 1985. The Middle Cambrian metazoan *Wiwaxia corrugata* (Matthew) from the Burgess Shale and Ogygopsis Shale, British Columbia, Canada. Philosophical Transactions Royal Society London B307:507-582.

FLOWER, R.H. 1961. Organisms attached to Montoya corals. Memoirs New Mexico Bureau Mines Mineral Resources 7:1-124.

GUENSBURG, T.E. & J. SPRINKLE. 1992. Rise of echinoderms in the Paleozoic evolutionary fauna: significance of paleoenvironmental controls. Geology 20:407-410.

HAVLICEK, V. 1972. Life habitat of some Ordovician inarticulate brachiopods. Vestnik Ustredniho Ustavu Geologickeho 47:229-233.

HESSELBO, S.P. 1989. The aglaspid arthropod *Beckwithia* from the Cambrian of Utah and Wisconsin. Journal of Paleontology 63:636-642.

KOBLUK, D.R. & N.P. JAMES.1979. Cavity-dwelling organisms in Lower Cambrian patch reefs from southern Labrador. Lethaia, 12:193-218.

LESCINSKY, H.L. 1995. Life orientation of concavo-convex brachiopods: overturning the paradigm. Paleobiology 21:520-551.

___(in press). Epibiont communities: recruitment and competition on Carboniferous brachiopods from North America. Journal of Paleontology.

LEWIS, R.D. 1982. Holdfasts. University of Kansas Paleontological Contributions, Monograph 1:57-64.

MAYORAL, E., J.C. GUTIERREZ MARCO & J. MARTINELL. 1994. Primeras evidencias de briozoos perforantes (Ctenostomata) en braquiopodos Ordovicicos de los Montes de Toledo (Zona Centroiberica Meridional, Espana). Revista Española Paleontologia 9:185-194.

McNAMARA, R.J. 1978. Symbiosis between gastropods and bryozoans in the Late Ordovician of Cumbria, England. Lethaia 11:24-40.

MORRIS, R.W. & S.H. FELTON. 1993. Symbiotic association of crinoids, platyceratid gastropods and *Cornulites* in the Upper Ordovician (Cincinnatian) of the Cincinnati Ohio region. Palaios 8:465-476.

RICHARDS, R.P. 1972. Autecology of Richmondian brachiopods (Late Ordovician of Indiana and Ohio). Journal of Paleontology 46:386-405.

RIDING, R. & A. Y. ZHURAVLEV. 1995. Structure and diversity of oldest sponge-microbe reefs: Lower Cambrian, Aldan River, Siberia. Geology 23:649-652.

SEPKOSKI, J.J. Jr. 1981. A factor analytic description of the Phanerozoic marine fossil record. Paleobiology 7:36-53.

___& P.M. SHEEHAN. 1983. Diversification, faunal change and community replacement during the Ordovician radiations, p.673-717. In, M.J.S. Tevesz & P.L. McCall (eds.), Biotic interactions in recent and fossil communities. Plenum Press, New York.

SIGNOR, P.W. & G.J. VERMEIJ. 1994. The plankton and the benthos: origins and early history of an evolving relationship. Paleobiology 20:297-319.

SPRINKLE, J, 1973. Morphology and evolution of blastozoan echinoderms. Museum of Comparative Zoology Harvard University Special Publication, 283. pg.

___1982 (ed.) Echinoderm faunas from the Bromide Formation (Middle Ordovician) of Oklahoma. University of Kansas Paleontological Contributions, Monograph 1:1- 369.

STEINHAUFF, D.M. & K.R. WALKER. 1995. Recognizing exposure, drowning, and 'missed beats': platform-interior to platform-margin sequence stratigraphy of Middle Ordovician limestones, East Texas. Journal Sedimentary Research 65B:183-207.

TAYLOR, P.D. 1984. *Marcusodictyon* Bassler from the Lower Ordovician of Estonia: not the earliest bryozoan but a phosphatic problematica. Alcheringa 8:177-186.

___& J. MICHALIK. 1991. Cyclostome bryozoans from the late Triassic (Rhaetian) of the West Carpathians, Czechoslovakia. Neues Jahrbuch Geologie Paläontologie, Abhandlungen 182:285-302.

___ & M.A. WILSON. 1994. *Corynotrypa* from the Ordovician of North America: colony growth in a primitive stenolaemate bryozoan. Journal of Paleontology 68:241-257.

WADDINGTON, J.B. 1980. A soft substrate community with edrioasteroids from the Verulam Formation (Middle Ordovician) at Gamebridge, Ontario. Canadian Journal Earth Sciences 17:674-679.

WILSON, M.A. 1985. Disturbance and ecologic succession in an Upper Ordovician cobble-dwelling hardground fauna. Science 228:575-577.

___& T.J. PALMER. 1992. Hardgrounds and hardground faunas. University of Wales, Aberystwyth, Institute of Earth Studies Publications 9:1-131.

___, ___, T.E. GUENSBURG, C.D. FINTON & L. E. KAUFMAN. 1992. The development of an Early Ordovician hardground community in response to rapid sea-floor calcite precipitation. Lethaia 25:19-34.

FUNCTIONAL MORPHOLOGY - PALEOECOLOGY OF PYGOPID BRACHIOPODS FROM THE WESTERN CARPATHIAN MESOZOIC

JOZEF MICHALIK

Geological Institute, Slovakian Academy of Sciences, 842 26 Bratislava, Slovakia

ABSTRACT- The Jurassic-Cretaceous pygopid 'key-hole' brachiopods, possessing a central perforation of the shell, were able to live in poorly oxygenated environments probably because of effective use of nutrients and wide dispersal of young individuals. Their shells appear to have been oriented on the substrate by a well-developed pedicle muscle. Although specifically adapted to deep-sea conditions, the pygopids might also have been opportunists able to live in shallow slope settings in competition with other benthos.

INTRODUCTION

The pygopids are a group of distinctive brachiopods crossing the Jurassic-Cretacreous boundary in the Carpathians, and other regions of the Tethyan Realm(Figure 1). Although their distribution and probable mode of life have been investigated in many precious studies, their peculiar morphology and occurrences remain fascinating speculation to paleobiologists and paleobiogeographers.

The function of the pygopid central perforation remained unresolved for a long time (Ager, 1963). Suess (1852), Pictet (1867), and Tawney & Davidson (1869) believed that the bilobation of *Pygope* was an ecologically rather than genetically induced feature. Bather (in Buckman, 1906) stressed the

importance of bilobation of brachiopod shells in poorly oxygenated environments, although his homeomorphic interpretation was rejected by Jarre (1962). Dacqué (1921) first suggested that the pedicle passed through the perforation to stabilize and strengthen the shell attachment to the substrate. Vogel (1966) studied pygopid functional morphology in detail and came to the conclusion that the central perforation functioned as a tube to drain the exhalant current from the mantle cavity and to separate the exhalant and inhalant currents. This, it was argued, enabled the pygopids to live on a deep sea bottom free of other benthos. Ager (1965) suggested that the central perforation was an adaptation to quiet and oxygen-poor conditions by extension of the mantle margin. Occur-rences of pygopids in many western Carpathian localities (Kotanski & Radwanski, 1959; Barczek, 1972; Siblík, 1979; Nekvasilová, 1980; Krobicki, 1994), however, also indicate distribution in shallow marine facies.

Relating the diameter of the pygopid central perforation to the level of nutrient supply, Kázmér (1990) suggested that forms with large perforations (*Pygope janitor*, *Pygites diphyoides*) lived on the nutrient-rich northern Tethyan ocean margin, and those with small perforations (*Pygope catulloi*, *P. diphya*) inhabited the nutrient-poor southern margin of the Tethys. This assumption, however, is not supported by pygopid distribution

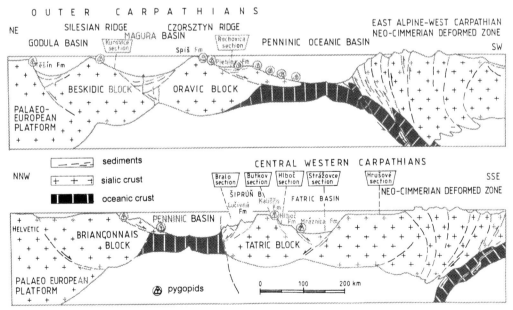

Figure 1. Interpreted geological paleocross-sections of the outer and central Carpathians during the early Hauterivian, showing the occurrence of pygopid brachiopods.

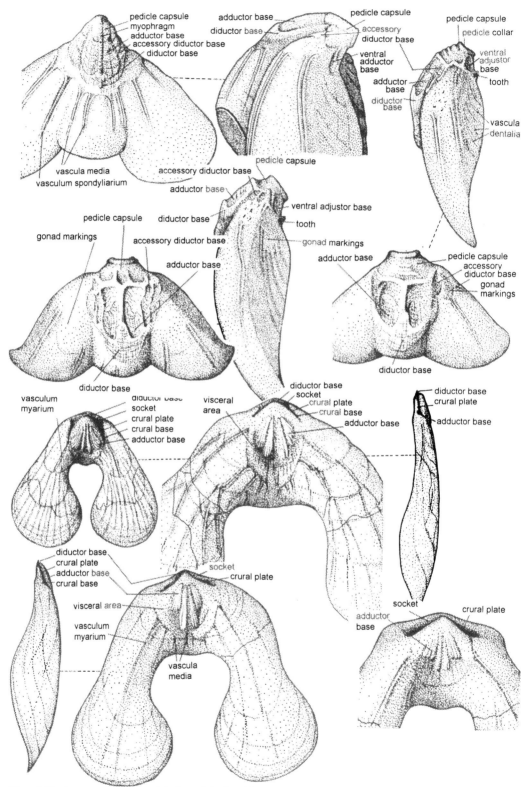

Figure 2. Muscle scars and mantle canals in the ventral valve of *Pygope janitor* at top, and dorsal valve below, Kruchy Hill, Pieniny Klippen Belt, Slovakia.

patterns. *P. diphya*, for example, is common in the Pieniny Klippen Belt of the paleoeuropean shelf, whereas *P. diphyoides* occurs in central Carpathian-, Austro-, and South Alpine areas located south of the Penninic Ocean.

SOFT BODY IMPRINTS

Moulds of pygopid shells from several localities show distinct imprints of soft body parts (muscle, pallial canals, and ovaria). The muscle scars are best preserved in the ventral valve (Figure 2). The posterior part of the muscle field was marked by a semilunate pedicle capsula, divided either by two short radial furrows, or by several rows of pits parallel to the foraminal margin. The muscle field proper is oval, separated from the pedicle capsula by wide furrows. Inconspicuous furrows at the anterior margin define the adductor, diductor, and accessory diductor scars. In the adult stage, the shape of ventral diductor scars varies from taxon to taxon (e.g., triangular in *Pygites diphyoides*, anchor-like in *Pygope catuloi*). The adductor scars bear radial striae, crossed by thin concentric lines and prominent concentric grooves formed by anterior shifting of the scar during ontogeny. Ventral adjustor scars are also present, with the most distinct ones being elongate in outline, striated, and located posterior to the teeth, parallel to the hingeline.

Ovarial markings (gonoglyphs) are located posterolaterally from the muscle field (Figures 2-3). Mantle canals (pallio-glyphs) are a pinnate system of the vasculum dentalium, vasculum spondyliarium, and vasculum medium. Each canal consisted of two parallel grooves separated by a medial ridge, which may have supported a row of ciliae. New branches of primary canals originate by false bifurcation (i.e., laterally inserted to existing canals). In some instances, new canals originated irrespective of the abandoned, old canal system. The anterior and anterolateral margins bear fine, parallel secondary canals, which probably supported sensory setae of the pallial margin. The vasculum dentalium was formed by two arched branches protruding from gonoglyphs near the teeth. The vasculum spondyliarium consisted of one pair of prominent, and two or three minor branches. The vasculum medium was formed by two pairs of inconspicuous branches.

Compared to deep-water forms, the pygopids from the Czorsztyn Ridge (*Pyope janitor*, *P. diphya*) show a 'juvenile' shape, characterised by short, diverging anterolateral lobes with a straight lateral commissure. These shells were regarded by Krobicki (1994) as opportunistic forms.

The lateral commissure is more or less straight in *Pygope janitor*, *Pygites diphyoides*, juvenile *Pygope diphya*, and *P. catulloi*. In mature forms of *P. diphya* and *P. catulloi*, however, the lateral commissure becomes sinuous, with overlap posteriorly. When the shell was open, the middle part of the strongly sinuous lateral commissure would remain closed in some varieties of *P. catulloi* (e.g., *P. sima* Zeuschner, 1860),

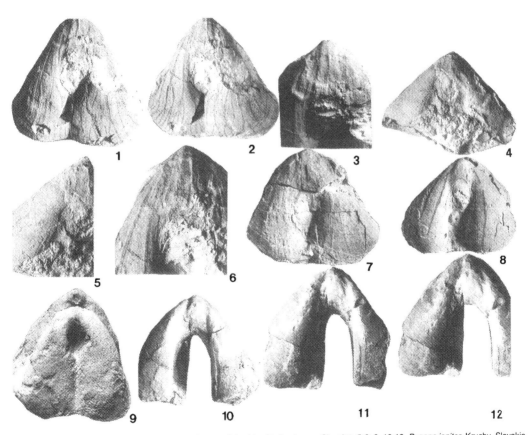

Figure 3. Mantle canal system in pygopids. 1-4, 7 *Pygites diphyoides*, Podhorie near Slavnica. 5-6, 8, 10-12, *Pygope janitor*, Kruchy, Slovakia (5-6, 8), Eperkeshegy, Hungary (10-12). 9, *Pygope catulloi*, Brvniste, Slovakia. All c. x1.

177

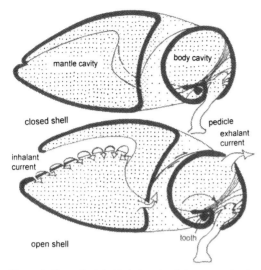

mantle cavity

body cavity

closed shell

pedicle

exhalant
current

inhalant
current

tooth

open shell

Figure 4. Reconstructed inhalant and exhalant gaps in the commissure of *Pygopes catulloi* when the shell was open.

as did the lateral part of the commissure covered by a squama of the opposite valve margin. This indicates that the postero-lateral exhalant current suggested by Vogel (1966) probably did not exist. The contact of the two valves in the central tube, however, is perpendicular to the direction of articulation and can open into two opposite gaps. The posterior gap could have functioned to channel the exhalant current. The function of the other gap is poorly understood, but it could have been used to expel body waste, reproductive cells, or larvae. A large space in the lateral lobes is retained in shallow-water pygopids and could have served as a larval brood pouch.

The commissure of the imperforate pygopids (*Triangope euganeensis, P. triangulus*) is largely straight, and only the anterior commissure in gerontic forms of *P. triangulus* is indented and strongly sulciplicate, and the posterior parts are covered by glotta. Thus inhalant and exhalant currents could well have been separated at the flat anterior margin. This implies a possible semi-infaunal mode of life, at least in gerontic individuals (Figure 4).

CONCLUSIONS

Strong bilobation of the perforated pygopids facilitated separation of the inhalant and exhalant currents (Vogel, 1966). Jet-like exhalant opening in the middle part of the shell could have served as a device for long-distance dispersal of reproductive cells. Complex musculature controlled the articulation of pygopid shells wihich had specific commissures. The myoglyphs indicate existence of strong muscles with accessory strands. A well-developed pedicle supported by a pedicle capsule could have moved the shell around to use nutrient-rich currents from any direction. Dense vascular canals maximized the circulation of oxygen and nutrients. Large gonads occupied the posterolateral part of the mantle.

Despite their specialized morphology, some pygopids, especially the neotenous forms of *Pygope janitor* and *P. diphya* from Czorsztyn Ridge, behaved as opportunists, inhabiting both basinal and elevated block facies (Krobicki, 1994). Pygopids could utilise nutrients from very weak

suspension currents. Their paleogeographic distribution in pelagic settings indicates that they could tolerate lowered oxygen levels and live on nutrients provided by either surficial or deep upwelling currents (Rehákavá & Michlik, 1994; Rehákavá, 1994).

Acknowledgments

The study was supported by Grant GA1080 from the Slovakian Academy of Sciences. The author thanks the Organizing Committee of the Third International Brachiopod Congress for a travel grant from the Natural Sciences and Engineering Research Council of Canada to present the paper in the conference.

REFERENCES

AGER, D.V. 1963. Principles of paleoecology. McGraw-Hill, New York, 371p.

___1965. The adaptation of Mesozoic brachiopods to different environments. Palaeogeography, Palaeoclimatology, Palaeoecology 1:143-179.

BARCZYK, W. 1972. Some representatives of the family Pygopidae (Brachiopoda) from the Tithonian of the Pieniny Klippen Belt. Acta Geologica Polonica 22:1-6.

BUCKMAN, S.S. 1906. Brachiopod homeomorphy: *Pygope, Antinomia, Pygites*. Quarterly Journal Geological Society London 62:433-454.

DACQUÉ, E.1921. Vergleichende biologische Formenkunde der fossilen niederen Tiere. Borntraeger, Berlin, 777p.

JARRE, P. 1962. Revision du genre *Pygope*. Travaux Labora-toire Géologie Grenoble 38:23-120.

KÁZMÉR, M. 1990. Tithonian-Neocomian Pygopidae (Brachio-poda) in the Alpine-Carpathian region. Általanos Földtani Szemle 25:327-335.

KOTANSKI, Z & A. RADWANSKI. 1959. Fauna with *Pygope diphya* and limburgites in the Tithonian Tatric Osobita sequence. Acta Geologica Polonica 9:519-534.

KROBICKI, M. 1994. Stratigraphic significance and paleoecology of the Tithonian-Berriasian brachiopods in the Pieniny Klippen Belt, Carpathians, Poland. Studia Geologica Polonica 106:89-156.

NEKVASILOVÁ, O.1980. Terebratulida (Brachiopoda) from the Lower Cretaceous of Stramberk (northeast Moravia), Czechoslovakia. Sborník Geologickych Vednik 23:49-80.

PICTET, F.1867. Étude monographique des Térebratules du groupe de la *Terebratula diphya*. Mélanges paléonto-logiques 3:135-202.

REHÁKOVÁ, D. 1994. Upper Jurassic-Lower Cretaceous carbonate microfacies and environmental models for the western Carpathians and adjacent paleogeographic units. Cretaceous Research 16:283-297.

___& J. MICHALÍK. 1994. Abundance and distribution of Late Jurassic-Early Cretaceous microplankton in western Carpathians. Geobios 27:135-156.

SIBLIK, M. 1979. Brachiopods of the Vrsatec Castel Klippe (Bajocian- ? Berriasian) near Llava (Slovakia). Zapadné Karpaty, Paleontologia 4:35-64.

SUESS, E. 1852. Über *Terebratula diphya*. Sitzungsberichte Kaiserlichen Akademie Wissenschaften 8:553-567.

TAWNEY, E. & T. DAVIDSON. 1869. On the occurrence of *Terebratula diphya* in the Alps of the Canton de Vaud. Quarterly Journal Geological Society London 25:305-309.

VOGEL, K. 1966. Eine funktionsmorphologische Studie an der Brachiopodengattung *Pygope* (Malm bis Unterkreide). Neues Jahrbuch Geologie Paläontologie 125:423-442.

ZEUSCHNER, L. 1860. Die Brachiopoden des Stramberger Kalkes. Neues Jahrbuch Mineralogie, Geognosie, Geologie 1860:678-691.

PRINCIPAL TRENDS IN EARLY ATHYRID EVOLUTION

TANYA L. MODZALEVSKAYA

All-Russian Geological Research Institute (VSEGEI), Srednii prospekt 74, St Petersburg 199026, Russia

ABSTRACT -- The most stable morphological features in early athyrid morphology appear to be the brachidium and cardinalium. The cardinalium allows division into at least three suborders: Meristelloidea, Athyroidea, and Retzioidea. Many other features (complex jugum, muscle attachment, hinge mechanisms, apical structures and external form) were more changeable and led to the development of morphologically distinct groups. Other athyrid evolutionary developments were related to combination of early morphological structures, and occurred irregularly. Parallel, and often isomorphic phylogenetic lineages, were typical for athyrid evolution (smooth and ribbed, punctate and impunctate shells). Despite this, each stock was characterised by a distinct trend of development, with simultaneous elaboration of existing and appearance of new structures.

INTRODUCTION

Athyrids are characterized by laterally directed spiralia, a simple to complex jugum, and complex internal cardinalia, combined with a comparatively monotonous external, normally astrophic shell. During their evolution from the Ordovician through the early Jurassic, all were benthic, attaching to the substrate at most ontogenetic stages. In the Early and Middle Paleozoic, athyrids occupied various niches of the shallow water shelf (Benthic Assemblage 2 or 3). The most important factor determining athyrid radiation and extinction appears to have been their adaptation to the environment. The cardinalium of athyrids is similar to that of rhynchonellids and atrypids, and the presence of a laterally-oriented, calcite spiral lophophore support makes them similar to spiriferids. There are different viewpoints not only concerning the nature of suprageneric taxa (for instance the rank of retziids, nucleospirids and athyrisinids, and the assignment of the Siluro-Devonian dayiids, coelospirinids, anoplothecinids and kayseriinids), but also on the systematic position of the group as a whole.

Beurlen (1952) was the first to suggest that athyrids originated from cyclospirids in the Late Ordovician (Caradoc), though *Cyclospira* lacks a jugum, has medially, instead of laterally directed spiralia, and occurs later than smooth athyrids with laterally-directed spiralia in the Caradoc. Following him, many researchers connected athyrid evolution and origins with atrypids (Boucot et al., 1964; Williams, 1965; Rudwick, 1970). The smooth-shelled Late Ordovican (Ashgill) hindellids, characterized by a 'primitive' cardinalium and simple jugum, were among the ancestral line of meristelloid and athyroid stocks, a viewpoint most common at present. Grunt (1989) analysed the common features of the lophophore, cardinalium, shell wall, and apical structures. She concluded that rhynchonellids could be the ancestors of athyrids, since they are rather extensively represented in the Ordovician, have a septalium and ribs, and could have evolved into athyrids by reversing the direction of the lophophore laterally, and by forming calcareous spiral supports and a jugum. The origin of nucleospirids, which appear with athyroids at the same time in the Early Silurian (late Llandovery), is not yet clear. Their

cardinalium is rather similar to that of the Athyroidea, but the brachidium has meristelloid features. Boucot et al. (1964) suggested that nucleospirids were related to athyroids because of similar internal and external structures.

EVOLUTION

Grunt (1989), comparing the morphology of athyrids with those of Recent brachiopods, presumed that structural features of rhynchonellids and terebratulids appeared to be well adapted, allowing athyrids to cross the Paleozoic-Mesozoic boundary. However, these features were apparently inadequate for subsequent survival. Among well-adapted characters were a complex cardinalium and astrophic shell (which ensured a

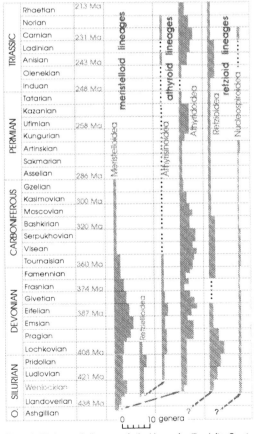

Figure 1. Phylogenetic lineages of athyrid superfamilies (after Grunt, 1989), with number of genera marking stocks [this excludes coelospirinids, dayiids].

179

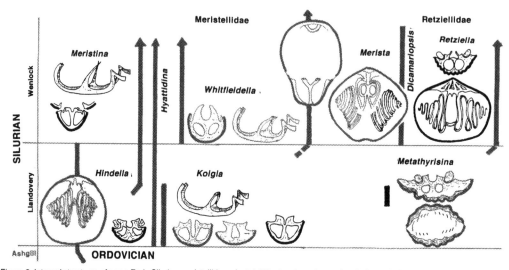

Figure 2. Internal structures of some Early Silurian meristellids and retziellids showing a burst of evolution at the Llandovery-Wenlock boundary.

strong hinge mechanism), a pedicle muscle throughout life, a calcite spiral lophophore and jugum, and strong shell wall (Figure 1).

Athyrids are generally classified into three groups, the meristelloids, athyroids and retzioids. The first (meristelloid) is characterised by a discrete cardinal plate and septum, the second (athyroids), by a complete cardinal plate, and open or overgrown visceral foramen and only slightly developed septum, and the third, (retzioids), by combinations of the previous. Punctation is viewed in this light as a secondary character at the superfamily level. The initial feature of athyrids was the structure of the cardinal plate with a sessile septalium (e.g. as seen in Early Silurian Hyattidina). Further evolution of cardinal plates proceeded along three trends: (1) Elevation of the inner cardinal plates above the valve bottom, formation of the septum-septalium, and junction or intergrowth of inner cardinal plates to form a single plate. Meristelloids display such changes in the cardinalium during evolution. This feature appears least stable in the Silurian genus Koigia (Figure 2), whose cardinalium evolved in several ways. One of these is characterised a deep sessile septalium, already typical of younger hindellids (e.g. Silurian Whitfieldella). In the other, a tendency towards elevation of inner cardinal plates is observed with an extension parallel to the hinge. (2) The fusion of the cardinal plate, and reduction of the brachial median septum, could be a step towards formation of the dorsal perforated plate near the dorsal beak, seen in didymothyridines (e.g. Glassina, Greenfieldia). (3) Athyroids evolved with a cardinalium similar to that of meristinids, in which the septum and septalium are the main feature. Meristella shells, on the contrary, had a wide open, flat septalium. In Whitfieldella, the septalium is covered by unusual horizontal plates. In athyroids in general, the cardinalium has a relatively consistent plan. Exceptions are the genus Pseudoprotathyris and the related Eifelian genus Leptathyris, which have a neotenous shape and hypertrophic outer cardinal plates (Figure 3).

The other main feature of athyrids is the structure of the brachidium. Their complex spiral lophophore has the apices of spiral cones laterally directed. The shape and number of spiral whorls depend, in part, on shape and size (age) of the shell. The spiralium takes up most of shell space, ensuring optimum use of the mantle cavity. The connecting structures between the hinge and spiralium are the crura and primary lamellae. In the centre of the shell, the jugum arises from primary lamellae in a postero-dorsal direction. Initially (e.g. in Ordovician Hindella, Whitfieldella), the jugum was simple, but later, evolution provided complications. In the meristelloids, generation of loops strengthened the brachidial apparatus. In athyroids, the jugal saddle was joined by means of a jugal stem and additional lamellae in the crural area, controlling the location of the junction with the primary spiral whorls (Figure 2). For instance, in Glassina the jugum was located near the beak, and lacks a jugal stem. In other genera (e.g. Didymothyris, Collarothyris, Greenfieldia), it protruded forward, towards the shell centre

ATHYRID PHYLOGENY

In athyrid evolution parallelism was especially striking, perhaps due to specialization in early forms. In the Llandovery, there was divergence of outer shell features and a number of near homeomorphic, parallel phylogenetic lineages appear. Characteristic are different external shell structures, but similar internal features. These include smooth and ribbed forms with similar cardinalia, e.g. in Meristelloidea and Athyridoidea. Smooth meristellids, including the Dayiidae (Copper, 1986) appeared at the close of the Ordovician or Early Silurian, continuing to the end of the Carboniferous (Figure 1). Modification of the cardinalium, appearance of additional diagnostic inner structures, and increasing diversity of meristellids occurred near the Llandovery-Wenlock boundary. Widespread Meristina, Merista, Whitfieldella, and Dicamariopsis, which appeared at close of the Llandovery are represented by a mozaic of primitive and advanced features, which became most pronounced in the Early and Middle Devonian. Such specialized structures as 'mystrochial plates' evolved very rapidly, complicating and strengthening not only the apical apparatus, but also the cardinalium. Meristina, in addition to mystrochial plates, acquired a thick, heavy prismatic shell layer. Large, thick, convex shells required stronger muscles, which were attached to a septalium elevated above the dorsal valve. Meristina obtusa reveals the overgrowth of septalium and outer hinge plates by secondary shell substance (by analogy with pentamerids, this may be

180

called a 'hinge process'). According to Grunt (1989), the appearance of prismatic shell delayed evolutionary progress in such athyrid lineages. Meristellids were mostly extinct by the Middle Devonian, with the exception of the widespread genus *Meristella* which survived the Pragian-Emsian, perhaps favoured by its complex jugum, which may have enhanced feeding. But at the end of the Carboniferous, only *Camarophorella* remained. An arched platform (the 'shoe lifter process') appeared in *Merista*, located above the bottom of the ventral valve, together with mystrochial plates. In the related genus *Dicamariopsis*, these mystrochial plates are also located in the dorsal valve.

Ribbed impunctate meristellids (retziellioids) appeared late, starting with Llandovery *Metathyrisina*. Having retained the primitive structure of the cardinalium, this group of brachiopods completed its evolution by the Early Devonian and was confined SE Asia and Australia (Rong et al.,1994).

Appearance of the athyroid stock, with characteristic cardinal plates, is confined to the end of the Wenlock. Therefore, their cardinal plate appears to be a late derived structure, compared to their septalium. Such a cardinalium is also recorded in punctate retziids, which started their evolution in the Wenlock (e.g. *Homeospira*). Along with appearance of punctae, a subhorizontal cardinal platform evolved as a 'superimposed', complete cardinal plate, on top of the septum and ,which strengthened the hinge. In *Nucleospira*, the strengthening of the apical apparatus was favoured by the cardinalium: the cardinal plate, being curved in the posterior-ventral direction, entered the umbonal cavity of the ventral valve. The structure

of the apical apparatus, on which the non-strophic shell growth depended, together with a complex cardinalium, ensured a strong union of both valves.

From Pridoli, and through the Early Devonian, retziids almost lasted to the Eifelian, when their evolution slowed, to continue, however, until the end of Triassic. A retziid bifurcate jugal stem formed jugal branches, which, in the latest Triassic genera became elongated, forming double spirals (as in *Hungarispira*: Dagys, 1974), repeating double spiral lamellae seen in Siluro-Devonian coelospirid-anoplothecid evolution.

Having become established in the Wenlock, smooth athyrids (*Glassina*), after undergoing a number of crises, underwent several periods of radiation until the end of Triassic. Marked diversification is associated with the multiple appearance of internal and external structures at the end Silurian. Athyrids had a smooth shell with fine growth lines, inherited by Late Paleozoic forms, e.g.*Composita*, *Spirigerella*. Starting with Devonian, *Protathyris* and other genera were characterized by the presence of growth lamellae (*Athyris*, *Atrythyris*), often covered by dense ribs or dissected into spines along the margin as in *Cleiothyridina*, *Delthachania*, and *Pinegathyris*. Devonian plicathyridines were characterised in the beginning by one or two lateral radial plicae (*Anathyris*, *Plicathyris*). Towards the end of their evolution the number of plicae on the surface reached about ten (*Flexathyris*), and, along with plicae, dense growth lamellae appeared. In addition to their convergent homeomorphy with spiriferids (transversely triangular outline, straight hinge) they were characterized by a thickened umbo, a pedicle collar, and highly elevated outer

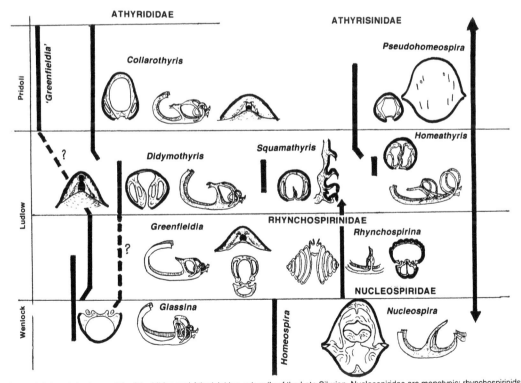

Figure 3. Internal structures of the Athyrididae and Athyrisinidae, primarily of the Late Silurian. Nucleospiridae are monotypic; rhynchospirinids have a complex cardinalium and brachidium.

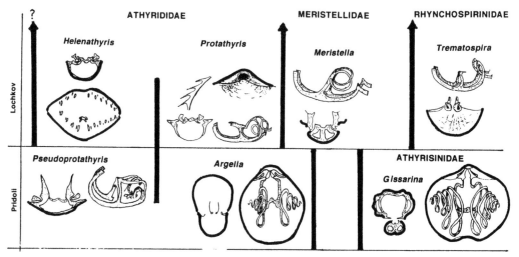

Figure 4. Variation in structures of athyrids evolving during the Devonian.

parts of the cardinal plate with a cardinal process (*Anathyris, Anathyrella, Camelicania, Pinegathyris*). These structures enabled plicathyridines, at the close of the Emsian, to fill many ecological niches in the Cantabrian-Armoricain region of Old World Realm and during the Devonian to migrate to Eastern Sibirea and the American continent. In N Eurasia, their mass extinction is recorded at the Frasnian-Famennian boundary (Alvarez, 1990: see also Rzhonsnitskaya & Modzalevskaya, this volume).

Ribbed athyrisinids, retaining a cardinalium and brachidium similar to smooth forms, are confined to the Late Silurian (Ludfordian: *Squamathyris, Homeathyris*, with maximum generic diversity in the Devonian (Eifelian), after which they disappeared, only to reappear in the Mid-Triassic. Two existing parallel lineages comprise the smooth athyrids, e.g. *Didymothyris-Collarothyris-Protathyris*, and ribbed athyrids, e.g. *Homeathyris-Pseudohomeospira-Athyrisina*. There appears to be a synchronous change of internal structures at the same boundary, e.g. the replacement in the Middle Ludfordian (Middle Ludlow) of the pedicle fulcrum by the pedicle collar, and disappearance of these structures by the end Silurian. Change in apical apparatus is here thought to be a significant evolutionary feature of early athyrids. The pedicle fulcrum has two plates, joined in the centre, formed by secondary shell, while the pedicle collar is a concave plate, resting with its ends on the delthyrium margin and sometimes supported by a short, low median septum. Between the bases of the dental plate are a pedicle fulcrum and collar, as seen in *Didymothyris, Collarothyris, Squamathyris*, and *Pseudohomeospira*. Extinction of athyrids with a pedicle fulcrum coincides with the beginning of a sedimentary cycle in the Arctic islands in the NE Russian Platform along the western slope of the polar to central Urals. Athyrids are also noted for a repetition of such elements in Late Devonian plicathyridines, and in the Late Permian genus *Spirigerella*. In the Triassic, the pedicle collar is characteristic of all Athyridoidea, and consists of a composite beak infilling (*Oxycolpella, Majkolpella*). In the punctate early rhynchospirinids, deltidial plates may be overgrown. Punctate retziids in the Triassic formed pedicle collars in the form of complete, tubes (*Hustedia, Hustediella, Neoretzia*).

CONCLUSIONS

Evolution of athyrids appears to display parallel synchronous generation and formation of morphologies. Since the set of morphological features was rather limited, evolutionary processes occurred by recombination of early features, with each lineage characterised by its own trends. Acquired specialized structures could perform additional functions.

REFERENCES

ALEKSEEVA, R. E. 1969. Novye svoeobraznye nizhne-devonskie spiriferidy (podotryad Athyridina). Doklady Akad.Nauk SSSR.187, 5:1157-1159.

ALVAREZ, F. 1990. Devonian athyrid brachiopods from the Cantabrian Zone (NW Spain). Biostratigraphie Paléozoique11:1-311.

BEURLEN, K.1952. Phylogenie und System der Brachiopoda Articulata. Neues Jahrbuch Geologie Paläontologie Monatshefte 3:111-125.

BOUCOT, A.J., J.G. JOHNSON & R.D. STATON. 1964. On some atrypoid, retzioid and athyroid Brachiopoda. Journal of Paleontolology 38(5):805-822.

___1965. Suborder Retziidina Boucot, Johnson & Staton, 1964, p. H649-H654. In, R. C. Moore (ed.), Treatise on Invertebrate Paleontology, Part H, Brachiopoda. Geological Society of America & University of Kansas Press, Lawrence.

COPPER, P. 1986. Filter-feeding and evolution in early spire-bearing brachiopods. In, P.R. Racheboeuf & C.C. Emig (eds.), Les brachiopodes fossiles et actuelles. Biostratigraphie du Paléozoique 4:219-230.

DAGYS, A.S. 1974. Triassovye brakhiopody (Morfologia, sistema, filogenia, stratigraphicheskoe znachenie i biogeografia).Izdatelstvo Nauka, Siberskoi otdelenie, 387 p.

GRUNT, T.A. 1989. Otryad Athyridida. Evolutsionaya morfologia i istoricheskoe razvitie. Akademia Nauk SSSR,Trudy Paleontologicheskii Institut 238:1-140 .

HALL, J. & J.M. CLARKE. 1893. An introduction to the study of the genera of Paleozoic Brachiopoda. New York Geological Survey 8(2):1-317.

___1894. An introduction to the Brachiopoda, 13th Annual Report New York State Geologist for the year 1893, Palaeontology, part 2: 751-943.

KOZLOWSKI, R. 1929. Les brachiopodes gotlandiens de la Podolie Polonaise. Palaeontologica polonica 1:1-254.

MENAKOVA, G.N. & O.I. NIKIFOROVA. 1986. Novye predstaviteli poznesiluriiskikh brakhiopod Zeravshano-Gissarskoi Gornoi oblasti. Paleontologiceskii Zhurnal 4:65-76.

MODZALEVSKAYA, T. L.1979. K sistematike paleozoiskikh atiridid. Paleontologicheskii Zhurnal 2:48-63.

---1981. Brakhiopody, p.146-157. In, L.V. Nekhorosheva (ed.). Obyasnitelnaya zapiska k skheme stratigrafii verkhne-siluriiskikh otlozhenii Vaigachsko-Yuzhnonovozemelskogo regiona. VNII Okeangeologia, 205p.

----1985. Brakhiopody silura i rannego devona evropeiskoi-chasti SSSR. Otryad Athyridida. Akademia Nauk SSSR:1-128.

NIKIFOROVA, O.I. 1970. Brakhiopody grebenskogo gorizonta Vaigacha (poznii silur), p.97-149. In, S.V. Cherkesova (ed.). Stratigrafia i fauna siluriiskikh otlozhenii Vaigacha (sbornik statej). NIIGA, 237p.

RONG JIAYU, D.L. STRUSZ, A.J. BOUCOT, LIPU FU, T.L. MODZALEVSKAYA & YANG-ZHENG SU. 1994. The Retziellidae (Silurian ribbed impunctate athyridine brachiopods). Acta Paleontologica Sinica 33:546-574.

___& XUECHANG YANG. 1981. Middle and late Early Silurian brachiopod faunas in Southwest China. Memoirs Nanjing Institute Geology 13:1673-770.

RUDWICK, M.J. 1970. Living and fossil Brachiopods, Hutchinson University Library, 199 p.

WILLIAMS, A.1956. The calcareous shell of the Brachiopoda and its importance to their classification. Biological Reviews 31(3): 243-287.

LATE CRETACEOUS RHYNCHONELLID ASSEMBLAGES OF NORTH BULGARIA

NEDA MOTCHUROVA-DEKOVA

National Museum of Natural History, 1 Tzar Osvoboditel Blvd, 1000 Sofia, Bulgaria

ABSTRACT- Five rhynchonellid assemblages are recognized in the epicontinental, marine Upper Cretaceous of northern Bulgaria, below the end-Cretaceous extinction event. Ages of the assemblages were determined by associated macrofossils and/or by correlation with occurrences of studied rhynchonellid taxa in other regions. Available data fail to meet criteria necessary to establish a biozonal scheme, nevertheless these assemblages provided an alternative for dating lithostrati-graphic complexes, where other fossils were absent. Described are the *Orbirhynchia mantelliana* (Cenomanian), *Cyclothyris difformis* (lower-middle Cenomanian), *Orbirhynchia reedensis* (middle-upper Turonian), *Cyclothyris globata* (uppermost upper Santonian), and *Septatoechia* aff. *baugasii* (upper Campanian) assemblages of Cretaceous age.

INTRODUCTION

Despite their relatively slow evolutionary rates compared to ammonites and belemnites, much attention during the last decades has been paid to Mesozoic brachiopods for their biostratigraphic significance. Their importance is appreciable in regions where other fossils are rare or absent. Among Late Cretaceous brachiopods, late Campanian and Maastrichtian micromorphs are considered to be of stratigraphic value in northwest Europe. On the basis of such brachiopods, Surlyk (1984) established twelve biozones in the uppermost Cam-panian and Maastrichtian. These formed the most refined zonation for this part of the Upper Cretaceous, although other zonal schemes

Campanian	U	*Septatoechia* aff. *baugasii* assemblage
	L	
Santonian	U	*Cyclothyris globata* assmbl.
	L	
Coniacian		
Turonian	U	*Orbirhynchia reedensis* assemblage
	M	
	L	
Cenomanian	U	
	M	*Cyclothyris difformis* assemblage / *Orbirhynchia mantelliana* assemblage
	L	

Figure 1. Chronostratigraphic range of the Late Cretaceous rhynchonellid assemblages in North Bulgaria.

based on belemnites and micro- and nanofossils have also been established. The possibility to establish a brachiopod zonation in northwest Europe was favoured by a number of factors, particularly the monotonous chalk facies and abundant brachiopods in many outcrops thoughout the uninterrupted upper Campanian-Maastrichtian stages.

Brachiopods are one of the most widespread fossil groups in Bulgarian, epicontinental marine, Upper Cretaceous rocks. They occur in abundance at several stratigraphic horizons. Brachiopods were collected from 36 Upper Cretaceous sections in North Bulgaria (Figure 2). A taxonomic study of the order Rhynchonellida was carried out as a first stage of the brachiopod investigations (Motchurova-Dekova, 1994, 1995). One of the final goals was to establish a brachiopod stratigraphy for these sections.

Some difficulties were encountered in an attempt to define a rhynchonellid biostratigraphy. The main problem was the sporadic occurrence of rhynchonellids in outcrop. In many Upper Cretaceous sections brachiopods were found at one level only, while underlying and the overlying strata were sterile for macrobrachiopods. Although macromorphic terebratulids and craniids were not included in the taxonomic study, they were also carefully collected, since they occurred together with rhynchonellids. The epicontinental Upper Cretaceous in Bulgaria is represented by a variety of alternating facies attesting to different environments, but only some favoured the distribution of brachiopods. The facies setting in which these macromorphic brachiopods were found, testify to shallow subtidal environments in relatively low latitudes. The main rockforming components were siliciclastics and carbonates, forming sandstones, calcareous sandstones, sandy limestones, and biodetrital limestones.

Another limiting factor to the distribution of brachiopods in the region was a decrease in the rate of sedimentation, or interrupted sedimentation, during the Late Cretaceous, manifested by hiatuses or hard grounds in the sequences. The third problem concerned the fact that some of the collected fossils were ap-parently not buried in situ. Some of these specimens were broken, and it was not clear whether this was a result of syn-chronous transport, or resedimentation. The rhynchonellids studied did not include unbroken phylogenetic series and it was difficult to determine relationships between successive taxa.

Notwithstanding the problems discussed above, five rhynchonellid assemblages, derived from 36 localities, were outlined in the epi-continental Upper Cretaceous strata of northern Bulgaria (Figure 1). The term assemblage is used here for any complex of rhynchonellid species occurring in the same strata. Such assem-blages may include autochthonous and allochthonous elements here. Generalized Upper Cretaceous sections were plotted to show the vertical distribution of these assemblages at different of localities (Figure 2).

185

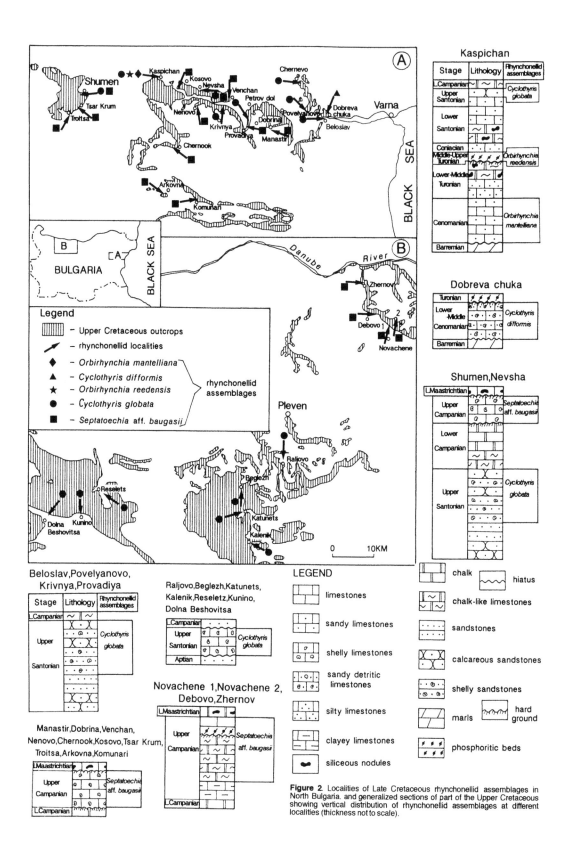

Figure 2. Localities of Late Cretaceous rhynchonellid assemblages in North Bulgaria. and generalized sections of part of the Upper Cretaceous showing vertical distribution of rhynchonellid assemblages at different localities (thickness not to scale).

186

Figure 3. Representative rhynchonellids from Bulgarian Cretaceous assemblages. 1. *Orbirhynchia mantelliana* (J. de C. Sowerby), #I-308, Kaspichan, Cenomanian, x1.2; 2. *Cyclothyris difformis* (Valenciennes, in Lamarck), #2758, Dobreva chuka, lower-middle Cenomanian, x1.2; 3. *Orbirhynchia reedensis* (Etheridge), #D-7, Kaspichan, middle-upper Turonian, x1.6; 4. *Cyclothyris globata* (Arnaud), #2449, Venchan, uppermost upper Santonian, x2; 5. *Septatoechia* aff. *baugasii* (d'Orbigny), #3656, Chernook, upper Campanian, x1.2.

The extensive studies of Jolkicev (1979-1980, 1989) served as a litho- and chronostratigraphic base for this paper. No single macrofossil group was employed, and no zonal macrofossil scheme was available for the investigated lithostratigraphic complexes. Ammonites, belemnites, inoceramids, and echinoids are most commonly used for dating the Upper Cretaceous in the region (Jolkicev, 1979-1980; 1989). The microfossil zonations available were only partially applicable because rhynchonellids occur in beds in which microfossils are absent or not characteristic. Brachiopod assemblages below cite frequent and rare species, and their distribution.

RHYNCHONELLID ASSEMBLAGES

Orbirhynchia mantelliana assemblage

Frequent species: *Orbirhynchia mantelliana* (J. de C. Sowerby) (Figure 3:1) and *O.* aff. *mantelliana*. Both species occur throughout the Cenomanian. *Orbirhynchia parkinsoni parkinsoni* Owen and *O. parkinsoni rencurelensis* (Jacob & Fallot) were found only in the lower Cenomanian. Rare species: *Orbirhynchia wiesti* (Quenstedt), *O.* cf. *boussensis* Owen, *O. ventriplanata* (Schloenbach), *O. lunpolaica* Ching & Ye, and *O. wilmingtonensis* Owen. These species occur in the upper Cenomanian. Distribution: This assemblage is recorded near Kaspichan, NE Bulgaria within sandy, detrital limestones of the Madara Fm., and is marked by single specimens of *Orbi-rhynchia mantelliana* near Petrov dol and Chernevo. Chrono-stratigraphic range, correlations: Cenomanian, determined by ammonites (Tzankov, 1982; Jolkicev, 1989). *Orbirhynchia mantelliana* is considered a good biostratigraphic marker within the middle Cenomanian in England (Owen, 1988). This species is also reported from the Cenomanian of Poland (Popiel-Barczyk, 1977), Serbia (Mitrovic-Petrovic & Radulovic,1994), Belgium and France (Davidson, 1852-1855). Other taxa occur widely in the Cenomanian or have a greater stratigraphic range.

Cyclothyris difformis assemblage

Frequent species: *Cyclothyris difformis* (Valenciennes in Lamarck) (Figure 3:2), and *C. compressa* (Valenciennes in Lamarck). Rare species:*Cyclothyris juigneti* Owen, *C.* cf. *contorta* (d'Orbigny), *Cretirhynchia* cf. *robusta* (Tate). Distribution: The assemblage is recorded at Dobreva chuka, near Beloslav, NE Bulgaria, in sandy limestones, Madara Fm., and with single specimens from Kaspichan and Chernevo (Figure 2). Chronostratigraphic range, correlations: Lower-middle Cenomanian. Since other characteristic fossils were not found at Dobreva chuka, age was determined by correlation with *Cyclothyris difformis* in other regions. This species is recorded in the lower and middle Cenomanian of France and South England (Owen, 1962, 1988). There are records of *C.* aff./cf. *difformis* from the lower Cenomanian of Bohemia (Nekvasilova, 1973) and Poland (Popiel-Barczyk, 1977). *C. compressa* is reported from the Cenomanian in France and South England (Owen, 1988). *C. juigneti* occurs in the middle Cenomanian of France (Owen, 1988), and *C. contorta* from the Cenomanian of Belgium and the Pyrénées (d'Orbigny,1847).

Orbirhynchia reedensis assemblage

Frequent species: *Orbirhynchia reedensis* (Etheridge) (Figure 3:3) and *O. extensa* Pettitt. Rare species: *O. lunpolaica* Ching & Ye, *O. wilmingtonensis* Owen, *O. cuvieri* Pettitt, *O. dispansa* Pettitt, *O.* cf. *praedispansa* Pettitt, and *Parthirhynchia* sp. Distribution: The assemblage is found near Kaspichan (Figure 2) within the condensed phosphoritic bed in the lowest part of the Dobrindol Fm. Chronostratigraphic range, correlations: Middle-upper Turonian. Condensation of the section and age of the phosphoritic bed is determined by ammonites (Jolkicev, 1989), and nannofossils (Sinnyovsky, 1988). *Orbirhynchia reedensis* is considered as characteristic for the upper Turonian in Donbass, the Crimea, Caspian Lowland, Manguishlak, Tuarkir (Katz, 1974;

Nechrikova, 1982), Bohemia (Nekvasilova, 1974), Saxony (Geinitz, 1873), and England (Pettitt, 1954). *O. extensa* is reported from the Cenomanian of Poland (Popiel-Barczyk, 1977) and the lower Turonian of England (Pettitt, 1954). *O. dispansa* is indicative of the upper Turonian in England (Pettitt, 1954).

Cyclothyris globata assemblage

Frequent species: *Cyclothyris globata* (Arnaud) (Figure 3:4), *Cyclothyris?* n.sp. and *Cretirhynchia?* n.sp. In NE Bulgaria *Cyclothyris globata* occurs in lower stratigraphic levels, whereas the latter two new species are found only in the topmost horizons, and in central N Bulgaria the assemblage consists only of the two new species. Distribution: In contrast to the first three assemblages this one is very widely distributed (Figure 2). In NE Bulgaria this assemblage occurs in the topmost Shumen Fm. (sandy limestones and calcareous sandstones) at Shumen, Provadiya, Beloslav, Povelyanovo, Chernevo, Krivnya, Nevsha, and Kaspichan. In central N Bulgaria, the assemblage is found within the Kalen Fm. (limestones) near Raljovo, Beglezh, Katunets, Kalenik, Reselets, Kunino, and Dolna Beshovitsa. Chronostratigraphic range, cor-relations: Uppermost upper Santonian, determined by associated ammonites, echinoids and bivalves (Jolkicev,1989). Ammonites are recorded only in NE Bulgaria, but the associated macrofossils correlate with Upper Santonian sediments from outcrops in central N Bulgaria, where ammonites are not found. Rhynchonellids suggest that the Kalen Fm. from central N Bulgaria corresponds only to the topmost Shumen Fm. from NE Bulgaria, where new species of *Cyclothyris?* and *Creti-rhynchia?* appear, because *Cyclothyris globata* from the lower beds of the Shumen Fm. is not found in the Kalen. It is difficult to correlate the rhynchonellids beyond the investigated area since two of the taxa are new. *Cyclothyris globata* is reported from the Coniacian-Santonian of the Caucasus (Aliev & Titova, 1988), and Campanian of France (Gaspard, 1983), Serbia, Macedonia, Croatia and Slovenia (Radulovic, personal commu-nication).

Septatoechia aff. *baugasii* assemblage

Frequent species: *Septatoechia* aff. *baugasii* (d'Orbigny) (Figure 3:5), and *S. baugasii* s.s. (d'Orbigny). Rare species: *Septatoechia eudesi* (Coquand), *S. inflata* Titova, *Cretirhyn-chia woodwardi* (Davidson), *C.* ex gr. *vespertilio* (d'Orbigny), *C. intermedia* Pettitt. Distribution: This assemblage is wide-spread, occurring in twelve localities in NE Bulgaria (Figure 2). In most sections the assemblage occurs throughout the Nikopol Fm. (detrital limestones), near Shumen, Manastir, Venchan, Nevsha, Nenovo, Kosovo, Chernook, Tsar Krum, and Troitsa. In outcrops near Dobrina, Arkovna and Komunari, the assemblage was also found in lower levels of the Mezdra Fm. (chalky limestones with chert nodules), but is not so diverse. In central N Bulgaria the assemblage was found near Novachene: at Novachene 1 (Nikopol Fm. and lowest Mezdra Fm.), Novachene 2 (Mezdra Fm.), Debovo (Nikopol Fm.), and Zhernov (Mezdra Fm.). Chronostratigraphic range, corre-lations: Upper Campanian, via ammonites, belemnites (Jolki-cev,1989). The most common species in this assemblage, *Septatoechia* aff. *baugasii*, is probably new. '*Rhynchonella*' *baugasii* is reported from the Coniacian, and '*Rhynchonella*' *vespertilio*, from the Coniacian-Santonian of France (Gaspard, 1983). *Cretirhynchia intermedia* and *C. woodwardi* occur res-pectively in the lower-middle Campanian and upper Campanian-lower Maastrichtian of England (Pettitt, 1954).

CONCLUSIONS

Five rhynchonellid assemblages have been outlined in the marine epicontinental Late Cretaceous of N Bulgaria. Each of the first three assemblages occurs only in one locality, while the last two are widely distributed. A coherent biostratigraphic scheme based on rhynchonellids is not yet possible. The age of the assemblages was determined by associated macrofossils and/or by correlation with occurrences of studied rhynchonellid taxa elsewhere. Rhynchonellid assemblages may be an alternative for dating strata, where other characteristic fossils are absent.

Acknowledgments

I sincerely thank P. Tchoumatchenco, N. Jolkicev and M. Verguilova (Sofia) for their continuous guidance, help, discussions and suggestions with research and manuscript preparation. I am especially indebted to I. Dieni (Padua) for his careful reading and comments on the manuscript. I highly appreciate the assistance of my husband V. Dekov (Sofia) in collecting and preparing material and in preparing figures. H. Stoynov (Vancouver) is thanked for improving the English text. Final preparation of the manuscript was carried out during tenure of a fellowship from the University of Padua. A travel grant was received from the Natural Sciences and Engineering Research Council of Canada to attend the Third International Brachiopod Congress.

REFERENCES

ALIEV, O.V. & M.V. TITOVA. 1988. Brachiopoda. Upper Cretaceous, p.220-240. In, A.A. Ali-zade (ed.), Cretaceous fauna of Azerbaidzhan. Elm, Baku.

ARNAUD, M.H. 1877. Mémoire sur le terrain Crétacé du Sud-Ouest de la France. Mémoires Société Géologique France, 2,10 (4):1-110.

DAVIDSON, T. 1852-1855. A monograph of the British fossil Brachiopoda, The Cretaceous Brachiopoda. Palaeontogra-phical Society Monographs, 2:1-117.

FAGE, G. 1934. Les rhynchonelles du Crétacé supérieur des Charentes. Bulletin Société Géologique France, 5(4):433-441.

GEINITZ, H.B. 1873. Das Elbthalgebirge in Sachsen, V, Brachiopoden und Pelecypoden. Palaeontographica, 20 (1):147-207.

GASPARD, D. 1983. Distribution des brachiopodes du Coniacien au Maastrichtien en France et pays limitrophes. Géologie Méditerranéenne, 10(3-4):229-238.

JOLKICEV, N.A. 1979-1980. Stratigraphie der Coniac-Maastrichtablagerungen in den Zentralteilen des Vorbalkans und der Mösischen Platte. Annuaire Université Sofia, 72 (1, Géologie):5-78.

___1989. Stratigraphy of the epicontinental Upper Cretaceous in Bulgaria. Kliment Ohridski University Press, Sofia, 184p.

KATZ, Yu. I. 1974. Type Brachiopoda, p.240-275. In, G.Ya. Krymgolts (ed.), Atlas of the Upper Cretaceous fauna of the Donbass. Nedra, Moscow.

MITROVIC-PETROVIC, J & V. RADULOVIC. 1994. Fossil fauna from Cenomanian tuffite beds of Grlja (Stara planina mountain, Eastern Serbia). Annales Géologiques Péninsule Balkanique, 58 (1):119-138.

MOTCHUROVA-DEKOVA, N. 1994. Brachiopods of order Rhynchonellida from the Northeuropean Upper Cretaceous in Bulgaria - taxonomy and stratigraphic significance. Unpublished Ph.D. thesis, University of Mining and Geology, Sofia, 227 p.

___(1995). Late Cretaceous Rhynchonellida (Brachiopoda) from Bulgaria, I, Genus *Cyclothyris* M'Coy. Geologica Balcanica, 25 (3-4):35-74.

NECHRIKOVA, N. I. 1982. Type Brachiopoda, p. 26-49. In, V.A. Sobetzkii (ed.), Atlas of the invertebrates from the Late Cretaceous seas of the Caspian Lowland, Nauka, Moscow.

NEKVASILOVA, O. 1973. The brachiopod genus *Bohemi-rhynchia* gen.n. and *Cyclothyris* M'Coy (Rhynchonellidae) from the Upper Cretaceous of Bohemia. Sbornik Geolo-gickych Ved, Paleontology, P15:75-117.

____1974. Genus *Cretirhynchia* and *Orbirhynchia* (Brachio-poda) from the Upper Cretaceous of North-West Bohemia. Sbornik Geologickych Ved, Paleontologie, 16:35-67.

ORBIGNY, A. d'. 1847. Paléontologie française. Terrains Crétacés, 4, Brachiopodes. Bertrand, Paris, 390 p.

OWEN, E.F. 1962. The brachiopod genus *Cyclothyris*. Bulletin British Museum Natural History, Geology 7(2):37-63.

____1988. Cenomanian brachiopods from the Lower Chalk of Britain and northern Europe. ibid. 44(2):65-175.

PETTITT, N.E. 1954. A monograph on the Rhynchonellidae of the British Chalk, II, Palaeontographical Society Monographs, 107:27-52.

POPIEL-BARCZYK, E. 1977. A further study of Albian-Cenomanian brachiopods from the environs of Annopol on the Vistula, with some remarks on related species from the Cracow Region, Poland. Prace Muzeum Ziemi, 26:25-54.

SINNYOVSKY, D. 1988. Upper Cretaceous nannoplancton zonation in North Bulgaria. Geologica Balcanica, 18(6):59-78.

SURLYK, F. 1984. The Maastrichtian Stage in NW Europe and its brachiopod zonation. Bulletin Geological Society Denmark, 33(1-2):217-223.

TZANKOV, V. 1982. Les fossiles de Bulgarie, Crétacé Supé-rieur, Cephalopoda (Nautiloidea, Ammonoidea) et Echino-dermata (Echinoidea), 5A:1-136.

RELATIONSHIPS BETWEEN SILICICLASTICS/CARBONATES AND SILURIAN BRACHIOPOD COMMUNITY DISTRIBUTION IN LITHUANIA

PETRAS MUSTEIKIS & D. KAMINSKAS

Department of Geology and Mineralogy, Vilnius University, Vilnius 2009, Lithuania

ABSTRACT- Silurian brachiopod communities from three cored boreholes in the Baltic Basin were discriminated by quantitative taxon analysis, and their distribution was compared to carbonate and siliciclastic rock content. From late Llandovery to early Ludlow time, terrigenous sediments dominated in this part of the Baltic Basin, and brachiopod communities tolerant to siliciclastic, dysaerobic settings developed. With reduced siliciclastic input and basin shallowing in the early Ludlow, these communities were replaced by *Atrypa* and *Spirigerina* communities adapted to aerobic conditions and reduced siliciclastic supply. In the late Ludlow and Pridoli, carbonate sedimentation prevailed and only four moderate to low diversity brachiopod communities existed. Results of this study are similar to the those obtained from evaluating the relationship between these communities and whole rock chemical analyses (for Si, Al, Ti, Fe, Ca, Mg-oxides). Whole rock standard analyses may thus be used for general quantitative evaluation of the association between sedimentation and brachiopod asemblages.

INTRODUCTION

Silurian strata in the Lithuanian part of the Baltic Basin attained a thickness of 40-800m, occurring at borehole depths from100-1200m. Silurian strata of the Baltic Basin outcrop in northern Estonia 400 km to the northwest of Lithuania, on the island of Saaremaa 600km to the north, and on the island of Gotland to the east. During the last decade, there have been several studies on the Silurian brachiopod communities of Lithuania. Brachiopod communities were discriminated in 16 boreholes more or less evenly distributed in the Baltic Sea in the west, across about 300km of the territory of Lithuania, and to the state of Byelarus in the east. The relationship between brachiopod community distribution and sedimentary environment has previously been established qualitatively.

Quantitative analyses of the brachiopod communities were carried out for 3 of 16 boreholes from central Lithuania, for which brachiopod communities have already been distin-

Figure 1. Map of the Baltic region and Lithuania showing the location of the boreholes analysed.

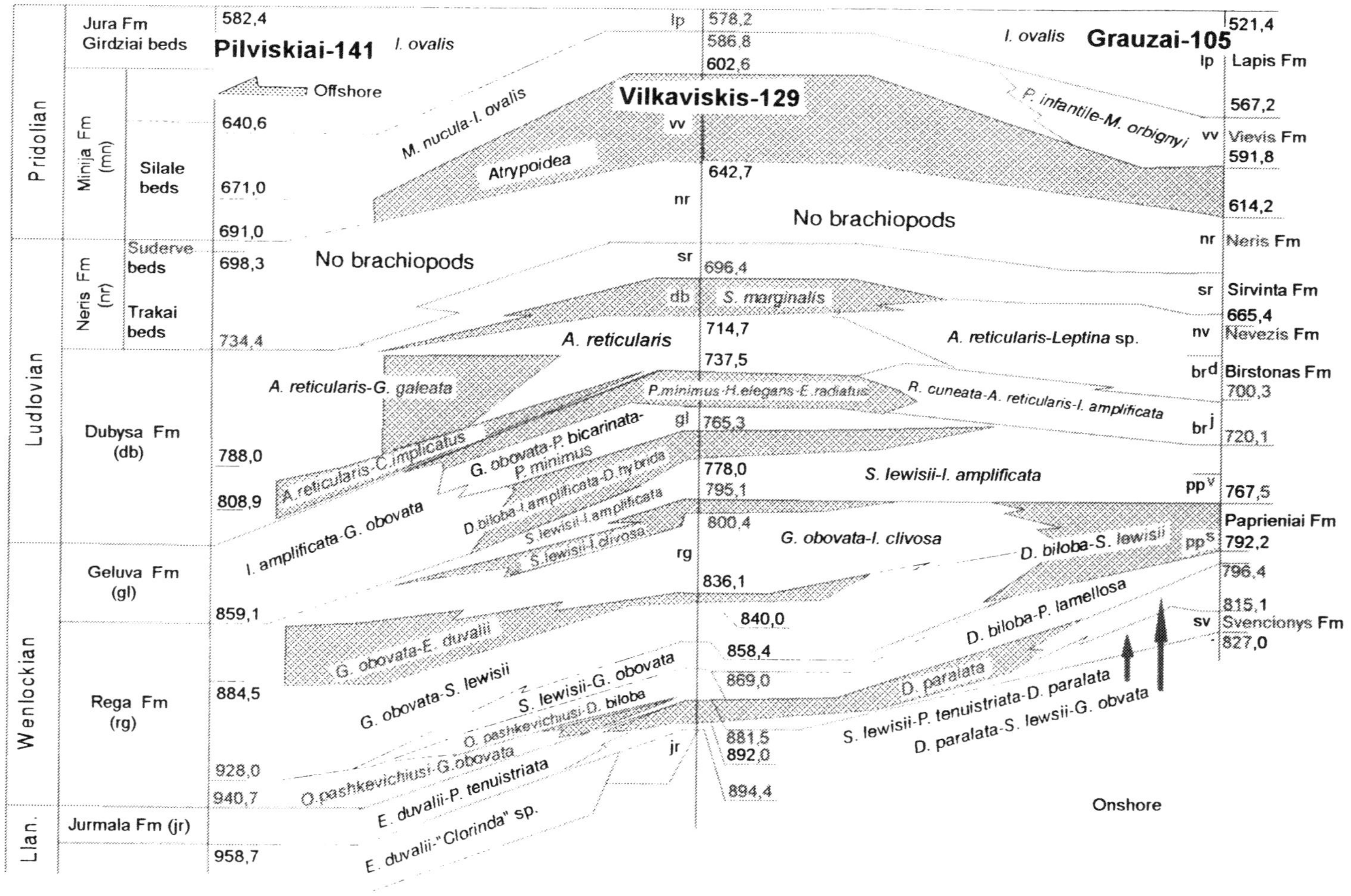

Figure 2. Distribution of brachiopod communities across the three boreholes. Key: jr, Jurmala Fm.; sv, Svencionys Fm.; rg, Riga Fm.; gl, Geluva Fm.; pp^s, Sutkai Beds, Paprieniai Fm.; pp^v, Vilkija Beds, Paprieniai Fm.; br^j, Jonava Beds, Birstonas Fm.; br^d, Dotnuva Beds, Birstonas Fm.; db, Dubysa Fm.; sr, Sirvinta Fm.; nr, Neris Fm.; nr^t, Trakai Beds, Neris Fm.; nr^s, Suderve Beds, Neris Fm.; mn^s Silale Beds, Minija Fm.; jr^d, Girdziai Beds, Jura Fm.; vv, Vievis Fm.; lp, Lapes Fm.

192

guished (Figures 1-2). In a recent study, sedimentary environment was evaluated by whole rock analyses (Si, Al, Ti, Fe, Ca, and Mg oxides) carried out at the Geological Survey of Lithuania (Musteikis & Kaminskas, in press). A total of 154 spot samples for this analysis were collected by the Geological Survey separately from brachiopod community studies. In the present companion study, sedimentary environment was evaluated by calcite, dolomite, and siliciclastic content of the host cored rock for the brachiopod communities sampled and studied previously. The relationship of brachiopod community distribution and sedimentary environments, as deduced from whole rock analyses, is here compared with the result of calcite, dolomite, and siliciclastic content analyses of the bore holes.

MATERIAL AND METHODS

A total of 384 core samples were taken from the Silurian sections of Grauzai-105, Vilkaviskis-129, and Pilviskiai-141 boreholes from central Lithuania (113, 129, and 142 samples respectively). There were 65 taxa (4,895 specimens) identified from Grauzai, 76 taxa (11,567 specimens) from Vilkaviskis, and 68 taxa (22,182 specimens) form Pilviskiai boreholes. The numbers of individuals of each taxon were counted in all samples and brachiopod communities were discriminated using a 'Radion 6' program (Ragelskiene, Computer Centre, Vilnius University). Discussion on the method of community discrimination and description of brachiopod communities can be found in Musteikis (1989,1991, 1993) and Musteikis & Kaminskas (in press).

Average values of carbonates (calcite and dolomite) and siliciclastics were estimated and plotted against stratigraphic ranges of brachiopod communities (Figure 3). Data were summarized into four time intervals: late Llandovery, early Wenlock, late Wenlock-early Ludlow, and late Ludlow-Pridoli, which correspond partly to six Silurian tectonosedimentary macrocycles of the Baltic Basin identified previously by Lapinskas (1987). To clarify position of communities within the basin and within the Benthic Assemblage suite of Boucot (1975), communities were plotted on diagrams for siliciclastic and carbonate content. If a community was recognized in more than one section, an average value was taken. Oxide content was evaluated according to the colour of the host rock. Greenish to greenish-grey, grey, and dark-grey to black colours were interpreted as aerobic, dysaerobic, and anaerobic conditions respectively. Water turbulence was evaluated using brachiopod articulation data, although disarticulation could also have been the result of bioturbation.

BRACHIOPOD COMMUNITY-SEDIMENT RELATIONSHIPS

Siliciclastics. Siliciclastic content showed a general decrease until the early Pridoli and increase to the end of Pridoli (Appendix: Figure 7). It ranged from 93% to 3% with changes from terrigenous to carbonate sedimentation at the Wenlock-Ludlow boundary and from carbonate to terrigenous sedimentation at the end of the Pridoli. Several episodes of terrigenous influx were detected: in the late Llandovery, at the Llandovery-Wenlock boundary, in the early Wenlock, in the late Wenlock, and at the end of the Pridoli.

Five brachiopod communities were recognized in the late Llandovery (Figures 2,3). Their positions estimated from siliciclastic and oxide content are the same as those estimated from SiO2 content (Musteikis & Kaminskas, in press). The Eoplectodonta duvalii - 'Clorinda' . community lived in an anaerobic, high siliciclastic environment; the Skenidioides lewisii-

Pentlandella tenuistriata - Dicoelosia paralata community in a quiet, dysaerobic environment below wavebase; the Dicoelosia paralata - Skenidioides lewisii - 'Glassia obovata' community in a dysaerobic, moderate energy environment above wave base; the Dicoelosia paralata community in an aerobic environment, and the Onniella pashkevichiusi - Dicoelosia biloba community also in an aerobic, but low siliciclastic environment.

In the early Wenlock, 14 brachiopod communities were distinguished from rocks with very similar siliciclastic content (60%-80%: Figure 4). The 'Glassia obovata' - Eoplectodonta duvalli and E. duvalli - 'G. obovata' communities lived on anaerobic, high siliciclastic substrates. Their positions estimated from siliciclastic content surprisingly contradict those estimated from SiO2 content (Musteikis & Kaminskas, in press). In the dysaerobic environment, a succession of 'G. obovata' dominated communities occurred in the following order of increasingly higher siliciclastic content: Skenidioides lewisii - 'Glassia obovata', 'Glassia obovata' - Skenidioides lewisii, 'Glassia obovata' - Isorthis clivosa, Onniella pashkevichiusi - 'Glassia obovata', Isorthis amplificata -'Glassia obovata' . In the dysaerobic environment above wave base (moderate turbulence) brachiopod communities occurred in the following order, ranked by decreasing siliciclastic content: Dicoelosia paralata - Skenidioides lewisii - 'Glassia obovata', Skenidioides lewisii - Isorthis clivosa, and Skenidioides lewisii - Isorthis amplificata. In aerobic environments, brachiopod communities also occurred within a very narrow siliciclastic range so that terrigenous supply was not the main regulating factor for distribution. In the moderately turbulent environment, they occurred in the following order by decreasing siliciclastic content: Onniella pashkevichiusi - Dicoelosia biloba, Dicoelosia biloba - Ptychopleurella lamellosa, and Dicoelosia biloba - Skenidioides lewisii. The Rhynchotreta cuneata - Atrypa reticularis (s.l.)- Isorthis amplificata community, perhaps existed in rough to moderately turbulent environments, according to articulation data.

Eleven communities were discriminated in the late Wenlock-early Ludlow (Figure 5). The general decrease in siliciclastics,

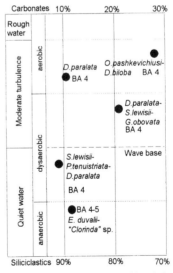

Figure 3. Late Llandovery brachiopod communities relative to energy-oxidation facies in Lithuania.

193

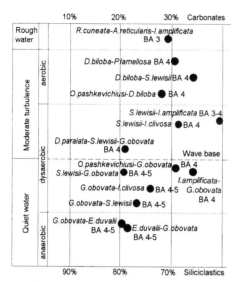

Figure 4. Early Wenlock brachiopod communities in Lithuania in relation to energy-oxidation facies.

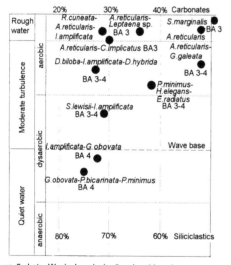

Figure 5. Late Wenlock-early Ludlow brachiopod communities in relation to energy-oxidation facies in Lithuania.

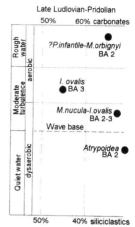

Figure 6. Late Ludlow-Pridoli brachiopod communities in Lithuania in relation to energy-oxidation facies.

Dalejina hybrida, Atrypa reticularis (s.l.) - Craniops implicatus, Protochonetes minimus - Howellela elegans - Eospirifer radiatus, and Atrypa reticularis (s.l.)- Gypidula galeata. The latter and the Atrypa 'reticularis' - Craniops implicatus community showed inverse positions, as estimated from SiO2 whole rock analysis (Musteikis & Kaminskas, in press). In aerobic rough water environments, communities occurred in the following order with decreasing siliciclastic content: Rhynchotreta cuneata - Atrypa reticularis (s.l.)- Isorthis amplificata, Atrypa reticularis (s.l.), and Spirigerina marginalis.

Only four moderate to low diversity communities have been identified from the upper Ludlow-Pridoli carbonate-dominated sequence (Figure 6). With decreasing siliciclastic content, they occurred in the following order: Isorthis ovalis (aerobic moderate turbulence), ?Pseudoprotathyris infantile - Morinorhynchus orbigny (aerobic rough water), Microsphaeridiorhynchus nucula - Isorthis ovalis (dysaerobic moderate turbulence), and Atrypoidea (dysaerobic, quiet water). According to these data, the Isorthis ovalis community appears more tolerant of siliciclastics than the Microsphaeridiorhynchus nucula - Isorthis ovalis community, as considered earlier (Musteikis, 1989,1991,1993). It contradicts their relative position estimated from SiO2 whole rock analyses (Musteikis & Kaminskas, in press).

Carbonates. In the Baltic Basin, carbonates show an inverse relationship with siliciclastics through time, i.e. overall increase from the late Llandovery to early Pridoli, and decrease to the end of the Pridoli (Figure 7). The change-over from terrigenous to carbonate sedimentation took place in early Ludlow time, and carbonate sedimentation dominated during the late Ludlow and Pridoli. The conversion of carbonate to terrigenous sediments occurred at the end of Pridoli. The position of brachiopod communities estimated from increase in carbonate content, predictably coincides with that estimated from decrease in siliciclastic content (Figures 3-6).

As expected, calcite change through time corresponds well with carbonate content. Dolomite content increased in the Ludlow lagoonal facies (lacking brachiopods), and in the Pridoli Isorthis ovalis community. Its increase just below lagoonal facies, and below Devonian dolomitic strata, is interpreted as secondary dolomitization, not influencing brachiopod

and increase in oxide content during this period was the reason for the continued existence of only three communities: Isorthis amplificata - 'Glassia obovata', Skenidioides lewisii - Isorthis amplificata, Rhynchotreta cuneata - Atrypa reticularis (s.l.) - Isorthis amplificata. In dysaerobic environments existed the 'Glassia obovata' - Protozeuga bicarinata - Protochonetes minimus, Isorthis amplificata - 'Glassia obovata' (both in quiet water below wave base), and Skenidioides lewisii -Isorthis amplificata (moderate turbulence, above wave base) communities. With decreasing siliciclastic content, in aerobic, moderately turbulent environments, communities occurred in following order: Dicoelosia biloba - Isorthis amplificata -

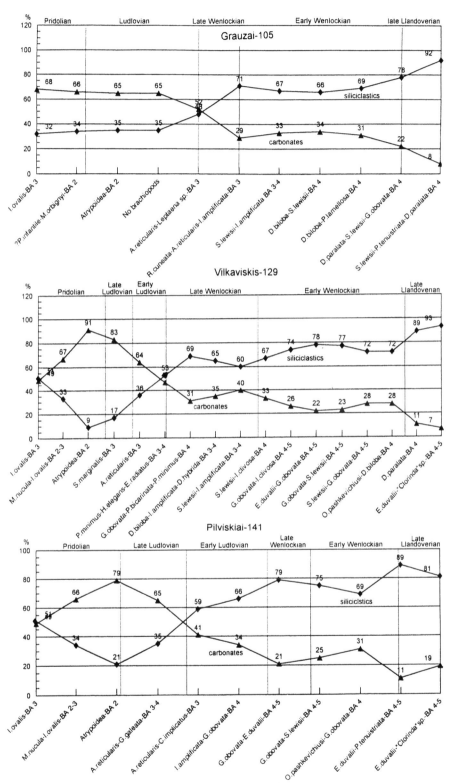

Figure 7. Average values of carbonates versus siliciclastics, calculated against brachiopod communities for three boreholes in Lithuania.

195

distribution. Secondary dolomitization has been confirmed by strontium and barium ratios, which show no change, although dolomite quantities increase. Thus the sum of calcite and dolomite content average values for brachiopod-bearing intervals has been used for evaluation of the relationship between sedimentary environment and community distribution (Figures 3-6).

CONCLUSIONS

This study shows a correlation between siliciclastic and carbonate content from bore hole data and brachiopod community distribution. From late Llandovery to early Ludlow time, terrigenous sedimentation dominated in the study area of the Baltic Basin, where brachiopod communities tolerant to high siliciclastic content developed. The *Skenidioides lewisii* - dominated communities lived in dysaerobic environments with an abundant siliciclastic supply, like communities dominated by *'Glassia obovata'* which lived in dysaerobic to anaerobic environments with a similar rate of siliciclastic input. This strongly confirms the relative position of the *Dicoelosia - Skenidioides* communities within Benthic Assemblage 4.

With inversion from terrigenous to carbonate sedimentation, and overall basin shallowing-upward in the early Ludlow, communities tolerant to high siliciclastic content and dysaerobic environment were replaced by *Atrypa reticularis* (s.l.) dominated communities, and the *Spirigerina 'marginalis'* community existed in an aerobic environment with lower siliciclastic influx. In the late Ludlow and Pridoli, carbonate sedimentation dominated, but this reverted to siliciclastic sedimentation at the end of the Pridoli. Only four moderate and low diversity communities existed in such environments. With regard to sediment-brachiopod community relationships, the results of this study appear to be very similar to, but more precise than, those derived from standard whole rock analyses (Musteikis & Kaminskas, in press). Whole rock analyses from borehole samples may be used for general quantitative evaluation of sediment-brachiopods relationships.

REFERENCES

BOUCOT, A.J. 1975. Evolution and extinction rate controls. Elsevier, Amsterdam, 427p.

LAPINSKAS, P. 1987. Formatsii silura Baltiiskoy sineklizy, p.103-116. In, R.G. Goretskii & P. Suveizdis (eds.), Tektonika, fatsii i formatsii zapada Vostochno-Evropeiskoi platformy. Nauka i tekhnika, Minsk.

MUSTEIKIS, P. 1989. Opyt vydeleniya brakhiopodovykh soobchschestva silura Litvy. Lietuvos aukstuju mokyklu mokslo darbai. Geologija 10:52-71.

___1991. Siluriiskie soobchschestva brakhiopod v razreze skv. Vilkaviskis-129. Lietuvos aukstuju mokyklu mokslo darbai. ibid. 12:47-66.

___1993. Silurian brachiopod communities in the section of the Pilviskiai-141 borehole. Lietuvos aukstuju mokyklu mokslo darbai. ibid.14 (1):118 -130.

MUSTEIKIS, P. & D. KAMINSKAS (in press). Geochemical parameters of the sedimentation and the distribution of Silurian brachiopod communities in Lithuania. Historical Biology.

SCHERBINA, V.N. 1958. O metodike massovogo opredeleniya karbonatonosnykh osadochnykh porod. Institut Geologii AN BSSR, Minsk 1:131-144.

ORDOVICIAN BRACHIOPOD BIOGEOGRAPHY IN THE IAPETUS SUTURE ZONE OF IRELAND: PROVINCIAL DYNAMICS IN A CHANGING OCEAN

M.A. PARKES & D.A.T. HARPER

Geological Survey of Ireland, Haddington Road, Dublin 4;
Department of Geology, University College Galway, Galway, Ireland

ABSTRACT-The Ordovician brachiopod faunas from terranes in the Iapetus suture zone of Ireland have been revised, permitting a better characterisation of their generic composition, and emphasising their disparate positions during the Early Ordovician. Mid-Ordovician faunas signal the development of the Anglo-Welsh faunal province across the Leinster Terrane, and highlight the marginal situations of faunas from Bellew-stown and Grangegeeth. The Ashgill *Foliomena* fauna from Ballyvorgal in the Central Terrane enhances the similarity with coeval faunas from Oriel Brook in the Grangegeeth Terrane. *Hirnantia* brachiopod faunas existing across the entire Iapetus suture zone are represented by assemblages at Kildare in the Leinster Terrane. Improved stratigraphic correlation, and refined discrimination of biogeographical affinity and flux within the Iapetus Ocean system is possible. Rapid cross-latitudinal migration of these terranes during the Ordovician is indicated by their changing provincial faunal signals.

INTRODUCTION

Harper & Parkes (1989) undertook a preliminary review of Ordovician brachiopod provinciality across Ireland, and also contributed to a synthesis of Irish Caledonide terranes (in Murphy et al.,1991), using the knowledge of brachiopod faunas at that time to discriminate the separate elements of a collage of crustal terranes, as they developed through the Early Paleozoic. In Ordovician sequences worldwide, brachiopods remain one of the most useful fossil groups in defining areas of disparate geological history, or discrete terranes.

This paper reports on recent and ongoing detailed work on the brachiopod faunas of terranes of the area of eastern and central Ireland, broadly the Iapetus suture zone trending NE to SW across Ireland. The area has been the subject of considerable contention in the last decade, and only proper reassessment of the faunas is likely to allow definitive judgments to be made. Some recent work on trilobites (Romano & Owen, 1993) has been important in definition of paleogeographical affinity of one segment, yet other work, on graptolites, has only added further opinion (Lenz & Vaughan, 1994).

The discussions of brachiopod faunas herein are based on recent work, on new collections in part, and also on re-examination of previously unavailable material in the 19th century collections of the Geological Survey of Ireland. The background to these collections is described by Sleeman (1992), and their modern curation is reported in Parkes et al. (1994). The paleoenvironmental significance of certain associations is now much better understood through work in areas with more diverse and better preserved assemblages in continuous successions. This improved knowledge can be reflected back to poor faunas in Ireland in isolated sections. An example is the *Foliomena* fauna from Ballyvorgal described originally by Harper (1980), which is discussed herein based on newly available material. Throughout this paper the Ordovician stratigraphical terminology used is that of Fortey et al. (1995), wherein redefinition of stages in Britain includes the division of the Llandeilo into a new Llandeilian (late Llanvirn) stage and an early Caradoc new Aurelucian stage including the old Costonian as a substage.

THE LEINSTER TERRANE

In the southern part of this terrane the very important fauna of the early Caradoc or older Tramore Limestone Formation unfortunately remains undescribed, but may prove critical in tracking interprovincial migrations during the climax of the *gracilis* transgression; it apparently contains a mixture of low and high latitude faunas (Carlisle, 1979). The Burrellian brachiopod faunas of the main Wexford development of the Late Ordovician, volcanic-dominated Duncannon Group have been recently described by Parkes (1992, 1994) documenting their Anglo-Welsh affinities, and broad Avalonian position in the Iapetus ocean. Recent work has been concentrated in the northern part of the terrane, perhaps the least well exposed, and the area now closest to the Iapetus suture.

The outermost arc volcanics in Ireland, from southeast subduction beneath a northward moving Eastern Avalonia occur at Kildare and Lambay Island/Portrane and at Balbriggan. Kildare has been known as a famous fossil locality, because of the Rawtheyan Kildare Limestone Formation. The brachiopods of this carbonate mound have still to be fully described, but the Hirnantia Mudstone Member contains a *Hirnantia* fauna identified by Wright (1968). This is the only example of this important Late Ordovician fauna in Ireland, but it seems likely that it was present across the Iapetus suture area (Rong & Harper, 1988). A very restricted variant of it is found in the deep water facies of the Tirnaskea Formation of Pomeroy in the Northwestern Terrane of Ireland (Mitchell, 1977). Two new collections from the formation have been recently documented by Harper et al. (1995). One is a Rawtheyan deep water assemblage with *Dedzetina*, *Sericoidea*, *Proboscisambon* and *Protozyga* ? The other confirms a Hirnantian age for the upper part of the formation, including the brachiopods *Dysprosorthis* and *Eostropheodonta*.

The andesitic volcanics at Grange Hill in the Kildare Inlier are tightly constrained by Soudleyan and Longvillian faunas below and above respectively (Parkes & Palmer, 1994). Whilst the geochemical affinity of the volcanics at Portrane and Lambay Island with those at Kildare has long been recognised, they have been only loosely dated by the fossils from the associated limestones as Late Ordovician in age. In the Ordovician correlation chart for Britain and Ireland, Williams noted a late Cautleyan age for the Portrane Limestone Formation, but the older rocks were only provisionally dated as no older than Streffordian (Williams et al., 1972).

THE PORTRANE LIMESTONE

Recent work here has correlated several horizons using a variety of fossil groups (Parkes et al., in prep.). The most reliable date is from a horizon of black slates with Burrellian graptolites grading up into a nodular limestone horizon with a brachiopod fauna. This limestone horizon and some overlying shell lags, or coquinas in the tuffaceous sediments contain *Sowerbyella sericea* and *Kjaerina*, both indicating a mid-Caradoc age. The package of graptolites, brachiopods and some other elements (bryozoans, orthocones) reasonably constrain the age to the Burrellian, and suggest the andesites at Portrane are probably contemporaneous with the volcanics at Kildare.

The succession below this horizon is reasonably continuous from the top of the andesites. Above this level, the structure is considerably more complex, although fossil evidence indicates possibly continuous horizons correlated with the Caradoc Streffordian Stage through Pusgillian to lower Cautleyan, below the high Cautleyan Portrane Limestone. However, brachiopods are sparse in these levels and not diagnostic. Those of the Portrane Limestone Formation mainly described by Wright (1963,1964) are comparable to many North European faunas, including those from the Baltic and Scotland, and reflect the substantial loss of provinciality in Ordovician brachiopod faunas, by the Ashgill through physical convergence of the terranes by ocean closure and broader climatic influences under similar latitudinal zones. Evidence of Hirnantian glaciation in the form of paleokarst and unconformable Silurian strata (Parkes, 1993) indicates emergence; no younger brachiopod faunas are known here other than those of the Cautleyan Portrane Limestone.

LAMBAY ISLAND

The Ordovician volcanic complex of Lambay Island has had little attention from paleontologists. Although richly fossiliferous horizons have long been known, their inaccessibility means little is known about them, and their age has been only crudely bracketed as Late Ordovician, usually suggested to be Ashgill. Much collecting has been done recently. The principal locality of Kiln Point, has bedded carbonates overlying andesites with a rich coral fauna, but few brachiopods. Faulted above these is a polymict conglomerate, interpreted as a possible forereef breccia. Rare siltstone clasts within this have yielded abundant *Sowerbyella sericea*, with few other genera. Much of the material in the GSI collections came from the cliffs north of Kiln Point (virtually inaccessible), but it is not clear whether the specimens were from the conglomerate or not.

A second new locality at Seal Hole has provided a nearly monospecific assemblage of *Rhynchotrema*. This is adjacent to a graptolite locality on Heath Hill, with a Burrellian, probably Soudleyan assemblage. Raven's Well, an old GSI locality in the immediate vicinity, has yielded a few *Platystrophia* specimens. Also nearby is a conglomerate which has provided a large limestone block, densely packed with the strophomenoid *?Macrocoelia*, and a few other genera, including *Sowerbyella*, *Plaesiomys* and indeterminate orthoids.

Although the age and field relationships of the various fossiliferous units are far from clear, as might be expected in the products of a large andesitic volcano, it is most likely that the eruption occurred during the mid-Caradoc. Portrane, whose rocks are the distal version of Lambay Island, supports this. The suggested Ashgill age from museum collections of Kiln Point (Lambay Island) limestone fossils and a few recent conodont samples, is a reflection of continued carbonate environments around an effectively inactive emergent volcanic island. The brachiopods probably represent opportunistic species, occupying unstable niches around a volcanic island during the mid-Caradoc.

BALBRIGGAN

The description in the Geological Survey Memoir to the Balbriggan area (Hull et al.,1871) of the shore west of Cardy Rocks, on Bremore headland 'where at two or three places the gray or brown slates are very fossiliferous' is somewhat extravagant. The preserved material in the GSI collections amounts to two or three specimens of each species listed in the memoir. There were five brachiopod species, but trilobites, an ostracode, and bryozoan are also indicated. Together they indicated a 'Bala' or Late Ordovician age. This material was re-examined by Mason (1961), whose determinations and records of some additional species were also recorded by France (1967) in a general paper on the volcanic rocks of Balbriggan. The fauna was suggested to be Longvillian, and the whole of the section to be Caradoc in age. Graptolitic mudrocks of the *clingani* biozone, near the top of the formation constrain the age of volcanism here. Murphy (1987) formalised the stratigraphy and noted that new material from the base of the Belcamp Formation was probably Harnagian to Soudleyan in age, i.e. older than suggested by Mason. The material (collected by Ian Mitchell) was unfortunately not described, but we have now been able to examine this. The list of fauna is based on examination of all the available material (Table 1).

Table 1. List of brachiopods from the Balbriggan succession at Bremore headland.

Bremore (opposite Cardy Rocks, N of Balbriggan)
Sowerbyella sericea
Chonetoidea cf. *abdita*
Nicolella sp.
Kjaerina sp.
Plaesiomys sp.
Hedstroemina sp.

Mason (1961) reported *Kjaerina* cf. *geniculata* and *Howellites* cf. *striata* from approximately the same locality. This material has not been traced, but is consistent with the overall indication of a Longvillian age for the sediments here. The trilobites are currently being re-assessed in detail by Dr. A. Owen, but are also consistent with a mid-Caradoc age. Also in the GSI collections are numerous very small lingulellid and paterulid? brachiopods. However, locality data is poor and lithology suggests they are probably from south of Balbriggan in graptolitic facies of either the *anceps* Biozone or else of Llandovery age.

HERBERTSTOWN

The Ordovician succession of Herbertstown was first described by Romano (1980a) and correlated with the rocks at Balbriggan. Later Harper et al. (1985) described a fauna of a mid-late Caradoc age, below the volcanics. Parkes (1994) identified a very similar fauna below the andesites at Kildare. The fauna is therefore an Anglo Welsh one from the northernmost margins of the Leinster Terrane. Murphy (1987), however, reversed the Herbertstown stratigraphy, with the ? Soudleyan fauna therefore largely overlying the volcanics, unlike at Kildare. Any new fossil localities may help determine which interpretation is correct, but exposure is poor.

BELLEWSTOWN TERRANE
HILLTOWN FORMATION (Llanvirn)

The significance of a new collection of *Paralenorthis*, *Jaanussonites* and '*Ahtiella*' cf.*paucirugosa* was reported by Harper et al. (1990, 1992). It suggested a correlation with parts of the Dunnage terrane of Newfoundland during the Early Ordovician. In other words, it emphasised the oceanic or intermediate affinity of the fauna, loosely associated with the high latitude Avalonian Celtic Province (Neuman & Harper, 1992), but not the low latitude Laurentian Toquima - Table Head Province. Earlier records (Harper & Rast, 1964) of brachiopods from Cooksgrove in the same formation were noted, but no material could be traced, and no new material could be collected. However, curation of the GSI collections has turned up some material which was obviously examined by J.C. Harper, but as previously noted, they are only shell exteriors not materially affecting earlier conclusions. In addition some indeterminate lingulids and other smaller inarticulates are present in black and grey slates of the Hilltown Formation.

CARNES FORMATION (Caradoc)

The base of this formation has been defined locally as a calcareous member, the Bellewstown Limestone. Contrasting ideas have still to be settled, but it would appear to represent a condensed deposit ranging from late Llanvirn to perhaps early Caradoc, mainly based on conodonts isolated from blocks. It may prove to be more closely associated with the Hilltown Formation, rather than an early deposit in the area of the Carnes Formation volcanics. A Mediterranean aspect was noted by Harper & Parkes (1989), mainly based on the occurrence of *Aegiromena*. Further work is required to fully assess the brachiopod fauna of this horizon, but the only forms listed by Harper & Rast (1964) were *Aegiromena* cf. *aquila*, *Kjaerina* , *Orthis*, and a geniculate strophomenoid.

The age determination for the rest of the formation of Actonian by Harper & Rast (1964) was revised by Brenchley et al. (1977) to uppermost Longvillian (Table 2). An Anglo-Welsh affinity invites comparisons to faunas of the Leinster Terrane (Parkes, 1992,1994). It is evident that by the mid-Caradoc the Bellewstown Terrane close to the Leinster Terrane in contrast to its earlier Ordovician history, allowing immigration for many Anglo-Welsh taxa. Murphy (1987) has suggested this was achieved by gradual amalgamation during late Ordovician transtensional movements with the Lowther Lodge fault separating the Bellewstown Terrane from the Balbriggan sequence of the Leinster Terrane.

Table 2. Composite list of brachiopods from the Carnes Formation (Caradoc) of the Bellewstown Terrane.

Bicuspina spiriferoides	Paterula sp.
Chonetoidea ?	Resserella sp.
Glyptorthis sp.	Salopia salteri
Gunnarella cf. rigida	Sowerbyella sericea
Kjaerina?	Strophomena cancellata
Leptestiina sp.	clitambonitids
Nicolella aff.actoniae	dalmanellids
Obolus cf. audax	

GRANGEGEETH TERRANE

Harper & Parkes (1989) noted the Scoto-Appalachian affinity of the early Caradoc brachiopod fauna from here, earlier described by Harper (1952) and Romano (1980b). Owen et al. (1992) elaborated on the biogeographical affinity of the faunas of the Grangegeeth terrane with emphasis on the trilobites, later described by Romano & Owen (1993). In Owen et al. (1992), a complex migrational history of brachiopod genera was discussed, and it was concluded that Grangeeth showed a strong Scoto-Appalachian signal. Brachiopods are comparable to the faunas of Pomeroy, and the Blount Belt of the Laurentian margin (Jaaunusson & Bergström,1980; Potter & Boucot, 1992).

The GSI collection has confirmed the occurrence of many of the brachiopod genera from the main locality at Grangegeeth (GSI J.C. Harper Loc. 1), itself now unavailable. The brachiopod fauna is dominated by *Hesperorthis inostrantzefi*, *Glyptomena* and *Leptestia*, but high diversity of brachiopods alone is evident, although there are other elements. Full taxonomic description of these brachiopods by D. Harper will be completed soon (Table 3). A Cautleyan fauna of *Glyptorthis*, *Skenidioides*, *Dicoelosia*, *Leptestiina*, *Chonetoidea* and *Christiania portlocki* was described from the Oriel Brook Formation, the youngest part of the Grangegeeth succession, by Harper & Mitchell (1982).

Table 3. Brachiopod genera from the early Caradoc of the Grangegeeth Terrane.

Orthisocrania	Anoptambonites
Oxoplecia	Anazyga
Palaeostrophomena	Cyrtonella
Plaesiomys	Dactylogonia
Platystrophia	Glyptomena
Productorthis	Hesperorthis
Salopia	Kullervo
Sowerbyella	Leptestia
Triplesia	Lingulella
Valcourea	Macrocoelia

CENTRAL TERRANE

The *Foliomena* fauna from Ballyvorgal in the Central Terrane, described by Harper (1980), consisted of only single specimens of *Dedzetina*, *Sericoidea*, *Foliomena*, *Christiania* and *Cyclospira*, together with an indeterminate leptellinid. Newly available material in the GSI collections has expanded this (Table 4). The new material also invites comparisons with other Ashgill faunas.There are some similarities with the brachiopods from the Oriel Brook Formation of the Grangegeeth Terrane (Harper & Mitchell, 1982), especially in terms of the small size of specimens reflecting deeper water environments, although the Oriel Brook fauna was possibly from shallower depths. Based on study of numerous *Foliomena* faunas worldwide, but especially in China, Rong et al. (1995) have established a *Kassinella-Christiania* Association in China, which is the shallower water equivalent of the *Foliomena* fauna in its normally deep water setting. The Ballyvorgal fauna is clearly an undifferentiated *Foliomena* fauna with all the key taxa, comparable to the many now known around the world (Cocks & Rong, 1988).

Table 4. Counts of *Foliomena* fauna brachiopods from Ballyvorgal in GSI collections.

Sericoidea	4	Cyclospira	3
Foliomena	1	Christiania	2
Dedzetina	2		

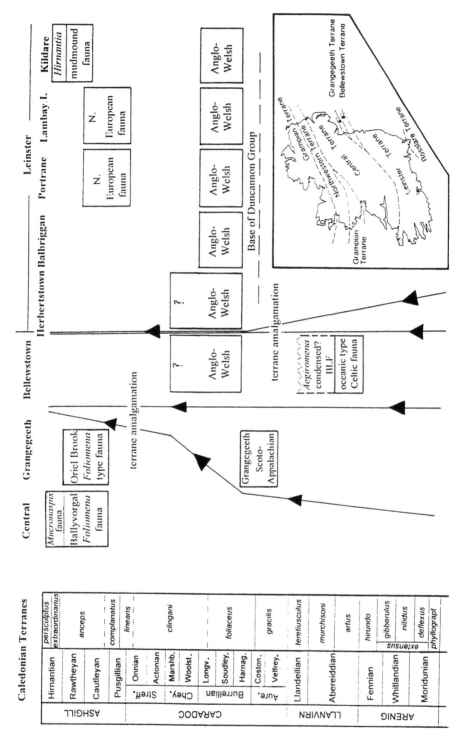

Figure 1. Key types of brachiopod faunas in the Iapetus Suture Zone of Ireland, showing provincial affinities and approximate times of amalgamation. BLF = Bellewstown Limestone Formation.

200

BIOGEOGRAPHY AND PROVINCIAL DYNAMICS

This review of new and existing data on brachiopod distributions in the Iapetus suture zone inevitably raises questions about the validity of this kind of analysis. A brief look at how interpretations of the various brachiopod faunas have changed through the years, and even today, are subject of contentious debate, and emphasises the need for new data and the continued revision of existing information. For example the Grangegeeth fauna was considered by Harper (1952) to have Baltic affinity, an opinion reinforced, according to Williams (1956), by the discovery of *Productorthis* in the assemblage. By 1989 enough data existed to allow us to discount the 'Baltic' significance, and compare it to Scoto-Appalachian assemblages. Parkes & Vaughan (1991) and Vaughan & Johnston (1992) clearly positioned Grangegeeth to the north of a single suture in the Caradoc. Most authors have agreed with this, especially when integrating information from other groups e.g. Owen et al. (1992). However, recently, Lenz & Vaughan (1994) have placed the Grangegeeth terrane on the Avalonian side of Iapetus, based on graptolite data from a recent borehole near Navan.

We prefer the reliability of benthic brachiopods and trilobites to the less well defined provinciality of the planktic graptolites, and maintain a Laurentian marginal position, to the north of the Iapetus suture during the mid Ordovician. However, the Arenig-Llanvirn position may well have been in higher latitudes, as a mid-oceanic island, since there are Llanvirn Atlantic province graptolites in the Slane Group. The very rapid drift rates of different continental blocks, derived mainly from paleomagnetic data is becoming apparent. Harper (1992) and Harper et al. (in press) have argued that changing provincial signals from intra-lapetan terranes result from rapid cross-latitudinal movements.

The distribution of different brachiopod faunas and their provinciality shows the approximate timing of amalgamation of terranes. What is obvious is the well defined mid-Caradoc Anglo-Welsh Province across the Leinster Terrane (northern localities only shown on Figure 1). The Bellewstown Terrane, with its oceanic or Celtic signature in the Early Ordovician must have approached or joined the Leinster Terrane by the mid-Caradoc. By the Ashgill, ecologically controlled faunas with stratigraphically restricted ranges, but widespread distribution, were the norm rather than provincial control of older Ordovician series. The ongoing study of Ordovician brachiopod faunas provides perhaps the best discriminant, independent test, helping to refine and improve our understanding of other paleogeographic reconstructions and models for development of the Iapetus Ocean.

Acknowledgments

M.A. Parkes is grateful to the Geological Survey of Ireland for financial support to attend the Third International Brachiopod Congress, and to present this paper. Ian Mitchell kindly provided new material from Balbriggan. Chris Stillman enabled fieldwork by Parkes on Lambay Island, with permission of Lord Revelstoke (deceased).

REFERENCES

BRENCHLEY, P.J., J.C. HARPER, W.I. MITCHELL, & M. ROMANO. 1977. A re-appraisal of some Ordovician successions in eastern Ireland. Proceedings Royal Irish Academy 77(B):65-85.

CARLISLE, H. 1979. Ordovician stratigraphy of the Tra-more area, County Waterford, with a revised Ordovician correlation chart for southeast Ireland, p.545-554. In, A.L. Harris, C.H. Holland & B.E. Leake (eds.). The Cale-donides of the British Isles - reviewed. Geological Society, London.

COCKS, L.R.M. & JIAYU RONG.1988. A review of the late Ordovician *Foliomena* brachiopod fauna with new data from China, Wales, and Poland. Palaeontology 31:53-67.

FORTEY, R.A., D.A.T. HARPER, J.K. INGHAM, A.W. OWEN, & A.W.A. RUSHTON. 1995. A revision of Ordovician series and stages from the historical type area. Geological Magazine 132:15-30.

FRANCE, D.S. 1967. The geology of Ordovician rocks at Balbriggan, County Dublin, Eire. Geological Journal 5(2):291-304.

HARPER, D.A.T., 1980. The brachiopod *Foliomena* fauna in the upper Ordovician Ballyvorgal Group of Slieve Bernagh, County Clare. Irish Journal of Earth Sciences, 2: 189-192.

___1992. Ordovician provincial signals from Appalachian-Caledonian terranes. Terra Nova 4:204-209.

___, C. MAC NIOCAILL, R.B. NEUMAN & S.H.WILLIAMS (in press). Distribution of early Ordovician circum and intra Iapetus brachiopods - the Celtic province revisited.

___& W.I. MITCHELL. 1982. Upper Ordovician (Ashgill) brachiopods from the Oriel Brook Formation, County Louth. Journal Earth Sciences Royal Dublin Society 5:31-35.

___,___& JIAYU RONG. 1995. New faunal data from the highest Ordovician rocks at Pomeroy, County Tyrone, Northern Ireland. Scottish Journal Geology 30:187-190.

___,A.W. OWEN & M. ROMANO. 1985. Upper Ordovician brachiopods and trilobites from the Clashford House Formation, near Herbertstown, Co. Meath, Ireland. Bulletin British Museum Natural History 38 (5):287-308.

___, F.C. MURPHY & M.A. PARKES. 1992. Intra-lapetus brachiopods from the Ordovician of eastern Ireland: implications for Caledonide correlation: reply. Canadian Journal Earth Sciences 29: 833-834.

___& M.A. PARKES. 1989. Palaeontological constraints on the definition and development of Irish Caledonide terranes. Journal Geological Society London 146: 413-415.

___,___,A.N. HOEY & F.C. MURPHY. 1990. Intra - Iapetus brachiopods from the Ordovician of eastern Ireland: implications for Caledonide correlation. Canadian Journal Earth Sciences 27: 1757-1761.

HARPER, J.C., 1952. The Ordovician rocks between Collon (Co. Louth) and Grangegeeth (Co. Meath). Scientific Proceedings Royal Dublin Society 26:85-112.

___& N. RAST. 1964. The faunal succession and volcanic rocks of the Ordovician near Bellewstown, Co. Meath. Proceedings Royal Irish Academy 64(B):1-23.

HULL, E., R.J. CRUISE & W.H. BAILY. 1871. Explanatory memoir to accompany sheets 91 and 92 of the maps of the Geological Survey of Ireland, 46p

JAANUSSON, V. & S.M. BERGSTRÖM. 1980. Middle Ordovician faunal spatial differentiation in Baltoscandia and the Appalachians. Alcheringa 4:89-110.

LENZ, A.C. & A.P.M. VAUGHAN. 1994. A Late Ordovician to middle Wenlockian graptolite sequence from a borehole within the Rathkenny Tract, eastern Ireland, and its relation to the paleogeography of the Iapetus Ocean. Canadian Journal Earth Sciences 31: 608-616.

MASON, C.F. 1961. The geology of the Lower Palaeozoic rocks around Balbriggan, north County Dublin. Unpublished M.Sc. Thesis, Queen's University Belfast, 60p.

MITCHELL, W.I., 1977. The Ordovician Brachiopoda from Pomeroy, Co. Tyrone. Palaeontographical Society Monographs, London, 138p.

MURPHY, F.C. 1987. Evidence for late Ordovician amalgamation of volcanogenic terranes in the Iapetus suture zone, eastern Ireland. Transactions of the Royal Society Edinburgh, Earth Sciences 78:153-167.

___, (&18 auctt.) 1991. Appraisal of Caledonian suspect terranes in Ireland. Irish Journal Earth Sciences 11:11-41.

NEUMAN, R.B. & D.A.T. HARPER. 1992. Palaeo-geographic significance of Arenig-Llanvirn Toquima- Table Head and Celtic brachiopod assemblages, p.241-254. In, B.D. Webby, B.D. & J.R. Laurie (eds.), Global Perspectives on Ordovician Geology, Balkema, Rotterdam.

OWEN, A.W., D.A.T. HARPER & M. ROMANO. 1992. The Ordovician biogeography of the Grangegeeth terrane and the Iapetus suture zone in eastern Ireland. Journal Geological Society London 149: 3-6.

PARKES, M.A. 1992. Caradoc brachiopods from the Leinster terrane (southeast Ireland) - a lost piece of the Iapetus puzzle? Terra Nova 4: 223-230.

___1993. Palaeokarst at Portrane, Co. Dublin: evidence for Hirnantian glaciation. Irish Journal Earth Sciences,12: 75-81.

___1994. The brachiopods of the Duncannon Group (Middle-Upper Ordovician) of southeast Ireland. Bulletin British Museum, Natural History 50 (2):105-174.

___& D. PALMER.1994. The stratigraphy and palaeontology of the Lower Palaeozoic Kildare inlier, Co. Kildare, Ireland. Irish Journal Earth Sciences 14: 65-81.

___& A.P.M. VAUGHAN. 1991. Discussion on sequence stratigraphy of the Welsh Basin. Journal Geological Society London 148: 1144.

___,A.G. SLEEMAN & R. MAHER. 1994. Geological Survey of Ireland: National Heritage Council funded curation project of 19th century collection. Geological Curator 6 (2): 70.

POTTER, A.W. & A.J. BOUCOT.1992. Middle and late Ordovician brachiopod assemblages of North America, p.307-323. In, B.D. Webby, B.D. & J.R. Laurie (eds.), Global Perspectives on Ordovician Geology. Balkema, Rotterdam.

ROMANO, M., 1980a. The Ordovician rocks around Herbertstown (County Meath) and their correlation with the succession at Balbriggan (County Dublin), Ireland. Journal Earth Sciences Royal Dublin Society 3:205-215.

___1980b. The stratigraphy of the Ordovician rocks between Slane (County Meath) and Collon (County Louth), eastern Ireland. Journal Earth Sciences Royal Dublin Society 3:53-79.

___& A.W. OWEN.1993. Early Caradoc trilobites of eastern Ireland and their palaeogeographical significance. Palaeontology 36: 681-720.

RONG, JIA-YU & D.A.T. HARPER. 1988. A global synthesis of the latest Ordovician Hirnantian brachiopod faunas. Transactions Royal Society Edinburgh, Earth Sciences 79:383-402.

___, ___, RENBIN ZHAN & RONGYU LI. 1994. Kassinella - Christiania Associations in the early Ashgill Foliomena brachiopod fauna of South China. Lethaia 27: 19-28.

SLEEMAN, A.G., 1992. The palaeontological collections of the Geological Survey of Ireland. Geological Curator 5 (7): 283-291.

VAUGHAN, A.P.M. & J.D. JOHNSTON. 1992. Structural constraints on closure across the Iapetus suture in eastern Ireland. Journal Geological Society London 149: 65-74.

WILLIAMS, A., 1956. Productorthis in Ireland. Proceedings Royal Irish Academy 57(B):179-183.

___, I. STRACHAN, D.A. BASSETT, W.T. DEAN, J.K. INGHAM, A.D. WRIGHT& H.B.WHITTINGTON. 1972. A correlation of Ordovician rocks in the British Isles. Geological Society London Special Report 3:1-74.

WRIGHT, A.D. 1963. The fauna of the Portrane Limestone, I, The inarticulate brachiopods. Bulletin British Museum Natural History, Geology 8:221-254.

___1964. The fauna of the Portrane Limestone, II, The articulate brachiopods. ibid. 9: 157-256.

___1968. A westward extension of the Upper Ashgillian Hirnantia fauna. Lethaia 1:352-367.

ORGANIC CONTENTS AND ELEMENTAL COMPOSITION OF BRACHIOPOD SHELL AND MANTLE TISSUES

LLOYD S. PECK & TRACY M. EDWARDS

British Antarctic Survey, High Cross, Madingley Rd, Cambridge CB3 0ET;
University of Wales, Swansea, Singleton Park, Swansea SA2 8PP, Wales, UK.

ABSTRACT- Shells and mantle tissues of six brachiopod species were analysed for elemental carbon, hydrogen and nitrogen (CHN), and ash-free dry mass, (AFDM) content. Four were punctate articulates (*Liothyrella neozelanica*, *L. uva*, *Terebratulina retusa* and *Kraussina rubra*), one was an impunctate articulate (*Notosaria nigricans*) and another was a punctate inarticulate (*Neocrania anomala*). All C:N ratios were between 3.8 and 5.8. These were low values and indicated a high protein content in all samples. Unexpectedly, elemental analysis revealed higher C:N ratios in shells than mantle tissues, suggesting that protein contents of mantle tissues were higher than shells, including the impunctate *N. nigricans*. Organic content assessments for shells using an HCl digest method were not significantly different from loss on ignition estimates for 2 of 3 species analysed. Similar HCl digest estimates for mantle tissue samples produced underestimates of around 50%, and the method is not appropriate for soft tissues. Shell AFDM estimates calculated from CHN assessments were lower than loss on ignition assessments in 5 of the 6 species studied (except *L. neozelanica*). The average estimates were 2.45% for calculations based on CHN content and 3.01% for loss on ignition. For mantle tissues the calculations based on elemental analysis produced estimates around 50% of those from the loss on ignition method. The major aim of this project, to apportion fractions of shell organic content to caeca and protein matrix was not possible because the C:N ratios obtained for mantle tissues were lower than for shell samples.

INTRODUCTION

Articulate brachiopods have been studied for more than a century in relation to their fossil remains and their importance in interpretation of evolutionary processes. Only in the last 20 to 30 years has work been carried out on brachiopod biology, and this has lead to a far from complete understanding of their life habits and ecological characteristics. We now know that, as well as having restricted distribution patterns (James et al., 1992), low energy characteristics are common attributes of the living members of the phylum. They exhibit slow rates of shell growth (Rosenberg et al., 1988), low metabolic rates (Peck, 1989; Peck et al., 1986a,b, 1987a, 1989; Shumway, 1982; Thayer, 1986), pump water through their mantle cavities under laminar flow conditions (LaBarbera, 1981), clear water slowly while feeding (Rhodes, 1990), and have very low tissue organic contents (Peck, 1992). The findings have lead to articulates being viewed as efficient species, minimising losses during energy transformation processes.

When studying ecology and physiology a variable of fundamental importance is the size of the organism under investigation. For studies such as assessments of metabolic or feeding rate, good estimates of the amount of organic tissue involved in the process are essential to the interpretation of functional processes. This is especially so for comparisons between taxa, such as those in the above studies which categorised articulate brachiopods as low energy species. Poor estimates of the amount of body tissue involved in a physiological process could lead to an inappropriate categorisation of a group in terms of its utilisation of energy.

In most invertebrates the estimation of organic content is obtained by dissecting out the metabolically active tissue and measuring the loss of mass on ignition, called the ash-free dry mass (AFDM). In punctate brachiopods the presence of mantle extensions (caeca) through the shell make the estimation of organic content more difficult. Previous studies have suggested that, in punctate species, around 50% of the total

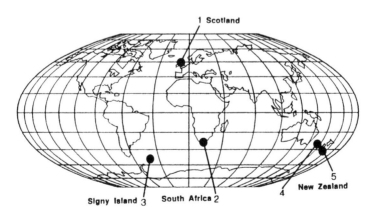

1 Scotland

Signy Island 3   South Africa 2   New Zealand 5   4

Figure 1. Sites where specimen collections were made. 1. *T. retusa* and *N. anomala* were collected at 170 m depth near Kerrera Island, Firth of Lorne, Scotland. 2. L. uva , a shallow location at Signy Island, Antarctica. 3. K. rubra , lowest intertidal level available from a beach on the west coast of South Africa. 4. *L. neozelanica*, Doubtful Sound, South Island, New Zealand, 4-8 m depth. 5. *N. nigricans*, 200 m depth off Karitane near Otago, South Island, New Zealand.

metabolically active tissue of the brachiopod may be located in the caeca (Curry & Ansell, 1986; Peck et al., 1987b). A major problem with these assessments has been the separation of caecal AFDM from the protein matrix of the shell. The contents of the caeca are cellular and should be included in assessments of metabolically active material. The protein matrix of the shell valves, on the other hand, is clearly not metabolically active and should be excluded.

In this study mantle tissues and shell organic contents were measured in four punctate articulate species (*Terebratulina retusa* (Linnaeus), *Liothyrella neozelanica* (Thomson, 1918), *Liothyrella uva* (Broderip, 1833) and *Kraussina rubra* (Pallas, 1776)), one impunctate articulate (*Notosaria nigricans* (Sowerby, 1846)) and one punctate inarticulate (*Neocrania anomala*, (Müller, 1776)). Three methods were used to estimate total organic content: the usual loss on ignition AFDM method; measuring loss on ignition after first removing $CaCO_3$ by dissolution in HCl; and by calculating organic content from CHN assessments. A major aim was to use C:N ratios in shell and mantle tissues to differentiate between organic matter located in caeca and the protein matrix of the shell.

MATERIALS AND METHODS

Brachiopods were collected from five sites around the world (Figure 1). Most species were collected by dredge from a boat, or by using SCUBA techniques (Table 1). In all cases, with the exception of *L. uva*, specimens were freeze dried after collection. They were then transported dry to the UK, and held in a desiccator until they were analysed.

Prior to analysis all shells were cleaned of epifauna. Mantle tissues were then carefully removed, making sure that there was no contamination from the gonads located in the mantle tissues. This was primarily done by taking samples from areas of mantle away from where the gonads were sited. After removing the mantle tissue, samples shells were cleaned of all internal tissues. Shells were then ground into powder, and mantle tissue samples divided into three so that all subsequent analyses were carried out on material from the same animals.

The first estimate of organic content was obtained from the commonly used loss on ignition method. Shell and mantle

Table 1. Collection dates and details for brachiopods in organic and elemental composition assessments (*specimens were collected from a site near the British Antarctic Survey station on Signy Island in March 1993 and held live in aquaria until analyses were carried out in August 1993).

Species	Type	Collection Date	Collection Method	Depth
Neocrania anomala	inarticulate punctate	June 1993	dredge	170 m
Terebratulina retusa	articulate punctate	June 1993	dredge	170 m
Liothyrella neozelanica	articulate punctate	October 1992	scuba	4-18 m
Liothyrella uva	articulate punctate	August 1993	scuba	10-20 m
Kraussina rubra	articulate punctate	February 1993	shore gathered	intertidal
Notosaria nigricans	articulate impunctate	October 1992	dredge	200 m

tissues were firstly dried to constant weight at 60°C, cooled to room temperature in a desiccator and weighed. They were then ignited in a muffle furnace at 475°C, again cooled to room temperature in a desiccator and reweighed. The difference in the two weighings gave the AFDM.

Secondly, AFDM was assessed after the removal of $CaCO_3$ from samples using HCl with a similar method to Rodhouse et al. (1984). Approximately 0.5 g of shell or mantle tissue was placed into stoppered 60 cm^3 boiling tubes and concentrated HCl added to excess. Tubes were left at room temperature for a minimum of 24 h to allow complete removal of all $CaCO_3$. The contents of the tubes were then filtered onto pre-ashed pre-weighed GF/F filter papers and washed with distilled water. The organic content of this material was obtained from loss on ignition as described above.

Thirdly, organic contents of brachiopod shell and mantle tissues were estimated via elemental analysis. The method used was similar to those of Kristensen & Andersen (1987) and Verardo et al. (1990). Samples of around 100 mg of shell or mantle tissue were placed in 14 cm^3 glass sample vials. Sulphurous acid was added to excess (when no further evolution of gas was obtained upon further addition) and vials left at room temperature for a minimum of 48 h to allow removal of all carbonate. Samples were then dried to constant weight at 60° C in the boiling tubes. The sulphurous acid reacted with any $CaCO_3$ present to produce $CaSO_4$, CO_2 and H_2O. Inorganic carbon was, therefore, removed as CO_2 when the samples were re-dried. Subsamples of around 1 mg were taken from the boiling tubes, transferred to pre-ashed aluminium boats and processed in a Carlo-Erba 1106 elemental analyser, using acetanilide as the standard. The data produced was in terms of the proportion (% dry mass) of each sample accounted for by the elements carbon, nitrogen and hydrogen. Organic contents (AFDM) were calculated from %C values using data given in Gnaiger and Bitterlich (1984) for fish muscle, liver, and gut tissues and weighted for the actual C:N ratio in each sample.

RESULTS

Ratios of carbon to nitrogen in mantle tissues ranged from 4.3 to 4.9, and from 3.8 to 5.8 for shells (Figure 2). For mantle tissues the highest value was obtained from *T. retusa*, and the lowest for *L. uva*, both punctate articulate species. The ratios for *N. nigricans* and *N. anomala* were within the range for the punctate articulates. Data for shells showed a wider spread, with *L. uva* again exhibiting the lowest value and *T. retusa* the highest. In all species, except *L. uva* shell C:N ratios were higher than mantle tissue values.

Shell AFDM estimates based on elemental analyses ranged from 1.8% of the dry mass, for *T. retusa*, to 4.5% for *N. anomala* (Figure 3). The value for *N. nigricans* was within the range for the punctate species, but the figure for *N. anomala* was significantly higher than all other species (F = 48.2, P < 0.001). Shell AFDM measured as loss on ignition showed the same pattern, with the value for the impunctate *N. nigricans* falling within those for the punctate species, and the inarticulate *N. anomala* having the highest shell organic content (Figure 3). The minimum content measured this way was 1.9% (*L. neozelanica*), and the maximum was 4.8%. In all species studied, except *L. neozelanica*, shell AFDM estimated from elemental analysis data was lower than the loss on ignition data (Figure 3). Values for *K. rubra*, *L. neozelanica* and *N. anomala* were not significantly different (P>0.05, 95% CI overlap), however, the loss on ignition method produced significantly higher estimates for *N. nigricans*, *T. retusa* and *L.*

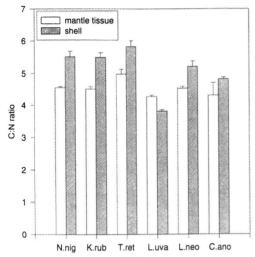

Figure 2. Mantle tissue and shell C:N ratios for 4 punctate articulate species (*K. rubra*, *T. retusa*, *L. neozelanica* and *L. uva*), 1 impunctate articulate (*N. nigricans*) and 1 punctate inarticulate (*N. anomala*). Values shown are means - SE; open bars indicate mantle tissue measures, hatched bars indicate data for shells; N.nig = *Notosaria nigricans*, K.rub = *Kraussina rubra*, T.ret = *Terebratulina retusa*, L.uva = *Liothyrella uva*. L.neo = *Liothyrella uva*, C.ano = *Neocrania anomala*. Ratios are based on carbon and nitrogen % dry mass measures.

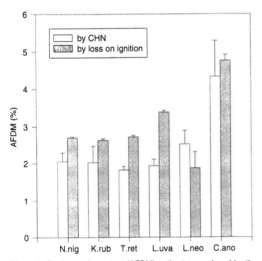

Figure 3. Shell organic content (AFDM) estimates produced by the standard loss on ignition method (hatched bars) and from calculations based on elemental CHN analyses (open bars). Data shown are means - SE, and are presented as % dry mass. (abbreviations as fig. 2).

uva. On average shell organic content estimated by the loss on ignition method was 3.01% (SE=0.39), and via elemental analysis it was 2.45% (SE=0.40). The results from the two methods were significantly different (ANOVA, F=8.62, P< 0.01). The two different methods produced more divergent data for mantle tissue AFDM estimates than for shells (Figure 4). For

both methods *N. anomala* tissues produced the highest values, and the data for *N. nigricans* were within the range for punctate species. The estimates calculated from CHN analyses were, however, lower for all species than loss on ignition assessments. On average data calculated from elemental analyses were only 53% of the values produced by the standard method.

The third method used to measure AFDM, by firstly removing skeletal $CaCO_3$, was only conducted on three species, *N. nigricans*, *L. neozelanica* and *L. uva* (Table 2). Data obtained for shells of *N. nigricans* and *L. uva* were not significantly different from the loss on ignition method (t=0.48, P=0.959, t=1.07, P=0.092, respectively). However, the analysis for *L. neozelanica* shell following HCl treatment indicated an AFDM content more than 3 times that produced by loss on ignition, a difference which was highly significant (t = 4.59, P = 0.004). Tissue AFDM estimates after HCl treatment were all dramatically lower than results from loss on ignition (Table 2). Data produced were between 4.5 and 6 times lower than the standard method, and all differences were highly significant.

DISCUSSION

Mass fraction (proportion of AFDM) C:N ratios were all in the range 3.8 to 5.8. Gnaiger & Bitterlich (1984) analysed muscle, liver, gut and fat of the silver carp, *Hypophtalmichthys molitrix*, and found that C:N ratios were inversely related to the protein content of those tissues. Muscle, liver, gut and fat were around 88%, 55%, 22% and 4% protein respectively, and their C:N ratios were 3.4, 5.6, 7.7 and 79. On this basis the organic contents of the shells and tissues analysed here were between 50% and 80% protein. C:N ratios in shells were higher than mantle tissues in 5 of the 6 species studied (Figure 2). This result was surprising, as it was expected that the protein matrix of the shell would tend to lower the C:N ratio, and cellular non-protein compounds would raise ratios in mantle tissues. Higher C:N ratios in shells compared with mantle tissues of the impunctate *N. nigricans* were especially surprising. Possible technical errors associated with the measurements which could have produced these data include contamination of shell

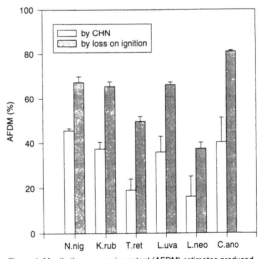

Figure 4. Mantle tissue organic content (AFDM) estimates produced by loss on ignition (hatched bars) and from elemental CHN analyses (open bars). Values shown are means -SE and are presented as % dry mass (abbreviations as Fig. 2).

Table 2. Organic contents of brachiopod shells and mantle tissues produced using the standard loss on ignition method, and from assessments made following the dissolution of skeletal $CaCO_3$ using HCl. Student's t and significance values refer to comparisons between methods used. In all cases n = 10.

Analysis	Species	Loss on ignition		HCl digest		t	P (<)
		mean	SE	mean	SE		
shell	N. nigricans	2.70	0.029	2.63	0.122	0.48	0.959
	L. neozelanica	1.87	0.428	5.54	0.640	4.59	0.004
	L. uva	3.38	0.050	3.51	0.120	1.07	0.092
mantle tissue	N. nigricans	67.4	2.60	12.0	1.15	20.5	0.001
	L. neozelanica	37.4	2.72	8.23	1.29	9.73	0.001
	L. uva	66.2	1.21	11.8	0.65	41.8	0.001

samples with inorganic carbon from shell carbonate, or the contamination of mantle tissue samples with high nitrogen compounds such as DNA.

The use of sulphurous acid to remove calcium carbonate from samples was developed for carbonate rich sediments, where a problem of contamination by inorganic carbon similar to that in the analysis of brachiopod shells is encountered. Several methods have been employed in sediment analysis (Byers et al., 1978, Kristensen & Andersen, 1987). The sulphurous acid digest method of Verardo et al. (1990) was chosen because of the advantages over other acid digest methods. Carbonate carbon is removed as CO_2 and excess acid volatilised off during drying as SO_2 and H_2O, which also removes problems associated with waters-of-hydration of calcium salts. The use of excess acid, and allowing samples to stand for 48 h prior to drying should have removed all inorganic carbonates, as was found by Verardo et al. (1990). If this was a problem large variation in the data obtained would also be expected, because of differing samples sizes and amounts of acid added. The small variation in C:N ratios obtained suggests that this was not a significant error.

Errors associated with contamination of mantle tissue samples with gonadal material of high DNA content were avoided by taking samples from mantle well away from the gonads. Contamination of shell samples with mantle tissue with high C:N ratios was also unlikely because the internal surfaces of the shells were cleaned rigorously prior to analysis, and also because of the low C:N ratios obtained for mantle tissues.

Higher C:N ratios in shell samples compared with mantle tissues would be obtained if the mantle samples contained high levels of protein with a high nitrogen content, or if shell samples incorporated some non-protein material. Jope (1965, 1969, 1977) defined the structure of many brachiopod shell proteins. She found that structural proteins in mantle tissues and shells were of very similar composition. Non-protein contents of shells were small, except in Neocrania, where there were significant amounts of chitin. The possibility therefore exists that the caecal contents in punctate articulates, and caecal contents combined with chitin in Neocrania raise C:N ratios in shell samples. However, this is not likely to be the major effect here, as the impunctate N. nigricans also had a higher C:N ratio in its shell. It is more likely that the storage layer located between the coelomic epithelia and outer mantle epithelium plays a major role in lowering mantle C:N ratios. This layer protrudes into the proximal sections of the caeca and contains protein and glycoprotein droplets in a collagen matrix (James, 1989; James et al., 1992). The contents of the storage layer vary seasonally (Curry et al., 1989; James et al., 1991), which may help to explain the generally lower C:N ratios in L. uva, the only species

sampled in late winter conditions and held for long periods at low ration levels in aquaria.

Shell organic content measurements using the loss on ignition method were criticised by Goulletquer & Wolowicz (1989), who produced data from the analysis of shells of the bivalve molluscs Ruditapes philipinarum, Cardium edule and Cardium glaucum which indicated that this method overestimated organic content by factors between 2.5 and 4.8 compared with an acid digest method. They explained their data in terms of losses of inorganic carbonate as CO_2 during ignition in a muffle furnace. We used a similar acid digest method here, trapping material on GF/F filter papers and rinsing in distilled water after digestion, however we did use HCl compared with their trichloroacetic acid. Our data indicate remarkably small differences between the methods for two of the three species investigated (Table 2). The value of 5.5% organic for the shell of L. neozelanica is much higher than any previous estimates for brachiopod shells (Curry et al., 1989), and is difficult to explain. However, there is clearly no gross overestimation by the loss on ignition method. This was also found for studies on organic assessments of high carbon content sediments. Byers et al. (1978) found that the loss on ignition method was much more accurate than acid digestion in such samples. The similar comparison of methods here on mantle tissues (Table 2) showed that an acid digest method produces massive underestimates of organic content and is inappropriate. This is due to organic compounds passing through the filters during the washing phase of the process. There must be similar losses when analysing shells, however, our data indicate that these losses are not significant. Because more carbonate is removed as temperature increases it would not necessarily be the case that such losses would be insignificant at ignition temperatures above 470°C.

The argument that carbonate losses cause an artificially high estimate of organic content in shells could also be put forward to explain the data in Figure 3, where AFDM estimates from the loss on ignition method are all higher than for estimates calculated from CHN assessments, with the exception of the results for L. neozelanica. However, the same arguments used in the discussion of the HCl digest method apply here, and the likely explanation of the differences is that the composition of articulate brachiopod shell organic material differs significantly from the 'standard' fish tissues analysed by Gnaiger & Bitterlich (1984), even when allowances are made for the C:N ratios obtained here. Clearly a better understanding of the stoichiometry associated with calculating AFDM from elemental data for brachiopod tissues would be necessary before a robust estimate of AFDM was produced by this method.

Differences between AFDM estimates calculated from CHN assessments compared with loss on ignition data were greater for mantle tissues than shells (Figures 3, 4). As for shells the underestimate in mantle tissue AFDM calculated using CHN data is probably produced by differences in composition from the tissues analysed by Gnaiger & Bitterlich (1984). Unlike the HCl digest method the extra underestimate for mantle tissues here could not have been caused by losses during a washing phase, as whole samples were dried prior to elemental analysis. The overall aim of using C:N ratio data from shell and tissue to allow a calculation of proportions of shell AFDM located in caeca and protein matrix was not possible because of the lower C:N ratios found in the mantle tissues.

Acknowledgments
We thank Alan Ansell, Norton Hiller, Daphne Lee and Simon Brockington for assistance in the collection of specimens. Elizabeth Prothero-Thomas gave advice and training on the use of the Carlo-Erba elemental analyser and Andrew Clarke provided invaluable discussion and comment. The work was supported by the Natural Environment Research Council of the UK.

REFERENCES

BYERS, S.C., E.L. MILLS &P.L. STEWART. 1978. A comparison of methods of determining organic carbon in marine sediments, with suggestions for a standard method. Hydrobiologia 58: 43-47.

CURRY, G.B. & A.D. ANSELL. 1986. Tissue mass in living brachiopods, p.231-241. In, P.R. Racheboeuf & C.C. Emig (eds.), Proceedings First International Congress on Brachiopods. Biostratigraphie Paléozoique 4.

___,___, M.A. JAMES & L.S. PECK. 1989. Physiological constraints on living and fossil brachiopods. Transactions Royal Society Edinburgh, Earth Sciences 80: 255-262.

GNAIGER, E. & G. BITTERLICH. 1984. Proximate biochemical composition and caloric content calculated from elemental CHN analysis; a stoichiometric concept. Oecologia 62: 1-10.

GOULLETQUER, P, & M. WOLOWICZ. 1989. The shell of Cardium edule, Cardium glaucum, and Ruditapes philipinarum: organic composition and energy value, as determined by different methods. Journal Marine Biological Assciation U.K. 69: 563-572.

JAMES, M.A. 1989. The reproductive biology of the Recent articulate brachiopod Terebratulina retusa (Linnaeus). PhD thesis, University of Glasgow.

___, A.D. ANSELL & G.B. CURRY. 1991. Repro-ductive cycle of the brachiopod Terebratulina retusa on the west coast of Scotland. Marine Biology 109: 441-451.

___,___, M.J. COLLINS, G.B. CURRY, L.S. PECK & M.C. RHODES. 1992. Biology of living brachiopods. Advances in Marine Biology 28: 175-387.

JOPE, H.M. 1965. Composition of brachiopod shell, p.156-164. In, R.C. Moore (ed.), Treatise on Invertebrate Palaeontology, H, Brachiopoda. The Geological Society of America and the University of Kansas Press, Lawrence, Kansas.

___1969. The protein of brachiopod shell-III. Comparison with structural protein of soft tissue. Comparative Biochemistry Physiology 30: 209-224.

___1977. Brachiopod shell proteins: their functions and taxonomic significance. American Zoologist 17:133-140.

KRISTENSEN, E. & F. ANDERSEN. 1987. Determination of organic carbon in marine sediments: a comparison of two CHN analyser methods. Journal Experimental Marine Biology Ecology 109:15-23.

LABARBERA, M. 1981. Water flow patterns in and around three species of articulate brachiopods. Journal Experimental Marine Biology Ecology, 55:185-206.

PECK, L.S. 1989. Temperature and basal metabolism in two Antarctic marine herbivores. Journal Experimental Marine Biology Ecology 127:1-12.

___1992. The tissues of articulate brachiopods and their value to predators. Philosophical Transactions Royal Society London, B., 339: 17-32.

___D.J. MORRIS, A. CLARKE & L.J. HOLMES. 1986a. Oxygen consumption and nitrogen excretion in the Antarctic brachiopod Liothyrella uva (Jackson, 1912) under simulated winter conditions. Journal Experimental Marine Biology Ecology 104: 203-213.

___, ___, ___&___ 1986b. Oxygen consumption and the role of caeca in the Recent Antarctic brachiopod Liothyrella uva notorcadensis (Jackson, 1912), p.349-355. In, P.R. Racheboeuf & C.C. Emig (eds.), Proceedings First International Congress on Brachiopods, Brest, France. Biostratigraphi

___, A. CLARKE & L.J. HOLMES. 1987a. Summer meta-bolism and seasonal changes in biochemical composition of the Antarctic brachiopod Liothyrella uva (Broderip, 1833). Journal Experimental Marine Biology Ecology 114: 85-97.

___, ___ &___ 1987b. Size, shape and the distribution of organic matter in the Recent Antarctic brachiopod Liothyrella uva. Lethaia 20: 33-40.

___, G.B. CURRY, A.D. ANSELL & M. JAMES. 1989. Temperature and starvation effects on the metabolism of the brachiopod Terebratulina retusa (L.). Historical Biology 2: 101-110.

RHODES, M.C. 1990. Extinction patterns and comparative physiology of brachiopods and bivalves. Unpublished PhD thesis, University of Pennsylvania, Philadelphia, USA.

RODHOUSE, P.G., C.M. RODEN, M.P. HENSEY & T.H. RYAN. 1984. Production of mussels, Mytilus edulis on the shore and in suspended culture. Marine Biology 84: 27-34.

ROSENBERG, G.D., W.W. HUGHES & R.D. TKACHUCK. 1988. Intermediatory metabolism and shell growth in the brachiopod Terebratulina transversa. Lethaia 21: 219-230.

SHUMWAY, S.A. 1982. Oxygen consumption in brachiopods and the possible role of punctae. Journal Experimental Marine Biology 58: 207-220.

THAYER, C.W. 1986. Respiration and the function of brachiopod punctae. Lethaia 19: 23-31.

VERARDO, D.J., P.N. FROELICH & A. McINTYRE. 1990. Determination of organic carbon and nitrogen in marine sediments using the Carlo-Erba NA 1500 analyser. Deep Sea Research 37: 157-165.

207

RADIATION OF THE EARLIEST CALCAREOUS BRACHIOPODS

LEONID I. POPOV, LARS E. HOLMER & MICHAEL G. BASSETT

All-Union Geological Research Institute (VSEGEI), Srednii Prospekt 74, 199026 St Petersburg, Russia;
Institute of Earth Sciences, Historical Geology and Palaeontology, Norbyvägen 22, S-752 36 Uppsala, Sweden;
Department of Geology, National Museum of Wales, Cardiff CF1 3NP, Wales, UK

ABSTRACT- The Cambrian radiation, and morphological variation, of five calcareous-shelled brachiopod orders, including taxa assigned to both articulates and inarticulates in the past, is reviewed, and analysed cladistically. The oldest known calcareous-shelled brachiopods are early Atdabanian obolellids. Chileides arose in the Botomian, and kutorginids in the Late Atdabanian or early Botomian. The oldest protorthoids ['articulates'] appear in the middle Early Cambrian, but the group did not radiate until Middle Cambrian times.

INTRODUCTION

Phylogenetic analysis suggests that calcareous and organophosphatic brachiopods had consistently separate shell chemistries from early in their evolution. Comparatively recent discoveries of Early and Middle Cambrian brachiopod faunas in Israel, Jordan, North Africa, Kirgizia, China and Australia show that both the taxonomic and morphological diversities of the earliest calcareous stocks are considerably greater than previously understood (Cooper, 1976; Roberts & Jell, 1990; Popov & Tikhonov, 1990; Geyer, 1994; see also summary by Popov, 1992). Among these finds, calcareous 'inarticulates' from China (Jin & Wang, 1992) have proved to be especially important for the understanding of the earliest phylogeny of this group; previously, the oldest known representatives of the non-pediculate, calcareous 'inarticulated' stocks (craniides, craniopsides, trimerellides) were from the Ordovician, and it was generally assumed that they had originated from organophosphatic ancestors at about that time (e.g. Williams & Rowell, 1965, fig. 141). This paper gives a brief account of the morphological variation within the five Orders of Cambrian calcareous brachiopods - the Obolellida, Chileida, Kutorginida, Orthida, and Craniopsida, and investigates their earliest radiation in the light of a phylogenetic (cladistic) analysis at the Family level. The detailed phylogenetic relationship between the calcareous and phosphatic brachiopods is outside the scope of this review; following our earlier studies, we consider that this divergence took place at an early stage in the history of the Phylum (Popov et al. 1993; Holmer et al. 1995).

MORPHOLOGY OF CAMBRIAN CALCAREOUS BRACHIOPODS

Obolellida. This Order comprises the oldest known calcareous brachiopods, from the early Atdabanian (Pelman, 1977). The group includes both forms that lack articulation, and those that have primitive articulatory structures, consisting of paired ventral denticles and dorsal sockets (Popov, 1992, fig.1). Such structures were acquired first within the Family Obolellidae, in which the Botomian genus Bicia has a pair of small, rudimentary denticles alongside the narrow, open delthyrium (e.g. Ushatinskaya, 1988, fig. 1, pl. 9). The denticles are direct extensions of the inner sides of the ventral proparea, as demonstrated by the fact that the growth lines can be traced continuously from the proparea to the distal tips of the denticular nubs. This suggests that the denticles were formed along the mantle margin and were composed partly of

primary shell; the latter conclusion requires further study for confirmation, because all available specimens of Bicia are either silicified or preserved as moulds. In Trematobolus and some genera of the Family Naukatidae, the denticles are composed entirely of secondary shell, and in the latter group they are supported by an arcuate plate, the so-called anterise (Popov, 1992, fig.1). The dorsal sockets are poorly developed to absent in most obolellides, but they are present in Naukatidae and some species of Trematobolus.

An open delthyrium is present only in the earliest Obolellidae (e. g. Obolella, Bicia and Ivshinella); in all other obolellides it is covered by a concave 'pseudodeltidium'. A morphologically similar structure is known otherwise only in the Eichwaldiidae; however, in this latter group the structure is part of the articulatory mechanism, and there is no evidence of homology with the obolellides. The obolellide muscle system is closely comparable with that of other 'articulated' brachiopods in having anterior and posterior adductor scars, which form a quadripartite muscle field in the dorsal valve of Trematobolus and some related genera, but are radially arranged in other obolellides; a single pair of oblique muscles (the internal oblique) was attached dorsally to a small area at the bottom of the notothyrial cavity. In some 'articulated' obolellides, the attachment scar of the internal oblique muscles is located posterior to the axis of rotation, suggesting that these may have served as diductors. In the Obolellidae and Trematobolidae, the mantle canal system is baculate with dorsal vascula media, but the mantle canals of Naukatidae are not known.

Chileida. The chileides first appeared in the Botomian, and the Order includes the earliest known calcareous brachiopods with a strophic shell; they lack any trace of articulatory structures along the posterior margin. The dorsal valve in all chileides lacks a pseudointerarea and is characterised invariably by hemiperipheral growth. This suggests the existence of a single generative zone along the posterior margin of the mantle, so that it is unlikely that the mantle lobes were completely separated. If this were so, then the axis of rotation was possibly fixed entirely by fused mantle lobes (Popov & Tikhonov, 1990; Popov 1992, fig.1).

Taxa in this group also have an unusually large ventral umbonal perforation, which is enlarged by resorption and sometimes covered posteriorly by a plate. The plate is more or less identical in morphology to the colleplax described by Wright (1981) in eichwaldiids. The function of the ventral per-foration is not understood fully, but it is unlikely that it served as a pedicle opening, both because of its anterior position and the fact that the chileides also have a delthyrium. If it is homologous with a colleplax, it is possible that it was the site of an organic pad secreted by the outer epithelium, as proposed by Wright (1981). Popov & Tikhonov (1990), and Popov (1992) speculated that the perforation may have served as part of an hydraulic shell-opening mechanism.

Chileide shell structure is known only from *Kotujella*, and according to Williams (1990), this may represent the oldest known endopunctate brachiopod with a fibrous secondary layer. The muscle system of the chileides is poorly understood; according to Popov & Tikhonov (1990), a set of internal oblique muscles may have attached posteromedially to the dorsal valve. Their location suggests that they are possibly homologous with the diductors of articulated brachiopods and obolellides, but because they are situated anterior to the axis of rotation they cannot have aided in shell opening. It is probable that chileides opened their shell by contraction of outside lateral muscles attached anteriorly to the body wall, comparable with the arrangement in Recent craniides. In the Matutellidae, both valves have pinnate mantle canals; in *Chile* there are three pairs of main trunks in the ventral valve and only one pair in the dorsal valve.

Kutorginida. This Order, as interpreted here, includes the three families Kutorginidae, Nisusiidae, and Agyrekiidae; the first known representatives are from the late Atdabanian or early Botomian. Kutorginides have a biconvex shell with a wide, straight posterior margin, and a large, widely triangular delthyrium covered by a convex pseudodeltidium (Popov, 1992, fig.1). The ventral apex is perforated by a small, rounded apical foramen. The dorsal diductor scars are located on the floor of the notothyrial cavity. According to Rowell & Caruso (1985), the Nisusiidae seem to have had an open digestive tract with a posteromedian anus.

Recent studies of *Kutorgina* and *Nisusia* (Rowell & Caruso, 1985; Popov & Tikhonov, 1990) indicate that their hinge mechanism was simple and quite different from that of most other Palaeozoic brachiopods. In *Kutorgina* it consists of two narrowly triangular propareas in the dorsal valve with paired ridges that border the margins of notothyrium, fitting into two deep, socket-like furrows that bound the margins of the pseudodeltidium in the ventral valve. In the Nisusiidae, the lateral extensions from the pseudodeltidium fit into 'sockets' on the inner side of the dorsal interarea; as noted by Rowell & Caruso (1985), 'the growth lines on the dorsal interarea are seemingly continous with those on the sockets; consequently the outer surface of the sockets, like the external surface of the remainder of the interarea, would have consisted of primary shell layer secreted by the dorsal mantle margin'. In similar fashion the growth characters of the dorsal hinge ridges suggest continuity with those on the interarea, and infer deposition as primary shell. As yet there is no direct confirmation of these interpretations from studies of shell structure because all available material is silicified. However, the available evidence as summarised here suggests that the kutorginide structures are not homologous with the sockets and socket ridges of the orthides and strophomenides, which comprise only secondary shell. Study of *Nisusia* (Popov & Tikhonov, 1990; Popov, 1992) suggests that a rudimentary kutorginide pattern of articulation is present in juveniles and that nisusiides are related closely to the kutorginides.

Orthida. A detailed account of the shell morphology of Cambrian orthides is outside the scope of this paper, but they are included in the analysis below in order to determine their possible phylogenetic relationship with other calcareous brachiopods. The Early Cambrian and early Middle Cambrian orthides belong mostly to the Superfamily Protorthoidea. In all protorthoids the ventral visceral area is situated on a highly elevated pseudospondylium; the diductor muscles were probably attached in the dorsal valve to the highly raised notothyrial platform. The earliest and, possibly, most primitive protorthoid genera, such as *Glyptoria* and *Israeleria* (Protorthoidea 1), have strophic shells and paired teeth on the

lateral sides of the delthyrial opening; sockets and true brachiophores are lacking (Popov, 1992, fig.1), appearing for the first time in early Middle Cambrian genera (Protorthoidea 2), such as *Arctohedra* (Roberts & Jell, 1990). Our analysis suggests that this latter group, including *Arctohedra*, may not be protorthids but may warrant familial or even higher level separation; it is beyond the scope of this review to establish new taxa, so we refer to the group as Protorthoidea 2 in order to differentiate it from the 'true' members of the Superfamily (Protorthoidea 1).

The oldest known orthides with deltidiodont teeth, rudimentary dental plates and brachiophores are eoorthides from the early Middle Cambrian (Ordian) of Australia. The Billingsellidae are also referred usually to the Orthida, but as they lack dental plates and have a laminar shell structure their detailed affinities require further investigation.

Craniopsida. The only possible Early Cambrian craniopside is *Heliomedusa* from China (Jin & Wang, 1992). This genus seems to have had a calcareous shell and is similar to other craniopsides in having a relatively large visceral cavity with a well developed posterior body wall, and in lacking any trace of a pedicle; it also has well developed marginal setae and, possibly, pinnate mantle canals with paired vascula lateralia in both valves. The only other possible Cambrian craniopside brachiopod is *Discinopsis* from the early Middle Cambrian of Canada, but it is very poorly known (Holmer & Popov, unpubl.).

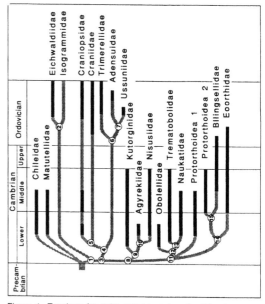

Figure 1. Topology for strict consensus tree for 19 calcareous Cambro-Ordovician early brachiopod families. The known stratigraphic distribution is indicated by black bars. Numbered nodes suported by the character states listed in Appendix 1.

PHYLOGENETIC ANALYSIS

Methods. The data matrix (Table 1) was analysed by means of the PAUP programme (Phylogenetic Analysis Using Parsimony 3.1.1; Swofford, 1993). The heuristic search (ACCTRAN optimization) using global swapping with the MULPARS option was used in the analysis of 42 unordered and unweighted

Table 1. Character State Matrix used in PAUP analysis of characters listed in Appendix 2. Characters were also coded as missing (?) or polymorphic in cases where the character is not applicable to the taxon.

	1	2	3	4	5	6	7	8	9	10	11	12	13	14	15	16	17	18	19	20	21	22	23	24	25	26	27	28	29	30	31	32	33	34	35	36	37	38	39	40	41	42
	prf	prt	gdv	orr	pmw	pmh	hod	pdl	nch	cav	iav	iad	pfo	af	col	thr	sor	fbd	ant	dpl	hso	cpl	bra	tdp	plf	spl	ols	ios	la	vvp	dvp	dam	cpr	nop	mcv	vvl	vvm	mcd	dvl	sld	por	shr
Chileidae	2	0	0	0	0	0	0	1	0	0	1	0	1	2	1	0	0	0	0	0	0	0	0	0	0	0	0	0	0	0	0	1	0	0	?	?	?	?	?	?	?	1
Craniopsidae	0	0	0	0	2	2	0	0	0	0	0	1	1	2	0	1	0	0	0	0	0	0	0	0	0	0	1	1	1	1	?	2	0	0	?	0	0	0	?	?	0	0
Obolellidae	0	0	0	1	1	0	0	0	0	0	1	1	0	0	0	1	0	0	0	0	0	0	0	0	0	0	0	2	0	0	0	1	0	1	0	1	?	1	1	?	1	1
Matutellidae	1	1	1	?	0	0	0	1	0	0	1	0	1	2	?	2	1	0	0	0	0	0	0	0	0	0	0	1	0	0	?	0	0	0	1	1	?	0	?	1	1	?
Trematobolidae	2	0	1	1	1	0	0	3	0	0	1	1	3	0	?	2	2	0	0	0	0	0	0	0	0	0	0	2	0	2	0	0	0	1	1	1	1	2	?	?	0	0
Naukatidea	0	?	0	0	1	0	0	3	0	0	1	1	2	0	0	2	1	1	1	0	0	0	0	1	0	0	1	2	0	2	0	1	0	1	?	?	?	?	?	?	?	0
Kutorgenidae	2	0	0	0	0	1	0	2	0	1	1	1	1	1	0	0	2	0	0	0	0	0	0	1	0	0	0	2	0	0	0	0	1	1	0	?	?	0	?	?	?	0
Agyrekiidae	2	0	0	0	0	1	0	2	1	0	1	1	1	1	0	0	2	1	1	0	0	0	0	0	0	1	0	2	0	0	0	1	1	1	?	?	?	0	?	1	0	0
Nisusiidae	2	?	0	2	0	1	0	2	0	0	1	1	1	1	0	0	2	0	0	0	0	0	0	0	0	0	0	2	0	0	0	?	1	1	0	0	?	?	1	?	?	0
Protorthidae1	2	?	0	?	0	0	0	0	0	0	1	1	1	0	0	3	0	0	0	0	0	0	1	0	0	1	0	2	4	0	?	0	1	1	0	0	?	0	?	1	?	0
Protorthidae2	?	?	0	?	0	0	0	0	0	0	1	1	1	0	0	3	3	0	0	0	0	0	2	0	0	0	0	2	4	0	0	0	1	1	2	0	1	2	?	0	0	0
Billingsellidae	1	0	0	2	0	1	0	2	1	0	1	1	1	1	0	3	3	0	0	0	0	0	2	0	0	0	0	2	0	0	0	0	1	1	2	0	2	0	?	0	0	0
Eoorthidae	2	?	0	2	0	0	0	0	0	0	1	1	1	0	0	3	3	0	0	1	0	1	0	0	0	0	1	1	0	0	0	0	0	0	2	0	?	?	0	?	0	0
Ussuniidae	0	0	0	0	2	2	1	0	0	0	0	0	0	0	0	0	0	0	0	0	0	1	0	0	0	0	0	1	1	0	0	0	0	0	0	1	1	?	?	2	?	0
Adensuidae	1	0	0	0	2	2	1	1	0	1	0	0	0	0	0	0	0	0	0	0	1	2	0	0	0	0	0	2	1	0	1	2	0	0	0	1	1	?	2	2	1	0
Trimerellidae	?	0	0	2	2	2	0	1	0	0	0	0	0	0	0	0	0	0	0	0	1	2	0	0	0	0	1	2	3	1	2	2	0	0	?	0	0	0	?	2	0	0
Cranlidae	?	0	0	?	2	2	0	0	0	0	0	0	0	0	0	0	0	0	0	0	0	0	0	0	0	0	0	0	0	0	0	2	0	0	0	?	?	0	?	0	1	0
Eichwaldiidae	2	1	1	0	2	0	0	0	0	0	1	0	0	2	1	0	0	0	0	0	0	0	0	0	1	0	0	2	1	0	?	?	2	2	?	?	?	0	?	1	1	1
Isogrammidae	2	?	1	?	0	0	0	0	0	0	1	1	1	2	1	0	0	0	0	0	0	0	0	0	1	0	0	2	0	0	?	?	2	0	?	?	?	0	?	?	?	1

characters selected from 19 family rank taxa, representing the Cambrian obolellides, kutorginides, chileides, and early orthides, but including also post-Cambrian craniiformean families (sensu Popov et al., 1993) within the Craniopsida, Trimerellida, and Craniida as well as the Order Dictyonellida and earliest articulated brachiopods within the Protorthidae, Eoorthidae and Billingsellidae. No formal outgroups were identified, because the potential candidates (that is, e.g., phoronids, bryozoans etc.) lack a bivalved shell and their characters are not useful in attempting to polarize any of the characters observable from fossil groups.

<u>Results and discussion</u>. The analysis produced three trees, each 87 steps long, with a consistency index of 0.759, of which only the strict consensus tree is shown (Figure 1); the three resulting trees differ only in the position of the Prot-orthoidea 1. Because of the problems of selecting outgroups, the polarities of the characters are uncertain. Although the Early Cambrian families Craniopsidae, Chileidae, Matutellidae and Obolellidae are ingroup taxa, they seem to have a rela-tively more 'simple' morphology by comparison with the other taxa and may hypothetically exhibit more primitive characters; any of them could be used as a potential 'basal reference taxon' for rooting the trees. Our analysis suggests that chileids and eichwaldiids are closely related phylogenetically, with the selection of the craniformeans or obolellids as basal reference taxa, the chileids and matutellids form a mono-phyletic group together with the eichwaldiids and isogrammids, but in the figured cladogram (Figure 1) the chileids were selected as the root.

Our cladogram (Figure 1) supports the recognition of the orders Obolellida (node 12), Dictyonellida (node 2), and Kutorginida (including Nisusiidae; node 9), as well as the craniformeans (node 4) as monophyletic groups. With the exception of dictyonellides, all these lineages already existed in the Botomian, and the main radiation and morphological diversi-fication of these groups clearly took place earlier in the early Cambrian (Tommotian and Atdabanian). Although the chileides and dictyonellides are morphologically quite different from the craniformeans, these taxa form a potential monophyletic group if the tree is rooted in the obolellides; however, it seems more likely that they are primitive by comparison with the articulated brachiopod taxa, including obolellides and kutorginides. The cladogram (Figure 1) also supports an unnamed monophyletic group (node 8) consisting of the kutorginides, obolellides, and orthides, characterized by having a dorsal interarea and notothyrial platform. The obolellides and orthides are also united in a monophyletic group by having paired teeth and saccate dorsal mantle canals, indicating that the obolellide denticles represent a derived state for this group, whereas the orthides supposedly retained the primitive condition.

The phylogenetic position of the craniformeans is uncertain; they may represent a stem group for the analysed taxa, but it is equally possible that they are derived in relation to the chileides as proposed tentatively here (Figure 1). If this assumption is correct, it indicates that the loss of a pedicle, ventral interarea and other characters represent derived conditions for the group.

REFERENCES

COOPER, G. A. 1976. Lower Cambrian brachiopods from the Rift Valley (Israel and Jordan). Journal Paleontology 50:269-289.

GEYER, G. 1994. A new obolellid brachiopod from the Lower Cambrian of Morocco. Journal Paleontology 68:995-1002.

HOLMER, L. E., L.E. POPOV, M.G. BASSETT & J. LAURIE (in press). Phylogenetic analysis and ordinal classification of the Brachiopoda. Palaeontology 38.

PELMAN, Yu.L. 1977. Ranne i srednekembriiskije bezzam-kovyje brakhiopody Sibirskoj Platformy [Early and Middle Cambrian inarticulate brachiopods of the Siberian Plate]. Trudy Instituta Geologii Geofiziki, Sibirskogo otdelenija 36:1-168.

POPOV, L.E. 1992. The Cambrian radiation of brachiopods, p.399-423. In, J.H. Lipps &P.W.Signor (eds.). Origin and early evolution of Metazoa. Plenum, New York.

___& Yu.A. TIKHONOV. 1990. Rannekembriiskie brakhiopodi iz yuzhnoi Kirgizii [Early Cambrian brachiopods from southern Kirgizia]. Paleontologicheskii Zhurnal 3(1990):33-46.

___, M.G. BASSETT, L.E. HOLMER & J. LAURIE. 1993. Phylo-genetic analysis of higher taxa of Brachiopoda. Lethaia 26:1-5.

ROBERTS, J. & P.A. JELL. 1990. Early Middle Cambrian (Ordian) brachiopods of the Coonigan Formation, western New South Wales. Alcheringa 14:257-309.

ROWELL, A.J. & N.E. CARUSO. 1985. The evolutionary signi-ficance of Nisusia sulcata, an early articulate brachiopod. Journal Paleontology 59:1227-1242.

SWOFFORD, D.L. 1993. Phylogenetic analysis using parsi-mony, Version 3.1.1. Computer program, Illinois Natural History Survey, Champaign, Illinois.

USHATINSKAYA, G. T. 1988. Obolellids (brachiopods) with articulate valve structures from the Lower Cambrian of Zabajkalie [Obolellidy (brakhiopody) s zamkovym sochle-neniem stvorok iz nizhnego kembriya Zabajkaliya]. Paleon-tologicheskij Zhurnal 1 (1988):34-39.

WILLIAMS, A. 1990. Biomineralization in the lophophorates, p.67-82. In, J.G. Carter (ed.). Skeletal biomineralization: patterns, processes and evolutionary trends. Van Nostrand Reinhold, New York.

___& A.J. ROWELL. 1965. Evolution and phylogeny, p.164-197. In, R.C. Moore (ed.). Treatise on invertebrate paleontology, H. Geological Society of America and University of Kansas Press, Lawrence.

WRIGHT, A. D. 1981. The external surface of Dictyonella and of other pitted brachiopods. Palaeontology 24:443-481.

APPENDIX 1

List of synapomorphies supporting the numbered nodes in Fig. 1. Characters and character states numbered as in Appendix 2. Homoplasies are marked by *:

(1)pseudodeltidium absent (8:0)*; internal oblique muscles functioning as diductors, attached posteromedially (28:2)*.
(2)anterior commissure uniplicate (2:1)*; cardinal process with shaft (33:2).
(3) dorsal valve with holoperipheral or mixoperipheral growth (3:0); ventral perforation absent (14:0)*; colleplax absent (15:0); laminar secondary layer (40:0)*; endopunctation absent (41:0)*; shell resorption absent (42:0).
(4)shell biconvex (1:0)*; posterior margin broadly convex (5:2)*; delthyrial opening absent (6:2)*; ventral interarea absent (11:0); pedicle opening absent (13:0)*; oblique lateral muscles absent (27:1)*; levator ani absent (29:1); dorsal adductor muscle scars not forming single field (32:2)*.
(5)diductors attached posterolaterally to dorsal valve (28:0)*; ventral vascula media absent (37:0).
(6)concave median plate present (7:1); cardinal areas present (10:1); hinge socket present (21:1); hinge plate present (22:1); ventral vascula lateralia present (36:1)*; shell aragonitic (40:2)*.
(7)shell dorsibiconvex (1:1)*; hinge plate well developed (22:2).
(8)shell costate or costellate (4:2)*; dorsal interarea present (12:1); notothyrial platform present (34:1).
(9)broad delthyrium and notothyrium (6:1)*; pseudodeltidium convex (8:2)*; ventral perforation small (14:1)*; sockets on margins of notothyrium, composed partly of primary shell (17:2); articulatory furrows on pseudodeltidium present (18:1)*; shell with fibrous secondary layer (40:1)*.
(10)shell smooth (4:0)*; triangular dorsal propareas present (24:1).
(11)paired teeth (16:3); dorsal mantle canals saccate (38:2).
(12) shell biconvex (1:0)*; shell striate (4:1)*; posterior margin straight short (5:1)*; denticles partly composed of primary shell (16:1)*; ventral mantle canals pinnate (35:1); ventral vascula lateralia present (36:1)*; dorsal vascula lateralia present (39:1).
(13)pseudodeltidium concave (8:3)*; pedicle foramen on interarea (13:2)*;

denticles composed of secondary shell (16:2); sockets on lateral sides of notothyrial cavity (17:1).

(14) paired sockets present (17:3); dorsal adductor scar forming quadripartite single field (32:0)*; cardinal process present (33:1); ventral mantle canals saccate (35:2).

(15) shell dorsibiconvex (1:1)*; sockets bounded by socket plates or brachiophores (23:2).

APPENDIX 2

List of coded characters used in cladistic analysis (full discussion of characters will be published elsewhere).

(1) prl-shell convexity: biconvex (0), dorsibiconvex (1),ventribiconvex (2).
(2) prt-anterior commissure: rectimarginate (0), uniplicate (1).
(3) gdv-growth of dorsal valve: holoperipheral, or mixoperipheral (0), hemiperipheral (1).
(4) orr-ornamentation: smooth (0), striate (1), costate or costellate (2).
(5) pmw-posterior margin: straight wide (0), straight short (1), broadly convex (2).
(pmh-delthyrial opening: narrow delthyrium (0), broad delthyrium and notothyrium (1), absent (2).
(7) hod-concave ventral median plate: absent (0), present (1).
(8) pdl-pseudodeltidium: absent (0), apical (1), convex (2), concave (3).
(9) nch-chilidium: absent (0), present, forming single plate (1).
(10) cav-cardinal areas: absent (0), present (1).
(11) iav - ventral interarea: absent (0), present (1).
(12) iad-dorsal interarea: absent (0), present (1).
(13) pfo-pedicle opening: absent (0), delthyrial (1), pedicle foramen on interarea (2), pedicle foramen umbonal or antrior to umbo (3).
(14) af-ventral perforation: absent (0), small, apical (1), extending anteriorly through resorption (2).
(15) col-colleplax: absent (0), present (1).
(16) thr - teeth: absent (0), denticles partly composed of primary shell (1), denticles composed of secondary shell (2), paired teeth (3).
(17) soc - sockets: absent (0), on lateral sides of notothyrial cavity (1), on margins of notothyrium, partly composed of primary shell (2), paired sockets (3).
(18) fbd-articulatory furrows on pseudodeltidium: absent(0), present(1).
(19) ant-anterise: absent (0), present (1).
(20) dpl-dental plates: absent (0), present (1).
(21) hso-hinge socket: absent (0), present (1).
(22) cpl-hinge plate: absent (0), vestigial (1), well developed (2).
(23) bra-sockets and brachiophores: absent (0), sockets absent, brachiophore nubs present (1), sockets bounded by socket plates or brachiophores (2).
(24) tdp-triangular dorsal propareas: absent (0), present (1).
(25) plf-ventral posterolateral furrows: absent (0), present (1).
(26) plf-sockets and socket ridges partly composed of primary shell: absent (0), present (1).
(27) ols-oblique lateral muscles: absent (0), present (1).
(28) ios-internal oblique muscles: diductors attached posterolaterally to dorsal valve (0), attached posteromedianly, not functioning as diductors (1), diductors attached posteromedianly (2).
(29) la - levator ani: absent (0), present (1).
(30) vvp-ventral muscle platform: absent (0), solid (1), free anteriorly (2), excavated or vaulted anteriorly (3), forming spondylium-like structure (4).
(31) dvp-dorsal muscle platform: absent (0), present (1).
(32) dam-dorsal adductor muscle scars: forming single field, quadripartite (0), forming single field, radially disposed (1), not forming single field (2).
(33) cpr-cardinal process: absent (0), present (1), present, with shaft (2).
(34) nop-notothyrial platform: absent (0), present (1).
(35) mcv-ventral mantle canals: pinnate (0), baculate (1), saccate (2).
(36) vvl-ventral vascula lateralia: absent (0), present (1).
(37) vvm-ventral vascula media: absent (0), present (1).
(38) mcd-dorsal mantle canals: pinnate (0), baculate (1), saccate (2).
(39) dvl-dorsal vascula lateralia: absent (0), present (1).
(40) skl-shell structure: with laminar secondary layer (0), with fibrous secondary layer (1), aragonitic (2).
(41) por-endopunctation: absent (0), present (1).
(42) shr-resorption: absent (0), present (1).

FACIES CONTROL OF JURASSIC BRACHIOPODS: EXAMPLES FROM CENTRAL ASIA

ELENA L. PROSOROVSKAYA

All-Russian Geological Research Institute (VSEGEI),
Sredni prospekt 74, 199026 St Petersburg, Russia

ABSTRACT- Abundant brachiopods are present in widespread Jurassic strata of the Great Balkhan, Tuarkyr, Copetdag, and Kugitang areas in Central Asia. These strata were formed in low latitudes during a major sedimentary cycle beginning with continental facies, through a gradual onset of marine transgression, and ending in regressive facies of the middle or late Oxfordian and Tithonian. The beginning and end of the Jurassic sequence in the region are characterized by heterogeneous, complex sedimentary settings, in contrast to rather homogeneous facies in the middle. Rhynchonellids, terebratulids, spiriferids and athyrids were a rich and diverse benthos, mainly confined to carbonates such as laminated limestones, reefal limestones (mid-Callovian to early Oxfordian), and black shales or dolomitic interbeds within siliciclastic regressive facies (late Jurassic). Strict confinement of brachiopods to specific facies is not always observed: brachiopod species ranges are plotted from Aalenian through Tithonian strata for four major regional sections.

INTRODUCTION

In latest Triassic through earliest Jurassic time, the African Gondwana and the Laurasia supercontinent collided, closing the Paleo-Tethys Ocean, as simultaneously the Atlantic ocean was opening to the west between Europe and North America. As a result, complex upland areas were formed in the Central Asia region, with formations of bauxite deposits. In the late Early Jurassic-early Middle Jurassic, coal-bearing deposits dominated in the coastal plains along the divergent margin (location of the Meso-Tethys Ocean) of the supercontinents. The Laurasia continental slope was gradually displaced and basinal shale, turbidites, and olistostromes were formed. In the Great Balkhan area of Central Asia (Figure 1), a small-shelled Bajocian rhynchonellid, *Striirhynchia dorsetensis* occurred exclusively in calcareous bands in deep water shales. The occurrence of this normally shallow water brachiopod in basinal deposits may be explained by post-mortem redeposition in deep water black shales. Brachiopods were abundant among the Jurassic benthic organisms of Central Asia, and were found in numerous settings (Prosorovskaya, 1968).

With continued transgression, brachiopods settled on shallow-shelf, black muds, and other siliciclastic substrates. Such examples can be found in basal, black shale distal facies of the Balkhan area, and onshore black shale-siliciclastic facies of

Figure 1. Jurassic outcrops of central Asia shown in black, with the four major sections indicated.

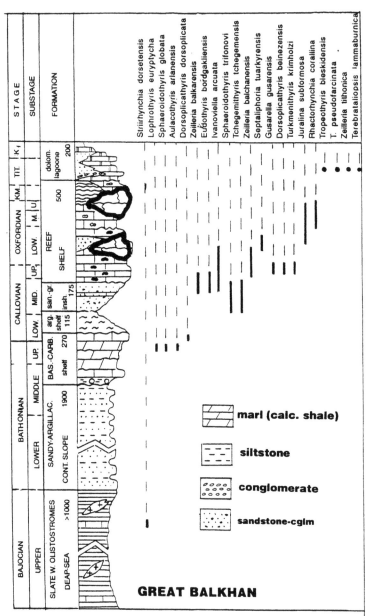

Figure 2. Jurassic brachiopod species in the Great Balkhan. Shales with olistostromes (basinal to lower continental slope facies), sandy-argillaceous sediments (upper continental slope to outer shelf), basal black shale (shelf), shales (shelf), sands-conglomerates, with black shale bands (inshore), reefal (shelf), dolomitic (nearshore-lagoonal).

the Kugitang area. Late Bathonian (reported previously as middle Bathonian, Prosorovskaya, 1986,1993a) brachiopods were nearly identical in both facies, but the association of *Lophrothyris euryptycha*, *Sphaeroidothyris globata*, and *Aulacothyris arlanensis* were confined to ferruginous, calcareous sandstones.

Maximum extent of the Meso-Tethys Ocean was reached in the middle Callovian, creating large shelf environments on the margins of Laurasia. During this time shales, sandstones and

conglomerates accumulated in the Great Balkhan, black shales around Tuarkyr, and black shales-siliciclastics in the Kugitang area. In the Great Balkhan, only very rare *Sphaeroidothyris trifonovi* (Moisseev) and *Tchegemithyris tchegemiensis* (Moisseev), more common *Euidothyris bordgakliensis* Prosorovskaya, *Zeilleria balkarensis* Moisseev, and abundant *Ivanoviella arcuata* (Rollier) occur in limestone beds within the shaly units. A similar but richer assemblage occurs in limestone interbeds of the overlying sandstones and conglomerates. In the Tuarkyr area, *Sphaeroidothyris trifonovi*

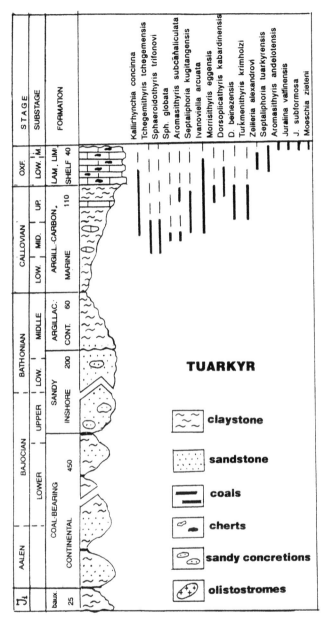

Figure 3. Jurassic brachiopods at Tuarkyr. Bauxite-bearing (laterites), coal-bearing (continental), sandy (inshore), argillaceous (continental), argillaceous-black shales (marine shelf), laminated limestones (shelf).

and *Tchegemithyris tchegemiensis* are found with *Septaliphoria kugitangensis* Moisseev, *Kallirhynchia concinna* (Sowerby), *Sphaeroidothyris globata* (Sowerby), and *Aromasithyris subcanaliculata* (Oppel), all occurring in middle Callovian normal shale and black shale facies. The absence of brachiopods in lower Callovian shales can probably be explained by unsuitable substrate conditions, and reduced salinity in the basin during the early Callovian. Almeras & Moulan (1983) have shown that Jurassic brachiopods in France are rare or absent on silty substrates, under turbulent bottom or emergent conditions, and under conditions of rapid muddy sedimentation.

In eastern Central Asia, at Kugitang, two different brachiopod assemblages were identified in the upper black shale-- siliciclastic sequence. The lower assemblage, of earliest Callovian age, consists only of small rhynchonellids like *Burmirrhynchia*, *Kutchirhynchia*, and *Tetrarhynchia*, and the upper assemblage, also of early Callovian age, contains only terebratulids such as *Aromasithyris subcanaliculata*, *A.. subingluviosa*, and *Gusarella longa*. It was observed by Ager (1965, 1986) that Jurassic brachiopods changed abruptly with lithologies, and by Johnson (1982) that Devonian brachiopods could change abruptly within the same facies. The Kugitang brachiopod assemblages conform to the latter abrupt type of change within the same facies.

In the middle Callovian, marine transgression accelerated and created space for brachiopod expansion. Widespread marine shelf areas created relatively homogeneous sedimentary facies, which led to similar, monotonous brachiopod associations throughout Central Asia. Nevertheless, each area within Central Asian basin, appears to have had its own characteristic fossil assemblages. This may have been caused by isolation of parts of the basin, preventing mixing of brachiopods, affinities for specific substrates, or establishment of specialist niches. This intra-basinal isolation is corroborated by the distribution of nektic ammonite species.

In the middle and latest Callovian, through earliest Oxfordian, carbonate platforms with reefs were widely distributed in Central Asia. In an onshore setting there were barrier reef formations in the Great Balkhan (Figure 2) and Kugitang areas, and in the late Oxfordian also at Copetdag. Abundant brachiopods, especially terebratulids, are confined to the lower part of these formations in the Great Balkhan, with such species as *Dorsoplicathyris beinezensis, Turkmenithyris krimholzi, Euidothyris bordgakliensis, Juralina subformosa, Zeilleria balchanensis, Gusarella gusarensis, Septaliphoria tuakyrensis,* and *Ivanoviella arcuata.* Brachiopods are rare in the upper part of the carbonate succession, including *Rhactorhynchia corallina, Septaliphoria pinguis, Lophrothyris subsella, Postepithyris, Juralina,* and *Moeschia zieteni* at Copetdag. A largely endemic association is confined to the lowermost reefal formation at Kugitang, with abundant large-winged *Septaliphoria guldaraensis, S. ismudica, S. lamelliformis, Kutchirhynchia kutchensis, Gusarella gusarensis,* and *G. subquadrata.* Brachiopods are rare in the upper reefal formation. Behind the barrier reef, lagoonal carbonate muds were deposited, which became laminated micrites in which many brachiopods occur at Tuarkyr (Figure 3), e.g. *Kallirhynchia concinna, Ivanoviella arcuata, Septaliphoria tuarkyrensis, Aromasithyris subcanaliculata, Dorsoplicathyris kabardinensis,* and *Zeilleria alexandrovi.* In the upper part of this lagoonal facies occur *Aromasithyris andelotensis, Juralina ex gr. valfinensis, J. subformosa,* and *Moeschia zieteni.*

At the end of the Jurassic, the Tethys ocean retreated from Central Asia, a period marked by regressions. Marine sedimentation continued in some onshore basins, where black calcareous muds and evaporites were deposited in a shallow, restricted shelf setting. These were subsequently transformed to black shales, dolomites, halites, gypsum and anhydrites in the Great Balkhan, Kugitang and Copetdag areas (Figures 4-5). In the Great Balkhan, a very specific brachiopod assemblage developed, consisting of *Tropeothyris ? bieskidensis, T. ? pseudofarcinata, T ? vialovi, Zeilleria tithonica,* and *Terebrataliopsis lammaburunica.* A striking feature of the brachiopods found in this Late Jurassic interval, from the Carpathians, the Crimea, northern Caucasus, Transcaucasus, and many other parts of Europe, is their much lower endemicity than for brachiopods from the underlying and overlying strata (Ager, 1975; Barczyk, 1979; Moisseev, 1934; Makridin & Kamyshan, 1964; Tkhorzevsky, 1989; Sandy, 1988, 1991; Prosorovskaya, 1993). This probably indicates a brief, but widespread, transgression during a long regressive cycle. The more cosmopolitan nature of the faunas, however, could also be attributed to a lack of detailed study of the brachiopods from this interval.

CONCLUSIONS

Some Jurassic brachiopods of Central Asia appear to have

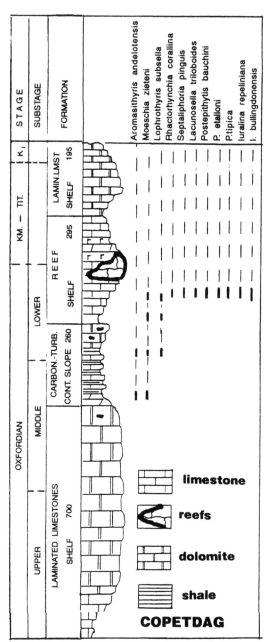

Figure 4. Jurassic brachiopods at Copetdag. Laminated limestone (shelf), black shales-turbidites (continental slope), reef (shelf).

been strongly endemic or to have been confined to specific facies. For example, *Striirhynchia dorsetensis* is known only from shales in the Great Balkhan. The assemblage of *Lophrothyris euryptycha, Sphaeroidothyris globata,* and *Aulacothyris arlanensis* occurs only in ferruginous-calcareous sandstones in rather widely separated areas of Balkhan and Kugitang. *Tropeothyris, Zeilleria,* and *Terebrataliopsis* are found only in dolomites of the Balkhan. However, other brachiopods, e.g., *Euidothyris bordgakliensis* and particularly

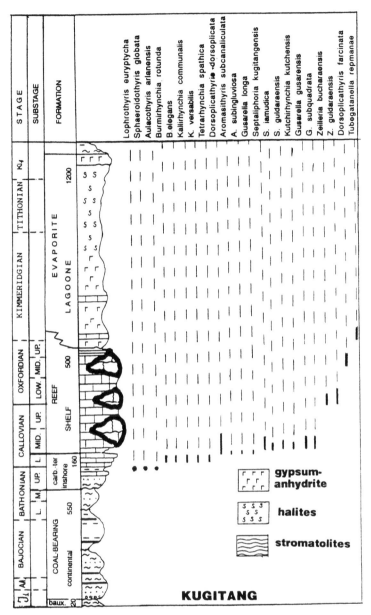

Figure 5. Jurassic brachiopods at Kugitang. Bauxite-bearing (laterites), coal-bearing (continental), carbonaceous-siliciclastic (inshore), reef (shelf), evaporite (lagoonal).

Ivanoviella 'arcuata', occur in several formations. In the Kugitang sequence, there are four stratigraphic horizons characterized by distinct brachiopod assemblages, each confined to a specific lithologic unit. The first horizon is the ferruginous sandstone mentioned above, the second contains extremely small rhynchonellids, the third shows a similar lithology to the second but contains only terebratulids; and in the fourth mainly large 'winged' rhynchonellids occur in reefal carbonates.

Ager (1986) mentioned that there appeared to be no phyletic gradualism within Mesozoic brachiopods. In Central Asia some sudden brachiopod faunal replacements occur within seemingly uniform sedimentary environments. The general observation is that species are crowded at particular horizons, with barren intervals between these. Jurassic brachiopods of Central Asia appear to useful for regional stratigraphic correlations, but are less useful for inter-regional correlations, a seen elsewhere, e.g.Ager, 1979; Almeras et al., 1991, 1994; Tchoumatchenco, 1984, 1986; Mancenido, 1990; Voros, 1993. Because of a close relationship of brachiopods to substrates, brachiopods may be useful in interpreting sedimentary environments.

Acknowledgments

The Natural Sciences and Engineering Research Council of Canada is thanked for providing travel support to attend the Third International Brachiopod Congress in Sudbury, Ontario September 2-5, 1995.

REFERENCES

AGER, D.V. 1965. The adaptation of Mesozoic brachiopods to different environments. Palaeogeography, Palaeoclimatology, Palaeoecology 1:143-172.

___1975. Brachiopods at the Jurassic-Cretaceous boundary. Mémoires Bureau Recherches Géologiques Minières 86:150-162.

___1986. The stratigraphical distribution of the Mesozoic Brachiopoda, p.33-41. In, P.R. Racheboeuf & C.C. Emig (eds.). Actes 1er Congrès International sur le Brachiopodes, Brest, 1985. Biostratigraphie Paléozoique 4.

ALMERAS, Y., A. BOULLIER & B. LAURIN. 1991. Les zones de brachiopodes du Jurassique en France. Annales Scientifique Université Franche Comte, Besançon, Geologie 4 (10):3-30.

___, S. ELMI, L. MEKAHLI, A. OUALI-MEHADGII, D. SADKI & M. TLILI. 1994. Biostratigraphie des brachiopodes du Jurassique moyen dans le domaine atlasique (Maroc, Algerie), p.219-241. Proceedings 3rd International Meeting on Aalenian and Bajocian Stratigraphy. Miscellanea Servizio Geologico Nazionale 5.

___& G. MOULAN. 1983. Influence des paléoenvironments sur la phylogenie des brachiopodes: exemples des terebratulides du Lias Provençal (France). Geobios 16 (2):243-248.

BARCZYK, W. 1979. Brachiopods from the Jurassic/Cretaceous boundary of Rogoznik and Czorztyn in the Pieniny Klippen Belt. Acta Geologica Polonica 29:207-214.

JOHNSON, J.G. 1982. Occurrence of phyletic gradualism and punctuated equilibria through geologic time, Journal Paleontology 56:1329-1331.

MAKRIDIN, V.P. & V.P. KAMYSHAN. 1964. Stratigraphical distribution of brachiopods at Jurassic of western and central parts of the Northern Caucasus. Trudy Geologii Poleznim Iskopaemim Severnogo Kavkaza 11:54-61.

MANCENIDO, M.O. 1990. The succession of Early Jurassic brachiopod faunas from Argentina: correlations and affinities, p. 397-404. In D.I. MacKinnon, D.E. Lee & J.D. Campbell (eds.), Brachiopods through Time, Proceedings 2nd International Brachiopod Congress. Balkema, Rotterdam.

MOISSEEV, A.S. 1934. The Jurassic Brachiopoda of Crimea and Caucasus. Trudy VSEGEI 203: 1-213.

PROSOROVSKAYA, E.L. 1968. Jurassic brachiopods of Turkmania. Nauka, Leningrad, 194p.

___1986. Brachiopod biozones in the Jurassic of southern USSR, p.357-361. P.R. Racheboeuf & C.C. Emigs (eds.), Brachiopodes fossiles et actuelles, Actes 1er Congrès International sur les brachiopodes. Biostratigraphie du Paléozoique 4:357-361.

___1993a. Brachiopod subdivisions in the Jurassic of southern USSR. Palaeogeography, Palaeoclimato-logy, Palaeoecology 100:183-188.

___1993b. Brachiopods at the Jurassic/Cretaceous boundary from the Ukrainian Carpathians, Crimea, Caucasus, and Transcaspian region, p.109-112. In, A. Vorös (ed.), Mesozoic Brachiopods of Alpine Europe. Hungarian Geological Society, Budapest.

SANDY, M. 1988. Tithonian Brachiopoda. Mémoires de Société Géologique France154:71-74.

TKHORZHEVSKII, E.S. 1989. Composition of shells and classification of Tithonian Terebratulida (Brachiopoda) of Penninian Cliffs of the Carpathians. Byulletin Moskovskogo Obshchestvo Ispytatelei Prirody 64:75-84.

TCHOUMATCHENCO, P. 1984. Les zones de brachiopodes du Jurassique d'Algérie du Nord et leur correlation avec les zones de brachiopodes en Bulgarie. Geological Survey Denmark Symposium 3:863-882.

___1986. Répartition paléoecologique des brachiopodes jurassique dans le Monts de Tiaret et l'Ouarsenis occidental (Algerie), p.389-398. In, P.R. Rachebeoeuf & C.C. Emigs (eds.), Brachiopodes fossiles et actuelles, Actes 1er Congrès International sur le Brachiopodes. Biostratigraphie Paléozoique 4.

VORÖS, A. 1993. Jurassic brachiopods of the Bakony Mts. (Hungary): global and local effects in changing diversity, p.179-187. In, A. Vorös (ed.) Mesozoic Brachiopods of Alpine Europe. Hungarian Geological Society, Budapest.

HYDRODYNAMIC STABILITY OF EMPTY SHELLS OF EXTANT TEREBRATULIDS AND RHYNCHONELLIDS: IMPLICATIONS FOR LIFE   HABIT OF EXTINCT BICONVEX TAXA

SONJA M. RODRIGUEZ & RICHARD R. ALEXANDER

Department of Geological and Marine Sciences
Rider University, Lawrenceville, New Jersey 08648, USA

ABSTRACT- Empty shells of the extant species *Dallina septigera, Hemithyris psittacea, Gryphus vitreus, Laqueus californianus, Liothyrella uva, Neothyris lenticularis, Notosaria nigricans, Terebratalia transversa, Terebratella sanguinea,* and *Terebratulina retusa* were placed alternately on fine gravel (2.0 to 4.0mm) and fine sand (0.20mm) in a recirculating flume in each of nine potential life orientations, including three with the commissural plane erect, three reclining on the dorsal valve, and three reclining on the ventral valve. Mean threshold current velocities to transport each species were determined in each orientation on each substrate. Velocities for shells resting on gravel ranged from 15 to 40cm/sec.Velocities for shells in the same orientation resting on sand usually averaged 5 to 10cm/sec lower. The most stable reclining orientation for all species on either substrate was with the posterior margin upcurrent. Velocities for erect shells averaged 1 to 4cm/sec higher when a lateral margin faced upcurrent for eight of ten species. Posteriorly weighted shells (*T. transversa*) were more stable relative to elongate (e.g., *L. uva*), globose (e.g., *L. californianus*), or conical (e.g.,*H. psittacea, T. sanguinea*) shells. Eddy-generating plicae (*D. septigera*) and costae (*T. sanguinea; N. nigricans*) decreased shell stability relative to similar sized costellate (*T. retusa; H. psittacea*) or smooth (*G. vitreus*) shells.

INTRODUCTION

Hydrodynamic stability of biconvex brachiopod shells in various orientations on sediment has been a subject of much speculation, but limited experimentation. It has been postulated that extinct biconvex brachiopods with broad interareas (spiriferids) and posteriorly weighted umbonal regions (pentamerids) to support the erect valves would be stable in current agitated habitats (Fürsich & Hurst, 1974). Alexander (1984) experimented on the stability of alate (winged), strophic (linear hinge line), and non-strophic (curved hinge line) extinct biconvex brachiopods in various feasible life orientations, but specimens were filled with lithified sediment which constrained the analysis and inferences. Furthermore, only six of nine feasible life orientations (Figure 1) were tested. Menard & Boucot (1951) experimented with transportation of articulated empty shells and single valves of nine extant terebratulid species in a flume, but also did not test each of the nine possible orientations a biconvex brachiopod may assume relative to the current and the substrate (Figure 1). Extant rhynchonellids, which have shapes contrasting with those of terebratulids, were not tested.

Whereas pedicle attachment of extant species provides stability to the shell in several orientations, the pedicle of many extinct biconvex families was lacking or atrophied during life, as evidenced by a pseudodeltidium or conjunct deltidial plates secreted across the foramen (Alexander, 1977). Stability of dispersed shells of non-pediculate adult spiriferids, atrypids, and pentamerids, which lacked or lost a functional pedicle during ontogeny, depended solely on hydrodynamic streamlining and center of gravity of articulated valves in each orientation. Many 'tetrahedral' or hemipyrami-

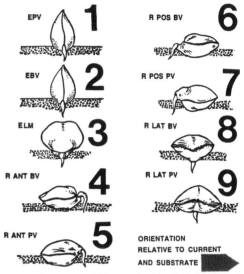

Figure 1. Orientations of biconvex brachiopod shells relative to current (black arrow) and substrate. Acronyms indicate valves either erect (E) or reclining (R). For erect orientations, pedicle/ventral (PV) or brachial /dorsal (BV) valve upcurrent. For reclining orientations, pedicle/ventral (PV) or brachial/dorsal (BV) valve uppermost. and anterior (ANT), or posterior (POS) or lateral margin (LAT) facing upcurrent. In subsequent figures, orientation number is given along horizontal axis of graph.

dal spiriferids and frilled, resupinate atrypids, respectively employed 'iceberg' and 'snowshoe' strategies for stabilization on soft substrates (Thayer, 1975). However, such a dichotomous adaptational classification does not necessarily indicate which biconvex shell shapes were most stable in which of nine feasible life orientations (Figure 1).

Although at least four biconvex orders of articulates are extinct, many of their biconvex shell profiles fortunately persist. The dorsal, lateral, or anterior profiles of many extant rhynchonellids and terebratulids resemble analogous profiles of their extinct relatives among the spiriferid, atrypid, orthid, and pentamerid orders. Experimentation on empty shells of extant terebratulids and rhynchonellids in different orientations provides insights into the hydrodynamic stability afforded by shells of varying biconvexity (dorsibiconvex to equibiconvex to ventribiconvex), which are well represented among the genera of extinct orders (McGhee, 1980). Comparable relief in the central fold of the valves exists between modern articulates and the sinus and sulcus of many extinct, non-pediculate biconvex taxa. Ornamentation in rhynchonellid and terebratulid shells range from plicate to costellate, similar to extinct biconvex taxa.

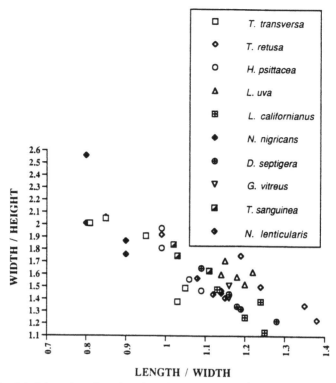

Figure 2. Representation of shell shape of experimental specimens in morphospace diagram with length/width plotted against width/height.

Additionally, the minute amount of tissue encapsulated between the valves marginally influences the center of gravity of the brachiopod. Thus the mass of unattached empty biconvex shells of extant species closely approximates the distributed biomass of living brachiopods except for the pedicle. Experimentation on these empty shells may therefore elucidate the stability afforded by similar shell shapes among extinct non-pediculate taxa in different feasible life orientation(s) and identify the most stable orientation(s) for unattached brachiopods. Comparative hydrodynamic stability of unattached modern shells may help explain why certain taxa of extinct biconvex brachiopods retained a functional pedicle throughout life whereas others shed their pedicle in late ontogenetic stages.

METHODS AND MATERIALS

The following ten species were used in the flume experiments: (1) *Dallina septigera*, (2) *Hemithyris psittacea*, (3) *Gryphus vitreus*, (4) *Laqueus californianus*, (5) *Liothyrella uva*, (6) *Neothyris lenticularis*, (7) *Notosaria nigricans*, (8) *Terebratalia transversa*, (9) *Terebratella sanguinea* and (10) *Terebratulina retusa*. Length, width, and height (to nearest 1.0mm) as well as shell mass (to nearest 1.0g) were recorded for each specimen. Specimens of these species provide a continuum of shell shapes from transversely compressed (*N. nigricans, T. transversa*), to globose (*L. californianus*), to elongate (*T. retusa*) (Figure 2). The species include smooth (e.g., *L. uva, N. lenticularis, G. vitreus*), costellate (e.g., *T. retusa, H. psittacea*), and costate (*N. nigricans, T. transversa, T, sanguinea*) shells. Species with rectimarginate (=no central fold; e.g. *G. vitreus*) and strongly unisulcate (*T. transversa*) to biplicate (e.g., *D. septigera*) anterior profiles are represented.

Also represented are species with triangular shaped beaks (*N. nigricans; T. sanguinea*), rounded posteriors with undersized pedicle openings (*N. lenticularis*), and nearly strophic (short linear) hinges (*T. transversa*). Finally, species with posteriorly weighted shells (*T. transversa*) to uniformly thick valves (*L. uva*) were included.

Empty shells of these species were placed on the sediment covered bottom of a plexiglass flume consisting of side by side, 3m long, 20cm wide channels. Two different grain sizes of sieved sediment alternately veneered the channel bottom, namely fine gravel (2 to 4mm) and fine sand (0.20mm), for tests on ten species. Flow through the channels was monitored immediately downcurrent of the specimen by a Marsh-McBirney 210B Current Meter. Current velocities were steadily increased through the flume until the specimen was placed in the traction load on the sediment. The velocity competent to transport the shells is hereafter referred to as the threshold velocity. A third sediment, silt (<.063mm), was abandoned as a test substrate because suspensed fines obscured visibility of the specimens.

Each specimen was placed in each of nine possible life orientations on each sediment grain size (Figure 1). In six of these orientations (Figure 1:4-9), the shell was reclining on one valve. Alternately, the dorsal and ventral valves were undermost, with the anterior margin facing upcurrent, or downcurrent, or with a lateral margin facing upcurrent. In three orientations the shell is erect with the commissural plane perpendicular to the sediment surface, and either the dorsal or ventral valve, or a lateral margin, facing upcurrent (Figure 1:1-3). A minimum of three replications of each specimen of ten species in each orientation were tested on sand; eight of

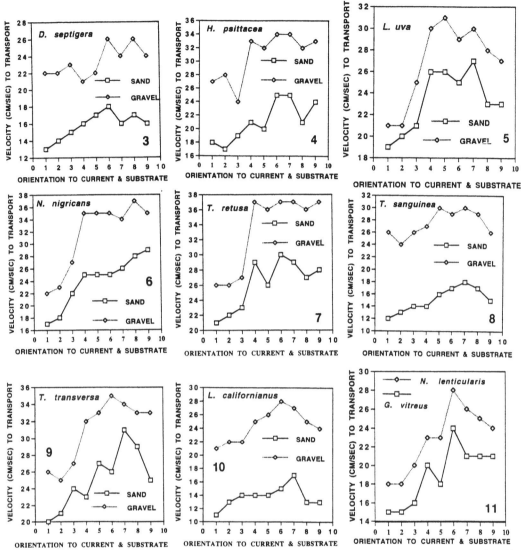

Figures 3-11. Mean current velocity (cm/sec) to transport shells from nine different initial orientations (see Figure 1) on gravel and sand for (3) *Dallina septigera*, (4) *Hemithyris psittacea*, (5) *Liothyrella uva*, (6) *Notosaria nigricans*, (7) *Terebratulina retusa*, (8) *Terebratella sanguinea*, (9) *Terbratalia transversa*, (10) *Laqueus californianus*, (11) *Neothyris lenticularis* and *Gryphus vitreus*.

these species were also tested on gravel (Figures 3-11). *Gryphus vitreus* and *Neothyris lenticularis* were not tested on gravel because of logistical and equipment problems. Mean threshold velocity to transport the shells from each initial orientation was calculated on the basis of a minimum of 12 tests per species. Differences in mean threshold velocity in each orientation were compared with every other orientation by t-tests.

EXPERIMENTAL RESULTS

All species in each orientation required higher threshold velocities on gravel in comparison to the same orientation on sand (Figures 3-10). All species were more stable in all reclining orientations (Figure 1:4-9) relative to any erect

orientation (Figure 1:1-3) on either gravel or sand (Figures 3-11), except for *Laqueus californianus* which was more stable on sand with the valves erect, and a lateral margin upcurrent, than reclining with a lateral margin upcurrent (Figure 10). Most species required a threshold velocity 8-15cm/sec faster to transport shells from the most stable reclining orientation relative to the least stable erect orientation. Exceptions are *Dallina septigera* (Figure 3) and *T. sanguinea* (Figure 8) for which threshold velocities were only 5-6cm/sec faster for the most stable reclining versus least stable erect orientation.

Species with compressed valves, i.e., width/height ratio>1.7 (*Notosaria nigricans, Terebratulina retusa, Terebratalia transversa, Hemithyris psittacea*; Figure 2), were particularly stable in reclining orientations on gravel. Threshold velocities of 32-38cm/sec were required for these species. In contrast, species with an inflated anterior profile, i.e., width/ height<1.5 (*Laqueus californianus, Dallina septigera*; Figure 2) required threshold current velocities of 20-28cm/sec for shells initially reclining on gravel. On sand, *L. californianus, D. septigera*, and *Terebratella sanguinea* (Figures 3, 8, 10), were the least stable shell shapes, requiring lower threshold velocities in each orientation relative to other species.

Among reclining orientations, the most stable orientation was with the posterior facing upcurrent (Figure 1:6-7). The exception was *Notosaria nigricans* on both substrates (Figure 6), and *Liothyrella uva* on gravel (Figure 5), which were more stable with the lateral margin facing upcurrent or the anterior facing upcurrent, respectively. Among erect orientations on either substrate, shells of all species were more stable if a lateral margin faced upcurrent than if either valve faced upcurrent, with the exception of *Hemithyris psittacea* on gravel (Figure 4), which was most stable with the brachial valve facing upcurrent .

DISCUSSION

Different mean threshold velocities to transport each species in different orientations resulted from the combination of differences in: (1) drag forces, both pressure and friction, (2) lift created by the pressure differential between upper and lower shell surfaces, and (3) acceleration reaction force (see Vermeij, 1993, p.68-73, for a succinct review of these forces on shelled molluscs). Furthermore, turbulence generated around erect or reclining valves created sediment scouring in the sand around the weight-bearing area which destabilized the shells. The size and depth of the scouring is related to the cross-sectional area of the shell intercepting the flow. Destabilization of the shell is also dependent on the distance between the center of gravity of the shell and the weight-bearing area, as well as the shell surface area in contact with the sediment.

Shells remained stable on gravel in their initial orientation at velocities which placed them in the traction load on sand. Not surprisingly, sand was eroded at a lower threshold velocity than gravel (Hjulstrom, 1939) and horseshoe-shaped depressions were initiated in the sediment upcurrent from the shell. These depressions expanded downcurrent, arching around each valve, as previously noted in similar experiments on fossil brachiopods (Alexander, 1984). As the sand on which the valves rested was progressively undermined, the shell listed, toppled, and tumbled in the traction load. Shells initially reclining on a sandy substrate also experienced under-sapping of sediment supporting one valve in the turbulent flow around the valves. Progressively larger area of the pedicle or brachial valve lost contact with the sand surface, and eventually the shell was flipped over as lift increased. At this same velocity on gravel, the undermost valve of shells maintained the same area of contact with the uneroded substrate.

Reclining orientations were more stable than erect orientations on the same substrate (Figures 3-11) because less cross-sectional surface area intercepted the flow when the valves were reclining relative to an entire brachial or pedicle valve intercepting the flow (Figure 1). Drag forces on the shell were less in reclining orientations, particularly with the anterior or posterior upcurrent, and the reduced turbulence generated less sediment-scour undermining the valves. In reclining orientations the center of gravity of the shell was closer to the sediment surface while the weight-bearing surface area in contact with the sediment increased, thereby affording the shell greater stability relative to erect orientations.

The difference in the threshold velocity for shells in the most stable reclining, versus the most stable erect orientation on gravel was minimal (3 cm/sec) for *Dallina septigera* (Figure 3), and maximal (10 cm/sec) for *Notosaria nigricans* (Figure 6) and *Terebratulina retusa* (Figure 7). Shells of *N. nigricans* and *T. retusa* are very compressed (width>>height) and elongated (length>>width), respectively (Figure 2). The costate, flattened, conical shells of *N. nigricans* generated considerable sediment-scouring turbulence around erect valves which were supported by minimum shell surface area, namely a pointed beak shallowly imbedded in the sand. Consequently the shells toppled at a comparatively low threshold velocity. However, these compressed shells had maximum surface area in contact with, and minimum shell relief above, the sediment surface when the valves reclined. Similarly, shells of *T. retusa* had minimum shell surface propping up the narrow valves in erect orientations, whereas a reclining orientation with the posterior upcurrent minimized the cross-sectional surface area intercepting the flow (Figure 1:6-7). Thus shells of these species were very stable in reclining relative to erect orientations on gravel (Figure 7). In contrast, *Dallina septigera* has inflated valves, and the dorsal, lateral and anterior profiles did not vary appreciably in the cross-sectional area intercepting the current. Consequently, lift and drag forces on the shell resting on gravel did not vary greatly between erect versus reclining orientations and threshold velocities in each orientation were similar (Figure 3).

The only exception where reclining orientations were less stable than erect orientations occurred on sand with specimens of *Laqueus californianus* (Figure 10). When a lateral margin of an erect shell faced into the current, a slightly higher threshold velocity was required to transport the shells than when the valves were reclining and the anterior margin upcurrent. This species has the most globose shape, with both length/width and width/height commonly<1.2 (Figure 2). Comparing the anterior, lateral and dorsal-ventral views, cross-sectional surface area perpendicular to the flow is not substantially different in any erect versus any reclining orientations, so that drag forces are not substantially different in these orientations. The acceleration reaction force acting on the shell, which is proportional to specimen volume (Denny, 1995), did not change in different orientations. What did change significantly was lift, which is maximum for inflated shells in a reclining orientation with a lateral margin upcurrent. In this orientation, the globose shell made contact with the sediment only at two points, namely the foramen and the mid-length of the undermost valve, less than when the shell rested on its posterior. Thus shells were less stable reclining with a lateral margin upcurrent than erect (Figure 10).

The most stable reclining orientation was with the posterior facing upcurrent and either the dorsal or ventral valve uppermost for all but two species. For species with length significantly exceeding width (Figure 2) this reclining orientation placed the minimum cross-sectional area perpendicular to the current. Furthermore, upcurrent scouring of sand around the shell posterior caused the valves to slump gradually into the upcurrent depression, buttressed by sediment piling up against the downcurrent anterior margin. The shell stabilized in the depression with its anterior-posterior axis inclined downcurrent, and much of its undermost valve resting against the sediment. Similar stabilized positions resulted in the experiments on empty terebratulid shells by Menard & Boucot (1951). Adult shells of most species are more likely to be buried in this orientation rather than transported away at current velocities below 18cm/sec.

One exception, *N. nigricans*, was more stable when a lateral margin of the reclining shell, rather than the posterior, faced upcurrent (Figure 6). This species has a flattened conical shape and often an asymmetric central fold that generated unequal sediment scouring around the weight-bearing area of the shell. Once the reclining valves were undermined, a shell with its posterior upcurrent was prone to slump onto a lateral margin, a very unstable orientation from which it was immediately transported. The other exception, *Liothyrella uva*, was very slightly more stable when the anterior faced upcurrent rather than downcurrent on gravel. This species has the greatest length, i.e., more than 45mm, of any tested species and has rectimarginate (no fold), smooth valves of uniform thickness. Consequently, there is little, if any difference in the cross-sectional area intercepting the flow when the anterior or posterior profile faced upcurrent. Drag, lift, and acceleration forces, as well as sediment scouring turbulence, differed insignificantly around reclining shells with either the anterior or posterior facing upcurrent.

The most stable orientation on sand with the valves erect had the lateral margin of the shell facing upcurrent. This orientation minimized the cross-sectional area intercepting the flow of water (Figure 1). Less drag was generated when a lateral margin faced upcurrent than with either the dorsal or ventral valve facing upcurrent. Furthermore, the more compressed the valves (Figure 2), the greater the difference in threshold velocity between orientations with either valve upcurrent versus a lateral margin upcurrent (Figure 3-11). Both *Notosaria nigricans* (Figure 6) and *Terebratalia transversa* (Figure 9) required a threshold velocity 4-5 cm/sec higher to transport shells with a lateral margin versus the dorsal or ventral valve initially upcurrent. Among species with more inflated shells, differences in mean threshold velocity to transport shells were only 2 to 3 cm/sec among the three erect orientations (Figures 3,10).

Shell shapes which were least stable in erect orientations were strongly equibiconvex, or globose, e.g., *Laqueus californianus*, or had well developed plicae (*Dallina septigera*) or costae *(Terebratella sanguinea)* ornamenting a well-defined central fold. Erect shells of *L. californianus* on sand were transported at the lowest threshold velocity of any species, i.e.,11cm/sec (Figure 10). Shells of *D. septigera* and *Neothyris lenticularis* not only have inflated valves, but lack the posterior weighting of the valves which characterize *Terebratalia transversa*. Although, *L. californianus* has shells of similar length to *T. transversa*, the latter species required a threshold velocity twice that which transported *L. californianus* from an erect orientation with a lateral margin upcurrent (Figure 9 vs.10).

The plicate fold in the shells of *Dallina septigera* generated considerable turbulence, thereby increasing sediment scour which destabilized the valves. The smooth valved species *Gryphus vitreus* has the same shell length and mass as *D. septigera*, yet the former species has no central fold and required a higher threshold velocity in each orientation on sand (Figure 11 vs. 3). The sulcate *Terebratella sanguinea* and *Hemithyris psittacea* have similar shapes, and their 'morphospace' (Figure 2) overlaps appreciably. They share triangular beaks, but differ in that *H. psittacea* is costellate, whereas *T. sanguinea* is costate. Sediment scouring turbulence, generated by coarser ornamentation of *T. sanguinea*, contributed to its lesser stability in all orientations on sand (Figure 4 vs. 8).

IMPLICATIONS FOR UNATTACHED BICONVEX PALEOZOIC BRACHIOPODS

More than 200 genera of biconvex brachiopods, including spiriferids, pentamerids, atrypids, and rhynchonellids lacked a functional pedicle in adulthood, if not throughout life (Alexander, 1977). The pedicle opening of many spiriferiid species was either sealed off by a pseudodeltidium throughout life or the pedicle atrophied as stegidial or deltidial plates accreted across the delthyrium. The obstructing incurving beak of the opposite valve progressively blocked off the delthyrium in pentamerids. Although the clustered life style that characterized some pentamerids (Ziegler et al.,1966), and rhynchonellids (Hallam, 1962), provided stability to erect individuals nestled within the cluster, many unattached biconvex brachiopods did not live so densely aggregated. For such dispersed,unattached brachiopods, hydrodynamic stability was afforded exclusively by the orientation of the shell to current and substrate (Alexander, 1984) and any subsidence of the valves into the sediment (Thayer, 1975).

The experiments on the empty shells of extant terebratulids and rhynchonellids, particularly on sandy substrates, offer insights into which orientations and shell shapes were probably hydrodynamically stable among Paleozoic biconvex taxa, even if they lived on muddy substrates. The experiments on extant species showed that threshold velocities increased when the substrate became coarser (gravel replaced sand), but the pattern of stable versus unstable orientations among the nine feasible life orientations remained virtually unchanged (Figures 3-11). The results suggest that the same stability relationships between erect and reclining orientations would occur if mud had been the substrate. Hjulstrom's experiments (1939) showed that the velocity to erode either consolidated clay and silt (.01 to .02mm) or fine gravel (2-4.0mm) is the same. Thus these experiments also have extrapolative value for assessment of stable orientations and shapes on fine grained substrates occupied by many Paleozoic biconvex taxa.

Among spiriferids which propped up the valves on their strophic (linear) hinge line, the experiments on the shells of *T. transversa* suggest that the erect orientation with a lateral margin facing into the current was the most stable (Figure 9). Supporting evidence is found in the asymmetrical distribution of epizoans across the valves of *Paraspirifer bownockeri*, interpreted as a response by suspension feeding encrusters to the food-bearing eddies impinging against the downcurrent lateral margin, the 'wake' of the flow around the shell (Kesling et al., 1980). Admittedly, a reclining orientation with a lateral margin facing upcurrent could also account for the asymmetrical distribution of filter-feeding encrusters. Asymmetrical distributions of epizoans on Ordovician biconvex shells (Alexander & Scharpf, 1990) are also congruent with a host life orientation with the shell erect or reclining and a lateral margin upcurrent.

The experiments on unattached shells of extant species also showed how unstable in erect orientations are species with flattened conical shapes, an architecture common to rhynchonellids. Paleozoic rhynchonellids with such unstable conical shapes, e.g., Ordovician *Rhynchotreta*, remained pediculate throughout life, whereas other rhynchonellids that became globose in gerontic life stages, e.g., Mesozoic *Septirhynchia* (Mancenido & Walley, 1979), lost a functional pedicle. Paleozoic and Mesozoic rhynchonellids commonly lived with their valves erect (Hallam, 1962; Richards, 1972; Westbroek, et al., 1975; Mancenido & Walley, 1979; Alexander, 1987; Alexander & Gibson, 1993). Unattached inflated shells in such orientations would have been hydrodynamically unstable, as evidenced from the experiments on globose *Laqueus californianus* (Figure 10), unless countered by the posterior-weighting that characterized these geniculate rhynchonellids (Mancenido & Walley, 1979). The brephic stage of many rhynchonellids, with its compressed valves, would also have been unstable in erect orientations, much like the unattached flattened shells of *Notosaria nigricans* (Figure 6), were it not for the anchoring pedicle. Retention of a functional pedicle in brephic stages of rhynchonellids was imperative for hydrodynamic stability in the absence of posterior weighting of the valves in early ontogenetic stages.

In contrast, many spriferids had very stable shapes for erect orientations (Alexander, 1984), much like the nearly strophic Recent *Terebratalia transversa* (Figure 9). Brephic stages in many spiriferids have lateral profiles similar to gerontic stages. Over geologic time, selective pressure for retention of the pedicle relaxed in this order, even on the posteriorly weighted brephic stages. Mid-Paleozoic spiriferid families evolved which lacked a pedicle in the earliest ontogenetic stage, as evidenced by the pseudodeltidium.

Many Paleozoic pentamerids have a shell outline similar to the thin valved Recent *Liothyrella uva*, a design which is not very stable for erect orientations (Figure 5). Yet pentamerids were very stable in erect orientations, as evidenced by abundantly in situ preservations (Ziegler et al., 1966). Pentamerids have extensively thickened shell posteriors that countered any instability in the globose shell after the pedicle atrophied. But, for most non-strophic Paleozoic brachiopods that did not have substantially thickened valves posteriorly, such as some atrypids and athyrids, reclining orientations provided stability, at least after death, if not in life. Inferred life positions based on dorsibiconvex to resupinate shell convexities (e.g., Copper, 1967), epibiont distributions (e.g., Cuffey et al. ,1995), taphonomic shell deformation (Alexander & Gibson, 1993), and flume experiments on fossil shells (Alexander, 1984), are now corroborated by flume experiments on empty shells of extant nonstrophic biconvex species. Reclining orientations are most stable for non-strophic brachiopods in the absence of pedicles to hold the shells erect.

CONCLUSIONS

Empty shells of extant terebratulids and rhynchonellids are more stable on gravel than sand. Scouring of sand from around weight-bearing areas of shells destabilizes the shell at lower competency velocities. Reclining orientations of these nonstrophic shells are most stable if the anterior faced downcurrent, a position that would require live individuals to filter water from the downcurrent wake (=leeside suspension feeding). Among valve-erect orientations, shells with the lateral margin upcurrent are most stable, an orientation complementary to the separation of lateral incurrents from medial excurrents of live terebratulids. In such an erect orientation, globose, equibiconvex valves are less stable than compressed valves, and posteriorly weighted shells are more stable than shells of uniform thickness. Plicae and costae decrease shell stability by increasing turbulence and sediment scour around the weight bearing area of ornamented shells.

Acknowledgments
We appreciate specimen donations from Gordon Curry, Michael LaBarbera, Mark James, Daphne Lee, and Melissa Rhodes for use in this investigation. Jackie Frizano assisted with the experiments and data analysis of certain species. Melissa Rhodes kindly made critical suggestions on manuscript drafts.

REFERENCES

ALEXANDER, R.R. 1977. Generic longevity of articulate brachiopods in relation to the mode of stabilization on the substrate. Palaeogeography, Palaeoclimatology, Palaeoecology 21:209-226.

___1984. Comparative hydrodynamic stability of brachiopod shells on current-scoured arenaceous substrates. Lethaia 17:17-32

___1987. Intraspecific selective survival within variably uniplicate Late Devonian brachiopods. Lethaia 20:315-325.

___& C.D. SCHARPF.1990. Epizoans on Late Ordovician brachiopods from southeastern Indiana. Historical Biology 4:179-202.

___& M.A. GIBSON.1993. Paleozoic brachiopod autecology based on taphonomy: example from the Devonian Ross Formation of Tennessee. Palaeogeography, Palaeoclimatology, Palaeoecology 100:25-35.

COPPER, P. 1967. Adaptations and life habits of Devonian atrypid brachiopods. Palaeogeography, Palaeoclimatology, Palaeoecology 3:363-379.

CUFFEY,C.A., A.J. ROBB, J.T. LEMBCKE & R.J. CUFFEY. 1995. Epizoic bryozoans and corals as indicators of life and post-mortem orientations of the Devonian brachiopod *Meristella*. Lethaia 28:139-154.

DENNY, M. 1995. Survival in the surf zone. American Scientist 83:166-173.

FÜRSICH, F.T. & J.M. HURST. 1974. Environmental factors determining the distribution of brachiopods. Palaeontology 17:879-900.

HALLAM, A.1962. Brachiopod life assemblages from the Marlstone rock-bed of Leicestershire. Palaeontology 4:653-659.

HJULSTROM, F. 1939. Transportation of detritus by moving water, p.159-184. In, P. D. Trask (ed.), Recent marine sediments. American Association Petroleum Geologists, Tulsa.

KESSLING, R.V., R. HOARE & D.K. SPARKS. 1980. Epizoans of the Middle Devonian brachiopod *Paraspirifer bownockeri*: their relationships to one another and to the host. Journal Paleontology 54:1141-1154.

McGHEE Jr., G. R. 1980. Shell geometry and stability strategies in the biconvex Brachiopoda. Neues Jahrbuch Geologie Paläontologie, Monatshefte 3:155-184.

MANCENIDO, M.O. & C.D. WALLEY. 1979. Functional morphology and ontogenetic variation in the Callovian brachiopod *Septirhynchia* from Tunisia. Palaeontology 22:317-338.

MENARD, H.W. & A.J. BOUCOT. 1951. Experiments on the movement of shells by water. American Journal Science 249:131-151.

RICHARDS, R.P. 1972. Autecology of Richmondian brachiopods (Late Ordovician of Indiana and Ohio). Journal Paleontology 46:386-405.

THAYER, C. W. 1975. Morphologic adaptations of benthic invertebrates to soft substrata. Journal Marine Research 33:177-189.

VERMEIJ, G.J. 1993. A natural history of shells. Princeton University Press, Princeton, 207 p.

WESTBROEK, P., F. NEIJNDORFF & J.H. STEL. 1975. Ecology and functional morphology of an uncinulid brachiopod from the Devonian of Spain. Palaeontology 18:367-375.

ZIEGLER, A.M., A.J. BOUCOT & R.P. SHELDON. 1966. Silurian pentameroid brachiopods preserved in position of growth. Journal Paleontology 40:1032-1036.

NOTES ON LIVING BRACHIOPOD ECOLOGY IN A SUBMARINE CAVE OFF THE CAMPANIA COAST, ITALY

EMMA TADDEI RUGGIERO

Dipartimento di Scienze della Terra, Università di Napoli 'Federico II', Napoli, Italy

ABSTRACT- Brachiopods living in a submarine cave off the Sorrento Peninsula included the inarticulate *Neocrania anomala*, and articulates *Megathiris detruncata*, *Argyrotheca cordata*, and *A. cuneata*. Population and individual analyses carried out on *Neocrania anomala* were followed during growth from 1991 to1995 by means of photographic surveys of areas on the cave wall. In this protected environment, the population of *N. anomala* remained virtually unchanged through five years of observation. Their growth, evaluated on basis of the shell areas occupied, was almost static, or very low. Analysis of their growth rate shows that these brachiopods could take more than 40 years to reach the maximum dimensions observed.

INTRODUCTION

In situ observation of brachiopods along the coasts of Campania was carried out to obtain data on environment and mode of life, which, in turn, could be useful for understanding present day distribution. A brachiopod community consisting of *Neocrania anomala*, *Megathiris detruncata*, *Argyrotheca cordata*, and *A. cuneata* was found living in submarine caves off the Campanian coast, from low-tide level to a depth of about 20m. A preliminary paper presented in 1993 dealt with brachiopods specifically from Isca Cave on the Sorrento Peninsula (Taddei Ruggiero, 1994), and in which wall biocoenoses (sampled by removing their substratum), thanatocoenoses from bottom sediments, and results of the first two years of observation of individuals in two fixed squares, were presented. In the present paper, observations on the same cave specimens, made from from 1991 to 1995, are provided.

THE CAVE SETTING

The cave studied is located on the island of Isca near Nerano, on the south side of the Sorrento Perninsula, flanking the Gulf of Naples (Figure 1). The cave is large, ca. 60m long (N-S) and 30m wide, with a broad submerged opening, 2m wide and 8m high (-14m to -8m deep), facing south. Its partially subaerial, dome-shaped vault reaches a level of +11m and its floor rises from -14m at the entrance, to up to -2m on the opposing side. Though the opening is broad, the light decreases rapidly and almost the whole cave is in total darkness. The cave has subvertical walls and is opened inside Cretaceous limestone.

The percentage of oxygen in the cave is normal, pH is between 8.03 and 8.18, and salinity measures 37.05 (with a slightly lower value than outside, 37.50). The half-day tidal fluctuations are within a range of 25cm. A feeble, though perceptible anti-clockwise current produces a good supply of water along the walls; there is no such current at the centre of the cave. Due to topographical characteristics of the cave, sediments from outside are prevented from being introduced inside. The sedimentation rate inside the cave is very low.

Endolithic organisms, such as *Lithophaga lithophaga* and *Cliona celata*, bore into the limestone cave walls up to 5-10cm deep. These borers are completely covered by epifauna in

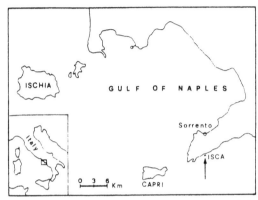

Figure 1. The Island of Isca, off the Campanian coast, Italy, on which the Isca brachiopod cave is located.

several layers. The distribution of the individuals is spotty and depends chiefly on light. In the cave part closer to the opening, red algae (*Peyssonnelia*), bryozoans, colonial corals and large sponges are most numerous. At 4-5m to the inside, algae disappear, and the wall population consists of sessile filter feeding benthos. Among the most numerous of these are four species of brachiopods (*Neocrania anomala*, *Megathiris detruncata*, *Argyrotheca cordata*, and *A. cuneata*), colonial and solitary corals, sponges (*Petronia ficiformis*, *Sycon elegans*, and *S. raphanus*), serpulid polychaetes (*Vermiliopsis* and *Semivermilia*), and the bryozoans *Miriapora* and *Sertella*. Less numerous are molluscs, represented almost exclusively by young individuals of *Arca noae*, and foraminiferans, mostly *Eponides repandus* and *Miniacina miniacea*.

METHODS

Due to entire cementation of the ventral valve of *Neocrania anomala* to the substrate, it is possible to trace variations through the years in the population by means of a photographic survey (Figure 2). This method permitted the study of thanatocoenoses of *N. anomala* as well, because, even when the organism died, its ventral valve adhered to the substrate for some years.

On the subvertical wall, in the zone richest in brachiopods, two squares, A (55x35cm) and B (20x30cm), were spotted and defined. Photographic documentation was made in July and August of each year from 1991 to 1995 by means of a Nikonos V, using 35 mm and macro lenses. Photographs were scanned, and image analysis carried out in digital format by employing specially designed, interactive software. This allowed us to determine the outline and area for each brachiopod shell in pixels. As the photo enlargement ratio is known, it was possible to calculate the real size of each individual. Analyses were of two types. The first was a population analysis by study of

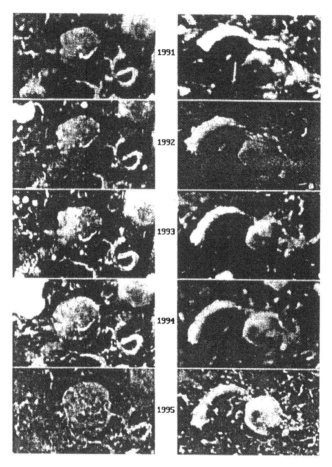

Figure 2. A sequence of five images of the inarticulate *Neocrania anomala* taken from 1991 to 1995 (top to bottom) of specimens # 194 (left) and # 204 (right), x2.6.

dimensions of a living population of *Neocrania anomala* (Figure 3a), and of its thanatocoenosis adhering to the wall (Figure 3b).

The second analysis required a study of the growth of individuals from 1991 to 1995. For this, 143 living specimens were measured, though not all appear in each of the 5 plots. In the analysis, aimed at emphasizing variations in area of specimens, only 85 were recorded for at least 4 years. The analysis consisted of evaluating average growth rate for each individual from 1991 to 1995 by means of regression study (Figure 4). Besides informing us on size variations, the photographic survey allowed us to follow the birth and death of individuals through the years

COMMUNITIES IN PLOTTED SQUARES

It was observed that most of the living organisms in the squares stuck to others, either living or dead. Associations mostly consisted of brachiopods and serpulids; less numerous were sponges, young individuals of *Arca noae*, *Miniacina*, bryozoans, and rare solitary corals. Among brachiopods, the inarticulate *Neocrania anomala* (Müller) dominated; less numerous were *Argyrotheca cuneata* (Risso) and *A. cordata* (Risso). *Megathiris detruncata* (Gmelin) was even rarer.

Neocrania anomala was present at up to 468 individuals per square meter in the two squares studied.

The most conspicuous variations in the community were amongst the sponges. The growth of serpulids was also variable: area occupied during the last year studied appeared much reduced. The bivalve *Arca noae* was usually observed only in its juvenile stages; possibly these later moved to the outside of the cave. Insofar as growth, one individual bivalve present in 1992 was still at the same place and almost the same in size in 1995.

Neocrania anomala was abundantly present as both living and dead individuals. Individuals which had died recently, showed white valve interiors; later on these valves were colonized by other organisms, often by young individuals of the same species. It was possible to record the death of particular individuals for each year of observation: specimen # 20 was found dead in 1992, # 5 in 1993, #152 in 1994, and #228 in 1995. The last was found bearing a shell of *Argyrotheca cuneata* on its valve. The fixation of two *Neocrania* was also observed, one in 1993 and the other in 1994; both were fixed inside a valve of dead *Neocrania*. Most valves of living *N. anomala* bore almost no epibionts. Serpulids were found only

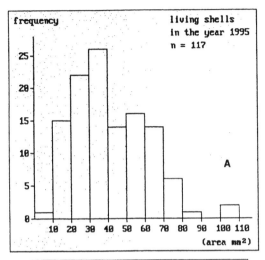

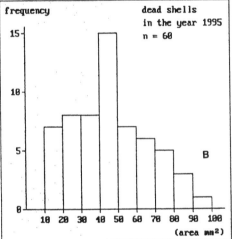

Figure 3. Population analyses of *Neocrania anomala* by photographic survey (1995), over an area of 2500 cm². (a) frequency histogram of the areas of living individuals; (b) frequency histogram of the areas of dead individuals. For the abscissae the dimensions are given in mm² (classes of 10), and on the ordinates the number of individuals in each class is given.

on a few megathyridids, and *Miniacina*.. No attachment of a juvenile individual of the same species was observed.

Photographs did not allow us to observe the growth of the terebratulid Megathyrididae. On account of their small size, *Megathiris detruncata* reaches a maximum width of 6mm and *Argyrotheca* 4mm, a greater enlargement of the negative is required. Moreover, their attachment, a short pedicle and the commissural plane almost at right angles to substratum, makes measurement extremely difficult. Only the presence and disappearance of these organisms could be observed. In 1995, on specimen # 98 of *Neocrania anomala* (in square A), there was an *Argyrotheca cuneata* of similar size present in 1991. We may deduce that its adult life spanned at least four years.

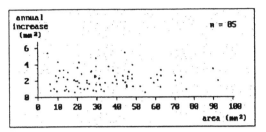

Figure 4. Scattergram of yearly increments of areas occupied by brachiopods (observations 1991-1995), as compared to total areas. Scale of ordinates is x4. Absolute increase remained almost unchanged independent of individual size, averaging about 2 sq. mm. Areas of specimens in first year (5.23-92.60 mm²) are shown on abscissae; yearly growth on ordinates (scale of y was magnified 4x to make measurements more legible).

ANALYSIS OF DATA

The population analysis carried out on *Neocrania anomala*, on data collected in 1995, has shown that for the total area of the two squares considered [2500cm2]; 117 living *N. anomala* occupied an area of 49.76 cm2 (2%), and 60 dead individuals occupied an area of 28 cm2 (1.1%). The biocoenosis histogram (Figure 3a) was slightly bi-modal, while the thanatocoenosis histogram (Figure 3b) was uni-modal, with a peak coinciding with the corresponding minimum of the biocoenosis. This minimum corresponds to class 40-50 mm2 (diameter 7-8 mm). This minimum appears to correspond with a crisis which took place when organisms reached a particular size. Analysis of the yearly growth of settlement areas in *N. anomala* shows that growth is fairly homogeneous and independent of individual size (Figure 4). This allows us to conclude that growth, estimated by absolute increase of attachment areas, follows a linear trend. Measurements of two specimens in their first year were taken in 1993 and 1994, and these covered 4 sq.mm and 2 sq.mm in 1995 respectively.

We hypothesize that growth of individuals took place with yearly increase of ca. 2 sq.mm, and followed a linear tren which passed approximately through the origin of the axes (Figure 4). Larger individuals could be more than 40 years old. These conclusions do not consider rapid growth under special circumstances. Nevertheless, it seems that rapid growth could not have taken place during the years of observation. The cave environment appears to have been conservative and stable, and to lie far from any source of anthropogenic pollution.

DISCUSSION

For *Neocrania anomala*, the most striking result was their very slow growth and apparently long life span. Variations observed in the area after 5 years, besides slow growth, were limited to the death of four shells and the new settlement of two shells. Rowell (1960), in his study of the first stages of growth of *N. anomala* in NW Scotland, found attached specimens with a diameter of 0.2mm and up. From a population analysis he inferred that there was a very high death rate among very young individuals, fast growth in the first year, and very slow growthj in the following years, with a short lifespan which could reach four or five years. Due to the method employed, and the type of cave substratum, shells smaller than 1mm diameter could escape attention. I have therefore no useful data for the first year of life in *Neocrania*, but later growth appears to have been very slow, and indicates that *N. anomala* in the Isca Cave has a long life span (Figure 5).

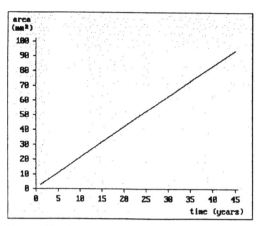

Figure 5. Virtual trend in the growth of individuals of *Neocrania anomala* (with reference to areas occupied by shells), based on results shown in Figure 4. Larger individuals (ca 100 sq. mm) ought to be more than 40 years old.

A long life span has not been hypothesized for living brachiopods. Paine (1969) believed that the articulate brachiopod *Terebratalia transversa* lived more than 10 years. Curry (1982) and Collins (1991) believed that *Terebratulina retusa* had very limited growth in its first six months of life, and that it then became considerably bigger, and reached ages of 7 to 8 years. These studies were made on populations obtained by dredging. Rickwood (1977), who studied living communities of *Waltonia inconspicua* from New Zealand, stade that growth was slow, unevenly distributed through life, and varied considerably between localities. In Otago Harbour *Waltonia inconspicua* was found to live up to 8, but rarely 15 years.

The population of *Neocrania* was stable, and variation due to death and settlement of new individuals was rare (Figure 3a). The stable cave environment probably led to a lack of sudden stress. The scarcity of newly settled individuals was perhaps due to insufficient suitable substrate available, with the consequence that larvae either did not succeed in fixation and died, were left to colonize other zones (even outside the cave), or were stuck to the proximity of existing adults. Rowell (1960) pointed out that *Neocrania* in Scotland died in large numbers in their first year, with only few survivors.

The problems connected with the presence or absence of juveniles in living and fossil populations have been faced by several authors (Paine,1969; Neall, 1969; Thayer,1975; Stewart,1981; Curry 1982; Collins,1991). Besides varying behaviour among different species, due to inherent genetic causes (number of eggs produced, presence of brood pouches, length of the larvae, larval photophoby or phototropy), and ecological constraints (type of attachment, type of substrate), it is possible that juveniles are lost from thanatocoenoses by currents, and that they therefore escape collection.

The fact that the Isca cave brachiopods were found only on the cave walls and not on boulders in the centralarea of the cave, was first explained by lower sedimentation or accretion rate on subvertical walls. Tunnicliffe & Wilson (1988) found that *Crania californica* from the fiords of British Columbia was most abundant on sloping surfaces (less than 90°), where sediments could accumulate. Rhodes & Thayer (1990) have shown experimentally that brachiopods can live in turbid waters, since

they are able to distinguish and expel detritic particles. This suggests that the distribution of *Neocrania anomala* is not exclusively due to greater sedimentation, but chiefly due to water and nutrient changes in the central part of the cave. Abundance of brachiopods in some dark caves may partially depend on their photophoby, or, even more, on the absence of competitors for space. In the case of the articulate megathyrids, low energy apparently makes attachment of larvae easier. Intense grazing in the photic zone, by chitons and regular echinoids, is contrasted by the total absence of herbivorous organisms in the aphotic cave setting (Asgaard & Stenton, 1984; Asgaard & Bromley, 1990; Thayer & Aiello Allmon, 1990).

CONCLUSIONS

The inarticulate *Neocrania anomala* in the Isca Cave is marked by very slow growth, with larger shells possibly reaching an age of more than 40 years. Populations were stable: death and attachment of new individuals were rare events. An adult lifespan of at least four years was observed in the articulate *Argyrotheca cuneata*. The restriction of brachiopods to vertical walls may be partially explained by lack of sedimentation, but also by inflow of nutrients in peripherical currents along cave walls. Abundance of brachiopods in this cryptic setting may be attributed to lack of light, low energy, low sedimentation, but also to absence of predators, grazing herbivores and competitors for space. Competition for space appears to play an important role, leading to complete cave wall cover, and stratification of both dead and living epibionts.

Acknowledgments

My most sincere thanks go to Domenico Fratta who assisted me in diving and photography, and my husband Roberto Taddei, who helped me in the statistical analyses.

REFERENCES

ASGAARD, U.& R.G.BROMLEY. 1990. Colonization by micromorph brachiopods in the shallow subtidal of the Eastern Mediterranean Sea, p.261-264. In, D.I. MacKinnon, D.E. Lee & J.D. Campbell (eds.), Brachiopods through time. Balkema, Rotterdam.

ASGAARD, U. & N. STENTOFT. 1984. Recent micromorph brachiopods from Barbados: palaeoecological and evolutionary implications. Geobios (Mémoire Speciale) 8:29-33.

COLLINS, M.J. 1991. Growth rate and substrate-related mortality of a benthic brachiopod population. Lethaia 24:1-11.

CURRY, G.B. 1982. Ecology and population dynamics of the Recent brachiopod *Terebratulina* from Scotland. Palaeontology 25:227-246.

NEALL, V.E. 1969. Notes on the ecology and paleoecology of *Neothyris*, an endemic New Zealand brachiopod. New Zealand Journal Marine Freshwater Research 4:117-125.

PAINE, R.T. 1969. Growth and size distribution of the brachiopod *Terebratalia transversa* Sowerby. Science 45:337-343.

RHODES, M.C. & C.W. THAYER. 1990. Effects of turbidity on suspension feeding: are brachiopods better than bivalves?, p.191-196. In, D.I. MacKinnon, D.E. Lee & J.D. Campbell (eds.), Brachiopods through time. Balkema, Rotterdam.

ROWELL, A.J. 1960. Some early stages in the development of the brachiopod *Crania anomala* (Müller). Annals Magazine Natural History 3:35-52.

RICKWOOD, A.E. 1977. Age, growth and shape of the intertidal brachiopod *Waltonia inconspicua* Sowerby from New Zealand. American Zoologist 17:63-73.

STEWART, I. R. 1981. Population structure of articulate brachiopod species from soft and hard substrates. New Zealand Journal Zoology 8:197-207.

TADDEI RUGGIERO, E. 1994. Brachiopods from bio- and thanatocoenoses of the Isca submarine cave (Sorrento Peninsula). In, R. Matteucci, M.G. Carboni & J.S. Pignatti (eds.), Studies on Ecology and Palaeoecology of Benthic Communities. Bollettino della Società Paleontologica Italiana, Special Volume 2: 313-323.

THAYER C.W. 1975. Size-frequency and population structure of brachiopods. Palaeogeography, Palaeoclimatology, Palaeoecology 17:139-148.

----& R. AIELLO ALLMON. 1990. Unpalatable thecideid brachiopods from Palau: ecological and evolutionary implications, p.253-260. In, D.I. MacKinon, D.E. Lee & J.D. Campbell (eds.), Brachiopods through time. Balkema, Rotterdam.

TUNNICLIFFE V. & K. WILSON. 1988. Brachiopod populations: distribution in fjords of British Columbia (Canada) and tolerance of low oxygen concentrations. Marine Ecology, Progress Series 47:117-128.

EVOLUTION OF DEVONIAN PLICATHYRIDINE BRACHIOPODS, NORTHERN EURASIA

M. A. RZHONSNITSKAYA & T .L. MODZALEVSKAYA

All-Russian Geological Research Institute (VSEGEI), St Petersburg, Srednii prospekt 74, Russia199026

ABSTRACT- Plicathyridines, a group of short lived, dynamically evolving, Old World and northern Eurasian athyrids of the Middle and Late Devonian, vanished in stepwise fashion during the Frasnian-Famennian mass extinction. Plicathyridine genera, as a rule, have short stratigraphic ranges, with regional differences in their occurrence, radiation and extinction. Two major innovations occurred, one in the Emsian-Eifelian (*serotinus-partitus* conodont zones) of the Rheno-Ardennian province, and the other in the Early to Late Frasnian (*transitans-linguiformis* conodont zones) of the Altai-Sayan province in the Kuznetsk Basin. Their disappearance coincides with the base of the Famennian lower *triangularis* Zone and is correlated westward with the Upper Kellwasser event.

INTRODUCTION

Plicathyridines constitute a distinctive and early diversified group within the Devonian shelly faunas of northern Eurasia, are well known in many regions and occur at many localities where Devonian brachiopod sequences are studied. In the European part, they are known from the Russian platform, middle and southern Timan, the Arctic Islands, Pai-Khoi, the west slopes of the southern Urals and in the Transcaucasus (Nalivkin,1941; Lyashenko, 1959, 1973; Cherkesova, 1973; Mamedov & Rzhonsnitskaya, 1985; Markovskii,1989). In Asian areas they occur on north Kotelny Island, in the Kuznetsk Basin, Gorny Altai and south Sette-Daban ranges (Khalfin, 1946; Rzhonsnitskaya, 1959, 1968; Gratsianova, Zinchenko & Kulkov,1960; Nikolaev & Rzhonsnitskaya,1967: Figure 1).

Since the first papers about the stratigraphic and geographic distribution of plicathyridines, a great deal of new information about these brachiopods, concerning their morphology, systematic, biostratigraphic and biogeographic occurrences, has been published (Grunt, 1980; Gratsianova & Dagys, 1983; Alvarez, 1990; Anonymous, 1992). Published and unpublished data on plicathyridines of Northern Eurasia allow us to review the morphologic diversity and stepwise nature of plicathyridine disappearance at the Frasnian-Famennian boundary.

LINEAGES OF PLICATHYRIDINE SPECIES

Plicathyridine genera, e. g. *Plicathyris* Khalfin 1946, *Anathyris* Peetz 1961, *Hexarhytis* Alvarez 1990, and *Anathyrella* Khalfin 1960, tend to be paucispecific and defined by species groups or lineages. They are characterized by a transversely triangular shape and long hinges (winged), similar to spiriferids (as seen in *Anathyris*), and large sizes, with straight or slightly curved hinge line (*Anathyrella, Anathyris*). There is a high angle to the ventral posterior umbo, a pedicle collar, and highly elevated outer part of the cardinal plate, or cardinal process. Available records of a wide range of variation in external morphology show a sometimes bizarre appearance. These genera are abundant in the Russian Platform, and especially in the Kuznetsk Basin, with many Emsian-Frasnian endemic species (*ezquerrai, phalaena, helmerseni, tschernyshevi, tarda, montzevi, ussovi,* and *monstrum* : Figure 2).

The plicathyridines probably evolved from *Hexarhytis ?undata* (Defrance), which appeared in the Late Lochkov: this species possessed a well-developed fold and sulcus, but no radial plicae nor median fold on the sulcus. The oldest reliable plicathyridine (*Hexarhytis*) is known from the Late Pragian-Early Emsian of the Rheno-Ardennes area, and in the Eifelian of the Transcaucasus (Figure 3). Such lineages as *Hexarhytis bonarensis - H. ferronosensis, Plicathyris ezquerrai* and *Anathyris phalaena* evolved during the Emsian-Eifelian. During the Emsian they were diverse and abundant in the Cantabrian region and Armorican Massif, even to their last occurrence in the western Pyrenées, north Africa and Arabia (Alvarez, 1990).

By contrast with European plicathyridines, Siberian forms are restricted to the uppermost Givetian and lowest Frasnian and the species *Plicathyris sibirica* (Figure 3) is distinct externally by its rounded plicae and small size.

Diagnostic features of the *Anathyris phalaena* lineage are mucronate ears and a well-developed cardinal area on both valves. The sinus is restricted by well-developed, rounded plicae and the sulcus consists of two large folds, divided by the fold (Figure 3). This group, which appeared in the late Emsian, and diversified modestly through this time, underwent striking diversification in the Frasnian of the Kuznetsk Basin. The *Anathyris phalaena* group is also well known in the Middle Devonian (Eifelian) of England (Hope's Nose, Devonshire) and in Missouri, USA. (Alvarez, 1990).

The *Anathyris helmerseni* lineage shows a trend from subequally shells with transverse outline and straight hinge line, to those that possess unequally convex shells and acute cardinal angles. Added to this are *Anathyris svinordensis* Nalivkin (Svinord Beds of the 'Main Devonian Field', Russian Platform) and *Anathyris* n. sp. 'V' (Figure 3), the transitional form between *Plicathyris* and *Anathyris*. These are characterized by lateral plicae only in the distal margin, as seen in *Anathyris* n. sp. 'H' (Figure 3).

The *Anathyris tschernyschewi* lineage is characterised by a rounded subrectangular outline and almost straight hinge, with sinus and sulcus sharing a longitudinal median fold (Figure 4). Among the species assigned to this group, there is only one other, *Anathyris* n. sp. 'L' (= '*Anathyris helmerseni*' of Lyashenko, 1959) from the Timanian Horizon of Middle Timan.

Anathyris tarda Markovski, 1989 (Askyn Horizon, west slopes of the Urals) is typical of its lineage, having smaller shells, a rounded or trapezoidal outline with slightly developed plicae, and a restricted sinus. There is a rather high sulcus enlarged by a median fold anteriorly. *Anathyris* n.sp. 'S'(Figure 4) is the last member within this group, occurring in the Strelna Horizon of the Kuznetsk Basin.

There are two species in the early Frasnian *Anathyris ? monzevi* group. Both species (from the Pskov Beds of the NW Russian Platform, and the Ustjrech Formation of south Timan)

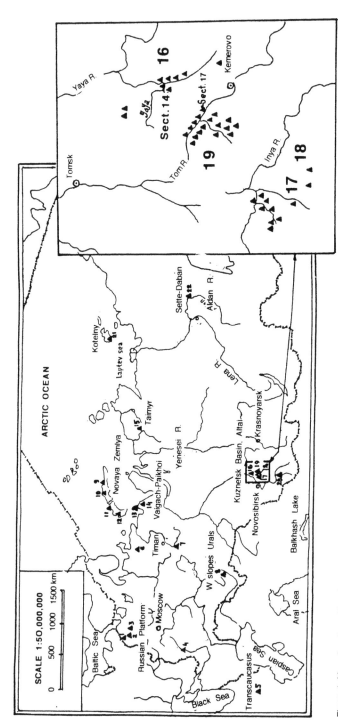

Figure 1. Map of northern Eurasia showing main localities of plicathyridine brachiopods, and inset map of the Tom and Yaya river areas of the Kuznetsk Basin, where many plicathyridine localities are found. Localities: 1-4, Russian Platform; 5, Transcaucasus; 6-7, Timan; 8, W slopes Urals; 9-12, Novaya Zemlya; 13-14, Vaigach-Paikhoi; 15, Taimyr; 16-20, Kuznetsk Basin, Altai; 21, Kotelny; 22, Sette-Daban.

234

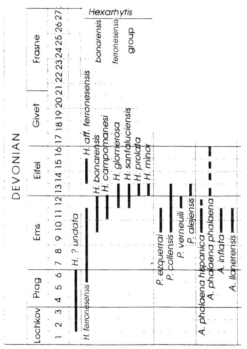

Figure 2. Ranges of Devonian plicathyridines, referred to numbered conodont zones, adapted from Alvarez (1990), and new data.

possess a shell that is not large, with rounded outline, a poorly defined sinus, and sulcus with a median fold. They represent, perhaps, a neotenous stage of other plicathyridine species.

The rapid evolution of the *Anathyrella ussovi* lineage is characterized by the possession of complex and unusual external and internal structures, which began in the Kuryak Horizon and Sergiev Formation of the NE and NW margins of the Kuznetsk Basin (e.g. *A. tyzhnovi* : Figure 4). Within this group, plicathyridines are characterised by rather large shells with hypertrophic, large sinus and sulcus, and there is usually no median fold. This species group perhaps lead to the development of the morphologically distinct *A. monstrum*

lineage, the last *Anathyrella* 'giants': their main feature is a distinct longitudinal median fold on the sulcus. These occurred mainly in the Kelbess Formation of the NW margin of the Kuznetsk Basin and Gorny Altai (Figure 4). Thus, plicathyridine groups form almost continuous ranges, pointing to a stepwise and final extinction pattern by the end of the Frasnian.

KUZNETSK BASIN BIOSTRATIGRAPHY

The Frasnian-Famennian sequence was studied in coastal-marine carbonate facies, well exposed on the right bank of the

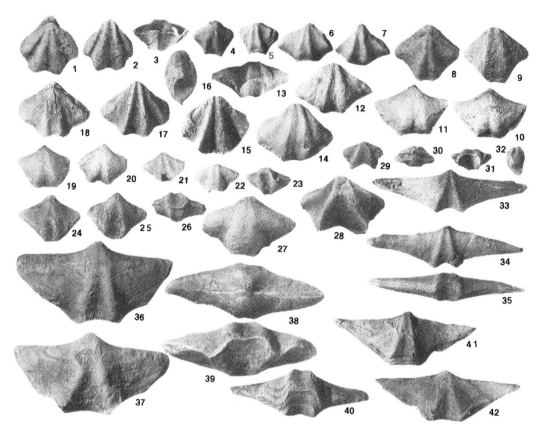

Figure 3. Selected growth forms of plicathyridines. 1-3, *Hexarhytis* ex. gr. *ferronesensis* (Verneuil & Archiac), Transcaucasus; 4-18, *Plicathyris sibirica* (Khalfina), Isylyl Horizon, Kuznetsk Basin; 19-27, *Anathyris* n.sp. `H', Isyly Horizon, Kuznetsk Basin; 28-32, *Anathyris* n.sp. `V', Novosibirsk beds, Sette-Daban; 33-42, *Anathyris phalaena supraphalaena* Khalfin, Vassino Horizon, Kuznetsk Basin. All approx. x0.8, except 27-28, c. x1.5.

Tom River and the left bank of the Yaya River. Here, during Upper Frasnian time, calkcareous sands and shales were deposited. The rich brachiopod communities (of Benthic Assemblages 2-3), dominated by plicathyridines, arose in the littoral zone alongside pentamerids, atrypids, spiriferids, and other athyrids. There are in sandy-shaly limestone coquinas with *Anathyrella* in the upper part of the Kelbess Formation. The plicathyridines, as well as other brachiopods, are found in the Solomino and Vassino horizons, which are correlated with the *linguiformis* Zone: *Anathyrella ussovi* can be treated as an index fossil for the uppermost Solomino Horizon. On the basis of plicathyridines, biozones are distinguished which usually correspond with the vertical range of the characteristic association in the sections studied. Changes in taxonomic composition with the beginning of each new sedimentory cycle have been used for drawing a boundary between the Solomino and Peshcherka horizons. The changes coincide with the appearence of Famennian *Icriodus* and other conodonts which characterise the Upper *triangularis* Zone of the Lower Famennian. Plicathyridines are absent near this boundary or several meters below it. The most noteworthy changes, however, are the appearances of other new taxa of athyrids, believed to be derived from the conservative and long ranging stocks. An example is *Athyris globularis* with its novel morphology.

BIOGEOGRAPHY

There are no plicathyridines in the Upper Devonian of the Old World Realm. At this time they occurred only on the Russian Platform and adjacent areas, and in the Altai-Sayan province. It can be shown that two different centres of plicathyridines distribution, with two major innovations occurred. The first was in Western Europe, the Cantabro-Armorican area of the Rheno-Ardennes province. Here the first appearance of plicathyridines is in Pragian time (*Hexarhytis, Plicathyris*). Their diversification was in the Late Emsian to Early Eifelian, and the first crises for plicathyridines, in terms of low diversity, was the late Eifelian and Givetian, when they became almost completely extinct. These bioevents coincide with what may have been global climatic cooling (Alvarez, 1990). The second was in Asia, in the Kuznetsk-Altai area of the Altai-Sayan province. Plicathyridines are known no earlier than the late Givetian. During the Early to Middle Frasnian, plicathyridines display great abundance and diversification, with many endemic taxa e.g., *Anathyris* and *Anathyrella*. The appearance of endemic genera and species resulted from their occurrence in a shallow, isolated sea. At the same time in other parts of northern Eurasia (the Russian Platform and adjacent areas), only the *Anathyris helmerseni* group developed during the early Frasnian. This platform was flooded, allowing relatively unrestricted migration. The distribution of plicathyridines shows a possible migration path between the Cantabro-

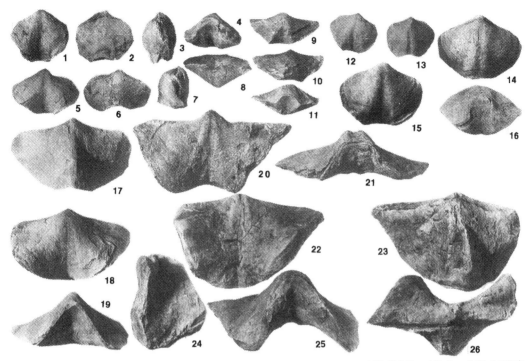

Figure 4. Selected species of plicathyridines from the Kuznetsk Basin. 1-8, *Anathyrella tyzhonovi* (Khalfin), Kuryak Horizon; 9-11, *Anathyris fimbriata* Khalfin, Isyly Horizon; 12-13, *Anathyris* n.sp. `S', Solomino Horizon; 17-19, *Anathyris tschernyschewi* Khalfina, Vassino Horizon; 20-21, *Anathyrella ussovi* (Khalfin), Solomino Horizon; 22-26, *Anathyrella* n.sp. `Su', Kelbes Formation. All c. x0.8.

Armorican area, the Russian Platform, Eastern Siberia and Laurentia during the Devonian (Alvarez, 1990). Perhaps plicathyridines migrated via England, the northwest Russian Platform, Timan, and the Arctic islands up to NE Asia.

Plicathyris and *Anathyris* are more globally distributed taxa than *Hexarhytis*, which is found only in the Transcaucasus.

CONCLUSIONS

The Frasnian-Famennian boundary event in the Siberian region resulted not only in the stepwise extinction of all plicathyridines, but also pentamerids, atrypids and other brachiopods, as well as tentaculitids and some ammonoids. Major extinctions and radiations in Siberian brachiopod evolution are commonly associated with sealevel fall or rise, and there is also a correlation between sealevel change and trends in plicathyridine survivorship throughout Upper Devonian time, with higher rates of extinction perhaps during colder climates.

Acknowledgments

Research was supported by the International Foundation 'Cultural Initiative', and by the Russian Academy of Natural Sciences, program 'Biodiversity'. Travel funds to the Third International Brachiopod Congress in Sudbury were provided to T.L. Modzalevskaya by the Natural Sciences and Engineering Research Council of Canada

REFERENCES

ALVAREZ, F. 1990. Devonian athyrid brachiopods from Cantabrian Zone (NW Spain) Biostratigraphie Paléozoique 11: 1-31.

ANONYMOUS. 1992. Tipovye razrezy pogranichnykh otlozhenii srednego i verkhnego devona. SNIIGGIMS, Kuzbass 1992:1-136.

CHERKESOVA, S.V. 1973. Novaya Zemlya i Vaigach, 1, p.316-335; Taimyr, 2, p.139-147. In, D. V. Nalivkin, M. A. Rzhonsnitskaya & B. P. Markovski (eds.). Devonskaya Sistema, Stratigrafiya, SSSR, Moskva.GRATSIANOVA, R.T. & A.S. DAGYS. 1983. Morfologia i voprosy sistematiki nekotorykh devonskikh atiridid Zapadnoi Sibir, p. 15-17. Trudy Institut Geologi, Sibirskoi Otdel AN SSSR 538:15-17.

KHALFIN, L. L. 1946. O paleozoiskoi faune Sibiri i zadachakh sibirskoi paleontologii. Izvestia ZSFAN SSSR, Seriya Geologiya 1: 44-69.

GRUNT, T.A. 1980. Atiridid Russkoi Platformy. Trudy Paleontologicheskii Institut 182: 1-164.

LYASHENKO, A.I. 1959. Atlas brakhiopod i stratigrafiya devonskikh otlozhenii tsentralnykh oblastei Russkoi Platformy. Gostoptekhizdat, Moskva, 450p.

____1973. Brakhiopody i stratigrafiya nizhnefranskikh otlozhenii Yuzhnogo Timana i Volgo-Uralskoi neftegazonoshoi provintsi. Trudy VNIGRI 134: 1-278.

MAMEDOV, A.B. & M.A. RZHONSNITSKAYA. 1985. Devonian of the south Transcaucasus. Courier Forschungsinstitut Senckenberg 75:135-136.

MARKOVSKII, B.P. 1989. Novy vidy franskikh brakhiopod

(Atrypida, Athyridida, Terebratulida) zapadnogo sklona Yuzhnogo Urala. Ezhegodnik 22:88-105.

NALIVKIN, D.V. 1941. Brakhiopody Glavnogo Devonskogo Polya 1: 139-226. In, R.F. Hekker (ed.). Fauna Glavnogo Devonskogo Polya, AN SSSR.

NIKOLAEV, A.A. & M.A. RZHONSNITSKAYA. 1967. Devonian of the northeastern USSR. International Symposium on the Devonian System, Calgary 1:

RZHONSNITSKAYA, M.A. 1959. K stratigrafii devonskikh otlozhenii Kuznetskogo Basseina. Sov Geologiya 9:

___1968. Biostratigrafiya devona okraina Kuznetskogo Basseina. Trudy VSEGEI 1: 1-285.

[p.s. readers are advised to write the authors for additional distribution data of species in the Tom and Yaya river sections. The term horizont is transliterated here as 'horizon', more or less replacing the western term 'member'].

PEREGRINELLA (BRACHIOPODA; RHYNCHONELLIDA) FROM THE EARLY CRETACEOUS, WRANGELLIA TERRANE, ALASKA

MICHAEL R. SANDY & ROBERT B. BLODGETT

Department of Geology, University of Dayton, Dayton, Ohio 45469-2364;
United States Geological Survey, Reston, Virginia 22092, USA.

ABSTRACT- Brachiopods from Bonanza Creek in the eastern Alaska Range of south-central Alaska had previously been considered to include pentamerid brachiopods of Devonian age. Re-investigation of this material indicates that these are rhynchonellids belonging to the enigmatic Early Cretaceous genus *Peregrinella*, and are here described as *Peregrinella chisania* n.sp., marking the first record of this genus from the Wrangellia Terrane, Alaska This supports more recent age-designations for the Chisana Formation, based on Early Cretaceous *Buchia* and *Inoceramus* bivalves.

INTRODUCTION

A collection of medium-sized brachiopods made during the early years of this century by S. R. Capps from Bonanza Creek in the eastern Alaska Range (Nabesna A-2 quadrangle) of south-central Alaska were referred to the pentamerid genus *Pentamerella* ? by Kirk (in, Capps, 1916) and used to date the rocks as Devonian. Later geologic mapping (Richter & Jones, 1973) in the region, however, disclosed that rocks from the Bonanza Creek area were rich in Late Jurassic and Early Cretaceous species of the bivalve *Buchia* . Moreover, this part of the eastern Alaska Range is considered part of the extensive Wrangellia Terrane (Poulton et al., 1992), which is not known to contain Devonian rocks. Re-investigation of these specimens, including the taking of serial sections, indicates that these are rhynchonellids referable to the genus *Peregrinella*.

SIGNIFICANCE OF *PEREGRINELLA*

Peregrinella has been considered a Tethyan rhynchonellid genus in its distribution by a number of authors (e.g., Ager 1967, 1986; Ager & Sun, 1986; Sandy 1989b, 1991), though it has also been reported earlier from California. A peregrinellid from Alaska, associated with bivalves such as *Buchia* and *Inoceramus*, suggests a broader latitudinal distribution for the brachiopod. Regarding *Buchia sublaevis*, one of the species found in the vicinity of Bonanza Creek, Jones et al., 1980 (p. A20) commented, 'This species of *Buchia* and mode of occurrence is common in other Lower Cretaceous rocks deposited in high paleolatitudes and may place a northern stamp on the enclosing rocks'. Their comments were based on material from the Upper Chulitna District, Alaska.

The Wrangellia plate has a geologic history that is generally considered to have indicated low-latitude settings during the Triassic. Low-latitude placement of Wrangellia is not considered likely during the Early Cretaceous. A broad latitudinal distribution for *Peregrinella* could be accounted for by deep water preferences in low-latitudes, and shallow-water environments in higher latitudes, as waters of specific temperature are commonly tracked by benthic organisms. Such a suggestion was made by Masse (1992) for the seemingly anomalous shallow-water occurrence of pygopid ('Tethyan') brachiopods in Greenland, that are typically found in low-latitude Tethyan settings, in what are usually considered deeper-water settings.

Is *Peregrinella* is useful as an indicator of paleolatitude (i.e., Tethyan low-latitudes)? It would seem that some consideration of relative bathymetry is necessary. However, the association of *Peregrinella* with sites identified as possibly chemosynthetic vents (see below) suggests that the distribution of such brachiopods may, in fact, not have any latitudinal control or association.

The highly disjunct distribution of *Peregrinella* has led to models of 'unusual' paleoecology to explain its occurrence, such as short-lived shallowing of deep-water basins (Thieuloy, 1972), or the transport of limestone blocks developed on rocky shorelines to deeper, basin slope and basinal environments (Ager,1965b). However, Lemoine et al. (1982) suggested that *Peregrinella* from Rottier, southern France, was associated with hydrothermal activity, and Campbell et al. (1993) suggested a cold seep association for Hauterivian *Peregrinella* and Tithonian *Cooperrhynchia* Sandy & Campbell (1994) from California. Such paleoecological associations could well explain the highly disjunct paleobiogeographic distributions of these, and related genera in isolated carbonates that formed in tectonically active regions during the Mesozoic. Further investigation would be necessary to determine whether the Alaskan occurrence is associated with a hydrothermal or cold seep setting.

SYSTEMATIC PALEONTOLOGY

Superfamily DIMERELLOIDEA Buckman, 1918
Family DIMERELLIDAE Buckman, 1918
Subfamily PEREGRINELLINAE Ager, 1965a
Genus *Peregrinella* Oehlert, 1887

Peregrinella chisania n. sp.

Diagnosis. Triangular tapering outline, biconvex profile, incurved ventral umbo; typically 15-20 strong ribs on both valves, non-bifurcating; anterior commissure rectimarginate; dental lamellae in ventral valve; well-developed median septum in dorsal valve forming septalium with hinge plates posteriorly; hinge plates flat but also curving anteriorly; two lobate swellings at hinge plate-median septum junction, developing into anteriorly projecting crura with laterally flattened outline in transverse section).

Description. Shells c.15-30mm long, 16-26mm wide, 10-18mm deep, slightly longer than wide, with 14-21 ribs. Dental lamellae present; well-developed median septum in dorsal valve, forming septalium with flat and anteriorly curved hinge plates posteriorly; at hinge plate-median septum junction, two lobate swellings developed, anteriorly changing into crura, which, in transverse section, have laterally flattened outline, and relatively consistent position with respect to the median septum (as far as traced: Figures 1-2).

Type locality and horizon. Bonanza Creek, just below the

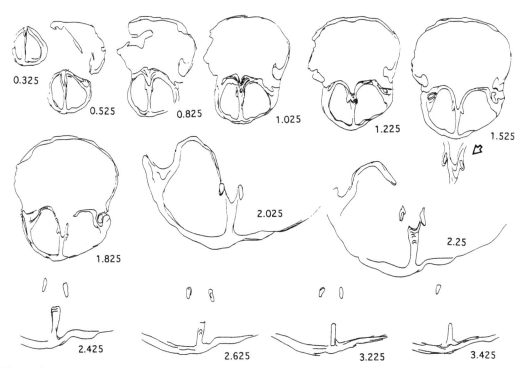

Figure 1. Seria; l sections of *Peregrinella chisania* n.sp., paratype USNM 487761, Magnification: 0.0-1.825, 2.425-3.435, x7.5; detail of 1.525 (arrowed), 2.025, 2.25, x15.

mouth of Little Eldorado Creek, USGS locality 743-SD, Nabesna A-2 Quadrangle, south-central Alaska, suspected Chisana Formation, Lower Cretaceous, probably Valanginian-Hauterivian [holotype and paratypes, a collection of 12 brachiopods, including 2 juveniles, USNM 487759-USNM 487770]. The possibility that the brachiopods may have been collected from an isolated carbonate lens or boulder should not be discounted until the original collecting site has been relocated. The infilling matrix is a black micritic limestone, out of character with the clastic and volcaniclastic-dominated sedimentary sequence. *Peregrinella* is generally interpreted as an indicator of Hauterivian age in southern Europe (Thieuloy, 1972). However, other occurrences have been interpreted to range from Berriasian to Hauterivian (Sun, 1986). A younger Cretaceous age for this Alaska material is also possible (Barremian?- see fossil localities numbers 1-4 near the mouth of Little Eldorado Creek and Bonanza Creek; localities 5-12 are upstream, listed by Richter & Jones,1973). Etymology: after the Chisana Fm.

Remarks. The internal structures of one specimen (Figure 1) were investigated by transverse serial sectioning (following Sandy, 1989a). In addition, one specimen in the collection had had its pedicle umbo ground down and the median septum and hinge plates were visible (Figure 2) in a similar configuration to that seen in the sectioned specimen. Dental lamellae in the ventral valve, were not seen in serial sections (due to umbonal damage), but determined by examination of other specimens (Figure 2).

The specimens from Alaska do not possess the rounded outline or broad hingeline that typify *Peregrinella* , such as the type *P. multicarinata* , or species described by Sun (1986, 1987) from Tibet. Compared to *P. whitneyi* from California, the

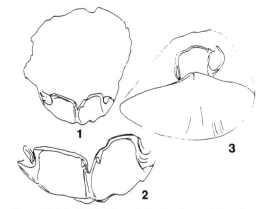

Figure 2. Polished detail of *Peregrinella chisania* n.sp. (1-2) umbones ground to reveal median septum in dorsal valve forming septalium; at hinge plate-median septum junction, two lobate swellings develop (incipient crura). 1 = x7.5; 2 = x15. USNM 487764; (3) Ventral valve (uppermost) showing dental lamellae; a median septum in dorsal valve (lowermost) hinted at by median line in umbonal area. USNM 487760.

Alaskan material is more coarsely ribbed and has a narrower hingeline. Despite these differences they are like *Peregrinella* in the presence of dental lamellae, a well-developed dorsal median septum with initially horizontal hinge plates, a deep septalium, bulbous swellings on the median septum that

240

Figure 3. *Peregrinella chisania* n.sp. 1-4, paratype, USNM 487760; 5-7, paratype USNM 487761 [specimen serially sectioned, Fig.1]; 8, paratype, USNM 487762; 9-12, holotype, USNM 487759. All x1.5

develop into crura, and anteriorly projecting crura. The complete crura may not be preserved in the Alaskan material. The acute inflection of the hinge plates towards the median septum, and the shape and depth of insertion of the teeth are features in common with *P. whitneyi* (see Ager, 1968). The Alaska specimens do not reach the large size typical of *Peregrinella* (e.g., Orbigny, 1847; Stanton, 1895; Thieuloy, 1972); the possiblity that the material represents juvenile specimens cannot be discounted. Where shell material is preserved, it is fibrous, suggestive of secondary wall prismatic calcite, supporting assignment to the Rhynchonellida rather than other Mesozoic articulates such as the Terebratulida. A number of specimens appear to have the trace of a median septum in the pedicle valve suggesting the presence of a spondylium. However, this may be because of slightly stronger development of the central rib or a slightly thicker shell, and there appears to be no ventral septal structure. This trace may have lead Kirk to refer the material to a pentamerid genus.

Acknowledgments

M.R.S. acknowledges the donors of The Petroleum Research Fund, administered by the American Chemical Society and the University of Dayton Student Fellows Program for support of this research. Thanks to Dr. Donald H. Richter, USGS, Anchorage, Alaska, and Dr. Ellis F. Owen, The Natural History Museum, London, for discussion, comments, and information that improved the manuscript, Lindsey Griffith, University of Dayton, for preparing the serial sections in Figure 1, and to Dean Paul Morman, College of Arts and Sciences, University of Dayton for continuing assistance. Specimens are deposited in the collections of the United States National Museum, Smithsonian Institution, Washington, D.C.

REFERENCES

AGER, D.V. 1965a. Mesozoic and Cenozoic Rhynchonellacea, p.597-625. In, R.C. Moore (ed.), Treatise on Invertebrate Paleontology, H, Brachiopoda. Geological Society of America, University of Lawrence Press, Kansas.

___1965b. The adaptation of Mesozoic brachiopods to different environments. Palaeogeography, Palaeoecology, Palaeoclimatology 1:143-172.

___1967. Some Mesozoic brachiopods in the Tethyan region, p.135-151. In, C.G. Adams & D.V. Ager (eds.), Aspects of Tethyan biogeography. Systematics Association Special Publication 7.

___1968. The supposedly ubiquitous Tethyan brachiopod *Halorella* and its relations. Journal Palaeontological Society India 5-9 (for 1960-64):54-70.

___1986. Migrating fossils, moving plates and an expanding earth. Modern Geology 10:377-390.

___& D.L. SUN. 1986. Distribution of Mesozoic brachiopods on the northern and southern shores of Tethys. Palaeontologia Cathaysiana 4:23-51.

BUCKMAN, S. S. 1918. The Brachiopoda of the Namyau Beds, northern Shan States, Burma. Palaeontologica Indica 3(2):1-299.

CAMPBELL, K.A., S. CARLSON & D.J. BOTTJER. 1993. Fossil cold seep limestones and associated chemosynthetic macroinvertebrate faunas, Jurassic-Cretaceous, Great Valley Group, Sacramento Valley, California, p.37-50. In, S.A. Graham & D.R. Lowe (eds.), Advances in the sedimentary geology of the Great Valley Group, Sacramento Valley, California. Pacific Section Society Economic Paleontologists Mineralogists 73.

CAPPS, S.R. 1916. The Chisana-White River District Alaska. United States Geological Survey Bulletin 630:130 .

DUMÉRIL, A.M.C. 1806. Zoologique analytique ou méthode naturelle de classification des animaux. Allais, Paris, 344p.

HUXLEY, T.H. 1869. An introduction to the classification of animals. Churchill and Sons, London, 147p.

KÜHN, O. 1949. Lehrbuch der Paläozoologie. E. Schweizerbart, Stuttgart, 326p.

JONES, D.L., N.J. SILBERLING, B. CSEJTEY Jr., W H. NELSON & C. D. BLOME. 1980. Age and structural significance of ophiolite and adjoining rocks in the Upper Chulitna District, south-central Alaska. U.S. Geological Survey Professional Paper 1121-A:1-21.

LEMOINE, M. A., A. ARNAUD-VANNEAU, H. ARNAUD, R. LETOLLE, C. MEVEL & J.P. THIEULOY. 1982. Indices possibles de paléo-hydrothermalisme marin dans le Jurassique et le Crétacé des Alpes occidentales (océan téthysien et sa marge continentale européenne): essai d'inventaire. Bulletin Société Géologique France 24:641-647.

MASSE, J. P. 1992. The Lower Cretaceous Mesogean benthic ecosystems - palaeoecologic aspects and palaeobiogeographic implications. Palaeogeography, Palaeoecology, Palaeoclimatology 91:331-345.

OEHLERT, D. P. 1887. Brachiopodes, p.1189-1334. In, J.C. Fischer (ed.), Manuel de Conchyliologie. Paris.

D'ORBIGNY, A. 1847. Paléontologie française. Terrain crétacé. Cinquième partie: Mollusques Brachiopodes. Paris. 390 p., pl. 490-599.

POULTON, T.P., R.L. DETTERMAN, R.L. HALL, D.L. JONES, J.A. PETERSON, P.L. SMITH, D.G. TAYLOR, H.W. TIPPER & G.E.G. WESTERMANN.1992. Western Canada and United States, p.29-92. In, G.E.G. Westermann (ed.), The Jurassic of the circum-Pacific. Cambridge University Press, Cambridge.

RICHTER, D.H. & D.L. JONES. 1973. Reconnaissance geologic map of the Nabesna A-2 Quadrangle, Alaska. United States Geological Survey Miscellaneous Geologic Investigations, Map I-749.

SANDY, M.R. 1989a. Preparation of serial sections, p.146-156. In, R.M. Feldmann, R.E. Chapman & J.T. Hannibal (eds.), Paleotechniques. Paleontological Society Special Publication 4.

___1989b. Jurassic and Cretaceous brachiopods as indicators of Tethys - Atlantic-Pacific Mesozoic gateways. Abstracts 28th International Geological Congress, Washington 3:17-18.

___1991. Biogeographic affinities of some Jurassic-Cretaceous brachiopod faunas from the Americas and their relation to tectonic and paleoceanographic events, p. 415-422. In, D.I. MacKinnon, D.E. Lee & J.D. Campbell (eds.), Brachiopods through time. Balkema, Rotterdam.

___& K.A. CAMPBELL. 1994. A new rhynchonellid genus from Tithonian (Upper Jurassic) cold seep deposits of California and its paleoenvironmental significance. Journal Paleontology, 68:1243-1252.

STANTON, T.W. 1895. Contributions to the Cretaceous paleontology of the Pacific Coast: the fauna of the Knoxville beds. U.S. Geological Survey Bulletin, 132p.

SUN, D.L. 1986. Discovery of Early Cretaceous *Peregrinella* (Brachiopoda) in Xizang (Tibet) and its significance. Palaeontologia Cathaysiana 2:211-227.

___1987. Early Cretaceous brachiopods from Baingoin and Xainza, northern Xizang. Bulletin Nanjing Institute of Geology Palaeontology1987(6):63-103.

THIEULOY, J.P. 1972. Biostratigraphie des lentilles à pérégrinelles (brachiopodes) de l'Hauterivien de Rottier (Drôme, France). Geobios 5:5-53.

[Further information on measurements for serial sections at 0.001mm intervals, and types, available from authors: editors].

EARLY DEVONIAN BRACHIOPODS FROM THE ZERAVSHAN MOUNTAIN RANGE, SOUTHERN TYAN-SHAN, CENTRAL ASIA

V. P. SAPELNIKOV & L. I. MIZENS

Institute of Geology and Geochemistry, Russian Academy of Sciences, Ekaterinburg, Russia

ABSTRACT- Early Devonian pentamerid, rhynchonellid, and atrypid brachiopods are described from the type area of the Pragian-Emsian boundary in the Kitab Geological Reserve, in the Zeravshan Mountain Range, southern Tyan-Shan (Uzbekistan). Stratigraphic ranges of common brachiopods are plotted. The Sangitovar horizon (upper Lochkovian) contains an extremely abundant and diverse brachiopod assemblage. On the basis of brachiopods, the two lower horizons of the Emsian, the Zinzilban and Norbonak, are subdivided into three parts, correlated with conodont zones. Three new, stratigraphically useful rhynchonellid species are described: *Corvinopugnax erinae*, *Isopoma triplicatum*, and *Sphaerirhynchia parvula*.

INTRODUCTION

There is a general lack of published information on brachiopods and their stratigraphic distribution in the upper Lochkovian and lower Emsian of the Kitab Geological Reserve (Uzkekistan), where the standard section was selected in 1989 by the International Subcommission on Devonian Stratigraphy (Oliver & Chlupac, 1991; Elkin et al., 1994a, b). Our study of the Devonian brachiopods of the Kitab Reserve began in 1987, when the section was recognized as a candidate for the Global Stratotype Section and Point (GSSP). Descriptions of Early Devonian pentamerids, rhynchonellids, and atrypids from the main sections in the Reserve (including the type section of the Pragian-Emsian boundary along the Zinzilban Gorge and others along the Khodzha-Kurgan Gorge) have now been completed (Figure 1). Lithological studies of these sections can be found in Kim et al. (1978, 1982, 1984) and Ziegler & Elkin (1995). Data on some of the brachiopods and their biostratigraphy have been published (Sapelnikov & Mizens, 1992, 1995a; Rybkina, 1989; Mizens & Sapelnikov, 1996).

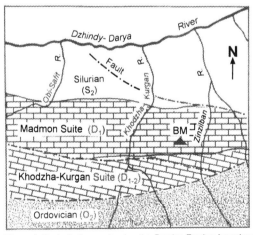

Figure 1. Geological map showing the Pragian-Emsian boundary section in the Kitab State Geological Reserve, Uzbekistan.

This paper provides a biostratigraphic synthesis for the brachiopods hitherto studied (Table 1), and describes three new stratigraphcally useful rhynchonellid species, based on specimens collected by the authors, E.A. Elkin, and I.A. Kim from 1988 to 1990. The fossil localities follow the Guide to Field Excursions (Kim et al., 1978), and stratigraphy of Ziegler & Elkin (1995).

BIOSTRATIGRAPHY

The upper Lochkovian Sangitovar Horizon contains an abundant and diverse brachiopod assemblage, distinct from those above and below. Correlation of the top of the Sangitovar Horizon with the Lochkovian-Pragian boundary (Elkin & Kim, 1988) on the basis of brachiopods is largely conditional. This is because the boundary is located within the massive reefal limestone sequence of the Madmon Suite, and many species that are confined to the Sangitovar horizon occur also in the Pragian elsewhere in Eurasia (Sapelnikov & Mizens, 1992, 1995; Sapelnikov et al., 1995; Rybkina, 1989). Marked by a major paleoevent, the Khukar/Zinzilban horizon boundary coincides with the Pragian/Emsian stage boundary (*dehiscens* Zone), which was drawn 35cm above the base of the Khodzhakurgan Suite at the International Standard Type Section located in Zinzilban Gorge. This boundary is marked by strata containing different brachiopod assemblages. For the 67 species described, only one species, *Gypidula* cf. *kayseri* (Peetz), crosses the boundary. Above the *dehiscens* Zone, genera characteristic of the Emsian and Eifelian occur at various levels in the Zinzilban and Norbonak horizons, e.g. *Leviconchidiella*, *Sieberelloides*, *Kransia*, *Asiatotoechia*, *Zeravshanotoechia*, *Oglu*, *Punctatrypa*, and *Crassipunctatrypa*. Several species are known elsewhere only from post-Pragian strata, e.g. *Gypidula boxitica* Khodalevich, *G. subvenetus* Khodalevich, *G. pseudoacutolobata* Rzhonsnitskaya, *Ivdelinia egorovi* Andronov, *I. intima* (Khodalevich), *Sieberella* cf. *ulanica* Malygina & Sapelnikov, *Atrypa markovskii* Rzhonsnitskaya, *Punctatrypa perpolita* (Khodalevich), and *Carinatina syrmatica* (Breivel).

A sharp turnover of the brachiopod fauna at the base of the Khodzhakurgan Suite is closely related to facies changes within the Kitab Basin at the end of the Pragian, but it probably also reflected local evolutionary changes. Thus the Pragian/Emsian boundary interval may be regarded as an important paleobiological event which marked a turning point in the evolution of Devonian biota of Eurasia and Australia. This 'event' has been interpreted to be closed related to eustatic sealevel rise, and corresponds to the 'Zinzilban sedimentological event' (the Madmon/Khodzhakurgan boundary) in the Zeravshan region (Ellkin et al., 1994), to the Zlichovian transgression in Western Europe (House, 1985), and to the "*Pirenea dehiscens* transgression-regression interlude" in Australia (Talent, 1989; Talent et al., 1993). Widespread brachiopod species that occur in the type boundary section provide an opportunity for precise global correlation based on both pelagic and benthic faunas (Sapelnikov & Mizens, 1992).

Table 1. Stratigraphic ranges of Early Devonian brachiopods in the type area of the Pragian/Emsian boundary.

BRACHIOPOD SPECIES DESCRIBED	Prag Sangitovar upper	Lochkov Khukar	Ems Zinzilban L	Zinzilban M	Zinzilban U	Norbonak L	Norbonak M	Norbonak U
Clorindina arataeformis (Nikiforova)	●							
Gypidula chodalevitchi Sapelnikov	●	●						
G. matutinalis (Khodalevich)	●	●						
G. nucalis (Khodalevich)	●							
G. nux Khodalevich	●							
G. vulgaris Breivel & Breivel	●							
G. problematica crassa Khodalevich	●							
G. cf. irgislensis Tyazheva	●							
G. cf. procerulaeformis Kulkov	●							
G. cf. kayseri (Peetz)	●	●	●					
G. cf. pseudoacutolobata Rzhonsnitskaya				●		●		
G. cf. subvenetus (Khodalevich)						●		
G. boxitica Khodalevich						●		
Leviconchidiella gyrifera (Malygina & Sapelnikov)						●		
L. vagranica (Khodalevich)						●		
Sieberelloides aff. weberi (Khodalevich)				●		●		
Ivdelinia procerula (Barrande)	●							
I. dzhausensis (Larin)	●							
I. cf. egorovi (Andronov)						●		
Sieberella cf. sieberi (Buch)			●					
S. cf. ulanica Malygina & Sapelnikov					●			
Latonotoechia latilinguata Sapelnikov & Mizens	●	●						
Hebetoechia nitidula (Barrande)	●							
Linguopugnoides kimi Rzhonsnitskaya	●							
L. magnoplicatus Sapelnikov &Mizens	●							
Corvinopugnax erinae n. sp.	●							
Isopoma triplicatum n. sp.	●							
Sphaerirhynchia parvula n. sp	●							
Kransia parallelepipela (Bronn)			●			●	●	
Asiatotoechia sinsilbanica Rzhonsnitskaya & Koss.				●				
Zeravshanotoechia zeravshanica Rzhonsnitskaya					●			
Atrypopsis pseudothetis Rzhonsnitskaya	●							
Spirigerina druida Havlicek	●	?						
Sibirispira inflata Alekseeva	●							
Tenuiatrypa tabuskaensis (Khodalevich)	●	●						
Carinatina praesignifera Rzhonsnitskaya	●	●						
C. arimaspa (Eichwald)	●	●						
C. syrmatica Breivel			●	●			●	
Totia cf. similis (Breivel & Breivel)		●						
Atrypa markovskii diligens Mizens				●	●		●	
Oglu cf. laticostatus Havlicek				●	●			
Punctatrypa perpolita (Khodalevich)				●	●			
Crassipunctatrypa crassiconcentrica Mizens								●
Quadrithyrina parvula Rybkina	●	●						
Q. crassa Larin	●	●						
Ambothyris larini Rybkina & Kim	●							

The Lochkovian/Pragian boundary is not as sharply defined as the Pragian/Emsian boundary, not only in southern Tyan-Shan, but also worldwide (Talent & Elkin, 1987; Talent et al., 1993; Sapelnikov & Mizens, 1995a; Sapelnikov et al., 1995; Chlupac et al., 1985; Chlupac & Oliver, 1989). Our recent survey of the geological and geographical distribution of the rhynchonellid Linguopugnoides Havlicek, and identification of three new species from the Zeravshan and Turkestan ranges (Table 1), made it possible to refine the Lochkovian/Pragian boundary limits. Linguopugnoides appears to be characteristic of the Lochkovian in Central Asia, the Prague Basin, Central Europe, Siberia, the Urals, and northwestern China. More than half the Linguopugnoides species do not extend beyond the late Lochkovian, which forms the basis for an international Linguopugnoides carens - L. subcarens - L. pararemissus brachiopod Biozone.

On the basis of Devonian brachiopod and conodont faunas in the Zeravshan region, the Zinzilban and Norbonak Horizons were each divided into three parts (Sapelnikov & Mizens, 1994a, b, 1995, in press; Elkin et al., 1994b). In the lower 26m of the Zinzilban Horizon, pentamerids, rhynchonellids, and atrypids are virtually absent but strophomenids are present and these extend up section (Kim et al., 1978). In the middle part of the Zinzilban Horizon (27-79m), pentamerids, rhynchonellids, and atrypids become dominant, including such genera as Sieberelloides, Multicosta, Asiatotoechia, Carinatoechia, Atrypa, Oglu, and Punctatrypa. The upper part contains an impoverished brachiopod fauna, with the zonal species Sieberella ulanica and Punctatrypa perpolita Khodalevich. The 'strophomenid beds' approximately correlate with the interval from the base of the *kitabicus* Conodont Zone (new name for *dehiscens* Zone) to the first appearance of *Polygnathus pannonicus* and *P. hindei* (Elkin et al., 1994). The middle part coincides with the conodont range of *P. hindei*, the lower range of *P. pannonicus*, and the last occurrence of *P. sokolovi*. The *Sieberella ulanica* beds correlate with the *Polygnathus pannonicus* and *P. kitabicus* Conodont Zones.

In the Norbonak Horizon, brachiopods are confined largely to the middle part (135-170m), including the zonal species *Leviconchidiella vagranica* (Khodalevich), *L. gyrifera* (Malygina & Sapelnikov), and the Emsian index species *Gypidula boxitica*, *G. subvenetus*, and *Ivdelinia egorovi*. The lower part (113-114m) contains only a few specimens of *Kransia parallelepipeda* (Bronn). Brachiopods are absent in the upper part (171-208m). The brachiopod horizon is correlative with the *excavatus* Conodont Zone. The change in brachiopod faunas across the Zinzilban/Norbonak boundary was likely controlled by a facies change, as indicated by the upward increase in grey detrital limestones in a generally dark, thinly bedded, muddy limestone sequence (Kim et al., 1984).

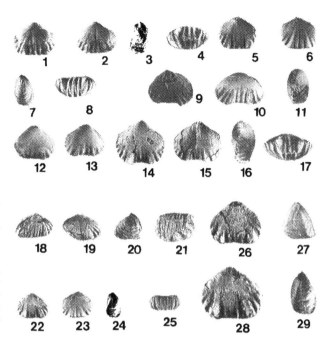

Figure 2. New species of Zeravshan rhynchonellids. 1-8, *Corvinopugnax erinae* n.sp. 1-4, holotype, locality B-867.5; 5-8, specimen CSGM 976/B165, locality B-853.5. 9-17, *Isopoma triplicatum* n.sp. 9-11, CSGM 976/B168, locality Z-874; 12-13, CSGM 976/B173, locality Z-874; 14-17, CSGM 976/B167, locality B783.5. 18-29, *Sphaerirhynchia parvula* n.sp. 18-21, CSGM 976/B137, locality B-839; 22-25, CSGM 976/B135, locality B-839; 26-27, CSGM 976B140, locality B839, x2; 28-29, holotype, x2. All Sangitovar horizon, x1, unless otherwise noted.

SYSTEMATIC PALEONTOLOGY
Order Rhynchonellida
Family Pugnacidae Rzhonsnitskaya, 1956
Genus **Corvinopugnax** Havlicek, 1961
Corvinopugnax erinae n. sp.
Figure 2:1-8

Etymology. In honour of the Uzbekistan paleontologist M. V. Erina. Holotype. CSGM 976/B164, from the left bank of the Dzhindy-Darya River, Bursykhirman Mountain, Zeravshan Range, Madmon Suite, Sangitovar horizon, Lochkovian. Material. 340 complete shells, >40vv, 28dv. Age and distribution. Sangitovar Horizon (late Lochkovian), Bursykhirman Mountain, Zeravshan Range.

Description. Shell small [l=8-10mm, w=11-12mm, d=5-7mm] suboval to subpentagonal, transverse, moderately dorsibiconvex; ventral valve uniformly convex or somewhat flattened anteriorly; hingeline curved; beak low, weakly incurved or suberect, raised slightly above hingeline; ventral interarea absent; delthyrium small, narrow, open; sulcus developed in anterior third of valve, broad, shallow, with flat floor, well defined near anterior margin, extending into broadly trapezoid tongue; dorsal valve strongly convex, with small beak appressed to hingeline; fold low, beginning at mid valve length, with flat or slightly convex top, becoming well defined only toward anterior margin, bearing narrow median furrow in most specimens; shell surface smooth in umbonal portion, but bearing subangular costae in anterior third, with 3 costae (rarely 2 or4) in sulcus, 4 (rarely 3 or 5) on fold, and 2 to 4 on each flank. Dental plates short, thin; teeth small, elongate; hinge plates relatively low, curved; crural plates low, thin; low myophragm present in some specimens.

Remarks. The new species differs from the type species, *Corvinopugnax corvinus* (Barrande, 1847; in Havlicek, 1961) in its smaller, less strongly transverse shell, narrower sulcus, less tumid dorsal valve, coarser costae in the sulcus and on the fold, and presence of a median furrow on the fold. *C.*

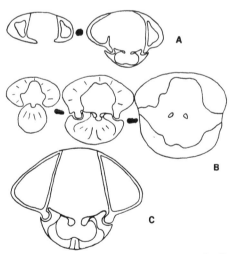

Figure 3. Serial sections (a), *Corvinopugnax erinae* n.sp., locality B-867.5, Bursykhirman Mtn, Zeravshan Range, x7; (b) *Isopoma triplicatum* n.sp., locality Z-874, Zinzilban Gorge, Zeravshan Range, x6; (c) *Sphaerirhynchia parvula* n.sp., locality B-839, Bursykhirman Mtn, Zeravshan Range, x18. All from Sangitovar Horizon, late Lochkovian.

tichiensis Alekseeva, 1967 from the Lower Devonian of NE Siberia has a smaller, elongate shell with more numerous lateral costae.

Genus *Isopoma* Torley, 1934
Isopoma triplicatum n.sp.
Figure 2:9-17

Etymology. Latin *tri- plicata*, meaning with three plicae.
Holotype. CSGM 976/B167, left bank of the Dzhingy-Darya River, Bursykhirman Mountain, Madmon Suite, Sangitovar Horizon, Lochkovian. Material. 60 shells, 42 vv, 30 dv. Age and distribution. Bursykhirman Mountain and Zinzilban Gorge, Zeravshan Range, Sangitovar Horizon, late Lochkovian.

Description. Shell medium-sized [l=9-13mm, w=11-15mm, d=6-8mm], broadly suboval, with greatest width at mid-length; hingeline curved.; anterior margin weakly emarginate; ventral valve weakly convex, anteriorly flattened in some specimens; beak low, incurved, sometimes touching dorsal apex; ventral interarea rudimentary or absent; sulcus broad, shallow, with flat floor, well defined anteriorly by pair of plicae; tongue broad, subtrapezoidal; dorsal valve uniformly convex, deeper than ventral valve; small beak appressed to hingeline or buried in delthyrial cavity; fold low, visible only toward anterior margin; shell surface smooth posteriorly, becoming coarsely ribbed in anterior third, with three in sulcus, four on fold, and two to four on each flank.; growth lines fine, averaging 8 or 9 per mm. Shell walls very thick, umbonal part filled by secondary thickening; dental plates absent; teeth massive; hinge plates crescentiform, curved; crural plates relatively low, thin, dorsolaterally directed; crura strong.

Remarks. The new species closely resembles *Isopoma transuralica* (Chernyshev, 1893; in Sapelnikov, 1987) but differs in its larger and generally broader shell. It can be distinguished from *Isopoma alecto* (Barrande, 1847; in Havlicek, 1961) from the Pragian Vinarice Limestone of the Prag Basin by its less elongate shell with a greater apical angle, costate shell flanks, less strong and more numerous ribs, especially in the sulcus and on the fold. *I. triplicatum* differs from *I.? postmodica* (Scupin, 1906; in Gratsianova, 1967) in having a broader shall with a less tumid dorsal valve, more rounded costae, lower tongue and well-developed growth lines, and in the absence of a median furrow on the dorsal fold.

Family Uncinulidae Rzhonsnitskaya, 1956
Genus *Sphaerirhynchia* Cooper & Muir-Wood, 1951
Sphaerirhynchia parvula n.sp.
Figure 2:18-29

Etymology. Latin *parvula*, small. Holotype. CSGM 976/B135, left bank Dzhindy-Darya River, Bursykhirman Mountain, Madmon Suite, Sangitovar Horizon, late Lochkovian. Material. 62 shells, 4vv, 2dv. Age and distribution. Bursykhirman Mountain, Zeravshan Range, Sangitovar Horizon, late Lochkovian.

Description. Shell small [l=6-7mm, w=8-12mm, d=5-8mm], subpentagonal, broad, uniformly convex, with greatest convexity attained at anterior margin.; hingeline short, curved; ventral valve weakly convex or flattened; beak small, pointed, erect or suberect; sulcus anteriorly delimited by pair of strong costae; tongue broad, truncated.; dorsal valve deepest at anterior margin, with low beak appressed to hingeline; median furrow developed from apex to anterior

margin; fold low, broad, with flat top, well defined in anterior half of valve.; shell surface smooth posteriorly, anteriorly covered by coarse, subangular costae, with 3 to 5 in sulcus, 4 to 6 on fold, and 2 to 5 on each flank; costae in sulcus sometimes bearing longitudinal grooves. Umbonal shell filled by secondary thickening; dental plates short, thin; teeth rounded; sockets shallow; crural plates subhorizontal; septalium absent; median septum absent/reduced to low ridge.

Remarks. The new species is most similar to *Sphaerirhynchia vijaica* (Khodalevich, 1951; Breivel & Breivel, 1977), but differs in having a much smaller shell with higher and better developed dental plates.

Acknowledgments
This study was funded by grants NM7000 and NM7300 from the International Science Foundation and the Russian Government. Type specimens are reposited in the Museum of the Kitab Geological Reserve (KSGR), Uzbekistan and in the Central Silurian Geological Museum (CSGM), Ekaterinburg, Russia.

REFERENCES

ALEKSEEVA, R. E. 1967. Brakhiopody i stratigrafiya nizhnego Devona severo-vostoka SSSR. Moskva, Nauka, 81 p.
BREIVEL, I.A. & M.G. BREIVEL. 1977. Brakhiopody, p.52-105. In, Ministerstvo Geologii RSFSR, Uralskoe Territorialnoe Geologicheskoe Upravlenie, Biostratigrafiya i fauna ran-nego Devona vostochnogo sklona Urala, Nedra, Moskva.
CHERNYSHEV, T.N. 1893. Fauna nizhnego Devona vostochnogo sklona Urala. Trudy geologicheskogo Komiteta 4(3):1-221.
CHLUPAC, I., P. LUKES, F. PARIS & H. P. SCHÖNLAUB. 1985. The Lochkovian-Pragian boundary stratotype in the Barrandian area, Czechoslovakia. Subcommission on Devonian Stratigraphy. Rennes, 11 p. [unpublished].
CHLUPAC, I. & W. A. OLIVER. 1989. Decision on the Lochkovian-Pragian boundary stratotype (Lower Devonian). Episodes 12:109-113.
ELKIN, E. A. & A. I. KIM. Devonian of the Asiatic part of the USSR: Recent achievements and problems. Memoirs Canadian Society Petroleum Geologists 14:663-667.
ELKIN, E.A., K. WEDDIGE, N. IZOKH & M.V. ERINA. 1994a. New Emsian conodont zonation (Lower Devonian). Courier Forschungsinstitut Senkenberg 168:139-157.
ELKIN, E. A., N.G. IZOKH, N.V. SENNIKOV, A.Yu. YAZIKOV, A.I. KIM & M.V. ERINA. 1994b. Vazhneishie globalnye sedimentologicheskie i biologicheskie sobytiya Devona Yuzhnogo Tyan-Shanya i yuga Zapadnoi Sibiri. Stratigrafiya, Geologicheskaya Korrelyatsia 2(3):24-31.
GRATSIANOVA, R.T. 1967. Brakhiopody i stratigrafiya nizhnego Devona Gornogo Altaya. Nauka, Moskva, 176p.
HAVLICEK, V. 1961. Rhynchonelloidea des böhmischen älteren Paläozoikums (Brachiopoda). Rozpravy Ustredního ústavu geologického 27: 1-211.
HOUSE, M.R. 1985. Correlation of mid-Palaeozoic ammonoid evolutionary events with global sedimentary perturbations. Nature 313:17-22.
KHODALEVICH, A.N. 1951. Nizhnedevonskie i eifelskie brakhiopody Ivdelskogo i Serovskogo raionov Sverdlovskoi oblasti. Gosgeolizdat, Moskva, 169p.
KIM, A.I., M.V. ERINA, L.S. APEKINA & A.I. LESOVAYA. 1984. Biostratigrafiya Devona Zeravshano-Gissarskoi gornoi oblasti. FAN, Tashkent, 95p.
KIM, A.I., E.A. ELKIN & M.V. ERINA. 1982. Biostratigrafiya nizhnego i srednego Devona Srednei Azii, p.85-92. In, Biostratigrafiya pogranichnykh otlozhenii nizhnego i srednego Devona. Nauka, Leningrad.

___,___,___& R.T. GRATSIANOVA. 1978. Type sections of the Lower and Middle Devonian boundary beds in Middle Asia, A guide to field excursions, Tashkent, 55 p.

MIZENS, L.I. & V.P. SAPELNIKOV. 1995. The genus *Linguopugnoides* Havlicek in space and time [abstract, p.59]. In, R. Mawson & J. Talent (eds.), First Australian Conodont Symposium, and Boucot Symposium, Macquarie University, Sydney.

___ &___(in press). The new species and biostratigraphic significance of the middle Palaeozoic brachiopod genus *Linguopugnoides* Havlicek, 1960. Historical Biology.

OLIVER, W.A. & I. CHLUPAC. 1991. Defining the Devonian: 1979-1989. Lethaia 24:119-122.

RYBKINA, N. L. 1989. Nekotorye brakhiopody verkhnei chasti madmonskoi svity Yuzhnogo Tyan-Shanya, p.101-106. In, Novye danye po ranne i srednepaleozoiskim brakhiopodam SSSR, Sverdlovsk.

SAPELNIKOV V.P. & L.I. MIZENS. 1992. Novye danye o brakhiopodakh pogranichnykh prazhskonizhneemskikh otlozhenii Zeravshanskogo Khrebta, p.42-53. In, Novye danye po stratigrafii i litologii Urala i Srednei Azii, Ekaterinburg.

___&___1994. Osobennosti stratigraficheskogo raspredeleniya brakhiopod v tipovom raione mezhdunarodnoi granitsy Pragian-Emsian na Yuzhnom Tyan-Shane. Ezhegodnik1993:9-11.

___&___1995. Novye danye po korrelyatsii nizhnedevonskikh obrazovanii Urala i tipovoi oblasti yarusnoi granitsy Pragian-Ems na Yuzhnom Tyan-Shane. Ezhegodnik1994:7-10.

___&___(in press). Brachiopods from the type locality of the Pragian-Emsian boundary. In, W. Ziegler & E.A. Elkin (eds.), Pragian/Emsian boundary stratotype. Courier Forschungsinstitut Senkenberg.

___,___& V.P. SHATOV.1987. Stratigrafiya i brakhiopody verkhnesiluriiskikh-srednedevoskikh otlozhenii Severa vostochnogo sklona Urala. Nauka, Moskva, 224p.

SAPELNIKOV V.P., M.P. SNIGIREVA, A.Z. BIKBAEV & L.I. MIZENS. 1995. Zonal subdivision of Early-Middle Devonian reef deposits of the Urals based on conodonts and brachiopods. In, R. Mawson & J. Talent (eds.), First Australian Conodont Symposium, and Boucot Symposium, Courier Forschungs-institut Senkenberg 182:399-419.

TALENT, J.A.1989. Transgression-regression patterns for the Silurian and Devonian of Australia, p.201-219. In, Edwin Sherborn Hills Memorial Volume, Blackwell, Carlton.

___& E.A. ELKIN. 1987. Transgression-regression patterns for the Devonian of Australia and southern West Siberia. Courier Forschungsinstitut Senkenberg 92:235-249.

___, R. MAWSON, A.S. ANDREW, P.J. HAMILTON & D.G. WHITFORD. 1993. Middle Paleozoic extinction events: faunal and isotopic date. Palaeogeography, Palaeoclimatology, Palaeoecology 104:139-152.

CLASSIFICATION OF PALEOZOIC RHYNCHONELLID BRACHIOPODS

NORMAN M. SAVAGE

Department of Geological Sciences, University of Oregon, Eugene 97403, USA

ABSTRACT- A major revision of Paleozoic rhynchonellid brachiopods is proposed. Seven new families include the Sphenotretidae, Niorhynicidae, Machaerariidae, Obturamentellidae, Aseptirhynchiidae, Petasmariidae, and Lambdarinidae. Diagnoses of 23 new subfamilies comprise the Hemitoechiinae, Ripidiorhynchinae, Sphaerirhynchiinae, Glossinulininae, Betterberglinae, Beckmanniinae, Corvinopugnacinae, Iowarhynchinae, Linguopugnoidinae, Fenestrirostrinae, Platyterorhynchinae, Gigantorhynchinae, Stenometoporhynchinae, Basilicorhynchinae, Axiodeaneiinae, Madarosiinae, Uncinulinellininae, Nipponirhynchiinae, Acolosiinae, Lambdarininae, Loborininae, Minyspheniinae, and Dzieduszyckiinae.

INTRODUCTION

A review of Paleozoic rhynchonellids for revision of the 1965 Brachiopod Treatise has resulted in a substantially differing classification. The total number of described genera has more than doubled during the three decades since the earlier treatise. Of 456 genera proposed in the literature as Paleozoic rhynchonellid genera, 360 are included in the classification outlined below, 34 are considered synonyms, 48 are set aside as nomina dubia, and 13 do not belong to the Order Rhynchonellida. Few comparative global taxonomic studies of rhynchonellids have been attempted since the 1965 Treatise. Classification has been hindered by differences in preservation and by various approaches to description. Comparing interiors described from silicified material with those that have been serially sectioned, or burnt and scraped, or described from molds, hampers detailed analysis, and some workers have assigned taxonomic significance to features readily seen in only one type of preservation. Information about deltidial plates, crura, and muscle fields is lacking for most genera, yet these appear to be of key taxonomical importance.

Uneven morphological information has obstructed a cladistic approach. Cladistic results were most useful where morphologic features of genera being compared were equally well known. Relationship of a genus of uncertain affinity was made after subjecting it to analyses that included genera from two or more possible families. However, efforts to arrive at a classification based primarily on a cladistic approach were thwarted by lack of a reliable morphological data base. Of over 50 shell character states used in these analyses, about one quarter to one third were missing from the information available for most genera. A classification based principally on cladistic analysis was adopted briefly, but was subsequently modified to the present classification, based partly on a phenetic methodology.

In Paleozoic rhynchonellids several external and internal features, such as fold, surface striae, costae patterns, marginal spines, dental plates, and cardinal process, appear to have evolved several times. Evolving features were followed stratigraphically, within a succession of somewhat similar genera, to avoid grouping unrelated homeomorphs. The break in recorded lineages at the Paleozoic-Mesozoic boundary appears to be mostly the result of global marine regression.

There are few continuous marine deposits across the Permian-Triassic boundary and thus few described rhynchonellids. From what is known of latest Permian and earliest Triassic rhynchonellids, it seems that only rare genera survived the extinction crisis.

SHELL CHARACTERS

External features. Size may be significant. Some genera, such as *Dorytreta*, are consistently small, others, such as *Plethorhyncha*, are consistently large, and most others, such as *Pugnax*, are more variable. Thus size should not be dismissed from generic diagnosis, allowing for environmental effects, especially temperature and factors causing stunting. Size differences have been included in diagnoses where this seems meaningful. Shape is an important generic feature, notwithstanding the frequency of homeomorphs. External features include outline, profile, and strength of fold. Rhynchonellids are usually dorsibiconvex, with varying amounts of dorsal valve inflation. Inflation may become extreme and distinctive, as in *Parapugnax*, but is otherwise not generically diagnostic. Degree of beak curvature is recorded for most genera: curvature often increases with age and is cautioned as a generic character.

Strength of the fold-sulcus in rhynchonellids is related to shape. Other external features useful in generic diagnosis include the commissure, marginal spines, interareas (relatively rare in Paleozoic rhynchonellids), deltidial plates, and foramen. Most rhynchonellids have a strongly uniplicate anterior commissure, though some are unisulcate; the latter, though rare, appears more in smooth genera. Shells usually have costae or stronger plicae superimposed on the fold-sulcus, resulting in a zig-zag margin. The extent of commissural deflection is often great enough to be termed a tongue or linguiform extension. This may be rectangular, as in *Uncinulus*, or more rounded as in *Leiorhynchus*.

Radial ornament is present in the earliest known rhynchonellids: smooth forms occur later, becoming increasingly frequent, but never prevalent. Ornament is of generic value, and changes occur during growth can be diagnostic, e.g. from smooth to costate, development of fold, and flanking plicae, and relate to gape. Ratio of body size and water inflow requirement to commissure length increased disproportionately with growth. Predators may have influenced development of the zig-zag margin and other flexures. Costae and plicae increase shell strength, and combinations of costae, fold-sulcus, and flanking plicae occur. Fine radial striae are often visible only in well preserved specimens, and may be superimposed on coarser costae. In others, fine costae merge to form coarse costae or plicae anteriorly. Superimposed striae may have arisen only once, to justify inclusion of these taxa in a single family.

Concentric ornament consists of fine growth lines and has little taxonomic value. In Ordovician *Lepidocyclus*, concentric growth features are strongly lamellose and may have generic significance, although this is not clear at the Lepidocyclinae subfamily level. Marginal spines (Westbroek ,1968; Rudwick, 1970) are most elongate in the Uncinuloidea, where some have geniculate dorsal and ventral anterior margins so that the spines are vertically disposed. In *Kransia,* the valve margins are straight, but spines long, fitting into grooves on the opposite valve to form a grill that probably restricted access to intruders. In *Uncinulus,* valve margins are moderately serrated and spines less elongate; shorter spines in*Glossinulus* and *Eoglossinulus* arise at the tips of more strongly serrated valve margins, and even shorter spines in *Sphaerirhynchia* arise from the tips of zig-zag valve margins. These differences were considered by Westbroek (1968) to be of greater than generic significance, and led him to divide the Uncinulidae into three lineages. Marginal spines are characteristic of several families that typically include genera of globular shape, thick shells, and a large cardinal process, grouped here within the Superfamily Uncinuloidea.

Interareas are rarely developed in Paleozoic, and partly obscured by the incurved beak, where developed. Interareas have been described in a few thin forms where umbones are not pronounced, generally in early genera, e.g. *Drepano-rhyncha , Oligorhynchia.* Delthyrium and deltidial plates are of uncertain taxonomic value. Many early rhynchonellids, such as *Ancistrorhyncha, Drepanorhyncha,* and *Orthorhynchula,* have an open delthyrium or very narrow marginal thickening that leaves the delthyrium open. Others, such as *Fenestri-rostra,* have narrow, elongate deltidial plates that do not meet. Most rhynchonellids, however, have well developed deltidial plates. Conditions include disjunct deltidial plates, as in *Cupularostrum,* conjunct as in *Hypsiptycha,* and duplex, as in *Rhynchotrema.* A rounded foramen to accommodate the pedicle is usually present near the posterior apex of the valve so that even when the deltidial plates are conjunct, they often lack their apical tips. Occasionally deltidial plates are alate, as in *Ptilotorhynchus,* but this condition may vary greatly within a genus. The position of the foramen relative to the apex of the delthyrium has often been used as a generic character when it departs significantly from the usual submesothyrid condition.

Internal features.   Internal diagnostic features in the ventral valve include dental plates, and associated lateral umbonal cavities, teeth, and platforms. Those in the dorsal valve include the hinge plates, cardinal process, median septum, and crura. Muscle scars are often present, but are inadequately known for many genera. They are indistinct in serial sections except in families where the muscle attachment areas are deeply impressed, such as the Leiorhynchidae. Dental plates are present in most Paleozoic rhynchonellids; they are reasonably constant within a genus, but are not easily where a genus has dental plates close to the shell wall or where lateral umbonal cavities became filled with callus. Usually dental plates are vertical, convergent toward the floor, convergent to form a spondylium, or otherwise constant. Umbonal cavities have not been used much in generic diagnosis and the feature is probably too dependent on infraspecific variation, age of individuals, and callus thickness. Teeth of rhynchonellids may be elongate and transversely corrugated in heavy thick-shelled genera with sturdy dental plates, as in *Trigonirhynchia,* or small and smooth in genera with short dental plates, or more delicate shells. A fossette may be developed on the inner side of the tooth, where it presses against the inner socket ridge. This may add to the rigidity of the articulation, and fit of the inner socket ridges into the fossettes may have become as important in holding the valves together as the fit of the teeth into the sockets. In this review of Paleozoic rhynchonellids the general the shape and surface of the teeth have not been

given much taxonomic significance, partly because they are infrequently described and partly because they vary with shell thickness and other age factors.

Muscle platforms are unusual in Paleozoic rhynchonellids (the Stenoscismatidae are described separately). Families with muscle platforms are the Camerorhorinidae, which have a spondylium, and the Tetracameridae and Rhynchotetradidae, which have a low or sometimes sessile spondylium. A few genera in other families have a spondylium. These muscle platforms are valued in generic and familial diagnosis.

Hinge plates include all the cardinalia except the cardinal process, thus comprising sockets, socket ridges, outer and inner hinge plates, crural bases and, if present, crural plates and septalium. Hinge plates performed two important unrelated functions in rhynchonellids, providing support for sockets, and attachment for crura. Outermost parts of hinge plates are connected to valve walls, and are often coincident with outer socket ridges. Outer hinge plates extend medially from inner socket ridges to crural bases, and inner hinge plates, or sometimes crural plates, extend medially from crural bases. The term 'inner hinge plates' is used for plates that are horizontal or subhorizontal, as in *Glossinotoechia.* If plates medial of crural bases are inclined to meet the valve floor, or form a V-shaped septalium supported by a median septum that joins the valve floor, they are generally termed 'crural plates'. Often crural bases may be seen as discrete structures, projecting anteriorly from hinge plates or crural plates. Occasionally a 'cover plate' caps the septalium for part or all its length. The terminology used here differs slightly from that of Westbroek (1968) who considered the cover plate over a septalium to be formed by inner hinge plates. Jin (1989) defined a septalium as a trough-shaped chamber encom-passed by a pair of hinge plates, crural bases, and septalial plates, and did not limit the composition of the septalium to any particular pair of these plates. The component features of hinge plates are variously developed within the rhynchonellids. More terms are needed to describe the relatively complex hinge plate of *Trigonirhynchia,* with its outer hinge plates, crural bases, septalium, and convex cover plate, than that of *Australirhynchia,* where the swollen inner socket ridges almost meet and the crura appear to arise directly from these ridges. Where the material is well preserved, as with most of the Anticosti and Wenlock rhynchonellids, it is possible to use these hinge plate distinctions. For Paleozoic rhynchonellids in general, the available information about hinge plate features is too uneven for it to form a satisfactory part of familial diagno-ses. Outer socket ridges may have some generic significance. Genera with a relatively wide ventral valve at the hinge line, e.g.*Glossinulus* and *Eucharitina,* usually have teeth directed inward so that the outer socket ridge may be low. Genera with a relatively narrow ventral valve at the hinge line, such as *Stege-rhynchus* and *Lepidocyclus,* usually have much larger outer socket ridges.

The socket floor may be smooth, as in *Glossinulus,* or crenulate, as in *Trigonirhynchia,* and this is used as a generic characteristic, where recognized. In some genera, crenulation may be most marked in gerontic specimens. The length and orientation of the sockets is often dependent on whether the shell is strophic, in which case the sockets are usually short and parallel to the hinge line, as in *Sphaerirhynchia,* or astrophic, which is normal in rhynchonellids and causes the teeth and sockets to be more anteriorly directed. As these anteriorly directed structures do not articulate significantly, they can become elongate, as in *Trigonirhynchia.* Inner socket ridges vary in size and may be low in more strophic shells, such as *Porostictia*, or very large in astrophic shells such as *Australirhynchia*, particularly for gerontic specimens. Medially from the inner socket ridges there is not much outer hinge plate present in earlier rhynchonellids (*Rhynchotrema*), but by about the Middle Devonian, hinge plates are often broader and more complex so that in *Uncinulus*, *Parapugnax*, and *Porostictia* there are distinct, horizontal outer hinge plates between inner socket ridges and crural bases.

Crural bases form a dense structure along the medial edge of the outer hinge plate. Inward from crural bases hinge plates are often divided, as in *Orthorhynchula* and *Callipleura*, and outer hinge plates and crural bases supported by crural plates that extend down to the valve floor, as in *Callipleura*. Or, crural plates may unite to form a septalium, supported by a median septum, with a cover plate, as in *Trigonirhynchia*, or without a cover plate, as in *Hebetoechia*. The septalium is a V-shaped or U-shaped structure, restricted to the hinge line, and apparently serving to to buttress the sockets. Genera without a septalium usually have alternative support for sockets, as with the crural plates of *Callipleura* and *Machaeraria*. Sockets may rest directly on the valve floor, as in *Sulcatina*, or be supported by thick shell below the hinge plate, as in *Pegmarhynchia*.. Rhynchonellids are characterized by these structures, but Jin (1989) has shown that in *Fenestrirostra* several different structures evolved in successive early Llandoverian species on Anticosti Island. In each of the species there is some development of a septalium posteriorly, but its importance as a support differs considerably, and thus it seems unwise to give too much generic significance to the septalium. The median septum is part of the buttress for the articulation, is variably developed, and not a reliable family character. In many families, such as the Ancistrorhynchidae, Aseptirhynchiidae, Allorhynchidae, and Pontisiidae, a dorsal median septum is rare or absent, usually because the sockets are small and supported by the valve walls. In other families, such as the Uncinulidae, the presence of a median septum is variable, but usually the hinge plates are close to the valve floor and supported by crural plates or callus. In other rhynchonellid families the median septum is consistently present, and a good generic character.

A cardinal process is absent from most Paleozoic rhynchonellid families. The process increased the area of diductor muscle attachment or otherwise made the attachment more secure, varying with the plan of the particular taxon and its stage of growth. It also permitted the diductor attachment surface to protrude into the ventral valve to permit specific valve articulation relationships and muscle arrangements, as discussed by Rudwick (1970). In families where it occurs, it is variable, but usually regarded as having generic significance. In the Rhynchotrematidae, the cardinal process is normally septiform, which seems to be a primitive form of the structure. In the Orthorhynchulidae, a cardinal process is generally present, but its shape varies from septiform, as in *Orthorhynchula*, to branched, as in *Orthorhynchyllion*. In the Uncinuloidea, several genera have a tongue-like cardinal process that protrudes well into the ventral valve (*Glossinulus*, *Markitoechia*), but others have a broader multi-lobed process (*Glossinotoechia*), or a bulbous process (*Uncinulus*). In the Eatoniidae, the cardinal process is bilobate, trilobate, or quadrilobate, and in *Eatonia* and *Pleiopleurina*, it is massive, bilobed, and deeply excavated, forming a structure that occupies much of the hinge plate.

Crura of fossil rhynchonellids are varied: those of most Paleozoic genera are poorly known. Crura appear to be of considerable taxonomic value. In living rhynchonellids, distal ends of crura are associated with proximal ends of soft spiral lophophores, but apparently do not support the lophophores. Crural bases enclose an angle within which food particles are directed from food grooves of the lophophore into the mouth. Dagys (1968) has described 13 different types of crura known in post-Paleozoic taxa, but most of these have not been recognized in Paleozoic genera. Cooper & Grant (1976), Baranov (1977, 1980), and Jin (1989) paid particular attention to crura in Paleozoic genera they described.

Endopunctation in Paleozoic rhynchonellids is rare and punctate genera do not appear to be otherwise closely related, though they are provisionally grouped in the separate superfamily Rhynchoporoidea. *Rhynchopora* and *Tretorhynchia* are well known: *Rariella* is regarded a nomen dubium.

Order RHYNCHONELLIDA Kuhn,1949.

Diagnosis. Rostrate, biconvex articulates usually with dorsal fold-ventral sulcus; interareas typically low, limited to ventral valve; commonly coarse, sometimes fine costae; mostly impunctate. Ventral interior typically with dental plates; spondylium normally absent. Dorsal interior commonly with medium septum supporting septalium or hinge plates; outer parts of hinge plates form bases for crura, that in Recent genera are attached to spirolophous lophophore. Range: Middle Ordovician-Recent. Paleozoic Superfamilies: Ancistrorhynchoidea, Rhynchotrematoidea, Uncinuloidea, Pugnacoidea, Camarotoechioidea, Stenoscismatoidea, Rhynchotetradoidea, Wellerelloidea, Lamdarinoidea, Rhynchoporoidea, Dimerelloidea.

Superfamily ANCISTRORHYNCHOIDEA Cooper, 1956a
Diagnosis. Rhynchonellida with subcircular, transverse, or elongate outline; dorsal fold usual, occasionally dorsal sulcus present; ventral interarea low; delthyrium open or with small deltidial plates anteriorly; costae fine to coarse, extending from beaks. Dental plates usually well developed, occasionally absent; hinge plates divided; dorsal median septum, cardinal process usually absent; crura long, ventrally curved. Range: Middle Ordovician-Lower Devonian.
Families: Ancistrorhynchidae, Oligorhynchiidae, Sphenotretidae, Niorhynicidae. For genera Figure 1.
Family ANCISTRORHYNCHIDAE Cooper, 1956a
Diagnosis. Ancistrorhynchoidea with subcircular outline; dorsal fold, ventral sulcus; costae fine to medium; delthyrium open. Dental plates usually short; hinge plates divided; dorsal median septum, cardinal process absent. Range: Middle Ordovician-Lower Devonian. Genera: Figure 1.
Family OLIGORHYNCHIIDAE Cooper, 1956a
Diagnosis. Ancistrorhynchoidea with elongate subtriangular outline; beak suberect; plicae coarse, few; weak dorsal fold masked by coarse plicae; delthyrium with small deltidial plates anteriorly. Dental plates distinct and vertical; hinge plates divided; dorsal median septum, cardinal process usually absent. Range: Middle-Upper Ordovician. Genera: Figure1.
Family SPHENOTRETIDAE n. fam.
Diagnosis. Small, elongate Ancistrorhynchoidea; weak dorsal sulcus, ventral fold; low ventral interarea; costae fine; foramen with small deltidial plates. Dental plates short; hinge plates small, divided; cardinal process absent. Range: Middle-Upper Ordovician. Genera: Figure 1.
Family NIORHYNICIDAE n. fam.
Diagnosis. Ancistrorhynchoidea with subcircular outline; low dorsal fold, ventral sulcus; costae coarse; delthyrium with small deltidial plates. Dental plates absent; dorsal median

septum short; septalium short, with or without cover plate; cardinal process absent. Range: Lower-Upper Silurian. Genera: See Figure 1.

Superfamily RHYNCHOTREMATOIDEA Schuchert, 1913
Diagnosis. Rhynchonellida with subtriangular to subpentagonal outline; dorsal fold, ventral sulcus; costae medium to coarse, usually extending from beaks; delthyrium open, or with deltidial plates variously developed. Dental plates present or fused to valve walls; dorsal median septum usually long, occasionally short or absent; hinge plates thick; septalium variably developed; cardinal process septiform to unilobed, rarely absent; crura long, ventrally curved. Range: Middle Ordovician-Lower Carboniferous. Families: Rhynchotrematidae, Orthorhynchulidae, Machaerariidae, Trigonirhynchiidae, Leptocoeliidae, Phoenicitoechiidae.

Family RHYNCHOTREMATIDAE Schuchert, 1913
Diagnosis. Early Rhynchotrematoidea with strong dorsal median septum; hinge plates short, thick; cardinal process septiform. Range: Middle Ordovician-Upper Silurian. Subfamilies: Rhynchotrematinae, Lepidocyclinae [Figure 2]

Subfamily RHYNCHOTREMATINAE Schuchert, 1913
Diagnosis. Rhynchotrematidae with strong, simple costae. Delthyrium usually open or with disjunct deltidial plates; dental plates short. Range: Middle Ordovician-Upper Silurian. Genera: Figure 2.

Subfamily LEPIDOCYCLINAE Cooper, 1956a
Diagnosis. Rhynchotrematidae with subcircular to elongate oval outline; strongly biconvex; costae coarse, crossed by imbricating growth lamellae. Delthyrium usually with conjunct deltidial plates; dental plates fused to thick shell walls, or obscured by callus. Range: Upper Ordovician-Upper Silurian. Genera: Figure 2.

Family ORTHORHYNCHULIDAE Cooper, 1956a
Diagnosis. Rhynchotrematoidea with delthyrium open or with incipient deltidial plates; ventral interarea commonly present; dental plates reduced; hinge plates sloping medially to join septalium, sessile or supported on low median septum; cardinal process septiform, lobed, or branching. Range: Middle Ordovician-Lower Silurian. Genera: Figure 2.

Family MACHAERARIIDAE n. fam.
Diagnosis. Rhynchotrematoidea with subpentagonal to triangular outline; dorsal fold, ventral sulcus strong; costae medium to coarse, extending from beaks; delthyrium with disjunct to conjunct deltidial plates. Dental plates short or fused with valve walls; teeth elongate and swollen; dorsal median septum, septalium incipient to absent; hinge plates undivided, sessile; inner socket ridges prominent to massive; cardinal process septiform or unilobed. Range: Lower Devonian-Middle Devonian. Genera: Figure 2.

Family TRIGONIRHYNCHIIDAE Schmidt, 1965a
Diagnosis. Rhynchotrematoidea with subtriangular outline; dorsal fold, ventral sulcus variable; costae often extending from beaks, but commonly umbones smooth; delthyrium open, or with disjunct-conjunct deltidial plates; anterior commissure often dentate, sometimes spinose. Dental plates, dorsal medium septum present; septalium with or without cover plate; cardinal process absent. Range: Middle Ordovician-Lower Carboniferous. Subfamilies: Trigonirhynchiinae, Hemitoechiinae, Rostricellulinae, Ripidiorhynchinae, Virginiatinae.

Subfamily TRIGONIRHYNCHIINAE Schmidt, 1965a
Diagnosis. Trigonirhynchiidae with covered septalium. Range: Lower Silurian-Lower Carboniferous. Genera: Figure 3.

Subfamily HEMITOECHIINAE n. subfam.
Diagnosis: Trigonirhynchiidae with uncovered septalium. Range: Middle Ordovician-Upper Devonian. Genera: Figure 3.

Subfamily ROSTRICELLULINAE Rozman, 1969
Diagnosis. Early Trigonirhynchidae with uncovered septalium and costae crossed by distinct concentric lamellae; delthyrium open or with incipient deltidial plates; dorsal median septum short. Range: Middle Ordovician-Lower Silurian. Genera: Figure 3.

Subfamily RIPIDIORHYNCHINAE n. fam.
Diagnosis. Late Trigonirhynchiidae with strong fold, sulcus, anterior tongue conspicuous; delthyrium with deltidial plates usually conjunct; costae numerous, straight, arising from beaks. Dental plates long; dorsal median septum short; septalium with/without cover plate anteriorly. Range: Middle Devonian-Lower Carboniferous. Genera: Figure 3.

Subfamily VIRGINIATINAE Amsden, 1974
Diagnosis. Trigonirhynchiidae with elongate subtriangular outline; apical angle acute; equibiconvex lateral profile; fold, sulcus weak to absent; anterior commissure rectimarginate to moderately unisulcate or uniplicate; costae coarse, straight, simple, arising at beaks or on umbones; ventral beak straight to erect; delthyrium with disjunct or conjunct deltidial plates. Dental plates well developed, close to valve walls, vertical; dorsal median septum short, low, thick; septalium open or with partial cover plate. Range: Upper Ordovician-Upper Silurian. Genera: Figure 3.

Family LEPTOCOELIIDAE Boucot & Gill, 1956
Diagnosis. Rhynchotrematoidea with subcircular to subpentagonal outline; lateral profile planoconvex to biconvex, valves meeting at acute angle laterally; fold, sulcus weak; costae simple, strong, extending from beaks; delthyrium open or with disjunct deltidial plates. Dental plates usually absent; dorsal median septum weak or absent; hinge plates sessile; cardinal process large to absent. Range: Lower Silurian-Middle Devonian. Genera: Figure 2.

Family PHOENICITOECHIIDAE Havlícek, 1990
Diagnosis. Rhynchotrematoidea with subcircular to subtrigonal outline; fold, sulcus moderate to strong; costae strong, simple, sometimes arising by bifurcation, intercalation; umbones generally smooth; delthyrium with disjunct to conjunct deltidial plates. Dental plates short, thin; hinge plates forming wide, open, ventrally convex septalium; cardinal process absent. Range: Lower Devonian. Genera: Figure 2.

Superfamily UNCINULOIDEA Rzhonsnitskaya, 1956
Diagnosis. Rhynchonellida with globular, cuboid or lenticular shape; dorsal fold, ventral sulcus weak; tongue high; costae numerous, flattened anteriorly on paries geniculatus; anterior, lateral margin spines usually present; foramen small, commonly with conjunct to disjunct deltidial plates anteriorly; squamae and glottae often developed. Dental plates weak or infilled with callus; dorsal median septum, septalium, cardinal process generally present. Range: Lower Silurian-Upper Devonian. Families: Uncinulidae, Glossinotoechiidae, Hebetoechiidae, Obturamentellidae, Hypothyridinidae, Eatoniidae, Hadrorhynchiidae, Innaechiidae.

Family UNCINULIDAE Rzhonsnitskaya, 1956
Diagnosis. Uncinuloidea with thick shell; anterior marginal spines long. Dental plates short or obscured by umbonal callus; dorsal median septum present; septalium usually infilled by callus; cardinal process wide, multilobed. Range: Lower-Upper Devonian. Genera: Figure 4.

Family GLOSSINOTOECHIIDAE Havlícek, 1992
Diagnosis. Uncinuloidea with cordiform to subtriangular outline, acuminate posteriorly; tongue distinct; costae numerous, flattened anteriorly; marginal spines developed; foramen mesothyrid with conjunct deltidial plates. Dental plates short, close to valve walls; septalium usually with cover plate; cardinal process high, linguiform, often striated. Range: Upper Silurian-Middle Devonian. Genera: Figure 4.

Family HEBETOECHIIDAE Havlícek, 1960

Diagnosis. Uncinuloidea strongly biconvex to globular; fold, sulcus weak; lateral-anterior margins vertical in mature specimens; tongue high; costae numerous, flattened, grooved on paries geniculatus; marginal spines developed; small foramen with disjunct to conjunct deltidial plates usually present; squamae, glottae frequently developed. Dental plates vertical to ventrally convergent; dorsal median septum distinct; septalium deep, without cover plate, often infilled with callus; cardinal process usually absent or incipient in early genera, larger and commonly multilobed or longitudinally striated in later genera.   Range: Lower Silurian-Middle Devonian. Subfamilies: Hebetoechiinae, Sphaerirhynchiinae, Glossulininae, Betterbergiinae, Beckmanniinae.

Subfamily HEBETOECHIINAE Havlícek, 1960

Diagnosis. Hebetoechiidae with few, relatively coarse costae; umbones smooth; marginal spines short. Dental plates short; dorsal median septum distinct; septalium present; low, callus-like cardinal process usually developed.   Range: Upper Silurian-Lower Devonian. Genera: Figure 4.

Subfamily SPHAERIRHYNCHIINAE n. subfam.

Diagnosis. Globular Hebetoechiidae with weak fold, high tongue, fine, distally grooved costae. Dental plates convergent ventrally; dorsal median septum well-developed; cardinal process absent or callus-like.   Range: Lower Silurian-Lower Devonian. Genera: Figure 4.

Subfamily GLOSSULININAE n. subfam.

Diagnosis. Transversely subpentagonal Hebetoechiidae with sharply delineated fold, sulcus, rectangular tongue; fine costae flattened, grooved on paries geniculatus; well-developed marginal spines. Dental plates short or obscured by callus; septalium reduced; cardinal process small to absent. Range: Lower Devonian. Genera: Figure 4.

Subfamily BETTERBERGIINAE n. subfam.

Diagnosis. Thick-shelled, rounded Hebetoechiidae with low fold, fine costae flattened, grooved anteriorly; long marginal spines. Dental plates short or obscured by callus; dorsal median septum, septalium short; cardinal process commonly present. Range: Lower-Middle Devonian. Genera: Figure 4.

Subfamily BECKMANNIINAE n. subfam.

Diagnosis. Subcircular Hebetoechiidae with broad fold, smooth umbones, rounded exterior features; marginal spines weakly developed. Dental plates short; dorsal median septum commonly prominent; callus-like cardinal process generally present. Range: Middle Devonian. Genera: Figure 4.

Family OBTURAMENTELLIDAE n. fam.

Diagnosis. Uncinuloidea with very thick shell; costae coarse; marginal spines short; dental plates rudimentary, mostly infilled with callus. Teeth massive, longitudinally grooved, arising from shell walls; ventral muscle field deeply impressed; dorsal median septum low, thick, long; septalium filled posteriorly with massive, linguiform, bilobed cardinal process. Range: Upper Silurian-Lower Devonian. Genera: Figure 4

Family HYPOTHYRIDINIDAE Rzhonsnitskaya, 1956

Diagnosis. Uncinuloidea with dorsal median septum and septalium weak or lacking. Cardinal process usually wide, with myophore consisting of numerous vertical ridges, but cardinal process sometimes poorly developed or absent.   Range: Lower-Upper Devonian. Genera: Figure 5.

Family EATONIIDAE Schmidt, 1965

Diagnosis. Uncinuloidea with subcircular to elongate outline; costae usually extending from beaks, often with superimposed striae; dorsal fold, ventral sulcus strong, but narrow; lateral margins of valves generally meeting at acute angle, never truncated; foramen with conjunct deltidial plates anteriorly; anterior and lateral valve margins commonly with stubby spines that project into notches in opposite valve. Dental plates absent or infilled with callus; ventral muscle field large

with strong border and distinct myophragm; large diductor scars enclosing small adductor scars, also with strong border and myophragm; dorsal median septum usually weak or absent; cardinal process very large, linguiform, bilobed to quadrulobed; hinge plates usually united by cardinal process; crural bases stout; crura short, laterally compressed. Range: Upper Silurian-Lower Devonian. Genera: Figure 4.

Family HADRORHYNCHIIDAE McLaren, 1965

Diagnosis. Uncinuloidea with coarse costae; commissure acutely denticulate with marginal grooves; fold, sulcus developed anteriorly; tongue high, rectangular to acute. Small septalium; hinge plates horizontal, divided; bilobed callus-like cardinal process commonly present. Range: Middle Devonian. Genera: Figure 5.

Family INNAECHIIDAE Baranov, 1980

Diagnosis. Uncinuloidea lacking septalium, cardinal process. Range: Upper Silurian-Middle Devonian. Subfamilies: Innaechiinae, Dogdoinae, Vladimirirhynchinae, Corvinopugnacinae.

Subfamily INNAECHIINAE Baranov, 1980.

Diagnosis: Innaechiidae with median septum; dental plates very short. Range: Lower Devonian. Genera: Figure 5.

Subfamily DOGDOINAE Baranov, 1982

Diagnosis. Innaechiidae lacking dental plates, dorsal median septum. Range: Lower Devonian. Genera: Figure 5.

Subfamily VLADIMIRIRHYNCHINAE Baranov, 1982

Diagnosis. Innaechiidae with short dental plates; median furrow dividing fold, corresponding ridge dividing sulcus; dorsal median septum commonly absent. Range: Lower-Middle Devonian. Genera: Figure 5.

Subfamily CORVINOPUGNACINAE n. subfam.

Diagnosis. Innaechiidae with fold, sulcus low, wide; costae numerous; short dental plates present. Range: Upper Silurian-Middle Devonian. Genera: Figure 5.

Superfamily PUGNACOIDEA Rzhonsnitskaja, 1956

Diagnosis. Rhynchonellida with transversely ovate to subtriangular outline; strongly inflated anteriorly; dorsal fold and ventral sulcus distinct; tongue high; costae few, simple, most pronounced anteriorly; umbones usually smooth; superimposed radial striae or costellae occasionally present; foramen generally present, with conjunct deltidial plates. Dental plates short to absent, or rarely long; dorsal median septum short to absent; hinge plates commonly divided, often sloping dorsally to merge with crural plates; septalium absent to rare, short when present; cardinal process absent. Range: Lower Devonian-Upper Permian. Families: Pugnacidae, Ladogiidae, Plectorhynchellidae, Rozmanariidae, Aseptirhynchiidae, Yunnanellidae, Petasmariidae, Camerophorinidae, Paranorellidae.

Family PUGNACIDAE Rzhonsnitskaya, 1956

Diagnosis. Pugnacoidea with high fold, tongue; umbones usually smooth; costae few; superimposed radial striae common; foramen generally present, with conjunct deltidial plates. Dental plates variable, occasionally lacking; dorsal median septum short to absent; hinge plates divided; crural plates inclined dorsally; septalium absent.   Range: Lower Devonian-Upper Permian. Genera: Figure 6.

Family LADOGIIDAE Lyashenko, 1973

Diagnosis. Pugnacoidea with high, rounded to acuminate tongue; fine, evenly spaced costae arising at beaks. Dental plates and dorsal median septum distinct; septalium short, often with cover plate. Range: Lower-Upper Devonian. Genera: Figure 6

Family PLECTORHYNCHELLIDAE Rzhonsnitskaya, 1958

Diagnosis. Small Pugnoidea with dorsal sulcus, ventral fold, longitudinally ovate outline; anterior commissure plicosulcate, smooth or with weak costae on fold, in sulcus. Dental plates short; dorsal median septum absent or very short; hinge plates divided. Range: Lower-Upper Devonian. Subfamilies: Plecto-rhynchellinae, Pygmaellinae.

Subfamily PLECTORHYNCHELLINAE Rzhonsnitskaya, 1958
Diagnosis: Plectorhynchellidae with crura long, closely set. Range: Middle-Upper Devonian. Genera: Figure 6.

Subfamily PYGMAELLINAE Baranov, 1977
Diagnosis. Early Plectorhynchellidae with crura lyre-shaped in cross-section. Range: Lower Devonian. Genera: Figure 6.

Family ROZMANARIIDAE Havlícek, 1982
Diagnosis. Pugnacoidea with transversely ovate to subcircular outline; ventral fold, dorsal sulcus; ventrally directed tongue pronounced in some genera; costae weak to absent; foramen with conjunct deltidial plates anteriorly. Dental plates short to absent; dorsal median septum weak or lacking; hinge plates divided; cardinal process absent. Range: Lower-Upper Devonian. Genera: Figure 6.

Family ASEPTIRHYNCHIIDAE n. fam.
Diagnosis. Pugnacoidea lacking dental plates, dorsal median septum, septalium; fold, sulcus developed anteriorly. Range: Lower-Upper Devonian. Genera: Figure 6.

Family YUNNANELLIDAE Rzhonsnitskaja, 1959
Diagnosis. Pugnacoidea with fine costae posteriorly merging into coarse costae anteriorly. Dental plates usually present; dorsal median septum supporting short septalium. Range: Middle-Upper Devonian. Genera: Figure 6.

Family PETASMARIIDAE n. fam.
Diagnosis Pugnacoidea with strong dental plates; hinge plates divided, horizontal; fine striae present on surface, rows of pits in well-preserved specimens. Dorsal median septum short but distinct; septalium short to absent. Range: Middle Devonian-Upper Permian. Genera: Figure 7.

Family CAMEROPHORINIDAE Rzhonsnitskaya, 1958.
Diagnosis Pugnacoidea with dental plates joined to form spondylium. Dorsal median septum, septalium absent; hinge plates undivided. Range: Middle Devonian. Genera: Figure 7.

Family PARANORELLIDAE Cooper & Grant, 1976
Diagnosis. Pugnacoidea with subcircular to elongate outline; fold and sulcus weak, commonly ventral fold, dorsal sulcus; costae weak to absent. Hinge plates divided; dental plates, dorsal median septum short. Range: Lower Carboniferous-Upper Permian. Subfamilies: Paranorellinae, Iowarhynchinae.

Subfamily PARANORELLINAE Cooper and Grant, 1976
Diagnosis. Subcircular Paranorellidae with dorsal sulcus. Range: Lower Carboniferous-Upper Permian. Genera: Figure 7.

Subfamily IOWARHYNCHINAE n. subfam.
Diagnosis. Small, smooth Paranorellidae with elongate oval outline; weak dorsal fold; mesothyrid foramen. Dental plates short; dorsal median septum, septalium absent; hinge plates divided. Range: Lower Carboniferous. Genera: Figure 7.

Superfamily CAMAROTOECHIOIDEA Schuchert, 1929
Diagnosis. Rhynchonellida with subcircular to transversely ovate outline; dorsal fold, ventral sulcus weak to moderate; tongue low to high; costae weak to strong, simple to bifurcating, more pronounced anteriorly, umbones often smooth; foramen small with conjunct deltidial plates anteriorly. Dental plates strong, usually convergent ventrally, occasionally infilled with callus; dorsal median septum long; septalium short, usually without cover plate; hinge plates divided anterior of septalium; cardinal process absent; crura long, ventrally curved; shell often thick, muscle fields deeply impressed. Range: Lower Silurian-Lower Carboniferous. Families: Camarotoechiidae, Leiorhynchidae, Septalariidae.

Family CAMAROTOECHIIDAE Schuchert, 1929
Diagnosis. Camarotoechioidea with short dental plates, low dorsal median septum. Range: Lower Silurian-Middle Devonian. Subfamilies: Camarotoechiinae, Linguopugnoidinae.

Subfamily CAMAROTOECHIINAE Schuchert, 1929
Diagnosis: Moderately biconvex Camarotoechiidae with costae sometimes bifurcate; dental plates nearly vertical. Range: Lower-Middle Devonian. Genera: Figure 8.

Subfamily LINGUOPUGNOIDINAE n. subfam.
Diagnosis. Early Camarotoechiidae with broad fold, sulcus; high tongue; strong, simple costae; weak dental plates; thin dorsal median septum. Range: Lower Silurian-Lower Devonian. Genera: Figure 8.

Family LEIORHYNCHIDAE Stainbrook, 1945
Diagnosis. Camarotoechioidea with strong dental plates convergent ventrally, and dorsal median septum long and high; profile strongly biconvex; shell commonly thick, with ventral muscle field deeply impressed. Range: Lower Silurian-Lower Carboniferous. Subfamilies: Leiorhynchinae, Innuitellinae, Fenestrirostrinae, Platyterorhynchinae, Gigantorhynchinae, Stenometoporhynchinae, Calvinariinae, Basilicorhynchinae.

Subfamily LEIORHYNCHINAE Stainbrook, 1945
Diagnosis. Leiorhynchidae with smooth umbones. Range: Lower Devonian-Lower Carboniferous. Genera: Figure 8.

Subfamily INNUITELLINAE Crickmay, 1968
Diagnosis. Leiorhynchidae with dental plates supported by buttresses; dorsal valve inflated; fold, costae weak. Range: Lower Devonian. Genera: Figure 8.

Subfamily FENESTRIROSTRINAE n. subfam.
Diagnosis. Early Leiorhynchidae with low costae, dental plates short to absent. Range: Lower Silurian-Lower Devonian. Genera: Figure 8.

Subfamily PLATYTERORHYNCHINAE n. subfam.
Diagnosis. Leiorhynchidae with lenticular profile, low fold; costae weak, uneven; dental plates short, almost meeting ventrally. Range: Middle-Upper Devonian. Genera: Figure 8.

Subfamily GIGANTORHYNCHINAE n. subfam.
Diagnosis. Large Leiorhynchidae, smooth; low fold, sulcus; cardinal process present; thick, united hinge plates; dental plates, septalium absent. Range: Middle Devonian. Genera: Figure 8.

Subfamily STENOMETOPORHYNCHINAE n. subfam.
Diagnosis. Leiorhynchidae with ventral valve flattened to concave, dorsal valve highly inflated; dental plates, dorsal median septum, hinge plates very short; septalium absent. Range: Upper Devonian. Genera: Figure 8.

Subfamily CALVINARIINAE Sartenaer, 1994
Diagnosis. Leiorhynchidae with transverse outline; smooth umbones, high tongue; few coarse costae; strongly convergent dental plates. Range: Upper Devonian. Genera: Figure 8.

Subfamily BASILICORHYNCHINAE n. subfam.
Diagnosis. Leiorhynchidae with subcircular outline; high tongue; strong costae; dental plates, dorsal median septum distinct. Range: Upper Devonian. Genera: Figure 8.

Family SEPTALARIIDAE Havlícek, 1960
Diagnosis. Camarotoechioidea with tongue generally extended into trail; hinge plates united; dorsal median septum high; cardinal process commonly developed. Range: Lower-Upper . Devonian. Genera: Figure 9.

Superfamily STENOSCISMATOIDEA Oehlert, 1887.
Diagnosis. Rhynchonellida with spondylium in ventral valve, camarophorium dorsal valve. Range: Lower Devonian-Upper Permian. Families: Stenoscismatidae, Atriboniidae.

Superfamily RHYNCHOTETRADOIDEA Likharev,
in Rzhonsnitskaya, 1956
Diagnosis. Rhynchonellida with elongate oval outline; apical angle acute; dorsal fold, ventral sulcus weak; planareas concave; costae coarse to very coarse, often with superimposed striae anteriorly; foramen with conjunct deltidial plates. Dental plates converge ventrally to form sessile

spondylium or spondylium duplex; spondylium with or without lateral buttresses; dorsal median septum strong; septalium present without cover plate; cardinal process absent; crural bases usually triangular in cross-section. Range: Upper Devonian-Upper Carboniferous. Families: Rhynchotetradidae, Tetracameridae.

Family RHYNCHOTETRADIDAE Likharev, in Rzhonsnitskaya, 1956

Diagnosis. Rhynchotetradoidea without lateral buttresses in ventral valve; shell surface with radial striae. Range: Upper Devonian-Upper Carboniferous. Subfamilies: Rhynchotetradinae, Axiodeaneiinae.

Subfamily RHYNCHOTETRADINAE Likharev, in Rzhonsnitskaya, 1956

Diagnosis. Rhynchotetradidae with sessile spondylium or spondylium duplex in ventral valve. Range: Lower-Upper Carboniferous. Genera: Figure 11.

Subfamily AXIODEANEIINAE n. subfam.

Diagnosis. Rhynchotetradidae with sessile spondylium; elongate outline; costae very coarse. Range: Upper Devonian-Lower Carboniferous. Genera: Figure 11.

Family TETRACAMERIDAE Likharev, in Rzhonsnitskaya, 1956

Diagnosis. Rhynchotetradoidea with lateral buttresses in ventral valve; shell surface without radial striae. Range: Lower Carboniferous. Genera: Figure 11.

Superfamily WELLERELLOIDEA Likharev, in Rzhonsnitskaya, 1956

Diagnosis. Rhynchonellida with subcircular to subpentagonal outline; dorsal fold, ventral sulcus strong to weak; costae usually simple; umbones commonly smooth; delthyrium with/without deltidial plates. Dental plates, dorsal median septum variously developed; hinge plates undivided, notched, or totally divided; cardinal process absent. Range: Lower Carboniferous-Upper Cretaceous. Paleozoic families: Wellerellidae, Pontisiidae, Allorhynchidae, Petasmatheridae, Amphipellidae.

Family WELLERELLIDAE Likharev, in Rzhonsnitskaya, 1956

Diagnosis. Wellerelloidea with subpentagonal outline; costae usually coarse. Dental plates often short, vertical, occasionally convergent ventrally, rarely absent. Range: Upper Carboniferous-Permian. Subfamilies: Wellerellinae, Tricoriinae, Trophisininae, Nipponirhynchiinae, Uncinunellininae, Strigirhynchiinae, Madarosiinae.

Subfamily WELLERELLINAE Likharev, in Rzhonsnitskaya, 1956

Diagnosis. Wellerellidae with short dental plates; costae arising anterior of beaks; delthyrium with conjunct deltidial plates; dorsal median septum present; hinge plates usually undivided. Range: Upper Carboniferous-Permian. Genera: Figure 10.

Subfamily TRICORIINAE Cooper & Grant, 1976.

Diagnosis. Wellerellidae with completely costate shells, rectimarginate anterior commissure, obsolete dental plates, strong dorsal median septum. Range: Lower Permian. Genera: Figure 10.

Subfamily TROPHISININAE Cooper & Grant, 1976

Diagnosis. Wellerellidae with finely costate shells and strongly convergent dental plates. Range: Lower Permian. Genera: Figure10.

Subfamily STRIGIRHYNCHIINAE Cooper & Grant, 1976

Diagnosis. Wellerellidae with costae arising at beaks; distinct dental plates; high dorsal median septum; hinge plates usually undivided. Range: Permian. Genera: Figure 10.

Subfamily MADAROSIINAE n. subfam.

Diagnosis. Wellerellidae with smooth surface, emarginate anterior, high dorsal median septum, open delthyrium. Range: Permian. Genera: Figure 10.

Subfamily UNCINUNELLININAE n. subfam.

Diagnosis. Wellerellidae with anteriorly flattened costae; marginal spines developed from intertroughs; median septum low to absent; septalium absent. Range: Permian. Genera: Figure 10.

Subfamily NIPPONIRHYNCHIINAE n. subfam.

Diagnosis. Wellerellidae with anteriorly flattened costae; marginal spines developed from intertroughs; dorsal median septum high; septalium well developed. Range: Lower Permian. Genera: Figure 10.

Family PONTISIIDAE Cooper & Grant, 1976.

Diagnosis. Wellerelloidea with dorsal median septum weak to lacking; costae usually coarse; dental plates commonly distinct; hinge plates undivided, but notched anteriorly; dorsal median septum absent. Range: Upper Carboniferous-Lower Triassic. Subfamilies: Pontisiinae, Acolosiinae.

Subfamily PONTISIINAE Cooper & Grant, 1976

Diagnosis. Pontisiidae with strong costae. Range: Upper Carboniferous-Lower Triassic. Genera: Figure 10.

Subfamily ACOLOSIINAE n. subfam.

Diagnosis. Pontisiidae with surface smooth except for very weak costae anteriorly. Range: Permian. Genera: Figure 10.

Family ALLORHYNCHIDAE Cooper & Grant, 1976

Diagnosis. Wellerelloidea with subcircular to subpentagonal outline; costae extending all or most of shell length; dorsal fold, ventral sulcus weak to absent; delthyrium with or without deltidial plates; dental plates, dorsal median septum weak to absent; hinge plates divided. Range: Lower Carboniferous-Upper Permian. Genera: Figure 10.

Family PETASMATHERIDAE Cooper & Grant, 1976

Diagnosis. Small Wellerelloidea with dorsal fold weak to absent; anterior commissure rectimarginate; costae strong, simple, from beaks; well developed interarea; delthyrium open, without deltidial plates; dental plates short; hinge plates widely divided. Range: Permian. Genera: Figure 10.

Family AMPHIPELLIDAE Cooper & Grant, 1976

Diagnosis. Welleroidea with subcircular outline, posterolateral pouches (apricatria) in both valves; very short dental plates. Range: Lower Permian. Genera: Figure 10.

Superfamily LAMBDARINOIDEA Brunton & Champion, 1974

Diagnosis. Very small Rhynchonellida with bilobed to cordiform outline; smooth surface; commonly with dorsal, ventral sulci, emarginate anterior; delthyrium with symphytium or conjunct deltidial plates or open; ventral median septum commonly present anteriorly; dorsal median septum absent posteriorly, rarely anteriorly; hinge plates short; cardinal process absent. Range: Upper Devonian-Upper Carboniferous. Families: Lambdarinidae.

Family LAMBDARINIDAE Brunton & Champion, 1974

Diagnosis. Lambdarinoidea with bilobed to cordiform outline; symphytium present or absent; ventral interarea long; foramen circular or delthyrium open; dental plates short or absent; ventral, dorsal median septa absent, or present anteriorly. Range: Upper Devonian-Upper Carboniferous. Subfamilies: Lambdarininae, Loborininae, Minyspheniinae.

Subfamily LAMBDARININAE Brunton & Champion, 1974

Diagnosis. Lambdarinidae with symphytium; ventral interarea long; foramen circular; dental plates short; ventral, dorsal median septa absent. Range: Lower Carboniferous. Genera: Figure 9.

Subfamily LOBORININAE n. subfam.

Diagnosis. Lambdarinidae lacking a symphytium; delthyrium elongate, open; dental plates fused to valve walls; teeth

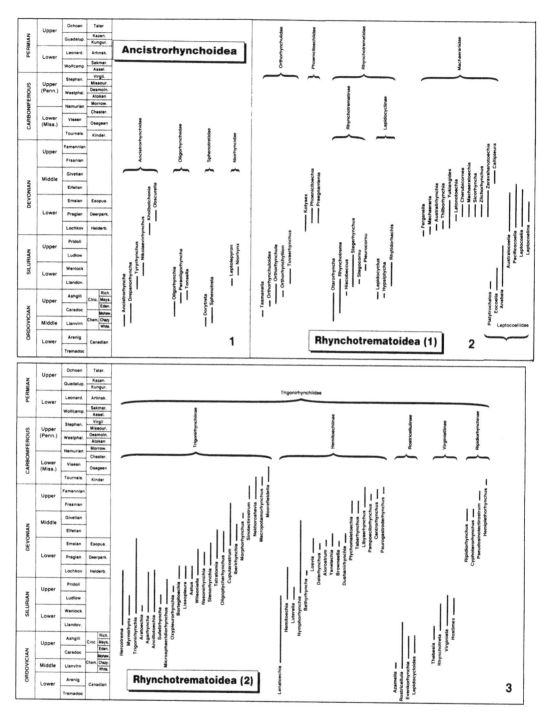

Ancistrorhynchoidea

1

Rhynchotrematoidea (1)

2

Rhynchotrematoidea (2)

3

massive; hinge plates poorly developed; ventral, dorsal median septa absent. Range: Upper Devonian. Genera: Fig. 9.

Subfamily MINISPHENIINAE, n. subfam.

Diagnosis. Lambdarinidae lacking symphytium, dental plates; dorsal, ventral median septa present anteriorly; teeth directed antero-medially. Range: Upper Carboniferous. Genera: Fig. 9.

Superfamily RHYNCHPOROIDEA Muir-Wood, 1955

Diagnosis. Rhynchonellida with endopunctate shell; costal interspaces commonly developed into marginal spines. Range: Lower Carboniferous-Upper Permian. Families: Rhynchoporidae.

Family RHYNCHOPORIDAE Muir-Wood, 1955

Diagnosis. Characters of superfamily. Range: As for superfamily. Genera: *Rhynchopora, Tretorhynchia*.

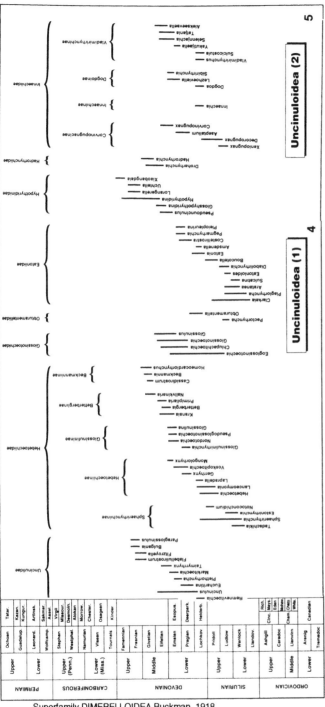

NOMINA DUBIA

The following genera are considered nomina dubia, in most cases because the type material is insufficiently well-preserved or insufficiently well-described to warrant generic status at this time: *Ancorhynchia, Areella, Asiarhynchia, Beichuanella, Beichuanrhynchus, Boreali-rhynchia, Camarotoechioides, Chivatschella, Corrugatimediorostrum, Donella, Dorsisinus, Ferganotoechia, Globidorsum, Glyptorhynchia, Hypoleiorhynchus, Irgislella, Laevorhynchia, Langkawia, Laosia, Leiorhynchoides, Linxiang-xiella, Miaohuangrhynchus, Mongolirhynchia, Nantanella, Neimongolella, Paratetratomia, Paryphorrhynchopora, Payuella, Perakia, Pla-tyglossariorhynchus, Plekonina, Protorhyn-cha, Pseudopugnax, Pugnacina, Rhynchotre-taoides, Rhynchotretina, Rhynoleichus, Salair-otoechia, Sichuanrhynchus, Straelenia, Tane-rhynchia, Togaella, Trilobatoechia, Uralotoe-chia, Wulungguia, Yanbianella, Yarirhynchia, Yingtangella.*

REFERENCES

AGER, D.V. 1965. Mesozoic and Cenozoic Rhynchonellacea, p.597-625. In, R.C. Moore (ed.), Treatise on Invertebrate Paleontology, H, Brachiopoda. Geological Society of America, University of Kansas Press, Lawrence.

AMSDEN, T.W. 1974. Late Ordovician and Early Silurian articulate brachiopods from Oklahoma, southwestern Illinois and eastern Missouri. Oklahoma Geological Survey Bulletin 111:1-154.

BARANOV, V.V. 1977. New Early Devonian rhynchonellids in the northeastern USSR. Paleontologicheskii Zhurnal 1977 (3):75-82.

___1980. Morphology of crura and new rhyn-chonellid taxa. ibid. 1980(4):75-90.

___1982. New Devonian rhynchonellids and athyridids from eastern Yakutia. ibid. 1982(2):41-51.

BOUCOT, A.J. & E.D. GILL. 1956. *Australo-coelia*, a new Lower Devonian brachiopod from South Africa, South America, and Australia. Journal Paleontology 30(5):1173-1178.

___,J.G. JOHNSON & R.D. STATON. 1965. Suborder Atrypi-dina, p.632-649. In, R.C. Moore (ed.), Treatise on Inverte-brate Paleontology. H, Brachiopoda. Geological Society of America, University of Kansas Press.

BRUNTON, C.H.C. & C. CHAMPION. 1974. A Lower Carboniferous brachiopod fauna from the Manifold Valley, Staffordshire. Palae-ontology 17(4):811-840.

BUCKMAN, S.S. 1918. The Brachiopoda of the Namyau Beds, Northern Shan States, Bur-ma. Geological Survey India Memoirs, Palae-ontologia Indica 3(2):1-299.

COOPER, G.A. 1956a. Chazyan and related brachiopods. Smithsonian Miscellaneous Collections127(1):1-1024;127(2):1025-1245.

___1956b. New Pennsylvanian brachiopods. Journal Paleontology 30(3):521-530.

Superfamily DIMERELLOIDEA Buckman, 1918
Family PEREGRINELLIDAE Ager, 1965
Subfamily DZIEDUSZYCKIINAE n. subfam.

Diagnosis. Large, transversely ovate Peregrinellidae with strong, simple, full costae; bisulcate; dental plates short, vertical; dorsal median septum short; crura long, thin, closely set. Range: Upper Devonian. Genera: Figure 11.

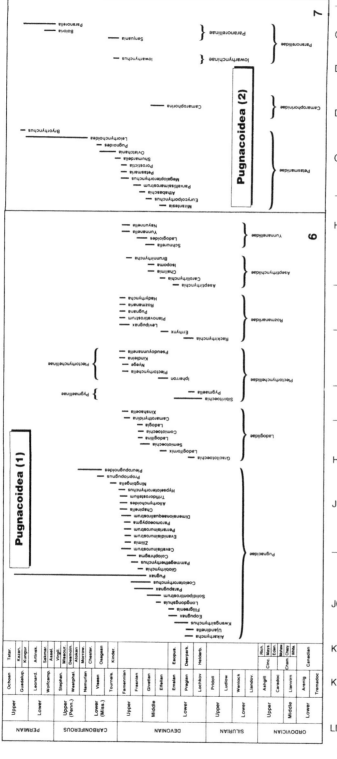

___& R.E. GRANT. 1976. Permian brachiopods of West Texas, Smithsonian Contributions Paleobiology 21(1/2):1923-2607.

CRICKMAY, C.H. 1968. Discoveries in the Devonian of Western Canada. DeMille Books, Calgary, 12p.

DAGYS, A.S. 1968. Jurassic and Lower Cretaceous brachiopods from northern Siberia (in Russian). Trudy Institut Geologii i Geofiziki 41:1-167.

DUNBAR, C.O. & G.E. CONDRA. 1932. Brachiopoda of the Pennsylvanian system in Nebraska. Nebraska Geological Survey Bulletin 2(5):1-377.

GRANT, R.E. 1965. The brachiopod superfamily Stenoscismatacea. Smithsonian Miscellaneous Collections 148(2):1-185.

___1976. Permian brachiopods from southern Thailand. Paleontological Society Memoir, Supplement 9:1-269.

HAVLICEK, V. 1960. Bericht über die Ergebnisse der Revision der böhmischen ältpaläozoischen Rhynchonelloidea. Ustredniho Ústavu Geologického Véstník 35:241-244.

___1961. Rhynchonelloidea des böhmischen älteren Paläozoikums (Brachiopoda). Ustredniho Ústavu Geologického Rozpravy 27:1-211.

___1982. New Pugnacidae and Plectorhynchellidae (Brachiopoda) in the Silurian and Devonian rocks of Bohemia. Ustredniho Ústavu Geologického Véstník 57(2):111-114.

___1990. New Lower Devonian (Pragian) Rhynchonellid brachiopods in the Koneprusy area (Czechoslovakia). ibid. 65 (4): 211-222.

___1992. New Lower Devonian (Lochkovian-Zlichovian) brachiopods in the Prague Basin. Sbornik Geologickych Ved (Paleontologie) 32:55-122.

HOARE, R.D. & R.H. MAPES. (in press). Cardiarina cordata Cooper, 1956, (Brachiopoda), terebratuloid or rhynchonelloid? Journal Paleontology.

JIN, J. 1989. Late Ordovician-Early Silurian rhynchonellid brachiopods from Anticosti Island, Quebec. Biostratigraphie Paléozoique 10: 1-127.

___, W.G.E. CALDWELL & B.S. NORFORD. 1993. Early Silurian brachiopods and biostratigraphy of the Hudson Bay lowlands, Manitoba, Ontario, and Quebec. Geological Survey Canada Bulletin 457:1- 221.

JOHNSON, J.G. 1975. Devonian brachiopods from the Quadrithyris Zone (upper Lochkovian), Canadian Arctic Archi-pelago. Geological Survey Canada Bulletin 235:5-57.

KÜHN, O. 1949. Lehrbuch der Paläozoologie. Schweizerbart'sche Verlagsbuchhandlung, Stuttgart, 326p.

KULKOV, N.P. 1963. Brakhiopody Solovikhinskikh Sloev Nizhnego Devona gornogo Altaya. Akademiya Nauk SSSR, Moscow, 135p.

LIKHAREV, B.K. 1956, p.125-126. In, M.A. Rzhonsnitskaya, Systematization of Rhynchonellida. International Geological Congress Mexico Report 20.

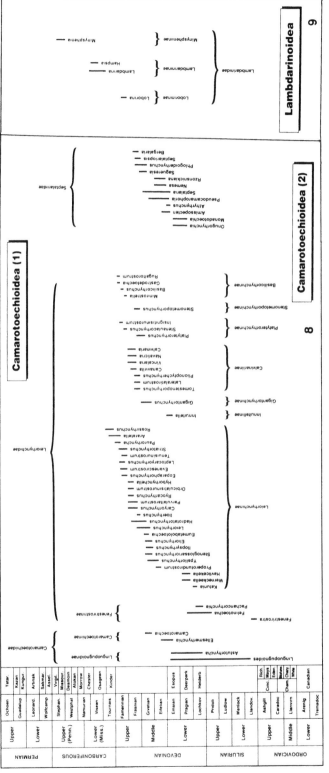

LYASHENKO, A.I. 1973. Brachiopods and stratigraphy of lower Frasnian deposits of South Timan and Volga-Ural oil-gas bearing province. Trudy Vsesoyuznyi Nauchno-Issledovatelskii Geologo-Razvedochnyi Neftianoi Institut (VNIGNI) 134:1-279.

McLAREN, D.J. 1965. Paleozoic Rhynchonellacea. In, R.C. Moore (Ed.), Treatise on Invertebrate Paleontology, H, Brachiopoda. Geological Society of America, University of Kansas Press.

MUIR-WOOD, H.M. 1955. A history of the classification of the phylum Brachiopoda. British Museum (Natural History), London, 124p.

OEHLERT, D.P. 1887. Brachiopodes. In, P. Fischer, Manuel de Conchyliologie. F. Savy , Paris.

ROZMAN, K.S. 1969. Late Ordovician brachiopods of the Siberian Platform. Paleontologicheskii Zhurnal 1969(3):86-108.

RUDWICK, M.J.S. 1970. Living and fossil brachiopods. Hutchinson, London, 199p.

RZHONSNITSKAYA, M.A. 1956. Systematization of Rhynchonellida. Resumenes de los Trabajos Presentados. International Geological Congress, Mexico, Report 20:125-126.

___1958. On rhynchonellid systematics. International Geological Congress Mexico Transactions 20(7):107-121.

___1959. On the systematics of the rhynchonellids. Paleontologickeskii Zhurnal 1959(1): 25-36.

SARTENAER, P. 1961. Étude nouvelle en deux parties, du genre *Camarotoechia* HALL et CLARKE, 1893. Première partie: *Atrypa congregata* CONRAD, espèce-type. Bulletin Institut Royal Sciences Naturelles Belgique 37:1-11.

___1966. *Ripidiorhynchus*, nouveau genre de brachiopode rhynchonellide du Frasnien. ibid. 42 (3):1-15.

___1994. *Canvirila*, nouveau genre calvinariide (Rhynchonellide, Brachiopode) de la partie moyenne du Frasnien. ibid., 64:97-108.

SCHMIDT, H. 1965. Paleozoic Rhynchonellacea, p.552-597 (partim). In, R.C. Moore (ed.), Treatise on Invertebrate Paleontology, H, Brachiopoda. Geological Society of America, University of Kansas Press.

SCHMIDT, H. 1965a. Neue Befunde an Paläozoischen Rhynchonellacea (Brachiopoda). Senckenbergiana Lethaea 46(1):1-25.

___1975. Septalariinae (Brachiopoda, Rhynchonellida) im Devon westlich und östlich des Rheins. Senckenbergiana Lethaea 56(2/3):85-121.

SCHMIDT, H. & D.J. MCLAREN.1965. Paleozoic Rhynchonellacea, p.552-597. In, R.C. Moore (ed.), Treatise on Inver-tebrate Paleontology, H, Brachiopoda. Geological Society of America, University of Kansas Press.

SCHUCHERT, C. 1913. Brachiopoda, p.355-420. In, K.A. von Zittel (ed.), Palaeontology, 2nd Edn.,1 (transl. C.R. Eastman), London.

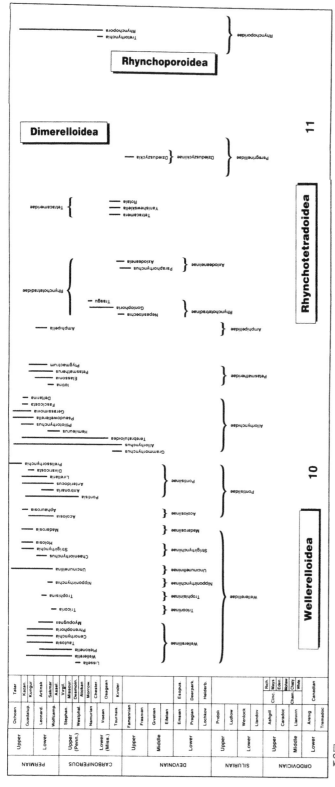

_____1929. In, C.Schuchert & C.M. Levene, Brachiopoda (Generum et genotyporum index et bibliographia). Fossilium Catalogus, 1. Animalia 42:1-140. Junk, Berlin.

STAINBROOK, M.A. 1945. Brachiopoda of the Independence Shale of Iowa. Geological Society America Memoir 14:1-74.

WESTBROEK, P. 1968. Morphological observations with systematic implications on some Paleozoic Rhynchonellida from Europe, with special emphasis on the Uncinulidae. Leidse Geologische Mededelingen 14:1-82.

XU, G.R. & R.E. GRANT. 1994. Brachiopods near the Permian-Triassic boundary in South China. Smithsonian Contributions Paleobiology 76:1-68.

XU, H.K. & Z.G. YAO. 1984. Paleozoic rhynchonellids without septum and septalium, and their classification. Acta Palaeontologica Sinica 23(5):554-567.

[*Readers are adivsed to request author for range charts: scale of reduction make these barely legible:editors]

SHELL ULTRASTRUCTURE OF MESOZOIC CRANIID BRACHIOPODS FROM THE UKRAINE

T.N. SMIRNOVA

Paleontology Department, Geological Faculty, Moscow State University, 119899 Moscow, Russia

ABSTRACT- Little known Late Jurassic and Early Cretaceous inarticulate craniid brachiopods from the Ukraine are examined for their shell ultrastructure. A new form, 'Craniscus' izyumensis n.sp. is described from the Upper Oxfordian (Late Jurassic) of eastern Ukraine, and a new Early Cretaceous (Berriasian) craniid genus, Conocrania, type C. spinacostata from Crimea, is diagnosed, and the type species redescribed in detail. SEM reveals a layer of coarse rhomboidal crystals between the primary layer of acicular calcite and the secondary laminar layer. C.spinacostata usually has a double-layered strongly recrystallized shell: a primary layer of small crystals, and a secondary layer of laminae.

INTRODUCTION

Jurassic and Cretaceous rocks are widespread in the eastern part of the Ukraine and the Crimea, and are repersented by shales and limestones deposited in shallow stable shelf settings, including sponge and microbial reef development. Locally brachiopods are abundant, with the new genus Conocrania coming from reefal deposits. This paper examines shell structure of Late Juras-sic and Early Cretaceous inarticulates belonging to the Craniida from the Ukraine (Figure 1). Late Cretaceous and Recent Craniidae craniids have been studied more extensively (Williams & Wright, 1970; Lee & Brunton,1986), thus the Ukrainian forms fill a gap in our knowledge of earlier extinct forms.

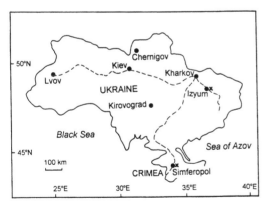

Figure 1. Map of Ukraine showing Early Cretaceous (Berriasian) localities in the Crimea, with village of Kuchki, 30 km E of Simferopol (Conocrania), and the Upper Jurassic (Oxfordian) at Izyum, east Ukraine, with Craniscus izyumensis. Key: fossil localities = x.

SYSTEMATIC PALEONTOLOGY

Order Craniida Waagen, 1885
Superfamily Craniacea Menke, 1828
Family Craniidae Menke, 1828
Genus Craniscus Dall, 1871
'Craniscus' izyumensis n.sp.

Type locality. Upper Jurassic, Upper Oxfordian; from near the town of Izyum, eastern Ukraine, near Kharkov [see name]. Material.The holotype N195/12942, and measured samples N196-199/12942 are housed in the VSEGEI, St Petersburg, Russia.

Diagnosis. Dorsal valve low, conical, with thin shell, usually subrectangular, elongated; posterior margin straight, anterior margin slightly rounded, length greater than width; costae variably pronounced on valve margins.

Description. Shells small, 2.7-3.2mm long, 2.2-3.0mm wide. Dorsal valve irregular, subrectangular in outline; posterior margin shorter than anterior; apex subcentral, or proximal to posterior margin. Protegulum composed of compact calcite devoid of punctae; very fine granular structure, granules c.1μm thick [at x7000]. Dorsal valve inner surface slightly concave, partly convex; subperipheral rim low on concave shell surface, flat on convex areas; tubercles barely protruding; subperipheral rim penetrated by punctae. Limbus flattened, thin, with irregular, often spinous margins; limbus and subperipheral rim arched in middle of posterior margin; feebly marked punctal rows situated in shallow furrows. Anterior adductor scars elongated, arched, with strongly curved lateral ends; inner sides wide, joined distally; horizontally oriented, posterior border parallel to posterior valve margin. Dorsal protractor scars divergent,slightly pronounced; elevation to support dorsal protractors, and median septum, lacking; posterior adductor scars rounded; inner margins elongated to valve centre.

Primary layer of dorsal valve composed of irregular, elongated, splinter-like crystals inclined at 70-80° to shell surface; crystal axes parallel; butting ends isometric 1-1.5μm across; thickness of layer 25μm; boundary with layer of coarse, rhomboidal crystals distinct, irregular, lacking transition between layers; crystal axes of both layers inclined at 45°; layer of coarse, rhomboidal crystals composed of elongated crystals with sharp ends; crystals radiating from apex, long axes slightly inclined to valve surface, or almost parallel to it; crystal bundles changing directions along valve surface, especially on costae, keeping orientation nearly parallel to valve surface; crystal sublayers overlap, due to differing orientation of crystals in each sublayer; crystals up to 10μm across, 50-60μm long; long axes perpendicular to growth lines;

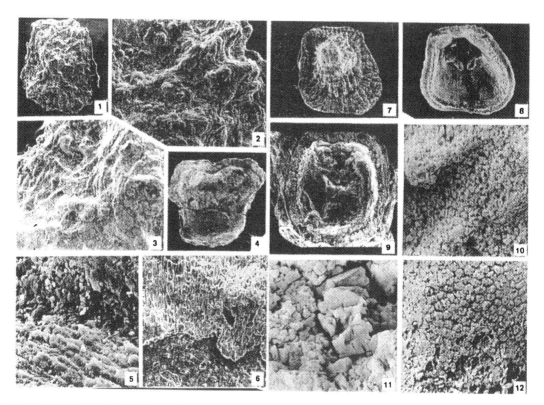

Figure 2. SEM micrographs 1-6: Holotype and paratypes of `Craniscus' izyumensis. 1, Holotype N195/12942, exterior surface of dorsal valve, x17. 2, fragment of #1, x100. 3, fragment of #1 with protegulum, x75. 4, N199/12942, internal surface of dorsal valve, x17. 5, N200/12942, boundary between primary layer of acicular calcite (above) and layer of coarse rhomboidal crystalline calcite (below), x1520. 6, N201/12942, boundary between layer of coarse rhomboidal crystalline calcite (above) and laminar layer (below), x1520. 7-12: Topotypes of *Conocrania spinacostata*. 7, N169/12942, exterior surface of dorsal valve, x25. 8, N168/12942, internal surface of dorsal valve, x25. 9, N174/12942, ventral valve interior surface, x30. 10, N182/12942, primary layer on the exterior surface of dorsal valve, x350. 11, N182/12942, laminar layer on the inner part of destroyed border of dorsal valve, x1300. 12, N183/12942, anterior adductor scars on the interior surface of dorsal valve, x35.

layer of coarse, rhomboidal crystals c.30µm thick; lowest layer represented by laminae parallel to valve floor.

Remarks. 'Craniscus' izyumensis is distinguished from *Craniscus barskovi* Smirnova (1972) by smaller size and elongated outline of the shell, rough ribs on the brachial shell margins, and by the absence of knobbed relief, a dorsal septum and platform for anterior adductor support. The new species has less elongated anterior adductor scars, and lacks a median bend in the subperipheral rim near the posterior margin.

Conocrania n. gen.

Type species. *Craniscus spinacostatum* Smirnova, 1972, from the Burulcha River, Crimea, Ukraine; Early Cretaceous, Berriasian. Etymology of genus: after the Latin *cono*.

Diagnosis. Shell small, usually no more than 3mm in length, costate, subtrapeziform or subrectangular in outline; dorsal valve highly conical; costae strongly marked; valve margin thick; anterior adductor scars crescentic, bean-shaped or oval; platform for anterior adductor scars support lacking; ventral valve slighly convex, cemented to substrate over entire surface; dorsal valve punctal canals unbranched.

Discussion. This new genus differs from *Craniscus* Dall, 1871, in having a small, costate, subtrapeziform shell, and highly conical dorsal valve with thick margins.

Conocrania spinacostata (Smirnova, 1972)

Craniscus spinacostatum Smirnova, 1972, p.23, pl.1, fig.2; Smirnova, 1990, p.4, pl.1, fig.1. From Early Cretaceous (Berriasian) sponge reefs; Kuchki, Burulcha river, 25 km from Simferopol, Crimea, Ukraine. Holotype formerly N136/92, Paleontology Department, Geological Faculty, Moscow State University, Russia, now N165/12942 in VSEGEI, St Petersburg, Russia. Addition material of 23 dorsal and two ventral valves from type locality.

Diagnosis. *Conocrania* with subtrapeziform outline, short posterior margin, distinct tubercles on the subperipheral rim, slender crescentic anterior adductor scars, absence of elevation for brachial protractor support, and anterior septum.

Description. Shells small, 1.7-3.0mm long, 1.8-3.2mm wide, with short, straight, posterior margin; wide, rounded anterior margin; width usually larger than length; apex near posterior margin; anterior margin ocasionally with median hollow; valve slopes high, convex, curved; maximum bend situated at 2/3 distance from apex; valve margins possibly elevated and with wave-like gaps not in one plane; triangular, flat, sharply inclined area of dorsal valve between posterior margin and apex; costae pronounced, weak or absent near apex; distinct costae appear at 1/3 distance from apex; most of valve surface covered by radial costae, in several stages; costae with flattened surface, vertical steps; new costae formed by intercalation; usually up to 4 steps of costae; first step from apex containing 25-30 costae; at valve margins 60-70 costae, rarely up to 90 costae; spacing of 10-12 costae per mm; distance between costae from 50-200μm; costae 40-50μm wide; costae of every step very distinct; costal ends sharply elevated, forming spine-like relief, especially at shell curvature, bend. Protegulum 0.3-0.5 mm in diameter, hemispherical in shape, with smooth surface devoid of sculpture and punctae.

Most concave part of valve between anterior and posterior adductor scars; valve margins thickened almost equally along perimeter; single row of high tubercles on subperipheral rim, with maximum width c.100μm; joining other tubercles along a belt. Punctae in hollows delimited by rows of small knobs. Prominent anterior adductor scars situated on flat inner valve surface, crescentic, with wide, rounded, inner part, sharp bill-shaped outer ends; anterior scars inclined at 90°, seldom at 75-80°; strongly convex posterior side of anterior adductor scars directed to posterior adductor scars; inner sides of anterior adductor scars proximal; dorsal elevator scars hemispherical or subtrigonal in outline, located in frontal grooves of anterior adductor scars; posterior adductor scars large, oval, joining brachial valve posterior margin; variably concave, inclined at >90°; tubercles absent on subperipheral rim near posterior adductor scars.

Dorsal protractor scars situated on flat inner surface of valve in front of anterior adductor scars; two tear-shaped scars c.200μm in length, divergent at sharp angle. Ventral valve subquadrate, slightly elongated in outline; posterior margin short, straight; anterior margin rounded; lateral margins subparallel, slightly convex; ventral valve inner surface strongly concave on lateral margins, flat-convex in middle. Limbus equally wide along perimeter; limbus surface concave on lateral part of valve, near posterior margin; limbus flat near anterior margin; subperipheral rim with row of tubercles high in posterior part of valve, flattened anteriorly. Two arched bends on subperipheral rim posterior margin, limited by posterior adductor scars; tubercles absent in middle part of valve posterior margin; subperipheral rim ending in gap deflected posteriorly; posterior adductor scars small, transversaly oval, deep; anterior adductor scars slender, crescentic, directed to each other at right angles; slightly convex, heart-shaped platform situated in front of anterior adductor scars, delimiting two fields with faintly expressed vascula lateralia.

The dorsal shell was strongly recrystallized. Primary layer of regular, finely-granular calcite composed of small, angular, rhomboidal crystals from 3-4μm to 7-8μm across; angles between crystals 75-80° and 110-115°; crystals oriented at c.90° to outer surface; primary layer 15-20μm thick; laminar layer 90-100μm thick; boundary with primary layer irregular, sharp; laminae very thin, <1μm, parallel to inner surface; laminae curved considerably near punctal canals, in upper part c.30μm thick; inclusions of massive prismatic calcite 20μm long; small, rectangular laminae on valve floor. Costae of fine granulated calcite in upper part, and laminar layer inner side. Anterior adductor scar surface distinctiv net-like; composed of hexagonal cells, forming honeycomb relief; distance between opposite sides of cells 12-15μm; cell walls composed of small laminae from 2-10μm across. Punctae not branched, diameters 10-20μm, located irregularly on valve floor; most of subperipheral rim devoid of punctae; row of large punctae 15-20μm in diameter limited to subperipheral internal rim; some punctae on low surface of subperipheral rim; numerous punctae on limbus, mostly circularly arranged or in rows perpendicular to shell margin; diameter of limbus punctae c.20μm, punctal rows with 3-6 punctae per row, situated in hollows; distance between limbus punctae from 20-100μm; dorsal valve floor punctae placed irregularly, forming rows parallel to valve margins; few irregularly disposed punctae on anterior, posterior adductor scars. Upper part of tubercles composed of fine granulated crystals of calcite, inner part of laminar layer, with boundary between layers well marked.

Remarks. Conocrania spinacostata differs from C. radegasti (Nekvasilova, 1982) by its subtrapeziform outline, shorter posterior margin, distinct tubercles on the subperipheral rim, more slender crescentic anterior adductor scars, and an absence of elevation for brachial protractor support, and septum in the anterior part of the dorsal valve.

Acknowledgments

I acknowledge the generous support of the Brian Mason Trust Fund, Geology Department, University of Canterbury, Christchurch, New Zealand, which enabled me to travel to New Zealand to begin this study. I gratefully acknowledgehelp from Dr D.I. MacKinnon, Neil Andrews (SEM), and Albert Downing (photography). Rex Doescher, of the Smithsonian Institute Brachiopod Information Center, provided complete bibliographic information. I am indebted to colleagues from the Paleontological Institute in Moscow, and Dr G.T. Ushatinskaya and E.A. Gegallo for the possibility to use SEM and consultations. The Natural Sciences and Engineering Research Council of Canada provided travel support to attend the Third International Brachiopod Congress in Sudbury, Ontario, September 1995.

REFERENCES

LEE, D.E., & C.H.C. BRUNTON.1986. Neocrania n. gen., and a revision of Cretaceous-Recent brachiopod genera in the family Craniidae. Bulletin British Museum Natural History (Geology) 40(4):141-160.

SMIRNOVA, T.N. 1972. Rannemelovye brakhiopody Kryma i Severnogo Kavkaza. Nauka, Moscow,143p.

___1990. Sistema rannemelovykh brakhiopod. Nauka, Moscow, 239p.

WILLIAMS, A. & A.D. WRIGHT. 1970. Shell structure of Craniacea and other calcareous inarticulate Brachiopoda. Special Papers Palaeontology 7:1-51.

NOTES ON THE TAXONOMY OF MESOZOIC RHYNCHONELLIDA

H. SULSER

Paläontologisches Institut und Museum, Universität Zürich, Karl Schmid-Str.4, CH-8006 Zürich, Switzerland

ABSTRACT- Among the genera of Mesozoic rhynchonellids more than 1600 species (including subspecies) are now placed in over 220 genera, about one fifth tentatively. The unequal size of genera is due partly to the working style of researchers and partly to the notorious problem of defining taxa unambiguously. In only a few instances does the size of the genus seem to reflect phylogeny. The life span of most Mesozoic rhynchonellid genera covers one to three stratigraphic stages. For long duration genera, up to 10 stages duration denotes their supposed conservative nature. Some paleogeographic implications, e.g. occurrence or migration pattern, are herein considered taxonomically. Mesozoic rhynchonellids assigned to a genus outside 'Rhynchonella'" sensu lato, contain up to 90-95% of the revised species worldwide, and are just a part of the species which have been named in the last 150 years. Rhynchonnelid genera herein are organised according to species density.

GENERA AND SPECIES

The order Rhynchonellida Kuhn comprises 1630 species (including most subspecies), placed in 224 genera (Sulser, 1993). This number includes also genera of questionable validity, but objective or rejected subjective synonyms, junior homonyms and nomina nuda are omitted. In 340 of the species/genus combinations (about 20% of the total) the proposed attributions are tentative and a question mark (?) must be placed behind the genus name. For 250 to 300 species strong similarities with other named species, or possible synonymy exists.

The incentive to establish new genera seems unbroken. Compared to genera, species have 'grown' in a more balanced way. At present there are three times more genera than in 1950 and these have doubled since 1968 (Table 1). The present relation between species and genera (tentative attributions included) results in a calculated 'species density' of 7.3 per genus. This average value implies an equal distribution which, although it does not exist in reality, enables a definition of density. Monospecific genera constitute 26% of Mesozoic rhynchonellids, with 'small' genera (2 to 5 sp.) forming 39%. The 'mid field' genera (6 to 10 sp.) represent 12%, whereas the 'large' genera (11 to over 50 species) claim 23% of the total (Table 2).

In a natural system of classification, the size of a genus, as a measure of abundance or rarity of assigned species, should be biologically significant. In the existing classification system of Mesozoic rhynchonellids, as probably in all systems concerning fossils, specific attributions to genera are influenced by difficulty in defining taxa without ambiguity. There are genera, e.g. Torquirhynchia Childs, which house all asymmetric rhynchonellids of Late Jurassic age. As a hypothesis for asymmetric growth is still speculative, Torquirhynchia is instead more a framework for a cluster of similar asymmetric forms than a genus in the true sense (Childs, 1969). Classification procedures reflect to a great extent the individual working methods of authors. Burmirhynchia Buckman, an ill-defined genus originally filled with a multitude of 56 Indian species, with only trivial differences, has surprisingly not been abandoned as one would expect. It has been used also for European species with comparable paleohabitats (Laurin 1984).

In the present species and genus distribution we find two prevailing trends: the creation of new genera with few species, and the use of large genera for new specific attributions. Most of the small genera (often with 1 or 2 species only) originated in Asiatic (Caucasus, Crimea, Pamir, Himalayas, China, Siberia) or European areas. A good portion has been erected within the last two decades.

Confidence in new genus names is sometimes hampered by lack of thorough discussion of affinities with established genera, or a lack of effort in amending known genera. The danger may arise that parallel taxonomies are developed, just enlarging the nomenclature, but without any real progress in knowledge. Praecyclothyris Makridin is almost certainly a junior synonym of Somalirhynchia Weir, which was not mentioned in the Russian Treatise (Sarycheva, 1960). Cyclothyris M'Coy, Belbekella Moisseev and Lamellaerhynchia Burri are also likely synonymous. Plicarostrum Burri and

Table 1. Taxa list showing Mesozoic rhynchonellid development from 1900 to 1992 (compiled from Ager, 1965; Ager et al., 1972; Doescher, 1981; Sulser, 1993).

Year	Genera	Species & Subspecies	Year	Genera	Species & Subspecies
1900	8	550	1964	104	1070
1910	8	610	1970	132	1220
1920	52	780	1979	176	1410
1930	54	820	1990	217	1590
1940	78	930	1992	224	1630
1950	78	950			

Table 2. Distribution of Mesozoic rhynchonellid species according to genera.

145 Small Genera	27 Medium Genera	52 Large genera
58 with 1 species	6 with 1 species	19 with 1 species
39 with 2 species	8 with 1 species	12 with 1 species
25 with 3 species	4 with 1 species	13 with 1 species
12 with 4 species	3 with 1 species	4 with 1 species
11 with 5 species	6 with 1 species	2 with 1 species
		2 with 1 species

Burrirhynchia Owen differ in minor internal details which are normally met at the species rank.

Looking at the larger genera, the reasons for having attained their present 'size' are often quite obvious. *Acanthothiris* d'Orbigny (Middle Jurassic) is composed of 27 species. For a long time it was the sole genus for all spiny rhynchonellids of that period. Later the genus was split, and some of its species were placed into new combinations of *Acanthorhynchia* Buckman, *Echinirhynchia* Childs, *Acanthothyropsis* Kamyshan, *Paraacanthothiris* Kamyshan, and others. The type species of *Acanthothiris*, *A. spinosa* (Linne),dates back more than 200 years, and it is impossible today to verify all specimens ever described as '*spinosa*', whether they belong to this species or not. In order to manage the difficulties with this old species, many authors have preferred to establish new species, with sound definitions, for material on which they work. Some of these species would probably fall into the variation range of '*spinosa*', if we had a clear idea of the species, and if we could finally define and decide what really is '*spinosa*'.

Cretirhynchia Pettitt, confined to the Late Cretaceous, houses some 30 species. Pettitt (1950) contributed 10 new English species, most showing only minor differences. Later, about the same number of *Cretirhynchia* species known before 1900, and new ones after 1950, were added. New species were mostly reported from Russia. Noticably no Cretaceous species set up between 1900 and 1950 were transferred to *Cretirhynchia*. It is to be expected that a critical review, assisted by statistical variability studies, would reduce English, and perhaps also Russian diversity, now dominating the genus *Cretirhynchia*.

Lacunosella Wisniewska is a genus defined by its highly diagnostic internal features, especially the falcifer crura. It ranges from the base of the Late Jurassic to the Early Cretaceous, and shows a rich assemblage of species, with regard to stratigraphic and geographic distribution. After separation of forms with a smooth shell, placed now in *Fortunella* Calzada (about 12 sp.), *Lacunosella* still comprises some 50 species. Some 15-20 of these species are provisional attributions and come mainly from horizons at the Jurassic/Cretaceous boundary. It is evident that *Lacunosella*, even after eventual elimination of possible synonyms, will retain the shape of a 'large', probably monophyletic genus.

The Late Middle Jurassic genera *Goniorhynchia* Buckman and *Robustirhynchia* Seifert represent two genera which differ greatly in their validity. *Goniorhynchia* is characteristic for its conspicuous thickening of internal structures, and it was speculated that this feature might signify a final stage in terms of phylogeny. But, local environmental adaptations are possible. It is difficult to judge if *Goniorhynchia* is just a late offshoot of the important, widespread genus *Cymatorhynchia* Buckman and therefore perhaps of questionable necessity, or if it embraces a group of biologically related forms. It is reasonable to keep these brachiopod species in a separate genus usable for future research. *Robustirhynchia*, on the other hand, occupies a weak positon. It was based on a few specimens with defect brachidia, all from one locality. At present it is no more than a collective name for some individuals of a somewhat unusual shape.

STRATIGRAPHIC ASPECTS

In 1972, Ager et al. noted the inequal weight assigned to the genera of Mesozoic rhynchonellids. Their dominance in the Jurassic, and sparsity in the Triassic and Cretaceous, did not

Table 3. Proportion of Mesozoic genera assigned to rhynchonellids and terebratulids (development 1965-1990).

	number of genera & percentage of total **Rhynchonellids**		number of genera & percentage of total **Terebratulids**	
	1965	1990	1965	1990
Cretaceous	14 (13%)	34(16%)	44 (32%)	130 (32%)
Jurassic	74 (71%)	126 (53%)	73 (53%)	213 (52%)
Triassic	17 (16%)	57 (26%)	21 (15%)	64 (16%)
TOTAL	105	217	138	407

Table 4. Mesozoic rhynchonellid genera and their stratigraphic ranges.

Number of Genus (-era)	Duration in Stage(s)	Number of Genus (-era)	Duration in Stage(s)
103	1	5	6
51	2	2	7
30	3	2	8
17	4	1	9
12	5	1	11

reflect the real diversity of the three faunas. This situation arose mainly through the work of Buckman (1918), who proposed a plethora of Jurassic genera, largely British and European forms. In the meantime, this disproportional distribution was smoothed out, mainly by the pioneering work of Dagys (1974) for the Triassic, and further by Russian and Chinese authors. For the Cretaceous the number of genera has kept abreast with the general increase of Mesozoic genera. It may be noted that in the other main group of Mesozoic articulate brachiopods, the terebratulids (with several superfamilies), the marked accumulation of new genera from 1965 to 1990 was homogenous, and did not change the proportion in the assemblage of Triassic, Jurassic and Cretaceous taxa (Table 3).

Nearly half the Mesozoic rhynchonellid genera are restricted to one single, and about 25% to two stratigraphic stages. The remainder extend from 3 to 11 (!) stages, but here most of the generic assignments are provisional. With regard to the present extreme of genus splitting, these cases should be reviewed carefully (Table 4). *Ptilorhynchia* Crickmay, with nine assigned species, was erected on the basis of a rare Callovian rhynchonellid from California. It was used further to describe Siberian forms ranging from the Middle Jurassic (Bajocian) to the Early Cretaceous (Hauterivian, 9 stages), and Late Jurassic specimens from British Columbia (Dagys, 1968; Owen, 1981). The generic uniformity of all these forms of different age was considered a possible argument for migration paths and their resulting paleogeographic distribution. More detailed knowledge of the true relationship of these rhynchonellids seems desirable. *Monticlarella* Wisniewska is another example of a long-ranging genus, from the base of the Middle Jurassic to Late Cretaceous. Its type species is from the Oxfordian of Poland, and is distinctive both externally and internally. *Monticlarella* was classified in the superfamily Dimerellacea Buckman, which is believed to share some apparently primitve characters with Paleozoic rhynchonellids. But some of these characters, like the cardinal process and long crura, cannot be observed in *Monticlarella*, and it is an

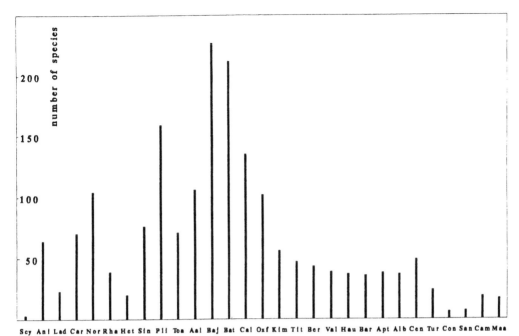

Figure 1. Mesozoic rhynchonellid species occurring by stratigraphic stages, with peak in the Bajocian-Bathonian.

open question whether these were lost by evolution from ancestral forms. A thorough study of all known *Monticlarella*, including those of questionable attributions, might help clarify the interesting question whether this genus really represents a conservative stock extended through 11 stratigraphic stages.

At the species level, the occurrence of rhynchonellids in each stage may be quantified, where tentative generic assignments can be included (Figure 1). As generally known, there were diversity maxima towards the end of the Triassic, in the Middle Liassic, and a pronounced peak in the Late Middle Jurassic, followed by almost continuous decline to the end Cretaceous. In the Bajocian-Bathonian we find 440 species, which constitute more than 20% of the total of all Mesozoic species (2000) counted by stage. If this is compared with the number of nominal species (1550), an average 'stage cover' of 1.3 species is calculated. In other words, two thirds show a life period limited to one stage, whereas one third persisted during two stages (the small number of consolidated species surviving even longer are not calculated). We should, however, remember the fact that the stratigraphic range of brachiopods is apt to vary from place to place. By their mode of colonization and life on the substrate, the duration span of a species may greatly depend on environmental conditons.

Misinterpretation in erecting species may arise through strict bed-by-bed sampling. Small differences in successive strata may blur the purely morphological approach and lead to the creation of neighbouring taxa which are at best local morpho-types.

PALEOGEOGRAPHIC IMPLICATIONS

Paleobiogeography has become an important aspect in brachiopod research. For the rhynchonellids from the Middle Jurassic of Saudi Arabia, 13 new genera were recently introduced by Cooper (1989). Most of these appear Ito be 'ordinary' rhynchonellids, but in the discussion of generic diagnoses there is scarcely any reference to known genera. The impression from such an isolated occurrence suggests an overestimated degree of endemism. One problem in compiling regional lists of brachiopods results from fragmentary information. More than half the c.100 'European' genera are described from England or France (Table 5). The question remains, if these are rare, only locally developed stocks, whether they are species different from elsewhere, or whether they simply have not yet been discovered outside England or France.

Only few rhynchonellids appear to have been tolerant and 'vital' enough to colonize remote regions, e.g. the boreal province and both shores of the Tethys ocean. *Stolmorhynchia* Buckman was considered an example of such a widespread genus from England, Italy, Spain, the Caucasus, and

Table 5. Rhynchonellid genera occurring in European countries. Key: parentheses () = in this country only.

England	59 (6)	Romania	25
France	55 (2)	Spain	24 (2)
Germany	40	Poland	18
Czechoslovakia	38 (2)	Bulgaria	8 (1)
Austria	33 (4)	Greece	5
Switzerland	32	Portugal	5
Italy	31	Belgium	3
Hungary	30	Denmark	2
Yugoslavia	27 (2)		

267

Morocco. A re-examination of the type species from the Middle Jurassic (Bajocian) of southern England yielded a clear picture and redefinition of the genus (Prosser 1993). From this, Caucasus species appear to be the only forms which, in addition to the English, can be assigned to *Stolmorhynchia*, and the presumed ubiquity of the genus should be redefined.

Considering the fact that evolution of any related group of brachiopods takes place both in space and in time, the definition of morphological limits to generic development, and the recognition of true monophyletic genera, become important questions, to which answers are still very difficult (Rousselle, 1984; Alméras, 1984).

REFERENCES

AGER, D.V. 1965. Mesozoic and Cenozoic Rhynchonellacea, p. H597-625. In, R.C. Moore (ed.), Treatise on Invertebrate Paleontology, Part H, Geological Society of America and Univiversity of Kansas Press, Lawrence.

AGER, D.V., A. CHILDS & D.A.B. PEARSON. 1972. The evolution of the Mesozoic Rhynchonellida. Géobios 5(2-3):157-235.

ALMÉRAS, Y. 1984. Comment traduire les enchaînements spécifiques dans la nomenclature générique. Exemples pris chez les Térébratulidés et les Rhynchonellidés (Brachiopodes) du Lias et du Dogger. Bulletin Société Géologique France 7 (26/4): 625-632.

BUCKMAN, S.S. 1918. The Brachiopoda of the Namyau Beds, northern Shan States, Burma. Memoirs Geological Survey India, Palaeontologia Indica 3(2): 1-299.

CHILDS, A. 1969. Upper Jurassic rhynchonellid brachiopods from northwestern Europe. Bulletin British Museum Natural History, Geological Supplement 6: 1-119.

COOPER, G.A. 1989. Jurassic brachiopods of Saudi Arabia. Smithsonian Contributions to Paleobiology 65:1- 213.

DAGYS, A.S. 1968. Jurassic and Early Cretaceous brachiopods of northern Siberia. Trudy Akademiya Nauk SSSR, Sibirskoi Otdelenie, Institut Geologii Geofiziki 41:1-167

___1974. Triassic brachiopods: morphology, classification, phylogeny, stratigraphic significance and biogeography. Izdatelstvo Nauk, Novosibirsk, 387p.

DOESCHER, R.A. 1981. Living and fossil brachiopod genera 1775-1979, Lists and Bibliography. Smithsonian Contributions Paleobiology 42:1- 238 .

LAURIN, B. 1984. Les rhynchonellides des plates-formes du Jurassique moyen en Europe occidentale: Dynamique des populations, évolution, systématique. Cahiers de Paléontologie (Invertébrés), 465 p.

OWEN, E.F. 1981. Distribution of some Mesozoic brachiopods in North America, p. 297-310. In, J. Gray, A.J. Boucot & W.B.N. Berry (eds.). Communities of the Past. Hutchinson Ross, Stroudsburg.

PETTITT, N.E. 1950, 1954. A monograph on the Rhynchonellidae of the British Chalk. Palaeontographical Society Monographs, 1:1-26 (1950); 2: 27-52 (1954).

PROSSER, C.D. 1993. The brachiopod *Stolmorhynchia stolidota* from the Bajocian of Dorset, England. Palaeontology 36(1): 195-200.

ROUSSELLE, L. 1984. Révision des espèces et diagnoses génériques (exemples pris chez les Brachiopodes du Jurassique). Bulletin Société Géologique France 7 (26/4): 621-623.

SARYCHEVA, T.G. (ed.).1960. Brakhiopody, p.112-324. In, Yu.A. Orlov (ed.), Osnovy paleontologii. Izdatelstvo Akademiya Nauk, Moskva.

SEIFERT, I. 1963. Die Brachiopoden des Oberen Dogger der schwäbischen Alb. Palaeontographica A 121(4-6): 156-203.

SULSER, H. 1993. Fossilium Catalogus, I, Animalia, Pars 132, Brachiopoda, Rhynchonellida Mesozoica. Kugler Publications, Amsterdam/New York, 281 p.

ZONATION AND PALEOECOLOGICAL DISTRIBUTION
OF BULGARIAN JURASSIC BRACHIOPODS

PLATON V. TCHOUMATCHENCO

Geological Institute, Bulgarian Academy of Sciences,
acad. G. Boncev Str. Bl. 24, 1113 Sofia, Bulgaria

ABSTRACT- Jurassic strata in Bulgaria are correlated by ammonites, and subdivided into thirteen brachiopod zones, from the base to the top as follows: (1) *Spiriferina walcotti* Zone - uppermost Hettangian-lower Sinemurian; (2) *Tetrarhynchia dunrobinensis* Zone - upper Sinemurian-lower lower Pliensbachian; (3) *Zeilleria quadrifida* Zone - upper lower Pliensbachian-lower upper Pliensbachian; (4) *Homoeorhynchia acuta* Zone - upper upper Pliensbachian-lower Toarcian; (5) *Homoeorhynchia cynocephala* Zone - Toarcian-lower Aalenian; (6) *Acanthothiris costata* Zone - Aalenian; (7) *A. sentosa* Zone - lower lower Bajocian; (8) *Sphaeroidothyris sphaeroidalis* Zone - upper lower Bajocian; (9) *Acanthothiris inflata* Zone - uppermost lower Bajocian and lower upper Bajocian; (10) *Wattonithyris wattonensis* Zone - uppermost Bajocian and lowermost lower Bathonian; (11) *Acanthothiris spinosa* Zone - lower and middle Bathonian; (12) *Lacunosella arolica* Zone - upper Callovian-Oxfordian; (13) *Lacunosella sparcicosta* Zone - Kimmeridgian-lower Tithonian. Barren intervals existed between the *A. spinosa* and *L. arolica* zones, and between the *L. arolica* and *L. sparcicosta* zones. Relationships of these brachiopods to substrates, biotopes, and paleogeography are reviewed.

INTRODUCTION

In the beginning of the Jurassic almost all of Bulgaria was continental. Later followed a marine transgression, along a series of grabens separated by horsts, later covering adjacent areas (Figure 1). These paleotectonic structures set the stage for subsequent Jurassic paleogeography and brachiopod distribution. Only in Holocene eastern Bulgaria exists a basin with siliciclastic turbidite sedimentation, and with a southern shelf region featuring carbonate sediments. Remnants of Jurassic shelf sediments are included in Middle Jurassic, deep water black shales as olistoliths, which repre-sent a source of information about paleoshelf environments and the brachiopods contained within (Tchoumatchenco, 1988). During the Callovian (i.e. late Middle Jurassic) Bul-garia, excepting the Rhodopes Massif, was covered by sea, with a trough filled by siliciclastic sediments in the south, and a carbonate platform to the north, the Moesian Platform (Tchoumatchenco & Sapunov, 1994). The Moesian Platform was in turn divided by a pelagic trough into west and east components. Today, Jurassic rocks crop out in central Bulgaria, but in the north, the Jurassic is covered by younger sediments, and information comes only from deep wells (Figure 1).

The first attempt to biostratigraphically subdivide Jurassic sediments based on brachiopods was made in 1967 (Tchou-matchenco, in Sapunov et al.,1967). The Lower Jurassic was then split into five zones. Later a biostratigraphical zonation of the Middle Jurassic into seven zones (Tchoumatchenco, 1977) was erected, and a further division of the Callovian and Upper Jurassic sediments into two zones and one barren interzone was made (Tchoumatchenco,1978). Taxonomy of the Bulgarian Jurassic brachiopods was undertaken by Tchoumatchenco (1978a,b,c;1983;1989), and an attempt was made to determine brachiopod paleoecologic distribution (Tchoumatchenco 1972b;1988,1993a-b, __et al.,1992).

EARLY JURASSIC BRACHIOPOD ZONES

Lower Jurassic strata of Bulgaria may be partitioned into a number of biostratigraphically useful brachiopod zones related to local substrate conditions (Tchoumatchenco, 1967, 1972; Tchoumatchenco, in Sapunov & Tchoumatchenco, 1988), which are listed from older to younger (Table 1):

Spiriferina walcotti Zone (Hettangian-lower Sinemurian). At this time of initial transgression, uniform conditions did not exist in Lower Jurassic basins of Bulgaria, and in the littoral biotope of the Izdremec Graben, near Komstiza, the coarse ribbed spire-bearer *Spiriferina* was common (Tchoumat-chenco,1972: Figure 1W). Shortly thereafter, conditions deepened, and sublittoral biotopes prevailed in the grabens. Still later, muddy bottoms were dominated by rhynchonellids (30%:*Gibbirhynchia*.), spiriferinids (30%: including 20% *Spiriferina walcotti* (Davidson), 10% *Liospiriferina tumida* (Buch)), and terebratulids (36%: 14%*Lobothyris grestenensis* (Radovanovic), 12% *L. punctata* (Sowerby), 10% *L. sub-ovoides* (Münster), followed by rare terebratellids (*Zeilleria indentata quiaoisensis* Choffat <10%). In the Mihailovgrad Graben (near Gaganitza), Izdremec Graben (Glojene), and at Jabalka agitated waters were dominated by terebratulids such as *Lobothyris punctata* and *L. grestenensis*. In the littoral zone (Beledie Han, Kremikovtzi brachiopods were rare, dis-articulated and broken. Elsewhere, conditions were apparent-ly unsuitable for brachiopods.

Table 1.

Stage & substage	Ammonite zone (Sapunov, 1988)	Brachiopod zone
Aalenian		
Toarcian	P. aalensis D. levesquei P. dispansus G. thouarsense H. variabilis H. bifrons D. tenuicostatum	H. cynocephala
Upper Pleinsbachian	P. spinatum A. margaritatus A. stokesi	H. acuta
Lower Pleinsbachian	O. figulinum A. capricornum A. maculatum B. luridum A. valdani barren interval	Z. quadrifida
Upper Sinemurian	E. raricostatum barren interval A. obtusum	T. dunrobinensis
Lower Sinemurian	barren interval A. bucklandi	S. walcotti
Hettangian	Fucinites sp barren interval	barren interval

269

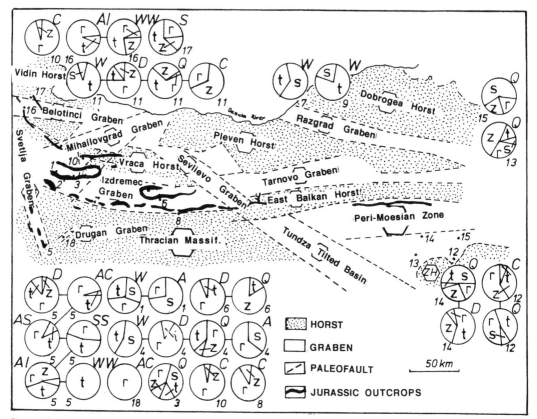

Figure 1. Principal tectono-paleogeographic structures in Bulgaria during Lower and Middle Jurassic time and brachiopod distribution. Key: W-*Spiriferina walcotti* Zone; D-*Tetrarhynchia dunrobinensis* Zone; Q-*Zeilleria quadrifida* Zone; A-*Homoeorhynchia acuta* Zone; C-*Homoeorhynchia cynocephala* Zone; AC-*Acanthothiris costata* Zone; AS-*Acanthothiris sentosa* Zone; SS-*Sphaeroidothyris sphaeroidalis* Zone; AI-*Acanthothiris inflata* Zone; WW-*Wattonithyris wattonensis* Zone; Sp-*Acanthothiris spinosa* Zone; s-Spiriferinida; r-Rhynchonellida; t-Terebratulidina; z-Terebratellidina. Sections: 1, Komstiza; 2, Ravna; 3 Gintzi; 4, Sarantzi-Kremikovtzi-Beledie Han; 5, Zabliano; 6, Teteven; 7, Glojene; 8, Troyan Balkan Mtns; 9, Jabalka; 10, Gorno Ozirovo; 11, Gaganitza; 12, Bilka-olistolith; 13, Djula-olistolith; 14, Cerkoviste-olistolithe; 15, Karaveljovo-olistolith; 16, Dolni Lom; 17, Belogradcik; 18-Drugan.

Tetrarhynchia dunrobinensis Zone (upper Sinemurian *A. obtusum* Zone- lower Carixian). In the central parts of the Mihajlovgrad Graben, near Gaganitza, and in the Izdremec Graben, near Teteven, an agitated sublittoral biotope carried a rich, diverse fauna dominated by rhynchonellids, e.g. *Tetrarhynchia dunrobinensis* Rollier (60-70%), and *Gibbirhynchia curviceps* (Quenstedt) (5-6%). Then followed up to 20% terebratulids, *Lobothyris subovoides* (Münster), with rare terebratellids, e.g. *Zeilleria darwini* (Deslongchamps) and *Z. numismalis* (Tchoumatchenco,1972). In peripheral parts of grabens (Sarantzi-Kremikovtzi, Glojene, Zabliano), a sublittoral-littoral biotope was dominated by *Tetrarhynchia dunrobinensis* (up to 85%), plus rare *Lobothyris subovoides* and *Zeilleria darwini*. In the deepest part of the Matoride Basin shelf, reconstructed from olistoliths at Stara Planina (Tchoumatchenco,1988), occurred the terebratulids *L. subovoides* and *Nucleata bodrakensis* (Moisseiev), terebratellids *Zeilleria darwini*, and *Z. numismalis* (Lamarck), but only rare *Homoeorhynchia prona* (Oppel), *Prionorhynchia* cf.*greppini* (Oppel), and *Gibbirhynchia amalthei* (Quenstedt).

Zeilleria quadrifida Zone (upper Carixian *A. maculatum* Zone-lower Domerian *A. stokesi* Zone). During this time brachio-

pods peaked and inhabited six different biotopes (Tchoumatchenco, 1972, 1994). In the central parts of the Mihailovgrad Graben (near Gaganitza), and at Izdremec, an agitated sublittoral biotope had a rich brachiopod fauna of terebratulids consisting of 50-60% *Lobothyris subpunctata* (Davidson), and 20% *L. edwardsii* (Davidson), followed by spiriferidines (13%*Spiriferina ascendes* Deslongchamps, 7% *Liospiriferina alpina falloti* (Conroy), and the terebratellids 7%*Zeilleria sarthacensis* (d'Orbigny); 6% *Z. subnumismalis* (Davidson), and 6% *Z. quadrifida* (Sowerby).In the deepest zones of the Izdremec graben, spiriferids prevailed, with 30% *Liospiriferina alpina falloti* (Corroy), 25% *Spiri-ferina oxyptera* (Buvignier), and 3% *S. haueri* (Suess). Rhynchonellids, *Cirpa langi* Ager, made up 31%, the terebratulids *L. subpunctata* 5%, and terebratellids *Z. quadrifida* 3%, and *Z. subnumismalis* 1.6%. At Teteven a special alternating habitat was occupied by shale-associated *Zeilleria subnumismalis* and marl-associated *L. subpunctata*.

The outer Mihailovgrad and Izdremec grabens were occupied by littoral or near-littoral biotopes, in which rhynchonellids, *Homoeorhynchia almaensis* (Moisseiev) formed 52-56% of the fauna, followed by the terebratulids *L. subpunctata* (28%), *L.*

270

edwardsii (4%), and terebratellids (46%), e.g. *Z. quadrifida* (8%), and *Z. subnumismalis* (6%). In the southern Matorides shelf (Tchoumatchenco, 1988: data from olistoliths included in black shales of the Middle Jurassic Kotel Fm. in the Stara Planina Mtns), there were quiet, shallow, lagoonal Bilka facies of red to pink micrites. Here occurred rhynchonellids such as *Cirpa borissiaki* (Moisseiev: 17%), *C.* cf. *langi* Ager (13%), and *Capillirostra* (17%), terebratulids *Lobothyris subpunctata* (43%), and rare *Liospiriferina alpina* (Oppel: 8%).

At Kotel, in sublittoral strongly agitated waters (grey-dark grey, red sandy, biodetrital limestones of Djula facial type) abundant brachiopods include the terebratellids (*Zeilleria quadrifida* (19%), *Z. subnumismalis* (15%), *Z. waterhousii* (10%), and *Aulacothyris resupinata* (6%). Others were the rhynchonellids *Homoeorhynchia almaensis* (Moisseiev: 10%), *H. acuta* (8%), and *Squamirhynchia squamiplex* (Quensted: 3%), terebratulids *Lobothyris subpunctata* (13%), and spiriferids *Spiriferina oxyptera* (5%), and *Liospiriferina alpina alpina* (5%). Most were disarticulated and fragmented. Under similar conditions, sediments of the Karaveljovo facies of alternating biodetritic limestones and marls, were formed. In these dominated the spiriferids *Liospiriferina alpina falloti* (32%), *Spiriferina haueri* (23%), and *L. alpina alpina* (3%). Others included *Zeilleria quadrifida* (18%), *Aulacothyris resupinata* (10%)), and *Homoeorhynchia almaensis* (Moisseiev: 9%), all also disarticulated and represented only by single valves, showing redeposition by strong currents on a sublittoral biotope with muddy bottom and calm waters.

The deepest biotope inhabited by brachiopods was a sublittoral to epibathyal, muddy, calm substrate, comprising the red shaly marls and clayey limestones of the Cerkoviste facies (at Kotel, Vessel-inovo, etc.). Shells include the terebratellids *Zeilleria cor* (38%), *Z. darwini* (2%), and spiriferids *Liospiriferina alpina falloti* (44%), rhynchonellids *Bodrakella bodrakensis* (Moisseiev:10%), *Homoeorhynchia acuta* (2%), and terebratulids *Nucleata bodrakensis* (Moisseiev: 4%). Brachiopods are small, whole shells, and belemnites almost vertical, showing that the fauna is autochtonous.

Homoeorhynchia acuta Zone(Domerian *A. margaritatus* Zone-lower Toarcian *D. tenuicostatum* Zone). The brachiopods belonging to this zone (Tchoumatchenco, 1972) have a limited distribution restricted to western Bulgaria. In the central Mihailovgrad Graben (Gaganitza), and in the deepest western Izdremec Graben (Komstitza), 'Spiriferina' villosa (Quenstedt) dominated (72%), with fewer thin-shelled rhynchonellids, *Homoeorhynchia acuta* (J. Sowerby: 28%), and rare *Quadratirhynchia quadrata* (Buckman). This formed in a sublittoral, muddy biotope. In the southern Izdremec Graben, a sublittoral agitated biotope, not far from the littoral zone at Beledie Han-Kremikovtzi, rhynchonellids with thick shells prevailed: *H. acuta* (47%), and *Quadratirhynchia quadrata* (17%), plus spiriferinids, e.g.'Spiriferina' villosa (36%).

Homoeorhynchia cynocephala Zone(Toarcian-*H. bifrons* Zone to Aalenian *L. opalinum* Zone). Conditions at this time were only favourable for brachiopods at isolated localities. In the central parts of the Mihailovgrad Graben (Gaganitza), and similarly in the Izdremec Graben (Beledie Han, Balkan), a habitat between sublittoral, agitated and muddy bottoms occurred. Dominant were terebratellids like *Zeilleria lycetti* (Davidson: 70%), with *Pseudogibbirhynchia moorei* (Davidson: 22%), and *Homoeorhynchia cynocephala* (Richard: 8%). In the outer graben (near Gorno Ozirovo a sublittoral, very agitated biotope was mainly inhabited by rhynchonellids such as *Homoeorhynchia cynocephala* (90%), and *Pseudogibbirhynchia moorei* (3%), with a few *Zeilleria lycetti* (7%).

Along the southern shelf of the Matorides Basin (Tchoumatchenco, 1988), in neptunian dykes, within karstic, red micrites of the Bilka facies, were brachiopods like those of the Izdremec Graben at Beledie Han, comprised of rhynchonellids like*Kallirhynchia platiloba* Muir-Wood (30%), *Homoeorhynchia cynocephala* (26%), and *Rhactorhynchia* (4%), terebratellids, e.g. *Aulacothyris blakei* (Davidson) (29%) and *Ornithella* (4%), and rare terebratulids, *Dundrothyris perovalis* (J. de C. Sowerby:5%). The brachiopods occur as coquinas with whole valves, which probably underwent short transport from a sublittoral, agitated biotope, and accumulated on the karst surface.

Table 2. Correlation of the Bulgarian Middle Jurassic (Aalenian-Bathonian) ammonite and brachiopod zones (after Tchoumatchenco, 1977).

Stage	Sub-stage	Ammonite zone	Brachiopod zone
Bathonian	Middle	T. subcontractus P. progracilis	A. spinosa
	Lower	O. fallax Z. zigzag	
Bajocian	Upper	P. parkinsoni G. garantiana S. subfurcatum	W. wattonensis A. inflata
	Lower	S. humphriesianum O. sauzei H. discites	S. sphaeroidalis A. sentosa
Aalenian		G. concavum L. murchisonae L. opalinum	A. costata H. cynocephala
Toarcian			

MIDDLE JURASSIC ZONES
(AALENIAN-MIDDLE BATHONIAN)

During the Middle Jurassic, land masses of the Rhodope Massif and Moesian Platform, representing horsts, were gradually transgressed (Tchoumatchenco,1993; Tchoumatchenco & Sapunov, 1994), which led to the formation of a shallow shoreline favouring brachiopods (between horst-grabens), and deeper straits in the central grabens, unsuitable for brachiopods. The following brachiopod biozones were defined (Tchoumatchenco, 1977: Table 2):

Acanthothiris costata Zone (upper Aalenian). Brachiopods from this zone were collected from the central parts of the small Svetlja Graben, near Zabljano in western Bulgaria (Tchoumatchenco, 1993). Here strongly agitated, sublittoral conditions showed many rhynchonellids, e.g. *Acanthothiris costata* d'Orbigny (83%), *Sphenorhynchia* cf. *matisconensis* (Lissajous, in Arcelin & Roche:12%), and the terebratulid *Sphaeroidothyris decipiens* (Deslongchamps: 4%). In another small bay of the Drugan Graben (near Staro Selo, W Bulgaria), not directly connected with the Svetlja Graben, was the endemic Bulgarian rhynchonellid *Druganirhynchia nevelinae* Tchoumatchenco, which inhabited a similar setting (Tchoumatchenco, 1983).

Acanthothiris sentosa Zone (lower lower Bajocian). These brachiopods inhabited only the Svetlja Graben (near Zablano: Tchoumatchenco, 1993), and were dominated by the rhynchonelids *A. sentosa* (Quenstedt: 49%) and *Druganirhynchia nevelinae* Tchoumatchenco (41%), and terebratulid *Milliothyris* cf. *inflata* (Roche:10%). These occurred in a very agitated sublittoral biotope.

Sphaeroidothyris sphaeroidalis Zone (upper lower Bajocian). This association was established in the Svetlja Graben (near Zabljano: Tchoumatchenco, 1993), in agitated, sublittoral waters in which the terebratulid *S. sphaeroidalis* (J. de C. Sowerby: 55%), and rhynchonellid, *Formosarhynchia* aff. *pugnacea* (Quenstedt: 45%) occur almost equally.

Acanthothiris inflata Zone(uppermost lower Bajocian-lower upper Bajocian). Brachiopods of this zone formed an association of *A. inflata* (Svetlja Graben, and an association of *A. elargata* near Dolni Lom, NW Bulgaria. The first was formed in agitated waters sublittoral conditions, with the rhynchonellids *Acanthothiris inflata* (Quenstedt: 50%), *A.elargata* Seifert (4%), and *Formosarhynchia subpugnacea* Seifert (4%), which were slightly more common than the terebratulids *Stiphrothyris cheltensis* (Buckman:34%), *Millythyris* cf. *milliensis* Almeras (4%), and terebratellid *Aulacothyris* cf. *carinata* (Lamarck: 4%). *Acanthothiris elargata* was found on the slopes of the Belotinci Graben in the vicinity of Dolni Lom, and inhabited strongly agitated, sublittoral zones, in which it dominated by 92% over terebratulids like *Cererithyris dorsetensis* Douglas & Arkell (8%).

Wattonithyris wattonensis Zone (upper upper Bajocian-lower lower Bathonian). This also contains two associations, that of *W. midfordensis* in SW Bulgaria (Svetlja Graben), and of *W. wattonensis* in NW Bulgaria (Belotinci Graben near Dolni Lom: Tchoumatchenco, 1993). *Wattonithyris midfordensis* occurs on its own in biodetritic, argillaceous limestones, formed in the quiet, sublittoral conditions. *Wattonithyris wattonensis* is found in biodetritic limestones, laid down in more agitated, sublittoral conditions with terebratulids, e.g. *W. wattonensis* Muir-Wood (46%), *W. roettingensis* Seifert (19%), and *Cererthyris fleischeri* (Oppel: 1%) dominated the terebratellids *Rugitella lomensis* Tchoumatchenco (19%), *R.* cf. *cadomensis* (Deslongchamps: 6%), and rhynchonellid*Kallirhynchia subsuperba* Tchoumatchenco (9%).

Acanthothiris spinosa Zone (lower-middle Bathonian). This zone contains brachiopods from oolitic-biodetritic limestones near Belogradcik, NW Bulgaria (Tchoumatchenco, 1993), formed in slightly agitated, sublittoral conditions. Here the terebratulids *Cererithyris intermedia* (Sowerby: 32%), *Sphaeroidothyris doultingensis* (Richardson & Walker: 30%), were found with the rhynchonellids *Acanthothiris spinosa* (Linnaeus: 23%) and *Acanthorhynchia panacanatina* (Buckman & Walker: 1%), and the terebratellid *Aulacothyris* cf. *mandelslohi* (Oppel: 14%).

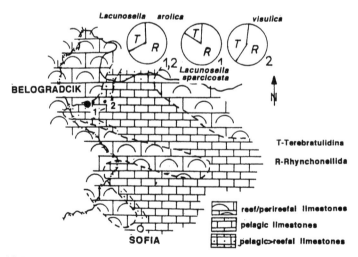

Figure 2. Distribution of Callovian-Tithonian brachiopods. Explanation of the circles: LA-*Lacunosella arolica* Zone; LS-*Lacunosella sparcicosta* Zone; LV-*Lacunosella sparcicosta-visulica* Zones. 1, Belogradcik-Oreshec railway station; 2, Necinska Bara (Gorno Belotinci).

272

Table 3. Correlation of Bulgarian Middle Jurassic (Bathonian)-Upper Jurassic ammonite and brachiopod zones.

Stage	Sub-stage	Ammonite zone (after Sapunov, 1976)	Brachiopod zone
Tithonian	Lower	T. vimineus S. swertschlageri H. hybonotum	
	Upper	H. beckeri	L. sparcicosta
Kimmeridgian	Middle	A. sesquinodosum Crussoliceras/A. sesqui- nodosum	
	Lower	C. divisum A. hypselocyclum H. desmoides	barren interval
Oxfordian	Upper	I. planula E. bimammatum P. bifurcatus	
	Middle	G. riazi P. antecedens P. episcopalis	L. arolica
	Lower	C. renggeri P. athletoides	
Callovian	Upper		
	Middle & Lower		barren interval
Bathonian	Upper		

MIDDLE JURASSIC (MIDDLE BATHONIAN-CALLOVIAN) - LATE JURASSIC

During the Callovian sedimentary changes affected the Bulgarian Moesian Platform and part of the Rhodope Massif (Tchoumatchenco, Sapunov, 1994). Shallow water carbonate platforms and deeper water pelagic basins were reconfigured (Figure 2). Conditions became favourable for brachiopods only on slopes of carbonate platforms, such a facies cropping out only in NW Bulgaria around Belogradcik (Tchoumatchenco, 1978). Here are two sections with many brachiopods, one near Belogradcik-Oreshec railway station, and another along the Necinska Bara River, near Gorno Belotinci, on the slope between the Vidin Complex Horst and the Belotinci Graben. In other transitional zones only single specimens occur. Brachiopod associations in NW Bulgaria are separated by two barren intervals, often with a sediment break, present in the Bathonian-Middle Callovian and lower Kimmeridgian (Table 3).

Lacunosella arolica Zone (upper Callovian-upper Oxfordian, up to I. planula Zone). This zone shows a slight increase in terebratulids at Necinska Bara (Tchoumatchenco, 1978). There, in a quiet sublittoral biotope, rhynchonellids comprise _Lacunosella arolica_ (Oppel: 44%), _Rhynchonella childsi_ Tchoumatchenco (8%), _Monticlarella_ cf. _czenstochawiensis_ (Roemer: 2%), _Lacunosella blanowicensis_ Wisniewska (5%), _Septocrurella multicostata_ Tchoumatchenco (7%), _Caucasella trigonella_ (Rothpletz: 2%), and '_Rhynchonella_' cf. _capilata_ Zittel (2%). The terebratulids included _Sellithyris engeli engeli_ (Rollier: 14%), _Nucleata euthymi_ (Pictet:12%), and _Sellithyris_ cf. _subsella_ (Leymerie: 3%). This zone is overlain by a barren interval.

Lacunosella sparcicosta Zone (middle Kimmeridgian- lower Tithonian). In this zone there are two associations (Tchoumatchenco, 1978), (a) L. sparcicosta , formed higher on the slope between Vidin Horst and Belotinci Graben, and (b) L. visulica -

on the lowest slope between the above (in Necinska Bara Valley, near Gorno Belotinci). In the _L. sparcicosta_ association, rhynchonellids such as _Lacunosella sparcicosta_ (Quenstedt: 42%), _Septocrurella sanctaeclarae_ (Roemer: 29%), _Monticlarella_ cf. _triloboides_ (Quenstedt: 3%), _L. selliformis_ (Lewinski: 6%), and _L. monsalvensis_ (Gillieron: 3%), reach up to 84% of the fauna. Then come the terebratulid _Nucleata nucleata_ (Schlotheim: 16%). In the _Lacunosella visulica_ assemblage, rhynchonellids include _Lacunosella visulica_ (Oppel: 45%), and _L. selliformis_ (16%), and there are terebratulids, e.g._Pygope janitor_ (Pictet: 33%), and _Juralina rauraca_ (Rollier: 6%). These differences reflect paleoecologic conditions, the first formed in a sublittoral relatively agitated biotope, while in the second, conditions were quieter, permitting an increase of the specialist _Pygope_. Terebratellids have not been found in Upper Jurassic strata of Bulgaria.

CONCLUSIONS

Rhynchonellida composed the greater part of the thanatocoenoses formed in shallow, highly turbulent water, but their percentage progressively diminished with relative deepening and reduction in energy. Terebratulidina took a more important part in biotopes with relatively calmer waters, inversely proportional to rhynchonellids. Terebratellidina and Spiriferidina became more abundant when Rhynchonellida and Terebratulidina declined. Terebratellidina and Spiriferidina were virtually absent under relatively high energy, but became dominant in calmer waters. The last spiriferinids disappeared in the lower Toarcian (Early Jurassic) tenuicostatum zone, c. 187 Ma.

Acknowledgments

The Natural Sciences and Engineering Research Council of Canada is thanked for assistance in attending the Third International Brachiopod Congress in Sudbury, Ontario.

273

REFERENCES

REVERT, J. & P. TCHOUMATCHENCO. 1973. Les zones de brachiopodes du Lias en Bulgarie et dans le Sud de la France (bassin des Causses): essai de corrélation. Documents Laboratoire Géologie Faculté Sciences, Lyon 56:181-191.

SAPUNOV, I. & J. STEPHANOV. 1964. The stages, substages, ammonite zones and subzones of the Lower and Middle Jurassic in the Western and Central Balkan Range (Bulgaria), p.705-718. Colloque du Jurassique, Luxembourg, 1962, Comptes rendus et mémoires, Luxembourg.

___,P. TCHOUMATCHENCO & V. SHOPOV. 1967. Biostratigraphy of the Lower Jurassic rocks near the village of Komstitsa, district of Sofia (Western Balkan Range). Bulletin Geological Institute16:125-143.

TCHOUMATCHENCO,P. 1972. Répartition stratigraphique des brachiopodes du Jurassique inférieur du Balkan central et occidental et du Kraïste (Bulgarie). ibid. 21; 63-84.

___1974. Note sur la répartition stratigraphique des brachiopodes du Jurassique inférieur dans les Balkanides centrales et occidentales (Bulgarie). Colloque du Jurassique, Luxembourg, 1967. Mémoires Bulletin Recherches Géologiques Minières, France 75:183-187.

___1972. Thanatocoenoses and biotopes of Lower Jurassic brachiopods in central and western Bulgaria. Palaeogeography, Palaeoclimatology, Palaeoecology 12:227-242.

___1977a. Les biozones de brachiopodes du Jurassique moyen en Bulgarie. Geologica Balcanica 7(1):97-108.

___1977b. Sur la stratigraphie des brachiopodes du Jurassique moyen dans la région de Belogradtchik. Reviews Bulgarian Geological Society 38(3):314-319.

___1978a. Brachiopodes du Jurassique moyen des environs du village de Dolni Lom, district de Vidin (Bulgarie du Nord-Ouest). Annales Université Sofia, Faculté Géologie Géographie 69(1976/1977):193-232.

___1978b. Callovian-Tithonian Brachiopoda from the northern limb of the Belogradcik Anticlinorium, Northwest Bulgaria. Palaeontology, Stratigraphy, Lithology 8:3-54.

___1978c. Middle Jurassic brachiopods from the Polatenska Formation near the Village of Zablyano, Radomir Area (West Bulgaria). ibid.9:27-56.

___1983. Druganirhynchia nevelinae gen. & sp. n. (Brachiopoda, Rhynchonellidae) and the partition of the Aalenian rhynchonellids in southwestern Bulgaria. Geologica Balcanica 13(6):69-78.

___1988. Réconstitution stratigraphique et paléogeographique du Jurassique inférieur et moyen à partir des olistolithes inclus dans la Formation de Kotel (Stara Planina orientale, Bulgarie). ibid.18(6):3-28.

___1989. Brachiopodes des olistolithes jurassiques inférieurs et moyens inclus dans la Formation de Kotel - Jurassique moyen (Stara Planina orientale, Bulgarie), Rhynchonellida. Palaeontology, Stratigraphy, Lithology 28:3-30.

___1990. Brachiopodes jurassiques inférieurs et moyens des olistolithes inclus dans la Formation de Kotel (Jurassique moyen) (Stara Planina orientale, Bulgarie), Spiriferida, Terebratulida. ibid. 28:3-40.

___,B. PEYBERNÈS, S. CERNJAVSKA, G. LACHKAR, J. SURMONT, J. DERCOURT, Z. IVANOV, J.P. ROLANDO, I. SAPUNOV & J. THIERRY. 1992. Étude d'un domaine de transition Balkan-Moésie: évolutions paléogéographiques et paléotectoniques du sillon du flysch jurassique inférieur et moyen dans la Stara Planina orientale (Bulgarie orientale). Bulletin Société Géologique France 163(1):49-61.

___1993a. Brachiopod thanatocoenoses in the Aalenian, Bajocian and Bathonian of western Bulgaria and their distribution. Palaeogeography, Palaeoclimatology, Palaeoecology 100:159-168.

___1993b. The horizontal distribution of brachiopods during the Zeilleria quadrifida Zone (Late Carixian-Early Domerian, Early Jurassic) in Bulgaria, p.143-150. In, Palfy, J. & Vörös, A. (eds.), Mesozoic brachiopods of Alpine Europe. Hungarian Geological Society, Budapest.

___& I. SAPUNOV. 1994. Intraplate tectonics in the Bulgarian part of the Moesian Platform during the Jurassic. Geologica Balcanica 24(3):3-12.

[*Editors' note: we have not attempted to translate the terms littoral, sublittoral, agitated, biodetritic, detritic, etc.]

BRACHIOPOD PALEOZOOGEOGRAPHY THROUGH THE CAMBRIAN

G. T. USHATINSKAYA

Paleontological Institute, Russian Academy of Sciences, Moscow, Russia

ABSTRACT- The oldest brachiopods, the 'inarticulates', appeared in the Tommotian, arriving with a number of other skeletal invertebrates at about the same time. Their outlined geographic distribution depended strongly on regional sedimentary settings and climates. Cambrian Lingulata consisted mostly of ubiquitous taxa inhabiting silici-clastic and carbonate environments on the continental shelf and slope. Long-ranging Calciata dispersed to several paleo-continents and preferred shallow water carbonate, and mixed siliciclastic-carbonate shelves. Short-ranging Calciata were confined to a single paleocontinent. The first 'articulates' (nisusiids) appeared in the Atdabanian. From the Atdabanian onwards, two large realms, the 'tropical' and the 'natal', became differentiated with regard to brachiopod distribution, and were characterised by differing lingulate/calciate ratios, with greater diversity of calciates in the tropical realm.

INTRODUCTION

Brachiopods acquired the capacity to build mineralised skeletons from nearly the earliest Cambrian [Tommotian, succeeding the brachiopod-barren basal Nemakit-Daldynian of the earliest Cambrian], and these left relatively common to abundant fossil remains useful for determining occurrence, distribution, migration, provincialism, and extinction patterns. So far, however, there have been few serious comprehensive studies on the paleozoogeography of Cambrian brachiopods. Previous work compared Cambrian brachiopod occurrences with the same zoogeographic units currently recognized for the trilobites (Lochman, 1956; Palmer, 1974; Ushatinskaya, 1986; Pelman et al., 1992). New data have made it possible to examine Cambrian brachiopod paleozoogeography in more detail.

Three outstanding general issues for the Cambrian remain:

Cambrian paleogeography. During the c.40-50 million years of the Cambrian Period, there were several major changes in plate tectonic movement. At the beginning of the Early Cambrian (Nemakit-Daldynian), breakup of the late Precambrian super-continent 'Paleopangea' created shallow epicontinental seas at low paleolatitudes (Kirschvink et al.,1984; Rozanov, 1984; Donnelly et al., 1990), and shallow carbonates and silici-clastics became widespread. In the Middle and Late Cambrian, the paleocontinents were more widely separated from one another. Shale and mixed carbonate-shale deposits accumulated over large areas on the outer shelves and continental slopes. Several different paleogeographic reconstructions for the Tommotian and following intervals, based on paleo-magnetic, lithologic and paleontologic data, have been attempted (Rozanov, 1984; Scotese & McKerrow, 1990; Kirschvink, 1992). This paper follows the reconstruction of Scotese & McKerrow (1990), with minor modifications matching our data.

Brachiopod systematics. This paper follows the classification proposed by Goryanskii & Popov (1985), modified by Popov et al. (1993), in which brachiopods are divided into two classes, Lingulata [with phosphatic shells], and Calciata [with calcareous shells, incorporating both 'inarticulate' and 'articulate' taxa]. Both classes are known from nearly the start of the Cambrian. Cambrian Calciata are represented by the orders Obolellida, Kutorginida, Trimerellida, some new groups described by Popov & Tikhonov (1990), and the articulate orders Orthida and Pentamerida. These made up only a small part of known calcareous brachiopods, as the main radiation of Calciata [the 'articulate' component] occurred in the Ordovi-

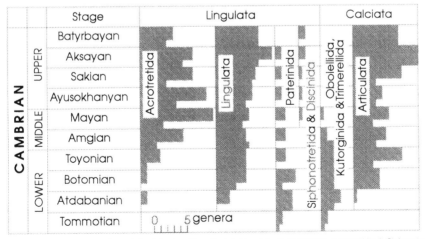

Figure 1. Range and relative abundance chart of Cambrian brachiopods. 1-4: Lingulata. 1, Acrotretida, 2, Lingulida, 3, Siphonotretida and Discinida, 4, Paterinida. 5-6: Calciata. 5, Obolellida, Kutorginida, and Trimerellida, 6, Articulata.

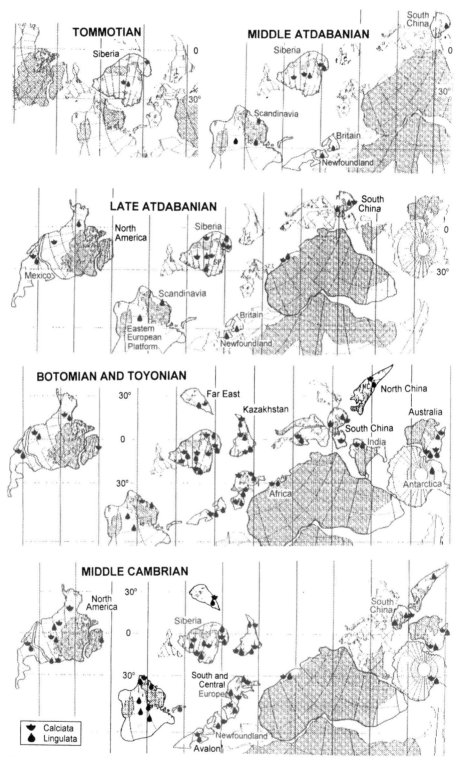

Figure 2. Cambrian brachiopod distribution for the Lingulata and Calciata with paleogeographic reconstructions of Cambrian oceans based on Scotese & McKerrow (1990). Land areas shown in grey.

cian, Silurian, and Devonian. All five orders of the Lingulata are known from the Cambrian, with the Lingulida, Acrotretida and Paterinida appearing in the Early Cambrian, the Siphonotretida in the Middle Cambrian, and the Discinida in the Late Cambrian. These made up about one third of all known lingulate genera (Figure 1).

Brachiopod distribution. Most brachiopods are benthic organisms and their major dispersal occurs primarily during the larval stage. The larvae of living Calciata are lecithitrophic, with a free swimming phase of several hours to two days before settling on the sea bottom. Recent Lingulata have planktotrophic larvae which swim for a period of several days to a month (Malakhov, 1976; Zezina, 1976), and thus have greater opportunity for wide dispersal. The larvae of Cambrian brachiopods probably had similar biological attributes. Some Cambrian lingulates are characterised by wide distribution, and quite possibly their long larval stage was one of the factors for this cosmopolitanism (Rowell, 1986). In addition, the larval surfaces of acrotretids, and some lingulids, have numerous small pits, which might have been an adaptation for a pelagic mode of life by reducing body weight (Biernat & Williams, 1970; Popov et al., 1982; Williams & Curry, 1991).

EARLY CAMBRIAN BIOGEOGRAPHY

Five brachiopod genera occur in the Tommotian Stage of the Siberian Platform and Mongolia, though none appear to have been present in the earliest, Nemakit-Daldynian, Stage of the Cambrian. The Siberia and Mongolia plates were covered mostly by shallow water, equatorial carbonate environments in the Tommotian (Figure 2). The oceans were apparently saturated with iron oxides and phosphorus (Rozanov, 1979; Rozanov & Zhuravlev, 1992). There were abundant archeocyath-microbial boundstones, and small archeocyath reefs were separated by depressions with clay-rich carbonates (Raaben, 1969). Brachiopods evolved in these diverse tropical habitats, occupied space between archeocyath-microbial reefs, and sometimes lived in such reefs. Paterinids (Family Cryptotretidae) appeared first, represented by *Aldanotreta* in the *Sunnaginicus* archeocyath Zone, and *Cryptotreta* in the *Regularis* archeocyath Zone. *Nochoroiella*, the earliest Calciata (order Obolellida), appeared in the *Regularis* zone in this region. The second oldest obolellid, *Obolella*, evolved near the end of the Tommotian (Pelman et al., 1992). *Khasagtina*, probably the earliest member of the Kutorginida, the made its appearance in the late Tommotian of Mongolia (Ushatinskaya, 1987).

Middle Atdabanian brachiopods, with 12 brachiopod genera, are known from the Siberian platform, Mongolia, the Altai-Sayan region, China, Great Britain, Newfoundland, the western part of the Eastern European [Russian] Platform, and Scandinavia (Figure 2). As for the Tommotian, brachiopod associations were most common on the Siberian Platform. Rich assemblages of *Cryptotreta* and *Obolella* are known in the middle Lena River area (Pelman et al., 1992), where the first articulate brachiopod, probably a nisusiid, was recorded by Ushatinskaya (1986). The distribution of cryptotretids and obolellids extended beyond the Siberian Plate. *Aldanotreta* and *Cryptotreta* are present in the upper Ust-Kundat and the lower Usa formations in the Altai-Sayan region (Pelman & Ermak, 1985; Pelman et al., 1992). *Aldanotreta* has been reported also from the Meishucun Formation of South China (Liu, 1979). Brasier (1986) noted that *Micromitra phillipsi* from the Home Farm Member of Nuneaton (UK), and in Bed Ab1 of Shropshire, was morphologically similar to *Aldanotreta*. In addition to cryptotretids, *Obolella,* and the earliest *Kutorgina* occur in the Altai-Sayan region (Pelman et al., 1992).

During the Atdabanian, the western part of the Eastern European Platform, Scandinavia, Great Britain and Newfoundland were probably located in higher latitudes, as indicated by predominant siliciclastic regimes (Figure 2). In the Eastern European Platform, the oldest brachiopods, *Micromitra* and *Mickwitzia*, are known from the *Holmia* Zone. These are paterinids (family Paterinidae), and probably aberrant forms (Jendryka-Fuglewicz, 1992). *Micromitra* is known also from the *Coleoloides typicalis* zone in Newfoundland (Bengtson & Fletcher, 1983), and *Mickwitzia* in the *Volbortella-Schmidtiellus mickwitzi* zone in Scandinavia (Bergström & Ahlberg, 1981). In the Middle Atdabanian, two other lingulate orders, the Lingulida and the Acrotretida, made their first appearance, which may have been associated with siliciclastic and mixed siliciclastic-carbonate facies in higher latitudes. Their representatives were reported from thinly bedded limestones interbedded with siliciclastics of the Lower Comley Formation in Great Britain (Hinz, 1987), from the Vergale horizon in the western Eastern European Platform (Jendryka-Fuglewicz, 1992), and from the *Holmia kjarulfi* zone of Scandinavia (Bergström & Ahlberg, 1981).

In younger Atdabanian time, brachiopods continued to be dispersed to new regions (Figure 2). Again, the richest assemblages are found in Siberia and adjacent regions, with such obolellid calciate brachiopods as *Obolella, Sibiria, Kutorgina*, and nisusiids being dominant. *Obolella* and *Sibiria* are particularly abundant in the Siberian Platform. Among the lingulates, the paterinids (*Cryptotreta, Micromitra*), small numbers of lingulids s.s.(*Botsfordia, Clivosilingula*), and acrotretids such as *Linnarssonia*, are known (Pelman et al., 1992). The obolellids *Bicia* , and *Magnicanalis*, were described in the Transbaikal (Ushatinskaya, 1988). *Dzunartzina* occurs in reefs near Dzun-Arts Mountain in western Mongolia (Ushatinskaya, 1993). In the siliciclastic and mixed carbonate-siliciclastic facies of western North America, the oldest brachiopods of the region, the lingulid *Paleoschmidtites*, were recorded in the Mackenzie Mountains, nisusiids and obolellids in Nevada, and nisusiids and *Mickwitzia* in the Caborca region of Mexico (Voronova et al., 1987; Rowell, 1977; Stewart et al., 1984). In the siliciclastics of the western Eastern European Platform and Scandinavia, rare but not new representatives of all three Early Cambrian lingulate orders have been found. Calciate remains are very scarce (Keller & Rozanov, 1979, 1980; Jendryka-Fuglewicz, 1992).

During the Tommotian and Atdabanian, there is differentiation in distribution of brachiopods both geographically and sedimentologically, although brachiopod zoogeographic realms cannot be clearly recognised. Calciates were dominant in the shallow shelf carbonate facies of the Siberian Platform and adjacent low latitudes. Lingulates were most numerous and diverse in siliciclastic facies at higher latitudes in shelf environments of the Eastern European Platform, Scandinavia, Great Britain, and Newfoundland. Later, acrotretids and paterinids became globally widespread both in siliciclastics and carbonates. The association of lingulids with terrigenous facies has been observed throughout their history. Although their remains are known in carbonates as old as the Botomian, lingulids are extremely abundant only in siliciclastics. Evidently, in shallow-water siliciclastic basins, where many brachiopods could not survive because of turbid, high energy conditions, lingulids were the only group to prosper because of tolerance to abundant suspended sediments and ability feed on phytoplankton.

In the Botomian and Toyonian, brachiopods became more diverse and abundant, are known from almost all regions, and in some areas are more abundant than any other fossils.

Although their distribution remained dependent on sedimentary environments and climates, this diminished later because lingulids, especially acrotretids, became widespread. Two major zoogeographic realms are distinguished provisionally. The first, the 'tropical realm', was very large and heterogeneous, and included regions mainly from low paleolatitudes, e.g. the Siberian Platform, Altai-Sayan, Mongolia, Kazakhstan, central Asia, South China, South Australia, and western North America, where carbonates prevailed. Adjoining these regions were probably southern and central Europe, North Africa, Arabia, and India, characterised by mixed carbonate-siliciclastics (Figure 2). Calciate brachiopods were especially diverse. Among the obolellids, *Obolella*, *Sibiria*, *Alisina*, and *Trematobolus* were widespread, especially in the Siberian Platform. *Kutorgina* became virtually ubiquitous, and more orthids made their appearance, e.g. *Nisusia*, *Matutella*, *Wimanella*, and *Arctohedra*. Lingulates also occupied an important place, with genera which had appeared in the Atdabanian becoming more numerous, such as *Botsfordia*, *Paleoschmidtites*, *Westonia*, *Clivosilingula*, *Micromitra*, *Paterina* and *Linnarssonia*. Their wide distribution was probably associated with the Botomian Transgression (Rozanov, 1984). New genera such as *Eothele*, *Homotreta*, *Hadrotreta*, and *Kyrshabactella* arose.

Most lingulates were cosmopolitan, though several territories with endemic brachiopod assemblages may be recognised, one of these comprising the present western and southern parts of the Siberian Platform, where carbonates at the margins of a vast evaporite basin contain a widespread *Kutorgina* fauna with 4-5 species (Pelman et al., 1992). *Kutorgina* was probably adapted to unusual life conditions. Two other territories with lingulates were South Kirgizia and the Dead Sea (Israel-Jordan), where mixed carbonate-silici-clastics, alternating with volcaniclastics, accumulated. A sig-nificant part of these assemblages was composed of calciates represented by endemic genera (Cooper, 1976; Popov & Tikhonov, 1990). Their diversity indicates a major radiation of calciates in these regions during the Botomian. Lingulates also occur, but are mostly cosmopolitan forms.

Jin et al. (1992, 1993) described well-preserved lingulids from shales of the Chiungchusu Formation of Yunnan (South China), including *Lingulepis*, *Lingulella*, and the first craniopsid (Jin et al.,1992). Moulds of very long pedicles were preserved in several pedicle valves of lingulids. This suggests that the pedicle, as an anchoring device in burrows, has not changed significantly through lingulid history, although some Cambrian lingulids were typical epibenthos (Goryanskii, 1969; Nikitin, 1989). Australia, the Northern Territory, western New South Wales, and South Australia, show rich brachiopod assemblages with about 25 genera of probable Toyonian age. About half of these brachiopods are calciates, including widespread genera and new forms. Most of the lingulate genera were cosmopolitan (Kruse, 1990; Roberts & Jell, 1990; Brock & Cooper, 1993).

The other Botomian-Toyonian brachiopod realm, here called the 'natal realm', includes the Eastern European Platform, Scandinavia, Great Britain, Avalonia, and Newfoundland, where siliciclastics, sometimes with thin carbonate units, are most common. Brachiopods show low diversity, with cosmopolitan lingulates dominant. *Micromitra*, *Paterina*, *Westonia*, *Lingulepis*, '*Lingulella*', *Eothele*, *Linnarssonia*, and *Acrotreta* occur mainly in siliciclastics (Keller & Rozanov, 1979, 1980; Jendryka-Fuglewicz, 1992). In the carbonates, and mixed siliciclastic-carbonates, lingulates and calciates co-occurred, including *Micromitra*, *Paterina*, *Lingulella*, *Botsfordia*, *Eothele*, *Linnarssonia*, *Obolella*, *Trematobolus*, *Kutorgina*, *Nisusia*, *Matutella*, and *Eoconcha* (Cooper, 1951; Shaw, 1955, 1962;

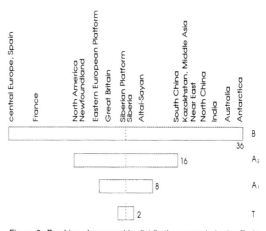

Figure 3. Brachiopod geographic distribution spread in the Early Cambrian (Tommotian-Botomian), showing increasing cosmopolitanism. T-Tommotian; A1-Early-Middle Atdabanian; A2-Late Atdabanian; B-Botomian. Numerals indicate numbers of localities in data base.

Rowell, 1962; Freyer, 1981; Elicki, 1994). These genera are known also from tropical realm, but the lingulates are most common, and the calciates erratic in distribution.

MIDDLE CAMBRIAN BIOGEOGRAPHY

During the Middle Cambrian the paleogeographic situation gradually changed. As in the Early Cambrian, many regions where brachiopods were common were located in low latitudes, but shallow epicontinental carbonate seas became restricted in extent. Limestones were commonly interbedded with, or even replaced by, siliciclastics, as in the Middle Cambrian of the Altai-Sayan region, Mongolia, and central Kazakhstan. Mixed carbonate-shales and shales accumulated on the outer shelf and slope of the Siberian Platform, Maly Karatau, Greenland, NE Australia, and New Zealand (Figure 2; see Henderson & MacKinnon, 1981; Zell & Rowell, 1988; Pelman et al., 1992; Ushatinskaya, 1994). Calciates were restricted largely to carbonate and mixed siliciclastic-carbonate facies of the shallow shelf. Muddy substrates were apparently not suitable for these thick-shelled calciates.

In the Middle Cambrian, the number of calciates decreased because of extinction in obolellids and kutorginids. The articulate brachiopods *Nisusia*, *Matutella*, *Wimanella* and *Diraphora*, which arose in the Early Cambrian, formed shell assemblages, and were widespread on two or three paleocontinents. *Engenella*, the first *Eoorthis*, and *Billingsella* appeared in the Middle Cambrian, but were generally rare. Taxonomic diversity of lingulates and especially acrotretids increased noticeably (Figure 1). The latter group was probably able to live on soft, muddy bottoms, as indicated by very small and light forms that probably lived on algal mats, forms with a high conical ventral valve that raised the shell above the bottom, and forms with apical nodes and thorns of similar function. By the end of Middle Cambrian, acrotretids comprised some 20 genera, including *Neotreta*, *Opistotreta*, *Rhondellina*, *Anabolotreta*, *Stilpnotreta*, and *Dactylotreta*. Most of these rapidly became widespread in subequatorial regions. For example, *Neotreta* is known from the Mayan Stage on the N Siberian Platform, central and S Kazakhstan, and in the Far East. *Anabolotreta* has also been found in these regions, in addition to Greenland, Australia, and New Zealand. A similar geographic distribution can be found for *Stilpnotreta*, *Opisto-*

treta, and *Angulotreta*. Lingulate assemblages in the late Middle Cambrian and early Late Cambrian are characterised by high taxonomic diversity, with up to 7-10 genera per location.

Assemblages in widely separated regions often show remarkable similarity, having about 50% of genera and species in common.

Siliciclastics, sometimes with carbonate interbeds, continued to accumulate in the 'natal realm' (Eastern European Platform, Scandinavia, Great Britain, Avalonia, and Newfoundland) of the temperate latitudes. Brachiopods were represented almost exclusively by lingulates. From the second part of the Middle Cambrian to the Early Ordovician, lingulates flourished in the western Eastern European Platform, forming shell beds known as the *Obolus* beds in many levels. These shell beds typically have low taxonomic diversity, with only one or two species at any locality. In Scandinavia and northern Europe, where fluctuating sedimentary environments produced alternating carbonates and siliciclastics, brachiopods showed high diversity of lingulates and some articulates, but these brachiopod assemblages were unlike those described from the west of the European Platform.

LATE CAMBRIAN BIOGEOGRAPHY

In the Late Cambrian, western North America, Greenland, the Siberian Platform, Altai-Sayan, Kazakhstan, Middle Asia, Far East, Australia, and Antarctica remained in low latitudes, with more widespread shelf carbonate and mixed siliciclastic-carbonate facies. Brachiopod assemblages from these regions were variable, with articulates dominant. During the Late Cambrian, articulate diversity grew steadily. Some genera, e.g., *Billingsella*, *Eoorthis*, and *Huenella*, were widespread, and represented by many species. According to Yadrenkina (1974), the Siberian platform, North America and Kazakhstan had 75% of the Late Cambrian brachiopod genera in common, in some instances even the species being the same. Taxonomic diversity of the lingulates was also high. Only towards the end of the Late Cambrian did lingulates decrease appreciably. Some genera, such as *Zhanatella*, *Disorhistus*, *Quadrisonia* and *Eurytreta*, were widespread. Others were restricted to one paleocontinent, such as *Boschakolia*, *Akmolina*, *Dicellomus*, and *Tropidoglossa*. Many endemic forms are known from the middle Late Cambrian of central Kazakhstan (Koneva et al., 1990; Holmer & Popov, 1994).

Siliciclastics of the Eastern European Platform contained mainly lingulates. In the shallow water facies, lingulids dominated and accumulated as shell beds. The *Ungula* community lived nearshore, the *Oepikites* community was distributed further out, and, in deeper waters, acrotretids and siphonotretids prevailed (Nikitin, 1989).

CONCLUSIONS

The oldest brachiopods appeared in the earliest Cambrian (Tommotian), and consisted of two lineages, the Lingulata and Calciata. Cambrian lingulates were mostly widespread genera and lived on siliciclastic and mixed siliciclastic-carbonate substrates in shelf and continental slope settings. In comparison, Cambrian calciates were less widespread than lingulates, and preferred shallow-water shelf carbonates and mixed siliciclastic-carbonate settings. From the middle Early Cambrian, two large realms, a 'tropical realm' and 'natal realm', may be distinguished on the basis of brachiopods, mainly by differing lingulate/calciate ratios, and greater diversity of calciates in the tropical realm.

Acknowledgments

I am grateful to Prof. A. Yu. Rozanov for helpful comments, to Y. Malakhovskaya for preparing paleogeographic maps on the computer, and to D. Gravestock (Australia) for comments on the language and scientific content of the manuscript. The work has been supported by RFFI Project 93-05-8503. A travel grant was received from the Natural Sciences and Engineering Research Council of Canada to attend the Third International Brachiopod Congress in Sudbury, Ontario.

REFERENCES

BENGTSON, S. & T. FLETCHER. 1983. The oldest sequence of skeletal fossils in the Lower Cambrian of southeastern Newfoundland. Canadian Journal Earth Sciences 20:525-536.

BERGSTRÖM, J. & Y. AHLBERG. 1981. Uppermost Lower Cambrian biostratigraphy in Scania, Sweden. Geologiske Foreningens Stockholm Förhandlingar 103:193-214.

BIERNAT, G. & A. WILLIAMS. 1970. Ultrastructure of the protegulum of some acrotretide brachiopods. Palaeontology 13:491-502.

BRASIER, M.D. 1986. The succession of small shelly fossils (especially conoidal microfossils) from English Precambrian-Cambrian boundary beds. Geological Magazine 123:237-256.

___1992. Background to the Cambrian Explosion. Journal Geological Society London149:585-587.

BROCK, G.A. & B.J. COOPER. 1993. Shelly fossils from the Early Cambrian (Toyonian) Wirrealpa, Aroona Creek, and Ramsay limestones of South Australia. Journal Paleontology 67:758-787.

COOPER, G.A. 1951. New brachiopods from the Lower Cambrian of Virginia. Journal Washington Academy Science 41:4-8.

___1976. Lower Cambrian brachiopods from the Rift valley (Israel and Jordan). Journal Paleontology 50:269-286.

DONNELLY, T.H., J.H. SHERGOLD, P.N. SOUTHGATE & C.J. BARNES. 1990: Events leading to global phosphogenesis around the Proterozoic/Cambrian boundary. In, A.J. Norholt & J. Jarvis (eds.), Phosphorite research and development. Geological Society Special Publications 52:273-287.

FREYER, G. 1981. Die unterkambrische Brachiopoden Fauna des Gorlitzer Schiefergebirges. Abhandlungen Berichten Naturkundemuseum Gorlitz 54(5):1-20.

ELICKI, O. & J. SCHNEIDER. 1992. Lower Cambrian (Atdabanian/Botomian) shallow carbonates of the Gorlitz Synclinorium (Saxony/Germany). Facies 26:55-66.

GORYANSKII, V.Yu.1969. Bezamkovie brakhiopody kembriiskikh i ordovikskikh otlozhenii severo-zapada Russkoi platformy. Nedra, Leningrad, 171p.

___& L.E. POPOV. 1985. Morfologia, systematicheskoe polozhenie i proiskhozhdenie bezamkovikh brakhiopod s karbonatnoi rakovinoi. Paleontologicheskii Zhurnal 3:3-14.

JENDRYKA-FUGLEWICZ, B. 1992. Analiza porownawcza ramienionogow z utworow Kambru gor Swietokrzyskich i platformy Prekembryjskiej w Polsce. Przeglad Geologiczny 3:150-155.

JIN YUGAN & HUAYU WANG. 1992. Revision of the Lower Cambrian brachiopod *Heliomedusa* Sun & Hou, 1987. Lethaia 25:35-49.

___, XUANGUANG HOU & HUAYU WANG. 1993. Lower Cambrian pediculate lingulids from Yunnan, China. Journal Paleontology 67:788-798.

HENDERSON, R.A. & D.I. MACKINNON. 1981. New Cambrian inarticulate Brachiopoda from Australia and the age of the Tasman Formation. Alcheringa 5:289-309.

HINZ, I. 1987. The Lower Cambrian microfauna of Comley and Rushton, Shropshire, England. Palaeontographica (A)196:1-100.

HOLMER, L.E. & L.E. POPOV. 1994. Cambrian-Ordovician lingulate brachiopods from Scandinavia, Kazakhstan, and South Ural Mountains. Fossils Strata 35:1-156.

KELLER, B.M. & A. YU. ROZANOV (eds.), 1979. Stratigrafiya verkhnedokembriiskikh i kembriiskikh otlozhenii zapada Vostochno-Evropeiskoi platformy. Nauka, Moskva, 234p.

___&___. 1980. Paleogeografiya i litologiya Venda i Kembriya zapada Vostochno-Evropeiskoi platformy. Nauka, Moskva, 118p.

KIRSCHVINK, J.L. 1992. A paleogeographic model for Vendian and Cambrian time, p.569-581. In, J.W. Schopf & C. Klein (eds.), The Proterozoic biosphere, a multidisciplinary study. Cambridge University Press.

___,T.D. BARR, A.YU. ROZANOV & A.YU. ZHURAVLEV. 1984. The destruction of Paleopangea (?) in the Early Cambrian. Abstracts 27th International Geological Congress, Moscow 4:113-114.

KONEVA, S.P., L.E. POPOV, G.T. USHATINSKAYA & N.V. ESAKOVA. 1990. Bezamkovie brakhiopody (akrotretidy) i mikroproblematiki iz verkhnego Kembriya Severo-Vostochnogo Kazakhstana, p.158-169. In, Biostratigrafiya i Paleontologiya Kembriya Severnoi Azii. Nauka, Novosibirsk.

KRUSE, P.D. 1990. Cambrian paleontology of the Daly basin. Northern Territory Geological Survey 7:1-58.

LIU DIYANG. 1979. Earliest Cambrian brachiopods from South West China. Acta Paleontologica Sinica 18:505-511.

LOCHMAN, C. 1956. Stratigraphy, paleontology and paleogeography of the Elliptocephala asaphoides strata in Cambridge and Hoosick quadrangles, New York. Bulletin Geological Society America 67:1331-1396.

MALAKHOV, V.V. 1976. Nekotorye stadii razvitia zamkovoi brakhiopod v Cnismatocentrum sakhalinensis parvum i problema evolutsii sposoba zakladki celomicheskoi mesodermi. Zoologicheskii Zhurnal 55:66-75.

NIKITIN, I.F. (ed.) 1989. Opornie razrezi i stratigrafiya kembro-ordovikskoi fosforitonosnoi obolovoi tolschi na severo-zapade Russkoi platformy. Nauka, Leningrad, 222p.

PALMER, A.R. 1974. Search for the Cambrian World. American Scientist 62:216-224

PELMAN, YU.L. & V.V. ERMAK. 1985. Novye dannye po stratigrafii ust-kundatskoi svity Kuznetskogo Alatau (r. Kiya, nizhnii Kembrii), p.16-32. Biostratigrafiya i biogeografiya paleozoya Sibiri. IGG, Novosibirsk.

___,N.A. AKSARINA, S.P. KONEVA, L.E.POPOV, L.P. SOBOLEV & G.T. USHATINSKAYA. 1992. Drevneishie brakhiopody territorii Severnoi Evrasii. Nauka, Novosibirsk, 145p.

POPOV, L.E., M.G. BASSETT, L.E. HOLMER & J. LAURIE. 1993. Phylogenetic analysis of higher taxa of Brachiopoda. Lethaia 26:1-5.

___,O.N. ZEZINA & J. NOVAK, 1982. Mikrostructura apikalnoi chasti rakoviny bezamkovikh brakhiopod i ee ecologicheskoe znachenie. Bullyutin MOIP, Biologicheskii Otdel 87:94-104.

___& Yu.A. TIKHONOV. 1990. Rannekembriiskie brakhiopody iz Yuzhnoi Kirgizii. Paleontologicheskii Zhurnal 3:33-45.

RAABEN, M.E. (ed.) 1969. Tommotskii Yarus i problema nizhnei granitsi Kembriya. Trudy Geologicheskogo Institita 206:1-380.

ROBERTS, J. & P.A. JELL. 1990. Early Middle Cambrian (Ordian) brachiopods of the Coonigan Formation, western New South Wales. Alcheringa 14:257-309.

ROWELL, A.J. 1962. The genera of the brachiopod superfamilies Obolellacea and Siphonotretacea. Journal Paleontology 36:136-152.

___1977. Early Cambrian brachiopods from the south-western Great Basin of California and Nevada. ibid. 51:68-85.

___1986. The distribution and inferred larval dispersion of Rhondellina dorei, a new Cambrian brachiopod (Acrotretida). ibid. 60:1056-1065.

ROZANOV, A.Yu. 1979. Nekotorye problemy izucheniya drevneishikh skeletnikh organizmov. Bullyutin MOIP, Geologia 54 (3):62-69.

___1984. Nekotorye aspekty izucheniya bio- i paleogeografii rannego Kembriya. Doklady 27 Mezhdunarodnii Geologicheskii Kongress, Paleontologia 2:85-93.

___& A.Yu. ZHURAVLEV. 1992. Lower Cambrian fossil records of the Soviet Union, p.205-281. In, J. Lipps & P. Signor (eds.), Origin and early evolution of Metazoa. Plenum Press, New York.

SCOTESE, C.R. & W.S. MCKERROW. 1990. Revised world maps and introduction. In, W.S. McKerrow & C.R. Scotese (eds.), Palaeozoic palaeogeography and biogeography. Geological Society Memoir 12:1-21.

SHAW, A.B. 1955. Paleontology of northwestern Vermont, The Lower Cambrian fauna. Journal Paleontology 29:775-805.

___1962. Paleontology of northwestern Vermont, VIII: Fauna of Hungerfold Slate; IX, Fauna of Monkton Quartzite. Journal Paleontology 36:314-345.

STEWART, Y.H., M. MCMENAMIN & J.M. MORALES-RAMIREZ. 1984. Upper Proterozoic and Cambrian rocks in the Caborca region, Sonora, Mexico. U.S. Geological Survey Professional Paper 1309:1-36.

USHATINSKAYA, G.T. 1986. Nakhodka drevneishei zamkovoi brakhiopody. Paleontologicheskii Zhurnal 4:102-103.

___1986. Brakhiopody Kembriya (obzor mestonakhozhdenii i nekotorye zakonomernosti geograficheskogo rasprostraneniya, p.7-34. In, Problemy paleogeografii Asii. Nauka, Moskva.

___1987. Neobichnye bezamkovye brakhiopody iz nizhnego Kembriya Mongolii. Paleontologicheskii Zhurnal 2:62-68.

___1988. Obolellidy (brakhiopody) s zamkovym sokhleneniem stvorok iz nizhnego Kembriya Zabaikaliya. ibid. :34-42.

___1993. Novyi rod paterinid (brakhiopody) iz nizhnego Kembriya zapadnoi Mongolii. ibid.1:115-118.

___1994. Novye sredne-verkhnekembriiskie akrotretidy (brakhiopody) severa Sibirskoi platformy. ibid., 4:38-54.

___& A.Yu. ZHURAVLEV. 1994. K probleme mineralizatsii skeleta (na primere brakhiopod). Doklady Akademii Nauk 337(2):231-234.

VORONOVA, L.G., N.A. DROZDOVA, E.A. ESAKOVA, E.A. ZHEGALLO, A.Yu. ZHURAVLEV, A.Yu. ROZANOV, T.A. SAYUTINA & G.T. USHATINSKAYA. 1987. Iskopaemye nizhnego Kembriya gor Makkenzi (Kanada). Trudy Paleontologicheskogo Instituta 224:1-88.

WILLIAMS, A. & G.B. CURRY. 1991. The microarchitecture of some acrotretide brachiopods, p.133-145. In, D.I. MacKinnon, D.E. Lee & J.D. Campbell (eds.), Brachiopods through time. Balkema, Rotterdam.

YADRENKINA, A.G. 1974. Biogeograficheskaya kharakteristika pozdnekembriiskikh i ordovikskikh brakhiopod severozapada Sibirskoi platformy. Trudy SNIIGGiMS 173:74-78.

ZEZINA O.N. 1976. Ekologiya i rasprostranenie sovremennikh brakhiopod. Nauka, Moskva, 136p.

ZELL, M.G. & A.J. ROWELL. 1988. Brachiopods of the Holm Dal Formation (late Middle Cambrian), central North Greenland. In, J.S. Peel (ed.), Stratigraphy and Palaeontology of the Holm Dal Formation (late Middle Cambrian), central North Greenland. Meddelelser Grønland, Geoscience 20:119-144.

NO SECOND CHANCES? NEW PERSPECTIVES ON BIOTIC INTERACTIONS IN POST-PALEOZOIC BRACHIOPOD HISTORY

JOSEPH A. WALSH

Department of Geophysical Sciences, University of Chicago
5734 South Ellis Avenue, Chicago, Illinois 60637 USA

ABSTRACT- Mesozoic brachiopods have long afforded the classic case study in biotic replacements. Brachiopod workers have generally addressed the issue in 'either/or' terms, either clades are 'ships that pass in the night', or direct competition between two clades is a major determinant of taxonomic diversity and ecological dominance. Here I outline an alternative perspective hypothesizing pre-emptive exclusion as an indirect effect of the community structure of Mesozoic benthic communities. The effects of interactions among incumbent clades suppresses the radiation of other clades, such as articulate brachiopods, without direct competitive displacement ever taking place. Brachiopods appear to be generally excluded from the present-day tropics. For rhynchonellides, this pattern originated as early as the Jurassic, and did not change subsequently. In the Jurassic there is also little variability in the higher taxonomic composition of assemblages in low paleolatitudes. While this pattern and its ecological correlates are highly consistent with a provisional model of pre-emptive exclusion, diversity data alone are not sufficient to eliminate other possible mechanisms. I outline preliminary criteria drawing on several types of data to distinguish pre-emptive exclusion from other processes that could generate the observed latitudinal pattern.

INTRODUCTION

Mesozoic brachiopods reiterate little of the clade's Paleozoic history. Following the mass extinction at the end of the Permian, they do not recapture earlier levels of taxonomic diversity, they no longer dominate benthic environments, and they show a remarkable sameness in morphology. Progress towards understanding why the brachiopods did not recover is of interest to ongoing debates about the role of biotic interactions in macroevolution. In the past, brachiopod workers have tended to couch their understanding of the situation in, that either clades are 'ships that pass in the night', with no effect on the taxonomic diversity of other clades (Gould & Calloway 1980), or direct competition is a major determinant of taxonomic diversity and ecological dominance. This limited range of alternatives may explain, in part, why post-Paleozoic brachiopods afford the classic case study in biotic interactions, yet the most rigorous research has been undertaken using data for other clades, e.g. Rosenzweig & McCord (1991) for turtles; Lidgard et al. (1993) for cyclostome vs. cheilostome bryozoans; Janis & Wilhelm (1993) for Tertiary pursuit predators; Roy (in press), for stromboidean gastropods.

One perspective on this problem comes from ecological patterns suggesting that present-day brachiopod distributions are somehow restricted relative to other clades or prior stratigraphic intervals. Rudwick (1970, Fig. 92) demonstrated that there are more Recent brachiopod species in temperate than tropical latitudes in both the northern and southern hemispheres, and my own analyses, incorporating data published since 1970, support this conclusion (Walsh, 1994; see Figure 1). The observed distribution is remarkable in departing both from the present-day norm (since most Recent clades show

maximum diversity in the tropics), and from the inferred Paleozoic pattern (a diversity gradient peaking at low latitudes, as was the case for brachiopods in the Carboniferous and Permian; Stehli et al. 1969; Humphreville & Bambach 1979; Kelley et al. 1990). The latitudinal pattern has not previously been reconstructed for any stratigraphic interval after the Permian.

This pattern is consistent with a large number of possible mechanisms, including overall 'exclusion' of brachiopods from the tropics through active maintenance of the latitudinal pattern in ecological time by direct competition with some other clade for some limiting resource. This mechanism might also explain why the brachiopod taxa that do occur in tropical latitudes are environmentally restricted (Jackson et al., 1971; Logan, 1977, 1981; Thayer, 1981; Asgaard & Stentoft, 1984; Thayer & Allmon, 1991). Previous workers have been explicit about who the interactors with brachiopods are thought to be (usually bivalves; Steele-Petrovic, 1979; Thayer, 1981,1985, 1986; Rhodes & Thayer, 1991; but sometimes predators; Donovan & Gale, 1990). Interference mechanisms, such as fouling by mussels, are observed in the present day (Thayer, 1985). The almost exclusive focus on bivalves as 'ecologically equivalent' to brachiopods (e.g. Thayer, 1981,1985) seems to invoke direct competition for a single discrete resource, usually substratum, in consequence of which tropical brachiopods fail to establish propagules outside of a few oligotrophic refuges.

Even if this is an accurate description of ecological interactions in the present day, it tells very little about the origins of the latitudinal pattern or its persistence in the geological past (Figure 1). Aside from those who (following Gould & Calloway, 1980) dismiss biotic interactions altogether, pre-vious workers have not systematically enumerated and eliminated competing hypotheses. The large number of mechanisms that are equally consistent with environmental restriction and low diversity in the tropics would include (but surely not be limited to) the following: (1) high-energy bivalves 'drove less well-adapted, low energy organisms into ecological refugia or 'safe places" (Rhodes & Thompson, 1993: p.323; this is 'displacive competition' of Hallam, 1987); or (2) Paleozoic representatives of the extant clades of articulate brachiopods were already restricted to 'refugial' environments during the Paleozoic, with other environments dominated by substrate-tolerant morphologies (e.g., free-lying spiriferides, quasi-infaunal strophomenides) that became extinct before the Mesozoic (Ager, 1965; Rudwick, 1970); or (3) low-paleolatitude brachiopods differentially survived biotic crises (such as the end-Permian, end-Triassic, and/or end-Cretaceous) in refugial environments ('safe places' of Vermeij, 1987); or (4) the radiation of predators, especially after the Jurassic, devastated brachiopod populations in all low-paleolatitude environments except safe places (Asgaard & Stentoft, 1984; Vermeij, 1987, 1990; Donovan and Gale 1990); or (5) low-paleolatitude brachiopods experienced pre-emptive exclusion (sensu Jablonski & Bottjer, 1990) by incumbent clades radiating in

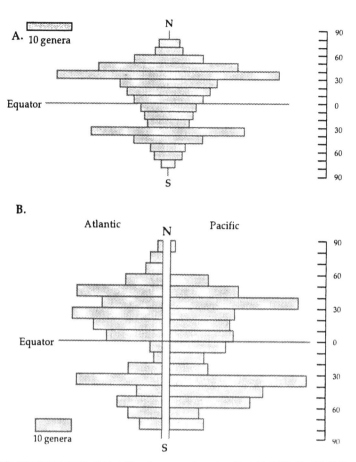

Figure 1. A) Recent latitudinal distribution of articulate brachiopod genera, redrawn from Rudwick (1970, Fig. 92). B) Recent latitudinal distribution of articulate genera, in the Atlantic and Pacific, incorporating post-1970 data.

non-refugial post-Paleozoic environments, without any sort of direct competitive displacement; or (6) a stochastic or directed trend towards miniaturization (Laurin & García-Joral, 1990), or origination or extinction bias in low-paleolatitude environments, has favored brachiopod lineages with small body sizes (such as those found in the tropics today); other parameters highly correlated with body size (e.g., fecundity, population size, geographic range; LaBarbera, 1986; Hanken & Wake, 1993; Taylor & Gotelli 1994) lead to the observed environmental distribution in the absence of biotic interactions; or (7) the Recent pattern originated through wholly abiotic mechanisms such as Tertiary changes in physical oceanography, rates of sedimentation and nutrient input, or primary productivity (Logan, 1977, 1979; Zezina ,1979; Bambach, 1993).

Not all of these alternatives specifically invoke competition, or even biotic mechanisms in general, and in many cases they need not be mutually exclusive. Furthermore, even the clade-specific direct competition of #1 (above) can unfold in at least five distinct ways (also not mutually exclusive) producing the same pattern in the fossil record (Sepkoski, in press).

ALTERNATIVE MODELS FOR BIOTIC INTERACTIONS

Such an array of hypotheses can best be addressed by relating diversity patterns to current macroevolutionary and ecological models in order to generate predictions about mechanisms testable with other data from the fossil record. One promising avenue of current research in ecology stresses indirect effects that arise through linked chains of direct species interactions or changes caused by interactions among many species (Wootton,1994a). Although most theoretical and empirical studies to date have focused on pairwise interactions between species, the total number of interactions among species in living communities is very high and not limited to direct physical confrontations such as interference or exploitative competition (Connell, 1983; Schoener, 1989; Fair-weather, 1990; Wootton, 1993,1994a,1994b; Heske et al., 1994; Tanner et al., 1994; Menge, 1995). Thus, patterns of 'exclusion' are likely to arise from the structure of entire communities rather than the presence or absence of any single clade.

In addition to presenting reconstructions of the latitudinal diversity gradient for Jurassic rhynchonellides, this paper summarizes macroevolutionary correlates of the pattern, and evaluates their consistency with patterns predicted to arise from pre-emptive exclusion of brachiopods by other clades incumbent in the structure of low-paleolatitude Mesozoic benthic communities. I view pre-emptive exclusion as a type of indirect effect that is a property of multi-species assemblages, with interactions among incumbent clades (rather than the presence or absence of any single clade) suppressing the radiation of non-incumbents. As currently understood, negative indirect effects such as the exploitative competition for substrate described for Recent tropical reef brachiopods would favor minimization of interaction by reduced coexistence with other clades (Wootton,1994a). Furthermore, simulation models predict long-term stabilization of assemblages (ibid.), such that the effects of pre-emption could persist on geolo-gical time scales. Patterns of exclusion consistent with this process would include the following:

Latitudinal diversity. Since the phenomenon of interest is low brachiopod diversity in the present-day tropics, pre-emption can be considered only if the latitudinal pattern is geologically old and originates at a time when the incumbency of brachiopods is overturned. This is most likely to occur following an episode of mass extinction (Jablonski, 1986; Rosenzweig & McCord, 1991), and thus a pattern that is congruent with survivorship biases following a mass extinction is consistent with pre-emptive exclusion rather than competitive displacement. Finally, a long-lived pattern cannot be due solely to the effects of ocean circulation or climate, which are both more highly variable on geological time scales.

Rates of morphological origination. There are significant geographical and environmental patterns in the historical record of origination of evolutionary innovations (Bottjer & Jablonski, 1988; Jablonski & Bottjer, 1990; Jablonski, 1993). The failure to generate novel morphologies may indicate the failure of a pre-empted clade to recapture a broad environmental distribution, either because it is unable to invade areas

and environments tending to produce morphological novelties, or because novelties are produced but perish without becoming established. The hypothesis that brachiopod mor-phologies are highly convergent within environments (and therefore low morphological variability in regional assemblages correlates with occurrence in a limited suite of environments), has long standing in the brachiopod literature. This is usually expressed in terms of homeomorphy, with contemporaneous, non-contemporaneous, sympatric and allopatric homeomorphs all having been documented. Recent rhynchonellide and tere-bratulide homeomorphs occur in the deep-sea (Muir-Wood, 1960; Cooper, 1972), and similar trends have been documented in the Jurassic and inferred to be driven by environmental pressures (Ager, 1965; Delance & Rollet, 1967 and references therein; Laurin, 1979; Boullier ,1981; Manceñido, 1983).

Rates of taxonomic origination. Low rates of taxonomic diversification (relative to the 'hyperexponential bursts' that typically follow mass extinctions; Miller & Sepkoski, 1988) that persist over geologically long intervals are consistent with differential survivorship in resource-limited environmental refugia. In contrast, most survivors of biotic crises would be geographically wide-ranging (Jablonski,1986; Jablonski & Raup, 1995), with survivors tending to diversify as their ranges subsequently fragment.

All of these predictions are based on extrapolations from the ecological correlates of diversity data and are appropriate only to a highly preliminary study such as this. More restrictive tests must be based on a data base of brachiopod occur-rences within assemblages across paleoenvironmental tran-sects (now in progress), as well as knowledge of geographical and environmental patterns in the survival of biotic crises. An outline of this approach is offered below.

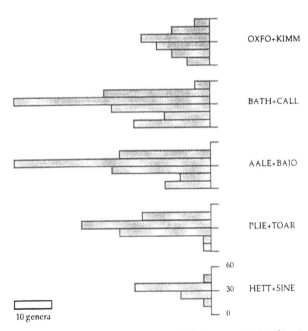

Figure 2. Distribution of rhynchonellide genera in low and intermediate paleolatitudes, grouped into 10° bands. Northern and southern hemi-sphere occurrences are combined and successive stratigraphic stages have been paired (except the Tithonian, which was excluded due to undersampling).

JURASSIC LATITUDINAL DIVERSITY

Methods. I compiled locality data (latitudes and longitudes on Recent coordinates) for most occurrences of all genera of Jurassic rhynchonellides reported in the paleontological literature (data are available from the author upon request). I excluded genera not reliably assigned to discrete localities or stratigraphic stages. To reconstruct the latitudinal diversity gradient in the Jurassic, these localities were assigned to lithospheric plates and rotated to inferred paleolatitudinal coordinates for each stratigraphic interval using computer programs developed by the Paleogeographic Atlas Project at the University of Chicago (Figure 2).

Because sample sizes for Jurassic rhynchonellides tend to be small, especially in low paleolatitudes, I have combined data for pairs of successive stratigraphic stages (for example, the Hettangian and Sinemurian are plotted together). To compare rates of diversification, I have divided occurrences for paired stages into low (0-20˚N and S) and intermediate (21˚-40˚N and S) paleolatitudes. I counted the number of genera whose first appearance in the fossil record falls in a given band of paleolatitude. I report both raw numbers of originations and, \because these appear to be proportional to diversity, rates of origination per taxon per million years. To appraise the distinctness of the evolving low paleolatitude fauna in morphological terms, I ordinated 10˚ paleolatitudinal bands for pairs of successive stages by their higher taxonomic composition using taxonomic subfamilies as proxies for novel morphologies (following the taxonomy of Shi & Grant, 1993). I used subfamilies because only one family of rhynchonellides originated in low latitudes during the Jurassic (the Callovian Septirhynchiidae Muir-Wood & Cooper 1951, considered a subfamily of the Rhynchonellidae by Manceñido & Walley 1979).

To evaluate the effect of possible differences in sampling intensity between latitudinal bands, I collected data on the number of genera of articulate brachiopods reported from 164 localities at low paleolatitudes (Saudi Arabia, Tunisia, Somalia, Kenya) and 161 localities in middle paleolatitudes (Paris Basin, French Jura, Causses, Gard). All localities were of Middle Jurassic age. I plotted mean numbers of genera per locality per individual stratigraphic stage in each of the monographs censused (Figure 3: see caption for references).

Results. A reconstruction of the latitudinal diversity gradient of Jurassic shows that there are fewer genera in low than middle paleolatitudes in every stage of the Jurassic (Figure 2). This pattern is present from the Hettangian/Sinemurian onward and is possibly congruent with a latitudinal bias in survivorship of the end-Triassic extinction. The overall pattern did not change in the Cretaceous (Walsh, 1994). Barring an as yet undocumented Tertiary radiation and extinction of low paleolatitude rhynchonellides, the latitudinal distribution observed since the earliest intervals of the Jurassic has endured to the present day. I suggest that latitudinal differences cannot be attributed to sampling intensity alone (Figure 3). Low-paleolatitude articulates have been as intensively collected as taxa from European localities, yet low-paleolatitude faunas do not exceed mid-paleolatitude faunas in mean numbers of genera per locality.

Total low-paleolatitude origination appears damped in comparison to origination at intermediate paleolatitudes (Figure 4A). Per-taxon rates of origination (Figure 4B) are comparable for low and middle paleolatitudes from the Pliensbachian through Bajocian, although not from the Bathonian onwards. Because meaningful error bars cannot be calculated for these rates for such small samples, the significance of the per-taxon rates is difficult to evaluate.

All well-sampled low-paleolatitude subfamily assemblages fall very close together (Figure 5). No brachiopods are reported from the 0-10˚ band for the Hettangian/Sinemurian, so this interval was not included in the analysis. Two 11˚-20˚ bands (Hettangian/Sinemurian and Pliensbachian/Toarcian) fall well outside the cluster of all other low paleolatitude assemblages due both to small samples (2 genera each), and to the absence of genera belonging to the Tetrarhynchiinae, the subfamily that dominates low paleolatitude faunas and largely accounts for the first principal component. The higher taxonomic composition of low paleolatitudinal faunas from the Pliensbachian through the Kimmeridgian shows virtually no change, whereas intermediate paleolatitude faunas are more highly variable in their subfamily composition. This is due to the waxing and waning of ten subfamilies that never occur in low paleolatitudes. Of eight subfamilies that do occur in low paleolatitudes, only two (the Tetrarhynchiinae and Cyclothyridinae) occur with more than a single genus in a given stratigraphic interval.

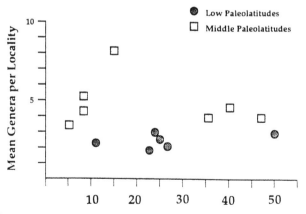

Figure 3. Sampling intensity and mean numbers of articulate genera per locality in low (North and East Africa) and middle (Paris Basin, Causses, and French Jura) paleolatitudes per stratigraphic stage of the Middle Jurassic. Data from Weir 1929, 1930, 1938; Dubar 1967; Alméras 1970; Alméras and Peybernès 1979; Walter and Alméras 1977, 1981; Cooper 1989; Garcia 1993.

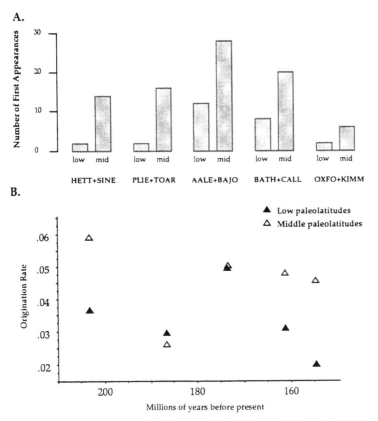

Figure 4. A) Number of stratigraphic first appearances of articulate genera in paired stages of the Jurassic, divided by occurrences in low (1-20°) and intermediate (21-40°) paleolatitudes. B) Rates of origination per taxon per million years for the same intervals.

DISCUSSION

Rhynchonellides in the Jurassic and Cretaceous were not diverse at low paleolatitudes, despite an appreciable global diversification in the Bajocian through Callovian. This pattern cannot be attributed to the effects of sampling alone. With one exception (discussed above), novel morphologies, as reflected in subfamilies, neither first appeared in low paleolatitudes, nor spread there in large numbers from intermediate or high paleolatitudes. This invariance of higher taxonomic structure in low-paleolatitude faunas suggests that rhynchonellides were environmentally restricted and pre-empted from benthic communities by other incumbents.

Although these conditions show little change during or after the Jurassic, it is not known whether they originate in the Jurassic, Triassic, or even Permian, and the relative importance of mass extinctions to changes in incumbency can therefore not be evaluated with these data. Furthermore, higher taxa cannot be regarded as true proxies for morphological diversity (or as units of novelty), since the equivalence of taxonomic ranks has not been demonstrated and the morphometric variance within a single subfamily could exceed that among other subfamilies. Finally, it must be acknowledged both that genera are not truly the units of incumbency (species are), and that 'low paleolatitudes' are not the ecological units from which species are pre-empted (specific environments are).

Ultimately, diversity data and biogeographical and morphological patterns may be sufficiently discordant that diversity data alone are not sufficient to test hypotheses of community-level phenomena (Foote, 1993; Lidgard and Roy 1994). The results presented here are not consistent with any single model of biotic interaction, but can be used to narrow alternatives and generate testable predictions about ecological mechanisms. Consistency with the pre-emptive model outlined above can be more rigorously demonstrated if diversity, environmental and assemblage data support the following predictions:

(1). The latitudinal diversity patterns (such as those described for the Recent and the Jurassic) will be highly congruent with patterns of survivorship from mass extinctions. Alternatives to this include the demonstration that Permian rhynchonellides (and/or terebratulides) also showed this pattern, or that the pattern arises slowly by a switch-over from a 'normal' to a 'reversed' gradient in the Triassic (possibly consistent with competitive displacement).

(2). Paleoenvironmental transitions on an onshore/offshore transect will be abrupt and congruent with survivorship, as above, at least in low paleolatitudes. Alternatives (such as the occurrence of Mesozoic patterns in the Permian, or slow transitions) may be consistent with competitive displacement rather than pre-emption.

(3). Morphological diversity (as measured in a multivariate empirical morphospace) will be limited in low relative to middle paleolatitudes, and congruent with survivorship. Environments and morphologies will be highly correlated.

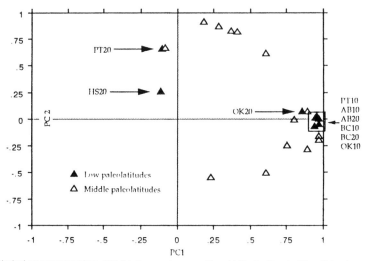

Figure 5. Results of principal component analysis of higher taxonomic composition of latitudinal bands. Closed triangles are low paleolatitude assemblages, open triangles intermediate paleolatitudes. Stages are paired and designated by initials of stage names and paleolatitudinal interval (thus, `HS20' indicates Hettangian/Sinemurian 11-20°, `PT10' is Pliensbachian/Toarcian 1-10°, etc.).

(4). The taxonomic composition of low paleolatitude brachiopod assemblages will be highly consistent between coeval localities. The transition to this condition will be correlated with an interval (or intervals) of biotic crisis. The composition of assemblages will be facies-specific. Gradual changes, or the presence of such patterns in the Permian, are alternatives.

CONCLUSIONS

(1). Mesozoic brachiopods do not recapture the high levels of ecological dominance and taxonomic diversity that characterized the group in the Paleozoic. Hypotheses advanced to explain the replacement of brachiopods have thus far been limited in scope and inadequately tested.

(2) The latitudinal diversity gradient of Recent brachiopods suggests 'exclusion' from the tropics, and this appears to be correlated with extreme environmental restriction. A similar latitudinal pattern existed for rhynchonellides in the Jurassic and Cretaceous, and similar environmental restrictions can tentatively be inferred from other observations. However, the congruence of Recent and ancient diversity patterns is not consistent with any single causal hypothesis, and does not favor biotic over abiotic mechanisms.

(3) The ecological correlates of such diversity patterns as rates of morphological origination can be used to generate predictions about changes in community structure. However, rigorous tests of these predictions must be based on observed patterns of survivorship from mass extinctions, paleoenvironmental transitions, and assemblage composition.

Acknowledgments

My thanks to the Paleogeographic Atlas Project, University of Chicago (and especially M.L. Hulver) for paleolatitudinal reconstructions. E. LeClair and M. Foote provided helpful discussion in the development of these ideas. The meticulous readings of M.M.R. Best, D. Jablonski, R. Lupia and K. Roy were instrumental in improving the manuscript. This research was supported in part by National Science Foundation Grant EAR93-17114 to D. Jablonski.

REFERENCES

AGER, D.V. 1965. The adaptation of Mesozoic brachiopods to different environments. Palaeogeography, Palaeoclimatology, Palaeoecology 1:143-172.

ALMERAS, Y.1970. Les Terebratulidae du Dogger dans le Mâconnais, le Mont d'Or lyonnais et le Jura méridional. Documents Laboratoires Géologie, Lyon 39:1-690.

___ & B. PEYBERNES. 1979. Les brachiopodes du Dogger des Pyrénées navarro-languedociennes. ibid. 76: 22-133.

ASGAARD, U. & N. STENTOFT. 1984. Recent micromorph brachiopods from Barbados: paleoecological and evolutionary implications. Géobios Memoire Spécial 8:29-38.

BAMBACH, R.K. 1993. Seafood through time: changes in biomass, energetics and productivity in the marine ecosystem. Paleobiology 19:372-397.

BOTTJER, D.J. & D. JABLONSKI. 1988. Paleoenvironmental patterns in the evolution of post-Paleozoic benthic marine invertebrates. Palaios 3:540-560.

CONNELL, J.H. 1983. Interpreting the results of field experiments: effects of indirect interactions. Oikos 41: 290-291.

COOPER, G.A.1972. Homeomorphy in Recent deep-sea brachiopods. Smithsonian Contributions Paleobiology 11: 1-25.

___1989. Jurassic brachiopods of Saudi Arabia. ibid. 65:1-213.

DELANCE, J.H. & A. ROLLET. 1967. Homéomorphie isochrone entre térébratulidés et zeilléridés (brachiopodes). Bulletin Société d'Histoire Naturelle Doubs 69(3):3-7.

DONOVAN, S.K. & A.S. GALE. 1990. Predatory asteroids and the decline of the articulate brachiopods. Lethaia 23:77-86.

DUBAR, G. 1967. Brachiopodes jurassiques du Sahara tunisien. Annales Paléontologie (Invertébrés) 53:33-101.

FAIRWEATHER, P.G. 1990. Is predation capable of interacting with other community processes on rocky reefs? Australian Journal Ecology 15: 453-464.

FOOTE, M. 1993. Discordance and concordance between morphological and taxonomic diversity. Paleobiology 19:185-204.

GARCIA, J.P. 1993. Les variations du niveau marin sur le Bassin de Paris au Bathonien-Callovien: impacts sur les communautés benthiques et sur l'évolution des ornithellidés

(Terebratulidina). Mémoires Géologiques Université Dijon 17:1-307.

GOULD, S.J. & C.B. CALLOWAY. 1980. Clams and brachiopods - ships that pass in the night. Paleobiology 6: 383-396.

HALLAM, A. 1987. Radiations and extinctions in relation to environmental change in the marine Lower Jurassic of northwest Europe. Paleobiology 14:364-369.

HANKEN, J. & D.B. WAKE. 1993. Miniaturization of body size: organismal consequences and evolutionary significance. Annual Review Ecology Systematics 24:501-519.

HESKE, E.J., J.H. BROWN & S. MISTRY. 1994. Long-term experimental study of a Chihuahuan Desert rodent community: 13 years of competition. Ecology 75:438-445.

HUMPHREVILLE, R. & R.K. BAMBACH. 1979. Influence of geography, climate and ocean circulation on the pattern of generic diversity of brachiopods in the Permian. Geological Society America Abstracts Programs 11(7):447.

JABLONSKI, D. 1986. Background and mass extinctions: the alternation of macroevolutionary regimes. Science 231:129-133.

___1993. The tropics as a source of evolutionary novelty through geological time. Nature 364:142-144.

___& D.J. BOTTJER. 1990. The ecology of evolutionary innovation: the fossil record, p.253-288. In, M.H. Nitecki (ed.), Evolutionary innovations. University Chicago Press, Chicago.

___& D.M. RAUP. 1995. Selectivity of end-Cretaceous marine bivalve extinctions. Science 268:389-391.

JACKSON, J.B.C., T.F. GOREAU & W.D. HARTMAN. 1971. Recent brachiopod-coralline sponge communities and their paleoenvironmental significance. Science 173:623-625.

JANIS, C.M. & P.B. WILHELM. 1993. Were there mammalian pursuit predators in the Tertiary? Dances with wolf avatars. Journal Mammalian Evolution 1:103-125.

KELLEY, P.H., A. RAYMOND & C.B. LUTKEN. 1990. Carboniferous brachiopod migration and latitudinal diversity: a new palaeoclimatic model, p.325-332. In, W.S. McKerrow & C.R. Scotese (eds.), Palaeozoic palaeogeography and biogeography. Geological Society Memoir 12, London.

LABARBERA, M. 1986. The evolution and ecology of body size, p. 69-98. In, D.M. Raup & D. Jablonski (eds.), Patterns and processes in the history of life. Springer Verlag, Berlin.

LAURIN, B. 1979. Convergence morphologique entre les brachiopodes sulqués: Septocrurella sanctae-clarae (Roemer) — Rhynchonellacea, Erymnariidae — et Nucelata nucleata (Schlotheim) — Terebratulacea, Pygopidae — dans l'Oxfordien de la région d'Aiglun (Alpes-maritimes). Géologie Méditerranéenne 6:417-422.

___& F. GARCIA-JORAL. 1990. Miniaturization and heterochrony in Homoeorhynchia meridionalis and H. cynocephala (Brachiopoda, Rhynchonellidae) from the Jurassic of the Iberian Range, Spain. Paleobiology 16:62-76.

LIDGARD, S., F.K. MCKINNEY & P.D. TAYLOR. 1993. Competition, clade replacement, and a history of cyclostome and cheilostome bryozoan diversity. Paleobiology 19:352-371.

___& K. ROY. 1994. Biotic replacements in the fossil record: new and neglected types of evidence. Geological Society America Abstracts Programs 26(7):52.

LOGAN, A. 1977. Reef-dwelling articulate brachiopods from Grand Cayman, B.W.I. Proceedings Third International Coral Reef Symposium, Miami, 1 (Biology):88-93.

___1979. The Recent Brachiopoda of the Mediterranean Sea. Bulletin Institut Océanographique, Monaco 72(1434):1-112.

___1981. Sessile invertebrate coelobite communities from shallow reef tunnels, Grand Cayman, B.W.I. Fourth Internaional Coral Reef Symposium, Proceedings 2:735-744.

MANCENIDO, M.O. 1983. A new terebratulid genus from Western Argentina and its homeomorphs (Brachiopoda, Early Jurassic). Ameghiniana 20:347-365.

___& C.D. WALLEY. 1979. Functional morphology and ontogenetic variation in the Callovian brachiopod Septirhynchia from Tunisia. Palaeontology 22:317-337.

MENGE, B.A. 1995. Indirect effects in marine rocky intertidal interaction webs: patterns and importance. Ecological Monographs 65:21-74.

MILLER, A.I. & J.J. SEPKOSKI. 1988. Modeling bivalve diversification: the effect of interaction on a macroevolutionary system. Paleobiology 14:364-369.

MUIR-WOOD, H.M. 1960. Homeomorphy in Recent Brachiopoda: Abyssothyris and Neorhynchia. Annals Magazine Natural History 13 (3):521-529.

RHODES, M.C.& C.W. THAYER. 1991. Mass extinctions: ecological selectivity and primary production. Geology 19:877-880.

___& R.J. THOMPSON. 1993. Comparative physiology of suspension-feeding in living brachiopods and bivalves: evolutionary implications. Paleobiology 19:322-334.

ROSENZWEIG, M.L. & R.D. McCORD. 1991. Incumbent replacement: evidence for long-term evolutionary progress. Paleobiology 17:202-213.

ROY, K. 1994. Effects of the Mesozoic marine revolution on the taxonomic, morphologic, and biogeographic evolution of a group: aporrhaid gastropods during the Mesozoic. Paleobiology 20:274-296.

___(in press). The roles of mass extinction and biotic interaction in large-scale replacements: a reexamination using the fossil record of stromboidean gastropods. Paleobiology.

RUDWICK, M.J.S. 1970. Living and fossil brachiopods. Hutchinson, London, 199p.

SCHOENER, T.W. 1989. Food webs from the small to the large. Ecology 70:1559-1589.

SEPKOSKI, J.J. (in press). Competition in macroevolution: the double wedge revisited. In, D.H. Erwin, D. Jablonski & J.H. Lipps (eds.), Evolutionary paleobiology: essays in honor of James W. Valentine.

SHI, X.-Y. & R.E. GRANT. 1993. Jurassic rhynchonellids: internal structures and taxonomic revisions. Smithsonian Contributions Paleobiology 73:1-190.

STEELE-PETROVIC, H.M. 1979. The physiological differences between articulate brachiopods and filter-feeding bivalves as a factor in the evolution of marine level-bottom communities. Palaeontology 22:101-134.

STEHLI, F.G., R.G. DOUGLAS & N.D. NEWELL. 1969. Generation and maintenance of gradients in taxonomic diversity. Science 164:947-949.

TANNER, J.E., T.P. HUGHES & J.H. CONNELL. 1994. Species coexistence, keystone species, and succession: a sensitivity analysis. Ecology 75:2204-2219.

TAYLOR, C.M. & N.J. GOTELLI. 1994. The macroecology of body Cyprinella - correlates of phylogeny, body-size, and geographical range. American Naturalist 144:549-569.

THAYER, C.W. 1981. Ecology of living brachiopods, p.110-126. In, T.W. Broadhead (ed.), Lophophorates: notes for a short course. University Tennessee Studies Geological Sciences 5.

___1985. Brachiopods versus mussels: competition, predation and palatability. Science 228:1527-1528.

___1986. Are brachiopods better than bivalves? Mechanisms of turbidity tolerance in articulates and their interaction with feeding. Paleobiology 12:161-174.

___& R.A. ALLMON. 1990. Unpalatable thecideid brachiopods from Palau: ecological and evolutionary implications, p. 253-260. In, D.I. MacKinnon, D.E. Lee, & J.D. Campbell (eds.), Brachiopods through time. Balkema, Rotterdam.

VERMEIJ, G.J. 1987. Evolution and escalation: an ecological history of life. Princeton University Press, Princeton.

WALSH, J.A. 1994. Latitudinal diversity gradients of post-Paleozoic articulate brachiopods: Recent patterns and their

Jurassic origins. Geological Society America Abstracts Programs 26(7):171.

WALTER, B. & Y. ALMERAS. 1977. Bryozoaires et brachiopodes des 'calcaires bajociens à bryozoaires' du Gard (France): paléontologie et paléoécologie. Géobios 10:907-955.

___&___1981. Bryozoaires et brachiopodes des 'calcaires bajociens à bryozoaires' des Causses (France) et leur paléoécologie. Géobios 14:361-387.

WEIR, J. 1929. Jurassic fossils from Jubaland, East Africa, collected by V.G. Glenday, and the Jurassic geology of Somaliland. Monographs Geological Department Hunterian Museum 3:1-63.

___1930. Mesozoic Brachiopoda and Mollusca from Mombasa. ibid. 4: 77-102.

___1938. The Jurassic faunas of Kenya with descriptions of some Brachiopoda and Mollusca. ibid. 5: 17-60.

WOOTTON, J.T. 1993. Indirect effects and habitat use in an intertidal community: interaction chains and interaction modifications. American Naturalist 141:71-89.

___1994a. The nature and consequences of indirect effects in ecological communities. Annual Review of Ecology Systematics 25:443-466.

___1994b. Predicting direct and indirect effects: an integrated approach using experiments and path analysis. Ecology 75:151-165.

ZEZINA, O.N. 1979. (The formation of the present-day brachiopod fauna on the shelves and slopes of the world ocean.) Byulletin Moskovskogo Obshchestva Ispytatelei Prirody, Otdel Biologicheskii 84:52-59.

DEGRADED INTRACRYSTALLINE PROTEINS AND AMINO ACIDS FROM FOSSIL BRACHIOPODS AND CONSIDERATIONS FOR AMINO ACID TAXONOMY

DEREK WALTON

Department of Earth Sciences, University of Derby,Kedleston Road, Derby DE22 1GB, UK

ABSTRACT- Four species of Recent and Plio-Pleistocene fossil brachiopods were collected from up to 20 horizons spanning the last 2.6Ma of sediment deposition in the South Wanganui Basin, New Zealand, and assayed for the preservation of intracrystalline proteins and/or amino acids. Results indicate that the proteins present in the shells of living and Recent brachiopods undergo rapid degradation through the decomposition of the peptide bond. Up to 80%of the constituent amino acids from the proteins are present in the free state by 0.2Ma, a rate of degradation higher than was originally expected for intracrystalline proteins. Quantitative analysis of concentrations of amino acids present within shells of fossil brachiopods indicate a range of reaction rates for degradation of individual amino acids after decomposition of proteins. Degradation of these amino acids may lead to total loss of compounds, to generation of non-standard amino acids, or to diagenetically-produced proteinaceous amino acids. Such degradation must be recognised for taxonomic considerations to be valid. Principal components analysis demonstrates that, within a horizon, taxonomic discrimination may be completed to at least subfamilial level. A greater understanding of degradative processes should allow a greater level of sensitivity for discrimination.

INTRODUCTION

Recent brachiopods, in common with other phyla, contain a complex mixture of intracrystalline biomolecules entombed within the inorganic phase. These include proteins (Curry et al., 1991; Cusack et al., 1992; Walton et al., 1993), lipids (Clegg, 1993) and carbohydrates (Clegg, 1993). Nucleic acids have not yet been detected. Fossil brachiopods also contain a range of intracrystalline molecules (Walton,1992), which represent the variously degraded remains of original organic material trapped during biomineralisation. As the fossil molecules are intracrystalline, they have not been degraded or contaminated by bacteria, and all degradation is due to the prevailing physico-chemical conditions. Degradation products remain trapped within the inorganic phase until demineralisation, allowing sampling of both preserved original biomolecules and the degradation products of less stable components.

Although the potential of proteins/amino acids for establishing molecular phylogenies was recognised in the early stages of their study, it was not clear to what extent compositional differences were due to genuine phylogenetic differentiation or were merely reflecting variable degradation of the original bio-molecules. Previous work has identified a number of possible degradative pathways, including hydrolysis of the protein and decarboxylation/deamination of amino acids.

Natural hydrolysis (the decomposition of peptide bonds by the action of time, heat or pressure) results in the release of amino acids from the original protein structure. Abelson (1955, 1956) noted that between 1% and 5% of all peptide bonds were broken in a period of 90ka. Van Kleef et al. (1975) demonstrated that the decay of proteins by non-enzymic processes took place in a relatively short time, and that the first stage in the breakdown was the natural hydrolysis of peptide bonds by the addition of of water. Goodfriend et al. (1992) demonstrated rapid hydrolysis of aspartic acid from the calcified tissues of corals over 350ka, and attributed this to the preferential cleavage of the peptide bond adjoining the molecule.

Once released from the protein, the thermal stability of the amino acid is clearly important. Abelson (1954) noted that the thermally unstable amino acids were either absent or present in very low concentrations in fossils. The thermal stability of alanine (a stable amino acid) in dilute aqueous solution was determined at elevated temperatures and extrapolated to normal temperatures and pressures. 10Ga would be required at 20°C in order for 63% of the molecule to degrade (Abelson, 1954), hence little or no thermal destruction of alanine should be expected at normal temperatures. Thermal reaction kinetics for the relatively unstable amino acids serine, threonine and phenylalanine show that at 20°C thermal decomposition of these compounds would take 0.1Ma, 1Ma and 10Ma respectively (Vallentyne, 1964), reflecting differences in their individual degradative reaction rates. A mixed solution of these amino acids did not alter the relative stability order of the amino acids as compared with pure solutions of individual amino acids, although the rate of decomposition was higher in a mixed solution than in pure solutions, indicating some inter-reaction of the samples (Vallentyne, 1964).

Molluscan shell powders (containing both inter- and intra-crystalline molecules) heated in the presence of water indicate that threonine, serine, cysteine, histidine and arginine are preferentially destroyed (Hare & Mitterer, 1969; Totten et al., 1972). In Miocene samples of the same species, the same amino acids had been lost. When compared to the analysis of pure compounds, the amino acids associated with the inorganic mineral phase appeared to be less stable.

Some amino acids may degrade into other amino acids by simple decarboxylation or deamination reactions (Vallentyne, 1964). The dehydration of the hydroxy-amino acids (serine and threonine) produces alanine and a-aminobutyric acid (Bada et al., 1978). The presence of non-standard amino acids such as a- and g-aminobutyric acid and b-alanine (Hare & Mitterer, 1967) has been used to indicate contamination of inter-crystalline samples (Schroeder, 1975), as the a-decarbox-ylation reaction is rare in pyrolysis experiments. However, in closed systems, such as for intracrystalline biomolecules, the presence of these amino acids will indicate the reaction and degradation of protein amino acids, rather than contamination by extraneous sources.

Clearly, differing rates of decomposition of the biomolecules explains the selective preservation of these biomolecules in the fossil record. The generation of proteinogenic amino acids from decomposition of other amino acids helps to explain the apparently anomalous concentration of some amino acids detected in fossils (e.g. Bada & Man, 1980). Examination of fossil molecules must include recognition of these reactions.

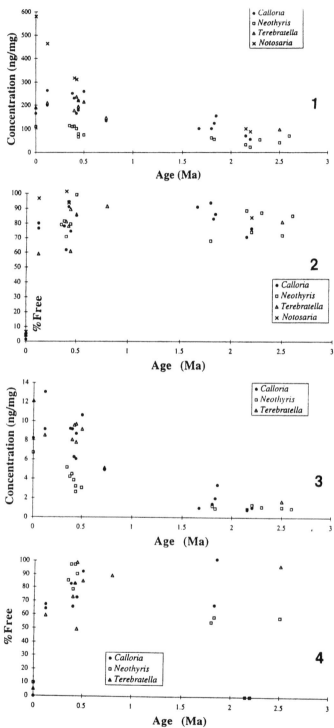

Figures 1-4. 1) Absolute concentration of acid soluble amino acids in four genera of brachiopods showing an overall decreasing concentration. Note initial rise (from Recent to youngest fossil) due to solubilisation of acid-insoluble proteins in the living organism; 2) Proportion of amino acids present in free state, showing a dramatic increase over first 0.12Ma and preservation of free amino acids in older samples indicating protection by mineral phase; 3) Concentration of aspartic acid/asparagine with an exponential decrease; 4) Proportion of aspartic acid/asparagine present in the free state. Overall rate is slower than the average for the other amino acids.

The aim of this study was to extract preserved amino acids, peptides and proteins from intracrystalline sites within the shells of brachiopods, to examine degradation of the biomolecules over time, and suggest some reaction pathways, prior to examining the role of amino acids in taxonomic analysis. Throughout this study, it has been assumed that there has been no significant evolution (i.e. change in composition) of the protein and that any change reflects diagenetic alteration.

METHODS AND MATERIALS

Recent and fossil brachiopods (*Neothyris*, *Calloria*, *Terebratella* and *Notosaria*) were collected in New Zealand from the localities given in Fleming (1953), Walton (1992) and Walton et al. (1993), and were prepared according to the methods of Walton & Curry (1994). Samples were scrubbed and encrusting epifauna removed by scraping. Articulated shells were disarticulated and body tissue removed (if present) before being soaked in an aqueous solution of bleach (10%v/v) for two hours at room temperature, washed with Milli RO™ water and air dried. Samples were ground and the powder incu-bated in an aqueous solution of bleach (10%v/v) under constant motion for 24hrs at room temperature, washed by repeated agitation with MilliQ™ water and centrifugation (typically 10 washes) and lyophilised. An aqueous solution of 2M HCl at a ratio of 11μL/mg was used in demineralisation. Finally, samples were centrifuged (20g.h.) to remove insoluble compounds.

Amino acid compositions were determined on the Applied Biosystems 420A amino acid analyser (Dupont et al., 1989), both with and without hydrolysis to ascertain the quantities of bound and unbound (free) amino acids which are present within the shell. Standard proteins and peptides were used during every analysis to ensure that hydrolysis proceeded to completion and blank analyses included to check for background levels of contamination. Analyses with hydrolysis were repeated at least three times, and those without at least twice to ensure reproducibility.

RESULTS

Bulk amino acid composition. The bulk amino acid composition can give a general guide to the state of preservation of the overall range of molecules (Figure 1). All samples analysed contain amino acid and show an overall decreasing trend in concentration with time. However, there appears to be an initial rise in the concentration of the amino acids (between the Recent samples and the youngest fossils), which differs in magnitude and duration, before the decreasing trend is initiated.

The proportion of amino acids which are present in the free state rises from negligible amounts in the Recent to greater than 58%by 0.12Ma (Figure 2), indicating rapid natural hydrolysis. The proportion of free amino acids also fluctuates with time; although remaining greater than 60% in all fossil samples. There is a corresponding variation in the proportion of amino acids which remain peptide bound. The lack of a consistent pattern in the percentage of free amino acids, and the corresponding concentration of peptide bound amino acids (either original or diagenetic bonding) indicates a complex system within the shell, whereby individual amino acids react at different rates in response to degradative factors.

Individual amino acids. Once the amino acids have been re-leased from the protein structure, they become essentially independent of the original protein and thus behave as chemical units and should be considered independently of

species. Aspartic acid shows an exponential decrease in con-centration over time (Figure 3), with correlation coefficients being highly significant. More than 80% of aspartic acid/asparagine has been lost from the acid soluble portion by 2.6Ma. The proportion of the amino acid present in the free state varies between samples. In *Neothyris*, there is a rapid rise to c.95% free (by 0.38Ma), followed by a maintenance of this proportion, even though the overall concentration of the biomolecule decreases over this time (Figure 4). In both *Calloria* and *Terebratella*, the rise in the rate of increase of the proportion of uncombined biomolecules is somewhat slower, reaching the same level by 0.5Ma.

Glutamic acid shows an exponential decrease in concentration (Figure 5). All samples show an increase in the concentration from Recent to youngest fossil, followed by rapid breakdown of up to 80% of the initial concentration by 2Ma. The proportion of free amino acids in the sample is low (Figure 6), in all cases being below 4%, and generally below 30%of the total present. The proportion shows a rapid increase to this level, followed by little fluctuation.

Serine decays rapidly, with c.80% of the original concentration lost by 0.12Ma (Figure 7). Following this rapid initial decrease, the concentration remains at a similar level. The proportion of serine present in the free state rapidly increases to a maxima (0.5Ma) and then decreases (Figure 8), until none remain in the uncombined state, indicating that any serine which remains in the sample after c.0.7Ma is bound via an HCl sensitive bond.

The concentration of alanine within the shells of the samples is an essentially random spread with respect to age (Figure 9). There is a large increase in concentration between the Recent and the fossil. In all cases, the concentration undergoes a large and rapid increase on transition from the living to the fossil, with a maxima at approximately 0.5Ma, followed by a decreasing trend, although in all cases the concentration of alanine in the oldest samples is similar to, or greater than, the initial concentration found in the Recent. The maxima of alanine concentration coincides with the highest proportion present in the free state. There is a rapid increase in the proportion of free amino acids with greater than 90% being free in most cases by 0.5Ma (Figure 10). A high percentage of free alanine remains throughout the samples studied, indicating that the increase in concentration of alanine must occur in the free state.

DISCUSSION

Natural hydrolysis of peptide bonds. The equilibrium between the condensation and hydrolysis reactions of a peptide bond is pushed towards the decomposition reaction, and promotes the natural hydrolysis of peptide-bound compounds in natural systems. The rate of this natural hydrolysis depends on temperature, water and the chemical characteristics of the amino acids on either side of the bond. The proportion of free amino acids rapidly increases from almost nothing in Recent samples (indicating that the amino acids are all bound into proteins) to c.80% in less than 0.5Ma, but does not proceed to completion. This reaction requires water to be included in the shell, either at the domain boundaries of the crystal or in fluid inclusions, and such water comprises up to 3% of the shell of articulate brachiopods (Gaffey, 1988), although the exact location is not known. After the initial rise to the maxima, the proportion of free amino acids tends to remain at a constant level, suggesting that the loss of free amino acids through degradative reactions, and the production of other free amino acids by natural hydrolysis reactions may be similar. If the rate of either were to fluctuate significantly, then the proportion of

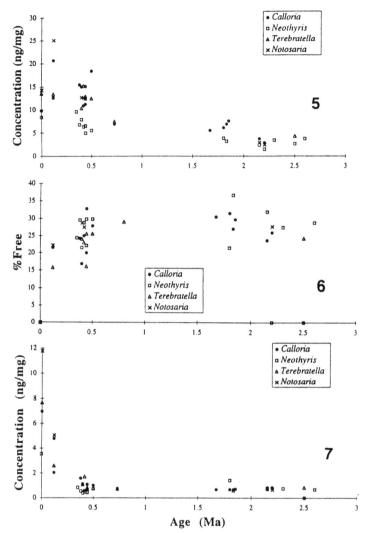

Figures 5-7. 5) Concentration of glutamic acid/glutamine showing a significant increase between Recent and youngest fossil; 6) Proportion of glutamic acid/glutamine present in free state. Note change in y-axis scale compared to other graphs. Level of free glutamicacid/glutamine is restricted by conversion to pyroglutamic acid; 7) Concentration of serine, showing preservation of this unstable amino acid in oldest samples at a low and constant level.

free amino acids present within the shell would also show a high rate of fluctuation.

Previous studies have shown that decomposition of peptide bonds will depend on the nature of the residues on either side of the bond. Protein sequencing (Cusack et al., 1992), immunology (Endo, 1992, 1994) and amino acid analysis (Walton et al., 1993) have all shown that the amino acid com-position of the species under study are different. The size of intracrystalline proteins also varies taxonomically (Curry et al., 1991; Walton et al., 1993). This will cause species level variation in the rate of natural hydrolysis between the samples as a product of the nature of the residues on either side of each peptide bond. Van Kleef et al. (1975) showed that susceptibility of bonds to natural hydrolysis was variable. As the primary sequence of intracrystalline proteins from brachio-

pods is not known, it is not possible to predict the effect of different residues on rates of natural hydrolysis.

Some amino acids remain peptide bound even in the oldest of the samples. There is no information regarding the size of these peptides, nor their primary sequence, as it has not proved possible to separate or purify fossil brachiopod peptides (Walton, 1992). An immunological investigation of Recent and fossil New Zealand brachiopods (Collins et al., 1991), recorded the presence of intracrystalline determinants in the fossil record, but at a much reduced level of reactivity (between 1% and 0.1% of the Recent signal). Natural hydrolysis will destroy the determinants, although the preserved determinants provide further evidence of the preservation of short peptides in the fossil record.

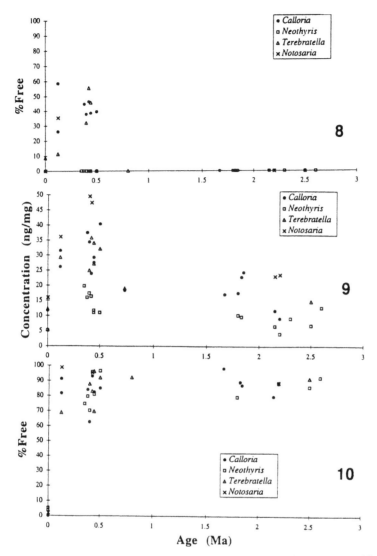

Figures 8 - 10.   8) Proportion of serine present in free state, which increases and decreases very rapidly. Older samples contain no serine in free state (c.f. Fig. 7) indicating that all remaining serine is bound to another molecule by an HCl sensitive bond; 9) Concentration of alanine over time. Dramatic rise in concentration is noted, probably due to diagenetic formation from serine and aspartic acid/asparagine. Final concentrations similar to those in the Recent; 10) Proportion of alanine present in free state.

Individual amino acids. Asparagine undergoes rapid and irreversible deamidation to produce aspartic acid and ammonia (Robinson & Rudd, 1974). Aspartic acid may decompose by either: (i), slow, reversible deamidation to fumaric acid and ammonia (Bada,1971), although this reaction cannot take place when aspartic acid is peptide bound (Bada & Man, 1980), or (ii), decarboxylation of the a- or b-carbons to form b-alanine or alanine. Bada & Miller (1969,1970) demonstrate less than 0.2% decarboxylation at between pH1 and 13, and 60-135°C, and conclude that this is unlikely in aqueous solution of the pure compound at elevated temperatures.

The amino acid b-alanine co-elutes with threonine, a relatively unstable amino acid. Notosaria contains a high concentration of aspartic acid, thus if a-decarboxylation occurred, the concentration of "threonine" would increase. This is observed with a dramatic increase between 0.12-0.5Ma, indicating that a-decarboxylation occurs in geological samples (Table 1). The coelution product is not observed in younger samples, indicating that b-alanine was probably formed in the free state, rather than when aspartic acid was peptide bound. This contrasts with estimates based on pyrolysis of the pure compound, suggesting a difference between pyrolysis reactions and fossil data, or an increase in the rate of reaction when other amino acids are present. The occurrence of a-decarboxylation may indicate that b-decarboxylation could also occur, forming alanine.

Glutamine undergoes rapid irreversible deamidation yielding glutamic acid (Robinson & Rudd, 1974). Glutamic acid is relatively stable amino acid (Vallentyne, 1964), although it may degrade by either (i), lactam formation to produce pyroglutamic acid, or (ii) g-decarboxylation to produce the non-protein amino acid, g-aminobutyric acid.

Pyroglutamic acid forms by the expulsion of water from glutamic acid (Wilson & Cannon, 1937), which may be converted back to glutamic acid through protein hydrolysis. Lactam formation may only take place when glutamic acid is in the free state. In fossil samples, the proportion of unbound glutamic acid is low (c.30%). This could be due to either a repeating sequence of glutamic acids, preferentially preserved, or formation of the lactam, preventing identification until after hydrolysis.

The concentration of glutamic acid decreases exponentially to c.20% of the Recent. There is no direct correlation between the age of the sample, the decrease in concentration of glutamic acid and the increase in the size of the peak at the position of g-aminobutyric acid. Lactam formation acts to preserve the biomolecule to some degree, although not extensively. This reaction, however, does not explain the decrease in the concentration of glutamic acid, which either occurs via decarboxylation to form g-aminobutyric acid, or via the decomposition of pyroglutamic acid. There is no direct evidence of a relationship between the decrease of glutamic acid and a corresponding increase of the g-aminobutyric acid, but this may be due, in turn, to the rapid destruction of this reaction product.

Serine decomposes by three possible pathways: (i), dehydration of the hydroxyl group in the side chain, resulting in the release of water and the eventual formation of alanine; (ii), aldol cleavage forming glycine and formaldehyde, or (iii), decarboxylation, resulting in the formation of ethanolamine, although dehydration is the prevalent decomposition reaction (Bada et al., 1978). The increase in concentration of alanine in fossil samples suggests dehydration reactions. Comparison of the data for serine and alanine show no direct correlation; while there is an increase in the concentration of alanine this cannot be explained solely by the decomposition of serine.

The proportion of free serine increases and decreases rapidly (Figure 8), from 0% in the Recent, to c.90% at 0.5Ma, and to 0% by c.0.7Ma, indicating rapid degradation whilst in the uncombined state. Older samples contain peptide bound serine. Uncombined serine decomposes predominantly by dehydration reactions, whereas aldol cleavage is dominant when the molecule is bound (Bada & Man, 1980). It is likely, therefore, that decomposition of serine will follow dehydration reactions when the amino acid is present in the free state. Decomposition by reversible aldol cleavage is negligible, as there is only a small rise in the corresponding concentration of glycine. It can thus be predicted that most of the serine in the shell has decomposed to alanine.

Alanine is thermally very stable (Abelson, 1954; Vallentyne, 1964), and decays via decarboxylation to form ethylamine and CO_2 (Abelson, 1954). In the presence of oxygen (the pyrolysis experiments were carried out under nitrogen), however, the rates of this pyrolysis reaction would increase dramatically. At room temperature, oxygen attacks alanine, causing the release of carboxyl carbon with a reaction half-life of 20ka (Conway & Libby, 1958). This indicates that the presence of oxygen is probably of vital importance in the decomposition of alanine over geological time.

The concentration of alanine present in the Recent samples is almost identical to that recovered from samples dated at c.2.6Ma. The intervening period, however, shows major variations in alanine concentration (Figure 9). All samples show a rapid increase in concentration, before decreasing to a level similar to that at which it started. Undoubtedly, this increase is due, at least in part, to its diagenetic formation from decomposition reactions, including the dehydration of serine. As there is no direct correlation between the decomposition of serine and the production of alanine, it is likely that other diagenetic reactions may produce alanine, for example the b-decarboxylation of aspartic acid, or the cleavage at the ring of phenylalanine and tyrosine. These reactions must play a part in the increase in concentration of alanine, as it cannot be explained by the decomposition of serine alone. The rate of decay of alanine appears, from both this and previous studies (Hare & Mitterer, 1967,1969), to be much more rapid than that recorded for pyrolysis of a solution of the pure amino acid. This phenomenon could have a number of causes, notably the presence of oxygen, surviving peptide bonds within the shell and the effect of mixtures of amino acids (Vallentyne, 1964).

Effect of carbohydrates on the destruction of amino acids. The reaction between carbohydrates and amino compounds is well documented (e.g. Hoering,1973,1980; Furth,1988), and results in condensation reactions and non-enzymic glycosylation. The reactions occur with interactions between the reducing groups of sugars and amino groups of other compounds to form glycosylamines. The end product of the reaction is a dark compound referred to as melanoidin, which has properties similar to that of humic acid (Hoering, 1973). Carbohydrates, including glucose (Clegg, personal commun.) are present within the shell (Collins et al., 1991), but it is not certain whether they are attached to the protein (Cusack, personal commun.).

Although not examined in this study, carbohydrates have an effect on amino acid decomposition. Pyrolysis reactions involving amino acids and glucose (Vallentyne, 1964) indicate that the higher the concentration of glucose in a standard solution of alanine, the faster the decomposition rate of the amino acid. The 0.37-life (63% destruction) of alanine at 167°C without glucose is just over 10 years, but is reduced to approximately two hours when 0.05M glucose is present, an increase in the rate of decomposition of some 40,000x. This interaction of carbohydrates with amino acids could well explain some of the accelerated degradation reactions detected in this study.

Molecular state of preservation. The proteins of fossil shells are almost completely decomposed, with only a limited number of peptide bonds surviving fossilisation. These bonds are associated with most of the amino acids (identified from the difference between the total and free amino acid concentration), although threonine and arginine are invariably absent from older samples. The presence of relatively unstable biomolecules such as serine in peptide bound compounds is likely to be due to the stabilising effect of the peptide bond, the stability of which is a function of the nature of the residues on the other side of the bond. The rate of natural hydrolysis then slows and reaches a plateau, corresponding to the complete destruction of the most labile peptide bonds and the preservation of less labile ones.

Individual amino acids also undergo degradative reactions, the majority of which take place when the biomolecule has been released from the protein and is present in the free state. Degradative reactions produce a range of reaction products, including the diagenetic generation of other amino acids which may be either proteinogenic or non-standard molecules. It is impossible to differentiate between the original and diagenetic amino acid, a factor which will distort any taxonomic relationship through time. For example, Jope (1967) found that the insoluble fraction of the intercrystalline protein from fossil bra-chiopods showed marked differences from the nearest living relatives. This study also found raised serine in the insoluble fraction, possibly indicating either post-mortem alteration or a sink for some serine. Both the present study, and that of Jope (1967) contrast with the results of Kolesnikov & Prosorov-skaya (1986), who recognised 'very familiar' compositions between Recent and fossil brachiopods (dating to the Juras-sic). Such results need to be treated with caution, as this study has shown that, at least in the soluble fraction, the amino acids are unstable to differing degrees, which results in changing amino acid ratios over time, and fossil samples, as a consequence, will have different amino acid ratios to those of extant species.

The concentration of amino acids in the shells increases in some cases between the Recent and the youngest fossils (Figure 1). Examination of the insoluble intracrystalline fraction of Recent samples (Walton, 1992) reveals that they contain acid-insoluble (but hydrolysable) proteins. These proteins will undergo natural hydrolysis, releasing amino acids into the soluble fraction, thus increasing the amino acid concentration. As acid insoluble samples from Recent samples age, they release peptides and free amino acids which are taken into solution by decalcification, effectively increasing the concen-tration of the amino acids.

Non-protein amino acids within the shell. b-alanine and g-aminobutyric acid are both present as intracrystalline molecules in fossil brachiopods. From kinetic studies and the examination of foraminifers, Schroeder (1975) concluded that the presence of these amino acids was indicative of conta-mination of the foraminifers' matrix from the sediment. The data presented here contradict this. Low concentrations of g-ami-nobutyric acid are present in all fossil samples, indicating low rates of conversion from glutamic acid. Notosaria, however, has a high concentration of aspartic acid, and older samples show an increase in the concentration of b-alanine, although this only accounts for c.45% of the decomposition of aspartic acid (Walton, 1992). As the amino acids have been released during decalcification from within the shell, it is unlikely that the non-protein amino acids were derived from the sediment in which the fossils are contained, and thus are indigenous to the fossil, formed by the diagenetic degradation of other amino acids. The presence of these amino acids should not neces-sarily be regarded as being indicative of contamination.

Amino acid taxonomy. Previous studies have used the amino acid content of Recent and fossil foraminiferans to complete chemotaxonomy (e.g. King & Hare, 1972; Haugen et al., 1989). These studies utilised principal components analysis to sum-marise the derived data from these samples, and group the samples with similar positions on principal components axes. These studies have analysed both Recent and fossil samples in the same analysis, which has led to a large spread of data. From the preceding discussion it may be seen that there are problems in this type of analysis as changes in the degradation pattern of the molecules will introduce further variation. For ex-ample, Haugen et al. (1989) recognised an increasing Gly/Ala ratio, indicative of the diagenetic production of amino acids.

King & Hare (1972) approached the problem in a different way, concentrating on the classification of Recent Foraminifera, but also examining fossil samples in order to determine whether the genetic differences are preserved in the fossil. It was concluded that there were changes in the amino acid composition of the fossil record, but that this did not affect the relationships between some of the more stable of the amino

acids. The study concluded that evolutionary changes could be traced through studies of the biochemical content of the fossils, although the authors were not able to say how far back in the fossil record such changes could be utilised.

It will be clear that the degradation of molecules over time will have a profound affect on amino acid composition of the shell and these changes must be considered in any investigation. Samples collected from the same horizon may be thought to have been subjected to similar geological processes through-out their history. The effect of this is to render the horizon as a time plane - a snapshot of geological time - whereby changes in the amino acid content due to diagenetic alteration of amino acids will be of approximately the same order in all samples, and hence differences in the compositions will be due to the initial biochemical difference of the species alone. This is obviously an oversimplification of possible relationships, and the amino acid composition of the fossils will be distorted over

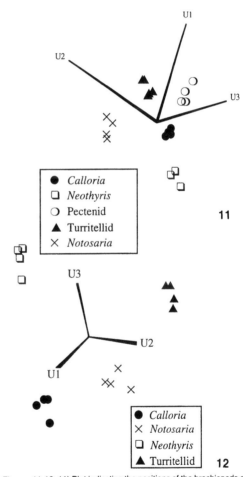

Figures 11-12. 11) Plot indicating the positions of the brachiopods on axes representing the first three principal components of the amino acid data from the Rapanui Marine Sand. The groupings are well separated indicating the preservation of a taxonomic signal; 12) Prin-cipal components plot for the amino acid data from the Upper Okiwa Group, demonstrating taxonomic discrimination at 2.15 Ma.

Table 1. Amino acid compositions of *Notosaria* demonstrating the apparent increase in the concentration of threonine which is due to the formation of b-alanine from aspartic acid.

Sample	Age (Ma)	Concentration (ng/mg)	
		Aspartic acid/ asparagine	Threonine
Recent	0	214.52	5.89
Rapanui Marine Sand	0.12	109.54	3.92
Tainui Shellbed	0.40	42.58	53.78
Pinnacle Sand	0.42	41.72	50.91
Upper Okiwa Group	2.15	1.98	22.29
Hautawa Shellbed	2.20	1.61	23.27

time by, for example, the rate and degree of diagenetic production of some amino acids, which will in turn depend on the initial concentration, and the effect of carbohydrates and of different mixture of amino acids in the sample.

Walton et al. (1993) demonstrated the taxonomic discrimination of Recent brachiopods through the use of amino acids utilising PCA. The same approach has been used to examine the fossil brachiopods in this study. Two examples are given. (I) The Rapanui Marine Sand member of the Rapanui Formation (Fleming, 1953) is the youngest formation studied and dates from c.0.12Ma. The first three principal components contain 86% of the total variation. Brachiopods can be separated to at least the subordinal level, and thus taxonomic discrimination is possible even though the amino acids have been degraded (Figure 11). In particular, this demonstrates the ability to utilise the intracrystalline component as the free amino acids are not leached from the system. (ii) The undifferentiated shell bed of the Upper Okiwa Group (Fleming, 1953) dates from c.2.15Ma. The first three principal com-ponents contain 91.8% of the total variation of the dataset. Brachiopods can be differentiated from each other and from the outgroup, again revealing a taxonomic signal (Figure 12).

The amino acid compositions of fossils are complex datasets, containing up to 17 variables. Information contained within datasets of this size are difficult to assimilate, and it is difficult to understand the relationships between amino acids, as these are between every member of the dataset. PCA has the advantage of summarising the data into fewer, derived variables which may then be used to differentiate the samples. The results from the PCA of the amino acid data extracted from fossils show that, within horizons, samples may be separated to at least the subordinal level in a range of samples, on the basis of their amino acid composition alone, and in some cases this separation is to the subfamilial level. These diagrams may be considered to be analogous to geochemical discrimination diagrams as the majority of the groupings described would be recognised as such, even if morphologically described groupings were not known.

Older samples show that even though significant degradation has taken place the brachiopods can clearly be differentiated from other organisms. Although the proximity of the groupings is not necessarily phylogenetic, each group can be seen to have a distinct amino acid signature and this must reflect the original taxonomic differences between the samples.

Acknowledgments

This work was completed at The University of Glasgow, whilst under the tenure of a UK NERC studentship. M. Cusack, M.J. Collins and G.B. Curry are thanked for their useful advice. The receipt of a Royal Society Travel Grant is also gratefully acknowledged.

REFERENCES

ABELSON, P.H. 1954. Organic constituents of fossils. Carnegie Institute Washington Yearbook 53:97-101.
___1955. ibid. 54: 107-109.
___1956. Paleobiochemistry. Scientific American 195:83-92.
BADA, J.L.1971. Kinetics of the non-biological decomposition and racemization of amino acids in natural waters. Advances Chemistry Series 106:309-331.
___& E.H. MAN.1980. Amino acid analysis in Deep Sea Drilling Project Cores: kinetics and mechanisms of some reactions and their applications in geochronology and in paleotemperature and heat flow determinations. Earth Science Reviews 16:21-55.
___& S L. MILLER. 1969. Kinetics and mechanism of the non-enzymatic reversible deamination of aspartic acid. Journal American Chemical Society 91:3946-3947.
___&___1970. Kinetics and mechanism of the reversible non-enzymatic deamination of aspartic acid. Journal American Chemical Society 92: 2774-2780.
___, M.Y. SHOU, E.H. MAN & R.A. SCHROEDER.1978. Decomposition of hydroxy amino acids in foraminifera tests: kinetics, mechanism and geochronological implications. EarthPlanetary Science Letters 41:67-76.
CLEGG, H. 1993. Biomolecules in Recent and fossil articulate brachiopods. Unpublished Ph. D. thesis, University of Newcastle Upon Tyne.
COLLINS, M.J., G. MUYZER, P. WESTBROEK, G.B. CURRY, P.A. SANDBERG, S.J. XU, R. QUINN & D. MACKINNON. 1991. Preservation of fossil biopolymeric structures: conclusive immunological evidence. Geochimica Cosmochimica Acta 55:2253 - 2257.
CONWAY, D. & W.F. LIBBY.1958. The measurement of very slow reaction rates; decarboxylation of alanine. Journal American Chemical Society 80:1077-1084.
CURRY, G.B., M. CUSACK, D. WALTON, K. ENDO, H. CLEGG, G. ABBOTT & H. ARMSTRONG.1991. Biogeochemistry of brachiopod intracrystalline molecules. Philosophical Transactions Royal Society London B, 333:359-366.
CUSACK, M., G.B. CURRY, H. CLEGG & G. ABBOTT. 1992. An intracrystalline chromoprotein from red brachiopod shells: implications for the process of biomineralisation. Comparative Biochemistry Physiology 102B:93-95.
DUPONT, D.R., P.S. KEIM, A. CHUI, R. BELLO, M. BOZZINI & K.J. WILSON. 1989. A comprehensive approach to amino acid analysis, p. 284-294. In, T.E. Hugli (ed.).Techniques in protein chemistry. Academic Press, New York.
ENDO, K. 1992. Molecular systematics of Recent and Pleistocene brachiopods. Unpublished Ph. D. thesis, University of Glasgow.
___, G.B. CURRY, R. QUINN, M.J. COLLINS & P. WEST-BROEK. 1994. Re-interpretation of terebratulide phylogeny based on immunological data. Palaeontology 37:349 - 374.
FLEMING, C.A. 1953. The geology of the Wanganui Subdivision. Department Scientific Industrial Research, Wellington, 362 p.
FURTH, A. 1988. Sweet peril for proteins. New Scientist 3 March: 58 - 62.
GAFFEY, S.J. 1988. Water in skeletal carbonates. Journal Sedimentary Petrology 58(3):397- 414.
GOODFRIEND, G.A., P.E. HARE & E.R.M. DRUFFEL. 1992. Aspartic acid racemization and protein diagenesis in corals over the last 350 years. Geochimica Cosmochimica Acta 56:3847 - 3850.
HARE, P.E. & R.M. MITTERER.1967. Non protein amino acids

in fossil shells. Carnegie Institute Washington Yearbook 65:362-364.

___&___1969. Laboratory simulation of amino acid diagenesis in fossils. ibid. 67:205-208.

HAUGEN, J.E., H.P. SEJRUP & N.B. VOGT. 1989. Chemotaxonomy of Quaternary benthic Foraminifera using amino acids. Journal Foraminiferal Research 19:38-51.

HOERING, T.C. 1973. A comparison of melanoidin and humic acid. Carnegie Institute Washington Yearbook 73:590-595.

___1980. The organic constituents of fossil mollusc shells, p.193-201. In, P.E. Hare, T.C. Hoering & K. King (eds.). Biogeochemistry of the amino acids. John Wiley, New York.

JOPE, M. 1967. The protein of brachiopod shell, II, Shell protein from fossil articulates: amino acid composition. Comparative Biochemistry Physiology 20: 601-605.

KING, K. & P.E. HARE.1972. Amino acid composition of the test as a taxonomic character for living and fossil planktonic Foraminifera. Micropaleontology 18:285-293.

KOLESNIKOV, C.M. & E.L. PROSOROVSKAYA. 1986. Biochemical investigations of Jurassic and Recent brachiopod shells, p.113-119. In, P.R. Racheboeuf & C.C. Emig (eds.). Les brachiopodes fossiles et actuels. Université de Bretagne Occidentale, Brest.

ROBINSON, A.B. & C.J. RUDD. 1974. Deamidation of glutaminyl and asparaginyl residues in peptides and proteins. Current Topics Cellular Regulation 8:247-294.

SCHROEDER, R.A.1975. Absence of b-alanine and g-aminobutyric acid in cleaned forminiferal shells: implications for use as a chemical criterion to indicate removal of non-indigenous amino acid contaminants. Earth Planetary Science Letters 25:274-278.

TOTTEN, D.K.J., F.D. DAVIDSON & R.W.G. WYCKOFF. 1972. Amino acid composition of heated oyster shells. Proceedings National Academy Sciences USA 69:784-785.

VALLENTYNE, J.R. 1964. Biogeochemistry of organic matter II: Thermal reaction kinetics and transformation products of amino compounds. Geochimica Cosmochimica Acta 32: 1353-1356.

VAN KLEEF, F.S.M., W.W. DEJONG & H.J. HOENDERS. 1975. Stepwise degradations and deamidation of the eye lens protein a-crystallin in ageing. Nature 258:264-266.

WALTON, D.I.1992. Biogeochemistry of brachiopod intracrystalline proteins and amino acids. Unpublished Ph. D. thesis, University of Glasgow.

___,M. CUSACK & G.B. CURRY.1993. Implications of the amino acid composition of Recent New Zealand brachiopods. Palaeontology 36:883-896.

___& G.B. CURRY.1994. Extraction, analysis and interpretation of intracrystalline amino acids from fossils. Lethaia 27:179-184.

WILSON, H. & R.K. CANNON. 1937. The glutamic acid-pyrrolidone carboxylic acid system. Journal Biological Chemistry 119:309-331.

TAXONOMIC IMPORTANCE OF BODY-MANTLE RELATIONSHIPS IN THE BRACHIOPODA

ANTHONY D. WRIGHT

School of Geosciences, The Queen's University of Belfast, Belfast BT7 1NN, Northern Ireland

ABSTRACT- Extant articulate brachiopods all fall within the telotremate grouping of Beecher (1891); nobody has ever seen the mantle of a protremate. Mantle setae in living brachiopods display wide variation in both form and function; some fossil shell details are interpreted in terms of setae and their implication for mantle distribution. Mantle canals, recently discovered on ventral interareas of procline clitambonitidines, and providing evidence of the mantle-coelomic cavity boundary posteriorly being at the margin of the delthyrium (Wright, 1994), are here recorded from orthidine, strophomenide, pentameride and spiriferide brachiopods. By contrast, the boundary, along with the fusion of the mantle lobes, is sited at the lateral ends of the hingeline in living rhynchonellides, terebratulides and thecideidines and presumably also in their fossil representatives. Apart from the straight-hinged spiriferides, this soft tissue distribution appears to tie in with either deltidiodont or cyrtomatodont dentition. The possibility of a mantle-lined area simply reflecting a strophic hinge would seem to be negated by the thecideidine study of Williams (1973); but the exact position of fusion of the mantle lobes, vis-à-vis delthyrial margin/end of hingeline (in close proximity in terebratulide-rhynchonellide stocks), has not been considered by zoologists. The retention of Beecher's Protremata and Telotremata for two readily recognisable, and moreover useful subclasses of articulate brachiopods, as advocated by Jaanusson (1971) is supported. The old arguments against Beecher's original diagnosis of the telotremates as having 'arms supported by calcareous crura, spirals or loops' continue to lose credence.

INTRODUCTION

One of the problems of brachiopod classification is that it has been largely established on extant stocks. Thus hinged forms have been conveniently placed in the Articulata, and those lacking a hinge in the Inarticulata of Huxley (1869). But, as pointed out previously (Wright, 1979), the Class Inarticulata, or more appropriately, the non-Articulata, is a simply a catch-all for anything that does not fit into the Articulata. Subsequently, attention has been given to these non-articulates by Gorjansky & Popov (1986) and Popov et al. (1993).

Looking then at the forms in present day seas which provided the basis for such division, one aspect of the non-articulates is that genera such as *Neocrania*, *Discinisca* and *Lingula* possess separate mantle lobes which extend all around the periphery in each valve. By contrast, the articulates, again represented by only three extant groups centered around *Hemithyris*, *Terebratulina* and *Lacazella*, have mantle lobes which fuse posteriorly at the lateral margins of the hinge line where they mark the edge of the body cavity. The problem is that this arrangement of the mantle in these extant Telotremata, to use the ordinal term of Beecher (1891), has been accepted as characterising all Articulata when in fact nobody has ever seen the mantle of a protrematous brachiopod.

The recent discovery of mantle canals lining the inner surface of the interareas in the ventral valves of certain clitambonitidine brachiopods with procline interareas (Wright, 1994), was interpreted as indicating a fundamentally different body-mantle distribution in the posterior of these extinct stocks compared with those of the extant articulate stocks. The current article presents further evidence concerning the mantle in other fossil articulate stocks that strengthens support for a two-fold division of these brachiopods which fits closely to that of Beecher (1891). Although Beecher established the Protremata and Telotremata as orders, these are best interpreted as subclasses as proposed by Jaanusson (1971, p. 43), the modern concept of orders as used in the brachiopod 'Treatise' (Williams & Rowell, 1965b) receiving widespread acceptance.

Critical to the present work is the recognition that the presence of mantle canals and related vascular markings across the interarea indicates that the interarea was lined with mantle, and hence was within the mantle cavity and not the body cavity as is the case with the areas of extant articulate brachiopods that are only represented by the orders Rhynchonellida, Terebratulida and the suborder Thecideidina. In these living, entirely telotrematous, stocks the mantle lobes of each valve fuse at the lateral ends of the hinge so that the areas are lined only by the shell secreting outer epithelium. Within the extinct protremate suborder Clitambonitidina, the evidence of mantle lined areas in some procline stocks means that fusion of the mantle lobes took place not at the margins of the hinge, but at the margins of the delthyrium, so that the body cavity was posteriorly restricted to the median part of the hinge region (Wright, 1994).

SETAL DEVELOPMENT

One feature that characterises the mantle of most brachiopods is that the mantle groove around the periphery of the shell houses setae located within setal follicles. These are best displayed in the non-articulate brachiopods which lack a hinge structure, although *Crania* (*Neocrania*) is a common form which lacks setae in the adult, certainly from the larval form with three pairs of cirri onwards (Rowell 1960, p.49). Setae are present in the early swimming larval stages (Neilsen, 1991, p.10), but are shed after settling. Setae are recorded in very young stages of *Discinisca* (down to the two pairs of cirri stage), and are regarded as embryonic organs with protective, stabilising and motile functions for the larva (Chuang 1977, p.49). In the adult discinids the setae occur around the entire periphery of each valve, extending in *Discinisca* to form an even fringe to the shell, although somewhat shorter along the posterior margin, and having the usual tactile and screening functions. The setae of *Pelagodiscus* are quite markedly differentiated. Foster (1974, p.40) records those along approximately three quarters of dorsal mantle edge of *P. atlanticus* as consisting

Figure 1. 1, *Eochonetes advena* Reed, Upper Ordovician (Ashgill), Drummock Group, Starfish Bed, Thraive Glen, Girvan, Scotland, ventral valve internal mould, BMNH B73914, x5. 2 and 5: *Tylothyris laminosa* (M'Coy), Lower Carboniferous (Asbian), Glencar Limestone, Sillees River, Bunnahone, County Fermanagh, Ireland. 2, anterior view of interior of silicified ventral valve, BMNH BB63601, x3; 5, anterior view of interior of silicified ventral valve, BMNH BB63600, x3. 3-4: *Retrorsirostra carleyi* (Hall), Upper Ordovician (Ashgill), Arnheim Formation, Lick Run, Morrow, Warren County, Ohio, U.S.A. 3, full view of interior of ventral valve, MUGM 29457, x1.5; 4, detail of mantle canals on the interarea to the left of the dental plate of ventral valve, MUGM 29457 (Fig. 3), x3.

of 'long (up to twice shell length) minutely spinose setae alternating with two to 10 short nonspinose setae, posterior quarter of mantle with only short type of setae; ventral mantle edge like dorsal, but all setae narrower, long setae interspersed with 20-40 short setae.'

Setae are well developed in the burrowing lingulids and, apart from the vicinity of the pedicle, occur around the periphery of each valve. The longer setae, antero-laterally and medianly, form tubes (Morse, 1902) facilitating the ingress and egress of water to the mantle cavity while at the same time screening out sand grains. The lateral setae are involved with the burrowing mechanism (Thayer & Steele-Petrovic,1975). An interesting observation by those authors was that in *Terebratulina unguicula*, in which setae have their more usual tactile function, about every tenth seta was larger than the rest, and that these do not retract but remain extended even when the valves are closed, presumably to maintain an early warning system. This may well account for the not uncommon phenomenon in some fossil orthides in which occasional relatively strong costellae of the exterior match unusually long and deep follicular embayments of the interior, such as are well illustrated by Cooper (1956) for *Laticrura* (pl.145B) and *Fascifera* (pl.152B). The presence of long, non-retractable setae occupying these sites would provide a rational explanation for their presence.

There seems no doubt that the follicular embayments of the crenulated valve margins in these early Palaeozoic articu-

late brachiopods accomodated the setal follicles (Williams & Wright 1963, p.19). From the anterior, the stature of the crenulations is typically reduced postero-laterally as the hinge line is approached, and could indicate that the setae are becoming shorter and finer. In *Rhipido-mella hessensis* these postero-lateral areas develop a small number of tubes to house a few setae in this region (Wright 1981, pl.69, fig.5) suggesting that although the setal length may be maintained, the density of the array is reduced. There is no evidence here of any setae along the posterior of the valves.

Although a few articulate brachiopods do have perforations along the posterior margin, these have not been interpreted as likely sites for projecting setae. Fine perforations are present along the posterior margin of the interarea in *Chonetoidea* and *Sentolunia* (Havlicek, 1967) and are similar to those noted by Reed (1917) in his genus *Eochonetes*, in which they continue internally as canals [Figure 1:1]. Here the tube-like structures of sediment on the mould represent the infill of canals through the shell tissue. Williams & Rowell (1965a) interpreted the perfo-rations of *Eochonetes* as containing anchoring strands of epithelium. The form and position of the canals is, as Brunton (1972, p. 21) has pointed out, essentially the same as the canals leading from the valve interior to the spines in chonetaceans, these latter being interpreted as an adaptation to stabilization of the shell (Brunton 1972, p. 23). These interpretations seem at first sight much more

300

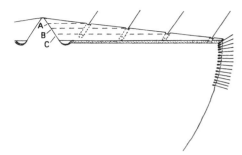

Figure 2. Envisaged periodic incorporation of postero-laterally orientated setae along the posterior margin of an *Eochonetes* ventral valve. A, B, C indicate successive growth stages of the interarea when a seta at the cardinal angle is incorporated into the shell, retaining contact with the mantle lining the interarea via a canal. Part of the lateral setal array is indicated in front of the cardinal angle.

plausible than any setal association, for in any event setae would be expected to emerge from the mantle at the growing edge of the hinge and not at the posterior of the interarea.

But it is worthwhile looking more closely at the disposition of the *Eochonetes* perforations. *Eochonetes* typically has 3 to 5 perforations symmetrically disposed along the posterior margin on either side of the ventral umbo. This suggests that at the genetically controlled instant of formation of each pair, the perforations would be located at the postero-lateral angles, and could therefore well mark the position of the most posteriorly situated setae of the array along the lateral and anterior margins (Figure 2). At this point the mantle edge, directed laterally on the main shell, would curve over to be directed anteriorly ready to form the growing edge of the ventral interarea.

The initial postero-lateral direction of the seta at the cardinal angle would be retained while the setal follicle maintained contact with the mantle as the shell thickened with growth, thus confining the seta within the perforation. This same orientation is incidentally typical of the spines along the posterior ventral margin of the chonetidines, which must again develop in a similar manner at the cardinal angles. One can speculate on the form and function of the postero-laterally directed eochonetid setae; the diameter of the perforations is larger than that of the coarser capillae of the differentiated radial ornament so that it is reasonable to suggest that these setae would be somewhat stouter that those of the main array and perhaps therefore non-retractable. The typically symmetrical pairing about the umbo is suggestive of a balancing function, additional to or replacing the sensory function, leading to stabilisation of the shell as suggested by Brunton (1972) for the chonetid spines as noted above.

This suggested incorporation of setae into the shell surface, and also the continuing contact of the setae with the epithelium, has been proposed recently for certain stocks of orthide brachiopods. The commonly occurring anteriorly directed ribs of such disparate genera as *Doleroides*, *Schizophoria* and *Tritoechia* have been interpreted as housing setae (Wright 1981, p. 472) which have been incorporated within the shell substance; the process of sealing into the shell has been demonstrated along the marginal crenulations of *Rhipidomella* (Wright 1981, fig.11). More recently, specimens of the clitambonitidine *Kullervo* in which the aditicule

passes through the shell to open well to the posterior of the anterior margin have been described by Wright & Rubel (1996, pl.1, figs. 1-6); the tube of *Eochonetes* indicates that this too maintained its connection with the mantle at a similar distance from the external opening.

One instance of a line of small pits occurring along the growing edge of the hinge itself is in *Unispirifer fluctuosus*, figured by Thomas (1971, pl. 28). However these pits, and the associated vertical flutings of the interarea, form part of a denticular development along the hinge. But there seems to be no case of setae developing along the growing margin of the interareas of protrematous brachiopods. This suggests that, although the interareas are lined with mantle and not simply outer epithelium, the mantle here was modified in that it did not possess setal follicles. This is perhaps not of major import for, as noted above, the setal density along the shell margins immediately in front of the cardinal angles in *Rhipidomella* is already quite sparse, while forms like *Lacazella* and *Neocrania* lack setae altogether in the adult stages.

GROUPS WITH VENTRAL AREAS LINED BY MANTLE

The presence of mantle canals has already been demonstrated on the ventral interareas of the clitambonitidine genera *Apomatella*, *Clinambon* and *Ilmarinia* (Wright,1994). Another feature that these three genera have in common is that their interareas are moderately long and are procline. This means that their inner surfaces are easily visible when examining the interiors of the valves, in contrast to the more usual apsacline forms where they are less obvious not only because of the restriction imposed by the acute angle between area and valve but also by additional shell secretion tending to fill and obscure this confined space. But, as was pointed out by Wright (1994, p. 224), this difference is only one of orientation and indeed vascular markings have now been observed in the rather more restricted space below the apsacline ventral area of *Estlandia* (Wright & Rubel, 1996, pl.2, fig.2).

There is therefore even less reason to doubt that the mantle-body disposition in these stocks was consistent throughout this suborder. On the basis of general shell morphology, and specifically the strophic, deltidiodont hinge, it was previously felt reasonable that the same soft tissue distribution would in addition characterise the other orthide stocks and also the strophomenides. The procline orthide *Retrosirostra carleyi* , from the Upper Ordovician Richmond Group of Ohio shows well preserved canals lining the interareas and the valve floor posterior to the gonocoeles (Figure 1:3-4), thus confirming the proposal that the disposition of mantle along the hinge region in the orthides generally is as previously described for the clitambonitidines. Regarding the strophomenides, Popov (person. commun.) has 'also observed similar mantle canals (to those of the figured clitambonitidines) along the posterior margin of the ventral valve of some Plectambonitoids.' This is particularly encouraging, as the ventral interareas in the strophomenides are commonly short and apsacline, and thus are not easily studied. Amongst the strophomenides, however, the Orthotetidina is a group which does contain genera with long (e.g. *Meekella*) or relatively long (e.g. *Schellwienella*) ventral interareas. In the orthotetidine material examined, consisting of silicified material from the Permian of Texas and the Carboniferous of Ireland, and reasonably well preserved mould material from the Devonian of the Falkland Islands, evidence of canals on the areas has yet to be found, and indeed only the vaguest traces have been observed on the valve floor beyond the muscle scars. But as was pointed out by Williams & Rowell (1965a, p.H132) the mantle canal patterns of the davidsoniaceans (orthotetidines), spiriferoids and pentameroids are rarely preserved.

The present search has however yielded some useful information on the latter two orders. Amongst the pentamerides, the widespread Lower Silurian *Stricklandia* is a wide hinged form which certainly possesses an area in the ventral valve, although the view that this is a true interarea has been challenged by Jaanusson (1971, p.36), who has demonstrated that in well preserved *Costistricklandia lirata* (ibid., pl.1, figs.1-2) the growth lines of the area are not straight. Examination of internal moulds of *Stricklandia lens* (Figure 3:1-2) reveals a series of pustules (corresponding to vascular pits on the calcareous inner shell surface) which not only occur on the main valve floor beneath the spondylium but extend across the break in slope on to the area. Thus the mantle, as in the orthides and strophomenides, again extends along the hinge line to the margins of the delthyrium, here defined by the edges of the spondylium.

Examination of the inner surfaces of the interareas of some straight hinged spiriferides has again revealed the presence of vascular markings and canals in suitably well preserved material. This includes some silicified specimens of the Carboniferous *Tylothyris laminosa* (Figure1:2,5) which show vascular markings beneath the interarea in the form of canals and pits. The latter are not as fine as the pitting of the gonocoeles, of which a little may be seen preserved in each valve along the frayed edges of the etched valve floor. The Silurian spiriferide *Cyrtia exporrecta* , in the form of internal moulds, some of which are sufficiently well preserved, show the moulds of vascular pits which are similar both on the interareas and on the valve floor lateral to the dental plates (Figure 3:3-5). No vascular markings have been found along the posterior margins of a small number of mouldic and silicified Upper Ordovician and Lower Carboniferous specimens of strophic athyridides so far examined, but the interareas of these narrow-hinged spire-bearers are much too restricted ever to be likely to show any vascular markings. The areas of some atrypides are even less well developed, being either minimal or non-existent.

DISCUSSION AND CLASSIFICATION

In summary, the evidence of the vascular markings on the internal surfaces of the areas indicates that they are underlain by mantle, thus forming part of the mantle cavity and indicating that fusion of the mantle lobes takes place at the edges of the delthyrium and not at the ends of the hingeline in the orders Orthida, Pentamerida, and Strophomenida.

In the order Spiriferida (Boucot, Johnson & Staton, 1965) there is also evidence of mantle below the interarea, but the non-strophic, narrow hinged spire-bearers, the Atrypida and Athyridida, are inferred to have mantle distribution similar to extant telotrematte stocks. It may be that in the spiriferides, the mantle distribution simply reflected a strophic hinge condition. Yet this seems unlikely as, according to Williams (1973), the fusion of the mantle lobes in extant thecideidines is at the ends of the hingeline so that, despite the strophic hinge, there can be no mantle canals or mantle beneath the area.

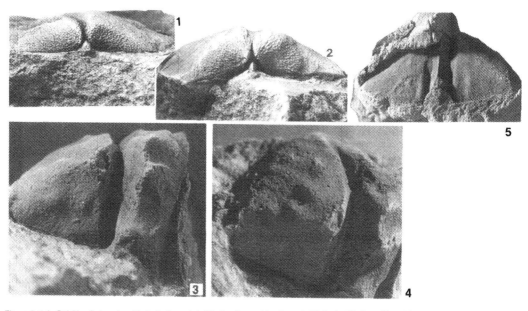

Figure 3. 1-2: *Stricklandia lens lens* (J. de C. Sowerby). Silurian (Lower Llandovery), A3 Beds, Allt Cwar Mawr, Llandovery, Wales. 1, posterior view of ventral valve internal mould, BMNH BB35011a, x4; 2, posterior view of ventral valve internal mould, BMNH BB35033, x3. 3-5: *Cyrtia exporrecta* (Wahlenberg). Lower Silurian (Telychian), Wether Law Linn Formation (lower member), Wether Law Linn, North Esk Inlier, Pentland Hills, Scotland. 3-4, posterior and postero-ventral views of ventral valve internal mould RMS Br136979 showing comparable moulds of vascular pits on interarea and shell floor, respectively, x5; 5, posterior view of ventral valve RMS Br136978, largely in the form of an internal mould, showing smooth interarea broken by moulds of vascular pits, x5.

Again, in the other extant articulate brachiopods, the Rhynchonellida and Terebratulida, all fusion of the mantle lobes purportedly takes place at the ends of the hingeline. However, a problem with the existing descriptions of living forms is that the exact position of fusion of the lobes has not been considered, and with the closeness of the teeth to the ends of what passes for the hinge-line in a non-strophic shell, any difference of disposition could have been overlooked. Investigations are in hand to establish the precise position of fusion in the living strophic *Megerlia* and *Kraussina*.

Jaanusson's (1971) advocacy for the retention of the two articulate divisions of Beecher (1891) was based on the characterisation of these two groups by their deltidiodont or cyrtomatodont dentitions. On this basis, Jaanusson (1971, p.43) felt that there was no need to replace Beecher's names despite the fact that they were established on other characters which were subsequently shown to be incorrect or poorly defined; indeed Beecher's 1891 criteria had already been modified by Schuchert (1897, 1913), and the emendation of diagnoses is a commonly accepted practice which equally well applies in this case. The view that Beecher's names Protremata and Telotremata for the clear divisions of the articulate brachiopods should be retained is fully endorsed herein, with the principal defining criteria being the deltidiodont-cyrtomatodont dentition of Jaanusson (1971) together with the differing disposition of the body and mantle posteriorly in the protremates as compared with that of extant telotremate groups.

But further, that part of Beecher's original diagnosis (1891, p.355) defining the Telotremata as having 'arms supported by calcareous crura, spirals or loops', which are lacking in the Protremata, is still a most useful guide to the two divisions. The discarding of the Beecher classification because of the sporadic presence of these calcareous supports in protrematous stocks is becoming increasingly unconvincing. Examples cited by Williams (1956, p. 271) as falling into this category were *Tropidoleptus*, *Enantiosphen*, *Thecospira*, *Leptaenisca* and *Enteletes*, etc. Although the work of Williams & Wright (1961) appeared to confirm that the first two genera were protremates, in *Tropidoleptus*, the recognition of the cyrtomatodont dentition by Jaanusson (1971) negates this. *Enantiosphen* remains enigmatic. *Thecospira*, as a thecideidine, is now regarded by some as a spiriferide (Baker, 1990). The spires of *Leptaenisca* are not the free-standing spires developed from crura, but simply ridges on the valve floor, and thus are no more true spires than the ridges in *Christiania* are true loops. The brachiophores of *Enteletes*, *Rhipidomella*, *Dicoelosia*, etc. are undoubtedly better developed and more complicated than in average orthides (Wright 1968, 1993), but apart from length they have little in common with rhynchonellide crura, so that the argument against the Beecher orders on this basis of lophophore support is becoming increasingly difficult to sustain.

On the other hand, those ordinal names proposed by Beecher (1891) for the inarticulate brachiopods, Atremata and Neotremata, cannot be similarly justified. The assignments of the non-articulates into the Class Inarticulata was itself misleading in its implied degree of unity (Wright, 1979), and, with the recognition of the fundamental importance of shell secretion, the two Beecher orders, which each contain a mixture of calcareous and chitinophosphatic shells, are irretrievably flawed. Gorjansky & Popov (1986) separated the chitinophosphatic brachiopods out as the Class Lingulata, and more recently Popov et al. (1993) have proposed the Class Calciata to contain the calcareous

brachiopods. This was taken by them to include the Subclass Craniformea and the unspecified Articulates. Following this classification, and supported by the mantle canal evidence, Wright (1994) suggested that the Calciata would consist of the Craniformea, Protremata and Telotremata; however, it may well be that differences between the calcareous non-articulates and the articulates require the retention of the Class Articulata to accommodate the Protremata and Telotremata.

But, in any event, the Subclasses Protremata and Telotremata, as redefined by Jaanusson (1971), form very useful and easily recognisable practical divisions, and should therefore stand from a valuable utilitarian as well as an evolutionary viewpoint.

Acknowledgments

I am grateful to Professor Valdar Jaanusson for much useful discussion; to Dr. Howard Brunton for arranging for material from his collection to be rephotographed at the Natural History Museum, London, and to Mr. Barrie Hartwell, School of Geosciences, Belfast, for photographic help with the other specimens; and to the Geological Museum at the Miami University, Oxford, Ohio, for the loan of the material from their collections figured herein. Dr. Stig Bergström at Ohio State University, Columbus, Ohio, very kindly arranged for me to receive a loan of a substantial sample of ventral valves of the procline orthide *Retrosirostra carleyi*. Specimens of the Silurian spiriferide *Cyrtia exporrecta* have very kindly been provided by Dr. Euan Clarkson of Edinburgh University. Repositories are the BMNH - The Natural History Museum, London; MUGM - Geological Museum, Miami University, Oxford, Ohio; RMS - Swedish Museum of Natural History (Riksmuseet), Stockholm.

REFERENCES

BAKER, P.G. 1990. The classification, origin and phylogeny of thecideidine brachiopods. Palaeontology 33:175-191.

BEECHER, C. E. 1891. Development of the Brachiopoda, Part 1, Introduction. American Journal of Science, Series 3, 41:343-357.

BOUCOT, A.J., J.G. JOHNSON & R.D. STATON. 1965. Order Spiriferida, p. H633-648. In, R.C. Mooore (ed.), Treatise on Invertebrate Paleontology, Part H, Brachiopoda. Geological Society of America, and University of Kansas Press, Lawrence.

BRUNTON, C.H.C. 1972. The shell structure of chonetacean brachiopods and their ancestors. Bulletin of the British Museum (Natural History), Geology Series 21:1-26.

CHUANG, S. H. 1977. Larval development in *Discinisca* (Inarticulate Brachiopod). American Zoologist 17:39-53.

COOPER, G.A. 1956. Chazyan and related brachiopods. Smithsonian Miscellaneous Collections 127 (I & II):1-1245.

FOSTER, M.W. 1974. Recent Antarctic and Subantarctic Brachiopods. American Geophysical Union, Antarctic Research Series 21:1-189.

GORJANSKY, W. J. & L.E. POPOV. 1986. On the origin and systematic position of the calcareous-shelled inarticulate brachiopods. Lethaia 19:233-240.

HAVLICEK, V. 1967. Brachiopoda of the suborder Strophomenidina in Czechoslovakia. Rozpravy Ustredniho Ustavu Geologickeho 33:1-235.

HUXLEY, T.H. 1869. An introduction to the classification of animals. John Churchill and Sons, London, 147 p.

JAANUSSON, V. 1971. Evolution of the brachiopod hinge. Smithsonian Contributions to Paleobiology, 3:33-46.

MORSE, E.S. 1902. Observations on living brachiopods.

Memoirs of the Boston Society of Natural History 5(8):313-386.

NEILSEN, C. 1991. The development of the brachiopod *Crania* (*Neocrania*) *anomala* (O.F. Müller) and its phylogenetic significance. Acta Zoologica 72:7-28.

POPOV, L.E., M.G. BASSETT, L.E. HOLMER & J. LAURIE. 1993. Phylogenetic analysis of higher taxa of Brachiopoda. Lethaia, 26:1-5.

REED, F.R.C. 1917. The Ordovician and Silurian Brachiopoda of the Girvan District. Transactions of the Royal Society of Edinburgh 51 (4):795-998.

ROWELL, A. J. 1960. Some early stages in the development of the brachiopod *Crania anomala* (Müller). Annals Magazine Natural History, Series 13 (3):35-52.

SCHUCHERT, C. 1897. A synopsis of American fossil Brachiopoda, including bibliography and synonomy. Bulletin US Geological Survey 87:1-464.

___1913. Class 2, Brachiopoda, p. 355-420. In, K.A. von Zittel, Text-book of Palaeontology, 1, Second Edition. Macmillan, London.

THAYER, C. W. & STEELE-PETROVIC, H. M. 1975. Burrowing of the lingulid brachiopod *Glottidia pyramidata* : its ecologic and paleoecologic significance. Lethaia 8:209-221.

THOMAS, G.A. 1971. Carboniferous and Early Permian brachiopods from Western and Northern Australia. Australian Bureau of Mineral Resources, Geology and Geophysics Bulletin, 56:1-277.

WILLIAMS, A. 1956. The calcareous shell of the Brachiopoda and its importance to their classification. Biological Reviews 31:243-287.

___1973. The secretion and structural evolution of the shell of thecideidine brachiopods. Philosophical Transactions of the Royal Society of London, Series B, 264:439-478.

___& A.J. ROWELL. 1965a. Evolution and Phylogeny, p. H164-199. In, R.C. Moore (ed.), Treatise on invertebrate paleontology, Part H, Brachiopoda. Geological Society of America and University of Kansas Press, Lawrence.

___&___1965b. Classification, p. H214-237. ibid.

___& A.D. WRIGHT. 1961. The origin of the loop in articulate brachiopods. Palaeontology 4:149-176.

___&___1963. The classification of the "*Orthis testudinaria*" group of brachiopods. Journal of Paleontology 37:1-32.

WRIGHT, A. D. 1968. The brachiopod *Dicoelosia biloba* (Linnaeus) and related species. Arkiv för Zoologi 20:261-319.

___1979. Brachiopod radiation, p.235-252. In, M.R. House (ed.), The origin of major invertebrate groups. Systematics Association Special Volume 12. Academic Press, London.

___ 1981. The external surface of *Dictyonella* and of other pitted brachiopods. Palaeontology 24:443-481.

___1993. A homoeomorph of the articulate brachiopod *Dicoelosia* from the Upper Ordovician Hulterstad fauna of Öland, Sweden. Geologiska Föreningens i Stockholm Förhandlingar 115:65-75.

___1994. Mantle canals on brachiopod interareas and their significance in brachiopod classification. Lethaia, 27:223-226.

___ & M. RUBEL. 1996. A review of the morphological features affecting the classification of clitambonitidine brachiopods. Palaeontology 39: 53-75.

REEF-DWELLING BRACHIOPODS FROM THE LATE PERMIAN OF THE CENTRAL YANGTZE RIVER AREA, CHINA

GUIRONG XU & RICHARD E. GRANT*

Department of Geology, China University of Geosciences, Yujiashan, Wuhan 430074, China;
National Museum of Natural History, Smithsonian Institution, Washington DC 20560, USA [*deceased]

ABSTRACT- Permian carbonate buildups of the central Yangtze area of South China (Hunan, Jiangxi, Hubei provinces) include sponge-microbial-algal and compound coral reefs. Reef-dwelling brachiopods were collected from the sponge-microbial-algal reef member of the Daluokeng Formation (Changxingian), and a few came from the Xiamidong Fm. (Wujiapingian). Some of these brachiopods, especially those from the Daluokeng section at Gaofeng (Hunan), are described here for the first time. Late Permian brachiopod-bearing reef sequences are known from four regions of China: Chenxian County (southern Hunan), Heshan County (Guangxi), and Longdonchuan and Cili counties (Shaanxi). Brachiopod faunas at these four reefal localities, (1) lived in a reefal framework, especially between algally-bound sponges, (2) were mostly endemics confined to their own narrow reef settings, with only two genera, *Araxathyris* and *Leptodus*, common to all faunas in the four regions, (3) lacked inarticulates, (4) included rare chonetids, (5) showed common spiriferids and rhynchonellids in the reef framework, and, (6) had productids, richthofenids and oldhaminids in reefal shell banks. A new reef-dwelling spiriferid genus, *Ladoliplica*, and two new species, the type *L. zigzagiformis*, and *L. platformia*, are described.

INTRODUCTION

Several localities exposing Upper Permian carbonate buildups have been found in the cebtral Yangtze area of South China, especially at Gaofeng (Cili County), Daping and Fanjiacun in Chenxi County (Hunan), Qingshuiyuan in Xiushui County (Jiangxi), Jingquanshan, in Puqi County, and Yushan in Chongyang County (Hubei). The area is located south of the Yangtze River, where Permian rocks are widespread, especially in the northern and southern parts. An ancient landmass, the Jiangnan Oldland, forms a geanticline in the middle part of this area. The basement of the geanticline is pre-Devonian, and middle Devonian rocks overlie the middle Silurian unconformably in this area. Carboniferous rocks are almost completely eroded in the northern part of Hunan. Two fault zones controlled sedimentation after the basement was formed. One was the Dayong-Cili fault zone, located on the north side of the area, the other the Hongjiang-Xupu fault zone. The Late Permian reef belt was developed along the north side of these two fault zones (Figure 1).

Carbonate buildups (reefs) in the central Yangtze area incorporate two associations, (1) sponge-algal-microbial and

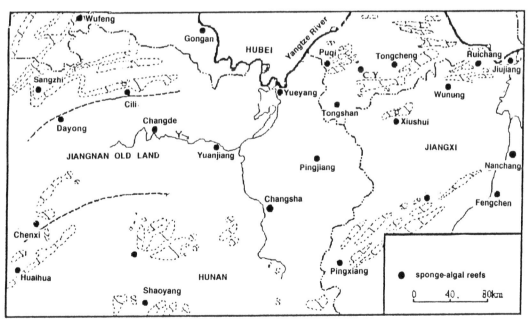

Figure 1. Map of the central Yangtze area, China, showing locations of Late Permian sponge-algal-microbial reefs (black circles) and associated brachiopiod localities.

305

(2) compound corals. The principal components of the Late Permian sponge-algal-microbial reefs in this area were reef building calcareous sponges, calcareous algae-microbial organisms, hydrozoans, and bryozoans. Coral reefs were characterised by only a few associates. Sponge- algal reefs were found in the lower Daluokeng Formation (lower Changxingian) at Gaofeng, Daping, Fanjiacun, Qingshuiyuang, Jingquanshan, and Yushan. Abundant reef-dwelling brachio-pods are associated with the sponge-algal frame builders, but only a few occurred in the coral reef community. The brachiopods in the sponge-algal reefs were difficult to find and collect because most were imbedded in very hard reef lime-stones (Figure 2:3-4). To separate brachiopod specimens from the matrix, the rock was heated, then cooled quickly so as to break up the matrix.

REGIONAL STRATIGRAPHY

The Upper Permian sequences in the central Yangtze area are different from those in South China. The following standard Permian sequenceat the Daluokeng section (near Daluokeng, c.3 km S of Gaofeng, Hunan) is describedfrom the top down: Daluokeng Formation (Upper Permian). Oolitic Limestone Mem-ber 3: greyish, medium-very thick bedded oolites, oncolitic dolostones, with crossbedding (middle), limestone breccias (lower), withSquamularia. Bioclastic Limestone Member 2: grey-dark grey, thick bedded wackestones-packstones, locally with intraclasts, dolomitization, with fora-minifers, gastropods, brachiopods [Uncinunellina, Spiri-gerella], bivalves, cephalopods, algae. Sponge-algal Frame-stone Member 1: greyish white, massive sponge-algal frame-stone, intercalated, lenticular beds of wackestones, shell beds containing corals, sponges, foraminifers, calcareous algae and the brachiopods Uncinunellina multicostioidea Xu & Grant, Spirigerella ovalioides Xu & Grant, Spirigerella shui-zhutangensis(Chan), Ladoliplica zigzagiformis, L. platformia.

Xiamidong Formation. Member 2: coral framestones, reef breccias, calcirudites, bioclastic wackestones-packstones; upper part light grey, thick bedded bioclastic wackestones-packstones, with foraminifers, echino-derms, brachiopods [Tyloplecta, Uncinunellina, Spirigerella guizhouensis Liao,Squamularia grandis Chao]. Lower part is a compound coral reef, with corals in life position, some associated with bryozoans, inozoan sponges, interspaces of corallites filled by lime mud, minor foraminifers, interbedded reef breccias. Member 1: chert nodules, bioclastic wackestones, grey-dark grey, thin-medium bedded, bioclastic, cherty wackestones, intercalated chert layers. Sponge-algal reefs show) abundant frame-building calcisponges (sphinctozoans, sclerosponges), binding algae; calcisponges preserved in growing position; framestones displaying unbedded or massive fabric; thick accumulations; reef builders mainly calcareous sponges, but calcareous algae, hydrozoans, bryozoans, corals; microfacies ranging from supratidal to marginal rim deposits. Reefs show features similar to those of other Permian reefs in South China,e.g. at Lichuan, a typical sponge-algal reef in northwest Hubei Province (Fan & Zhang, 1985).

BRACHIOPOD ASSEMBLAGES AND THEIR DISTRIBUTION

Most reef dwelling brachiopods were found in the sponge-algal reefs of the Daluokeng Formation, but a few were found to occur in coral reefs of the Xiamidong Formation. Three brachiopod assemblages are recognized:

Table 1. Stratigraphic occurrences of Late Permian brachiopods in the middle Yangtze area. Key: f = fossils intact; v = vv; d = dv.

	D III	D II	D I	X II	X I
Anidanthus sinosus				D (1)	
Araxathyris subpentagulata			D (1)		
Araxathyris sp.			D (1)		
Asioproductus bellus				D(1)	
Crurithyris pusilla			D (4)		
Crurithyris speciosa			D (1)		
Crurithyris sp.			D (4)		
Ladoliplica platformia			D (3)		
Ladoliplica zigzagiformis			D (8v,5d)		
Ladoliplica sp.			D (29v, 3d)		
Leptodus sp.		DX (2)	D (f)		
Martinia sp.			D (1)		
Orthothetina sp.	D (f)		D (f)		
Prelissorhynchia pseudoutah			D (1)		
Prelissorhynchia triplicatioides			D (4)		
Rostranteris ptychiventria			D (3)		
Spirigerella discurella		LD (1)	D (1)		
Spirigerella guizhouensis				D (1)	
Spirigerella ovalioides			D (17)		
Spirigerella shuizhutangensis			D (5)		
Spirigerella sp.		X (1)	D (5), Y (f)		
Squamularia elegantuloides			L (1)		X (1), Q (3)
Squamularia formilla			D (1), C (f)		
Squamularia grandis				D (3), Q (f), X (f)	
Squamularia sp.		Dx (1)	D (2)		
Uncinunellina multicostioides			D (2)	D (1)	
Uncinunellina timorensis			Ld (1), C (1)		
Uncinunellina sp.		D (1)	D (2), Z (1)	D (1)	

SECTIONS
D, Daluokeng
X, Xiamidong
Z, Zhuojiapo
L, Luiwangmiao
Dx, Daxi
Ld, Luiwangmiaodong
C, Daping
Y, Yushan
Q, Qingshuiyuan

Table 2. Comparison of four Late Permian reef-dwelling brachiopod faunas from the Yangtze area, China.

	LDC	DLK	HT	LB		LDC	DLK	HT	LB
ORTHIDA					Camarophorinella			2 *	
Acosarina			1	1	Hybostenoscisma			1 *	
Enteletes	1		2 *	1	Prelissorhynchia		2	1 *	1
Meekella	1		2		Terebratuloidea			1	
Orthothetina		1	2	1	Uncinunellina		3	2 *	
Orthotichia			1		SPIRIFERIDA				
Peltichia	1		1		Araxathyris	2	2	3 *	3
Perigeyerella	1		2	1	Cartorhium	1	1		
STROPHOMENIDA					Crurithyris		3	1 *	
Anidanthus		1	1		Eolaballa			1 *	
Asioproductus		1			Hustedia			1	
Chenxianoprodustus			2		Ladoliplica		3		
Compressoproductus			1		Martinia			1*	
Falafer			1		Rectambitus	1			
Gubleria			1		Semibrachythyrina			1 *	
Haydenella			1 *		Speciothyris			1 *	
Huatangia			1		Spirigerella		5 *		
Incisius			1		Squamularia	2	4	5 *	
Leptodus	1	1	3 *	2	TEREBRATULIDA				
Richthofenia	1	1	2		Dielasma		3 *	2	
Spinomarginifera	1		4 *	4	Hemiptychina			1 *	1
Strophalosiina			1 *		Notothyris			5 *	3
Tchernyschevia			2		Qinglongia			2 *	
Tyloplexta			1		Rostranteria	1	1		
RHYNCHONELLIDA					TOTAL GENERA	12	15	40	16
Allorhynchus			1 *		TOTAL SPECIES	14	30	69	24

LDC, *Longdonchuan section, Xikou county*; DLK, *Daluokeng section, Cili County*; HT, *Huatang section*; HS, *Matan section, Heshan Coal Mine*; ∗ *>10 specimens in one species*; 1.. *number of species present*.

Spirigerella guizhouensis - Squamularia grandis assemblage. This assemblage occurs in the upper part of the Xiamidong Fm. The main components include *Spirigerella guizhouensis* Liao, *Spirigerella shuizhutangensis* (Chan), *Asioproductus bellus* Zhan, *Uncinunellina multicostioidea* Xu & Grant, and *Squamu-laria grandis* Chao.

Prelissorhynchia triplicatioides - Ladoliplica zigzagiformis assemblage. This assemblage typically occurs in sponge-algal reefs in the lower Daluokeng Fm., and also at the Luiwangmiao and Luiwangmiaodong sections, Cili County, Daping section oin Chenxi County, and Yushan section, Chongyang County. Other elements include *Uncinumellina multicostioidea* Xu & Grant, *Ladoliplica platformia*, *L. zigzagiformis* [see below], *Araxathyris subpentagulata* Xu & Grant, *Spirigerella ovalioides* Xu & Grant, *Spirigerella shuizhutangensis* (Chan), *Prelissorhynchia pseudoutah* (Huang), *Rostranteris ptychiventria* Xu & Grant. Spiriferids and athyrids were the dominant reef dwellers in the central Yangtze area, with 7 genera and 19 species. Rhynchonellids make up 2 genera and 5 spp., and productoids 2 genera and 2 spp. Orthotetoids, richthofenids, oldhaminids, and terebratulids are represented by only one species each. No chonetids were found in this area.

Spirigerella discurella assemblage. A very small number of brachiopod fossils in this assemblage from the upper part (mbrs 2 and 3) of the Daluokeng Fm. occur at the type locality in Gaofeng. Representative were found also in the Xiamidong Fm. in the Daluokeng and Xiamidong sections. They are not associated with coral reef builders, but occur above the reef framestone or in the intercalated lenticular bioclastic wackestones. The assemblage is preserved in reefs at the same horizon in the Qingshuiyan section, Xiushui County.

COMPARISON WITH OTHER REEF DWELLING FAUNAS OF SOUTH CHINA

A brachiopod assemblage from Late Permian sponge-algal reefs at Huatang (Chenxian Co., S Hunan) described by Liao & Meng (1986) contained 40 genera and 69 brachiopod spp., many of them endemic (Table 2). This brachiopod fauna included the dominant Productida (Oldhaminoidea) with 23 spp. attributed to 14 genera, among which only three species, *Huatangia sulcatifera*, *Rugosomarginifera chengyaoyenensis*, and *Leptodus nobilis*, were represented abundantly. Spiriferids and athyrids included 18 spp. in 11 genera, with abundant *Martinia*, *Squamularia*, *Araxathyris*, common *Semibrachythyrina*, *Phricodithyris*, *Crurithyris*, and lesser *Eliva ? depressa* and *Eolaballa pristina*. Rhynchonellids, terebratulids, and dalmanellinids totaled 26 spp. in 15 genera, with *Stenoscismatoidea*, *Dielasma*, *Notothyris* and *Enteletes* represented abundantly. There were no inarticulates, very few Chonetoidea, and only a single shell of *Waagenites*.

307

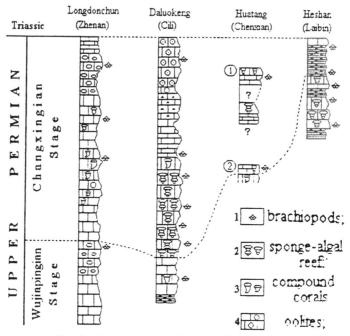

Figure 2. Correlation of four sections of Late Permian reef sequences of the central Yangtze region.

Most of the Late Permian brachiopods in the Huatang section were found in beds intercalated with sponge-algal frame-stones forming coquinas that might indicate formerly in situ brachiopod shell banks. Liao (1987) described a Late Permian silicified brachiopod fauna from Heshan, Laibin County (Guangxi). Some of the brachiopods were associated with sponge reefs, especially in the second member of the Heshan Fm., which yielded 16 genera and 24 spp. (Table 2, Figure 2). The brachiopod fauna of the Longdonchuan Fm. at Longdon-chuan, Xikou County (Shaanxi) was described by Xu & Grant (1994). The brachiopods were associated with compound rugose corals in the upper part of the Longdonchuan Fm., which contains 14 brachiopod spp. in 12 genera (Table 2).

The four Permian brachiopod faunas show several features in common: (1) they all inhabited a reef framework, especially between sponges bound by algae or cyanobacteria, (2) most of the brachiopods were endemic, confined to niches in the reefs with only two genera, *Araxathyris* and *Leptodus*, occur in all four faunas, (3) no inarticulates lived in the reefs; (4) chone-toids were found only occasionally, (5) spiriferids, athyrids and rhynchonellids were common in the reef framework, and (6) productids, richthofenids and oldhaminids formed shell banks in reef settings. To show the relationship between the four faunas, we used binary cluster analysis at the generic level (Figure 3). The Heshan fauna and the Huatang fauna were most closely related and clustered first. The Daluokeng and Longdonchuan faunas formed a second cluster. This is interpreted as caused by an old highland (the Jiangnan Oldland) being present at that time, but not completely separating the northern from southern central Yangtze Sea (refer to Figure 1).

Endemic genera and species were dominant in each of the four faunas. *Ladoliplica platformia*, *L. zigzagiformis*, *Spirigerella ovalioides*, and *Spirigerella shuizhutangensis* were abundant

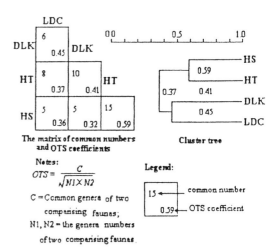

Figure 3. Cluster analysis of the four brachiopod faunas from the Late Permian carbonate buildups of the Yangtze region.

the Cili fauna. In the Longdonchuan Fm., *Rostranteria ptychi-ventria*, *Rectambitus bisulcata*, and *Cartorhium twifurcifer* were more abundant than others. *Notothyris minuta*, *N. depressus*, *Dielasma zhijinense*, *D. nummulum*, and *Prelissorhynchia subrotunda* were principal elements associated with sponge beds. In the Huatang fauna, brachiopods were abundant, formed shell mounds, and *Huatangia sulcastigera*, *Chenxiano-productus nitens*, *Hybastenoscisma bambusoides*, and *Eolaballa pristina* were typical endemic elements.

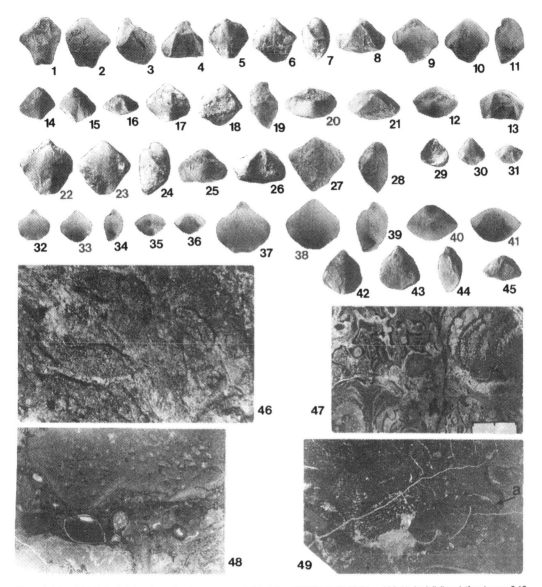

Figure 4. 1-8 and 14-28: *Ladoliplica zigzagiformis* n.gen., n.sp. 1-4, holotype S4F020. 9-13, 29-31 and 32-41: *Ladoliplica platformia* n.sp. 9-13, holotype S4F024. 46: sponge-algal framestone horizon 10. 47: polished surface, horizon 12. 48: brachiopod in sponge-algal framestone, horizon 6. 49: richthofeniid shell, horizon 2, all from Daluokeng section. Shells c. x0.75.

PALEOECOLOGY

Reef-dwelling brachiopods were found either within the reef framework, commonly with sponges and other frame builders, or in channels on reef flats. The reefal framework brachiopods showed (i) relatively small shells compared with counterparts in non-reef environments, (ii) a strong pedicle tfor attachment (Figure 4:3), (iii) smooth or simply ornamented shells (e.g. *Prelissorhynchia*, *Crurithyris*, *Spirigerella*), and, (iv) capacity also to inhabit non-reef environments. Channel-dwelling brachiopods lived at the bottom of reef flat channels. The spiriferid *Ladoliplica*, and productid *Richthofenia* were attached to the substrate by a strong pedicle or spines, and had relatively

wide cardinal areas to prevent the shells from being swept away by currents (Figure 4:4). A submarine micro-erosion sur-face existed beneath these fossils, indicating that they lived on a soft substrate in the reef environment. In the shell interior, structures of the apically cystose shell was preserved, indicating that the shells were not transported far from their living place before burial.

Reefs or banks were rarely formed by brachiopods alone, but some shell beds were intercalated between sponge framestone. In such shell beds large numbers of transported brachiopod shells accumulated, with few other organisms. These

309

shell beds were formed on a short-lived reef-flat environment after the reef 'partially died' [sic, ed.].

SYSTEMATIC PALEONTOLOGY
Order Spiriferida Waagen 1883
Family Martiniidae Waagen 1883
Ladoliplica n. gen.

Type species. *Ladoliplica zigzagiformis* n.sp. Age and distribution. Wujiapingian, Upper Permian, South China.

Diagnosis. Medium-sized, moderately biconvex, smooth shell, subpentagonal outline, strongly incurved beak, short hingeline, small interarea on both valves; dorsal interior with converging hinge plate at umbonal area; short, small, sometimes bifurcated cardinal process; lateral simple spiralia with only two coils.

Comparison. Some species of *Martinia* have a greatly extended tongue on the anterior fold, e.g. *Martinia rhomboidalis* (Girty, 1909, in Cooper & Grant, 1976), which is rather comparable with *Ladoliplica zigzagiformis*, but the American species has a conspicuously rounded fastigium. *Ladoliplica* is distinguished from *Martinia* in its subpentagonal outline, strongly incurved beak, zigzag lateral commissure, and extended, pointed tongue.

Ladoliplica zigzagiformis n.sp.
Figures 4:1-8, 14-21, 25-31; fig.5.

Type locality. Unit D1, Daluokeng Formation, beds 7, 9 and 16, Daluokeng section, Hunan.

Diagnosis. Medium-sized, lateral commissure zigzag, tongue of pedicle valve long, slightly rounded platform.

Description. Shell equally wide as long, widths 14-19 mm, moderately ventribiconvex; beaks strongly incurved, sharp; triangular interareas on both valves, open delthyrium, notothyrium; ventral valve deeper at mid-posterior, greatest swelling in front of umbonal region; dorsal valve less convex, longitudinally forming flat line along fold except umbonally; profile strongly curved posteriorly, flattened at middle, gradually sloping anteriorly; commissure zigzag laterally, uniplicate anteriorly, fold low at mid-length, gradually heightening anteriorly, forming slightly rounded fastigium; sulcus shallow, beginning near or slightly anterior to mid-length, moderately to greatly extended anteriorly, slightly rounded; costae absent, but with fine lines or fibres sparsely spaced on exfoliated or abraded shells, growth lamellae weak, visible only anteriorly.

Ventral interior with strong hinge teeth, triangular in outline, blunt; dental ridges extended toward midline, converging near umbonal region, forming short platform; dental plates absent; median septum umbonally. Dorsal interior with shallow hinge sockets; downward, short plate not reaching valve floor, beneath hinge socket; hinge plates converging umbonally; short, small cardinal process, sometimes bifurcated; spiralia simple, two coils.

Ladoliplica platformia n. sp.
Figures 4:9-13, 22-24, 32-41; fig.6.

Type locality. Unit D1, Daluokeng Formation, from beds 7, 15, Daluokeng section, Hunan.

Diagnosis. Flat, square tongue at anterior ventral valve, lateral commissures angularly zigzag.

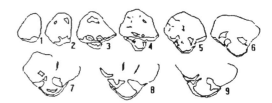

Figure 5. Serial sections of *Ladoliplica zigzagiformis* n.gen., n.sp., at 0.03 mm intervals, x3.

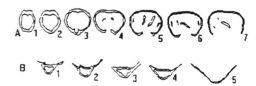

Figure 6. *Ladoliplica platformia* n.sp. A-ventral valve, sections at 0.04 mm intervals; B-dorsal valve, sections at 0.05 mm intervals, x2.

Description. Lengths 10-16mm, widths11-17mm, outline regularly subpentagonal in maturity, slightly transverse; moderately biconvex; ventral valve deepest at mid-posteriorly; hinge ends slightly protruding, sharply angled at angular, zigzag lateral commissure; uniplicate anteriorly; fold low at mid-length, gradually becoming very high anteriorly; dorsal valve steeply sloping anterolaterally, forming square tongue; ventral valve generally strongly convex, greatest swelling in front of umbonal region; valve profile strongly curved at umbonal region, flattened at middle, gradually sloping anteriorly; dorsal profile less convex, forming flat line along fold except in umbonal region; strongly convex transversely; fastigium at fold gradually sloping posteriorly, except for umbonal region, steeply sloped anteriorly.

Comparison. The new species is similar to *Ladoliplica zigzagiformis* in shape, incurved beak, short hinge line, and zigzag lateral commissure, but differs in its generally smaller size, more angular and narrow fold, zigzag lateral commissure, and square, flat anterior commissure, square tongue. The two species co-occur in bed 7 of the type species locality.

Acknowledgments

The publication of this paper is supported by the National Natural Scientific Foundation Committee of China, the China Geology and Mineral Resources Professional Foundation, and the Smithsonian Institution (USA). The Natural Sciences and Engineering Research Council of Canada, Ottawa, provided travels and accommodation support to attend the Third International brachiopod Congress. Other workers on the project were Luo Xinmin, Lin Qixiang, Wang Yongbiao, Chen Linzhou, and Xiao Shiyu of the China University of Geosciences. The author also thanks R.E. Grant (deceased) for his help.

REFERENCES

CHAO, Y.T. 1929. Carboniferous and Permian spiriferids of China. Palaeontologica Sinica B11(1):1-133.
COOPER, G.A. & R.E. GRANT.1969. Permian brachiopods of west Texas. Smithsonian Contributions Paleobiology 1:1-20.
___&___1974. Permian brachiopods of west Texas, II. ibid.14:233-793.

___&___1975. Permian brachiopods of west Texas, III. ibid.19:749-1920.

___&___1976a. Permian brachiopods of west Texas, IV. ibid. 21:1923-2607.

___&___1976b. Permian brachiopods of west Texas, V. ibid. 24:2609-3159.

FAN J.S. & W. ZHANG. 1985. Sphinctozoans from late Permian reefs of Lichuan, western Hubei, China. Facies 13:1-44.

GIRTY, G.H. 1909. The Guadalupian fauna. U.S. Geological Survey Professional Paper 58:1-651.

GRANT, R.E. 1971. Brachiopods in the Permian reef environment of West Texas. Proceedings North American Paleontological Convention 1969 J:1444-1481.

Li X.J., L.Z. Chen, X.M. Luo. 1993. The reefs of Changxing Formation in southern Hunan Province. Geologica Sinica 28:317-326.

LIAO, Z. T.1987 Paleoecological characters and stratigraphic significance of silicified brachiopods of the Upper Permian from Heshan, Laibin, Guangxi, p.81-125. In, Stratigraphy and paleontology of systemic boundaries in China, Permian-Triassic Boundary. Geological Publishing House, Beijing.

___& F.Y. MENG. 1986. Late Changxingian brachiopods from Huatang of Chen Xian County, southern Hunan. Memoirs Nanjing Institute Geology Palaeontology, Academia Sinica 22:71-94.

CHERNYSHEV, T.N. 1902. Die obercarbonischen Brachiopoden des Ural und des Timan. Mémoires Comité Géologique 16(2):1-749.

XU, G.R. 1987. Brachiopods, p.215-235. In, Z.Y. Yang, et al., Permian-Triassic boundary stratigraphy and fauna of South China. Geological Publishing House, Beijing.

___& R.E. Grant. 1994. Brachiopod faunae near the Permian-Triassic boundary in South China. Smithsonian Contributions Paleobiology 76:1-68.

YANG, Z.Y., S.B. WU, H.F. YIN, G.R. XU & K.X. ZHANG. 1991. Permo-Triassic events of South China. Geological Publishing House, Beijing, 183p (Chinese)

___,___,___,___ &___1993. ibid.,153p. (English).

PERMIAN BRACHIOPODS FROM THE TSUNEMORI FORMATION, SW JAPAN, AND THEIR PALEOBIOGEOGRAPHIC IMPLICATION.

JUICHI YANAGIDA

Department of Earth and Planetary Sciences, Faculty of Science, Kyushu University 33, Fukuoka 812-81, Japan

ABSTRACT-The Permian Tsunemori Formation of SW Japan, composed of a series of trenchfill deposits, is considered to have been formed by collisional collapse and accretion of the Akiyoshi Limestone at the end of the Mid-Permian. Brachiopods in the mudstones have been examined for the purpose of elucidating the last stage of the history of the Akiyoshi Limestone Group [=the Akiyoshi Seamount with a cap of thick reefs].With 16 species and 15 genera, faunal elements are strongly related to those of the Middle to Upper Permian of S and E China. Some cool water elements are also recognized: habitat temperatures may have fallen about the end of the Mid-Permian. Collison and accretion of the Akiyoshi Seamount is considered to have occurred at the end of the Guadalupian (=Midian), or at the beginning of the Dzhulfian.The depositional site of the Tsunemori Fm might originally have been close to the northern margin of the Yangtze Platform.

INTRODUCTION

A huge, exotic limestone mass called the Akiyoshi Limestone Group is distributed in the 'Inner Zone' of SW Japan. It is composed of oceanic carbonates ranging from the Early Carboniferous (Tournaisian) to the Middle Permian (=Midian, Guadalupian). It was developed as a reef complex (Ota, 1968) built on a seamount (Kanmera & Nishi, 1983) of basaltic and pyroclastic rocks (Yanagida et al, 1971). This limestone was developed mostly in the lower paleolatitudes, shifted westwards with plate movement. At about the end of the Mid-Permian, the Akiyoshi Seamount collapsed to form a microcontinent. Part of the seamount and its limestone cap accreted to terrigenous sediment to form the Tsunemori Fm. (Sano & Kanmera, 1991a-d), comprising black mudstones, sandstones,, and pebble-bearing mudstones with a total thickness of 350m (Toriyama, 1954). Lithofacies analysis shows that these sediments may be considered as trench-fill deposits (Kanmera & Nishi, 1983). These poorly sorted black mudstones contain fossils that are transported to the deep sea by turbidity flows. These include brachiopods, fusulinid forams, rare bivalves and cephalopods. Mixed fusulines of Carboniferous to Permian age appear in limestone blocks of different sizes, which belong to the Akiyoshi Limestone. Radiolarians include *Pseudoalbaiella globosa* and *Folliculus monacanthus*, reported by Kanmera & Sano (1986) also to be of mid-Permian

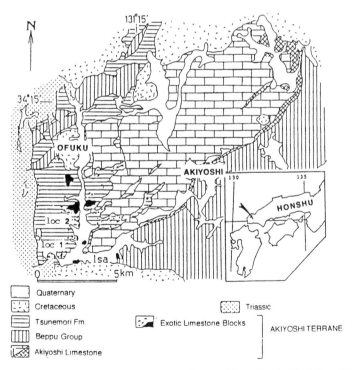

Figure 1. Index map of Middle Permian brachiopod localities 1 and 2 from the Tsunemori Fm. on Honshu Island, Japan (after Sano & Kanmera, 1991a).

313

Figure 2. Representative brachiopods of the Tsunemori Fm. 1, *Cathayspiriferina* aff. *fenshuijiangensis* Liang (x2); 2, 4, *Alispiriferella* sp. (x1.5); 3, *Spiriferella* aff. *qubuecensis* Chang (x1.5); 5-7, *Derbyoides* sp. (x2); 8, *Goniarina* cf. *subulata* Grant (x3); 9, *Waagenites* cf *wongiana* (Chao) (x3); 10, *Waagenites* aff. *striata* Liao (x3); 11, *Megousia* aff. *definita* Cooper & Grant (x2); 12, *Kozlowskia* sp. (x2); 13, Gubleria sp. (x1.5); 14, *Leptodus* sp. (x2); 15, *Enteletes* cf. *wannanensis* Zhang & Ching (x2); 16, *Stenocisma* sp. (x2).

age. The brachiopod fauna appears to provide additional data to the history of the Akiyoshi Limstone.

OCCURRENCE OF BRACHIOPODS, OTHER FAUNA

There are two brachiopod localities consisting of small road-side exposures (Figure 1). Both are characterised by pebble-bearing black mudstones, with grain sizes from well-rounded to sub-rounded pebbles to subangular granules, and are poorly sorted. Rocks include sandstones, mudstones, limestones and volcanics with brachiopods, crinoid ossicles, bryozoans, and fusulinids. Fusulinids include *Lepidolina multiseptata shiraiwensis* (Ozawa), *Yabeina* sp., *Sumatrina* sp., *Afghanella* cf. *schenki* Thompson, *Neoschwagerina* sp., *Chusenella* sp. *Paraschwagerina* sp., and *Pseudofusulina ambigua* (Deprat).

Most brachiopod shells were disarticulated, dissolved and preserved as moulds, which, however, show fine ornamentation examined as latex replicas. It is assumed that transport from the original habitat, and subsequent burial preserved the shells.

CORRELATION

Most of the fauna comes from Locality 1, dominated by spiriferids, but with low abundance (rarely more than 10 specimens). Taxa include *Spiriferella qubucensis* Chang, *Cathayspirina* aff. *fengshuijiangensis* Liang, *Waagenites* aff. *striata* Liao, *Alispiriferella* sp., *Leptodus* sp., *Gubleria* sp., *Waagenites* cf. *wongiana* (Chao), *Hustedia* sp., *Enteletes* cf.

wannanensis Zhang & Ching, *Derbyoides* sp., *Stenocisma* sp., *Goniarina* cf. *subulata* Grant, *Megousia* aff. *definita* Cooper & Grant, *Kozlowskia* sp., and *Tyloplecta* sp. At locality 2, only the first three taxa occur. The most remarkable character is the boreal nature of the fauna, e.g. with *Spiriferella*, *Alispiriferella*, *Megousia* and *Stenocisma*. Simultaneously there are warm water elements such as *Leptodus*, *Gubleria*, *Goniarina* and *Tyloplecta*. Mabuti (1937) probably discovered *Gemmellaroia ozawa* at Locality 1.

Spiriferella qubucensis is close to the species found by Zhang & Jin (1976) from the Permian Selong Group in the Jolmo Lungma region (Mount Everest), where seven species of the genus are described. *Alispiriferella* sp. may be the youngest species of the genus, known normally from the Arctic (Waterhouse et al., 1978). The *Cathayspiriferina* form, common in the Tsunemori Fm., is close to that reported by Liang from the Lengwu Fm. of Zhejiang province, China. *Leptodus* and *Gubleria* reveal Tethyan affinities. The orthotetids, *Derbyoides* and *Goniarina* are also small, like those described from the late Permian of Hydra, Greece (Grant, 1995). The small chonetids, e.g. *Waagenites*, and *Enteletes* are similar to those from the Lontang Fm of S China. The productid *Megousia* is very close to the species from the Permian Leonard Fm. and Cherry Canyon Fm. of Texas (Cooper & Grant, 1975). *Tyloplecta* is doubtfully identified by densely arranged spine bases and posterior strong rugae.

FAUNAL CHARACTER

The brachiopod fauna of the Tsunemori Fm is of Middle

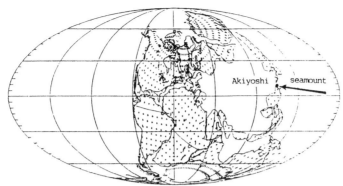

Figure 3. Inferred location of the Akiyoshi Seamount about the end of the Middle Permian, with arrow indicating direction of plate movement (map from Scotese & McKerrow, 1990).

Permian (Guadalupian) age. However, there are Late Permian characters as seen in *Enteletes* cf. *wannanensis, Waagenites* cf. *wongiana*, and *W. striata*. like those of the Lontang Fm. of S China Furthermore, *Goniarina* is close to the Late Permian form from Greece. *Cathayspiriferina* is close to the late Middle Permian species from Zhejiiang, E China. The Tsunemori fauna includes forms similar to the warm water fauna of S and E China, with elements of the cold water fauna totally absent in China.

CONCLUSIONS

The uppermost fusulinid zone in the Akiyoshi Limestone is the late Middle Permian *Lepidolina multiseptata shiraiwensis* Zone. The Akiyoshi Seamount accumulated reefal carbonates from the Early Carboniferous onwards.At the end of the Mid-Permian, these limestones collided with siliciclastic (terrigenous) facies, and subduction brought about the end of the reef complex, collapse of the limestone succession, and appearance of trench-fill deposits. Brachiopods and fusulinids were transported from shallow habitats to the deeper bottom facies by turbidity currents. The Tsunemori fauna is related to that of the Yangtze Platform, but with cooler elements added. This suggests that the Akiyoshi Limestone accumulated in lower latitudes and drifted towards, and ultimately collided with, the northern margin of the Yangtze Platform at the end of the Mid-Permian (Guadalupian) or in the early Late Permian (Dzhulfian).

Acknowledgments

The author wishes to thank the late Mr Goro Okafuji, Professor Emeritus Michihoro Kawano, Gentaro Naito (Mione City Museum of History and Folklore), Akihiko Sugimura, Takayosho Fukutomi, Fumio Takahashi, Hiseho Ishida, and Koichiro Takeuchi, who supplied many interesting specimens for study, and Masamichi Ota for long years of collaboration. Financial support came from the Ministry of Education, Science and Culture of Japan.

REFERENCES

COOPER, G.A. & R.E. GRANT. 1975. Permian brachiopods of West Texas. Smithsonian Contributions to Paleobiology 19: 795-1921.

GRANT. R.E. 1995. Upper Permian brachiopods of the Superfamily Orthotetoidea from Hydra Island, Greece. Journal of Paleontology 69: 755-670.

KANMERA, K. & H.NISHI. 1983. Accreted organic reef complex in Southwest Japan, p.195-206. In, M. HASHIMOTO & S. UEDA (eds.). Accretion tectonics in the circum-Pacific regions, Terra Publishing Co, Tokyo.

KANMERA, K. & H. SANO. 1986. Straigraphy and structural relationships among pre-Jurassic accretionary and collisional system in Akiyoshi Terrane, p.51-88. In, Guidebook for Excursion, International Geological Correlation Project 224. Osaka, Japan.

LIANG, W. 1990. Lengwu Formation of Permian and its brachiopod fauna in Zhejiang. Memoirs Ministry of Geology and Mineral Resources 2(10): 1-521.

LIAO, Z. 1980. Upper Permian brachiopods from western Guizhou, p. 241-277. In, Stratigraphy and Paleontology of Upper Permian coal-bearing formations in western Guizhou and Yunnan, Science Press, Beijing.

MABUTI, S. 1937. On a Permian brachiopoid *Gemmelaroia* (*Gemmelaroiella*) *ozawai* subgen, et sp. nov. from Japan. Proceedings Tokyo Imperial Academy 13: 16-19.

OTA, M. 1968. The Akiyoshi Limestone Group: a geosynclinal organic reef complex. Bulletin Akiyoshi-dai Museum of Nautral History 5: 1-44.

SANO, H. & K. KANMERA. 1991a. Collapse of ancient oceamnic reef complex - what happened during collision of Akiyoshi Reef Complex? Geologic setting and age of Akiyoshi terrane rocks on western Akiyoshi-dai plateau. Journal Geological Society of Japan 97: 113-133.

___1991b. Broken Limestone as collapse product. Ibid., 217-229.

___1991c. Limestone breccias, redeposited limestone, and mudstone injections. Ibid., 297-309.

___1991d. Sequence of collisional collapse and generation. Ibid., 631-644.

SCOTESE, C.R. & W.S. MCKERROW. 1990. Revised world map and introduction. In, W.S. MCKERROW & C.R. SCOTESE (eds.), Paleozoic paleogeography and biogeography. Geological Society Memoir 12: 435p.

TORIYAMA, R. 1954. Geology of Akiyoshi, Part II. Memoirs faculty of Science Kyushu University, Geology 5(1): 1-46.

WATERGOUSE, J.B. B.J. WADDINTON, & N. ARCHBOLD. 1978. The evolution of the Middle Carboniferous to Late Permian brachiopod family Spiriferellinae. Special papers Geological Association of Canada 18: 415-443.

YANAGIDA, J. , M. OTA, S. SUGIMURA, & T. HAIKAWA. 1971. On the geology of the northeastern party of the Akiyoshi Limestone Plateau. Science Reports, Department of Geology, Kyushu University 11: 105-114.

ZHANG, S. & Y. JIN. 1976. Late Paleozoic brachiopods from the Mount Jolmo Lungma region. In, Report of the Scientific expedition in the Mount Jolmo Lungma region, Paleontology, fascicle II, Science Press, Beijing.

FRASNIAN ATRYPOID BRACHIOPODS FROM SOUTH TIMAN

Yu. A. YUDINA

Timan Pechora Department, All-Russian Petroleum Scientific Research Geological Exploration Institute, Pushkin St 2, 169400 Ukhta, Komi Republic, Russia

ABSTRACT- Atrypids are present in all Frasnian brachio-pod faunas of South Timan: most of the fauna is assignable to the family Atrypidae, except possibly for 'Carinatina'. They are represented by *Pseudoatrypa*, *Desquamatia*, *Atryparia*, *Iowatrypa*, *Pseudogruene-waldtia*, *Radiatrypa*, *Rugosatrypa*, *Spinatrypa*, *Spin-atrypina*, and 'Carinatina'. Most atrypoids inhabited quiet water, muddy substrates; a few dwelled in reefs, e.g. 'Carinatina', *Radiatrypa*, and *Desquamatia*.

INTRODUCTION

Brachiopods are important tools for Devonian biostratigraphy in south Timan. They are the commonest, most numerous and diverse Frasnian megafossils, and include many atrypoids. The earliest Frasnian atrypids are known from the Timan suite. The upper part of this suite is correlated with the lowermost *asymmetricus* conodont Zone, but conodonts are not found in the lower Timan rocks. The Lower Timan atrypid brachiopods are represented by *Pseudoatrypa* and *Desquamatia*, accompanied by *Devonoproductus*, *Uchtospirifer*, *Komispirifer* and *Cyrtina*. In South Timan, *Pseudoatrypa*, *Desquamatia*, *Iowatrypa*, *Spinatrypina*, and *Spinatrypa* (*Spinatrypa*) were present during the early Middle Frasnian [Lower *asymmetricus* conodont Zone]. These are associated with *Ladogia*, *Comiotoechia*, *Cupularostrum* ?, *Elytha*, *Nordella*, and *Eleutherokomma* in the Ustyarega Suite.

Relatively deep water sediments were formed in sediment-starved depressions in south Timan during the late Middle Frasnian (M+U *asymmetricus* and *A. triangularis* Zone). This 'Domanik Suite' is composed mainly of bituminous carbonates and cherts, lacking articulate brachiopods. Various atrypid genera were characteristic for Late Frasnian time (*gigas* Conodont Zone). The highest number of atrypid genera (seven) occurred in middle Late Frasnian time. All the Middle Frasnian atrypid genera persisted when 'new' genera, such as *Radiatrypa*, *Rugosatrypa*, *Pseudogruenewaldtia*, and 'Carinatina' appeared in the area. In addition, the Vetlasyan, Syrachoy, and Ukhta Suites contain representatives of *Nervostrophia*, *Schuchertella*, *Productella*, *Gypidula*, *Theodossia*, *Adolfia*, *Cyrtospirifer*, and *Cryptonella*. In the Late Frasnian Liyayel and Sedyu Suites, atrypids are accompanied by *Monelasmina*, *Ryocarhynchus*, *Calvinaria*, *Warrenella*, *Cyrtospirifer*, and *Biernatella*. Late Frasnian reef brachiopod assemblages with atrypids include *Gypidula*, *Hypothyridina*, *Crurithyris*, and *Cryptonella*. The end of the Frasnian is marked by total extinction of the atrypid brachiopod fauna (Figure 1).

DISTRIBUTION OF SPECIES

In South Timan, the genus *Pseudoatrypa* Copper (1973) includes early Frasnian species from the Timan Suite, e.g. *P. grossheimi* (Lyashenko), *P. nefedovae* (Lyashenko), *P. velikaja* (Nalivkin). Middle Frasnian species from the Ustyarega suite include *P. martynovae* (Lyashenko), *P. philippovae* (Lyashenko), and *P. cf. richthofeni* (Kayser). The

late Frasnian species from the Syrachoy suite is *P. symmetrica* (Lyashenko). All are from shallow water siliciclastics and carbonates. *Desquamatia* Alekseeva (1960) is represented by two species, one, *D. nalivkini* (Lyashenko), occurs with *Pseudoatrypa* in assemblages from the lower and middle Frasnian. Shells of the second species, *D. poljanica* (Lyashenko) have been discovered in late Frasnian reefs. *Rugosatrypa* Rzhonsnitskaya (1975) and *Radiatrypa* Copper (1978) are present in Late Frasnian brachiopod fauna. *Rugosatrypa* is represented by *R. sp. 'E'*, which is abundant in argillaceous limestones of the Sedyu Suite. *Radiatrypa* is also represented by one species, *R. magnitica* (Nalivkin) occurring in reef complexes. Perhaps, the genus *Desquamatia* was an ancestor of both *Radiatrypa* and *Rugosatrypa*. Externally, *Radiatrypa* differs from *Desquamatia* in being more finely costate, with thickening of shell wall in umbonal region. This may possibly be an adaptation to a high energy reef environment. *Rugosatrypa*, in comparison with *Desquamatia*, is less coarsely ribbed and has fine lamellose growth lines. Possibly this is a result of *Rugosatrypa*'s adaptation to live in deeper and quieter waters, with more clays.

The genus *Atryparia* Copper (1964) has two species in south Timan during Late Frasnian *gigas* Zone. Shells of *A. vetlasjanica* (Lyashenko) have been encountered in marls and argillaceous limestones of the Vetlasyan Suite, infilling sediment-starved depressions. Single valves of *Atryparia* sp.

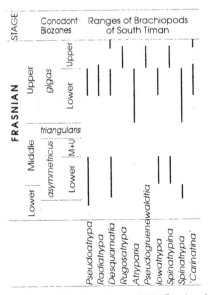

Figure 1. Ranges of brachiopod genera in the Frasnian of south Timan.

'N1' have been found in the upper Liyayel Suite (a slope facies of the Syrachoy reefs). *Iowatrypa* Copper (1973) is widespread in the lower Frasnian deposits of south Timan (lower *asymmetricus* Zone). Here, shells of *Iowatrypa timanica* (Markovskii) have been
collected from the Ustyarega shallow water limestones. The Late Frasnian marked a second occurrence of this genus: *Iowatrypa keranica* (Lyashenko) is characteristic of the Liyayel assemblage. *Pseudogruenewaldtia* Rzhonsnitskaya (1960) is known so far only from the Sedyu Suite. Interbeds of argillaceous limestone of this suite contain abundant shells of *P. tschernyschewi* Rzhonsnitskaya. The probable ancestor of *Pseudo-gruenewaldtia* is *Iowatrypa*.

The genus *Carinatina* Nalivkin (1941) may be represented by 'Carinatina' sp. 'B', rare in the Syrachoy and Ukhta reefs (*gigas* Zone). *Spinatrypa* Stainbrook (1951) occurs in nearly the whole Frasnian sequence of South Timan. A single indeterminate *Spinatrypa* (*Spinatrypa*) shell has been obtained from the Lower Frasnian Timan Suite. *Spinatrypa* (*S.*) *semilukiana* Lyashenko is characteristic of the Late Frasnian Vetlasyan Suite. The late Frasnian Syrachoy beds yield shells of *S.* (*S.*) ex gr. *semilukiana* Lyashenko and *S.*(*S.*) aff. *planosulcata* (Webster). *Spinatrypa* (*Isospinatrypa* ?) sp. 'N' is known so far only from the Sedyu Suite. Shells of *Spinatrypa* (*Spinatrypa*) are commonly present in shallow argillaceous limestones.

Spinatrypina Rzhonsnitskaya (1964) is represented by four species: *S. ninae* (Lyashenko) is known from the Ustyarega beds, and other species have been found in the upper part of Late Frasnian deposits, e.g. *S.* ex gr. *tubaecostata* (Paeckelmann) present in the Ukhta suite, *S. aschensis* (Markovskii), and *S. sikasa* (Markovskii) from the Sedyu Suite.

CONCLUSIONS

Frasnian deposits of south Timan are remarkable for their variety of brachiopods. Especially atrypoids are present in all complexes, often in great numbers. Atrypoids may be used successfully for detailed subdivision of Frasnian
sediments. Most atrypoids preferred low energy environments and soft, muddy bottoms. A few atrypoids moved to the high energy reef environment, e.g. *Radiatrypa*, 'Carinatina' and some *Desquamatia*.

REFERENCES

ALEKSEEVA, R.E. 1960. O novom podrode *Atrypa* (*Desquamatia*) subgen. nov., semeistva Atrypidae Gill (Brakhiopody). Doklady Akademiya Nauk SSSR 131(2): 421-424.

COPPER, P. 1964. European Mid-Devonian correlations. Nature, London 204: 363-364.

___ 1973. New Siluro-Devonian atrypoid brachiopods. Journal of Paleontology 47: 484-500.

LYASHENKO, A.I. 1973. Brakhiopody I stratigrafiya nizhne-franskiikh otlozhenii yuznogo Timana i Volgo-uralskoi Neftegazonosmoi Provintsii. Izdatelstvo NEDRA, 279p.

NALIVKIN, D.V. 1941. Brakhiopody glavnogo Devonskogo Polya, p.139-226. In, M.A. Batalina et al. (eds.), Fauna glavnogo Devonskogo Polya. Paleontologicheskii Institut, Akademiya Nauk SSSR.

STAINBROOK, M.A. 1951. Substitution for the pre-occupied name *Hystricina*. Journal Washington Academy of Sciences 41:196.

RZHONSNITSKAYA, M.A. 1964. O devonskikh atripidakh Kuznetskogo Basseina. Trudy VSEGEI 93: 91-112.

___1975. Biostratigrafiya devona okraina Kuznetskogo Basseina, 2, Opisanie brakhiopod. VSEGEI, Leningrad, 232 p.

FRASNIAN-FAMENNIAN BRACHIOPOD FAUNAL EXTINCTION DYNAMICS: AN EXAMPLE FROM SOUTHERN POLAND

A. BALINSKI

Institut Paleobiologii PAN, Al. Zwirki i Wigury 93, 02-089 Warszawa, Poland

ABSTRACT- Brachiopod species diversity in the Frasnian of southern Poland reveals that the fauna was most diversified in the early part of the stage. Following this, diversity gradually decreased, reaching a minimum just above the Frasnian-Famennian boundary. Recovery started in the late *P. triangularis* Chron. During the *P. crepida* Chron, surviving brachiopods seem to have fully recovered, reaching highest taxonomic diversification for the Famennian. The same data plotted as turnover rates show a sharp negative pulse expressing high extinction rates, and very low numbers of newly originating species. The second negative turnover pulse occurred during the later part of the *P. crepida* to *P. rhomboidea* chrons. This event was associated with strong regression in the basin, and migration of the brachiopod fauna out of the area. Co-occurring conodonts reveal a similar pattern of evolutionary dynamic.

INTRODUCTION

Devonian strata of the Debnik Anticline represent the southern facies of the Polish epicontinental basin and were deposited near the 'Pre-Carpathian Land' area (Figure 1). Today these strata are usually poorly exposed in a small isolated area. Nevertheless, Devonian rocks of the Debnik Anticline are important in understanding the geology and paleogeography of Poland. Fossils from the area have been studied since Roemer (1863) and Gürich (1903) (for reviews see Siedlecki, 1954; Narkiewicz & Racki, 1984). Recent studies of the Upper Devonian brachiopods, with their biostratigraphic framework, are based on conodonts (Balinski, 1979, 1995). This paper analyses brachiopod faunal dynamics during the Late Devonian with special emphasis on the Frasnian--Famennian biotic extinction crisis, and compares this with contemporaneous conodonts.

DIVERSITY OF THE BRACHIOPOD FAUNA

The Frasnian brachiopod succession (see Balinski, 1979; Figure 2) starts with an impoverished fauna represented by a single atrypid species *Spinatrypina* (*Spinatrypina*) sp. This part of the section represents a rather short episode of improved water circulation during the latest Givetian and earliest Frasnian, which was otherwise characterized by poorly oxygenated, stagnant water conditions (Racki & Balinski, 1981).

In the following early Frasnian *Cyrtospirifer bisellatus* interval, brachiopods are more diverse, which suggests better environmental conditions. A transgressive pulse recorded in the study area , and many other places of the world at that time, drowned the extensive Middle Devonian carbonate platform of southern Poland (Narkiewicz & Racki, 1987; see also Schlager 1981; Johnson et al., 1989). The brachiopod assemblage is dominated by spire-bearers (spiriferoids and atrypoids). This assemblage displays some similarity to the faunas of the early part of Frasnian of the East European Platform. The common species for these areas are

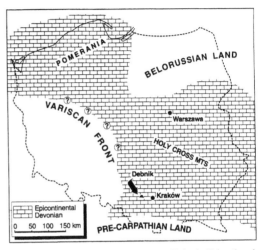

Figure 1. Schematic paleogeographic map of Poland with location of the Debnik Anticline Outcrops.

Desquamatia (*Neatrypa*) *velikaja* (Nalivkin), and *Spinatrypa* (*Spinatrypa*) *semilukiana* Lyashenko. The genus *Eleutherokomma* Crickmay is represented at Debnik by a characteristic endemic species *E. zarecznyi* (Gürich). In Russia this genus includes several related early Frasnian species described by Lyashenko (1959, 1978) as *Lamellispififer* (=*Mucrospirifer*) and *Dmitrispirifer* (=*Eleutherokomma*). A recent study of the distribution of *Eleutherokomma* has shown that the genus migrated to Europe from North America in the early Frasnian (Brice et al. 1994).

During the mid-Frasnian *Plionoptycherhynchus cracoviensis* interval (*punctata* to *hassi* conodont zones), the brachiopod fauna was the most diverse of the whole Frasnian (20 species). Its most important feature is the appearance of rhynchonelloids (7 species), which immediately became the dominant group in the association. The characteristic rhynchonelloid genera for the interval are *Hypothyridina*, *Coelotherorhynchus*, *Plionoptycherhynchus*, *Lateralatirostrum* and *Flabellisinurostrum*. Other brachiopods are represented by strophomenoids (4 spp.), spiriferoids (2 spp.), inarticulates (4 spp.), and atrypoids, orthids and pentameroids (each represented by a single species). Most characteristic for the assemblage are endemic *P. cracoviensis* (Gürich) and *Flabellisinurostrum guerichi* (Balinski). The Canadian species *Lateralatisinurostrum athabascense* (Kindle) and *Coelotherorhynchus schucherti* (Stainbrook) also occur in the studied area. There are several other species which are closely related to, or identical with, Russian forms, e.g. *Hypothyridina ascendoides* Nalivkin, *Atryparia* (*Costatrypa*) aff. *uralica* (Nalivkin), and *Chonetipustula* aff. *petini* (Nalivkin). The co-occurring conodont fauna documents an increasing proportion of

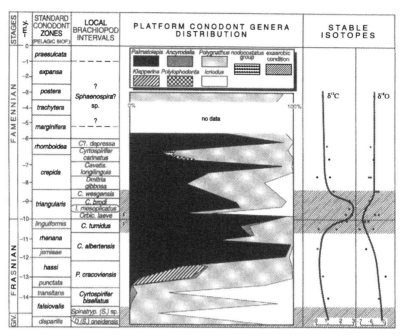

Figure 2. Comparative percentage distribution of platform conodont genera from the Late Devonian of the Debnik Anticline (southern Poland).

palmatolepids and deep-water, wide-platform polygnathids of the genus *Klapperina* (see Figure 2).

During the late Frasnian *Calvinaria albertensis* interval, the brachiopod assemblage was still diversified, and consisted of 13 species. Most characteristic are *Calvinaria albertensis albertensis* (Warren), *C. a.* minor Balinski, *Ryocarhynchus tumidus* (Kayser), *Radiatrypa alticola* (Frech), *Iowatrypa markowskii* (Lyashenko), and *Biernatella polonica* Balinski. The conodont fauna from this interval is dominated by palmatolepids and polygnathids and suggests the palma-tolepid-polygnathid to palmatolepid biofacies (Figure 2). These indicate maximum Frasnian transgression in the studied area. This event can be correlated with Event 6 of Sandberg et al. (1988), and with the T R cycle IId of Johnson et al. (1985). Generally, the assemblage of the *C. albertensis* interval displays considerable affinity with North American faunas from the Independence Shale, Iowa, and the Ireton Formation (and its equivalents) in Alberta. This affinity is indicated by such taxa as *C. albertensis albertensis* (Warren), *Iowatrypa markowskii* [similar to *I. americana* (Stainbrook)], *Theodossia* cf. *hungerfordi* (Hall), and *Tenticospirifer cyrtinaformis* (Hall & Whitfield).

This greatest Late Devonian (late Frasnian) eustatic rise drowned bioherms and biostromes and thus led to at least local extermination of both frame-building and reef-dwelling faunas (Sandberg et al., 1988, p. 196; Narkiewicz & Hoffman 1989, p. 20). In the latest part of the interval a distinct regressive pulse occurred, suggested by an admixture of quartz sand in the sediment and a shift in conodont biofacies from deep-water palmatolepid to slightly shallower, palmatolepid-polygnathid biofacies. The brachiopod fauna occurring in these strata is dominated by dwarfed forms, e.g. *Iowatrypa markowskii*, *Bier-natella polonica*, *Calvinaria albertensis minor*, and small linguloids.

The Frasnian sequence ends with the *Ryocarhynchus tumidus*

interval. This part of the section marks the beginning of stagnant and poorly-oxygenated water conditions. Generally, the brachiopod fauna is sparse, although in some layers the index species is abundant. Other brachiopods are represented by linguloids, linguliporids, *Radiatrypa alticola*, *Biernatella polonica*, *Athyris concentrica* (Buch), *Cyrtospirifer minor* (Gürich), and *Tenticospirifer cyrtinaformis*. Most characteristic is the western European species *Ryocarhynchus tumidus* (Kayser) which is known, both at Debnik and Senzeille, Belgium, in the same stratigraphic position, i.e. directly below the Frasnian-Famennian boundary (Bouckaert, 1972). The last atrypids and biernatellid athyridids died out in this interval in Poland. Co-occurring conodonts are the dominant genera *Palmatolepis* and *Polygnathus* , whereas *Icriodus* is totally absent (Figure 2).

Following the extinction events, during the early Famennian, stratified, poorly oxygenated conditions continued. The sequence starts with the *Orbiculatisinurostrum laeve* interval. Brachiopods are very rare and document the evident crisis at the F/F boundary (Figure 3). Most typical is the spiriferid *Cyrtospirifer minor* (Gürich) and the leiorhynchid rhynchonellid *O. laeve* (Gürich). Extremely rare are the productid *Praewaagenoconcha* cf. *speciosa* (Hall) and *Athyris concentrica*. Thus, this post-event relict assemblage exhibits very low taxic diversity. Brachiopods are characterized by small phenotype illustrating the 'Liliput effect' (Urbanek 1993), which occurs frequently in post-event assemblages. Small body size of organisms living in physiologically highly stressed environments seems to be a result of high juvenile mortality and possibly stunting (Kammer et al., 1986). The oldest Famennian specimens of *Cyrtospirifer minor* usually attain a lenmgth of 10mm (Balinski, 1979, p. 67, text-fig. 22). This species seems to be the only articulate which also occurs below the F-F boundary, and thus survived the extinction. It may represent an ecological generalist (see Harries et al., 1992), whose broad adaptive range enabled it to cope with perturbations associated with mass extinctions. The

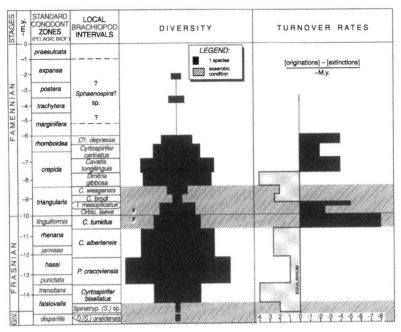

Figure 3. Species diversity pattern and turnover rates of Late Devonian brachiopods from Debnik Anticline (southern Poland), calculated at the resolution of local brachiopod intervals. Standard conodont zonation after Ziegler & Sandberg (1990).

leiorhynchid *O. laeve* represents a group of brachiopods which probably is well adapted to poorly oxygenated waters. Bowen et al. (1974) indicate that *Leiorhynchus* occurs as the only fossil in black shales, which suggest a dysaerobic environment. According to Alexander (1994), leiorhynchid brachiopods were adapted to dysoxic environments by living attached to other shells and thus perching a centimeter or more above the oxygen-depleted sediment surface.

Low diversity of the brachiopod fauna, and unfavourable, poorly oxygenated water conditions are characteristic also for the two succeeding Famennian intervals. In the *Iloerhynchus mesoplicatus* interval, *Cyrtospirifer minor* is still present, but *O. laeve* is replaced by another leiorhynchid, *Iloerhynchus mesoplicatus* Balinski. Very rare in this interval are lingulids and two other rhynchonellids, e.g. *Minirostrella rara* Balinski and *Colophragma* sp. In the *Cyrtospirifer brodi* interval, the brachiopod assemblage consists of only two species: *Nigerinoplica* sp. and *Cyrtospirifer brodi* (Venyukov), the successor to *C. minor*. A total absence of rhynchonellids is characteristic here.

During the early Famennian *Cyrtospirifer wesgensis* interval, a distinct facies change took place, expressed in sediments mainly by the presence of fine quartz sands which was shed from the emergent pre-Carpathian Land to the south (Figure 1). Brachiopods, though generally still rare, are abundant in some layers. The most characteristic brachiopod is *Cyrtospirifer wesgensis* Zheiba, which possesses a long hinge margin, very variable ventral area and asymmetric, twisted ventral beak (Balinski, 1995, p.60). These unusual features of the shell seem to be adaptations to more turbulent, unstable conditions on the substrate. The two coeval productids, *Nigerinoplica* sp. and *Sentosia profunda*, have long spines on the ventral valve and were also well adapted to these environments.

The *Dmitria gibbosa* interval (*crepida* conodont Zone) records a

termination of basinal dysoxia which continued during probably the whole *Palmatolepis triangularis* Chron. The brachiopod assemblage from this interval is one of the most diverse in the Famennian of Debnik and comprises 13 species belonging to Productida, Spiriferida, Athyridida, Rhynchonellida, and Orthida (Figure 3). It indicates a return of more favourable environmental conditions and a strong, post-extinction recovery of brachiopod faunas in southern Poland.

During the *Cavatisinurostrum longilinguis* interval [*crepida* Zone], brachiopods flourished and reached their greatest species diversity (14 species) in the regional Famennian. Most characteristic for the assemblage are spiriferids, athyridids, and rhynchonellids. These brachiopods are represented by a wide spectrum of adult shell dimensions, e.g. from 2mm length in the cardiarinid *Loborina lobata* Balinski, to more 40mm in *Dmitria globosa* (Gürich). The co-occurring fauna is also rich and diverse and is represented by sponges, gastropods, holothuroids, ophiuroids, echinoids, and fish remains. The conodonts are relatively rich and dominated by palmatolepids and polygnathids (Figures 2-3). They indicate a palmatolepid to palmatolepid-polygnathid biofacies and document the probable maximum of the Famennian transgression. This episode coincides with sealevel rise and a maximum of Famennian transgression (T-R cycle IIe) recognized in Euramerica (Johnson et al. 1985, 1989). Thus, the limestones of the *C. longilinguis* interval were deposited in an offshore, oxygenated setting rich in nutrients, favourable for brachiopods and other fauna.

Brachiopod diversity in the mid-Famennian *Cyrtospirifer carinatus* interval is slightly lower. The assemblage is dominated by medium to large-sized species, e.g. *Cyrtospirifer carinatus* Balinski, *Athyris sulcifera* Nalivkin, and *Mesoplica costata* Balinski. Less frequent are linguloids, *Schizophoria shubarica* Martynova, *Eoschuchertella* sp., *Evanescirostrum seversoni* (McLaren), *Athyris tau* Nalivkin, and *A.* sp. Almost

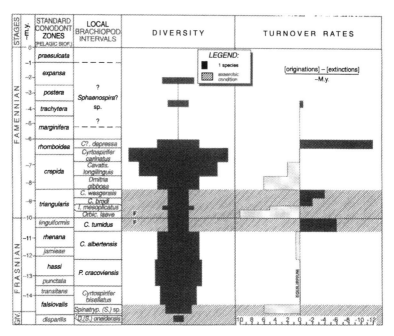

Figure 4. Species diversity pattern and turnover rates of Late Devonian conodonts from the Debnik anticline (southern Poland), calculated at the resolution of local brachiopod intervals. Standard conodont zonation after Ziegler & Sandberg (1990).

all these brachiopods possessed a strong, well-developed pedicle, and were adapted to turbulent water conditions. *M. costata* did not possess a pedicle but instead was equipped with a very long trail and strong spines enabling it to rest firmly on the seafloor. Characteristic for the interval is an increasing percentage of polygnathids and icriodontids (Figure 4), as well as common oncoids. Both strongly suggest a sealevel drop, and more vigorous conditions during the interval.

In the succeeding *Cyrtiorina* ? *depressa* interval [*rhomboidea* conodont Zone], the brachiopods again became rare and less diverse. Spiriferoids and productoids were the most important elements in the assemblage. Most typical are *Cyrtiorina* ? *depressa* Balinski, *Mesoplica* cf. *praelonga* (Sowerby), *M.* sp., *Leioproductus pauperculus* Cooper & Dutro, and *Eoschuchertella*. Totally absent are rhynchonellids. Interesting is, however, a reappearance of stromatoporoids (laminar forms) in this interval. Above the interval a strong regressive trend is accelerated and brachiopods disappear from the area. The only fossils are rare gastropods, massive stromatoporoids, and extremely rare conodonts. The latter are dominated by polygnathids and *Pelekysgnathus inclinatus*, suggesting a shallow, turbulent water, nearshore setting. There are also intervals with barren light grey, pure intrasparudites (calcareous sands).

The latest Famennian brachiopods come from the *Sphaenospira* ? sp. interval. These are poorly preserved, extremely rare, and represented by only three species. Typical are *Sphaenospira* ? sp. and *Sentosia* sp. Other fossils are represented by massive stromatoporoids and rare foraminifers, crinoids, and conodonts. The last occur in some layers only, and are represented by a probably mixed, latest Famennian to earliest Carboniferous *Polygnathus, Palmatolepis, Omolonognathus*, and *Pelekysgnathus*. These layers with conodonts, crinoids and rare brachiopods suggest that the prevailing regressive conditions were probably briefly interrupted in the

late Famennian by a transgressive pulse which brought more favourable conditions.

The late Famennian section is very poorly exposed. Thus, the data are incomplete and fundamental problems concerning late Famennian to earliest Carboniferous sedimentation, biostratigraphy, and faunal dynamics remain an unsolved question. However, there are suggestions that during latest Famennian time, a platform rimmed by shoals developed with sandy barriers and stromatoporoid patch reefs (Narkiewicz & Racki 1987). Probably several emersion episodes then took place (Paszkowski, 1995).

BRACHIOPOD SPECIES DIVERSITY VERSUS TURNOVER RATES

During the Late Devonian in southern Poland (Figure 3: left side), total standing diversity represents ecological diversity, with the number of species calculated for each brachiopod interval separately. Turnover rates are compared for the same set of data (Figure 3, right side). Numbers were calculated following the method given by Bayer et al. (1986) and McGhee (1988). According to this method, the turnover rate is the difference between the rate at which new species originate, and the rate at which species become extinct. Data were obtained at the resolution of local brachiopod intervals, correlated with the standard conodont zonal scheme (see Balinski, 1979, 1995). Duration of brachiopod intervals have been estimated in the range from 0.2my (*I. mesoplicatus* interval) to 1.5my (*C. albertensis* interval).

The Frasnian brachiopod fauna was most diversified in the early part of the stage (Figure 3). Diversity then gradually declined, reaching a minimum several meters above the Frasnian-Famennian boundary, i.e. during the *Cyrtospirifer brodi* interval. This decline in diversity is gradual, not catastrophic. The same data plotted as turnover rates (Figure

3), show that during most of the Frasnian origination rates were higher than the extinction rates. At the end of the Frasnian, however, there was a sharp negative pulse in species turnover rate. In this case the extinction rate became very high while origination rates dropped suddenly. The catastrophic extinction rate slightly lowered, but remained high, in the earliest Famennian. Thus, the sharp negative turnover rate in brachiopod faunas spanned some 2my of the latest Frasnian and earliest Famennian. The process of recovery started as early as the late *P. triangularis* Chron. During the *P. crepida* Chron, the brachiopod fauna seems to have fully recovered both in terms of species diversity and high origination rates. The regressive trend in the basin in the late part of *P. crepida* Chron, however, caused a return of negative turnover rates and emigration of brachiopod fauna from the area.

CONODONT FAUNAL DYNAMICS

The diversity of the conodont fauna and its turnover rate were compared to brachiopods (Figure 4). Generally, the pattern of species diversity is very similar for both groups. In the early Frasnian, diversity of conodonts was low; then it increased, reaching a maximum in the *P. cracoviensis* interval. In later intervals, diversity gradually declined and stayed low during the *C. tumidus* through *C. wesgensis* intervals. Beginning from the *D. gibbosa* brachiopod interval (*P. crepida* conodont Chron), diversification of the conodont fauna quickly increased, with highest values during the *C. carinatus* interval. During the *C . ? depressa* Chron, species diversity dropped again as regression in the basin progressed, and detrital stromatoporoid-rich carbonates prevailed.

Turnover rates of the conodont fauna show two major negative pulses. The first took place during the *C. tumidus* interval and thus coincided with the extinction event at the end of the Frasnian. The second occurred during the *S .? depressa* interval, emphasizing a major facies change in the basin. A strong pulse in turnover rates occurred at the base of the Famennian. This may suggests that the extinction event for conodonts was completed at the end of the Frasnian, while among brachiopods it continued for some time during the early Famennian.

DISCUSSION

Three stages can be recognized during the recovery from the F-F extinction of brachiopod faunas in southern Poland. The first is characteristic for the basal Famennian, i.e. the *O. laeve* and *I. mesoplicatus* intervals. The brachiopod assembages were then dominated by cyrtospiriferids and rhynchonellids (frequently rerpresented by stunted phenotypes), and exhibited very low species deversity. The second stage is characteristic for the late part of the *P. triangularis* Chron. During that time the brachiopod fauna was still poor in individuals and species and dominated by cyrtospiriferids and productids, with absence of rhynchonellids. Similar expansion of cyrtospiriferids and productids in the middle part of the *P. triangularis* Chron was noted in the Holy Cross Mountains (Racki, 1990). The third stage of brachiopod recovery took place during the *P. crepida* Chron (*D. gibbosa* and *C. longilinguis* brachiopod intervals). It is characterized by successful recovery of the shelf ecosystem which was seriously disturbed at the Frasnian-Famennian boundary. Brachiopods completely recovered at that time (but for extinct groups like atrypids) and quickly attained their greatest Famennian diversity (Figure 3). Among the flourishing groups of articulate brachiopods there were spire-bearers (spiriferids and athyridids), rhynchonellids, and to a lesser extent, productids. Racki (1990) described a similar recovery of the shelf habitats during the *P. crepida* Chron in the Holy Cross Mountains.

The stable carbon and oxygen isotopes in brachiopod shells from the Debnik anticline supplement the picture of environmental changes at the Frasnian-Famennian boundary in southern Poland. It shows substantial change in the ∂C13 isotopic composition (Halas et al., 1992; Figure 2). In the latest Frasnian, the carbon isotope curve shifted by 1.5 permil toward more positive values, and then remained high until the latest part of *P. triangularis* Chron. In the *P. crepida* Chron the curve began to fall to the previous level. These data suggest that an upwelling system appeared in this stagnant and stratified marine basin, spilling anoxic waters onto the shallow shelf (Halas et al., 1992, p.135). Correlation of the carbon isotope curve and the pattern of faunal dynamics is very suggestive (Figures 2-4). A similar positive shift of carbon isotope curve during late *P. linguiformis* to *P. triangularis* Chrons was observed in German and Belgian sections (Joachimski & Buggisch, 1992).

Diversity and turnover rates of brachiopod and conodont faunas of southern Poland can be compared with data for Appalachian, New York, and Ural faunas given by Bayer & McGhee (1986) and McGhee (1988, 1989). In all those cases, a sharp negative pulse in turnover rates occurred at the end of the Frasnian. This pulse, however, was not geologically instantaneous, instead, a period of high negative turnover rates spanned some 1.3my (for the New York State brachiopods) to 2my (for the Appalachian marine invertebrates, and southern Ural and southern Polish brachiopods: see McGhee, 1988, 1989).

After the extinction event, the faunas of S Poland recovered in the early Famennian, but then returned to negative turnover rates. A very similar pattern of faunal dynamics is recognizable among such widespread geographic regions as the New York Sate, the Appalachians, S Poland, and, to some extent, the S Urals. The decline in species diversity and negative turnover rates among brachiopods from Poland in the *S. ? depressa* interval was associated with a strong regression in the basin. Similar prolonged Famennian regressions were observed from the end of the *P. crepida* Chron up to the Early or Middle *P. expansa* Chron (T-R cycle IIe) in western Pomerania, NW Poland (Matyja 1993), Belgium, and the US (Johnson et al., 1985). Thus, as in Poland, the negative pulse in turnover rates during the Famennian in the New York and Appalachian faunas, is connected with regressions of the IIe cycle, and emigration of marine faunas. This event is of a different nature than that at the Frasnian Famennian boundary, recognized as a worldwide biotic crisis.

Acknowledgments

Sincere appreciation is extended to Dr G. Racki (Silesian University, Sosnowiec) for his useful comments and encouragement during the preparation of this paper. I thank Dr. A. Pisera for revising the typescript. The Natural Sciences and Engineeering Research Council of Canada provided travel funds to attend the Third International Brachiopod Congress in Sudbury, Canada.

REFERENCES

ALEXANDER, R. R. 1994. Distribution of pedicle boring traces and the life habit of Late Paleozoic leiorhynchid brachiopods from dysoxic habitats. Lethaia 27:227-234.

BALINSKI, A. 1979. Brachiopods and conodonts from the Frasnian of the Debnik anticline, southern Poland. Palaeontologia Polonica 39:1-95.

___1995. Brachiopods and conodont biostratigraphy of the Famennian from the Debnik anticline, southern Poland. Palaeontologia Polonica 54:1-88.

BAYER, U. & G.R. MCGHEE. 1986. Cyclic patterns in the Paleozoic and Mesozoic: implications for time scale calibrations. Paleoceanography 1:383-402.

BOUCKAERT, J., A. MOURAVIEFF, M. STREEL, J. THOREZ, & W. ZIEGLER. 1972. The Frasnian-Famennian boundary in Belgium. Geologica Palaeontologica 6:87-92.

BOWEN, Z. P., D.C. RHOADS, & A.L. MCALESTER. 1974. Marine benthic communities in the Upper Devonian of New York. Lethaia 7:93-120.

BRICE, D., B. MILHAU, & B. MISTIAEN. 1994. Affinités nord-américaines de taxons dévoniens (Givétien-Frasnien) du Boulonnais, nord de la France- migrations et diachronismes. Bulletin Société Géologique France165:291-306.

GÜRICH, G. 1903. Das Devon von Debnik bei Krakau. Beiträge Paläontologie Geologie Österreich-Ungarns Orients 15:127-164.

HALAS, S., A. BALINSKI, M. GRUSZCZYNSKI, A. HOFFMAN, K. MALKOWSKI, & M. NARKIEWICZ. 1992. Stable isotope record at the Frasnian/Famennian boundary in southern Poland. Neues Jahrbuch Geologie Paläontologie, Monatshefte, 1992:129-138.

HARRIES, P. J. & E.G. KAUFFMAN. 1992. Patterns of survival and recovery following the Cenomanian-Turonian (Late Cretaceous) mass extinction in the Western Interior Basin, United States. Fifth International Conference Global Bioevents Göttingen 278-298.

JOACHIMSKI, M. M. & W. BUGGISCH. 1992. Carbon isotope shifts at the Frasnian/Famennian boundary: evidence for worldwide Kellwasser events? Fifth International Conference on Global Bioevents Göttingen 58-59.

JOHNSON, J. G., G. KLAPPER, G. & C.A. SANDBERG. 1985. Devonian eustatic fluctuation in Euramerica. Geological Society America Bulletin 96: 567-587.

___& C.A. SANDBERG. 1989. Devonian eustatic events in the western United States and their biostratigraphic responses. Canadian Society Petroleum Geologists Memoir 14:171-178.

KAMMER, T.W. 1986. Ecologic stability of the dysaerobic biofacies during the Late Paleozoic. Lethaia 19:109 -121.

LYASHENKO, A.I. 1959. Atlas brakhiopod i stratigrafii devona Russkoy platformy. Gostoptehizdat 1-451.

___1973. Brakhiopody i stratigrafiya nizhnefranskikh otlozhenie yuzhnogo Timana i Volgo-Uralskoy neftegazo-nosnoy provintsii. Trudy Vsesoyuznogo Nauchno-Issledo-vatelskogo Geologorazvedochnogo Neftiannogo Instituta 134:1-129.

MATYJA, H. 1993. Upper Devonian of Western Pomerania. Acta Geologica Polonica 43:27-94.

MCGHEE, G.R. 1988. The Late Devonian extinction event: evidence for abrupt ecosystem collapse. Paleobiology 14:250-257.

___1989. Evolutionary dynamics of the Frasnian-Famennian extinction event. Canadian Society Petroleum Geologists Memior 14:23-28.

NARKIEWICZ, M. 1989. Turning points in sedimentary develop-ment in the Late Devonian in southern Poland. Canadian Society Petroleum Geologists Memoir 14:619-635.

___& A. HOFFMAN. 1989. The Frasnian/Famennian transition: the sequence of events in southern Poland and its implications. Acta Geologica Polonica 39:13 28.

___& G. RACKI. 1984. Stratygraphy of the Devonian of the Debnik anticline. Kwartalnik Geologiczny 28:513-546.

___&___.1987. Correlation and sedimentary development of the Upper Devonian between Debnik and Zawiercie in southern Poland. Kwartalnik Geologiczny 31:341-356.

PASZKOWSKI, M. 1995. Strunian in Raclawka valley. In, M. Szulczewski & J. Dvorak (eds), Evolution of the Polish-Moravian carbonate platform in the Late Devonian and Early Carboniferous: Holy Cross Mts., Krakow Upland, Moravian karst. XIII International Congress Carboniferous-Permian, Guide to Excursion B4:24-26 .

RACKI, G., 1990. Frasnian/Famennian event in the Holy Cross Mts, central Poland: stratigraphic and ecologic aspects. ?

___& A. BALINSKI. 1981. Environmental interpretation of the atrypid shell beds from the Middle to Upper Devonian boundary of the Holly Cross Mts and Cracow Upland. Acta Geologica Polonica 31:177-211.

ROEMER, F. 1863. Die Altersbestimmung des schwarzen Marmors von Debnik im Gebiete von Krakau. Zeitschrift Deutschen Geologischen Gesellschaft 15:708-712.

SANDBERG, C. A., F.G. POOLE, & J.G. JOHNSON. 1989. Upper Devonian of Western United States. Canadian Society Petroleum Geologists Memior, 14:183-220.

SCHLAGER, W. 1981. The paradox of drowned reefs and carbonate platforms. Geological Society America Bulletin 92:197 211.

SIEDLECKI, S. 1954 Utwory paleozoiczne okolic Krakowa (zagadnienia stratygrafii i tektoniki). Biuletyn Instytutu Geologicznego,73:1 415.

URBANEK, A. 1993. Biotic crises in the history of Upper Silurian graptoloids: a palaeobiological model. Historical Biology 7:29 50.

ZIEGLER, W. & C.A. SANDBERG. 1990. The Late Devonian standard conodont zonation. Courier Forschungsinstitut Senckenberg 121:1 115.

A TO Z WORLD LIST OF BRACHIOPOD GENERA-SUBGENERA TO 1995

REX DOESCHER

Smithsonian Institution, National Museum, Washington, DC 20560, USA

COMMENTS- This alphabetic list of 4827 genera and subgenera is as complete as available to the end of 1995, and has been loaded down from ASCII-2 files via e-mail. Synonyms and homonyms are not separated. The list has not been verified. Transliteration of the cyrillic alphabet provided by Doescher uses German conversion, and has not been altered to fit the scheme generally employed in this volume (see cyrillic alphabet transliteration in preface, using GSA format). We hope it proves useful to those searching for publication dates and authorship. Our list count shows that the world centres for individual brachiopod taxonomic diversity are Washington, Prag, Brussels, and Brisbane. [*Editors.]

Aalenirhynchia Shi & Grant 1993
Aberia Melou 1990
Aboriginella Koneva 1983
Abrekia Dagys 1974
Absenticosta Lazarev 1991
Abyssorhynchia Zezina 1980
Abyssothyris Thomson 1927
Acambona White 1865
Acanthalosia Waterhouse 1986
Acanthambonia Cooper 1956
Acanthatia Muir-Wood & Cooper 1960
Acanthobasiliola Zezina 1981
Acanthocosta Roberts 1971
Acanthocrania Williams 1943
Acanthoglypha Williams & Curry 1985
Acanthoplecta Muir-Wood & Cooper 1960
Acanthoproductus Martynova 1970
Acanthorhynchia Buckman 1918
Acanthorthis Neuman 1977
Acanthospirifer Menakova 1964
Acanthospirina Schuchert & Levene 1929
Acanthothiris Orbigny 1850
Acanthothyropsis Kamyschan 1973
Acanthotoechia Williams & Curry 1985
Acareorthis Roberts 1990
Acculina Misius 1977
Achunoproductus Ustritsky 1971
Acidotocarena Liu 1979
Acolosia Cooper & Grant 1976
Acosarina Cooper & Grant 1969
Acritosia Cooper & Grant 1969
Acrobelesia Cooper 1983
Acrobrochus Cooper 1983
Acrosaccus Willard 1928
Acrospirifer Helmbrecht & Wedekind 1923
Acrothele Linnarsson 1876
Acrothyra Matthew 1901
Acrothyris Hou 1963
Acrotreta Kutorga 1848
Acrotretella Ireland 1961
Actinoconchus M'Coy 1844
Actinomena Opik 1930
Acuminothyris Roberts 1963
Acutatheca Stainbrook 1945
Acutella Ljaschenko 1973
Acutilineolus Amsden 1978
Acutoria Cooper & Dutro 1982
Adairia Gordon, Henry & Treworgy 1993

Adaptatrypa Struve 1980
Adectorhynchus Henry & Gordon 1985
Adensu Popov & Rukavischnikova 1986
Adiaphragma Xian 1978
Adminiculoria Waterhouse & Gupta 1979
Admixtella Rozman 1978
Admodorugosus Brunton, Mundy & Lazarev 1993
Adnatida Richardson 1991
Adolfia Gurich 1909
Adolfispirifer Krylova 1962
Adrenia Chatterton 1973
Adriana Gregorio 1930
Advenina Sandy 1986
Adygella Dagys 1959
Adygelloides Dagys 1959
Adygellopsis Sun & Shi 1985
Aegiria Opik 1933
Aegiromena Havlicek 1961
Aegironetes Havlicek 1967
Aemula Steinich 1968
Aenigmastricklandia Ziegler 1966
Aenigmastrophia Boucot 1971
Aenigmathyris Cooper 1971
Aequspiriferina Yang & Yin 1962
Aerothyris Allan 1939
Aesopomum Havlicek 1965
Aetheia Thomson 1915
Aethirhynchia Shi 1990
Afghanospirifer Plodowski 1968
Afilasma Stehli 1961
Agalatassia Popov & Holmer 1994
Agarhyncha Havlicek 1982
Agastapleura Xu 1978
Agelesia Cooper & Grant 1969
Agerinella Pajaud & Patrulius 1965
Agerithyris Radulovic 1991
Agramatia Sokolskaja 1948
Agulhasia King 1871
Agyrekia Koneva 1979
Ahtiella Opik 1932
Aidynkulirhynchia Ovtsharenko 1983
Aikarhyncha Havlicek 1990
Airtonia Cope 1934
Ajukuzella Ovtsharenko 1983
Akatchania Klets 1988
Akelina Severgina 1967
Akmolina Popov & Holmer 1994
Akopovorhynchia Smirnova 1990
Aksarinaia Koneva 1992
Aktassia Popov 1976
Ala Nalivkin 1979
Alabushevothyris Smirnova 1994
Alaskospira Kirk & Amsden 1952
Alatiformia Struve 1963
Alatochonetes Liang 1990
Alatoproductus Ching & Zhu 1978
Alatorthotetina He & Zhu 1985
Alatothyris Waterhouse & Gupta 1986
Aldanispirifer Alekseeva 1967
Aldanotreta Pelman 1977
Aldingia Thomson 1916

Alekseevaella Baranov 1980
Alexenia Ivanova 1935
Alichovia Gorjansky 1969
Aliconchidium St.joseph 1942
Alimbella Andreeva 1960
Alipunctifera Waterhouse 1975
Aliquantula Richardson 1991
Alisina Rowell 1962
Alispira Nikiforova 1961
Alispirifer Campbell 1961
Alispiriferella Waterhouse & Waddington 1982
Alitaria Cooper & Muir-Wood 1967
Alithyris Sun 1981
Allanella Crickmay 1953
Allanetes Boucot & Johnson 1967
Allorhynchoides Savage, Eberlein & Churkin 1978
Allorhynchus Weller 1910
Almerarhynchia Calzada 1974
Almiralthyris Calzada 1994
Almorhynchia Ovtsharenko 1983
Alorostrum Savitskii 1992
Alphachoristites Gatinaud 1949
Alphacyrtiopsis Gatinaud 1949
Alphaneospirifer Gatinaud 1949
Altaeorthis Severgina 1967
Altaestrophia Bublichenko 1956
Altaethyrella Severgina 1978
Altajella Kulkov 1962
Altiplecus Stehli 1954
Altoplicatella Xu & Liu 1983
Altorthis Andreeva 1960
Altunella Liu, Zhang & Di 1984
Alvarezites Struve 1992
Alwynella Spjeldnaes 1957
Alwynia Stehli 1961
Ambardella Andreeva 1987
Ambikella Sahni & Srivastava 1956
Ambocoelia Hall 1860
Amboglossa Wang & Zhu 1979
Ambonorthella Bassett 1972
Ambothyris George 1931
Amesopleura Carter 1967
Ametoria Cooper & Grant 1975
Amictocracens Henderson & MacKinnon 1981
Amissopecten Havlicek 1960
Amoenirhynchia Siblik 1986
Amoenospirifer Havlicek 1957
Amosina Boucot 1975
Amphiclina Laube 1865
Amphiclinodonta Bittner 1888
Amphigenia Hall 1867
Amphipella Cooper & Grant 1969
Amphiplecia Wright & Jaanusson 1993
Amphistrophia Hall & Clarke 1892
Amphistrophiella Harper & Boucot 1978
Amphithyris Thomson 1918
Amphitomella Bittner 1890
Amsdenella Tillman 1967
Amsdenina Boucot 1975
Amsdenostrophiella Harper & Boucot 1978
Amurothyris Koczyrkevicz 1976
Amydroptychus Cooper 1989
Amygdalocosta Waterhouse 1967
Anabaia Clarke 1893
Anabaria Lopushinskaja 1965
Anabolotreta Rowell & Henderson 1978
Anadyrella Dagys 1974
Anakinetica Richardson 1987

Anaptychius Hoover 1981
Anarhynchia Ager 1968
Anastrophia Hall 1867
Anathyrella Khalfin 1960
Anathyris Peetz 1901
Anatreta Mei 1993
Anatrypa Nalivkin 1941
Anazyga Davidson 1882
Anchigonites Opik 1939
Ancillotoechia Havlicek 1959
Ancistrocrania Dall 1877
Ancistrorhyncha Ulrich & Cooper 1942
Ancorellina Mancenido & Damborenea 1991
Ancorhynchia Ching & Ye 1979
Ancylostrophia Harper, Johnson & Boucot 1967
Andalucinetes Racheboeuf 1985
Andobolus Kozlowski 1930
Andreaspira Abramov & Grigorjewa 1986
Aneboconcha Cooper 1973
Anechophragma Neuman 1977
Anelotreta Pelman 1986
Anemonaria Cooper & Grant 1969
Aneuthelasma Cooper & Grant 1976
Angarella Asatkin 1932
Angiospirifer Legrand-Blain 1985
Angulotreta Palmer 1954
Angusticardinia Schuchert & Cooper 1931
Angustothyris Dagys 1972
Aniabrochus Cooper 1983
Anidanthus Hill 1950
Animonithyris Cooper 1983
Anisactinella Bittner 1890
Anisopleurella Cooper 1956
Annuloplatidia Zezina 1981
Anomactinella Bittner 1890
Anomalesia Cooper & Grant 1976
Anomaloglossa Percival 1978
Anomaloria Cooper & Grant 1969
Anomalorthis Ulrich & Cooper 1936
Anoplia Hall & Clarke 1892
Anopliopsis Girty 1938
Anoplotheca Sandberger 1855
Anoptambonites Williams 1962
Ansehia Termier & Termier 1970
Antelocoelia Isaacson 1977
Anteridocus Cooper & Grant 1976
Antezeilleria Xu & Liu 1983
Anthracospirifer Lane 1963
Anthracothyrina Legrand-Blain 1984
Antigaleatella Sapelnikov & Al 1975
Antigonambonites Opik 1934
Antigoniarcula Elliott 1959
Antinomia Catullo 1851
Antiptychina Zittel 1880
Antiquatonia Miloradovich 1945
Antirhynchonella Oehlert 1887
Antispirifer Williams 1916
Antistrix Johnson 1972
Antizygospira Fu 1982
Antronaria Cooper & Grant 1976
Anulatrypa Havlicek 1987
Anx Havlicek 1980
Aorhynchia MacFarlan 1992
Aparimarhynchia MacFarlan 1992
Apatecosia Cooper 1983
Apatobolus Popov 1980
Apatomorpha Cooper 1956
Apatorthis Opik 1933

Apatoskenidioides Liu, Zhu & Xue 1985
Aperispirifer Waterhouse 1968
Apertirhynchella Siblik 1986
Aphanomena Bergstrom 1968
Aphaurosia Cooper & Grant 1976
Apheathyris Fu 1982
Aphelesia Cooper 1959
Aphelotreta Rowell 1980
Apheoorthina Havlicek 1949
Apheoorthis Ulrich & Cooper 1936
Aphragmus Cooper 1983
Apicilirella Su 1976
Apistoconcha Conway Morris 1990
Apletosia Cooper 1983
Apodosia Smirnova & MacKinnon 1995
Apollonorthis Mitchell 1974
Apomatella Schuchert & Cooper 1931
Apopentamerus Boucot & Johnson 1979
Aporthophyla Ulrich & Cooper 1936
Aporthophylina Liu 1976
Apothyris Cooper 1989
Apousiella Carter 1972
Apringia Gregorio 1886
Apsilingula Williams 1977
Apsocalymma McIntosh 1974
Apsotreta Palmer 1954
Arabatia Cooper 1989
Arabicella Cooper 1989
Araksalosia Lazarev 1989
Araktina Koneva 1992
Araneatrypa Havlicek 1987
Arapsopleurum Cooper 1989
Arapsothyris Cooper 1989
Araratella Abramian & Al 1975
Araspirifer Havlicek 1987
Aratanea Schmidt 1967
Aratoechia Havlicek 1982
Araxathyris Grunt 1965
Araxilevis Sarytcheva 1965
Arbizustrophia Garcia-Alcalde 1972
Arcelinithyris Almeras 1971
Arceythyris Rollet 1964
Archaeorthis Schuchert & Cooper 1931
Archaiosteges Carter 1991
Archeochonetes Racheboeuf & Copper 1986
Arcosarina Ching, Sun & Ye 1979
Arcticalosia Waterhouse 1986
Arctispira Smith 1980
Arctitreta Whitfield 1908
Arctochonetes Ifanova 1968
Arctohedra Cooper 1936
Arctomeristina Amsden 1978
Arctosia Cooper 1983
Arctospirifer Stainbrook 1950
Arctothyris Dagys 1965
Arcualla Walmsley & Boucot 1975
Arcuatothyris Popiel-Barczyk 1972
Arcullina Waterhouse 1986
Ardiviscus Lazarev 1986
Arduspirifer Mittmeyer 1972
Areella Erlanger 1992
Arenaciarcula Elliott 1959
Arenorthis Havlicek 1970
Areostrophia Havlicek 1965
Argella Menakova & Nikiforova 1986
Argentiproductus Cooper & Muir-Wood 1951
Argorhynx Havlicek 1992
Argovithyris Rollet 1972

Argyrotheca Dall 1900
Arionthia Cooper & Grant 1976
Arktikina Grunt 1977
Aromasithyris Almeras 1971
Aroonia Bengtson 1990
Arquinisca Radwanska & Radwanski 1994
Artimyctella Liu 1979
Artiotreta Ireland 1961
Ascanigypa Havlicek 1990
Aseptagypa Brice 1982
Aseptalium Hou & Xian 1975
Aseptella Martinez & Winkler 1977
Aseptirhynchia Soja 1988
Aseptonetes Isaacson 1977
Asiacranaena Kaplun 1991
Asiarhynchia Su 1980
Asiomeekella Liang 1990
Askepasma Laurie 1986
Asperlinus Waterhouse & Piyasin 1970
Aspidothyris Diener 1908
Aspinosella Waterhouse 1982
Astamena Roomusoks 1989
Astegosia Cooper & Grant 1969
Astraborthis Williams 1974
Astua Havlicek 1992
Asturistrophia Garcia-Alcalde 1992
Astutorhyncha Havlicek 1961
Asymmetrochonetes Smith 1980
Asymphylotoechia Ross 1970
Asyrinx Hudson & Sudbury 1959
Asyrinxia Campbell 1957
Atactosia Cooper 1983
Atelelasma Cooper 1956
Atelelasmoidea Zeng 1987
Atelestegastus Cooper & Grant 1975
Atelithyris Smirnova 1975
Athabaschia Crickmay 1963
Athyrhynchus Johnson 1973
Athyris M'Coy 1844
Athyrisina Hayasaka 1922
Athryisinoidea Chen & Wang 1980
Athyrisinoides Jiang 1978
Athyrisinoides Chen & Wang 1978
Athyrisinopsis Zhang 1983
Athyrorhynchia Xu & Liu 1983
Atlanticocoelia Koch 1981
Atlantida Havlicek 1971
Atribonium Grant 1965
Atrypa Dalman 1828
Atryparia Copper 1966
Atrypella Kozlowski 1929
Atrypellina Menakova & Nikiforova 1986
Atrypina Hall & Clarke 1893
Atrypinella Khodalevich 1939
Atrypinopsis Rong & Yang 1981
Atrypoidea Mitchell & Dun 1920
Atrypopsis Poulsen 1943
Atrypunculus Havlicek 1987
Atrythyris Struve 1965
Attenuatella Stehli 1954
Auchmerella Struve 1964
Aucklandirhynchia MacFarlan 1992
Aulacatrypa Havlicek 1987
Aulacella Schuchert & Cooper 1931
Aulacophoria Schuchert & Cooper 1931
Aulacorhyna Strand 1928
Aulacothyris Douville 1879
Aulacothyroides Dagys 1965

Aulacothyropsis Dagys 1959
Aulidospira Williams 1962
Aulie Nikitin & Popov 1984
Aulites Richardson 1987
Aulonotreta Kutorga 1848
Auloprotonia Muir-Wood & Cooper 1960
Aulosteges Helmersen 1847
Auriculispina Waterhouse 1975
Austinella Foerste 1909
Australiarcula Elliott 1960
Australina Clarke 1912
Australirhynchia Savage 1968
Australispira Percival 1991
Australocoelia Boucot & Gill 1956
Australosia McKellar 1970
Australospirifer Caster 1939
Australostrophia Caster 1939
Austriellula Strand 1928
Austrirhynchia Ager 1959
Austrochoristites Roberts 1971
Austrohedra Roberts 1990
Austronoplia Isaacson 1977
Austrospirifer Glenister 1956
Austrothyris Allan 1939
Aviformia Xian 1988
Avisyrinx Martinez 1975
Avonia Thomas 1914
Avonothyris Buckman 1918
Axiodeaneia Clark 1917
Azamella Laurie 1991
Azygidium Waterhouse 1983
Babinia Racheboeuf & Branisa 1985
Babukella Dagys 1974
Backhausina Pajaud 1966
Bactrynium Emmrich 1855
Badainjarania Zhang 1981
Baeorhynchia Cooper 1989
Bagnorthis Levy & Nullo 1974
Bagrasia Nalivkin 1960
Bailliena Nelson & Johnson 1968
Bailongjiangella Fu 1982
Baissalosteges Kotljar 1989
Bajanhongorella Rozman 1977
Bajanorthis Andreeva 1968
Bajkuria Ustritsky 1963
Bajtugania Grunt 1980
Bakonyithyris Voros 1983
Balakhonia Sarytcheva 1963
Balanoconcha Campbell 1957
Balatonospira Dagys 1974
Balkhasheconcha Lazarev 1985
Bancroftina Sinclair 1946
Bandoproductus Jing & Sun 1981
Barbaestrophia Havlicek 1965
Barbarorthis Opik 1934
Barbarothyris Wang & Rong 1986
Barbatulella Williams & Lockley 1983
Barkolia Zhang 1981
Barrandina Booker 1926
Barroisella Hall & Clarke 1892
Barzellinia Gregorio 1930
Bashkiria Nalivkin 1979
Basilicorhynchus Crickmay 1952
Basiliola Dall 1908
Basiliolella Hondt 1987
Batenevotreta Ushatinskaia 1992
Bateridium Su, Rong & Li 1985
Baterospirifer Su & Li 1984

Bathymyonia Muir-Wood & Cooper 1960
Bathynanus Foster 1974
Bathyrhyncha Fuchs 1923
Baturria Carls 1974
Bazardarella Ovtsharenko 1983
Beachia Hall & Clarke 1893
Beckmannia Mohanti 1972
Becscia Copper 1995
Beecheria Hall & Clarke 1893
Begiarslania Titova 1992
Beichuanella Chen 1978
Beichuanrhynchus Chen 1990
Beitaia Rong & Yang 1974
Bejrutella Tchorszhewsky 1972
Bekkerella Reed 1936
Bekkerina Roomusoks 1993
Bekkeromena Roomusoks 1963
Belbekella Moisseiev 1939
Beleutella Litvinovich 1967
Bellaclathrus Winters 1963
Bellimurina Cooper 1956
Belothyris Smirnova 1960
Benignites Havlicek 1952
Bergalaria Schmidt 1975
Beschevella Poletaev 1975
Betachoristites Gatinaud 1949
Betacyrtiopsis Gatinaud 1949
Betaneospirifer Gatinaud 1949
Betterbergia Schmidt 1981
Biarea Torbakova 1959
Biarella Markowskii 1988
Bibatiola Grant 1976
Bicarinatina Batrukova 1969
Bicepsirhynchia Shi 1990
Bicia Walcott 1901
Biconostrophia Havlicek 1956
Biconvexiella Waterhouse 1983
Bicuspina Havlicek 1950
Bidentatus Khodalevich & Breivel 1972
Biernatella Balinski 1977
Biernatia Holmer 1989
Biernatium Havlicek 1975
Bifida Davidson 1882
Bifolium Elliott 1948
Bihendulirhynchia Muir-Wood 1935
Bihenithyris Muir-Wood 1935
Bilaminella Babanova 1964
Billingsella Hall & Clarke 1892
Bilobia Cooper 1956
Bilotina Reed 1944
Bimuria Ulrich & Cooper 1942
Biparetis Amsden 1974
Biplatyconcha Waterhouse 1983
Biplicatoria Cooper 1983
Birchsella Clarke 1987
Biseptum Khodalevich & Breivel 1959
Bisinocoelia Havlicek 1953
Bispinoproductus Stainbrook 1947
Bistramia Hoek 1912
Bisulcata Boucot & Johnson 1979
Bisulcina Titova 1977
Bittnerella Dagys 1974
Bittnerula Hall & Clarke 1894
Blasispirifer Kulikov 1950
Bleshidimerus Havlicek 1990
Bleshidium Havlicek 1982
Blochmannella Leidhold 1921
Blyskavomena Havlicek 1976

Bobinella Andreeva 1968
Bockelia Neuman 1989
Bodrakella Moisseiev 1936
Bohemiella Schuchert & Cooper 1931
Bohemirhynchia Nekvasilova 1973
Boicinetes Havlicek & Racheboeuf 1979
Bojarinovia Aksarina 1978
Bojodouvillina Havlicek 1967
Bojothyris Havlicek 1959
Bolgarithyris Titova 1986
Bolilaspirifer Sun 1981
Boloria Grunt 1973
Bomina Korovnikov & Ushatinskaia 1994
Boonderella Percival 1991
Booralia Campbell 1961
Boreadocamara Curry & Williams 1984
Boreadorthis Opik 1934
Borealirhynchia Su 1976
Borealis Boucot, Johnson & Rubel 1971
Borealispirifer Hou & Su 1993
Boreiospira Dagys 1974
Boreiothyris Dagys 1968
Bornhardtina Schulz 1914
Bortegitoechia Erlanger 1994
Borua Williams & Curry 1985
Bosquetella Smirnova 1969
Bothrionia Cooper & Grant 1975
Bothrostegium Cooper & Grant 1974
Bothrothyris Cooper 1983
Boticium Havlicek & Mergl 1982
Botsfordia Matthew 1891
Boubeithyris Cox & Middlemiss 1978
Bouchardia Davidson 1850
Bouchardiella Doello-Jurassicado 1922
Boucotella Bowen 1966
Boucotia Gill 1969
Boucotides Amsden 1968
Boucotinskia Brunton & Cocks 1967
Boucotstrophia Jahnke 1980
Bouskia Havlicek 1975
Bowanorthis Percival 1991
Bozshakolia Ushatinskaia 1986
Brachymimulus Cockerell 1929
Brachyprion Shaler 1865
Brachyspirifer Wedekind 1926
Brachythyrina Fredericks 1929
Brachythyrinella Waterhouse & Gupta 1978
Brachythyris M'Coy 1844
Brachyzyga Kozlowski 1929
Bracteoleptaena Havlicek 1963
Bradfordirhynchia Shi & Grant 1993
Brahimorthis Havlicek 1971
Branconia Gagel 1890
Brandysia Havlicek 1975
Branikia Havlicek 1957
Branxtonia Booker 1930
Brasilina Clarke 1921
Brasilioproductus Mendes 1959
Brevicamera Cooper 1956
Brevilamnulella Amsden 1974
Brevipelta Geyer 1994
Breviseptum Sapelnikov 1960
Brevispirifer Cooper 1942
Brochocarina Brunton 1968
Broeggeria Walcott 1902
Bronnothyris Popiel-Barczyk & Smirnova 1978
Brooksina Kirk 1922
Browneella Chatterton 1973

Brunnirhyncha Havlicek 1979
Bruntonites Struve 1992
Bryorhynchus Cooper & Grant 1969
Buceqia Havlicek 1984
Buchanathyris Talent 1956
Buckmanithyris Tchorszhewsky 1990
Bulahdelia Roberts 1976
Bulgania Erlanger 1994
Bullarina Jing & Sun 1981
Bullothyris Sucic-Protic 1971
Buntoxia Lazarev 1986
Burmirhynchia Buckman 1918
Burovia Ustritsky 1980
Burrirhynchia Owen 1962
Butkovia Baranov 1987
Buxtonia Thomas 1914
Buxtoniella Abramov & Grigorjewa 1986
Buxtonioides Mendes 1959
Bynguanoia Roberts 1990
Bystromena Williams 1974
Cacemia Mitchell 1974
Cactosteges Cooper & Grant 1975
Cadomella Munier-Chalmas 1887
Cadudium Havlicek 1985
Caenanoplia Carter 1968
Caenotreta Cocks 1979
Caeroplecia Williams 1974
Calcirhynchia Buckman 1918
Caledorhynchia MacFarlan 1992
Callaiapsida Grant 1971
Callicalyptella Boucot & Johnson 1972
Calliglypha Cloud 1948
Calliomarginatia Ching 1976
Callipentamerus Boucot 1964
Callipleura Cooper 1942
Calliprotonia Muir-Wood & Cooper 1960
Callispirifer Perry 1984
Callispirina Cooper & Muir-Wood 1951
Calloria Cooper & Lee 1993
Callospiriferina Rousselle 1977
Callytharrella Archbold 1985
Calvinaria Stainbrook 1945
Calvustrigis Carter 1987
Calyptolepta Neuman 1977
Calyptoria Cooper 1989
Camarelasma Cooper & Grant 1976
Camarium Hall 1859
Camarophorella Hall & Clarke 1893
Camarophorina Licharew 1934
Camarophorinella Licharew 1936
Camarospira Hall & Clarke 1893
Camarothyridina Linnik 1966
Camarotoechia Hall & Clarke 1893
Camarotoechioides Rzonsnitzkaja 1978
Cambrotrophia Ulrich & Cooper 1937
Camerella Billings 1859
Camerisma Grant 1965
Camerophorina Schmidt 1941
Camerothyris Bittner 1890
Campages Hedley 1905
Campylorthis Ulrich & Cooper 1942
Canadospira Dagys 1972
Canalilatus Pelman 1983
Canavirila Sartenaer 1994
Cancellospirifer Campbell 1953
Cancellothyris Thomson 1926
Cancrinella Fredericks 1928
Cancrinelloides Ustritsky 1963

Candispirifer Havlicek 1971
Caninella Liang 1990
Cantabriella Martinez & Rio Garcia 1987
Canthylotreta Rowell 1966
Capelliniella Strand 1928
Capillarina Cooper 1983
Capillirhynchia Buckman 1918
Capillirostra Cooper & Muir-Wood 1951
Capillispirifer Zhang 1983
Capillithyris Katz 1974
Capillomesolobus Pecar 1986
Capillonia Waterhouse 1973
Caplinoplia Havlicek & Racheboeuf 1979
Carapezzia Tomlin 1930
Carbocyrtina Ivanova 1975
Cardiarina Cooper 1956
Cardinirhynchia Buckman 1918
Cardinocrania Waagen 1885
Cardiothyris Roberts 1971
Caricula Grant 1976
Carinagypa Johnson & Ludvigsen 1972
Carinastrophia Gratsianova 1975
Carinatina Nalivkin 1930
Carinatinella Gratsianova 1967
Carinatothyris Tchorszhewsky 1990
Carinatrypa Copper 1973
Cariniferella Schuchert & Cooper 1931
Carinokoninckina Jin & Fang 1977
Carlinia Gordon 1971
Carlopsina Reed 1954
Carneithyris Sahni 1925
Carolirhynchia Havlicek 1992
Carpatothyris Smirnova 1975
Carpinaria Struve 1982
Carringtonia Brunton & Mundy 1986
Cartorhium Cooper & Grant 1976
Caryogyps Johnson, Boucot & Murphy 1976
Caryona Cooper 1983
Caryorhynchus Crickmay 1952
Casquella Percival 1978
Cassianospira Dagys 1972
Cassidirostrum McLaren 1961
Castellaroina Boucot 1972
Catacephalus Yang 1983
Catazyga Hall & Clarke 1893
Cathaysia Ching 1966
Cathayspirina Liang 1990
Caucasella Moisseiev 1934
Caucasiproductus Lazarev 1987
Caucasoproductus Kotljar 1989
Caucasorhynchia Dagys 1963
Caucasothyris Dagys 1974
Causea Wiman 1905
Cavatisinurostrum Sartenaer 1972
Cedulia Racheboeuf 1979
Celdobolus Havlicek 1982
Celebetes Grant 1976
Celidocrania Liu, Zhu & Xue 1985
Celsifornix Carter 1974
Celtanoplia Racheboeuf 1981
Cenorhynchia Cooper & Grant 1976
Centronella Billings 1859
Centronelloidea Weller 1914
Centrorhynchus Sartenaer 1970
Centrospirifer Tien 1938
Ceocypea Grant 1972
Ceramisia Cooper 1983
Cerasina Copper 1995

Ceratreta Bell 1941
Cerberatrypa Havlicek 1990
Cererithyris Buckman 1918
Chaeniorhynchus Cooper & Grant 1976
Chaganella Nikitin 1974
Chakassilingula Ushatinskaia 1992
Chalimia Baranov 1978
Chalimochonetes Baranov 1980
Chaloupskia Neuman 1989
Changshaispirifer Zhao 1977
Changtangella Xian 1983
Changyangrhynchus Yang 1984
Chaoella Licharew 1931
Chaoiella Fredericks 1933
Chaoina Ching 1974
Chapadella Greger 1920
Chapadella Quadros 1981
Chapinella Savage, Eberlein & Churkin 1978
Charionella Billings 1861
Charionoides Boucot, Johnson & Staton 1964
Charltonithyris Buckman 1918
Chascothyris Holzapfel 1895
Chathamithyris Allan 1932
Chattertonia Johnson 1976
Chatwinothyris Sahni 1925
Chaulistomella Cooper 1956
Cheirothyris Rollier 1919
Cheirothyropsis Makridin 1964
Chelononia Cooper & Grant 1974
Chemungia Caster 1939
Cheniothyris Buckman 1918
Chenxianoproductus Liao & Meng 1986
Cherkesovaena Havlicek 1987
Cherubicornea Havlicek 1992
Chianella Waterhouse 1975
Chile Popov & Tikhonov 1990
Chilianshania Yang & Ting 1962
Chilidiopsis Boucot 1959
Chilidorthis Havlicek 1972
Chimaerothyris Paulus, Struve & Wolfart 1963
Chivatschella Zavodowsky 1968
Chlidonophora Dall 1903
Chlupacina Havlicek & Racheboeuf 1979
Chlupacitoechia Havlicek 1992
Chnaurocoelia Johnson, Boucot & Murphy 1976
Choanodus Cooper & Grant 1974
Chondronia Cooper & Grant 1976
Chonetella Waagen 1884
Chonetes Fischer 1830
Chonetina Krotow 1888
Chonetinella Ramsbottom 1952
Chonetinetes Cooper & Grant 1969
Chonetipustula Paeckelmann 1931
Chonetoidea Jones 1928
Choniopora Schauroth 1854
Chonopectella Sarytcheva 1966
Chonopectoides Crickmay 1963
Chonopectus Hall & Clarke 1892
Chonosteges Muir-Wood & Cooper 1960
Chonostegoidella Lee & Yang 1986
Chonostegoides Sarytcheva 1965
Chonostrophia Hall & Clarke 1892
Chonostrophiella Boucot & Amsden 1964
Choperella Ljaschenko 1969
Choristitella Ivanov & Ivanova 1937
Choristites Fischer 1825
Choristothyris Cooper 1942
Christianella Liang 1983

Christiania Hall & Clarke 1892
Chrustenopora Havlicek 1968
Chrustenotreta Havlicek 1994
Chuanostrophia Xian 1988
Chynistrophia Havlicek 1977
Cilinella Havlicek 1970
Cimicinella Schmidt 1946
Cimicinoides Anderson, Boucot & Johnson 1969
Cincta Quenstedt 1868
Cinctifera Muir-Wood & Cooper 1960
Cinctopsis Sucic-Protic 1985
Cinerorthis Havlicek 1974
Cingolospiriferina Pozza 1992
Cingulodermis Havlicek 1971
Cirpa Gregorio 1930
Cistellarcula Elliott 1954
Claratrypa Havlicek 1987
Clarkeia Kozlowski 1923
Clarkella Walcott 1908
Clathrithyris Smirnova 1974
Clavigera Hector 1879
Clavodalejina Havlicek 1977
Cleiothyridellina Waterhouse 1978
Cleiothyridina Buckman 1906
Cliftonia Foerste 1909
Clinambon Schuchert & Cooper 1932
Clintonella Hall & Clarke 1893
Clistotrema Rowell 1963
Clitambonites Agassiz 1847
Clivosilingula Ushatinskaia 1993
Clivospirifer Ljaschenko 1973
Clorinda Barrande 1879
Clorindella Amsden 1964
Clorindina Khodalevich 1939
Clorindinella Rzonsnitzkaja 1975
Cloudella Boucot & Johnson 1963
Cloudothyris Boucot & Johnson 1968
Cnismatocentrum Dall 1920
Coelospira Hall 1863
Coelospirella Su 1976
Coelospirina Havlicek 1956
Coeloterorhynchus Sartenaer 1966
Coenothyris Douville 1879
Colaptomena Cooper 1956
Coledium Grant 1965
Colinella Owen 1981
Collarothyris Modzalevskaja 1970
Collemataria Cooper & Grant 1974
Colletostracia Farrell 1992
Collinithyris Middlemiss 1981
Collumatus Cooper & Grant 1969
Colongina Breivel & Breivel 1970
Colophragma Cooper & Dutro 1982
Colosia Cooper 1983
Colpotoria Cooper 1989
Columellithyris Tchorszhewsky & Radulovic 1984
Comatopoma Havlicek 1950
Comelicania Frech 1901
Comiotoechia Ljaschenko 1973
Companteris Lazarev 1981
Composita Brown 1849
Compositella Xu & Liu 1983
Compressoproductus Sarytcheva 1960
Compsoria Cooper 1973
Compsothyris Jackson 1918
Comuquia Grant 1976
Conarosia Cooper 1989
Conarothyris Cooper 1983

Concaviseptum Brock, Engelbretsen & Dean-Jones 1995
Conchidiella Khodalevich 1939
Conchidium Oehlert 1887
Concinnispirifer Boucot 1975
Concinnithyris Sahni 1929
Condrathyris Minato 1953
Conispirifer Ljaschenko 1985
Connectoproductus Donakova 1974
Conodiscus Ulrich & Cooper 1936
Conomimus Johnson, Boucot & Gronberg 1968
Conotreta Walcott 1889
Contradouvillina Gratsianova 1975
Convexothyris Smirnova 1994
Coolinia Bancroft 1949
Coolkilella Archbold 1993
Cooperea Cocks & Rong 1989
Cooperina Termier, Termier & Pajaud 1966
Cooperithyris Tchorszhewsky 1988
Cooperrhynchia Sandy & Campbell 1994
Coptothyris Jackson 1918
Corbicularia Ljaschenko 1973
Cordatomyonia Boucot, Gauri & Johnson 1966
Corineorthis Stubblefield 1939
Coriothyris Ovtsharenko 1983
Cornwallia Wilson 1932
Coronalosia Waterhouse & Gupta 1978
Corrugatimediorostrum Sartenaer 1970
Cortezorthis Johnson & Talent 1967
Corvinopugnax Havlicek 1961
Corylispirifer Gourvennec 1989
Coscinarina Muir-Wood & Cooper 1960
Coscinophora Cooper & Stehli 1955
Costachonetes Waterhouse 1975
Costachonetina Waterhouse 1981
Costacranaena Johnson & Perry 1976
Costalosiella Waterhouse 1983
Costanoplia Xu 1977
Costatispirifer Archbold & Thomas 1985
Costatrypa Copper 1973
Costatumulus Waterhouse 1983
Costellarina Muir-Wood & Cooper 1967
Costellirostra Cooper 1942
Costellispirifer Boucot 1973
Costicrura Hoover 1981
Costiferina Muir-Wood & Cooper 1960
Costinorella Dagys 1974
Costirhynchia Buckman 1918
Costirhynchopsis Dagys 1977
Costisorthis Havlicek 1974
Costispinifera Muir-Wood & Cooper 1960
Costispirifer Cooper 1942
Costispiriferina Dagys 1974
Costisteges Liao 1980
Costistricklandia Amsden 1953
Costistrophomena Sheehan 1987
Costistrophonella Harper & Boucot 1978
Costitrimerella Rong & Li 1993
Costoconcha Ching, Sun & Ye 1979
Craigella Reed 1935
Cranaena Hall & Clarke 1893
Cranaenella Fenton & Fenton 1924
Crania Retzius 1781
Craniops Hall 1859
Craniotreta Termier & Monod 1978
Craniscus Dall 1871
Craspedalosia Muir-Wood & Cooper 1960
Craspedelia Cooper 1956
Craspedona Cooper & Grant 1975

Crassatrypa Mizens 1977
Crassiorina Havlicek 1950
Crassipunctatrypa Mizens & Rzonsnitzkaja 1979
Crassispirifer Archbold & Thomas 1985
Crassitestella Baarli 1995
Crassumbo Carter 1967
Cratispirifer Archbold & Thomas 1985
Cratorhynchonella Tong 1982
Cratospirifer Tong 1986
Cremnorthis Williams 1963
Crenispirifer Stehli 1954
Cretirhynchia Pettitt 1950
Cricosia Cooper 1973
Crinisarina Cooper & Dutro 1982
Crinistrophia Havlicek 1967
Criopus Poli 1791
Cristicoma Popov 1980
Cristiferina Cooper 1956
Cromatrypa Havlicek 1987
Crossacanthia Gordon 1966
Crossalosia Muir-Wood & Cooper 1960
Crossiskenidium Williams & Curry 1985
Crozonorthis Melou 1976
Cruralina Smirnova 1966
Cruratula Bittner 1890
Cruricella Grant 1976
Crurirhynchia Dagys 1961
Crurispina Goldman & Mitchell 1990
Crurithyris George 1931
Cryptacanthia White & St.john 1867
Cryptatrypa Siehl 1962
Cryptonella Hall 1861
Cryptopora Jeffreys 1869
Cryptoporella Bitner & Pisera 1979
Cryptorhynchia Buckman 1918
Cryptospira Laurie 1991
Cryptospirifer Grabau 1931
Cryptothyrella Cooper 1942
Cryptothyris Bancroft 1945
Cryptotreta Pelman 1977
Ctenalosia Cooper & Stehli 1955
Ctenochonetes Racheboeuf 1976
Ctenokoninckina Xu 1978
Cubacula Lazarev 1984
Cubanirhynchia Kamyschan 1968
Cubanothyris Dagys 1959
Cudmorella Allan 1939
Cuersithyris Almeras & Moulan 1982
Cumberlandina Boucot 1975
Cuneirhynchia Buckman 1918
Cuparius Ross 1971
Cupularostrum Sartenaer 1961
Curticia Walcott 1905
Curtirhynchia Buckman 1918
Cuspidatella Modzalevskaja 1991
Cyclacantharia Cooper & Grant 1969
Cycladigera Havlicek 1971
Cyclocoelia Foerste 1909
Cyclomyonia Cooper 1956
Cyclorhynchia Baranov 1994
Cyclospira Hall & Clarke 1893
Cyclothyris M'Coy 1844
Cydimia Chatterton 1973
Cymatorhynchia Buckman 1918
Cymbidium Kirk 1926
Cymbistropheodonta Harper & Boucot 1978
Cymbithyris Cooper 1952
Cymbricia Roberts 1990

Cymoproductus Xu 1987
Cymostrophia Caster 1939
Cyndalia Grant 1993
Cyphomena Cooper 1956
Cyphomenoidea Cocks 1968
Cyphotalosia Carter 1967
Cyphoterorhynchus Sartenaer 1964
Cyranoia Cooper 1983
Cyrbasiotreta Williams & Curry 1985
Cyrolexis Grant 1965
Cyrtalosia Termier & Termier 1970
Cyrtella Fredericks 1924
Cyrtia Dalman 1828
Cyrtina Davidson 1858
Cyrtinaella Fredericks 1916
Cyrtinaellina Fredericks 1926
Cyrtinoides Yudina & Rzonsnitskaja 1985
Cyrtinopsis Scupin 1896
Cyrtiopsis Grabau 1923
Cyrtiorina Cooper & Dutro 1982
Cyrtoniscus Boucot & Harper 1968
Cyrtonotella Schuchert & Cooper 1931
Cyrtonotreta Holmer 1989
Cyrtospirifer Nalivkin 1924
Cyrtothyris Middlemiss 1959
Cystothyris Sun 1981
Dabashanospira Fu 1982
Dactylogonia Ulrich & Cooper 1942
Dactylotreta Rowell & Henderson 1978
Daghanirhynchia Muir-Wood 1935
Dagnachonetes Afanasjeva 1978
Dagysorhynchia Smirnova 1994
Dagyspirifer Hoover 1991
Dalaia Plodowski 1968
Dalejina Havlicek 1953
Dalejodiscus Havlicek 1961
Dalerhynchus Bai 1978
Dalinuria Lee & Gu 1976
Dalligas Steinich 1968
Dallina Beecher 1893
Dallinella Thomson 1915
Dallithyris Muir-Wood 1959
Dalmanella Hall & Clarke 1892
Dalmanellopsis Khalfin 1948
Danella Pajaud 1966
Danocrania Rosenkrantz 1964
Danzania Pavlova 1988
Dareithyris Siblik 1991
Darvasia Licharew 1934
Dasyalosia Muir-Wood & Cooper 1960
Dasysaria Cooper & Grant 1969
Datangia Yang 1977
Davanirhynchia Ovtsharenko 1983
Davidsonella Munier-Chalmas 1880
Davidsonia Bouchard-Chantereaux 1849
Davidsoniatrypa Lenz 1968
Davidsonina Schuchert & Levene 1929
Daviesiella Waagen 1884
Davoustia Racheboeuf 1976
Dawsonelloides Boucot & Harper 1968
Dayia Davidson 1881
Dearbornia Walcott 1908
Decoropugnax Havlicek 1960
Decurtella Gaetani 1966
Dedzetina Havlicek 1950
Delepinea Muir-Wood 1962
Deltachania Waterhouse 1971
Deltarhynchia Cooper 1989

Deltarina Cooper & Grant 1976
Deltatreta Ulrich 1926
Delthyris Dalman 1828
Deltospirifer Wang & Rong 1986
Demonedys Grant 1976
Dengalosia Manankov & Pavlova 1981
Densalvus Carter 1991
Densepustula Lazarev 1982
Denticuliphoria Licharew 1956
Dentospiriferina Dagys 1965
Deothossia Gatinaud 1949
Derbyaeconcha Licharew 1934
Derbyella Grabau 1931
Derbyia Waagen 1884
Derbyina Clarke 1913
Derbyoides Dunbar & Condra 1932
Dereta Elliott 1959
Desatrypa Copper 1964
Desistrophia Talent 1988
Desmoinesia Hoare 1960
Desmorthis Ulrich & Cooper 1936
Desquamatia Alekseeva 1960
Destombesium Havlicek 1971
Devonalosia Muir-Wood & Cooper 1960
Devonamphistrophia Harper & Boucot 1978
Devonaria Biernat 1966
Devonochonetes Muir-Wood 1962
Devonogypa Havlicek 1951
Devonoproductus Stainbrook 1943
Diabolirhynchia Drot 1964
Diambonia Cooper & Kindle 1936
Diambonioidea Zeng 1987
Diandongia Rong 1974
Diaphelasma Ulrich & Cooper 1936
Diaphragmus Girty 1910
Diazoma Durkoop 1970
Dicamara Hall & Clarke 1893
Dicamaropsis Amsden 1968
Dicellomus Hall 1871
Diceromyonia Wang 1949
Dichacaenia Cooper & Dutro 1982
Dichospirifer Brice 1971
Dichotomosella Tchoumatchenco 1987
Dichozygopleura Renouf 1972
Dicoelosia King 1850
Dicoelospirifer Zhang 1989
Dicoelostrophia Wang 1955
Dicondylotreta Mei 1993
Dicraniscus Meek 1872
Dicrosia Cooper 1978
Dictyobolus Williams & Curry 1985
Dictyoclostoidea Wang & Ching 1964
Dictyoclostus Muir-Wood 1930
Dictyonathyris Xu 1978
Dictyonina Cooper 1942
Dictyonites Cooper 1956
Dictyostrophia Caster 1939
Dictyothyris Douville 1879
Dictyothyropsis Barczyk 1969
Dicystoconcha Termier & Al 1974
Didymelasma Cooper 1956
Didymoparcium Lenz 1977
Didymothyris Rubel & Modzalevskaja 1967
Diedrothyris Richardson 1980
Dielasma King 1859
Dielasmella Weller 1911
Dielasmina Waagen 1882
Dielasmoides Weller 1911

Dienope Cooper 1983
Dierisma Ching, Sun & Ye 1979
Diestothyris Thomson 1916
Digitia Gregorio 1930
Dignomia Hall 1872
Digonella Muir-Wood 1934
Dihelictera Copper 1995
Diholkorhynchia Yang & Xu 1966
Dilophosina Cooper 1983
Dimegelasma Cooper 1942
Dimensionaequalirostrum Sartenaer 1980
Dimerella Zittel 1870
Dinapophysia Maillieux 1935
Dinarella Bittner 1892
Dinarispira Dagys 1974
Dinobolus Hall 1871
Dinorthis Hall & Clarke 1892
Diochthofera Potter 1990
Dioristella Bittner 1890
Diorthelasma Cooper 1956
Diparelasma Ulrich & Cooper 1936
Diplanus Stehli 1954
Diplonorthis Mitchell 1977
Diplospirella Bittner 1890
Dipunctella Liang 1982
Diraphora Bell 1941
Dirinus M'Coy 1844
Discina Lamarck 1819
Discinisca Dall 1871
Discinolepis Waagen 1885
Discinopsis Matthew 1892
Discomyorthis Johnson 1970
Discotreta Ulrich & Cooper 1936
Discradisca Stenzel 1964
Disculina Deslongchamps 1884
Disphenia Grant 1988
Dispiriferina Siblik 1965
Dissoria Cooper 1989
Disulcatella Fu 1980
Ditreta Biernat 1973
Divaricosta Cooper & Grant 1969
Dixonella Gourvennec 1989
Djindella Menakova 1991
Dmitria Sidiachenko 1961
Dmitrispirifer Ljaschenko 1973
Dnestrina Nikiforova & Modzalevskaja 1968
Doescherella Abramov & Grigorjewa 1987
Dogdathyris Baranov 1994
Dogdoa Baranov 1982
Doleroides Cooper 1930
Dolerorthis Schuchert & Cooper 1931
Dolichobrochus Cooper 1983
Dolichomocelypha Liu 1987
Dolichosina Cooper 1983
Dolichozygus Cooper 1983
Doloresella Sando 1957
Domokhotia Abramov & Grigorjewa 1983
Donalosia Lazarev 1989
Donella Rotai 1931
Dongbaella Ching & Ye 1979
Dongbeiispirifer Liu 1977
Dorashamia Sarytcheva 1965
Dorsirugatia Lazarev 1992
Dorsisinus Sanders 1958
Dorsoplicathyris Almeras 1971
Dorsoscyphus Roberts 1971
Dorytreta Cooper 1956
Dotswoodia McKellar 1970

Douvillina Oehlert 1887
Douvillinaria Stainbrook 1945
Douvillinella Spriestersbach 1925
Douvillinoides Harper & Boucot 1978
Douvinella Ljaschenko 1985
Dowhatania Waterhouse & Gupta 1979
Drabodiscina Havlicek 1972
Draborthis Marek & Havlicek 1967
Drabovia Havlicek 1950
Drabovinella Havlicek 1950
Dracius Steinich 1967
Drahanorhynchus Havlicek 1967
Drahanostrophia Havlicek 1967
Drepanorhyncha Cooper 1956
Droharhynchia Sartenaer 1985
Druganirhynchia Tchoumatchenco 1983
Drummuckina Bancroft 1949
Duartea Mendes 1959
Dubaria Termier 1936
Dubioleptina Havlicek 1967
Dulankarella Rukavischnikova 1956
Dundrythyris Almeras 1971
Durranella Percival 1979
Duryeella Boucot 1975
Dushanirhynchia Wang & Zhu 1979
Dyoros Stehli 1954
Dyschrestia Grant 1976
Dyscolia Fischer & Oehlert 1890
Dyscritosia Cooper 1982
Dyscritothyris Cooper 1979
Dysedrosia Cooper 1983
Dysoristus Bell 1944
Dysprosorthis Rong 1984
Dyticospirifer Johnson 1966
Dzhagdicus Sobolev 1992
Dzhangirhynchia Ovtsharenko 1983
Dzharithyris Ovtsharenko 1989
Dzhebaglina Misius 1986
Dzieduszyckia Siemiradzki 1909
Dzirulina Nutsubidze 1945
Dzunarzina Ushatinskaia 1993
Eatonia Hall 1857
Eatonioides McLearn 1918
Eccentricosta Berdan 1963
Echinalosia Waterhouse 1967
Echinaria Muir-Wood & Cooper 1960
Echinauris Muir-Wood & Cooper 1960
Echinirhynchia Childs 1969
Echinocoelia Cooper & Williams 1935
Echinocoeliopsis Hamada 1968
Echinoconchella Lazarev 1985
Echinoconchus Weller 1914
Echinospirifer Ljaschenko 1973
Echinosteges Muir-Wood & Cooper 1960
Echyrosia Cooper 1989
Ecnomiosa Cooper 1977
Ectatoglossa Chu 1974
Ectenoglossa Sinclair 1945
Ectochoristites Campbell 1957
Ectoposia Cooper & Grant 1976
Ectorensselandia Johnson 1973
Ectorhipidium Boucot & Johnson 1979
Ectyphoria Cooper 1989
Edreja Koneva 1979
Edriosteges Muir-Wood & Cooper 1960
Ehlersella Boucot & Johnson 1966
Eichwaldia Billings 1858
Eifelatrypa Copper 1973

Eifyris Struve 1992
Eilotreta Amsden 1968
Ejnespirifer Fu 1982
Elasmata Waterhouse 1982
Elasmothyris Cooper 1956
Elassonia Cooper & Grant 1976
Elderra Richardson 1991
Elegesta Vladimirskaya 1985
Elenchus Aleksandrov 1973
Eleutherocrania Huene 1899
Eleutherokomma Crickmay 1950
Elinoria Cooper & Muir-Wood 1951
Eliorhynchus Sartenaer 1987
Elita Fredericks 1918
Eliva Fredericks 1924
Elivella Fredericks 1924
Elivina Fredericks 1924
Elkania Ford 1886
Elkanisca Havlicek 1982
Ella Fredericks 1918
Ellesmerhynchia Brice 1990
Elliottella Stehli 1955
Elliottina Pajaud 1963
Ellipsothyris Sahni 1925
Elliptoglossa Cooper 1956
Elliptostrophia Havlicek 1963
Elmaria Nalivkin 1947
Elsaella Alichova 1960
Elymospirifer Wang 1974
Elythyna Rzonsnitzkaja 1952
Emanuella Grabau 1923
Embolosia Cooper 1983
Emeithyris Xu 1978
Enallosia Cooper & Grant 1976
Enallothecidea Baker 1983
Enantiosphen Whidborne 1893
Enantiosphenella Johnson 1974
Enchondrospirifer Brice 1971
Endospirifer Tachibana 1981
Engenella Andreeva 1987
Enigmalosia Czarniecki 1969
Enodithyris Smirnova 1986
Ense Struve 1992
Entacanthadus Grant 1993
Enteletella Licharew 1926
Enteletes Fischer 1825
Enteletina Schuchert & Cooper 1931
Enteletoides Stuckenberg 1905
Eoamphistrophia Harper & Boucot 1978
Eoanastrophia Nikiforova & Sapelnikov 1973
Eoantiptychia Xu & Liu 1983
Eobiernatella Balinski 1995
Eobrachythyrina Lazarev & Poletaev 1982
Eobrachythyris Brice 1971
Eochonetes Reed 1917
Eochoristitella Qi 1983
Eochoristites Chu 1933
Eocoelia Nikiforova 1961
Eoconcha Cooper 1951
Eoconchidium Rozman 1967
Eoconulus Cooper 1956
Eocramatia Williams 1974
Eocymostrophia Baarli 1995
Eodallina Elliott 1959
Eodalmanella Havlicek 1950
Eodevonaria Breger 1906
Eodictyonella Wright 1994
Eodinobolus Rowell 1963

Eodiorthelasma Xu & Liu 1984
Eodmitria Brice 1982
Eoglossinotoechia Havlicek 1959
Eogryphus Hertlein & Grant 1944
Eohemithiris Hertlein & Grant 1944
Eohowellella Lopushinskaja 1976
Eokarpinskia Rzonsnitzkaja 1964
Eokirkidium Khodalevich & Sapelnikov 1970
Eolaballa Liao & Meng 1986
Eolacazella Elliott 1953
Eolissochonetes Hoare 1960
Eolyttonia Fredericks 1924
Eomaoristrophia Ushatinskaia 1983
Eomarginifera Muir-Wood 1930
Eomarginiferina Brunton 1966
Eomartiniopsis Sokolskaja 1941
Eomegastrophia Cocks 1967
Eonalivkinia Vladimirskaya 1985
Eoobolus Matthew 1902
Eoorthis Walcott 1908
Eoparaphorhynchus Sartenaer 1961
Eoperegrinella Ager 1968
Eopholidostrophia Harper, Johnson & Boucot 1967
Eoplectodonta Kozlowski 1929
Eoplicanoplia Boucot & Harper 1968
Eoplicoplasia Johnson & Lenz 1992
Eoproductella Rzonsnitzkaja 1980
Eopugnax Baranov 1991
Eoreticularia Nalivkin 1924
Eorhipidomella Hints 1971
Eorhynchula Liang 1983
Eoscaphelasma Koneva & Al 1990
Eoschizophoria Rong & Yang 1980
Eoschuchertella Gratsianova 1974
Eoseptaliphoria Ching & Sun 1976
Eosericoidea Zeng 1987
Eosiphonotreta Havlicek 1982
Eosophragmophora Wang 1974
Eosotrematorthis Wang 1955
Eospinatrypa Copper 1973
Eospirifer Schuchert 1913
Eospiriferina Grabau 1931
Eospirigerina Boucot & Johnson 1967
Eostrophalosia Stainbrook 1943
Eostropheodonta Bancroft 1949
Eostrophomena Walcott 1905
Eostrophonella Williams 1950
Eosyntrophopsis Yadrenkina 1989
Eosyringothyris Stainbrook 1943
Eothecidellina Baker 1991
Eothele Rowell 1980
Eousella Ovtsharenko 1976
Epacrosina Cooper 1983
Epelidoaegiria Strusz 1982
Ephippelasma Cooper 1956
Epicelia Grant 1972
Epicyrta Deslongchamps 1884
Epithyris Phillips 1841
Epithyroides Xu 1978
Epitomyonia Wright 1968
Equirostra Cooper & Muir-Wood 1951
Erbotreta Holmer & Ushatinskaia 1994
Erectocephalus Xian 1978
Eremithyris Brugge 1973
Eremotoechia Cooper 1956
Eremotrema Cooper 1956
Ericiatia Muir-Wood & Cooper 1960
Eridmatus Branson 1966

Eridorthis Foerste 1909
Erinostrophia Cocks & Worsley 1993
Eripnifera Potter 1990
Erismatina Waterhouse 1983
Eristenosia Cooper 1983
Errhynx Havlicek 1982
Erymnaria Cooper 1959
Erymnia Cooper 1977
Eschatelasma Popov 1981
Esilia Nikitin & Popov 1985
Espella Nilova 1965
Estlandia Schuchert & Cooper 1931
Estonirhynchia Schmidt 1954
Estonomena Roomusoks 1989
Etheridgina Oehlert 1887
Etherilosia Archbold 1993
Etymothyris Cloud 1942
Eucalathis Fischer & Oehlert 1890
Eucharitina Schmidt 1955
Eudesella Munier-Chalmas 1880
Eudesia King 1850
Eudesites Rioult 1966
Eudoxina Fredericks 1929
Euidothyris Buckman 1918
Eumetabolotoechia Sartenaer 1975
Eumetria Hall 1863
Eunella Hall & Clarke 1893
Euorthisina Havlicek 1950
Euractinella Bittner 1890
Eurekaspirifer Johnson 1966
Euroatrypa Oradovskaja 1982
Eurycolporhynchus Sartenaer 1968
Eurypthyris Ovtsharenko 1983
Eurysina Cooper 1983
Eurysites Cooper 1989
Eurysoria Cooper 1983
Euryspirifer Wedekind 1926
Eurytatospirifer Gatinaud 1949
Eurythyris Cloud 1942
Eurytreta Rowell 1966
Euxinella Moisseiev 1936
Evanescirostrum Sartenaer 1965
Evanidisinurostrum Sartenaer 1987
Evargyrotheca MacKinnon & Smirnova 1995
Evenkina Andreeva 1961
Evenkinorthis Yadrenkina 1977
Evenkorhynchia Rozman 1969
Exatrypa Copper 1967
Excavatorhynchia Jin & Fang 1977
Exceptothyris Sucic-Protic 1971
Expellobolus Havlicek 1982
Experilingula Koneva & Popov 1983
Fabulasteges Liang 1990
Faksethyris Asgaard 1971
Falafer Grant 1972
Falciferula Tchoumatchenco 1987
Fallax Atkins 1960
Fallaxispirifer Su 1976
Fallaxoproductus Lee, Gu & Li 1982
Falsatrypa Havlicek 1956
Famatinorthis Levy & Nullo 1973
Fanichonetes Xu And Grant 1994
Fardenia Lamont 1935
Farmerella Clarke 1992
Fascicoma Popov 1980
Fascicosta Stehli 1955
Fascicostella Schuchert & Cooper 1931
Fasciculina Cooper 1952

Fascifera Ulrich & Cooper 1942
Fascistropheodonta Harper & Boucot 1978
Fascizetina Havlicek 1975
Fehamya Mergl 1983
Felinotoechia Havlicek 1961
Fenestrirostra Cooper 1955
Fengzuella Li & Han 1980
Fenxiangella Wang 1978
Ferganella Nikiforova 1937
Ferganoproductus Galitzkaja 1977
Fernglenia Carter 1992
Ferrax Havlicek 1975
Ferrobolus Havlicek 1982
Ferrythyris Almeras 1971
Fezzanoglossa Havlicek 1973
Ffynnonia Neuman & Bates 1978
Fibulistrophia Garcia-Alcalde 1972
Fidespirifer Ljaschenko 1973
Filiatrypa Chen 1983
Filiconcha Dear 1969
Filigreenia Soja 1988
Fimbrinia Cooper 1972
Fimbriothyris Deslongchamps 1884
Fimbrispirifer Cooper 1942
Finkelnburgia Walcott 1905
Finospirifer Yin 1981
Fissirhynchia Pearson 1977
Fissurirostra Orbigny 1847
Fistulogonites Neuman 1971
Fitzroyella Veevers 1959
Flabellirhynchia Buckman 1918
Flabellitesia Zhang 1989
Flabellocyrtia Chorowicz & Termier 1975
Flabellothyris Deslongchamps 1884
Flabellulirostrum Sartenaer 1971
Fletcherithyris Campbell 1965
Fletcherithyroides Dagys 1977
Flexaria Muir-Wood & Cooper 1960
Flexathyris Grunt 1980
Floweria Cooper & Dutro 1982
Fluctuaria Muir-Wood & Cooper 1960
Foliomena Havlicek 1952
Fordinia Walcott 1908
Formosarhynchia Seifert 1963
Fortunella Calzada 1985
Fossatrypa Mizens & Rzonsnitzkaja 1979
Fossuliella Popov & Ushatinskaia 1992
Fosteria Zezina 1980
Foveola Gorjansky 1969
Frankiella Racheboeuf 1983
Franklinella Lenz 1973
Frechella Legrand-Blain 1986
Fredericksia Paeckelmann 1931
Frenula Dall 1871
Frenulina Dall 1895
Frieleia Dall 1895
Fulcriphoria Carls 1974
Furcirhynchia Buckman 1918
Furcitella Cooper 1956
Fusella M'Coy 1844
Fusichonetes Liao 1981
Fusiproductus Waterhouse 1975
Fusirhynchia Dagys 1968
Fusispirifer Waterhouse 1966
Gacella Williams 1962
Gacina Stehli 1961
Gagriella Moisseiev 1939
Galateastrophia Harper & Boucot 1978

Galeatagypa Sapelnikov 1981
Galeatathyris Jin 1983
Galeatella Muir-Wood & Cooper 1960
Galeatellina Sapelnikov & Al 1976
Galinella Popov & Holmer 1994
Galliennithyris Rollet 1966
Gamdaella Miloradovich 1947
Gamonetes Isaacson 1977
Gamphalosia Stainbrook 1945
Gannania Fu 1982
Gasconsia Northrop 1939
Gashaomiaoia Rong, Su & Li 1985
Gaspesia Clarke 1907
Gastrodetoechia Sartenaer 1965
Gatia Archbold 1993
Gefonia Licharew 1936
Gegenella Lee & Gu 1976
Gelidorthina Havlicek 1974
Gelidorthis Havlicek 1968
Gemerithyris Siblik 1977
Geminisulcispirifer Sartenaer 1982
Gemmarcula Elliott 1947
Gemmellaroia Cossmann 1898
Gemmellaroiella Mabuti 1937
Gemmulicosta Waterhouse 1971
Geniculifera Muir-Wood & Cooper 1960
Geniculina Roomusoks 1993
Geniculomclearnites Harper & Boucot 1978
Genuspirifer Liang 1990
Georgethyris Minato 1953
Geranocephalus Crickmay 1954
Gerassimovia Licharew 1956
Gerkispira Carter 1983
Gerolsteinites Struve 1990
Gerothyris Struve 1970
Gerrhynx Baranov 1991
Geyerella Schellwien 1900
Gibberostrophia Harper & Boucot 1978
Gibbirhynchia Buckman 1918
Gibbithyrella Ovtsharenko 1989
Gibbithyris Sahni 1925
Gibbochonetes Aisenverg 1971
Gibbospirifer Waterhouse 1971
Gigantoproductus Prentice 1950
Gigantorhynchus Sapelnikov & Malygina 1977
Gigantothyris Seifert 1963
Gilledia Stehli 1961
Giraldibella Havlicek 1977
Giraldiella Bancroft 1949
Girlasia Gregorio 1930
Girtyella Weller 1911
Gisilina Steinich 1963
Gissarina Menakova & Nikiforova 1986
Gjelispinifera Ivanova 1975
Glabrichonetina Waterhouse 1978
Glabrigalites Struve 1992
Glaciarcula Elliott 1956
Gladiostrophia Havlicek 1967
Glaphyrorthis Roberts 1990
Glassia Davidson 1881
Glassina Hall & Clarke 1893
Glazewskia Pajaud 1964
Glendonia McClung & Armstrong 1978
Globatrypa Mizens & Sapelnikov 1985
Globidorsum Jin 1988
Globiella Muir-Wood & Cooper 1960
Globirhynchia Buckman 1918
Globispirifer Tachibana 1964

Globithyris Cloud 1942
Globosobucina Waterhouse & Piyasin 1970
Globosochonetes Brunton 1968
Globosoproductus Litvinovich & Vorontsova 1983
Globulirhynchia Brice 1980
Glossella Cooper 1956
Glosseudesia Lobatscheva 1974
Glosshypothyridina Rzonsnitzkaja 1978
Glossinotoechia Havlicek 1959
Glossinulina Johnson 1975
Glossinulinirhynchia Baranov 1991
Glossinulus Schmidt 1942
Glossoleptaena Havlicek 1967
Glossorthis Opik 1930
Glossothyropsis Girty 1934
Glottidia Dall 1870
Glyphisaria Cooper 1983
Glyptacrothele Termier & Termier 1974
Glyptambonites Cooper 1956
Glypterina Boucot 1970
Glyptias Walcott 1901
Glyptoglossella Cooper 1960
Glyptogypa Struve 1992
Glyptomena Cooper 1956
Glyptorhynchia Shen & He 1994
Glyptoria Cooper 1976
Glyptorthis Foerste 1914
Glyptospirifer Hou & Xian 1975
Glyptosteges Cooper & Grant 1975
Glyptotrophia Ulrich & Cooper 1936
Gmelinmagas Marinelarena 1964
Gnamptorhynchos Jin 1989
Gnathorhynchia Buckman 1918
Goleomixa Grant 1976
Goliathyris Feldman & Owen 1988
Gonambonites Pander 1830
Gonathyris Baranov 1994
Gondolina Ching & Liao 1966
Goniarina Cooper & Grant 1969
Goniobrochus Cooper 1983
Goniocoelia Hall 1861
Goniophoria Janischewsky 1910
Goniorhynchia Buckman 1918
Goniothyris Buckman 1918
Goniothyropsis Ovtsharenko 1983
Goniotrema Ulrich & Cooper 1936
Gorchakovia Popov & Khazanovitch 1989
Gorgostrophia Havlicek 1967
Gorjanskya Tenjakova 1980
Gotatrypa Struve 1966
Goungjunspirifer Zhang 1983
Grabauellina Licharew 1934
Grabauicyrtiopsis Gatinaud 1949
Grabauispirifer Gatinaud 1949
Gracianella Johnson & Boucot 1967
Gracilotoechia Baranov 1977
Grammetaria Cooper 1959
Grammoplecia Wright & Jaanusson 1993
Grammorhynchus Roberts 1971
Grandaurispina Muir-Wood & Cooper 1960
Grandiproductella Lazarev 1987
Grandirhynchia Buckman 1918
Grandispirifer Yan 1959
Grantonia Brown 1953
Granulirhynchia Buckman 1918
Grasirhynchia Owen 1968
Grayina Boucot 1975
Grebenella Modzalevskaja & Beznosova 1992

Greenfieldia Grabau 1910
Greenockia Brown 1952
Greira Erlanger 1993
Grorudia Spjeldnaes 1957
Gruenewaldtia Tschernyschew 1885
Grumantia Ustritsky 1963
Gruntallina Waterhouse & Gupta 1986
Gryphus Megerle 1811
Guangdongina Mou & Liu 1989
Guangjiayanella Yang 1984
Guangshunia Xian 1978
Guangxiispirifer Xian 1978
Guangyuania Sheng 1975
Guaxa Garcia-Alcalde 1986
Gubleria Termier & Termier 1960
Guicyrtia Wang & Zhu 1979
Guilinospirifer Xu & Yao 1988
Guistrophia Wang & Rong 1986
Gunnarella Spjeldnaes 1957
Gunningblandella Percival 1979
Gusarella Prosorovskaja 1962
Guseriplia Dagys 1963
Guttasella Neuman 1977
Gwynia King 1859
Gwyniella Johansen 1987
Gypidula Hall 1867
Gypidulella Khodalevich & Breivel 1959
Gypidulina Rzonsnitzkaja 1956
Gypiduloides Savage & Baxter 1995
Gypospirifer Cooper & Grant 1976
Gyroselenella Li 1985
Gyrosina Cooper 1983
Gyrosoria Cooper 1973
Gyrothyris Thomson 1918
Habrobrochus Cooper 1983
Hadrorhynchia McLaren 1961
Hadrosia Cooper 1983
Hadrotatorhynchus Sartenaer 1986
Hadrotreta Rowell 1966
Hadyrhyncha Havlicek 1979
Hagabirhynchia Jefferies 1961
Hallina Winchell & Schuchert 1892
Hallinetes Racheboeuf & Feldman 1990
Halorella Bittner 1884
Halorellina Xu 1978
Halorelloidea Ager 1960
Hamburgia Weller 1911
Hamletella Hayasaka 1953
Hamlingella Reed 1943
Hampsia Morris 1994
Hamptonina Rollier 1919
Hanaeproductus Ficner & Havlicek 1978
Hansotreta Krause & Rowell 1975
Hanusitrypa Havlicek 1967
Haplospirifer Lee & Gu 1976
Harjumena Roomusoks 1993
Harknessella Reed 1917
Harmatosia Cooper 1983
Harpidium Kirk 1925
Harpotothyris Smirnova 1990
Harringtonina Boucot 1972
Harttella Bell 1929
Harttina Hall & Clarke 1893
Haupiria MacKinnon 1983
Havlicekella Amsden 1985
Havlicekia Boucot 1963
Haydenella Reed 1944
Haydenoides Chan 1977

Hebertella Hall & Clarke 1892
Hebetoechia Havlicek 1959
Hedeina Boucot 1957
Hedeinopsis Gourvennec 1990
Hedstroemina Bancroft 1929
Hefengia Xu 1991
Heimia Haas 1890
Helaspis Imbrie 1959
Helenaeproductus Lazarev 1991
Helenathyris Alekseeva 1969
Heligothyris Middlemiss 1991
Heliomedusa Sun & Hou 1987
Helmersenia Pander 1861
Helvetella Owen 1977
Hemichonetes Lee & Su 1980
Hemileurus Cooper & Grant 1976
Hemiplethorhynchus Peetz 1898
Hemipronites Pander 1830
Hemiptychina Waagen 1882
Hemistringocephalus Smirnov 1985
Hemithiris Orbigny 1847
Hemithyropsis Katz 1974
Hemitoechia Nikiforova 1970
Heosomocelypha Liu 1987
Herangirhynchia MacFarlan 1992
Hercosestria Cooper & Grant 1969
Hercosia Cooper & Grant 1969
Hercostrophia Williams 1950
Hercothyris Cooper 1979
Hercotrema Jin 1989
Hergetatrypa Havlicek 1987
Hesperinia Cooper 1956
Hesperithyris Dubar 1942
Hesperomena Cooper 1956
Hesperonomia Ulrich & Cooper 1936
Hesperonomiella Ulrich & Cooper 1936
Hesperorhynchia Warren 1937
Hesperorthis Schuchert & Cooper 1931
Hesperosia Cooper 1983
Hesperotrophia Ulrich & Cooper 1936
Hessenhausia Struve 1982
Heteralosia King 1938
Heteraria Cooper & Grant 1976
Heterelasma Girty 1909
Heterelasmina Licharew 1939
Heterobrochus Cooper 1983
Heteromychus Cooper 1989
Heterorthella Harper, Boucot & Walmsley 1969
Heterorthina Bancroft 1928
Heterorthis Hall & Clarke 1892
Hexarhytis Alvarez 1990
Hibernodonta Harper & Mitchell 1985
Himalairhynchia Ching & Sun 1976
Himathyris Waterhouse 1986
Hindella Davidson 1882
Hinganella Su 1980
Hingganoleptaena Zhu 1985
Hipparionix Vanuxem 1842
Hircinisca Havlicek 1960
Hirnantia Lamont 1935
Hirsutella Cooper & Muir-Wood 1951
Hiscobeccus Amsden 1983
Hisingerella Henningsmoen 1948
Hispanirhynchia Thomson 1927
Hispidaria Cooper & Dutro 1982
Histosyrinx Massa, Termier & Termier 1974
Holcorhynchella Dagys 1974
Holcorhynchia Buckman 1918

Holcospirifer Bassett, Cocks & Holland 1976
Holcothyris Buckman 1918
Hollardiella Drot 1967
Hollardina Racheboeuf & Al 1982
Holorhynchus Kiaer 1902
Holosia Cooper & Grant 1976
Holotricharina Cooper & Grant 1975
Holtedahlina Foerste 1924
Holynatrypa Havlicek 1973
Holynetes Havlicek & Racheboeuf 1979
Homaliarhynchia Shi 1990
Homeocardiorhynchus Sartenaer 1985
Homevalaria Waterhouse 1986
Homoeorhynchia Buckman 1918
Homoeospira Hall & Clarke 1893
Homoeospirella Amsden 1968
Homotreta Bell 1941
Hontorialosia Martinez 1979
Hopkinsirhynchia Shi & Grant 1993
Horderleyella Bancroft 1928
Horridonia Chao 1927
Hoskingia Campbell 1965
Hostimena Havlicek 1990
Hostimex Havlicek 1982
Howellella Kozlowski 1946
Howellites Bancroft 1945
Howelloidea Su 1980
Howittia Talent 1956
Howseia Logan 1963
Huacoella Benedetto & Herrera 1993
Huananochonetes Wang, Boucot & Rong 1981
Huatangia Liao & Meng 1986
Hubeiproductus Yang 1984
Huenella Walcott 1908
Huenellina Schuchert & Cooper 1931
Hulterstadia Wright 1993
Humaella Zhu 1982
Hunanoproductus Hou 1965
Hunanospirifer Tien 1938
Hunanotoechia Ma 1993
Hungarispira Dagys 1972
Hungaritheca Dagys 1972
Hustedia Hall & Clarke 1893
Hustedtiella Dagys 1972
Hyattidina Schuchert 1913
Hyborhynchella Cooper 1955
Hybostenoscisma Liao & Meng 1986
Hynniphoria Suess 1859
Hyperobolus Havlicek 1982
Hypoleiorhynchus Linnik 1976
Hypolinoproductus Liang 1982
Hyponeatrypa Struve 1966
Hypopsia Cooper & Grant 1974
Hypothyridina Buckman 1906
Hypselonetes Racheboeuf 1981
Hypseloterorhynchus Sartenaer 1971
Hypsiptycha Wang 1949
Hypsomyonia Cooper 1955
Hysterohowellella Carls 1985
Hysterolites Schlotheim 1820
Hystriculina Muir-Wood & Cooper 1960
Iberirhynchia Drot & Westbroek 1966
Iberithyris Kvakhadze 1972
Iberohowellella Carls, Meyn & Vespermann 1993
Iberomena Villas 1985
Icodonta Bell 1941
Idioglyptus Northrop 1939
Idiorthis McLearn 1924

Idiospira Cooper 1956
Idiostrophia Ulrich & Cooper 1936
Iheringithyris Levy 1961
Ikella Tjazheva 1972
Iliella Rukavischnikova 1980
Ilmarinia Opik 1934
Ilmenia Nalivkin 1941
Ilmeniopsis Xian 1983
Ilmenispina Havlicek 1959
Ilmospirifer Ljaschenko 1969
Iloerhynchus Balinski 1995
Ilopsyrhynchus Sartenaer 1988
Ilyinella Jassjukevitch 1973
Imacanthyris Grunt 1991
Imatrypa Havlicek 1977
Imbrexia Nalivkin 1937
Imbricatia Cooper 1952
Imbricatospira Fu 1980
Imdentistella Grunt 1991
Imperiospira Archbold & Thomas 1993
Impiacus Lazarev & Suursuren 1988
Implexina Poletaev 1971
Inaequalis Sucic-Protic 1971
Incisius Grant 1976
Incorthis Havlicek & Branisa 1980
Indaclor Havlicek 1990
Independatrypa Copper 1973
Indigia Barchatova 1973
Indorhynchia Ovtsharenko 1975
Indospirifer Grabau 1931
Inflatia Muir-Wood & Cooper 1960
Infurca Percival 1979
Ingelarella Campbell 1959
Ingria Opik 1930
Iniathyris Besnossova 1963
Iniproductus Lazarev 1990
Innaechia Baranov 1980
Innuitella Crickmay 1968
Inopinatarcula Elliott 1952
Insignitisinurostrum Sartenaer 1987
Institella Cooper 1942
Institifera Muir-Wood & Cooper 1960
Institina Muir-Wood & Cooper 1960
Inversella Opik 1933
Inversithyris Dagys 1968
Invertina Copper & Chen 1995
Invertrypa Struve 1961
Iotina Cooper & Grant 1976
Iowarhynchus Carter 1983
Iowatrypa Copper 1973
Ipherron Havlicek 1982
Iphidella Walcott 1905
Irboskites Bekker 1924
Irenothyris Pojariskaja 1966
Irgislella Tjazheva 1972
Irhirea Havlicek 1971
Iridistrophia Havlicek 1965
Iru Opik 1934
Ishimia Nikitin 1974
Isjuminelina Makridin 1960
Ismenia King 1850
Isochonetes Aisenverg 1985
Isocrania Jaekel 1902
Isogramma Meek & Worthen 1870
Isophragma Cooper 1956
Isopoma Torley 1934
Isorthis Kozlowski 1929
Isospinatrypa Struve 1966

Isovella Breivel & Breivel 1970
Israelaria Cooper 1976
Issedonia Popov 1980
Istokina Breivel & Breivel 1988
Isumithyris Hatai 1948
Iubagraspirifer Gatinaud 1949
Iugrabaspirifer Gatinaud 1949
Ivanothyris Havlicek 1957
Ivanoviella Makridin 1955
Ivdelinella Brice 1982
Ivdelinia Andronov 1961
Ivdeliniella Kim 1981
Ivshinella Koneva 1979
Iwaispirifer Tachibana 1964
Jaanussonites Neuman 1977
Jacetanella Racheboeuf, Batet & Magrans 1993
Jaffaia Thomson 1927
Jaisalmeria Sahni 1955
Jakutella Abramov 1970
Jakutochonetes Afanasjeva 1977
Jakutoproductus Kaschirzev 1959
Jakutostrophia Manankov 1991
Jarnesella Walcott 1905
Janiceps Frech 1901
Janiomya Havlicek 1967
Janius Havlicek 1957
Japanithyris Thomson 1927
Jarovathyris Havlicek 1987
Jevinellina Liu 1983
Jezercia Havlicek & Mergl 1982
Jiangdaspirifer Chen, Rao 1986
Jielingia Zeng 1987
Jiguliconcha Lazarev 1990
Jilinia Liu 1977
Jilinmartinia Lee & Gu 1980
Jilinospirifer Su 1980
Jipuproductus Sun 1983
Jisuina Grabau 1931
Jivinella Havlicek 1949
Johnsonathyris Savage, Eberlein & Churkin 1978
Johnsonetes Racheboeuf 1987
Johnsoniatrypa Zhang 1989
Jolkinia Breivel & Breivel 1988
Jolonica Dall 1920
Jolvia Sapelnikov 1960
Jonesea Cocks & Rong 1989
Joviatrypa Copper 1995
Juralina Kyansep 1961
Juresania Fredericks 1928
Juvavella Bittner 1888
Juvavellina Bittner 1896
Juxathyris Liang 1990
Juxoldhamina Liang 1990
Kabanoviella Smirnova 1973
Kachathyris Smirnova 1975
Kadraliproductus Galitzkaja 1977
Kafirnigania Katz 1962
Kahlella Legrand-Blain 1995
Kajnaria Nikitin & Popov 1984
Kalitvella Lazarev & Poletaev 1982
Kallirhynchia Buckman 1918
Kamoica Hatai 1936
Kampella Baarli 1988
Kaninospirifer Kulikov & Stepanov 1975
Kansuella Chao 1928
Kantinatrypa Havlicek, 1995
Kaplex Ficner & Havlicek 1975
Kaplicona Havlicek 1987

Karadagella Babanova 1965
Karadagithyris Tchorszhewsky 1974
Karakulina Andreeva 1972
Karathele Koneva 1986
Karavankina Ramovs 1969
Karbous Havlicek 1985
Karlicium Havlicek 1974
Karnotreta Williams & Curry 1985
Karpatiella Sucic-Protic 1985
Karpinskia Tschernyschew 1885
Kasakhstania Besnossova 1968
Kasetia Waterhouse 1981
Kassinella Borissiak 1956
Katastrophomena Cocks 1968
Katunia Kulkov 1963
Kavesia Lazarev 1987
Kawhiarhynchia MacFarlan 1992
Kayserella Hall & Clarke 1892
Kayseria Davidson 1882
Kedrovothyris Smirnova 1990
Keilamena Roomusoks 1993
Kelamelia Zhang 1983
Kelsovia Clarke 1990
Kendzhilgithyris Ovtsharenko 1983
Kentronetes Racheboeuf & Herrera 1994
Keokukia Carter 1990
Keratothyris Tuluweit 1965
Kericserella Voros 1983
Kerpina Struve 1961
Keskentassia Popov & Holmer 1994
Kestonithyris Sahni 1925
Keyserlingia Pander 1861
Keyserlingina Tschernyschew 1902
Khangaestrophia Mendbayar 1994
Khasagtina Ushatinskaia 1987
Khinganospirifer Su 1976
Khodalevichia Boucot & Johnson 1979
Kholbotchonia Baranov 1988
Kiaeromena Spjeldnaes 1957
Kiangsiella Grabau & Chao 1927
Kijanina Aksarina 1992
Kikaithyris Yabe & Hatai 1941
Kindleina Savage, Eberlein & Churkin 1978
Kinelina Ljaschenko 1969
Kingena Davidson 1852
Kingenella Popiel-Barczyk 1968
Kinghiria Litvinovich 1966
Kinnella Bergstrom 1968
Kintathyris Shi & Waterhouse 1991
Kirkidium Amsden, Boucot & Johnson 1967
Kirkina Salmon 1942
Kisilia Nalivkin 1979
Kitakamithyris Minato 1951
Kjaerina Bancroft 1929
Kjerulfina Bancroft 1929
Kleithriatreta Roberts 1990
Klipsteinella Dagys 1974
Klipsteinelloidea Sun 1981
Klukatrypa Havlicek 1987
Kochiproductus Dunbar 1955
Koeveskallina Dagys 1965
Koigia Modzalevskaja 1985
Kokomerena Misius 1986
Kolhidaella Moisseiev 1939
Kolihium Havlicek 1982
Kolymithyris Dagys 1965
Komiella Ljaschenko 1985
Komispirifer Ljaschenko 1973

Komukia Waterhouse 1982
Koninckella Munier-Chalmas 1880
Koninckina Suess 1853
Koninckodonta Bittner 1893
Konstantia Pajaud 1970
Korjakirhynchia Smirnova 1990
Kosirium Ficner & Havlicek 1975
Kosoidea Havlicek & Mergl 1988
Kosomena Havlicek 1990
Kotlaia Grant 1993
Kotujella Andreeva 1962
Kotujotreta Ushatinskaia 1994
Kotylotreta Koneva 1990
Kotysex Havlicek 1990
Kozhuchinella Severgina 1967
Kozlenia Havlicek 1987
Kozlowskia Fredericks 1933
Kozlowskiellina Boucot 1958
Kozlowskites Havlicek 1952
Kransia Westbroek 1968
Krattorthis Jaanusson & Bassett 1993
Kraussina Davidson 1859
Krejcigrafella Struve 1978
Krimargyrotheca MacKinnon & Smirnova 1995
Kritorhynchia Rong & Yang 1981
Krizistrophia Havlicek 1992
Krotovia Fredericks 1928
Kubanithyris Tchorszhewsky 1989
Kueichowella Yang 1978
Kullervo Opik 1932
Kulumbella Nikiforova 1960
Kumbella Khodalevich 1975
Kundatella Aksarina 1978
Kungaella Solomina 1988
Kunlunia Wang 1983
Kuntella Ovtsharenko 1991
Kurakithyris Hatai 1946
Kurnamena Roomusoks 1989
Kurtomarginifera Xu 1987
Kutchirhynchia Buckman 1918
Kutchithyris Buckman 1918
Kutorgina Billings 1861
Kutorginella Ivanova 1951
Kuvelousia Waterhouse 1968
Kuzgunia Klenina 1984
Kvania Havlicek 1994
Kvesanirhynchia Kvakhadze 1976
Kwangsia Grabau 1931
Kwangsirhynchus Hou & Xian 1975
Kymatothyris Struve 1970
Kyrshabaktella Koneva 1986
Kyrtatrypa Struve 1966
Labaia Licharew 1956
Laballa Moisseiev 1962
Labriproductus Cooper & Muir-Wood 1951
Lacazella Munier-Chalmas 1880
Lachrymula Graham 1970
Lacunaerhynchia Almeras 1966
Lacunarites Opik 1934
Lacunites Gorjansky 1969
Lacunosella Wisniewska 1932
Ladjia Veevers 1959
Ladogia Nalivkin 1941
Ladogiella Opik 1934
Ladogifornix Schmidt 1964
Ladogilina Ljaschenko 1973
Ladogilinella Ljaschenko 1973
Ladogioides McLaren 1961

Laevicyphomena Cocks 1968
Laevigaterhynchia Wisniewska-Zelichowska 1978
Laevirhynchia Dagys 1974
Laevispirifer Ushatinskaia 1977
Laevithyris Dagys 1974
Laevorhynchia Shen & He 1994
Laima Gravitis 1981
Laioporella Ivanova 1975
Lakhmina Oehlert 1887
Lamanskya Moberg & Segerberg 1906
Lamarckispirifer Gatinaud 1949
Lambdarina Brunton & Champion 1974
Lamellaerhynchia Burri 1953
Lamelliconchidium Kulkov 1968
Lamellokoninckina Jin & Fang 1977
Lamellosathyris Jin & Fang 1983
Lamellosia Cooper & Grant 1975
Laminatia Muir-Wood & Cooper 1960
Lamiproductus Liang 1990
Lammellosathyris Ching & Fang 1983
Lamnaespina Waterhouse 1976
Lamnimargus Waterhouse 1975
Lamniplica Waterhouse & Rao 1989
Lampangella Waterhouse 1983
Lancangjiangia Jin & Fang 1977
Lanceomyonia Havlicek 1960
Landonella MacKinnon 1993
Langella Mendes 1961
Langkawia Hamada 1969
Langshanthyris Sun 1987
Lanipustula Klets 1983
Laosia Mansuy 1913
Lapradella Baranov 1989
Laqueus Dall 1870
Larispirifer Enokjan & Poletaev 1986
Larium Gregorio 1930
Lateralatirostrum Sartenaer 1979
Laterispina Wang & Jin 1991
Lathamella Liu 1979
Laticrura Cooper 1956
Latiflexa Koczyrkevicz 1984
Latiplecus Lee & Gu 1976
Latiproductus Sarytcheva & Legrand-Blain 1977
Latirhynchia Fu 1980
Latispirifer Archbold & Thomas 1985
Latonotoechia Havlicek 1960
Lazella Radulovic 1991
Lazithyris Radulovic 1986
Lazutkinia Rzonsnitzkaja 1952
Leangella Opik 1933
Lebediorthis Severgina 1984
Leiochonetes Roberts 1976
Leiolepismatina Yang & Xu 1966
Leioproductus Stainbrook 1947
Leiorhynchoidea Cloud 1944
Leiorhynchoides Dovgal 1953
Leiorhynchus Hall 1860
Leioria Cooper 1976
Leioseptathyris Wu 1974
Leiothycridina Grunt 1980
Lenatoechia Nikiforova 1970
Lengwuella Liang 1990
Lenothyris Dagys 1968
Lenzia Perry, Boucot & Gabrielse 1981
Leontiella Yadrenkina 1982
Lepidocrania Cooper & Grant 1974
Lepidocycloides Nikiforova 1961
Lepidocyclus Wang 1949

Lepidoleptaena Havlicek 1963
Lepidomena Laurie 1991
Lepidorhynchia Burri 1956
Lepidorthis Wang 1955
Lepidospirifer Cooper & Grant 1969
Lepismatina Wang 1955
Leptaena Dalman 1828
Leptaenella Fredericks 1918
Leptaenisca Beecher 1890
Leptaenoidea Hedstrom 1917
Leptaenomendax Garcia-Alcalde 1978
Leptaenopoma Marek & Havlicek 1967
Leptaenopyxis Havlicek 1963
Leptagonia M'Coy 1844
Leptalosia Dunbar & Condra 1932
Leptathyris Siehl 1962
Leptella Hall & Clarke 1892
Leptellina Ulrich & Cooper 1936
Leptelloidea Jones 1928
Leptembolon Mickwitz 1896
Leptestia Bekker 1922
Leptestiina Havlicek 1952
Leptobolus Hall 1871
Leptocaryorhynchus Sartenaer 1970
Leptochonetes Havlicek & Racheboeuf 1979
Leptocoelia Hall 1857
Leptocoelina Johnson 1970
Leptodonta Khalfin 1955
Leptodontella Khalfin 1948
Leptodus Kayser 1883
Leptolepyron Jin 1989
Leptoptilum Opik 1930
Leptoskelidion Amsden 1974
Leptospira Boucot, Johnson & Staton 1964
Leptostrophia Hall & Clarke 1892
Leptostrophiella Harper & Boucot 1978
Leptothyrella Muir-Wood 1965
Leptothyrellopsis Bitner & Pisera 1979
Lercarella Mascle & Termier 1970
Lespius Gregorio 1930
Lessiniella Voros 1993
Lethamia Waterhouse 1973
Leurosina Cooper & Grant 1975
Levenea Schuchert & Cooper 1931
Levibiseptum Xian 1975
Leviconchidiella Rzonsnitzkaja 1960
Levigatella Andronov 1961
Levigypa Sapelnikov & Mizens 1985
Levipugnax Pushkin 1986
Levipustula Maxwell 1951
Levisapicus Tong & Al 1990
Levispira Mizens 1975
Levitusia Muir-Wood & Cooper 1960
Leymerithyris Calzada 1988
Lezhoeviella Baranov 1978
Lialosia Muir-Wood & Cooper 1960
Liberella Liang 1990
Libyaeglossa Havlicek 1973
Libyaerhynchus Mergl & Massa 1992
Libys Massa, Termier & Termier 1974
Licharewia Einor 1939
Licharewiella Ustritsky 1960
Lichnatrypa Wang & Rong 1986
Lievinella Boucot 1975
Ligatella Martinez 1978
Liljevallia Hedstrom 1917
Limbatrypa Copper 1982
Limbella Stehli 1954

Limbifera Brunton & Mundy 1988
Limbimurina Cooper 1956
Limstrophia Fu 1980
Lindinella Mergl & Slehoferova 1990
Lindstroemella Hall & Clarke 1890
Lineirhynchia Buckman 1918
Lingatrypa Mizens 1985
Lingshanella Xu & Yao 1986
Linguithyris Buckman 1918
Lingula Bruguiere 1797
Lingulapholis Schuchert 1913
Lingularia Biernat & Emig 1993
Lingulasma Ulrich 1889
Lingulella Salter 1866
Lingulellotreta Koneva 1983
Lingulepis Hall 1863
Lingulipora Girty 1898
Lingulobolus Matthew 1895
Lingulodiscina Whitfield 1890
Lingulops Hall 1872
Linguopugnoides Havlicek 1960
Linnarssonella Walcott 1902
Linnarssonia Walcott 1885
Linoporella Schuchert & Cooper 1931
Linoproductus Chao 1927
Linoprotonia Ferguson 1971
Linostrophomena Rong, Xu & Yang 1974
Linterella Amsden 1988
Linxiangxiella Yang 1984
Liocoelia Schuchert & Cooper 1931
Lioleptaena Xian 1978
Liolimbella Li 1991
Liosomena Liang 1990
Liosotella Cooper 1953
Liospiriferina Rousselle 1977
Liostrophia Cooper & Kindle 1936
Liothyrella Thomson 1916
Lipanteris Briggs 1986
Liralingua Graham 1970
Liramia Cooper 1983
Liraplecta Jing & Sun 1981
Liraria Cooper & Grant 1975
Liraspirifer Stainbrook 1950
Lirellaria Cooper & Grant 1976
Lirellarina Cooper 1989
Liricamera Cooper 1956
Liriplica Campbell 1961
Lissajousithyris Almeras 1971
Lissatrypa Twenhofel 1914
Lissatrypella Sapelnikov & Mizens 1982
Lissatrypoidea Boucot & Amsden 1958
Lissella Campbell 1961
Lissidium Lenz 1989
Lissochonetes Dunbar & Condra 1932
Lissocoelina Schuchert & Cooper 1931
Lissocrania Williams 1943
Lissoleptaena Havlicek 1992
Lissomarginifera Lane 1962
Lissopleura Whitfield 1896
Lissorhynchia Yang & Xu 1966
Lissosia Cooper & Grant 1975
Lissostrophia Amsden 1949
Lissothyris Smirnova 1987
Lissotreta Amsden 1968
Litocothia Grant 1976
Litothyris Roberts 1971
Liveringia Archbold 1987
Lixatrypa Havlicek 1987

Ljaschenkovia Batrukova 1969
Ljudmilispirifer Tcherkesova 1976
Llanoella Boucot 1975
Loboidothyris Buckman 1918
Loboidothyropsis Sucic-Protic 1971
Loborina Balinski 1982
Lobothyris Buckman 1918
Lobothyroides Xu 1978
Lobvia Breivel & Breivel 1977
Lochengia Grabau 1929
Lochkothele Havlicek & Mergl 1988
Loczyella Frech 1901
Loganella Boucot & Amsden 1958
Loilemia Reed 1936
Lokutella Voros 1983
Lomatiphora Roberts 1971
Lomatorthis Williams & Curry 1985
Longdongshuia Hou & Xian 1975
Longipegma Popov & Holmer 1994
Longispina Cooper 1942
Longithyris Katz & Popov 1974
Longvillia Bancroft 1933
Longxianirhynchia Fu 1982
Longyania Zhu 1990
Lopasnia Ilkhovsky 1990
Loperia Walcott 1905
Lophrothyris Buckman 1918
Lopingia Chan 1979
Lorangerella Crickmay 1963
Lordorthis Ross 1959
Loreleiella Racheboeuf 1986
Loriolithyris Middlemiss 1968
Losvia Breivel & Breivel 1976
Lotharingella Laurin 1984
Lowenstamia Stehli 1961
Loxophragmus Cooper & Grant 1974
Luanquella Garcia-Alcalde & Racheboeuf 1978
Ludfordina Kelly 1967
Luhaia Roomusoks 1956
Luhotreta Mergl & Slehoferova 1990
Lunaria Ching, Sun & Ye 1979
Lunoglossa Xu & Xie 1985
Lunpolaia Ching & Ye 1979
Luofuia Xu 1977
Luppovithyris Lobatscheva 1990
Lurgiticoma Popov 1980
Luterella Amsden 1988
Lutetiarcula Elliott 1954
Lychnothyris Voros 1983
Lycophoria Lahusen 1886
Lyonia Archbold 1983
Lyra Cumberland 1818
Lysidium Havlicek 1985
Lysigypa Havlicek 1990
Lytha Fredericks 1924
Lyttonia Waagen 1883
Maakina Andreeva 1961
Mabella Klenina 1984
Macandrevia King 1859
Machaeraria Cooper 1955
Machaeratoechia Havlicek 1992
Machaerocolella Li & Han 1980
Mackerrovia Cocks 1968
Maclarenella Stehli 1955
Macrocoelia Cooper 1956
Macroplectane Cossmann 1909
Macropleura Boucot 1963
Macropotamorhynchus Sartenaer 1970

Madarosia Cooper & Grant 1976
Madoia Sun & Ye 1982
Magadania Ganelin 1977
Magadina Thomson 1915
Magadinella Thomson 1915
Magas Sowerby 1818
Magasella Dall 1870
Magella Thomson 1915
Magellania Bayle 1880
Magharithyris Farag & Gatinaud 1962
Magicostrophia Zhu 1985
Magnicanalis Rowell 1962
Magniderbyia Ting 1965
Magniplicatina Waterhouse 1983
Magnithyris Sahni 1925
Magniventra Harper & Boucot 1978
Magnumbonella Carter 1968
Majkopella Moisseiev 1962
Makridinirhynchia Sucic-Protic 1969
Makridinithyris Tchorszhewsky 1989
Malayanoplia Hamada 1969
Malinella Andreeva 1982
Malleia Thomson 1927
Malloproductus Tachibana 1981
Maltaia Cooper 1983
Malurostrophia Campbell & Talent 1967
Malwirhynchia Chiplonker 1938
Mamatia Popov & Holmer 1994
Mametothyris Smirnova 1969
Mammosum Gregorio 1930
Mamutinetes Havlicek 1990
Manespira Copper 1986
Mangkeluia Xu 1991
Manithyris Foster 1974
Mannia Davidson 1874
Manosia Zeng 1983
Mansina Andreeva 1977
Maorielasma Waterhouse 1964
Maorirhynchia MacFarlan 1992
Maoristrophia Allan 1947
Mapingtichia Lee 1986
Marcharella Andreeva 1968
Margaritiproductus Lazarev 1986
Marginalosia Waterhouse 1978
Marginatia Muir-Wood & Cooper 1960
Marginicinctus Sutton 1938
Marginifera Waagen 1884
Marginirugus Sutton 1938
Marginoproductus Tan 1986
Marginorthis Liu, Zhu & Xue 1985
Marginovatia Gordon & Henry 1990
Mariannaella Sapelnikov & Al 1975
Mariaspirifer Tcherkesova 1991
Marinurnula Waterhouse 1964
Marionites Cooper & Muir-Wood 1951
Maritimithyris Smirnova 1986
Markamia Jin & Shi 1985
Markitoechia Havlicek 1959
Marklandella Harper, Boucot & Walmsley 1969
Martellia Wirth 1936
Martellispirifer Gatinaud 1949
Martinia M'Coy 1844
Martiniella Grabau & Tien 1931
Martiniopsis Waagen 1883
Martinothyris Minato 1953
Matanoleptodus Liao 1983
Matutella Cooper 1951
Mauispirifer Allan 1947

Maxillirhynchia Buckman 1918
Mayaothyris Sun 1987
Maydenella Laurie 1991
Mcewanella Foerste 1920
Mclearnites Caster 1945
Mclearnitesella Harper & Boucot 1978
Medessia Andreeva 1960
Mediospirifer Bublichenko 1956
Mediterranirhynchia Sucic-Protic 1969
Meekella White & St. John 1867
Megachonetes Sokolskaja 1950
Megakozlowskiella Boucot 1957
Megalopterorhynchus Sartenaer 1965
Megalosia Waterhouse 1988
Megamyonia Wang 1949
Meganterella Boucot 1959
Meganteris Suess 1855
Megaplectatrypa Zhang 1981
Megapleuronia Cooper 1952
Megasalopina Boucot, Gauri & Johnson 1966
Megaspinochonetes Yang & Rong 1982
Megasteges Waterhouse 1975
Megastrophia Caster 1939
Megastrophiella Harper & Boucot 1978
Megathiris Orbigny 1847
Megatschernyschewia Sremac 1986
Megerlia King 1850
Megerlina Deslongchamps 1884
Megousia Muir-Wood & Cooper 1960
Megumatrypa Harper 1973
Meifodia Williams 1951
Mendacella Cooper 1930
Mendathyris Cloud 1942
Mennespirifer Ljaschenko 1973
Mentzelia Quenstedt 1871
Mentzelioides Dagys 1974
Mentzeliopsis Trechmann 1918
Meonia Steinich 1963
Merciella Lamont & Gilbert 1945
Merista Suess 1851
Meristella Hall 1860
Meristelloides Isaacson 1977
Meristina Hall 1867
Meristorygma Carter 1974
Meristospira Grabau 1910
Merophricus Cooper 1983
Merospirifer Reed 1949
Mesochorispira Carter 1992
Mesodalmanella Havlicek 1950
Mesodouviella Harper & Boucot 1978
Mesodouvillina Williams 1950
Mesoleptostrophia Harper & Boucot 1978
Mesolissostrophia Williams 1950
Mesolobus Dunbar & Condra 1932
Mesonomia Ulrich & Cooper 1936
Mesopholidostrophia Williams 1950
Mesoplica Reed 1943
Mesoseptina Munoz 1994
Mesotreta Kutorga 1848
Metabolipa Godefroid 1974
Metacamarella Reed 1917
Metaplasia Hall & Clarke 1893
Metathyrisina Rong & Yang 1981
Metorthis Wang 1955
Metriolepis Cooper & Grant 1976
Mexicaria Cooper 1983
Mezounia Havlicek 1967
Miaohuangrhynchus Yang 1977

Micella Ovtsharenko 1976
Micidus Chatterton 1973
Mickwitzia Schmidt 1888
Micraphelia Cooper & Grant 1969
Microcardinalia Boucot & Ehlers 1963
Micromitra Meek 1873
Microrhynchia Muir-Wood 1952
Microsphaeridiorhynchus Sartenaer 1970
Microthyridina Schuchert & Levene 1929
Microtrypa Wilson 1945
Mictospirifer Johnson 1995
Millythyris Almeras 1971
Mimaria Cooper & Grant 1976
Mimatrypa Struve 1964
Mimella Cooper 1930
Mimikonstantia Baker & Elston 1984
Mimorina Cooper 1983
Minatothyris Vandercammen 1957
Mingenewia Archbold 1980
Miniplanus Waterhouse & Piyasin 1970
Miniprokopia Havlicek 1971
Minirostrella Balinski 1995
Minispina Waterhouse 1982
Minithyris Radulovic 1991
Minororthis Ivanov 1950
Minutilla Crickmay 1967
Minutostropheodonta Harper & Boucot 1978
Minutulirhynchia Smirnova 1992
Minysphenia Grant 1988
Minythyra Brunton 1984
Miogryphus Hertlein & Grant 1944
Mirantesia Mohanti 1972
Mirifusella Carter 1971
Mirilingula Popov 1983
Mirisquamea Sucic-Protic 1971
Mirorthis Zeng 1983
Mirtellaspirifer Gatinaud 1949
Misolia Seidlitz 1913
Mistproductus Yang 1991
Mitchellella Strusz 1984
Miyakothyris Hatai 1938
Mjoesina Spjeldnaes 1957
Mobergia Redlich 1899
Moderatoproductus Litvinovich & Vorontsova 1983
Modestella Owen 1961
Moeschia Boullier 1976
Mogoktella Andreeva 1968
Mogoliella Ischnazarov 1972
Moisseevia Makridin 1964
Moisseievia Dagys 1963
Molongcola Percival 1991
Molongella Savage 1974
Molongia Mitchell 1921
Monadotoechia Havlicek 1960
Monelasmina Cooper 1955
Mongolella Alekseeva 1976
Mongolina Grabau 1931
Mongoliopsis Grunt 1977
Mongolirhynchia Hou & Zhao 1976
Mongolochonetes Afanasjeva 1991
Mongolorhynx Erlanger 1992
Mongolosia Manankov & Pavlova 1976
Mongolospira Alekseeva 1977
Mongolostrophia Rozman & Rong 1993
Monobolina Salter 1866
Monoconvexa Pelman 1977
Monomerella Billings 1871
Monorthis Bates 1968

Monsardithyris Almeras 1971
Montanella Ovtsharenko 1976
Monticlarella Wisniewska 1932
Monticulifera Muir-Wood & Cooper 1960
Montsenetes Racheboeuf 1992
Moorefieldella Girty 1911
Moorellina Elliott 1953
Moquellina Ching, Sun & Ye 1979
Moraviaturia Sahni 1960
Moravilla Havlicek 1953
Moravostrophia Havlicek 1962
Morganella McKellar 1970
Morinatrypa Havlicek 1990
Morinorhynchus Havlicek 1965
Morphorhynchus Cooper & Dutro 1982
Morrisia Davidson 1852
Morrisina Grabau 1931
Morrisithyris Almeras 1971
Mosquella Makridin 1955
Moumina Fredericks 1924
Moutonithyris Middlemiss 1976
Mucroclipeus Goldman & Mitchell 1990
Mucrospirifer Grabau 1931
Mucrospiriferinella Waterhouse 1982
Muirwoodella Tchorszhewsky 1974
Muirwoodia Licharew 1947
Multicorhynchia Chen 1983
Multicosta Khodalevich & Breivel 1977
Multicostella Schuchert & Cooper 1931
Multiridgia Zeng 1987
Multispinula Rowell 1962
Multispirifer Kaplun 1961
Munhella Neuman 1971
Munieratrypa Mizens & Rzonsnitzkaja 1979
Muriferella Johnson & Talent 1967
Murihikurhynchia MacFarlan 1992
Murinella Cooper 1956
Murjukiana Severgina 1967
Murravia Thomson 1916
Murrinyinella Kruse 1990
Musculina Schuchert & Levene 1929
Mutationella Kozlowski 1929
Mycerosia Cooper 1989
Mylloconotreta Williams & Curry 1985
Myodelthyrium Thomas 1985
Myopugnax Glushenko 1975
Myotreta Gorjansky 1969
Myriospirifer Havlicek 1978
Myrmirhynx Havlicek 1982
Mystrophora Kayser 1871
Nabarredia Havlicek & Racheboeuf 1979
Nabiaoia Xu 1979
Nadiastrophia Talent 1963
Najadospirifer Havlicek 1957
Najdinothyris Makridin & Katz 1964
Nakazatothyris Minato & Kato 1977
Nalivkinaria Rzonsnitzkaja 1968
Nalivkinia Bublichenko 1928
Nanacalathis Zezina 1981
Nanambonites Liu 1976
Nanatrypa Sapelnikov & Mizens 1982
Nannirhynchia Buckman 1918
Nanorthis Ulrich & Cooper 1936
Nanospira Amsden 1949
Nanothyris Cloud 1942
Nantanella Grabau 1936
Nanukidium Jones 1979
Naradanithyris Tokuyama 1958

Narynella Andreeva 1987
Nasonirhynchia Havlicek 1992
Nastosia Cooper 1989
Naukat Popov & Tikhonov 1990
Navalicria Sartenaer 1989
Navispira Amsden 1983
Nayunnella Sartenaer 1961
Neaguithyris Georgescu 1991
Neatrypa Struve 1964
Nebenothyris Minato 1953
Negramithyris Prosorovskaja 1985
Neimongolella Zhang 1981
Nekhoroshevia Bublichenko 1956
Nekvasilovela Calzada 1987
Nematocrania Grant 1976
Nemesa Schmidt 1941
Neoancistrocrania Laurin 1992
Neoathyrisina Chen 1988
Neobolus Waagen 1885
Neobouchardia Thomson 1927
Neochonetes Muir-Wood 1962
Neocirpa Prosorovskaja 1985
Neocoelia McKellar 1966
Neocramatia Harper 1989
Neocrania Lee & Brunton 1986
Neocyrtina Yang & Xu 1966
Neodelthyris Hou 1963
Neoedriosteges Liang 1990
Neofascicosta Xu 1978
Neoglobithyris Havlicek 1984
Neogypidula Licharew 1961
Neohemithyris Yabe & Hatai 1934
Neokarpinskia Mizens 1977
Neokjaerina Levy & Nullo 1974
Neoliothyrina Sahni 1925
Neometabolipa Godefroid 1974
Neomunella Ozaki 1931
Neophricadothyris Licharew 1934
Neoplicatifera Ching, Liao & Fang 1974
Neopugilis Li 1991
Neoretzia Dagys 1963
Neorhynchia Thomson 1915
Neorhynchula Liang 1983
Neorichthofenia Shen, He & Zhu 1992
Neoschizophoria Yanagida 1983
Neospirifer Fredericks 1924
Neospirigerina Rzonsnitzkaja 1975
Neostrophia Ulrich & Cooper 1936
Neothecidella Pajaud 1970
Neothyris Douville 1879
Neotreta Sobolev 1976
Neowellerella Dagys 1974
Neoyanguania Shi 1988
Nepasitoechia Havlicek 1979
Nereidella Wang 1955
Nerthebrochus Cooper 1983
Nervostrophia Caster 1939
Neumanella Harper & Boucot 1978
Neumania Harper 1981
Neumayrithyris Tokuyama 1958
Nicolella Reed 1917
Nicoloidea Zeng 1987
Nicolorthis Havlicek 1981
Nigerinoplica Lazarev 1986
Nigeroplica Nalivkin 1975
Nikiforovaena Boucot 1963
Nikolaevirhynchus Baranov 1988
Ningbingella Roberts 1971

Ninglangothyris Jin & Fang 1977
Niorhynx Havlicek 1982
Nipponirhynchia Yanagida & Nishikawa 1984
Nipponithyris Yabe & Hatai 1934
Niquivilia Benedetto & Herrera 1993
Nisusia Walcott 1905
Niutoushania Liao 1984
Niviconia Cooper & Grant 1974
Nix Easton 1962
Nochoroiella Pelman 1983
Nocturnellia Havlicek 1950
Nodaea Tachibana 1981
Noetlingia Hall & Clarke 1893
Nondia Boucot & Chiang 1974
Nordathyris Grunt 1977
Nordella Ljaschenko 1973
Nordispirifer Ljaschenko 1973
Nordotoechia Tcherkesova 1965
Norella Bittner 1890
Notanoplia Gill 1950
Nothokuvelousia Waterhouse 1986
Nothopindax Cooper & Grant 1974
Nothorthis Ulrich & Cooper 1938
Notiobolus Popov 1981
Notiochonetes Muir-Wood 1962
Notoconchidium Gill 1950
Notoleptaena Gill 1951
Notolosia Archbold 1986
Notoparmella Johnson 1973
Notorthisina Havlicek & Branisa 1980
Notorygmia Cooper 1972
Notosaria Cooper 1959
Notoscaphidia Williams & Curry 1985
Notosia Cooper 1983
Notospirifer Harrington 1955
Notostrophia Waterhouse 1973
Notothyrina Licharew 1936
Notothyris Waagen 1882
Notozyga Cooper 1977
Novellinetes Havlicek & Racheboeuf 1979
Novozemelia Tcherkesova 1973
Nubialba Neuman 1994
Nucleata Quenstedt 1868
Nucleatina Katz 1962
Nucleatula Bittner 1888
Nucleorhynchia Liang 1983
Nucleospira Hall 1859
Nucleusorhynchia Sun & Ye 1982
Nudauris Stehli 1954
Nudirostralina Yang & Xu 1966
Nudispiriferina Yang & Xu 1966
Nudymia Lazarev 1990
Nugnecella Levy & Nullo 1974
Nuguschella Tjazheva 1960
Numericoma Popov 1980
Nurataella Larin 1973
Nuratamella Kalashnikov 1992
Nuria Misius 1986
Nurochonetes Ushatinskaia 1977
Nushbiella Popov 1986
Nyalamurhynchia Ching, Rong & Sun 1976
Nyege Veevers 1959
Nymphorhynchia Rzonsnitzkaja 1956
Oanduporella Hints 1975
Obesaria Havlicek 1957
Obliqunsteges Liang 1990
Oblongarcula Elliott 1959
Obnixia Hoover 1979

Obolella Billings 1861
Obolellina Billings 1871
Obolopsis Saito 1936
Obolorugia Zhang 1983
Obolus Eichwald 1829
Obovothyris Buckman 1927
Obscurella Baranov 1991
Obsoletirhynchia Shi 1992
Obturamentella Amsden 1958
Ochotathyris Dagys 1974
Ochotorhynchia Dagys 1968
Ocnerorthis Bell 1941
Ocorthis Melou 1982
Ocruranus Liu 1979
Odarovithyris Tchorszhewsky 1971
Odontospirifer Dunbar 1955
Odoratus Zhu 1985
Oehlertella Hall & Clarke 1890
Oepikina Salmon 1942
Oepikinella Wilson 1944
Oepikites Khazanovitch & Popov 1984
Ogbinia Sarytcheva 1965
Ogilviella Lenz 1968
Oglu Havlicek 1987
Oglupes Havlicek 1987
Ogmoplecia Wright & Jaanusson 1993
Ogmusia Cooper 1983
Oina Popov & Tikhonov 1990
Oiosia Cooper & Dutro 1982
Okathyris Smirnova 1975
Oldhamina Waagen 1883
Oldhaminella Wanner & Sieverts 1935
Oleneothyris Cooper 1942
Olentotreta Koneva & Al 1990
Oleorthis Havlicek 1968
Olgerdia Grigorjewa 1977
Oligomys Schuchert & Cooper 1931
Oligoptycherhynchus Sartenaer 1970
Oligorhachis Imbrie 1959
Oligorhynchia Cooper 1935
Oligorhytisia Cooper 1983
Oligorthis Ulrich & Cooper 1936
Oligothyrina Cooper 1956
Oliveirella Oliveira 1934
Ombonia Caneva 1906
Omnutakhella Lopushinskaja 1976
Omolonella Moisseiev 1936
Omolonia Alekseeva 1967
Omolonothyris Dagys 1968
Oncosarina Cooper & Grant 1969
Onegia Andreev 1993
Onniella Bancroft 1928
Onnizetina Havlicek 1974
Onopordumaria Waterhouse 1971
Onugorhynchia Havlicek 1992
Onychoplecia Cooper 1956
Onychotreta Ulrich & Cooper 1936
Opikella Amsden 1968
Opisthotreta Palmer 1954
Oppeliella Tchorszhewsky 1989
Opsiconidion Ludvigsen 1974
Orbicoelia Waterhouse & Piyasin 1970
Orbicula Cuvier 1798
Orbiculatisinurostrum Sartenaer 1984
Orbiculoidea Orbigny 1847
Orbiculopora Batrukova 1969
Orbiculothyris Wolfart 1968
Orbinaria Muir-Wood & Cooper 1960

Orbirhynchia Pettitt 1954
Orbithele Sdzuy 1955
Orenburgella Pavlova 1969
Orhoria Havlicek 1990
Oriensellina Smirnova 1986
Orientospira Dagys 1965
Orientospirifer Hou & Xian 1975
Orientothyris Katz & Popov 1974
Origostrophia Mitchell 1977
Oriskania Hall & Clarke 1893
Orlovirhynchia Dagys 1968
Ornatothyrella Dagys 1974
Ornatothyris Sahni 1929
Ornithella Deslongchamps 1884
Ornithothyris Sahni 1925
Ornothyrella Havlicek 1971
Orthambonites Pander 1830
Orthidiella Ulrich & Cooper 1936
Orthidium Hall & Clarke 1892
Orthiella Ljaschenko 1985
Orthis Dalman 1828
Orthisocrania Rowell 1963
Orthocarina Fu 1980
Orthoidea Friren 1876
Orthokopis Baarli 1995
Ortholina Calzada 1984
Orthonomaea Hall 1893
Orthopleura Imbrie 1959
Orthorhynchula Hall & Clarke 1893
Orthorhynchuloides Williams 1962
Orthorhynchyllion Jin 1989
Orthospirifer Pitrat 1975
Orthostrophella Amsden 1968
Orthostrophia Hall 1883
Orthotetella King 1931
Orthotetes Fischer 1829
Orthotetoides Lazarev 1984
Orthothetina Schellwien 1900
Orthothrix Geinitz 1847
Orthothyris Cooper 1955
Orthotichia Hall & Clarke 1892
Orthotoma Quenstedt 1869
Orthotropia Hall & Clarke 1893
Orulgania Solomina & Tschernjak 1961
Orusia Walcott 1905
Oslogonites Opik 1939
Oslomena Spjeldnaes 1957
Osmarella Pearson 1977
Otariella Waterhouse 1978
Otariella Popov & Holmer 1994
Otarorhyncha Nikiforova & Popov 1981
Otospirifer Hou & Xian 1975
Ottadalenites Harper 1981
Ottenbyella Popov & Holmer 1994
Otusia Walcott 1905
Ovalella Walmsley & Boucot 1975
Ovalospira Fu 1980
Ovatathyris Owen 1988
Ovatia Muir-Wood & Cooper 1960
Overtonia Thomas 1914
Overtoniina Grunt 1973
Ovidiella Nikitin & Popov 1984
Ovlatchania Abramov & Grigorjewa 1986
Owenirhynchia Calzada 1980
Oxlosia Ulrich & Cooper 1936
Oxoplecia Wilson 1913
Oxycolpella Dagys 1962
Oxypleurorhynchia Plodowski 1973

Ozora Carter 1990
Pachancorhynchia Brice 1986
Pachyglossella Cooper 1960
Pachymagas Ihering 1903
Pachymoorellina Baker 1989
Pachyplax Alvarez & Brunton 1990
Pachythyris Boullier 1976
Pacificocoelia Boucot 1975
Pacifithyris Hatai 1938
Paeckelmanella Licharew 1934
Pahlenella Schuchert & Cooper 1931
Pajaudina Logan 1988
Palaeobolus Matthew 1899
Palaeochoristites Sokolskaja 1941
Palaeoglossa Cockerell 1911
Palaeoldhamina Liang 1982
Palaeoleptostrophia Rong & Cocks 1994
Palaeoschizophoria Fu 1982
Palaeoschmidtites Koneva 1979
Palaeospirifer Martynova & Sverbilova 1968
Palaeostrophia Ulrich & Cooper 1936
Palaeostrophomena Holtedahl 1916
Palaeotrimerella Li & Han 1980
Palaferella Spriestersbach 1942
Paldiskia Gorjansky 1969
Paldiskites Havlicek 1982
Palinorthis Ulrich & Cooper 1936
Palmerhytis Brunton & Mundy 1986
Pamirorhynchia Ovtsharenko 1983
Pamirotheca Dagys 1974
Pamirothyris Dagys 1974
Pamirothyropsis Ovtsharenko 1979
Pammegetherhynchus Sartenaer 1977
Pampoecilorhynchus Sartenaer 1968
Panderina Schuchert & Cooper 1931
Panderites Roomusoks 1993
Pantellaria Dall 1919
Papiliolinus Waterhouse & Gupta 1977
Papillostrophia Havlicek 1967
Papodina Voros 1983
Paraacanthothyris Kamyschan 1973
Parabifolium Pajaud 1966
Parabornhardtina Sun & Hou 1964
Paraboubeithyris Middlemiss 1980
Parabuxtonia Yang & Zhang 1982
Paracapillithyris Katz & Popov 1974
Parachonetella Liao 1980
Parachonetes Johnson 1966
Parachoristites Barchatova 1968
Paracomposita Ding & Yao 1985
Paraconchidium Rong & Yang 1974
Paracostanoplia Xu 1977
Paracraniops Williams 1963
Paracrothyris Wu 1974
Paracrurithyris Liao 1981
Paracyrtiopsis Gatinaud 1949
Parademonedys Yang 1984
Paraderbyia Sun 1983
Paradinobolus Li & Han 1980
Paradolerorthis Zeng 1987
Paradoxothyris Xu 1978
Paradygella Liao & Sun 1974
Paraemanuella Yang 1977
Parageyerella He & Zhu 1985
Paraglossinulus Martynova 1988
Paragusarella Shi 1990
Parahemiptychina Chen, Rao 1986
Parajuresania Lazarev 1982

Parakansuella Tan 1987
Parakarpinskia Zhang 1983
Parakeyserlingina Fredericks 1916
Parakinetica Richardson 1987
Parakingena Sun 1981
Paralaballa Sun 1981
Paralazutkinia Jiang 1978
Paraldingia Richardson 1973
Paralenorthis Havlicek & Branisa 1980
Paralepismatina Yang & Xu 1966
Paraleptodus Lee & Gu 1980
Paraleptostrophia Harper & Boucot 1978
Parallelelasma Cooper 1956
Parallelora Carter 1974
Paralyttonia Wanner & Sieverts 1935
Paramarginatia Yang 1978
Paramarginifera Fredericks 1916
Paramartinia Reed 1949
Parameekella He & Zhu 1985
Paramentzelia Xu 1978
Paramerista Su 1976
Paramesolobus Afanasjeva 1975
Paramonticulifera Tong 1978
Paramuirwoodia Zhang 1983
Paranaia Clarke 1913
Paranisopleurella Zhang 1989
Paranorella Cloud 1944
Paranorellina Dagys 1974
Parantiptychia Xu & Liu 1983
Paranudirostralina Sun & Ye 1982
Paraoligorhyncha Popov 1981
Paraonychoplecia Percival 1991
Paraorthotetina He & Zhu 1985
Parapholidostrophia Johnson 1971
Paraphorhynchus Weller 1905
Paraplatythyris Sun 1987
Paraplicanoplia Xu 1977
Paraplicatifera Zhao & Tan 1984
Parapugnax Schmidt 1964
Parapulchratia Chan 1979
Paraquadrithyris Yang 1983
Parareticularia Lee & Gu 1976
Pararhactorhynchia Shi 1990
Pararhipidium Boucot & Johnson 1979
Pararhynchospirina Zhang 1981
Paraschizophoria Lazarev 1976
Paraspirifer Wedekind 1926
Paraspiriferina Reed 1944
Parastringocephalus Struve 1965
Parastrophina Schuchert & Levene 1929
Parastrophinella Schuchert & Cooper 1931
Parastrophonella Bublichenko 1956
Parasulcatinella Xu & Liu 1983
Paratetratomia Yang 1977
Parathecidea Backhaus 1959
Parathyridina Schuchert & Levene 1929
Parathyrisina Wang 1974
Paratreta Biernat 1973
Paratribonium Sapelnikov & Mizens 1991
Parazyga Hall & Clarke 1893
Parenteletes King 1931
Paritisteges Liang 1990
Parmephrix Brunton, Racheboeuf & Mundy 1994
Parmorthina Havlicek 1975
Parmorthis Schuchert & Cooper 1931
Parmula Menakova & Breivel 1987
Paromalomena Rong 1984
Paromoeopygma Sartenaer 1968

Parthirhynchia Titova 1980
Paruncinella Waterhouse & Gupta 1979
Parvaltissimarostrum Sartenaer & Xu 1991
Parvirhynchia Buckman 1918
Parvulaltarostrum Sartenaer 1979
Paryphella Liao 1981
Paryphorhynchopora Simorin 1956
Patagorhynchia Allan 1938
Paterina Beecher 1891
Paterorthis Havlicek 1971
Paterula Barrande 1879
Patriaspirifer Johnson 1995
Paucicostella Cooper 1956
Paucicrura Cooper 1956
Paucispinauria Waterhouse 1983
Paucispinifera Muir-Wood & Cooper 1960
Paucistrophia Wang & Zhu 1979
Paulinella Boucot & Racheboeuf 1987
Paulonia Nalivkin 1925
Paurogastroderhynchus Sartenaer 1970
Paurorhyncha Cooper 1942
Paurorthina Rubel 1961
Paurorthis Schuchert & Cooper 1931
Pavdenia Breivel & Breivel 1988
Payuella Grabau 1934
Pectorhyncha McLearn 1918
Peculneithyris Smirnova 1972
Peetzatrypa Rzonsnitzkaja 1975
Pegmarhynchia Cooper 1955
Pegmathyris Hatai 1938
Pegmatreta Bell 1941
Pelagodiscus Dall 1908
Pelaiella Martinez 1991
Peleicostella Havlicek 1971
Pelmania Koneva 1992
Pelonomia Cooper 1956
Peltichia Jing & Liao 1981
Pembrostrophia Bassett 1971
Pemphixina Cooper 1981
Peniculauris Muir-Wood & Cooper 1960
Pennospiriferina Dagys 1965
Pentactinella Bittner 1890
Pentagonia Cozzens 1846
Pentamerella Hall 1867
Pentamerifera Khodalevich 1939
Pentameroides Schuchert & Cooper 1931
Pentamerus Sowerby 1812
Pentithyris Cooper 1983
Pentlandella Boucot 1964
Pentlandina Bancroft 1949
Penzhinella Solomina 1985
Penzhinothyris Smirnova 1969
Perakia Hamada 1969
Peratos Copper 1986
Perditocardinia Schuchert & Cooper 1931
Peregrinella Oehlert 1887
Peregrinellina Hou & Wang 1984
Peregrinelloidea Dagys 1968
Periallus Hoover 1979
Perichonetes Xu 1979
Peridalejina Havlicek 1973
Perigeyerella Wang 1955
Perihustedia Xu 1991
Perimecocoelia Cooper 1956
Perissothyris Carter 1967
Peristerothyris Mancenido 1983
Peritrimerella Liang 1983
Peritritoechia Xu, Rong & Liu 1974

Permasyrinx Waterhouse 1983
Permianella He & Zhu 1979
Permicola Koczyrkevicz 1976
Permochonetes Afanasjeva 1977
Permophricodothyris Pavlova 1965
Permorthotetes Thomas 1958
Permospirifer Kulikov 1950
Permundaria Nakamura, Kato & Choi 1970
Perrarisinurostrum Sartenaer 1984
Perrierithyris Almeras 1971
Perryspirifer Jones & Boucot 1983
Peshiatrypa Xian 1978
Pesterevatrypa Rzonsnitzkaja 1975
Petalochonetes Afanasjeva 1988
Petalothyris Cooper 1983
Petasmaia Cooper & Grant 1969
Petasmaria Cooper & Dutro 1982
Petasmatherus Cooper & Grant 1969
Petria Mendes 1961
Petrocrania Raymond 1911
Petroria Wilson 1926
Petshorospirifer Fotieva 1985
Pexidella Bittner 1890
Phaceloorthis Percival 1991
Phaneropora Zezina 1981
Phapsirhynchia Pajaud 1976
Pharcidodiscus Roberts 1976
Pharetra Bolten 1798
Phaseolina Gaspard 1988
Phenacozugmayerella Hoover 1991
Philhedra Koken 1889
Philhedrella Kozlowski 1929
Philippotia Racheboeuf 1982
Phlogoiderhynchus Sartenaer 1970
Phoenicitoechia Havlicek 1960
Pholidostrophia Hall & Clarke 1892
Phragmophora Cooper 1955
Phragmorthis Cooper 1956
Phragmostrophia Harper, Johnson & Boucot 1967
Phragmothyris Cooper 1955
Phrenophoria Cooper & Grant 1969
Phricodothyris George 1932
Phyllolasma Liang 1990
Phyllonia Su 1980
Phymatothyris Cooper & Muir-Wood 1951
Physemella Godefroid 1974
Physetorhyncha Sartenaer & Rozman 1968
Physotreta Rowell 1966
Piarorhynchella Dagys 1974
Piarorhynchia Buckman 1918
Piarothyris Sahni 1925
Piatnitzkya Taboada 1993
Picnotreta Henderson & MacKinnon 1981
Pictothyris Thomson 1927
Pidiobolus Mergl 1995
Pilkena Richardson 1991
Piloricilla Carter 1987
Pinaxiothyris Dagys 1968
Pinegathyris Grunt 1980
Pinghuangella Jiang 1978
Pinguaella Boucot & Johnson 1979
Pinguispirifer Havlicek 1957
Pionodema Foerste 1912
Pionomena Cooper 1956
Pionopleurum Cooper 1989
Pionorthis Schuchert & Cooper 1931
Pionothyris Cooper 1983
Pirgulia Cooper & Muir-Wood 1951

Pirgumena Roomusoks 1993
Pirithyris Sun & Ye 1982
Pirotella Sucic-Protic 1985
Pirothyris Thomson 1927
Pirotothyris Sucic-Protic 1971
Pisirhynchia Buckman 1918
Pizarroa Hoek 1912
Placocliftonia Havlicek 1990
Placothyris Westphal 1970
Placotriplesia Amsden 1968
Plaesiomys Hall & Clarke 1892
Plagiorhyncha McLearn 1918
Planalvus Carter 1971
Planatrypa Copper 1967
Planatrypa Struve 1966
Plancella Amsden 1985
Planicardinia Savage 1968
Planidorsa Schuchert & Cooper 1931
Planihaydenella Chang 1987
Planirhynchia Sucic-Protic 1969
Planispina Stehli 1954
Planodouvillina Harper & Boucot 1978
Planoharknessella Havlicek 1950
Planoproductus Stainbrook 1947
Planothyris Glushenko 1975
Planovatirostrum Sartenaer 1970
Platidia Costa 1852
Platycancrinella Waterhouse 1983
Platyglossariorhynchus Sartenaer 1970
Platymena Cooper 1956
Platymerella Foerste 1909
Platyorthis Schuchert & Cooper 1931
Platyrachella Fenton & Fenton 1924
Platyselma Gordon 1966
Platyspirifer Grabau 1931
Platystrophia King 1850
Platyterorhynchus Sartenaer 1970
Platytoechia Neuman 1964
Platytrochalos Jin 1989
Playfairia Reed 1917
Plebejochonetes Boucot & Harper 1968
Plectambonites Pander 1830
Plectatrypa Schuchert & Cooper 1930
Plectelasma Cooper & Grant 1969
Plectella Lamansky 1905
Plectocamara Cooper 1956
Plectoconcha Cooper 1942
Plectodonta Kozlowski 1929
Plectodontella Havlicek 1953
Plectoglossa Cooper 1956
Plectoidothyris Buckman 1918
Plectorhynchella Cooper & Muir-Wood 1951
Plectorthis Hall & Clarke 1892
Plectospira Cooper 1942
Plectospirifer Grabau 1931
Plectosyntrophia Fu 1982
Plectothyrella Temple 1965
Plectothyris Buckman 1918
Plectotreta Ulrich & Cooper 1936
Plectotrophia Ulrich & Cooper 1936
Pleiopleurina Schmidt 1964
Plekonella Campbell 1953
Plekonina Waterhouse 1986
Plesicarinatina Mizens 1977
Plesiothyris Douville 1879
Plethorhyncha Hall & Clarke 1893
Pleuraloma Cooper 1989
Pleurelasma Cooper & Grant 1976

Pleurochonetes Isaacson 1977
Pleurocornu Havlicek 1961
Pleurodium Wang 1955
Pleurohorridonia Dunbar 1955
Pleuropugnoides Ferguson 1966
Pleurorthis Cooper 1956
Pleurothyrella Boucot, Caster 1963
Plicaea Aisenverg 1992
Plicanoplia Boucot & Harper 1968
Plicanoplites Havlicek 1974
Plicarostrum Burri 1953
Plicathyris Khalfin 1946
Plicatifera Chao 1927
Plicatiferina Kalashnikov 1980
Plicatocyrtia Gauri 1965
Plicatoderbya Thomas 1937
Plicatolingula Liu 1979
Plicatoria Cooper 1983
Plicatospiriferella Waterhouse & Waddington 1982
Plicatosyrinx Minato 1952
Plicidium Rong & Yang 1981
Plicirhynchia Allan 1947
Plicochonetes Paeckelmann 1930
Plicocoelina Boucot & Johnson 1966
Plicocyrtia Boucot 1963
Plicocyrtina Havlicek 1956
Plicodevonaria Boucot & Harper 1968
Plicogypa Rzonsnitzkaja 1975
Plicoplasia Boucot 1959
Plicoproductus Ljaschenko 1969
Plicostricklandia Boucot & Ehlers 1963
Plicostropheodonta Sokolskaja 1960
Plicotorynifer Abramov & Solomina 1970
Plionoptycherhynchus Sartenaer 1979
Pliothyrina Van Roy 1980
Ploughsharella Liang 1982
Pocockia Lazarev 1976
Podolella Kozlowski 1929
Podtsheremia Kalashnikov 1966
Poikilosakos Watson 1917
Poloniproductus Biernat & Lazarev 1988
Polylasma Popov 1980
Polymorpharia Cooper & Grant 1975
Polystylus Klets 1993
Polytoechia Hall & Clarke 1892
Pomatospirella Bittner 1892
Pomatotrema Ulrich & Cooper 1932
Pomeraniotreta Bednarczyk 1986
Pomeromena Mitchell 1977
Pompeckium Havlicek 1970
Pondospirifer Waterhouse 1978
Pontielasma Waterhouse & Piyasin 1970
Pontisia Cooper & Grant 1969
Porambonites Pander 1830
Porambonitoides Xu 1978
Poramborthis Havlicek 1949
Porostictia Cooper 1955
Portneufia Hoover 1979
Portranella Wright 1964
Posicomta Grunt 1986
Postamartinia Wang & Yang 1993
Postepithyris Makridin 1960
Pradochonetes Pardo & Garcia-Alcalde 1984
Pradoia Comte 1938
Praeangustothyris Koczyrkevicz 1984
Praeargyrotheca Smirnova 1983
Praecubanothyris Dagys 1974
Praecyclothyris Makridin 1955

Praegibbithyris Sun 1987
Praegnantenia Havlicek 1961
Praegoniothyris Ovtsharenko 1983
Praehorridonia Ustritsky 1962
Praekirkidium Breivel & Breivel 1988
Praelacazella Smirnova 1969
Praelacunosella Wisniewska-Zelichowska 1978
Praelongithyris Middlemiss 1959
Praemagadina MacKinnon 1993
Praemonticlarella Garcia Joral 1993
Praeneothyris Katz 1962
Praerhaetina Radulovic & Urosevic 1988
Praeudesia Hegab & Tchorszhewsky 1991
Praewaagenoconcha Sokolskaja 1948
Prantlina Havlicek 1949
Prelissorhynchia Xu And Grant 1994
Primipilaria Struve 1992
Primorewia Licharew & Kotljar 1978
Prionorhynchia Buckman 1918
Prionothyris Cloud 1942
Proanadyrella Xu & Liu 1983
Proatribonium Gratsianova 1967
Probolarina Cooper 1959
Probolionia Cooper 1957
Proboscidella Oehlert 1887
Proboscidina Isaacson 1977
Proboscisambon Havlicek & Mergl 1982
Procarinatina Mizens & Sapelnikov 1982
Procerulina Andronov 1961
Prochlidonophora Thomson & Owen 1979
Prochoristitella Legrand-Blain 1969
Proconchidium Sapelnikov 1969
Prodavidsonia Havlicek 1956
Productella Hall 1867
Productellana Stainbrook 1950
Productellina Reed 1943
Productelloides Kotlyar 1985
Productina Sutton 1938
Productorthis Kozlowski 1927
Productus Sowerby 1814
Progonambonites Opik 1934
Prokopia Havlicek 1953
Prolazutkinia Hou & Xian 1983
Promarginifera Shiells 1966
Pronalivkinia Rukavischnikova 1977
Propatella Grubbs 1939
Properotundirostrum Sartenaer 1986
Propriopugnus Brunton 1984
Propygope Bittner 1890
Prorensselaeria Raymond 1923
Proreticularia Havlicek 1957
Prorichthofenia King 1931
Prorugaria Waterhouse 1982
Proschizophoria Maillieux 1912
Prosoponella Li 1984
Prospira Maxwell 1954
Prosserella Grabau 1910
Prostricklandia Rukavischnikova & Al 1973
Prosyringothyris Fredericks 1916
Protambonites Havlicek 1972
Protanidanthus Liao 1979
Protathyris Kozlowski 1929
Protathyrisina Chu 1974
Protatrypa Boucot, Johnson & Staton 1964
Proteguliferina Licharew 1960
Protegulorhynchia Owen 1980
Proteorhynchia Owen 1981
Proteorthis Havlicek 1974

Protoanidanthus Waterhouse 1986
Protobolus Liu 1979
Protochonetes Muir-Wood 1962
Protocortezorthis Johnson & Talent 1967
Protocymostrophia Harper & Boucot 1978
Protodouvillina Harper & Boucot 1978
Protogusarella Perry & Chatterton 1979
Protohesperonomia Williams & Curry 1985
Protoleptostrophia Caster 1939
Protomegastrophia Caster 1939
Protomendacella Havlicek 1970
Protoniella Bell 1929
Protophragmapora Alekseeva 1967
Protoreticularia Su 1980
Protorhyncha Hall & Clarke 1893
Protorthis Hall & Clarke 1892
Protoshaleria Harper & Boucot 1978
Protosiphon Matthew 1897
Protoskenidioides Williams 1974
Prototegulithyris Almeras, Elmi & Benshili 1988
Prototreta Bell 1938
Protozeuga Twenhofel 1913
Protozyga Hall & Clarke 1893
Psamathopalass Liu 1979
Psebajithyris Tchorszhewsky 1974
Pseudoanisopleurella Xu 1978
Pseudoantiquatonia Zhan & Wu 1982
Pseudoathyrisina Chen 1979
Pseudoatrypa Copper 1973
Pseudoaulacothyris Smirnova 1990
Pseudoavonia Wang 1983
Pseudobolus Cooper 1956
Pseudobornhardtina Yang 1977
Pseudocamarophoria Wedekind 1926
Pseudocamarotoechia Kulkov 1974
Pseudochonetes Su 1976
Pseudoconchidium Nikiforova & Sapelnikov 1971
Pseudocrania M'Coy 1851
Pseudocyrtina Dagys 1962
Pseudoderbyia Licharew 1934
Pseudodicellomus Bell 1962
Pseudodicoelosia Boucot & Amsden 1958
Pseudodielasma Brill 1940
Pseudodouvillina Stainbrook 1945
Pseudogibbirhynchia Ager 1962
Pseudoglossinotoechia Tcherkesova 1967
Pseudoglossothyris Buckman 1901
Pseudogruenewaldtia Rzonsnitzkaja 1960
Pseudohalorella Dagys 1965
Pseudoharttina Licharew 1934
Pseudohaydenella Liang 1990
Pseudohomeospira Nikiforova 1970
Pseudojisuina Liang 1990
Pseudokingena Bose & Schlosser 1900
Pseudokymatothyris Chen 1979
Pseudolabaia Ching & Ye 1979
Pseudolaballa Dagys 1974
Pseudoleiorhynchus Rozman 1962
Pseudolepismatina Ching & Sun 1976
Pseudoleptaena Miloradovich 1947
Pseudoleptellina Andreev 1992
Pseudoleptodus Stehli 1956
Pseudolingula Mickwitz 1909
Pseudomarginifera Stepanov 1934
Pseudomendacella Zhang 1989
Pseudomeristina Grunt 1991
Pseudometoptoma Huene 1899
Pseudomimella Xu & Liu 1984

Pseudomonticlarella Smirnova 1987
Pseudomonticulifera Zhao & Tan 1984
Pseudoorthotetes Sokolskaja 1963
Pseudoparazyga Johnson 1970
Pseudopentagonia Besnossova 1963
Pseudopholidops Bekker 1921
Pseudoporambonites Zeng 1977
Pseudoprotathyris Modzalevskaja 1979
Pseudopugnax Licharew 1956
Pseudopygoides Xu 1978
Pseudorhaetina Sandy 1994
Pseudorostranteris Glushenko 1975
Pseudorugitela Dagys 1959
Pseudosieberella Godefroid 1972
Pseudosinotectirostrum Yudina 1991
Pseudospiriferina Yang & Xu 1966
Pseudospondylospira Hoover 1991
Pseudostrophalosia Clarke 1970
Pseudostrophochonetes Racheboeuf 1981
Pseudostrophomena Roomusoks 1963
Pseudosyringothyris Fredericks 1916
Pseudosyrinx Weller 1914
Pseudotubithyris Almeras 1971
Pseudouncinulus Rzonsnitzkaja 1968
Pseudoundispirifer Zhang 1987
Pseudowattonithyris Almeras 1971
Pseudowellerella Licharew 1956
Pseudoyunnanella Chen 1978
Psilocamara Cooper 1956
Psilocamerella Fu 1980
Psiloria Cooper 1976
Psilothyris Cooper 1955
Psioidea Hector 1879
Psioidiella Campbell 1968
Pteroplecta Waterhouse 1978
Pterospirifer Dunbar 1955
Pterostrophia Garratt 1985
Ptilorhynchia Crickmay 1933
Ptilotorhynchus Cooper & Grant 1976
Ptychoglyptus Willard 1928
Ptychomaletoechia Sartenaer 1961
Ptychopeltis Perner 1903
Ptychopleurella Schuchert & Cooper 1931
Ptyctorhynchia Buckman 1918
Ptyctothyris Buckman 1918
Ptygmactrum Cooper & Grant 1976
Puanatrypa Xian 1978
Puanospirifer Jiang 1978
Pugilis Sarytcheva 1949
Pugnacina Zuong & Rzonsnitzkaja 1968
Pugnaria Biernat & Racki 1986
Pugnax Hall & Clarke 1893
Pugnoides Weller 1910
Pulchratia Muir-Wood & Cooper 1960
Pulchrithyris Sahni 1925
Pulsia Ivanov 1925
Pumilus Atkins 1958
Punctatrypa Havlicek 1953
Punctocyrtella Plodowski 1968
Punctolira Ulrich & Cooper 1936
Punctopatella Grubbs 1939
Punctoproductus Liang 1990
Punctospirella Dagys 1974
Punctospirifer North 1920
Punctothyris Hyde 1953
Punctspinatrypa Rzonsnitzkaja 1975
Purdonella Reed 1944
Pusillagutta Misius 1986

Pustula Thomas 1914
Pustulatia Cooper 1956
Pustuloplica Waterhouse 1968
Pustulospiriferina Waterhouse 1983
Pycnobrochus Cooper 1983
Pycnoria Cooper 1989
Pygites Buckman 1906
Pygmaella Baranov 1977
Pygmochonetes Jing & Hu 1978
Pygope Link 1830
Pyraeneica Sucic-Protic 1971
Pyramidalia Nalivkin 1947
Pyramidathyris Hu 1983
Pyramina Ljaschenko 1969
Pyriusina Aksarina 1992
Qianjiangella Liang 1983
Qianomena Rong & Yang 1981
Qiansispirifer Yang 1977
Qilianoconcha Ching, Sun & Ye 1979
Qilianotryma Xu 1979
Qinghaispiriferina Sun & Ye 1982
Qinglongia Liao 1980
Qingthyris Xu & Liu 1983
Qingyenia Yang & Xu 1966
Qinlingia Rong, Zhang & Chen 1987
Qinlingotoechia Fu 1983
Qispiriferina Xu & Liu 1983
Quadratia Muir-Wood & Cooper 1960
Quadratirhynchia Buckman 1918
Quadrifarius Fuchs 1923
Quadrikentron Boucot & Gauri 1966
Quadrisonia Rowell & Henderson 1978
Quadrithyrina Havlicek 1987
Quadrithyris Havlicek 1957
Quadrochonetes Stehli 1954
Quangyuania Sheng 1975
Quasiavonia Brunton 1966
Quasidavidsonia Havlicek 1987
Quasimartinia Havlicek 1959
Quasistrophonella Harper & Boucot 1978
Quasithambonia Bednarczyk & Biernat 1978
Quebecia Walcott 1905
Quinquenella Waterhouse 1975
Quiringites Struve 1992
Quizhouspirifer Xian 1979
Quondongia Percival 1991
Rackirhynchia Havlicek 1990
Radiatrypa Copper 1978
Radimatrypa Havlicek 1990
Radiomena Havlicek 1962
Rafanoglossa Havlicek 1980
Rafinesquina Hall & Clarke 1892
Rahouiarhynchia Tchoumatchenco 1987
Railtonella Laurie 1991
Rakverina Roomusoks 1993
Ralfia Popov & Khazanovitch 1989
Ralia Lazarev 1987
Rallacosta Cooper & Grant 1976
Ramavectus Stehli 1954
Ramovsina Sremac 1986
Ranorthis Opik 1939
Raridium Sapelnikov & Mizens 1987
Rariella Zhang 1981
Rarithyris Tchorszhewsky 1989
Rastelligera Hector 1879
Ratburia Yanagida 1971
Raunites Opik 1939
Ravozetina Havlicek 1974

Rawdonia Peou 1979
Rebrovia Popov & Khazanovitch 1989
Rectambitus Xu And Grant 1994
Rectigypidula Zhang 1987
Rectirhynchia Buckman 1918
Rectithyris Sahni 1929
Rectotrophia Bates 1968
Redlichella Walcott 1908
Reedoconcha Kotljar 1964
Reeftonella Boucot 1959
Reeftonia Allan 1947
Reflexia Rotai 1931
Regelia Crickmay 1952
Reinversella Bates 1968
Remnevitoechia Gratsianova 1970
Renaudia Racheboeuf 1976
Rensselaeria Hall 1859
Rensselaerina Dunbar 1917
Rensselandia Hall 1867
Rensselandioidea Yang 1983
Replicoskenidioides Potter 1990
Resserella Bancroft 1928
Retaria Muir-Wood & Cooper 1960
Retichonetes Muir-Wood 1962
Reticularia M'Coy 1844
Reticulariina Fredericks 1916
Reticulariopsis Fredericks 1916
Reticulatia Muir-Wood & Cooper 1960
Reticulatochonetes Bublichenko 1956
Reticulatrypa Savage 1970
Retimarginifera Waterhouse 1970
Retroplexus Brunton & Mundy 1988
Retrorsirostra Schuchert & Cooper 1931
Retzia King 1850
Retziella Nikiforova 1937
Retzispirifer Kulkov 1960
Reuschella Bancroft 1928
Reveroides Sapelnikov 1976
Reversella Liang 1983
Rhabdostrophia Neuman 1989
Rhactomena Mitchell 1977
Rhactorhynchia Buckman 1918
Rhactorthis Williams 1963
Rhaetina Waagen 1882
Rhaetinopsis Yang & Xu 1966
Rhamnaria Muir-Wood & Cooper 1960
Rhapidothyris Tuluweit 1965
Rhenorensselaeria Kegel 1913
Rhenospirifer Mittmeyer 1972
Rhenostrophia Boucot 1960
Rhenothyris Struve 1970
Rhinotreta Holmer 1986
Rhipidium Schuchert & Cooper 1931
Rhipidomella Oehlert 1890
Rhipidomelloides Boucot & Amsden 1958
Rhipidomena Cooper 1956
Rhipidothyris Cooper & Williams 1935
Rhizothyris Thomson 1915
Rhombaria Cooper 1983
Rhomboidella Sucic-Protic 1985
Rhombospirifer Duan & Li 1985
Rhombothyris Middlemiss 1959
Rhondellina Rowell 1986
Rhynchatrypa Siehl 1962
Rhynchocamara Schuchert & Cooper 1931
Rhynchoferella Spriestersbach 1942
Rhynchonella Fischer 1809
Rhynchonellina Gemmellaro 1871

Rhynchonelloidea Buckman 1918
Rhynchonelloidella Muir-Wood 1936
Rhynchonellopsis Vincent 1893
Rhynchopora King 1865
Rhynchora Dalman 1828
Rhynchorina Oehlert 1887
Rhynchorthis Bates 1968
Rhynchospirifer Paulus 1957
Rhynchospirina Schuchert & Levene 1929
Rhynchotetra Weller 1910
Rhynchotrema Hall 1860
Rhynchotreta Hall 1879
Rhynchotretaoides Severgina 1967
Rhynchotretina Khalfin 1948
Rhynobolus Hall 1871
Rhynoleichus Abramov & Grigorjewa 1983
Rhyselasma Yadrenkina 1972
Rhysostrophia Ulrich & Cooper 1936
Rhysotreta Cooper 1956
Rhyssochonetes Johnson 1970
Rhytialosia Lazarev 1989
Rhytibulbus Li 1991
Rhytidorhachis Jin & Caldwell 1990
Rhytiophora Muir-Wood & Cooper 1960
Rhytirhynchia Cooper 1957
Rhytisia Cooper & Grant 1975
Rhytisoria Cooper 1983
Rhytistrophia Caster 1939
Richthofenia Kayser 1881
Rictia Gregorio 1930
Rigauxia Brice 1988
Rigbyella Stehli 1956
Rimirhynchia Buckman 1918
Rimirhynchopsis Dagys 1963
Riograndella Kobayashi 1937
Rionirhynchia Kamyschan & Kvakhadze 1980
Riorhynchia Jassjukevitch 1974
Rioultina Pajaud 1966
Ripidiorhynchus Sartenaer 1966
Robertorthis Havlicek 1977
Robinsonella Moisseiev 1936
Robustirhynchia Seifert 1963
Rochatorhynchia Katz 1962
Rocheithyris Almeras 1971
Rochtex Havlicek 1990
Roemerella Hall & Clarke 1890
Romingerina Hall & Clarke 1893
Rorespirifer Waterhouse & Piyasin 1970
Rosella Andreeva 1972
Rosobolus Havlicek 1982
Rossirhynchus Gaetani 1964
Rossithyris Owen 1980
Rostranteris Gemmellaro 1899
Rostricellula Ulrich & Cooper 1942
Rostrirhynchia Sucic-Protic 1969
Rostrospirifer Grabau 1931
Rotaia Rzonsnitzkaja 1959
Rotundostrophia Gratsianova 1960
Rouillieria Makridin 1960
Rowellella Wright 1963
Rowleyella Weller 1911
Rozmanaria Weyer 1972
Rudirhynchia Buckman 1918
Ruegenella Owen 1977
Rufispirifer Havlicek 1987
Rugaltarostrum Sartenaer 1961
Rugaria Cooper & Grant 1969
Rugatia Muir-Wood & Cooper 1960

Rugauris Muir-Wood & Cooper 1960
Rugia Steinich 1963
Rugicostella Muir-Wood & Cooper 1960
Rugitela Muir-Wood 1936
Rugithyris Buckman 1918
Rugivestis Muir-Wood & Cooper 1960
Rugoclostus Easton 1962
Rugoconcha Jing & Sun 1981
Rugodavidsonia Copper, 1995
Rugoleptaena Havlicek 1956
Rugolepyros Lenz 1989
Rugomena Roomusoks 1993
Rugosatrypa Rzonsnitzkaja 1975
Rugosochonetes Sokolskaja 1950
Rugosomarginifera Xu 1987
Rugosothyris Zhang 1987
Rugosowerbyella Mitchell 1977
Rugostrophia Neuman 1971
Rugulatia Sokolskaja 1952
Rurambonites Cocks & Rong 1989
Russiella Makridin 1964
Russirhynchia Buckman 1918
Rustella Walcott 1905
Ruthiphiala Carter 1988
Rutorhynchia Sun 1981
Rutrumella Harper 1981
Ryocarhynchus Sartenaer 1984
Rzonsnickiana Mamedov 1976
Saccogonum Havlicek 1971
Saccorhynchia Ching, Sun & Ye 1979
Sacothyris Ching, Sun & Ye 1979
Sacothyropsis Chen, Rao 1986
Saetosina Waterhouse 1986
Sagueresia Mohanti 1972
Saharonetes Havlicek 1984
Sajakella Nasikanova 1968
Sakawairhynchia Tokuyama 1957
Salacorthis Williams 1974
Salairella Severgina 1984
Salairina Alekseeva 1970
Salairotoechia Rzonsnitzkaja 1968
Salanygolina Ushatinskaia 1987
Salgirella Moisseiev 1936
Salonia Cooper & Whitcomb 1933
Salopia Williams 1955
Salopina Boucot 1960
Salopinella Yang & Rong 1982
Sampo Opik 1933
Sandia Sutherland & Harlow 1973
Sandrella Waterhouse 1986
Sanjuanella Benedetto & Herrera 1987
Sanjuanetes Racheboeuf & Herrera 1994
Sanjuania Amos 1958
Sanqiaothyris Yang & Xu 1966
Sanxiaella Rong & Chang 1981
Sardope Dieni, Middlemiss & Owen 1975
Sardorhynchia Ruggiero & Ungaro 1983
Sarganostega Cooper & Grant 1969
Sarytchevinella Waterhouse 1983
Sasyksoria Popov & Holmer 1994
Satpakella Koneva & Al 1990
Saucrobrochus Cooper 1983
Saucrorthis Xu, Rong & Liu 1974
Saughina Bancroft 1949
Saukrodictya Wright 1964
Savageina Boucot 1975
Scacchinella Gemmellaro 1891
Scalpellirhynchia Muir-Wood 1936

Scambocris Liu 1979
Scamnomena Bassett 1977
Scapharina Cooper & Grant 1975
Scaphelasma Cooper 1956
Scaphiocoelia Whitfield 1891
Scaphorthis Cooper 1956
Sceletonia Cooper & Grant 1974
Scenesia Cooper & Grant 1976
Schachriomonia Nikiforova 1978
Schalidomorthis Bassett 1981
Schedophyla Neuman 1971
Schellwienella Thomas 1910
Schistochonetes Roberts 1971
Schizambon Walcott 1884
Schizobolus Ulrich 1886
Schizocrania Hall & Whitfield 1875
Schizopholis Waagen 1885
Schizophorella Reed 1917
Schizophoria King 1850
Schizopleuronia Liao 1983
Schizoramma Foerste 1912
Schizoria Cooper 1989
Schizospirifer Grabau 1931
Schizostrophia Fu 1980
Schizotreta Kutorga 1848
Schizotretinia Havlicek 1994
Schizotretoides Termier & Monod 1978
Schmidtites Schuchert & Levene 1929
Schmidtomena Roomusoks 1989
Schnurella Schmidt 1964
Schrenkiella Barchatova 1973
Schuchertella Girty 1904
Schuchertellopsis Maillieux 1939
Schuchertina Walcott 1905
Schwagerispira Dagys 1972
Scissicosta Lazarev 1992
Scoloconcha Gordon 1966
Sculptospirifer Su 1980
Scumulus Steinich 1968
Scutepustula Sarytcheva 1963
Securina Voros 1983
Securithyris Voros 1983
Sedenticellula Cooper 1942
Sedjulina Ljaschenko 1985
Selenella Hall & Clarke 1893
Selennjachia Baranov 1982
Sellithyris Middlemiss 1959
Selloproductus Termier & Al 1974
Semenewia Paeckelmann 1930
Semibrachythyrina Yang & Ting 1962
Semicostella Muir-Wood & Cooper 1960
Semigublerina Liang 1990
Semileptagonia Karapetov 1971
Semilingula Popov 1990
Semilunataproductus Han 1987
Seminucella Carter 1987
Seminula M'Coy 1844
Semiotoechia Ljaschenko 1973
Semiplanella Sarytcheva & Legrand-Blain 1977
Semiplanus Sarytcheva 1952
Semiproductus Bublichenko 1956
Semitreta Biernat 1973
Sendaithyris Hatai 1940
Senokosica Sucic-Protic 1971
Sentolunia Havlicek 1967
Sentosia Muir-Wood & Cooper 1960
Sentosioides Lazarev 1992
Septacamarella Glushenko 1975

Septacamera Stepanov 1937
Septachonetes Chatterton 1973
Septalaria Leidhold 1928
Septalariopsis Chen 1978
Septaliphoria Leidhold 1921
Septaliphorioidea Yang & Xu 1966
Septamphiclina Jin & Fang 1977
Septaparmella Su 1976
Septarinia Muir-Wood & Cooper 1960
Septasteges Waterhouse & Piyasin 1970
Septathyris Boucot, Johnson & Staton 1964
Septatoechia Lobatscheva & Titova 1977
Septatrypa Kozlowski 1929
Septemirostellum Roberts 1971
Septicollarina Zezina 1981
Septiconcha Termier & Al 1974
Septirhynchia Muir-Wood 1935
Septocrurella Wisniewska 1932
Septocyclothyris Xu 1978
Septomena Roomusoks 1989
Septoproductus Frech 1911
Septorthis Hints 1973
Septospirifer Waterhouse 1971
Septospirigerella Grunt 1965
Septosyringothyris Vandercammen 1955
Septothyris Cooper & Williams 1935
Septulirhynchia Almeras 1966
Seratrypa Copper 1967
Serbarinia Morozov 1985
Serbiorhynchia Radulovic 1991
Serbiothyris Sucic-Protic 1971
Sergospirifer Ivanova 1952
Sergunkovia Nalivkin 1979
Sericoidea Lindstrom 1953
Serratocrista Brunton 1968
Serrulatrypa Havlicek 1977
Seseloidia Grant 1993
Sestropoma Cooper & Grant 1969
Setigerites Girty 1939
Settedabania Abramov 1970
Severella Sapelnikov 1963
Severginella Rozman 1981
Shagamella Boucot & Harper 1968
Shaleria Caster 1939
Shaleriella Harper & Boucot 1978
Shanxiproductus Duan & Lee 1985
Sharpirhynchia Shi & Grant 1993
Shiqianella Xian 1978
Shiragia Kobayashi 1935
Shishapangmaella Yang 1983
Shlyginia Nikitin & Popov 1983
Shoshonorthis Jaanusson & Bassett 1993
Shouxianella Liu 1983
Shovdolella Rozman 1992
Shrockia Boucot & Smith 1978
Shumardella Weller 1910
Siberiothyris Dagys 1968
Siberistrophia Astashkina 1970
Sibiratrypa Rzonsnitzkaja 1975
Sibiria Gorjansky 1977
Sibirirhynchia Rzonsnitzkaja 1968
Sibirispira Alekseeva 1968
Sibiritoechia Alekseeva 1966
Sicelia Gortani & Merla 1934
Sichuanothyris Shen, He & Zhu 1992
Sichuanrhynchus Tong 1978
Sicorhyncha Havlicek 1961
Sicularia Grant 1993

Sicyusella Liang 1990
Sieberella Oehlert 1887
Sieberelloides Rzonsnitzkaja 1975
Sigmelasma Potter 1990
Sigopallus Liu & Liu 1985
Sikasella Nalivkin 1979
Silesiathyris Brugge 1977
Similoleptaena Roomusoks 1989
Simplicarina Cooper & Grant 1975
Simplicithyris Zezina 1976
Sinochonetes Wang, Boucot & Rong 1981
Sinocyrtiopsis Gatinaud 1949
Sinoglossa Mei 1993
Sinoproductella Wang 1955
Sinopunctatrypa Wang, Copper & Rong 1983
Sinorhynchia Yang & Xu 1966
Sinorthis Wang 1955
Sinoshaleria Xian 1978
Sinospirifer Grabau 1931
Sinostrophia Hamada 1971
Sinotectirostrum Sartenaer 1961
Sinothyris Minato 1953
Sinotrimerella Li & Han 1980
Sinuatella Muir-Wood 1928
Sinucosta Dagys 1963
Sinucostella Xu & Liu 1983
Sinuplicorhynchia Dagys 1965
Sinusella Sucic-Protic 1985
Siphonobolus Havlicek 1982
Siphonosia Cooper & Grant 1975
Siphonotreta Verneuil 1845
Siphonotretella Popov & Holmer 1994
Sivorthis Jaanusson & Bassett 1993
Skelidorygma Carter 1974
Skenidioides Schuchert & Cooper 1931
Skenidium Hall 1860
Slavinithyris Tchorszhewsky 1986
Slovenirhynchia Siblik 1967
Smeathenella Bancroft 1928
Smirnovaena Nekvasilova 1985
Smirnovina Calzada 1985
Snezhnorhynchia Smirnova 1990
Soaresirhynchia Almeras 1994
Socraticum Gregorio 1930
Sogxianthyris Sun 1981
Sokolskya Aisenverg 1980
Solidipontirostrum Sartenaer 1970
Solitudinella Godefroid 1991
Somalirhynchia Weir 1925
Somalitela Muir-Wood 1935
Somalithyris Muir-Wood 1935
Sommeriella Archbold 1982
Sonculina Misius 1986
Songzichonetes Yang 1984
Soudleyella Bancroft 1945
Sowerbina Fredericks 1928
Sowerbyella Jones 1928
Sowerbyites Teichert 1937
Spanodonta Prendergast 1935
Spasskothyris Smirnova 1975
Sphaerathyris Baranov 1994
Sphaerirhynchia Cooper & Muir-Wood 1951
Sphaerobolus Matthew 1895
Sphaeroidothyris Buckman 1918
Sphenalosia Muir-Wood & Cooper 1960
Sphenarina Cooper 1959
Sphenophragmus Imbrie 1959
Sphenorhynchia Buckman 1918

Sphenorthis Grubbs 1939
Sphenospira Cooper 1954
Sphenosteges Muir-Wood & Cooper 1960
Sphenotreta Cooper 1956
Sphriganaria Cooper 1989
Spinarella Cooper & Grant 1975
Spinatrypa Stainbrook 1951
Spinatrypina Rzonsnitzkaja 1964
Spinauris Roberts 1971
Spinella Talent 1956
Spinifrons Stehli 1954
Spinilingula Cooper 1956
Spinocarinifera Roberts 1971
Spinochonetes Rong, Xu & Yang 1974
Spinocyrtia Fredericks 1916
Spinocyrtina Fredericks 1916
Spinolepismatina Dagys 1974
Spinolyttonia Sarytcheva 1964
Spinomarginifera Huang 1932
Spinomartinia Waterhouse 1968
Spinoplasia Boucot 1959
Spinorthis Wright 1964
Spinorugifera Roberts 1976
Spinospirifer Martynova 1961
Spinosteges Liang 1990
Spinostrophia Jiang 1978
Spinulicosta Nalivkin 1937
Spinuliplica Campbell 1961
Spinulothele Rowell 1977
Spinulothyris Antostschenko 1973
Spirelytha Fredericks 1924
Spirifer Sowerby 1818
Spiriferella Tschernyschew 1902
Spiriferellaoides Lee, Gu & Li 1980
Spiriferellina Fredericks 1924
Spiriferelloides Lee, Gu & Li 1985
Spiriferina Orbigny 1847
Spiriferinaella Fredericks 1926
Spiriferinoides Tokuyama 1957
Spirigerella Waagen 1883
Spirigerellina Dagys 1974
Spirigerina Orbigny 1847
Spirinella Johnston 1941
Spirisosium Gregorio 1930
Spiropunctifera Ivanova 1971
Spitispirifer Waterhouse & Gupta 1986
Spitzbergenia Kotljar 1977
Spondyglossella Havlicek 1980
Spondylobolus M'Coy 1851
Spondylopyxis Johnson, Boucot & Murphy 1976
Spondylospira Cooper 1942
Spondylospiriferina Dagys 1972
Spondylostrophia Kulkov 1967
Spondylothyris Su 1980
Spondylotreta Cooper 1956
Spuriosa Cooper & Grant 1975
Spurispirifer Havlicek 1971
Spyridiophora Cooper & Stehli 1955
Squamaria Muir-Wood & Cooper 1960
Squamathyris Modzalevskaja 1981
Squamatina Havlicek & Racheboeuf 1979
Squamilingulella Xu 1978
Squamiplana Sucic-Protic 1971
Squamirhynchia Buckman 1918
Squamularia Gemmellaro 1899
Squamulariina Fredericks 1916
Stainbrookia Cooper & Dutro 1982
Stauromata Hoover 1981

Stegacanthia Muir-Wood & Cooper 1960
Stegerhynchops Amsden 1978
Stegerhynchus Foerste 1909
Stegocornu Durkoop 1970
Stegorhynchella Rzonsnitzkaja 1959
Stegospira Fu 1982
Steinhagella Goldring 1957
Stelckia Crickmay 1963
Stenaulacorhynchus Sartenaer 1968
Stenobrochus Cooper 1983
Stenocamara Cooper 1956
Stenoglossariorhynchus Sartenaer 1970
Stenogmus Cooper 1983
Stenometoporhynchus Sartenaer 1987
Stenopentamerus Boucot & Johnson 1979
Stenorhynchia Brice 1981
Stenorhynchites Havlicek 1990
Stenorina Cooper 1989
Stenosarina Cooper 1977
Stenoscisma Conrad 1839
Stepanoconchus Lazarev 1985
Stepanoviella Zavodowsky 1960
Stepanoviina Zavodowsky 1968
Stereochia Grant 1976
Stethothyris Thomson 1918
Stichotrophia Cooper 1948
Stictozoster Grant 1976
Stilpnotreta Henderson & MacKinnon 1981
Stiphrothyris Buckman 1918
Stipulina Muir-Wood & Cooper 1960
Stita Gregorio 1930
Stolmorhynchia Buckman 1918
Stolzenburgiella Bittner 1903
Straelenia Maillieux 1935
Streptaria Cooper 1959
Streptis Davidson 1881
Streptopomum Havlicek 1967
Streptorhynchus King 1850
Striarina Cooper 1973
Striatifera Chao 1927
Striatochonetes Mikrjukov 1968
Striatoproductella Krylova 1962
Striatoproductus Nalivkin 1947
Striatopugnax Chen 1978
Striatorhynchus Pushkin 1986
Striatospica Waterhouse 1975
Stricklandia Billings 1859
Stricklandiella Sapelnikov & Al 1973
Stricklandistrophia Sapelnikov & Al 1975
Strigirhynchia Cooper & Grant 1969
Strigospina Liao 1979
Striirhynchia Buckman 1918
Striirichthofenia Lu 1982
Striispirifer Cooper & Muir-Wood 1951
Striithyris Muir-Wood 1935
Stringocephalus Defrance 1825
Stringodiscus Struve 1982
Stringomimus Struve 1965
Striochonetes Waterhouse & Piyasin 1970
Strixella Boucot & Amsden 1958
Strongylobrochus Cooper 1983
Strongyloria Cooper 1989
Strophalosia King 1844
Strophalosiella Licharew 1935
Strophalosiina Licharew 1935
Strophochonetes Muir-Wood 1962
Strophodonta Hall 1850
Strophomena Rafinesque 1824

Strophonella Hall 1879
Strophonellites Havlicek 1967
Strophonelloides Caster 1939
Strophopleura Stainbrook 1947
Strophoprion Twenhofel 1914
Strophoproductus Nalivkin 1937
Strophorichthofenia Termier & Al 1974
Stroudithyris Buckman 1918
Struveina Boucot 1975
Struvethyris Radulovic 1991
Stuartella Belanski 1928
Sturtella Savage 1971
Styxorthis Mergl 1991
Subansiria Sahni & Srivastava 1956
Subcuspidella Mittmeyer 1965
Subglobosochonetes Afanasjeva 1976
Sublepida Mizens & Sapelnikov 1982
Subquadriangulispirifer Sartenaer 1982
Subrensselandia Cloud 1942
Subriana Sapelnikov 1960
Subsinucephalus Struve 1992
Subspirifer Shan & Zhao 1981
Substriatifera Kotljar 1964
Subtaeniothaerus Solomina 1988
Suessia Deslongchamps 1855
Sufetirhyncha Havlicek 1982
Suiaella Moisseiev 1956
Sulcataria Cooper & Grant 1969
Sulcathyris Durkoop 1970
Sulcatina Schmidt 1964
Sulcatinella Dagys 1974
Sulcatorthis Zeng 1987
Sulcatospira Xu 1979
Sulcatospirifer Maxwell 1954
Sulcatostrophia Caster 1939
Sulcatothyris Dagys 1974
Sulcicosta Waterhouse 1983
Sulcicostula Baranov 1989
Sulciphoria Smirnova 1992
Sulciplica Waterhouse 1968
Sulciplicatatrypa Zhang 1983
Sulcirhynchia Burri 1953
Sulcirostra Cooper & Muir-Wood 1951
Sulcirugaria Waterhouse 1983
Sulcispiriferina Waterhouse & Gupta 1981
Sulcorhynchia Dagys 1974
Sulcupentamerus Zeng 1987
Sulevorthis Jaanusson & Bassett 1993
Sunacosarina Liang 1990
Supertrilobus Boucot & Johnson 1979
Surugathyris Yabe & Hatai 1934
Svalbardia Afanasjeva 1977
Svalbardoproductus Ustritsky 1962
Svaljavithyris Tchorszhewsky 1974
Svetlania Baranov 1982
Svobodaina Havlicek 1950
Swaicoelia Hamada 1968
Swantonia Walcott 1905
Symmatrypa Mizens & Sapelnikov 1975
Symphythyris Smirnova 1966
Synatrypa Copper 1966
Syndielasma Cooper 1956
Syntomaria Cooper 1982
Syntrilasma Meek & Worthen 1865
Syntrophia Hall & Clarke 1893
Syntrophina Ulrich 1928
Syntrophinella Ulrich & Cooper 1934
Syntrophioides Schuchert & Cooper 1931

Syntrophodonta Struve 1982
Syntrophopsis Ulrich & Cooper 1936
Sypharatrypa Copper 1982
Syringopleura Schuchert 1910
Syringospira Kindle 1909
Syringothyris Winchell 1863
Systenothyris Cooper 1983
Tabarhynchus Baranov 1989
Tabellina Waterhouse 1986
Tacinia Glushenko 1975
Tadschikia Nikiforova 1937
Taemostrophia Chatterton 1973
Taeniothaerus Whitehouse 1928
Taffia Ulrich 1926
Tafilaltia Havlicek 1970
Taimyrella Ustritsky 1963
Taimyropsis Ustritsky 1963
Taimyrothyris Dagys 1968
Taimyrrhynx Havlicek 1983
Tainotoechia Tcherkesova 1967
Tainuirhynchia MacFarlan 1992
Talasoproductus Litvinovich & Vorontsova 1983
Talentella Johnson 1990
Taleoleptaena Havlicek 1967
Tallinnites Roomusoks 1993
Talovia Severgina 1975
Talovkorhynchia Smirnova 1994
Tamarella Owen 1965
Tanakura Hatai 1936
Tanerhynchia Allan 1947
Tanggularella Shi 1990
Tangshanella Chao 1929
Tangxiangia Xu 1977
Tannuspirifer Ivanova 1960
Tanyoscapha Cooper 1983
Tanyothyris Cooper 1989
Taoqupospira Fu 1982
Tapajotia Dresser 1954
Taphrodonta Cooper 1956
Taphrorthis Cooper 1956
Taphrosestria Cooper & Grant 1975
Tarandrospirifer Simakov 1970
Tarfaya Havlicek 1971
Tarutiglossa Havlicek 1984
Tashanomena Zhan & Rong 1994
Tasmanella Laurie 1991
Tasmanorthis Laurie 1991
Tastaria Havlicek 1965
Tatjanaspirifer Tcherkesova 1991
Tatjania Baranov 1982
Tauromenia Seguenza 1885
Taurothyris Kyansep 1961
Tautosia Cooper & Grant 1969
Tazzarinia Havlicek 1971
Tchadania Kulkov 1985
Tcharella Andreeva 1987
Tchegemithyris Tchorszhewsky 1972
Tcherskidium Nikolayev & Sapelnikov 1969
Tecnocyrtina Johnson & Norris 1972
Tectarea Licharew 1928
Tectatrypa Mizens 1973
Tegulella Holmer 1989
Teguliferina Schuchert & Levene 1929
Tegulithyris Buckman 1918
Tegulocrea Carter 1992
Tegulorhynchia Chapman & Crespin 1923
Teichertina Veevers 1959
Teichostrophia Harper, Johnson & Boucot 1967

Telaeoshaleria Williams 1950
Teleoproductus Lee 1986
Telothyris Almeras & Moulan 1982
Templeella Rozman & Rong 1993
Tenellodermis Havlicek 1971
Teneobolus Mergl 1995
Tenerella Liang 1990
Tenisia Martynova 1970
Tenticospirifer Tien 1938
Tenuiatrypa Rzonsnitzkaja 1975
Tenuichonetes Jing & Hu 1978
Tenuicostella Mittmeyer & Geib 1967
Tenuisinurostrum Sartenaer 1967
Teratelasma Cooper 1956
Teratelasmella Laurie 1991
Terazkia Havlicek 1990
Terebratalia Beecher 1893
Terebrataliopsis Smirnova 1962
Terebratella Orbigny 1847
Terebratula Muller 1776
Terebratulina Orbigny 1847
Terebratuloidea Waagen 1883
Terebratulopsis Gregorio 1930
Terebrirostra Orbigny 1847
Terrakea Booker 1930
Tesuquea Sutherland & Harlow 1973
Tethorotes Manankov 1979
Tethyrete Havlicek 1994
Tethyrhynchia Logan 1994
Tethyspira Siblik 1991
Tetjuchithyris Smirnova 1986
Tetracamera Weller 1910
Tetractinella Bittner 1890
Tetragonetes Cooper & Grant 1975
Tetralobula Ulrich & Cooper 1936
Tetraodontella Jaanusson 1962
Tetraphalerella Wang 1949
Tetrarhynchia Buckman 1918
Tetratomia Schmidt 1941
Texarina Cooper & Grant 1970
Texathyris Carter 1972
Thadiqithyris Nazer 1987
Thaerodonta Wang 1949
Thamnosia Cooper & Grant 1969
Thaumatosia Cooper 1973
Thaumatrophia Wang 1955
Thebesia Amsden 1974
Thecidea Defrance 1822
Thecidella Munier-Chalmas 1887
Thecidellella Hayasaka 1938
Thecidellina Thomson 1915
Thecidiopsis Munier-Chalmas 1887
Thecocyrtella Bittner 1892
Thecocyrtelloidea Yang & Xu 1966
Thecospira Zugmayer 1880
Thecospirella Bittner 1900
Thecospiropsis Dagys 1974
Thedusia Cooper & Grant 1976
Theodossia Nalivkin 1925
Thiemella Williams 1908
Thliborhynchia Lenz 1967
Thomasaria Stainbrook 1945
Thomasella Fredericks 1928
Thuleproductus Sarytcheva & Waterhouse 1972
Thurmannella Leidhold 1921
Thyratryaria Xu & Liu 1983
Thysanobolus Havlicek 1982
Thysanotos Mickwitz 1896

Tianzhushanella Liu 1979
Tiaretithyris Tchoumatchenco 1986
Tibetatrypa Copper & Hou 1986
Tibetospirifer Liu & Wang 1990
Tibetothyris Ching & Sun 1976
Tichirhynchus Baranov 1989
Tichosina Cooper 1977
Tilasia Holmer 1991
Timalina Batrukova 1969
Timaniella Barchatova 1968
Timanospirifer Ljaschenko 1973
Timorhynchia Ager 1968
Timorina Stehli 1961
Tingella Grabau 1931
Tiocyrspis Sartenaer 1994
Tipispirifer Grant 1976
Tiramnia Grunt 1977
Tisimania Hatai 1938
Tissintia Havlicek 1970
Titanambonites Cooper 1956
Titanaria Muir-Wood & Cooper 1960
Titanomena Bergstrom 1968
Titanothyris Ching & Hu 1982
Tityrophoria Waterhouse 1971
Tivertonia Archbold 1983
Tobejalotreta Koneva & Al 1990
Togaella Severgina 1960
Togatrypa Havlicek 1987
Tolmatchoffia Fredericks 1933
Tomasina Hall & Clarke 1892
Tomestenoporhynchus Sartenaer 1993
Tomilia Sarytcheva 1963
Tomiopsis Benediktova 1956
Tomiproductus Sarytcheva 1963
Tonasirhynchia Lobatscheva & Smirnova 1994
Tongluella Liang 1990
Tongzithyris Ching, Liao & Fang 1974
Tonsella Amsden 1988
Toquimaella Johnson 1967
Toquirnia Ulrich & Cooper 1936
Tornquistia Paeckelmann 1930
Torosospirifer Gourvennec 1989
Torquirhynchia Childs 1969
Torynechus Cooper & Grant 1962
Torynelasma Cooper 1956
Torynifer Hall & Clarke 1893
Toryniferella Weyer 1967
Tosuhuthyris Sun & Ye 1982
Totia Rzonsnitzkaja & Mizens 1977
Tourmakeadia Williams & Curry 1985
Toxonelasma Cooper 1989
Toxorthis Temple 1968
Transennatia Waterhouse 1975
Transversaria Waterhouse & Gupta 1983
Trasgu Martinez 1979
Trautscholdia Ustritsky 1967
Treioria Neuman & Bates 1978
Trematis Sharpe 1848
Trematobolus Matthew 1893
Trematorthis Ulrich & Cooper 1938
Trematosia Cooper 1976
Trematospira Hall 1859
Treptotreta Henderson & MacKinnon 1981
Tretorhynchia Brunton 1971
Triadispira Dagys 1961
Triadithyris Dagys 1963
Triangope Dieni & Middlemiss 1981
Triasorhynchia Xu & Liu 1983

Triathyris Comte 1938
Trichochonetes Roberts 1976
Trichorhynchia Buckman 1918
Trichothyris Buckman 1918
Tricoria Cooper & Grant 1976
Tridensilis Su 1976
Trifidarcula Elliott 1959
Trifidorostellum Sartenaer 1961
Trigeria Bayle 1878
Trigonatrypa Havlicek 1990
Trigonellina Buckman 1907
Trigonirhynchella Dagys 1963
Trigonirhynchia Cooper 1942
Trigonirhynchioides Fu 1982
Trigonithyris Muir-Wood 1935
Trigonoglossa Dunbar & Condra 1932
Trigonosemus Koenig 1825
Trigonospirifer Wang, Rong & Chen 1987
Trigonotreta Koenig 1825
Trigonotrophia Andreeva 1972
Trigrammaria Wilson 1945
Trilobatoechia Xian 1990
Trilobostrophia Schischkina 1983
Trimerella Billings 1862
Trimurellina Mitchell 1977
Triplecella Wilson 1932
Triplesia Hall 1859
Triseptata Bondarev 1965
Triseptothyris Xu 1978
Tritoechia Ulrich & Cooper 1936
Trochalocyrtina Wright 1975
Trochifera Ching & Ye 1979
Trondorthis Neuman 1974
Tropeothyris Smirnova 1972
Trophisina Cooper & Grant 1976
Trophidelasma Cooper & Grant 1969
Tropidoglossa Rowell 1966
Tropidoleptus Hall 1857
Tropidothyris Cooper 1956
Tropiorhynchia Buckman 1918
Trotlandella Neuman 1974
Trucizetina Havlicek 1974
Truncalosia Imbrie 1959
Truncatenia Liao 1980
Tsaganella Oleneva 1993
Tschatkalia Nikiforova 1964
Tschernovia Kalashnikov 1993
Tschernyschewia Stoyanow 1910
Tshemsarythyris Ovtsharenko 1983
Tubaria Muir-Wood & Cooper 1960
Tubegatanella Prosorovskaja 1968
Tuberculatella Waterhouse 1982
Tuberculatospira Xian 1988
Tuberella Lee 1986
Tubersulculus Waterhouse 1971
Tubithyris Buckman 1918
Tubulostrophia Havlicek 1967
Tudiaophomena Xian 1978
Tufoleptina Havlicek 1961
Tulathyris Grunt 1976
Tulcumbella Campbell 1963
Tulipina Smirnova 1962
Tuloja Severgina 1983
Tulungospirifer Ching & Sun 1976
Tumarinia Solomina & Grigorjewa 1973
Tunarites Cooper & Muir-Wood 1951
Tunethyris Calzada 1994
Tungussotoechia Lopushinskaja 1976

Tunisiglossa Massa, Havlicek & Bonnefous 1977
Tuotalania Hu 1983
Turarella Andreeva 1987
Turgenostrophia Alekseeva 1983
Turkmenithyris Prosorovskaja 1962
Turriculum Gregorio 1930
Tuvaechonetes Kulkov 1984
Tuvaella Tschernyschew 1937
Tuvaerhynchus Kulkov 1985
Tuvaestrophia Kulkov 1985
Tuvinia Andreeva 1982
Twenhofelia Boucot & Smith 1978
Tyersella Philip 1962
Tylambonites Percival 1991
Tyloplecta Muir-Wood & Cooper 1960
Tylospiriferina Xu 1978
Tylothyris North 1920
Tyrganiella Kulkov 1970
Tyronella Mitchell 1977
Tyrothyris Opik 1953
Tyryrhynchus Baranov 1988
Tythothyris Zezina 1979
Uchtella Ljaschenko 1973
Uchtospirifer Ljaschenko 1957
Uexothyris Struve 1992
Ufonicoelia Havlicek 1992
Ujandinella Baranov 1977
Ujukella Andreev 1993
Ujukites Andreeva 1985
Ukoa Opik 1932
Ulbospirifer Gretchischnikova 1965
Uldziathyris Grunt 1977
Umboanctus Waterhouse 1971
Uncina Kaplun 1991
Uncinella Waagen 1883
Uncinulus Bayle 1878
Uncinunellina Grabau 1932
Uncisteges Jing & Hu 1978
Uncites Defrance 1825
Uncitispira Fu 1982
Undaria Muir-Wood & Cooper 1960
Undatrypa Copper 1978
Undellaria Cooper & Grant 1975
Undiferina Cooper 1956
Undispirifer Havlicek 1957
Undispiriferoides Xian 1978
Undithyrella Havlicek 1971
Undulella Cooper & Grant 1969
Undulorhyncha Amsden 1988
Ungula Pander 1830
Uniplicatorhynchia Sun & Ye 1982
Unispirifer Campbell 1957
Uralella Makridin 1960
Uralina Schuchert & Levene 1929
Uraloconchus Lazarev 1990
Uraloproductus Ustritsky 1971
Uralorhynchia Dagys 1968
Uralospira Mizens 1977
Uralospirifer Havlicek 1959
Uralotoechia Sapelnikov 1963
Urbanirhynchia Katz 1974
Urbimena Havlicek 1976
Urella Rzonsnitzkaja 1960
Urushtenia Licharew 1935
Urushtenoidea Jing & Hu 1978
Ushkolia Martynova & Sverbilova 1969
Usinia Aksarina 1992
Ussunia Nikitin & Popov 1984

Ussuricamara Koczyrkevicz 1969
Ussurirhynchia Koczyrkevicz 1976
Vaculina Koneva 1992
Vadimia Boucot & Rong 1994
Vadum Strusz 1982
Vaga Sapelnikov & Al 1973
Vagranella Sapelnikov 1960
Vagrania Alekseeva 1959
Valcourea Raymond 1911
Valdaria Bassett & Cocks 1974
Valdiviathyris Helmcke 1940
Vallomyonia Johnson 1966
Vandalotreta Mergl 1988
Vandercammenina Boucot 1975
Vandergrachtella Crickmay 1953
Vandobiella Pojariskaja 1966
Vaniella Kvakhadze 1974
Variatrypa Copper 1966
Vassilkovia Popov & Khazanovitch 1989
Vectella Owen 1965
Vediproductus Sarytcheva 1965
Veeversalosia Lazarev 1989
Veghirhynchia Dagys 1974
Vehnia Neuman 1989
Veliseptum Popov 1976
Vellamo Opik 1930
Velostrophia Havlicek 1965
Venezuelia Weisbord 1926
Verchojania Abramov 1970
Verkhotomia Sokolskaja 1963
Vermiculothecidea Elliott 1953
Verneuilia Hall & Clarke 1893
Vex Hoover 1979
Viallithyris Voros 1978
Viarhynchia Calzada 1974
Victorithyris Allan 1940
Viligella Dagys 1965
Viligothyris Dagys 1968
Villicundella Levy & Nullo 1974
Vincalaria Sartenaer 1989
Vincentirhynchia MacFarlan 1992
Vincia Gregorio 1930
Virbium Gregorio 1930
Virgiana Twenhofel 1914
Virgianella Nikiforova & Sapelnikov 1971
Virginiata Amsden 1968
Virgoria Neuman 1977
Viruella Roomusoks 1959
Visbyella Walmsley & Al 1968
Vitiliproductus Ching & Liao 1974
Vitimetula Hoover 1991
Vjalovithyris Tchorszhewsky 1989
Vladimirella Popov 1994
Vladimirirhynchus Baranov 1982
Vltavothyris Havlicek 1956
Voiseyella Roberts 1964
Volborthia Moller 1870
Volgathyris Smirnova 1987
Volgospirifer Schevtchenko 1970
Volirhynchia Dagys 1974
Voskopitoechia Havlicek 1992
Vosmiverstum Breivel & Breivel 1970
Waagenites Paeckelmann 1930
Waagenoconcha Chao 1927
Waconella Owen 1970
Wadiglossa Havlicek 1984
Waikatorhynchia MacFarlan 1992
Waiparia Thomson 1920

Wairakiella Waterhouse 1967
Wairakirhynchia MacFarlan 1992
Waisiuthyrina Beets 1943
Walcottina Cobbold 1921
Walkerithyris Cox & Middlemiss 1978
Waltonia Davidson 1850
Wangyuia Zhang 1989
Wardakia Termier & Al 1974
Warrenella Crickmay 1953
Warrenellina Brice 1982
Warsawia Carter 1974
Waterhouseiella Archbold 1983
Wattonithyris Muir-Wood 1936
Wattsella Bancroft 1928
Webbyspira Percival 1991
Weberithyris Smirnova 1969
Weibeia Fu 1980
Weiningia Ching & Liao 1974
Weizhouella Chen 1983
Weldonithyris Muir-Wood 1952
Wellerella Dunbar & Condra 1932
Wellerellina Shen, He & Zhu 1992
Wenxianirhynchus Zhang 1983
Werneckeella Lenz 1971
Werriea Campbell 1957
Westbroekina Savage & Baxter 1995
Westonia Walcott 1901
Westonisca Havlicek 1982
Westralicrania Cockbain 1967
Whidbornella Reed 1943
Whitfieldella Hall & Clarke 1893
Whitfieldia Davidson 1882
Whitspakia Stehli 1964
Whittardia Williams 1974
Wilberrya Yancey 1978
Wilsoniella Khalfin 1939
Wimanella Walcott 1908
Wimanoconcha Waterhouse 1983
Winterfeldia Spriestersbach 1942
Wiradjuriella Percival 1991
Wittenburgella Dagys 1959
Worobievella Dagys 1959
Wulongella Zhu 1985
Wulungguia Zhang 1981
Wutubulakia Hou & Zhang 1983
Wyatkina Fredericks 1931
Wyella Khodalevich 1939
Wyndhamia Booker 1930
Wynnia Walcott 1912
Wysogorskiella Hints 1975
Xana Garcia-Alcalde 1972
Xenambonites Cooper 1956
Xenelasma Ulrich & Cooper 1936
Xenelasmella Rozman 1964
Xenelasmopsis Rozman 1968
Xeniopugnax Havlicek 1982
Xenizostrophia Su 1976
Xenobrochus Cooper 1981
Xenocryptonella Zhang 1983
Xenomartinia Havlicek 1953
Xenorina Cooper 1989
Xenorthis Ulrich & Cooper 1936
Xenosaria Cooper & Grant 1976
Xenospirifer Hou & Xian 1975
Xenosteges Muir-Wood & Cooper 1960
Xenostrophia Wang 1974
Xenothyris Ching, Rong & Sun 1976
Xerospirifer Havlicek 1978

Xerxespirifer Cocks 1979
Xestosia Cooper & Grant 1975
Xestosina Cooper 1983
Xestotrema Cooper & Grant 1969
Xiangzhounia Ni & Yang 1977
Xiaobangdaia Wang 1985
Xinanorthis Xu, Rong & Liu 1974
Xinanospirifer Rong, Xu & Yang 1974
Xinjiangiproductus Yao & Fu 1987
Xinjiangochonetes Xu 1991
Xinjiangospirifer Hou & Zhang 1983
Xinjiangthyris Sun 1984
Xinshaoella Zhao 1977
Xinshaoproductus Tan 1986
Xizangostrophia Rong 1976
Xizispirifer Liang 1990
Xysila Copper 1995
Xystostrophia Havlicek 1965
Yabeithyris Hatai 1948
Yagonia Roberts 1976
Yakovlevia Fredericks 1925
Yakutijaella Baranov 1977
Yalongia Xu & Liu 1983
Yanbianella Tong 1978
Yanetechia Baranov 1980
Yangkongia Xu & Liu 1983
Yangtzeella Kolarova 1925
Yanguania Yang 1977
Yanishewskiella Licharew 1957
Yanospira Dagys 1977
Yaonoiella Waterhouse 1983
Yarirhynchia Jing & Sun 1981
Yatsengina Semichatova 1936
Yekerpene Struve 1992
Yeosinella Reed 1933
Yeothyris Struve 1992
Ygera Havlicek 1961
Ygerodiscus Havlicek 1967
Yichangorthis Zeng 1987
Yidunella Ching, Sun & Ye 1979
Yidurella Zeng 1987
Yingtangella Bai & Ying 1978
Yingwuspirifer Rong, Xu & Yang 1974
Yochelsonia Stehli 1961
Yongjia Liu 1977
Yorkia Walcott 1897
Ypsilorhynchus Sartenaer 1970
Yrctospirifer Gatinaud 1949
Yuanbaella Fu 1982
Yuezhuella Ching & Ye 1979
Yukiangides Havlicek 1992
Yulongella Sun 1981
Yunnanella Grabau 1923
Yunnanoleptaena Jahnke & Shi 1989
Yunshannella Lee & Gu 1982
Zaissania Sokolskaja 1968
Zanclorhyncha Xu 1979
Zdimir Barrande 1881
Zdimirella Tcherkesova 1973
Zeilleria Bayle 1878
Zeillerina Kyansep 1959
Zejszneria Siemiradzki 1922
Zellania Moore 1855
Zeravshania Menakova 1983
Zeravshanotoechia Rzonsnitzkaja 1977
Zeugopleura Carter 1988
Zeuschneria Smirnova 1975
Zhanatella Koneva 1986

Zhejiangella Liang 1982
Zhejiangellina Liang 1990
Zhejiangoproductus Liang 1990
Zhejiangorthis Liang 1983
Zhejiangospirifer Liang 1982
Zhenania Ding 1983
Zhexichonetes Liang 1982
Zhidothyris Ching, Sun & Ye 1979
Zhinania Liang 1990
Zhonghuacoelia Chen 1978
Zhongliangshania Shen, He & Zhu 1992
Zhongpingia Yang 1983
Zhuaconcha Liang 1990
Zia Sutherland & Harlow 1973
Ziganella Nalivkin 1960
Zilimia Nalivkin 1947
Zittelina Rollier 1919
Zlichopyramis Havlicek 1975
Zlichorhynchus Havlicek 1963
Zonathyris Struve 1992
Zophostrophia Veevers 1959
Zugmayerella Dagys 1963
Zugmayeria Waagen 1882
Zygatrypa Copper 1977
Zygonaria Cooper 1983
Zygospira Hall 1862
Zygospiraella Nikiforova 1961

ABSTRACTS PRESENTED AT THE THIRD INTERNATIONAL BRACHIOPOD CONGRESS, SEPTEMBER 2-5, 1995

SAKMARIAN (PERMIAN) BRACHIOPODS FROM SOUTHERN OMAN

LUCIA ANGIOLINI, A. BAUD, J. BROUTIN, H.A.HASHMI, J. MARCOCKS, J.P. PLATEL, A. PILLEVUIT & J. ROGER
Dipartimento di scienze della Terra, Via Mangiagalli 34, 20133 Milano, Italy
Located in the SE margin of the Arabian plate (Sultanate of Oman), the Haqf area is a region marked by gently deformed, uplifted Paleozoic formations. The Early Permian succession represents the last phase of a mega-sequence which began with deposition of Late Westphalian to ? Sakmarian Al Khlata tillites, succeeded by the transgressive marine Saiwan Fm. This unit contains a rich brachiopod fauna, together with bivalves, gastropods, crinoids, cephalopods and bryozoans. Brachiopods of the Saiwan Formation have been collected from the type section, close to the Saiwan 1 oil well and from a section located on the left bank of wadi Hanshi. The first fossiliferous level of the Saiwan Formation consists exclusively of *Cyrtella nagmargensis*. Succeeding fossiliferous levels show higher taxonomic diversity consisting of *Derbyella*, *Streptorhynchus* cf. *biani*, *Reedoconcha permixta*, *Subansiria*, *Permospirifer* cf. *wardakensis*, *Neospirifer* aff. *hardmani*, *Trigonotreta*, *Punctospirifer*, *Gjelispinifera*, and *Fletcherthyris*. Conularids, bivalves, cephalopods and gastropods are also present. The highest fossiliferous level of the Saiwan Formation is characterized by abundant *N.* aff. *hardmani*, together with bivalves. According to Uralian zonation, the Sakmarian is usually subdivided into Sterlitamakian and Tastubian. The first level of the Saiwan Formation is probably early Sterlitamakian, whereas the higher levels are probably Late Sterlitamakian on the basis of the following considerations: 1) the faunas of the Saiwan Formation lack such typical Asselian-Tastubian forms as *Brachythyrinella*, *Tomiopsis*, *Globiella* (Australia, Garwhal Himalaya, Unaria or the Indian shield, Kashmir). These genera possibly occur also in the Early Sterlitamakian (Australia, Garwhal Himalaya); 2) *Trigonotreta* is widespread in the Asselian Tastubian of Australia, Karakorum and Central Afghanistan but it is also present in the Sterlitamakian and in higher levels (Himalaya, Australia); 3) *Cyrtella* occurs both in the Asselian-Tastubian (Australia, Karakorum, Central Afghanistan) and in the Early Sterlitamakian (Australia, India, Himalaya); 4) *Subansiria* is an endemic genus of the Early Sterlitamakian of Peninsular India and Himalaya; 5) the *Cyrtella-Subansiria* fauna may straddle the Tastubian-Sterlitamakian boundary; and 6) *Taeniothaerus*, *Neospirifer* and *Streptorhynchus* seem to be of Late Sterlitamakian age (Himalaya, Central Afghanistan). The Saiwan faunas show affinity to the faunas of Himalaya, Peninsular India, central Afghanistan, and to a slightly lesser degree, of western Australia.

DELTHYRIAL STRUCTURES AND FUNCTIONAL MORPHOLOGY IN *CONCHIDIUM* FROM THE SILURIAN OF GOTLAND

MICHAEL G. BASSETT
Department of Geology, National Museum of Wales, Cardiff, CF1 3NP, UK
Some pentameride brachiopods have a true deltidium, but in *Conchidium biloculare* from the Klinteberg Limestone Formation (late Wenlock, Homerian) of Gotland, Sweden, the delthyrium is closed partially by a thick, concave to trough-like deltidial cover homologous with a pedicle collar. As in some spiriferides, the structure was formed by epithelium of an atrophying pedicle, but unlike the spiriferides the structure in *C. biloculare* retained a subapical foramen throughout most of ontogeny. The base of the delthyrium was sealed from early in ontogeny by the incurvature of the dorsal valve against the ventral palintrope. In gerontic shells, complete closure of the delthyrium was caused by increased shell thickening and incurvature to a point where the ventral beak rested on the dorsal umbo. Achievement of this state was accompanied by final pedicle atrophy and adoption of a free-lying life strategy.

BIOGEOCHEMICAL RESEARCH ON BRACHIOPOD SHELLS: OPPORTUNITIES AND PITFALLS

Y. A. BORISSENKO
Faculty of Geology and Geography, University of Kharkov, Kharkov, Ukraine
Studies of the mineralogy of brachiopod shells are at present based on too few observations with respect to the quantity of modern and fossil brachiopods. The results do not seem to be on firm enough ground. In types of mineralization, brachiopod shells are divided into francolite and calcite compositions that correspond to the Lingulata and Calciata classes. In the state of Mg-calcite, the Calciata can be divided into high- and low-Mg subtypes that correspond to subclasses. As to micro-impurity contents, particularly Mg and Sr (associated mainly with development of shell porosity), two groups of articulate brachiopods can be recognized, one with rather impure, and the other with relatively clean shell material. The assumption of complete or partial aragonite mineralization in some brachiopod groups has not yet been substantiated. The difference between francolite and calcite in micro-elemental shell composition is explained by many causes, the main ones being phylogenetic and physiological factors which reflect systematic positions of organisms. The essence of these factors is internal control of shell formation, which depends upon the metabolic activity of organisms and the function of mantle epithelial cells. These factors control the quantity of micro-elements, parts of which can replace isomorphously some components in composition of base materials. Mineral purity of shell material is determined largely by the rate of ontogenetic accretion of skeleton. Certain quantity of micro-elements penetrates passively depending on their concentration in the environment and physico-geographic conditions. Finally, the role of chemical and microstructural transformation of mineral and organic components is essential during and after shell formation. In general, one can suppose that a specific brachiopod shell composition depends on simultaneous or consecutive influence of different factors. In origin, these factors can be classified into four groups: genetic influence on the organism's physiology, environmental factors, secondary processes, and technical causes such as various degrees of shell cleaning from impurities of rock matrices and systematic errors in laboratory analyses. The technical causes also include different representations of analyzed objects from different investigated sections or areas. When making general biogeochemical calculations, the final result depends on the number of representations. Especially chosen and statistically calculated results of numerous biogeochemical investigations can guarantee to some extent the probability of reflection of the most typical, frequently repeated situations. On the basis of this, one can make a conclusion about possible changes of bionomic and phylogenetic indicators.

BIOMINERALIZATION OF THE CALCAREOUS-SHELLED, INARTICULATED BRACHIOPOD *NEOCRANIA ANOMALA*

KAREN BROWN
Geology Department, University of Glasgow,
Glasgow, G12 8QQ, UK

The inarticulated brachiopod *Neocrania anomala* is atypical of other inarticulated brachiopods. The species has no hinge between the two valves and so is classified as inarticulated, yet it and other members of its superfamily, the Cranioidea, possess a carbonate mineralized shell typical of the articulated brachiopods. This anomaly is illustrated also in the protein composition of the valves. The location of the protein is dissimilar to that of both classes and the constituent animo-acids can be seen to be a mixture of inarticulated and articulated levels, presenting problems for precise classification. To address this problem, the proteins which are enclosed and protected within the calcite crystals were examined for their amino-acid composition. From this information it will be possible to determine the primary sequence of the main protein and the subsequent protein sequence. Examination of this molecule and its interactions with the surrounding shell minerals will, in future, indicate the biomineralization process particular to this species. This will allow the subsequent assignment of the species and its superfamily to an existing class, or will provide evidence for the formation of a new one.

THE CARDINALIA OF ARTICULATED BRACHIOPODS

C.H.C. BRUNTON, F. ALVAREZ & D. I. MACKINNON
Paleontology Department, Natural History Museum, Cromwell Road, London, SW7 5BD, UK; Departamento de Geologia, Universidad de Oviedo, Oviedo, Spain; Geology Department, University of Canterbury, Christchurch, New Zealand

The vast output of paleontological literature in recent decades has had a profound effect on the mode of study and classification of the Brachiopoda. Along with the many publications has grown an extensive and complicated terminology. Many established terms are, of course, justifiable because each conveys in one or two words the subtle complexity of a particular structure which otherwise would require lengthy description in a variety of languages. But others, however, are either synonyms, in that they are used by different authors for structures of the same origin in different fossil groups, or they have been coined for unimportant growth variants of a well-known feature. Yet further unnecessary proliferation of terminology may be attributed to partitioning of research as a result of geographic, linguistic, geopolitical or stratigraphic boundaries or specialisations. The terminology used in recent years to describe the cardinalia of athyridides displays divergent opinions and differing usages, despite there being general agreement on other morphological terms. Different techniques used to observe internal structures, such as direct observation of disarticulated and cleaned valves, internal moulds, etching of silicified specimens or serial sectioning (grinding) and the reconstruction of internal structures, have contributed to a multiplicity of morphological terms, many of which are synonyms. These differing terminologies are misleading for comparative work and can play havoc when constructing cladograms. The problem of cardinalia terminology has been tackled by Brunton, Alvarez and MacKinnon (in press) who utilize Recent and fossil examples to illustrate structures and indicate which are homologous; in an attempt to reduce the number of terms used, they list those recommended which are suitable in a wide spectrum of brachiopods.

EXAMINATION OF THE EOCENE-OLIGOCENE TEMPERATURE DROP USING STABLE ISOTOPES IN NEW ZEALAND FOSSIL BRACHIOPODS

NANCY BUENING
Department of Geology, University of California,
Davis, CA 95616, USA

A global decline in temperature between the late Eocene and early Oligocene (38-33 Ma) has been documented by isotopic variations observed in foraminiferal calcite and by biogeographic shifts in the distribution of warm-water marine invertebrates. Paleotemperature estimates reconstructed from oxygen isotope values from foraminifera from deep sea cores in high southern latitudes suggest SST declined to 4˚C by the early Oligocene. In contrast, paleotemperature estimates derived from the biogeographic distribution of subtropical marine organisms from New Zealand indicate that ocean temperatures dropped to 16˚C. Determining the magnitude of the Eocene-Oligocene temperature drop is essential to document the history of Antarctic glaciation and the accompanying dramatic changes in climatic patterns. The objective of my study is to provide a quantitative record of paleotemperature estimates based on oxygen isotope values from brachiopods in a New Zealand shelf environment, that will document the magnitude of the decline in temperature at the Eocene-Oligocene boundary. Preliminary results indicate that shelf temperatures of New Zealand during the early Oligocene dropped to a low of 6˚C and continued to rise to a high of 12˚C during the middle Oligocene. By the early Miocene, New Zealand shelf temperatures were 9˚C. Results of ontogenetic sampling show temperature variations of 1˚C during the lifetime of an individual. This suggests very little seasonality during the Oligocene and early Miocene. The New Zealand shelf mean temperature of 8˚C for the early Oligocene obtained in this study agrees with paleotemperature reconstructions from DSDP site 277 on the Campbell Plateau south of New Zealand and from $18\partial O$ values of surface dwelling planktic foraminifera from 40˚S latitude (the approximate latitude of New Zealand during the early Oligocene) compiled from 10 DSDP and ODP sites. However, the estimated mean paleotemperature of 8˚C is approximately 8˚C lower than paleotemperatures from southern Australian and northeastern New Zealand early Oligocene shelf environments. This difference may be due to the relative positions of equatorial and polar currents circulating near New Zealand during the early Oligocene. Paleotemperature estimates from this study are considerably lower than those derived from New Zealand paleontological evidence. The reasons for the discrepancy is unclear, however, and it may be that our assumptions of temperature tolerances, based on modern distributions, are inaccurate.

STRATOCLADISTICS AND PENTAMERIDE PHYLOGENETIC INFERENCE

SANDRA J. CARLSON
Department of Geology, University of California,
Davis, CA 95616, USA

Stratocladistics (Fisher, 1980) provides a method for evaluating alternative phylogenetic hypotheses with respect to their concordance with both character data and patterns of stratigraphic occurrence. In order to evaluate the efficacy of this method empirically, I constructed a data matrix for the type genera of 19 pentameride families recognized in the revision of the Treatise on Invertebrate Paleontology, plus two genera of uncertain familial affiliation and two rhynchonellide genera, using four orthide genera as outgroups. I coded all genera for 72 characters of shell morphology, and analyzed the character data using the heuristic search option (random addition of taxa, 10 replicates) in PAUP 3.1.1. Twenty-two cladograms of length 308 resulted (Retention index = 0.53). The consensus diagram of these 22 trees contained two poorly resolved regions, indicating variation among tree topology in resolving relationships among several genera. I then added to the data matrix the observed stratigraphic range of each genus, as a multistate polymorphic stratigraphic character in MacClade 3.0, and examined the revised length (and topology) of each of the 22 cladograms obtained in the PAUP analysis. The revised lengths ranged from 359 to 367; completing a full search on each cladogram in MacClade 3.0, now including both stratigraphic range and morphological data, resulted in convergence toward two topologies (length 355) closely represented in the cladograms of the shortest revised length. Results are sensitive to the time scale of the stratigraphic coding; coding by period, epoch, or stage yield somewhat different results. For comparison, I included stratigraphic range as an ordered and irreversible character in a PAUP analysis. Twelve cladograms of length 402 resulted; their consensus topology differs significantly (and curiously) from the stratocladistic results. These results demonstrate that stratocladistics can be very useful in several facets of phylogenetic inference involving fossil or Recent taxa. It can provide a criterion for selecting a subset of the equally most parsimonious cladograms obtained from morphological data alone, when such data are ambiguous or in conflict. It can also be used in conjunction with morphological data to obtain a pattern of relationship that simultaneously minimizes both morphological and stratigraphical "parsimony debts". Fossils contain valuable temporal and anatomical information; now both can be used in computer-aided phylogenetic inference.

ORDER PENTAMERIDA

Sandra J. CARLSON and Arthur J. BOUCOT
Department of Geology, University of California, Davis, CA 95616, USA; Department of Zoology, Oregon State University, Corvallis, OR 97331, USA

Pentamerides comprise a small, but significant group of early and middle Paleozoic brachiopods. Interesting changes in shell size, shape, convexity and plication, hinge line length, nature of hinge structures and development of various muscle platforms in both dorsal and ventral valves characterize pentameride evolution, and illustrate transformations from more primitive orthide-like morphologies to more derived, rhynchonellide-like morphologies. Two suborders are recognized: the older, more primitive paraphyletic Syntrophiidina and younger, more derived, monophyletic Pentameridina. Rhynchonellida originates within the Syntrophiidina, between the two superfamilies recognized; the older Porambonitoidea and the younger Camerelloidea are both paraphyletic. No revolutionary changes in the older classification were deemed necessary in this revision; three family groups were elevated to superfamilies (Camerelloidea, Stricklandioidea, Gypiduloidea) and two small new families are recognized. Since the publication of the 1965 Treatise on Invertebrate Paleontology, the number of pentameride genera has approximately doubled. While many of the earlier-named genera (and higher taxa) originated in North America with North American brachiopodologists, most of the recently named genera originated in Russia and China, greatly expanding our knowledge of the global distribution of the group. The phylogenetic relationships among the pentamerides, and of pentamerides to other brachiopods are somewhat clearer than before, but are not unproblematic, particularly at the level of genus and family. Extensive homoplasy (homeomorphy) in

such an ancient group that survives to the present only in highly derived descendants will almost ensure generally unstable phylogenetic results. Emphasizing (or weighting) certain selected characters over others can produce quite different hypotheses of phylogeny. These difficulties underscore the problems associated with erecting a fully phylogenetic classification; the phylogeny must be robust to provide the classification with stability. Phylogenetic analysis does not solve all the problems of brachiopod classification, but it does clarify the nature of the problems, making them easier to test, and it provides the most defensible method available thus far for reconstructing brachiopod evolutionary history.

BRACHIOPOD EXTINCTIONS ACROSS THE ORDOVICIAN-SILURIAN BOUNDARY, ANTICOSTI ISLAND, CANADA

PAUL COPPER
Department of Earth Sciences, Laurentian University, Sudbury, Canada P3E 2C6

The Late Ordovician extinction on Anticosti Island was a two stage, downstep event for the very rich and well-preserved brachiopod fauna. The first major extinction saw the elimination of the Richmondian fauna within the upper part of the Vaureal Formation, but no abrupt departure of any taxa is seen at a single level. The last of the Richmondian taxa vanished in the Schmitt Creek Mbr of the Vaureal Fm. (late Rawtheyan, Ashgill), but a number of hold-overs carried into the overlying Ellis Bay Formation (c.60m thick). The Ellis Bay carries a largely Hirnantian fauna typified by *Eospirigerina* and *Hindella*, but also having *Hirnantia* in the upper sandy facies to the east: pentamerids include only *Parastrophinella*. The Ellis Bay fauna disappeared below, and within the upper part of the formation: this is the second extinction event, recording the final demise of many hold-over Ordovician taxa. No erosional breaks or hiatuses are known at the boundary. The new Silurian fauna above the last reefal and oncolitic limestones of the Ellis Bay Fm. is marked by dwarfed or stunted (*Becscia*, *Cryptothyrella*), as well as normally sized taxa, but a number of hold-overs in the earliest Silurian (e..g *Eospirigerina*, *Leptaena*, etc.) survived withou major changes. No major brachiopod families or subfamilies disappeared at the O/S boundary, and the extinction event is limited compared to the end Devonian extinction.

CENTRAL NORTH AMERICAN MIDDLE-UPPER DEVONIAN (EIFELIAN-FRASNIAN) BRACHIOPOD BIOEVENTS

JED DAY
Department of Geography-Geology, Illinois State University, Normal Illinois, 61790-4400, U.S.A.

Secular changes in the structure of Devonian benthic shelly biofacies, of which brachiopods were a dominant component, were primarily effected by eustatic sea level fluctuations and their consequent global and regional environmental changes (Johnson & Sandberg, 1989; Johnson 1990; Racki, 1989, 1992: Day, 1992 in press). In central North American, the most complete and best documented Middle-Upper Devonian (Eifelian-Frasnian) brachiopod sequence is developed in the Iowa Basin, as outlined in recent studies by Day (1989, 1992, 1994, in press). and Day & Koch (1994a, 1994b). The Devonian succession in the Iowa Basin consists of marine carbonate platform strata included in the Wapsipinicon and Cedar Valley groups, and mixed carbonates and clastics of the Lime Creek Formation. These strata yield at least 250 spp. of brachiopods whose ranges are now well understood and tied directly to the Iowa Basin Devonian conodont sequence. The Iowa Basin Devonian benthic and pelagic faunal sequence is divided into nineteen composite brachiopod-conodont Faunal Intervals comprising a high-resolution temporal framework that permits detailed study of secular changes in Middle and Upper Devonian shelly faunas in the region. Iowa Basin Eifelian-Frasnian strata comprise nine major sequences or Transgressive-Regressive (T-R) cycles. The timing of deposition of seven of nine Iowa Basin Eifelian-Frasnian T-R cycles corresponds directly to Euramerican eustatic T-R cycles of Ie-IId of Johnson and others (1985), Johnson & Klapper (1992), and subdivisions of T-R cycles of IIa (IIa-1 and IIa-2) and IIb (IIb-1 and IIb-2) defined by Day et al. (1994; in press). Significant migrations, phyletic transitions and radiations within numerous brachiopod lineages occurred during the initial deepening or sea level highstand stages of eight of nine major Eifelian-Frasnian transgressions recorded by the Iowa Basin Devonian succession. The timing of the most significant brachiopod migrations and radiations recorded by the Iowa Basic Devonian faunal sequence correspond to transgressions that initiated Devonian T-R cycles Ie (upper *kockelianus* Zone), IIa-1 (upper Middle *varcus* Subzone), IIa-2 (lower part *disparilis* Zone = lower *subterminus* Fauna), IIb-1 (*norrisi* Zone), IIb-2 (early Frasnian Montagne Noire Zone 3), IIc (M.N. Zones 5-6), and IId (M.N. Zone 9). Five major species-level extinction events are recorded in the brachiopod sequence. Their timing corresponds to eustatic sea level lowstands that terminated five of nine Iowa Basin Eifelian-Frasnian T-R cycles. Major brachiopod extinctions events correspond directly to regressions (and associated environmental events) that terminated Devonian T-R cycles Ie (early Givetian, *ensensis* Zone), IIa-2 (late Givetian, upper part *disparilis* Zone), IIb-2 (early Frasnian, M.N. Zone 4), IIc (late Frasnian - M.N. Zone 8?), and IId (latest Frasnian - M.N. Zone 13 or the Kellwasser II event).

ENDOSCOPIC INVESTIGATIONS OF FEEDING STRUCTURES AND MECHANISMS IN TWO RECENT PLECTOLOPHIC BRACHIOPODS

SHOVA R. DHAR , ALAN LOGAN ,
B. A. MACDONALD & J. E. WARD
Centre for Coastal Studies, University of New Brunswick, Saint John, New Brunswick, E2L 4L5 Canada; Salisbury State University, Salisbury, Maryland, 21807-6837, USA.
Endoscopic investigation permits direct *in vivo* observations of the lophophore in relatively undisturbed specimens. Using this method, the feeding structures and mechanisms of the Recent species *Terebratulina septentrionalis* (Couthouy) from the Bay of Fundy and *Terebratalia transversa* (Sowerby) from western Canada have been examined. A diet consisting of various-sized, light-reflective particles and natural seston was given to both species. Particle velocities in flow patterns through the lophophore were determined by means of video image analysis. Particle capture and transport of material to the food groove as well as rejection mechanisms and other types of behaviour were observed and add insight to feeding efficiency in typical plectolophic brachiopods. This work is part of an ongoing study which will make comparisons between these typical plectolophes and the spirolophe rhynchonellid *Hemithyris psittacea* from eastern Canada.

RELATIONSHIP OF SPINOSITY AND SHELL SIZE OF WAAGENOCONCHID BRACHIOPODS TO SEDIMENT GRAIN SIZE: PERMO-PENNSYLVANIAN, YUKON.

LAING FERGUSON & ROBERT M. WILLIAMS
Mount Allison University, Sackville, New Brunswick, Canada, E0A 3C0; Department of Zoology, University of British Columbia, Vancouver, BC, V6T 1Z4, Canada
Collections of several hundred productid brachiopod shells belonging to the genus *Waagenoconcha* Chao were examined in order to determine if any relationship existed between the mature shell size and the relative size and spacing of the quincuncially arranged spine bases which cover the shell and the coarseness of the sediment in which the shells had lived and were entombed. The shells came from three stratigraphic sections in the southwestern Yukon (two from the Peel River Rapids and one from the Tatonduk River) and were from the Pennsylvanian Ettrain Formation and the lower Permian Jungle Creek Formation. Immature shells were found to be plano-convex. After the initiation of the development of a trail at maturity, the brachial valve becomes markedly geniculate and the shells become concavo-convex. The flat part of the brachial valve is here referred to as the visceral disc. It was found that there is a strong positive correlation between the size (length) of the visceral disc and the grain size of the containing sediment (as indicated by terrigenous clastic material such as quartz and chert grains). The spine bases on the trail are much smaller and more closely spaced than those on the earlier formed part of the shell. This is possibly because they functioned to filter out only suspended sediment at the elevated commissure whereas the earlier formed (pre trail) spines had to contend with coarse sediment moving at the sediment-water interface. In this study it was thus the size and density of the spine bases on the early-formed part of the shell that were measured. A strong negative correlation was found between the coarseness and relative spacing of spine bases and the coarseness of the sediment. Large, mature shells with coarse, widely spaced spine bases were found in coarse-grained, high energy (near shore?) sediments. Smaller, mature shells with finer, more closely spaced spine bases were found in finer grained, low energy sediments. Within each of the three stratigraphic sections it was found that these associations are not unidirectional with time, but appear to be tied to coarsening-upward and fining-upward trends in the lithology. They thus appear to be phenotypical in nature and environmentally controlled rather than genetically controlled. Caution must be exercised in the use of the size and spacing of spine bases in discriminating species within the genus *Waagenoconcha*.

COMPARATIVE EMBRYOLOGICAL STUDIES ON *GLOTTIDIA* (LINGULATA), *TEREBRATALIA* (ARTICULATA) AND *PHORONIS* (PHORONIDA)

GARY FREEMAN
Department of Zoology, University of Texas at Austin, Texas 78712-1064, USA.
Fate maps have been prepared for *Glottidia pyradimata*, *Terebratalia transversa* and *Phoronis vancouverensis*. Fate maps indicate the origins in the uncleaved egg and early cleavage stage embryo of the different larval regions of these animals and the translocation of a given region of the early embryo to a new position at a later time during embryogenesis. In each of these clades, gastrulation occurs at the vegetal pole of the embryo. During gastrulation, mesodermal and endodermal cells invaginate into the embryo. In *Phoronis* and *Glottidia* the blastopore becomes the mouth of the larva. In *Terebratalia* the blastopore persists until late embryogenesis when the demarcation between the apical and mantle lobes of the future larva have appeared. It occupies the same position as the blastopore in *Glottidia* and *Phoronis*. The region that will form the apical lobe in *Glottidia* has a lateral position in the early embryo while the region that will form the apical lobe in

Terebratalia or the epistome (the apical lobe homologue) in *Phoronis* is found at the animal pole of the embryo. During gastrulation in *Terebratalia* and *Phoronis*, the apical lobe and epistome-forming regions move through 90 degrees to their definitive anterior lateral positions. The timing and mode of regional specification in these three clades have been studied by isolating animal, vegetal or lateral halves from embryos at different developmental stages. In all three groups the differentiation of an apical lobe or epistome depends on inductive signals that originate from the vegetal region of the embryo. If the presumptive apical lobe-forming region is removed from a cleavage stage *Glottidia* embryo, the remaining lateral half cannot regulate by forming an apical lobe. When the same experiment is done on *Phoronis* or *Terebratalia*, the remaining half will regulate by forming the missing apical lobe or epistome. The developments of articulate brachiopods and phoronids through late gastrulation are remarkably similar. The early developments of these two clades differ in significant ways from the development of the Linguloidea.

LATE EOCENE TEREBRATELLID RADIATION

NORTON HILLER

University of Canterbury / Canterbury Museum,
Christchurch, New Zealand

The Late Eocene in Australia and New Zealand records the almost coeval first appearance of several members of the Terebratellidae, a predominantly southern hemisphere family of brachiopods. Most of these forms belong to the subfamily Terebratellinae (in currently used classification) and all possess teloform loops. The genera involved include: *Diedrothyris* - an Australian form that is probably represented in New Zealand by the species previously identified as '*Neothyris*' *esdailei* Thomson, 1918. *Cudmorella* and *Victorithyris* - two forms known only from Australia. *Aliquantula* - an Australian form originally included in *Stethothyris*. *Stethothyris* - a form that presently includes species from both Australia and New Zealand but which probably can be divided into three separate genera. In addition to these is a new taxon from New Zealand, bringing to eight the number of genera making their first appearance in Late Eocene strata. It is likely that these forms share a common ancestor that underwent an adaptive radiation in Late Eocene times to exploit new ecological opportunities brought about by changing tectonic, oceanographic and climatic conditions. The nature of the ancestral stock is not known for certain. Cretaceous - Early Paleogene terebratellids are rare and poorly understood but it is believed that a form not unlike the Paleocene '*Campages*' *chathamensis* Allan, 1932 would make a suitable ancestor for the later Terebratellinae.

SHELL STRUCTURE OF DEVONIAN ATRYPID BRACHIOPODS FROM NAKHICHOVAN, AZERBAIDZHAN

V. N. KOMAROV

Moscow State Geological Prospecting Academy, 23 Miklukho-Marklay St., 117485 Moscow, Russia

The evidence is based on a collection of about 6530 Devonian atrypid specimens from 77 localities. The material was assigned to some 39 species. The wall of the atrypid shell is composed of three layers; a primary, a secondary (fibrous) and a tertiary (columnar) layer. The primary layer is frequently worn away. Its visible thickness varies from 10-18μ in *Atryparia dispersa* (Struve) to 130μ in *Gruenewaldtia latilinguis* (Schnur). Usually the primary layer is not differentiated. In that case it can be granular or more often cryptocrystalline. In *Atrypa norashenensis* Mamedov, *A. culminigera* Struve, *A. collega* Struve, *Atryparia dispersa* (Struve), *Spinatrypina comitata* Copper, *S. robusta* Copper, *Desquamatia subindependensis* Komarov, and *Carinatina arimaspa* (Eichwald), the cryptocrystalline layer consists of needle-shaped or suboval crystals. Their visible length usually varies from 6.1 to 34μ and their width from 0.8 to 11.5μ. In *Spinatrypa aspera araxica* Komarov, the cryptocrystalline layer consists of wedge-shaped crystals of variable visible length from 10 to 18μ and width from 1.8 to 3.2μ. Fine-grained primary layer (grain dimensions 0.5-5μ) was observed in *Pseudatrypa gjumuschlugensis* (Mamedov). Coarse-grained primary layer (grain dimensions 2.2-16.0μ) was observed in *Gruenewaldtia latilinguis* (Schnur). Differentiate primary layer was found in *Punctatrypa olgae* Nalivkin for the first time. It consists of two sublayers - the outer large-crystal (40-50μ in thickness) and inner fine-crystal (5-18.3μ in thickness). Micropores and longitudinal ridges can be related to the ultrastructural features of the primary layer crystals. The secondary layer shows great variation in the dimension and outline of the fibrous crystals even in a specimen. Micropores, ridges and different shading can be related to the ultrastructural features of the fibrous crystals. The tertiary layer varies in total thickness from about 0.5 to 5.3mm, as in *Atrypa culminigera*, *Atryparia dispersa*, *Pseudoatrypa gjumuschlugensis* and *Gruenewaldtia latilinguis*. In *Atrypa culminigera*, *A. norashenensis*, *Desquamatia subindependensis*, and *Punctatrypa olgae*, two or three layers of fibrous and columnar crystals may interfinger. Different shading, round or three-cornered depressions and very fine knobs can be related to the ultrastructural features of the columnar crystals. The dental plates show a two-layered structure, consisting of homogeneous or heterogeneous fibrous calcite and columnar calcite. Cryptocrystalline layer was not observed. The exposed structural diversity of dental plates is dependent upon the heterogeneity of their fibrous component. Great structural diversity is typical of the subfamily Spinatrypinae.

LIVING BRACHIOPODS DREDGED AROUND NEW CALEDONIA DURING THE 1985-1989 FRENCH CRUISES

BERNARD LAURIN

Centre des Sciences de la Terre, Université de Bourgogne,
21000 Dijon, France

Nine cruises explored the bathyal fauna around New Caledonia and Chesterfield Islands during the years 1985 to 1989. On a total of 4006 dredgings and trawlings at depths from 180 to 2660 m (half at 280 to 600 m), 178 provided brachiopods; more than 2900 living specimens were collected. The maximum richness of the brachiopod fauna occurs in depths of 200 to 800 m. The number of specimens is generally greater than 10 in most of the stations, and up to more than 200 in some. The fauna consists of 26 species belonging to 19 genera and 14 families. Two new genera (*Neoancistrocrania* and *Kanakythyris*) and five new species (*N. norfolki*, *K. pachyrhynchos*, *Stenosarina globosa*, *S. lata*, and *Fallax neocaledonensis*) were described. Four species (the rhynchonellide *Neorhynchia strebelei*, the terebratulides *Abyssothyris wyvillei*, *Kanakythyris pachyrhynchos*, and the dallinide *Nipponithyris afra*) show a strongly sulcate anterior commissure, a feature that is usually considered as typical of deep-sea brachiopods. Nevertheless, that feature occurs in the brachiopods of New Caledonia at shallower depths than previously recorded. Some species, such as the dallinid *F. neocaledonensis*, exhibit features that might be related to various taxa: dental plates typical of *Falla*, a hood in the adult brachidium as in *Campages*, and general shape homeomorphic with *Dallina floridana*. This results in the necessity of improving the characters used in more accurate studies, namely on development, to deduce significant homologies. In geographical distribution, the New Caledonian bathyal fauna includes both restricted and widely distributed species. The ranges of several species or genera are extended noticably. Each of the 5 species of the rhynchonellide genus *Basiliola* had a distinct geographical distribution. For example, around New Caledonia both the small (formerly only Japanese) *B. lucida,* and the larger (formerly only Hawaiian) *B. beecheri,* were collected. The genus *Grammetaria,* previously known off South America (1 specimen of *G. africana)* and off the Philippine Islands (3 specimens of *G. bartschi*), expands its range, as well as the number of collected specimens. *Stenosarina,* formerly known from the Caribbean and from the Mozambique Channel (under another name) expands its range greatly with the discovery of three species. *Nipponithyris afra*, previously known from the Japanese seas and from the Mozambique Channel, occurs in New Caledonia. It is noticeable that this sulcate species was dredged in the same station as the sulcate *Abyssothyris wyvillei.*

HISTORY OF THE GENUS *TEREBRATULA* (BRACHIOPODA: TEREBRATULIDA)

DAPHNE E. LEE , M. CALDERA & O. SIMONE
Geology Department, University of Otago, P. O. Box 56,
Dunedin, New Zealand; Dipartimento di Geologia e Geofisica,
Universita degli Studi di Bari, Italy

Terebratula terebratula (Linnaeus 1758), the name-bearer for the Order Terebratula, Phylum Brachiopoda, has a long, complex and illustrious history. The type specimen from "tophaceâ *Concretione"* near the town of Andria in southern Italy was first illustrated in a scientific treatise by Fabio Colonna in 1616. Colonna's specimen was refigured by Klein (1753), and the species was described (on the basis of the Colonna/Klein illustrations) as *Anomia terebratula* by Linnaeus (1758). The genus *Terebratula* was formally established by Müller (1776), who did not, however, include *Anomia terebratula* in his list of species. *Anomia terebratula* Linnaeus was subsequently designated as the type species of the genus *Terebratula* by Lamarck (1799). In spite of these distinguished antecedents, the type species of the genus has never been officially ratified; the whereabouts of the type is unknown; and the age and exact position of the type locality was uncertain. This paper summarises the history of *Terebratula terebratula* (Linnaeus) from 1616 to the present day. The type locality from which Colonna collected the specimen which he figured in 1616 has been rediscovered, the age established, a neotype selected, and an application made to the I.C.Z.N. to ratify the selection of the type species of the genus *Terebratula.*

VARIATION IN THE LOOPS OF TWO RECENT SPECIES OF *LIOTHYRELLA* (BRACHIOPODA: TEREBRATULIDA) FROM NEW ZEALAND AND SOUTH ORKNEY ISLANDS

DAPHNE E. LEE & C. R. SAMSON
Geology Department, University of Otago,
Box 56, Dunedin, New Zealand

Within three of the Mesozoic to Recent superfamilies included in Suborder Terebratulidina (Terebratuloidea, Dyscolioidea and Loboidothyridoidea), distinctions between genera are based principally on differences in the loop and cardinalia, as perhaps half of the 230 genera so far described are externally similar, being of medium size, oval in outline, biconvex, smooth, and

with moderately large permesothyrid to epithyrid foramens. This study investigates the variation present in the loops and cardinalia of two species of Recent terebratuloid brachiopods, *Liothyrella neozelanica* Thomson from New Zealand, and *Liothyrella uva* (Broderip) from Signy Island, South Orkneys, in order to determine the typical range of variation within and between clearly defined species within a single genus. Eleven parameters were measured for 140 individual brachiopods, and the variation quantified and presented visually and statistically. Variation in many features of the loop and cardinalia is shown to be significant. It is suggested that many genera of short-looped brachiopods placed in Suborder Terebratulidina which have been discriminated on the basis of minor differences in loop morphology should be re-examined, and some placed in synonymy.

ORDER TEREBRATULIDA

DAPHNE E. LEE & ARTHUR J. BOUCOT
Geology Department, University of Otago, P. O. Box 56,
Dunedin, New Zealand; Department of Zoology,
Oregon State University, Corvallis, Oregon, USA

The Order Terebratulida includes some 600 brachiopod genera placed in three suborders and 15 superfamilies which range in age from Devonian to Recent, a span of about 400 million years. Order Terebratulida includes the majority of living articulate brachiopods (around 75% of the genera and 90% of the species). The three existing brachiopod suborders, Centronellidina, Terebratulidina and Terebratellidina, will be used in the revised Treatise, but it is recognised that future work, especially on brachiopod genetics, may necessitate major reorganisation of the suborders and superfamilies within the Order Terebratulida. Several new superfamilies of Mesozoic to Recent age are recognised on the basis of new studies of loop ontogeny and presence or absence of characters such as dental plates and spiculation. A summary account of each superfamily lists the number of included families and genera, the stratigraphic range, the major diagnostic features, and areas which require further research. A number of enigmatic living genera such as *Fallax,* *Ecnomiosa,* and *Macandrevia* are shown to be closely related to Mesozoic progenitors. The often complex loop ontogeny in Order Terebratulida, the considerable degree of internal and external variation exhibited by many species and genera, together with the possibility of external homeomorphy, mean that recognition of terebratulide genera is fraught with problems. Careful examination of ontogenetic stages, preparation of loops and internal features by dissection as well as serial sectioning, and synthesis of genetic information with morphologic studies should, in due course, result in a more accurate understanding of the relationships between genera now included in Order Terebratulida.

MORPHOLOGY, PHYLOGENY AND TAXONOMIC AFFINITIES OF THE RECENT MICROMORPHIC BRACHIOPOD *GWYNIA CAPSULA* (JEFFREYS)

ALAN LOGAN , D. I. MACKINNON & J.E. PHORSON
Centre for Coastal Studies, University of New Brunswick,
Saint John, NB. E2L 4L5, Canada Department of Geology,
University of Canterbury, Christchurch, New Zealand; Fellside
Gardens, Balmont, Durham, DH1 1AR, UK

The skeletal and soft-part morphology and shell microstructure of the poorly-known Recent micromorphic brachiopod *Gwynia capsula* (Jeffreys) have been elucidated from specimens collected from the Northumberland coast and the Menai Strait, as well as from USNM material from Jersey, using the scanning electron microscope. The species is compared to the diminutive Recent brachiopod *Argyrotheca cistellula* (Searies-Wood), with which it often occurs, and also to the Middle Jurassic micromorph *Zellania davidsoni* Moore. On the basis of a morphological comparison between *Gwynia* and *Zellania* there is a strong case for a close phylogenetic link between these two taxa, despite the considerable time gap. Their taxonomic affinities are more difficult to assess, however, and may have to await periostracal studies on the megathyroid *Argyrotheca bermudana* Dall and molecular biology studies on *Gwynia capsula* soon to be initiated by researchers at the University of Glasgow.

DIAGNOSING THE DALLININAE (ORDER TEREBRATULIDA): A BRACHIOPOD SUBFAMILY WITH A LONGSTANDING IDENTITY PROBLEM

DAVID L. MACKINNON
Department of Geological Sciences, University of Canterbury, Christchurch, New Zealand.

Largely on the basis of loop development, Beecher (1893) subdivided Recent long-looped brachiopods of the Family Terebratellidae into three subfamilies, the Dallininae of the northern hemisphere, the Magellaninae of the southern hemisphere and the Megathyrinae inhabiting temperate waters on both sides of the equator. The nominate genus of Beecher's original subfamily Dallininae is the genus *Dallina* (type species *Terebratula septigera* Loven). Over the past 100 years, the taxonomic concept of the Dallininae has endured, and indeed prospered. Allan (1940) elevated the taxon to family-level status and Dagys (1968) further elevated the taxon to superfamily status, although the latter has not been universally accepted. Thus there currently exists in brachiopod literature the anomaly of a brachiopod subfamily (the Dallininae) that is assigned by various authors to one of two different superfamilies (the Terebratelloidea and Dallinoidea). Of longstanding importance in family level diagnosis of long-looped brachiopods has been patterns of loop ontogeny, but until comparatively recently, few complete ontogenetic sequences were known. This deficiency is now being addressed through an ongoing research programme involving scanning electron microscopy investigations of very small juveniles of all available living species of long-looped brachiopods. In the course of these investigations it has been found that existing descriptions of ontogenetic development in various Dallininae (as defined in the 1965 Treatise), including *Dallina septigera* (Loven), are inadequate and misleading, and that new data now becoming available on early ontogeny has the potential to resolve longstanding uncertainties regarding the systematic status and phylogenetic relationships of Dallininae and other higher level, long-looped taxa. Studies of juvenile *Dallina septigera* and a sympatric homeomorph *Fallax dalliniformis* Atkins (with which *D. septigera* has in the past been confused) reveal much greater morphological disparity than previously recognised. Despite some cautionary words by Atkins (1960), the notion that dental plates are present in young *Dallina* has persisted to the present day. However, a new examination of early growth stages of *D. septigera* confirms that dental plates are absent in juvenile as well as adult stages of that species. In *Fallax dalliniformis*, on the other hand, dental plates are present at all stages of growth. Compared to the early loop ontogenies of all other long-looped brachiopods studied to date, the early ontogeny of *D. septigera* most closely resembles that of austral Terebratellidae, such as *Calloria inconspicua* (Sowerby) and *Magellania flavescens* (Lamarck). Characteristics shared by *D. septigera* and the terebratellids include a septal pillar that is initially blade-like and non-bifurcate, the absence of dental

plates throughout all growth stages, the shortness of the pedicle collar and the absence of spiculation. The early ontogenetic motif of *Fallax dalliniformis*, by contrast, is much closer to that of Laqueidae.

SIMILARITY OF EARLY LOOP ONTOGENY IN THE RECENT BRACHIOPODS *MACANDREVIA* KING AND *ECNOMIOSA* COOPER: TAXONOMIC AND PHYLOGENETIC IMPLICATIONS

DAVID L. MACKINNON & DANIÈLE GASPARD
Department of Geological Sciences, University of Canterbury, Christchurch, New Zealand; Département des Sciences de la Terre, Université de Paris-Sud, F-91405 Orsay, France.

A recent immunological study by Endo et al. (1994) draws attention to a hitherto unsuspected linkage between the long-looped terebratulid brachiopod *Ecnomiosa* Cooper and a previously recognized anomalous subset of *Macandrevia*, Kraussinidae and Megathyrididae. Relying almost exclusively on previously published morphological data of other authors, Endo et al. also attempted to identify shared characteristics of *Ecnomiosa* and *Macandrevia* that might lend support to their unexpected findings. Given the unusual and controversial nature of such findings, we decided to re-examine the morphological basis for the *Ecnomiosa-Macandrevia* kraussinid-megathyrid linkage proposed by Endo et al., paying particular attention to early ontogenetic phases of loop development. While our investigations confirm strong morphological similarities in the early ontogenies of *Ecnomiosa* and *Macandrevia*, we have found certain aspects of the previously published data set used by Endo et al. to be either misleading or inaccurate. In this paper we present new morphological data that correct previous misinformation and that help clarify relationships. Unlike all other long-looped brachiopods (except megathyroids), but like short-looped brachiopods, the descending branches of juvenile *Macandrevia* and *Ecnomiosa* grow only from the crura. Whereas the crura themselves are built up from normal secondary layer fibres, these first-formed, very slender branches, which extend from the distal ends of the crura, are essentially non-fibrous. The branches continue to grow in two arcs which converge and eventually meet on a low septal pillar located mid-valve. Shortly thereafter, the septal pillar undergoes rapid upward (ventral) growth to form a hood with a bifurcate anterior margin composed of strong, obliquely interdigitating, spinose projections. Further growth and localised resorption with the appearance of two lateral lacunae in the ascending lamellae, sees the development of the bilacunar phase in both species. From this stage onwards both species undergo radically different ontogenetic development of the loop. In *Macandrevia*, the narrow waist of the septal pillar is completely resorbed so that the loop is once again supported only by the crura; initially the descending branches are still united medially, but later they separate to form a teloform loop. In *Ecnomiosa*, by contract, the loop remains firmly anchored to the septal pillar, now transformed into a full median septum, by four supports - two lateral connecting bands to the descending branches, and two medioventical connecting bands to the transverse band uniting the distal (posteroventral) ends of the ascending lamellae. In the terminal adult loop stage of *Ecnomiosa*, the lateral connecting bands are resorbed, leaving only two mediovertical bands uniting the transverse band of the hood to the median septum. On the basis of strong similarities in the loop ontogeny between *Macandrevia* and *Zeilleria leckenbyi* (Baker, 1972) we refer *Macandrevia* to the hitherto Mesozoic superfamily Zeillerioidea. Likewise, on similarities in adult loop morphology between *Ecnomiosa* and forms like *Belothyris*, *Kingena*, and

Aldingia, we refer *Ecnomiosa* to the new superfamily Kingenoidea. A recent study of Lower Cretaceous Praeargyrothecidae (MacKinnon & Smirnova 1995) suggests a zeilleriid origin for Megathyridoidea, thus lending credence to the suggestion from immunotaxonomy of an as yet unspecified linkage between *Macandrevia* and Megathyrididae. However, we can find no morphological evidence to support the proposed linkage (Endo et al., 1994) between Kraussinidae and either *Macandrevia*, *Ecnomiosa* or Megathyrididae.

POST-PALEOZOIC RHYNCHONELLIDES: AN OVERVIEW

MIGUEL MANCENIDO & ELLIS F. OWEN

La Plata Natural Sciences Museum, La Plata 1900, Argentina; Natural History Museum, London, SW7 5BD, UK

The number of Mesozoic and Ceinozoic nominal rhynchonellide genera has been dramatically increased to more than double during the last 30 years. The main results attained in the course of revisionary work on current classificatory/ evolutionary framework (to be adopted for the forthcoming updated edition of the Treatise), are outlined in this overview. Among progress beyond the classical scheme developed by Ager, Childs & Pearson (over two decades ago), the relationships between principal crural types have been reappraised and several lineages surviving from the Paleozoic have been recognized. Some of these (such as wellerellids, allorhynchids or rhynchotetradids) did not linger beyond Early/Middle Jurassic times and have thus made a comparatively minor contribution to the overall diversity of the group. The mainstream of post-Paleozoic rhynchonellides was channeled along relatively few, long-lasting, major Baupläne: (a) a dimerelloid stock, characterized by fold-less, subcircular, smooth to radially ribbed shells, with often sulcate dorsal valves (and sometimes ventrals, too), having reduced deltidial plates and laterally compressed (often long) crura, seems traceable back to the Devonian at least, and may reach the present via deep-sea cryptoporids; (b) a norelloid stock characterized by small, smooth to capillate (or gently fluted), ovoid to subtriangular shells, generally sulcate dorsal valves, minute beak and foramen, and crura arcuiform (or a derivation thereof; it apparently stemmed from likewise sulcate, late or perhaps even mid-Palaeozoic ancestors and has frieleiids as Recent deep-sea survivors); (c) a pugnacoid stock characterized by subtriangular to subpentagonal, smooth to pauci-semicostate shells, typically bearing a central plication, and divided hinge plates, but lacking septalium and dorsal median septum, and whose crura may be falciform, septiform or a modification thereof (it probably evolved from rather generalized mid Paleozoic ancestors, splitting into two main branches, the falcifer basiliolids with extant, off-shore representative and the septifer erymnariids which became extinct by mid-Tertiary); (d) a rhynchonelloid stock characterized by subtriangular, often cynocephalous, partly or fully costate shells, with well developed central uniplication, dorsal median septum and uncovered septalium, but squama and glotta wanting (like all previous ones), having raduliform and calcariform crura; with likely mid to late Paleozoic forerunners, yet without direct post-Mesozoic descendants known; (e) most 'ordinary-looking' rhynchonellides may be grouped in a hemithiridoid stock, characterized by sub-triangular to subpentagonal (sometimes globose), commonly strongly costate shells, with moderate beak, dorsal fold and squama and glotta usually well defined, dorsal median septum and uncovered septalium always present (though variably developed) and having raduliform crura (or a variation thereof); seemingly issued from Paleozoic ancestors with a covered septalium, the prevailing Jurassic Tetrarhynchinae may have left derivatives among living shallow-water hemithirids of (mainly) northern seas, whereas those from the southern hemisphere may represent descendants of the dominant Cretaceous cyclothyridine branch. In addition, a number of very specialized, short-lived offshoots, such as acantho-thirids (with peculiar, spiny exteriors) or septirhynchiids (with unique cardinal process and pentameroid outline) appear restricted to particular segments of the Mesozoic Era. Subjective opinion is often involved in the interpretations of certain lineages as either 'Lazarus' or 'Elvis' taxa. Nevertheless, various aspects of the improved broad picture obtained seem to reflect the role played by different heterochronic episodes, as well as the influence of the Mesozoic marine revolution on the evolutionary history of the whole group.

BRACHIOPOD RESEARCH AND ITS IMPACT UPON INVERTEBRATE PALEONTOLOGISTS: AN EXPOSTULATION

LAURA PEARSALL

Institute of Invertebrate Psychoanalysis, Sudbury, Canada

This study examines the symbiotic relationship which must inevitably occur between the invertebrate paleontologist and his/her objects of study, and more specifically, to the "enmeshment" risk involved in close examination of the intimate details of live entities limited by Ordovician boundaries and the stresses imposed by membership in a superfamily taxon (with its obvious attendant 'attachment' dilemma) by entities limited by 1990's singles bar boundaries and the stresses imposed by publish-or-perish identity crises. In brief, one runs the risk of emotional identification with one's collection (the 'special lampshell in one's life' syndrome), the obvious extension of which is the conferring upon a 400-million-year-old fossil one's own last name. It is by now a well-understood phenomenon (hinted at in the 1913 Farfegnugen Treatise) that the phylum Brachiopoda possesses an alarming capacity to project its characteristics upon the vulnerable researcher. The susceptible scientist is most likely to identify with the following characteristics: long-loop envy, short septum anxiety, monotonous external habitus, failure to generate novel morphologies, fixation upon the Upper Kellwasserian bioevent, all manner of biotic crises, intermittent transgressive/regressive episodes, an immoderate attachment to the substrate and, of course, the secret dread within all of us that one may be a closet homeomorph. Several colleagues, whom we were unable to treat in time, tragically went on to serial sectioning, and were heard to plead, when ultimately discovered, 'For the love of God, stop me before I grind again!'. For the brachiopods' part, interviews with exceptionally articulate specimens yielded the pitiable apology, ' Try living with my phylogenetic relationships and see how you'd evolve!'. Recommendations for rehabilitation of the afflicted paleobiologist/brachiopodologist include: leaving the hammer and sample bag at home when on a date, resolving one's own suspended sediments, slipping in a little poetry amidst the journals, being intense without being in tents, and most importantly, never falling in love with anything more than 1,000,000 years old.

THE MUSCLE SYSTEM OF ACROTRETOIDEAN BRACHIOPODS

LEONID E. POPOV AND LARS E. HOLMER
VSEGEI, Srednii Prospekt 74, 199026 St. Petersburg, Russia;
Institute of Earth Sciences, Norbyvagen 22,
S-752 36 Uppsala, Sweden.

The musculature of fossil linguloideans is fairly well understood, and appears to have been very conservative from the early Cambrian to the Recent. By contrast, it is more difficult to homologize the acrotretoidean muscle system with that of other lingulates. The most recent attempt at reconstructing acrotretoidean musculature was based mainly on the assumption that it is similar to that of Recent discinoideans. An alternative model involves an interpretation of the conical acrotretoidean ventral valve as representing a linguloidean ventral valve that is 'rolled up' along the posterior margin. Botsfordiides would then appear to have a morphology intermediate between that of the Obolidae and Acrotretidae, in having an incompletely 'rolled up' ventral valve. The acrotretoidean pedicle foramen would have been formed by the linguloidean pedicle groove. As a result of the acquisition of a conical valve shape, the ventral posterolateral muscle fields, with the equivalents of the linguloid transmedian and anterior lateral muscles, migrated up on to the posterior slope of the acrotretoid valve to form the central 'cardinal scars'; the dorsal 'cardinal scars' sometimes show possible traces of two muscles and may have been the sites of equivalents of either the transmedian, outside lateral, and/or middle lateral. The central, umbonal muscle scars then formed the acrotretoidean 'apical pits' posterolateral to the pedicle foramen, and this muscle was probably attached to the dorsal 'median buttress'. The acrotretoidean apical process is situated invariably between the proximal ends of the ventral vascula lateralia and undoubtedly represents a muscle platform with up to two pairs of muscle scars, which are possibly homologous with the linguloidean outside lateral, middle lateral and central muscles, whilst the dorsal 'antero-central' muscle scar may have served as the attachment site for the central and anterior lateral muscles. The acrotretoideans undoubtedly had a smaller number of muscles by comparison with the linguloids, and it is unlikely that, for example, the central muscles were absent in most forms, especially in extremely highly conical taxa that lack or have a reduced apical process; moreover the middle lateral and outside lateral muscles might have been united into a single muscle running from the dorsal cardinal muscle fields to the apical process. Assuming that this interpretation of acrotretoid muscles is correct, it is clear that the same muscle was attached to areas of very unequal size in opposite valves; thus it is unlikely that they had columnar muscles like those of all recent linguloideans and discinoideans, but instead the muscles might have been tendonized.

STRATEGIES OF PHYLOGENETIC CLASSIFICATION OF THE CLITAMBONITIDINE BRACHIOPODS

MADIS RUBEL
Institute of Geology, Estonian Academy of Sciences,
EE-0100 Tallinn, Estonia

In the revised classification of the suborder Clitambonitidina, the phylogenetic (cladistic) approach has been preferred before such classificatory procedures as formation of keys, directed towards decrease in the number of characters, or dendrograms, based on overall but phenetic similarity. From among several algorithms, PAUP has been chosen to clarify the phylogenetic relationships between the studied genera. A number of conceptions and transformations of their characters are discussed to apply the corresponding parsimony criteria.

It is concluded that introduction of any constraints in the character evolution leads to a drastic decrease in the number of equally parsimonious cladograms and, therefore, possible solutions.

EARLY LOOP ONTOGENY OF SOME RECENT LAQUEID BRACHIOPODS

MICHIKO SAITO
Geological Institute , University of Tokyo, 7-3-1 Hongo,
Bunkyo-ku, Tokyo 113, Japan

Early loop ontogeny of nine Recent laqueid species from Japanese waters were examined using scanning electron microscopy. Although juvenile specimens representing certain early loop stages were not available for some species, all the examined species appear to share roughly the same early stages of loop development as described by Richardson (1975). However, some differences and characteristic features exist in the early stages of loop ontogeny, such as the direction of the flanges, the presence or absence of division of the median septum and the timing of resorption of the projection on the median septum, all of which appear to be invariable at the genus level. A prime example is the case of the affinity between *Laqueus* and *Frenulina*, which have sometimes been included in the same subfamily. Inspections of loop morphology of early growth stages , especially of the above three characters, indicate that these two genera may have had different origins. Comparison of relative timing of loop development as standardized by shell length reveals some variation within a species, but no major variation among species. However, heterochrony is evident in the aberrant laqueid species '*Frenulia* ' sp., of which supposedly adult individuals have a loop of juvenile morphology of other species.

'TEREBRATULA' LIARDENSIS WHITEAVES 1898 (BRACHIOPODA, TEREBRATELLIDINA) - THE FIRST LONG-LOOPED TEREBRATULID RECORDED FROM THE TRIASSIC OF N AMERICA - AND ITS PALEOBIOGEOGRAPHIC-STRATIGRAPHIC SIGNIFICANCE

MICHAEL R. SANDY
Department of Geology, University of Dayton,
Dayton, Ohio 45469-2364, USA.

'*Terebratula*' *liardensis* Whiteaves, 1898 was originally described from the Liard Formation (Ladinian-Karnian, Middle to Upper Triassic) of the Liard River, northeastern British Columbia. The species has often been recorded from the Liard Formation of this region. In fact three terebratulid species have been described from this stratigraphic interval: '*Terebratula*' '*liardensis* Whiteaves 1998; *Coenothyris petriana* McLearn 1937; *Coenothyris silvana* McLearn 1937. It is believed that '*Terebratula*' *liardensis* is the senior synonym by priority of publication date of these three species. McLearn's species may be worthy of consideration as subspecies if some stratigraphic significance is apparent. Serial sections have been prepared through one specimen of "*Terebratula*" *liardensis* and the species is now referred to the genus *Aulacothyroides* Dagys. The genus is known from the Ladinian-Karnian (late Middle to early Late Triassic) of Siberia and also now from northeastern British Columbia. The Canadian record is from in place (autochthonous) sedimentary sequences on the western margin of the Canadian Craton. Therefore it appears that *Aulacothyroides* had at least a circum-Panthalassa (Pacific) distribution indicating that some brachiopod genera had significant longitudinal distributions in the Early Mesozoic. *Aulacothyroides liardensis* is not uncommon in the Liard Formation and it may have some biostratigraphic utility.

OLDEST RECORD OF PEDUNCULAR ATTACHMENT OF BRACHIOPODS TO CRINOID STEMS (U ORDOVICIAN, OHIO, USA)

MICHAEL R. SANDY
Department of Geology, University of Dayton,
Dayton, Ohio 45469-2364, USA

A bedding-plane slab of limestone from the Upper Ordovician (Waynesville Formation) of southwestern Ohio contains numerous brachiopod specimens (at least 60) exposed on its surface. The brachiopods are the atrypid *Zygospira modesta* and are aligned in an area measuring approximately 9cm x 2cm. A wide range of specimen-sizes from juvenile to adult are present. The crinoid and attached brachiopods are believed to represent a ciocoenosis. An articulated crinoid stem is exposed centrally and parallel to the long axis of this brachiopod concentration. It is clear here that the brachiopods encircle the crinoid column. This slab records the paleoecological phenomenon of brachiopods attached by pedicles to crinoid stems. This is perhaps no surprise - *Zygospira modesta* has been recorded attached to bryozoans in the Upper Ordovician of southwestern Ohio. However, direct evidence of such a relationship is rarely glimpsed in the fossil record. No evidence of the brachiopod pedicle attachment scar *Podichnus* was seen on the crinoid ossicles. The crinoid is believed to have lived as part of the benthos, attached to the sea floor. Subsequently *Zygospira* spat settled on the crinoid stem and grew, probably allowing successive spawnings of brachiopods to settle. The crinoid stem is largely intact and was most likely knocked over by a storm-generated current and rapidly buried by sediment transported during the same event, allowing the crinoid stem and surrounding brachiopods to be preserved close to their living configuration (except that the crinoid stem is now horizontal). The crinoid calyx or roots were not seen.There appear to be at least three strategies by which articulate brachiopods attach to crinoids:- direct cementation of the pedicle valve;- by curved spines along the hingeline that wrap around crinoid stems as in some Late Paleozoic productid brachiopods; - by pedicle attachment. This example of pedicle attachment is, I believe, the oldest record of articulate brachiopods attaching directly to crinoid stems, elevating them above their cousins 'stuck in the mud' or on the sea bottom - thereby fortuitously developing tiering in brachiopods.

REVISION OF SOME LOWER-MIDDLE JURASSIC BRACHIOPODS FROM THE CANADIAN CORDILLERA - PALEOBIOGEOGRAPHY AND DISTRIBUTION SIGNIFICANCE

MICHAEL R. SANDY , ALFRED DULAI , ATTILA VOROS & JOZSEF PALFY
Department of Geology, University of Dayton, Dayton, Ohio 45469-2364, USA; Hungarian Natural History Museum, Muzeum korut 14-116, 1088 Budapest, Hungary; Department of Geological Sciences, University of British Columbia, Vancouver, BC, V6T 2B4, Canada

A number of Jurassic brachiopod species were first described from the Canadian Cordillera, primarily the Queen Charlotte Islands (QCI), British Columbia, by Whiteaves (1876-1903) and Burwash, (1913). Collections made in recent years by Sandy and Palfy with re-investigations of original type material has allowed revision of some of these brachiopods and the addition

of new records. The following rhynchonellids are recorded: *Gibbirhynchia maudensis* (Whiteaves) from the Upper Pliensbachian of Maude Island, QCI; *Piarorhynchia* cf. *juvensis* (Quenstedt) is a new record from the Upper Pliensbachian of Graham Island, QCI; *Kallirhynchia nonsinuata* (Burwash) Bathonian, Moresby Group, Maude Island, QCI. The following terebratulids are recorded: *Lobothyris subpunctata* (Davidson), recorded from the Upper Sinemurian of the Telkwa Range, Smithers area, British Columbia; *Lobothyris?* *skindegatensis* (Whiteaves) from the Bathonian, Moresby Group, Maude Island, QCI. These represent the first well-documented (with serial sections) occurrences of the genera *Gibbirhynchia*, *Kallirhynchia* and *Lobothyris* from the western Cordillera of Canada. These genera together with *Piarorhynchia* are also well-represented in contemporaneous European faunas. These occurrences serve to emphasize the widespread longitudinal distribution of certain Jurassic brachiopod genera. This is further underlined by the recognition of species comparable or nonspecific with forms originally described from Europe, namely *Piarorhynchia* cf. *juvensis* and *Lobothyris subpunctata*. Additionally, from a stratigraphical point of view, some of these genera appear long-ranging, e.g. *Piarorhynchia* and *Lobothyris*. All of these records could reflect connections with European faunas via the Hispanic Corridor in the Early Jurassic and subsequently in the Middle Jurassic via the continued opening of the Central Atlantic Ocean. Only the Sinemurian occurrence of *Lobothyris* pre-dates the commonly accepted Pliensbachian date for the initiation of the Hispanic Corridor. Previous confusion regarding the possible Cretaceous age of '*Rhynchonella*' *nonsinuata* Burwash, "*Terebratula*" *skidegatensis* and '*Terebratula*' *grahamensis* can be discounted, although certainly Cretaceous brachiopods do occur locally in the Canadian Cordillera.

UNCINULID ECOLOGY AND DISTRIBUTION IN THE DEVONIAN OF THE CANTABRIAN MOUNTAINS (NORTHERN SPAIN)

DIETRICH SCHUMANN
Geologisch-Paläontologisches Institut, Technische Hochschule, 64287 Darmstadt, Germany

The uncinulids are characterised by their aberrant structure. Anatomical investigations by Schmidt (1937), revealed the inner spicular grille. This was suggested to function as a particle filter. This interpretation was principally accepted by Schumann (1963), Rudwick (1964), Westbroek, (1967) and Westbroek et al. (1975). The hypothesis put forward by Schmidt (1937) assumes that ontogenetic growth occurs in cycles; a period of longitudinal growth is followed by a period of vertical growth. This process is repeated until the adult stage is reached (= adult individual). This concept is presumably wrong. Growth probably occurs in two cycles only. Vertical growth follows longitudinal growth and there is no repetition of the process. This implies that small shells with vertical flanks are adult shells. Extensive investigations of Emsian/Eifelian sections in the Cantabrian Mountains (northern Spain) reveal clear facies dependencies of certain phenotypes of *Uncinulus orbgnyanus* (de Verneuil). The method of fixosessile attachment by numerous uncinulids is unknown. The foramen of some species is extremely small and apparently open throughout their lifetime. In other species, the foramen is covered completely by the umbo, thus making a fixosessile habit unlikely. Possible alternatives are discussed in analogy to recent Neothyrida.

GENERAL MORPHOLOGY, MORPHOLOGICAL PATTERNS AND SYSTEMATIC UNITS OF POST-PALEOZOIC CRANIID BRACHIOPODS

ANDREW V. SHAPOVALOV

Department of Paleontology, Moscow State University, Moscow 119899, Russia

This paper presents the analysis of features used for defining the diagnosis of post-Paleozoic craniid brachiopod species. For both valves, the following characters were considered: outlines, shell shape, muscle impressions (including supports and pendants), margin formation, ornamentation (including attachment scar) and vascular marks. Differences in these features are connected with the skeletal growth process determined by both the soft body of the organisms and the environment. The vascular impressions are dependent upon the preservation of the inner surface of the shell. If we view the morphology as the result of growth modes we can separate the characters determined by the inner organization of the organisms from the growth modifications induced by the environment. This enables us to understand numerous morphotypes realized among the craniid brachiopods and sheds new light on their taxonomy.

EARLY DIAGENETIC ALTERATION IN THE PHOSPHATIC BRACHIOPOD SHELLS AND THEIR CAUSES

GALINA T. USHATINSKAYA

Paleontological Institute, Russian Academy of Sciences, 117868 Moscow, Russia

Brachiopods with phosphatic shells, the Lingulata, are known from the beginning of the Cambrian. In the Early Paleozoic, the group consists of five orders and more than one hundred and fifty genera. Five Recent genera are descendants of two Early Paleozoic orders. Shells of Recent lingulates contain about 50% organic matter. The ancient shells contained the same or a higher amount. According to Emig (1990), the loss of organic matter in the *Lingula* shells after death leads to full disappearance of the shells over several days. In comparison, the disappearance of the calcarious shells of Recent articulate brachiopods in the same conditions take six to seven months. A comparative study of Recent and earliest lingulate shell microstructure and differences in the microstructure within same taxa from different facies allows us to suggest that the shell microstructure of the Cambrian lingulates is not always primary. It could have been formed at some stage during early diagenesis. The transformations included decomposition, transfer and redeposition of calcium phosphate within the valves. In the process, microorganisms played the most important role. The resulting microstructure was often similar to the primary one, but sometimes it was an invert reflection of a primary structure; there are transitional types as well. The process was probably very fast (hours or days). Sometimes phosphatized parcels of soft body, muscle tissues or cells of the outer epithelium were preserved. In the case where the microbial activity was suppressed, the post-mortem rebuilding in shells was minor or did not occur at all.

A MIDDLE TRIASSIC BRACHIOPOD BANK, NIDANG, XINYI COUNTY, GUIZHOU, CHINA

GUIRONG XU, XIAOQING YANG & QIXIANG LIN

Department of Geology, China University of Geosciences, Wuhan, 430074, China

Brachiopod coquinas of the Middle Triassic were found in the Nidang area, southwestern Guizhou Province, China. Brachiopod coquinas occur in the upper part of the Falang Formation. The Falang Formation contains abundant fossils with various taxa, including brachiopods, bivalves, ammonoids, gastropods, hexacorals, crinoids, bryozoans, calcareous algae, foraminifers and ostracodes. On the basis of index fossils, The Falang Formation is early Ladinian in age and three biozones: 1) *Protrachyceras deprati* Biozone; 2) *Daonella bulogensis bifurcata -Halobia kuei* assemblage zone; 3) *Sanqiaothyris subcircularis - Rhaetinopsis ovata* assemblage zone (or *Septaliphoria xingyiensis* acrozone). Six facies are identified: 1) Shell bank facies. There are two kinds of shell bank facies: a) brachiopod and (b) bivalve bank. 2) Crinoid bank facies of storm wave deposits. 3). Ooids, oncoids and alga-coated nodule bank facies. 4) Fore reef facies. 5) Silty bioclasts facies. 6) Dolomitized micritic facies. Three shallowing-up cycles can be divided in terms of the facies sequences. The brachiopod coquinas extend eastward for about 300m and change to bivalve coquinas with a few brachiopod. The thickness of brachiopod coquinas is more than 5m, but become less than 3m eastward, and change into thick bivalve coquinas. The brachiopod community is dominated by *Septaliphoria xingyiensis*, making up about 89% of the total individual number of brachiopods. Specimens in growth position indicate that the species might have a short pedicle attached to bioclasts.The Nidang area was situated at the continental margin between the Yangtze Platform and the South China Basin. A Ladinian compound hexacoral reef may be present in the Nidang area as inferred from abundant boulders in calcirudite consisting of compound hexacorals. Faults played an important role in reef complex evolution. The brachiopod bank developed near reef tops in the late stage of reef evolution, which provided abundant nutrition and calcium carbonate, as well as favourable substrate for the brachiopods. In the same conditions, bivalves were prosperous at the time, but in a deeper area with a softer substrate. In the late Falang stage, brachiopod banks were replaced by bivalve banks.

AUTHOR INDEX

Milton Keynes UK
Ingram Content Group UK Ltd.
UKHW051945071024
449327UK00026B/2170